# BUILDING YOUR OWN HOME

# BUILDING YOUR OWN HOME

## A comprehensive guide for owner-builders

**FULLY UPDATED EDITION INCLUDING NEW SECTIONS
EASY ACCESS & BUILDING IN BUSHFIRE PRONE AREAS**

### George Wilkie

This updated edition first published in Australia in 2021 by
New Holland Publishers
Sydney

Level 1, 178 Fox Valley Road, Wahroonga, NSW 2076, Australia

Copyright © 2021 text: George Wilkie
Copyright © 2021 illustrations: George Wilkie and Stuart Arden
Copyright © 2021: New Holland Publishers (Australia) Pty Ltd

This book includes text and illustrations from the following publications:

*Building Your Own Home*
George Wilkie and Stuart Arden
First published by Lansdowne-Rigby Publishers in 1984
Reprinted 1985, 1988, 1989, 1990, 1992, 1993 (twice), 1995
Revised edition published in 1997, reprinted 2001.
First published by New Holland Publishers in 2003. Reprinted in 2007, 2009, 2011, 2012 (twice), 2013, 2014, 2016, 2017, 2020, updated edition 2021

*Alterations and Additions to Your Home*
George Wilkie
First published by Lansdowne Publishing Pty Ltd in 1998

All rights reserved. No part of this publication may be reproduced, stored in a retrieval system or transmitted, in any form or by any means, electronic, mechanical, photocopying, recording or otherwise, without the prior written permission of the publishers and copyright holders.

National Library of Australia Cataloguing-in-Publication Data:

Wilkie, George.

Building your own home : a comprehensive guide for owner-builders. Rev. ed.

ISBN 9781742572161 (pbk.).

1. House construction - Amateurs' manuals. 2. House construction - Handbooks, manuals, etc. I. Title.

690.8

Managing Director: Fiona Schultz
Project Editors: Sophie Church and Claire de Medici
Designers: Kariman Roper, Andrew Davies
Production Director: Arlene Gippert
Printed in China

10 9 8 7 6 5 4 3

The information in this book is true and complete to the best of our knowledge. All recommendations are made without guarantee on the part of the author and the publisher. The author and publisher disclaim any liability for damages or injury resulting from the use of this information.

Keep up with New Holland Publishers:
NewHollandPublishers
@newhollandpublishers

# CONTENTS

| # | Title | Page |
|---|---|---|
| 1. | Contents | 5 |
| 2. | Introduction | 6 |
| 3. | A history of owner-building | 8 |
| 4. | Domestic architecture | 11 |
| 5. | A history of building materials | 15 |
| 6. | A history of building construction | 18 |
| 7. | Buy, build or extend? | 21 |
| 8. | Pre-design | 23 |
| 9. | Designing the home | 24 |
| 10. | Consultants | 29 |
| 11. | Authorities | 33 |
| 12. | Design drawings | 37 |
| 13. | Building drawings | 39 |
| 14. | Specifications | 48 |
| 15. | Schedules | 50 |
| 16. | Costing | 52 |
| 17. | Tendering, contracting and legal | 57 |
| 18. | Subcontractors | 60 |
| 19. | Suppliers | 64 |
| 20. | Administration | 68 |
| 21. | Achieving completion | 72 |
| 22. | Where to now? | 73 |
| 23. | Sites | 74 |
| 24. | The base | 80 |
| 25. | Site preparation | 81 |
| 26. | Footings | 84 |
| 27. | Sub-surface or agricultural drainage | 90 |
| 28. | Retaining walls | 92 |
| 29. | Ground-to-floor systems | 96 |
| 30. | Timber floor frames | 98 |
| 31. | Steel flooring systems | 100 |
| 32. | Concrete slabs | 102 |
| 33. | Walls—introduction | 109 |
| 34. | Timber wall framing | 110 |
| 35. | Metal wall framing | 115 |
| 36. | Brick veneer construction | 117 |
| 37. | Autoclaved Aerated Concrete (AAC) | 120 |
| 38. | Powerpanel® | 122 |
| 39. | Sheet and board claddings | 123 |
| 40. | Sheet and board products | 126 |
| 41. | Solid masonry construction | 128 |
| 42. | Interior linings | 138 |
| 43. | Windows | 140 |
| 44. | Doors | 149 |
| 45. | Roofs | 152 |
| 46. | Conventional roof frames | 155 |
| 47. | Trussed roofs | 161 |
| 48. | Roofs—flat, skillion and vaulted | 166 |
| 49. | Roof tiles | 169 |
| 50. | Roof sheets | 172 |
| 51. | Ceilings | 176 |
| 52. | Special roof details | 180 |
| 53. | Services | 183 |
| 54. | Drainer | 184 |
| 55. | Plumber | 189 |
| 56. | Electrician | 193 |
| 57. | Major appliances (whitegoods) | 196 |
| 58. | Fixing out | 198 |
| 59. | Trims | 199 |
| 60. | Kitchens | 201 |
| 61. | Bathrooms | 204 |
| 62. | Robes and presses | 210 |
| 63. | Fireplaces | 213 |
| 64. | Stairs and entry | 217 |
| 65. | Cabinet joints | 221 |
| 66. | Options | 223 |
| 67. | Double-storey construction | 225 |
| 68. | Car accommodation | 228 |
| 69. | Thermal comfort | 230 |
| 70. | Floor surface finishes | 233 |
| 71. | Painting | 235 |
| 72. | Building material properties and uses | 238 |
| 73. | Builders | 241 |
| 74. | City conditions | 242 |
| 75. | The green supplement | 243 |
| 76. | Enhanced access and livability | 259 |
| 77. | Building in bushfire prone areas | 266 |
| 78. | Glossary | 272 |
| 79. | Index | 277 |

# 2 • INTRODUCTION

The concept for Building Your Own Home materialised when Stuart Arden and George Wilkie were teaching architectural drafting and building construction classes with TAFE NSW.

They realised that the available published information available to students at that time had not kept pace with the changing construction methods emerging from the housing boom in Australia in the 1980s.

At this time the Sydney Building Information Centre (SBIC) was visited by a growing number of families interested in being directly involved in designing and building their own home. Stuart and George discussed this increasing interest with SBIC and an agreement was reached to offer a short evening course to provide information to budding owner builders.

The course notes developed for the SBIC course filled a large lever-arch binder and the weekly photocopying was absorbing many hours and many dollars every week to prepare. This problem could be solved if a more formally published set of bound notes could be produced and distributed to the course participants.

Feelers were put out to a few publishers in the Sydney metropolitan area. A response came from a small publisher that specialised in printing glossy annuals for the Australian motor sports fraternity, Motor Sport Press. They suggested that there could be a market for a publication that combined sound information for Owner Builders and was a useful textbook for students of architecture, building construction and drafting.

The motor sport publications by Berghouse Media Services Pty Ltd were hardcover glossy full colour books, that were partially funded by the inclusion of large format advertisements. This became the format for the first edition of Building Your Own Home, A Comprehensive Guide for Australian Owner Builders. Only the first edition contained advertisements and the hard cover, all editions that have followed have the red brick bond cover design and no advertisements.

House building has evolved since the first Edition, but surprisingly more slowly than some predicted.

Houses are more likely to be assembled rather than built-that is, components are made off-site then assembled into a whole on-site. This productivity improvement has attempted to maintain the price of the average project home within the funding potential of median income Australia's families.

Builders working in the 1970s would be amazed at the lack of hammer and saw work on a house building site, but the materials concrete, timber, bricks, tiles, plasterboard, aluminium -would be familiar and photos of the building progress would show little change in over 30 years.

The availability of materials, the fickleness of the weather, the intransigence of local government, the imposition of environmental issues, the concern for waste management, the cost of borrowing money, the skills of the site trades, and so on still provide reality for this revised book.

The format of BYOH has remained constant through all editions, the large A4 page size and the sound logical layout continue to allow easy access to its information.

People have suggested that the book would be better if it was sold as an interactive CD-ROM, with a blow-by-blow video.

These people cannot see the weary couple boning up on the finer points of, when should they install the windows in their dream home; or the young building student, slaving away at midnight after a day on the tools, tracing over illustrations to satisfy a lecturer's demand for sectional drawings through aluminium and timber windows.

Building your own home remains a great emotional and creative experience for many Australians, there will always be people who reject the rigid brick veneer tents, mass produced by the project home industry- they want their own individual 'Gloria Some' [1].

This new enlarged edition has been extensively revised and contains more information to assist in understanding designing for people of all abilities and a section on reducing bushfire impact.

Owner-Builders (OBs) remain the focus of the book but some changes have been introduced to improve its value as an educational text. It is impossible, in a book of manageable size, to provide every possible correct trade description. To that extent this is not the definitive text. Information, however, is given on the majority of trade descriptions used in Australian cottage construction, in a manner that is clear and as free as possible from anomalies.

If every individual building process was investigated and all the known methods and conventions currently being applied in Australia, there is every chance that new methods would be devised between the writing of the book and its subsequent publication. Even then, the tradespeople on your job would probably have their own special method that did not conform to any of our described methods.

Therefore, the book aims to communicate the most common methods and conventions used in the Australian residential building industry but notes that local conditions and idiosyncrasies may have evolved different methods. The building industry is not a uniform entity and, like all freely structured organisations, individual parts adapt to local influences and traditions.

These local variations are part of the character of the industry and are why it has developed in a steady evolutionary manner.

There have been very few revolutionary leaps forward in residential building construction.

Many methods and systems promoted as revolutionary have, in practice, proved to be less suitable than the methods they sought to replace.

The introduction to the first edition of this book noted that people had to be mad to think that they could be successful OBs.

This was based on the belief that the hardest part of building a home was the organisation of the tradespeople and the suppliers.

Times have changed and it is even more complicated to be an OB. It is not that building has become any more complicated (it remains much as it was in the 1980s), what have changed are the rules. Consent authorities have progressively moved from a set of weighty, but understandable, restrictive regulations, generally based on quantitative limitations, to qualitative, performance based regulations.

Many consent authorities require the applicant to prove that the house they want to build is suited to the environment. Building a cottage on a block of land is, in many locations, not a right but a privilege. Neighbours have rights to make representations to the controlling authorities, trees are protected, shadows are analysed and environmental impacts studied. Today's OBs are not mad but they certainly have to be courageous.

Dealing with modern environmental planning legislation has severely changed the emphasis of being an OB and has inevitably increased the stress levels. Once the greatest satisfaction an OB could achieve was to finish the house on time and on budget, now it is to start building on time and on budget.

The perennial problems of being Owner Building remain.

- It is difficult to maintain friendships. There is little time for socialising and some friends will go into hiding, fearing that they may asked them to cart wheelbarrow loads of concrete.
- Your personal relationships with loved ones will be strained. Few families believe that pouring many cubic metres of concrete on a Saturday afternoon is a bonding experience. Family additions begat during the building period, however, are far from uncommon-just build another bedroom.
- The local authorities becomes adversaries. OBs will begin to believe that they have been singled out to have the toughest building surveyor overseeing their project.
- OBs will learn every trick to cajole your bank manager to keep advancing funds, even when your project is still incomplete and has exceeded the original cost estimates.
- Tradespeople treat OBs as subhuman. They often want money in advance but are slow to return to the project to complete critical unfinished work.
- Suppliers seldom want to give OBs the same credit and discount terms they provide to real builders. Deliveries

---

1  Ref: Afferbeck Lauder, *Let stalk Strine*

are not always made when the OB is on site, so OBs are doomed to sit at work worrying about how many timber flooring boards have been pilfered from the footpath.
- Neighbours will often have an idealised image of the house that should be built. When the home doesn't fit that image, the dust from the building works will become a life-threatening hazard.

Potential OBs who are still smiling after reading that list may have the courage to be an OB. It is not an easy task and there will be problems along the way but the final result belongs to the OB. Time generally heals all but the most disastrous financial and interpersonal wounds.

A few OBs do successfully complete and occupy a house they build themselves without confronting any crisis worth noting.

But these people are far and few between, and generally are those who build in mud brick on a remote commune and are totally unaware of the disasters happening around them anyway.

OBs have ups and downs, but many have succeeded in building fine homes within budget. A project home may appear a much easier alternative, yet many families have lost the battle against escalating costs where the elements of the house that are covered by the primary cost do not meet their needs and so the casually approved variations that add thousands of dollars to the final account.

Kit homes are a middle-ground alternative and often appear economical, but careful worked forward cost estimates seldom show any financial advantage. Like project homes, kit homes are pre-designed, reducing immediately one of the most satisfying components of being an OB-the participation in the design process.

### How do OBs gather information?

When the first edition of this book was published, there was a serious gap in information available for OBs. Since then, tens of thousands of copies of this book have been sold, so the gap should have closed a little.

There are magazines featuring OB projects, books to assist in financial management, stories of people who have built their own homes and an assortment of other useful works.

Many of the books are Australian and provide important practical information.

Some books, however, are imported from the United States and Britain and, although they may contain worthwhile advice, the Australian cottage is distinctive and there can be real problems with information that comes from overseas.

### Why is this book necessary?

A survey in 1981 indicated that people said they chose to be OBs because they wanted:
- more house for the same money
- a better house for the same money
- better control of quality
- better control of quantity
- better control of time.

Today little has changed. In an ever more mass-produced world, the challenge to create a home from a vacant site (and the satisfaction of facing that challenge) is one of the few great opportunities still available.

Notice no prospective OB seems to aim to achieve the same house for less money. However, this is not strictly true as many OBs, when first asked the question 'Why are you going to build your own home?' did want to save money.

Further discussion determines that the ability to reproduce a project-type home for less money was seldom, if ever, the real reason for owner-building. It should be noted, though, that project display homes form the benchmark from which OBs judge gains or losses.

This book recognises that OBs generally begin with a fair amount of knowledge gleaned from some or all of the above sources, but they need a framework into which that knowledge can be placed. This book attempt to set down the collections of facts, applications and general information of residential house building to assist prospective OBs plan an OB project.

This book is divided into two parts. The first part outlines the building industry, its participants and operation as applicable to OBs. It is important that OBs understand the complex system of interaction that is the building industry. Residential house building is not inherently a complex activity but the rules, written and unwritten must be understood by prospective OBs if they are to be successful with their project.

The second part concentrates on specific building operations and is laid out for the reader to follow these operations in a roughly chronological manner through the normal construction process.

Although the book is divided, the following themes are applied generally:
1. Information is presented in a manner directly applicable to OBs and constraints and benefits are explored wherever appropriate.
2. Current building systems are dissected and presented in a manner that can be easily understood.

# 3 • A HISTORY OF OWNER-BUILDING

What were the factors that allowed owner-building to develop in Australia?

The phenomenon of owner-building probably happens in Australia to a greater extent than in any other western country. The reasons for this are influenced by two significant factors:

1. Australian cities are particularly suburban in nature and are predominantly subdivided into individual family-held residential allotments, each containing a freestanding single-family cottage. This predominance of freestanding cottages means that OBs can carry on a building program with little or no effect on neighbours because no part of any house is in contact with another, as they are in many parts of Europe.
2. The cottage building industry is structured in such a way that most of the on-site assembly of building components is carried out by specialist sub-trades. This means that an owner builder can immediately save the cost of a builder's overhead and profit, by direct employment of sub-trades. This saving is in the vicinity of 15–25 per cent of the estimated contract value if the work was carried out by a regular builder.

Once, quite recently, the homes built by OBs and those built by traditional builders were of much the same quality. This changed significantly during the 1990s, when the majority of homes were built by project home companies. The homes they built and continue to build, are of such a quality that it is unlikely that an OB would be bothered replicating them.

Project homes are universally constructed with:
- basic concrete slab floors, often very low to the ground as to appear to be below the ground
- standardised ceiling heights of 2400+ millimetres to allow cheap flat ceiling trussed roofs
- basic powercoated aluminium windows of standardised proportions
- plasterboard internal linings to walls and ceilings
- one light fitting and two power points per room
- extensive use of faux period trims and details in moulded medium density fibre-board (MDF), gypsum and plastic.

These are seldom the materials or methods used by OBs; therefore, care must be taken when attempting to compare the cost of owner-building with the cost of a builder-built home—compared like for like the OB still should be able to make significant savings.

Are you the first OB? The answer is obviously no. There have been OBs since the beginning of constructed shelter. The question should be: when did owner-building cease to be a normal practice? This is a difficult question to answer, certainly on an international scale, but if we examine the Australian scene, we can set down a potted history of owner-building in this country.

| **Chronology of Owner Building*** | | | |
|---|---|---|---|
| | Workers | Middle | Affluent |
| 1788–1820s | All self. No architects/builders. | Convict labour. | Some prefabrication in England. |
| 1820–1850s | Mainly self-built cottages. | Building trade groups began to emerge to satisfy a growing wealth. | Architects, fashion books from Europe. Builders, tradespeople become the norm. |
| 1850–1890s | Mainly self-built freestanding cottages. Introduction of terrace houses in inner city areas. | Builders and tradespeople. Use of pattern books. Move to suburban rings around cities. | |
| 1890–1914 | Builders, skilled tradespeople, manufactured components, prefabrication in Australia. Late Victorian/Federation (Queen Anne revival) housing styles almost universally builder-constructed. The size, degree of embellishment and quality of material varied with income group. | | |
| 1915–1920s | World War I—building industry at virtual standstill. | | |
| 1920–1930s | War service housing schemes; in an effort to house the veterans of the war as they married and commenced having children, the Commonwealth Government developed a scheme of lending money at a low interest, but required strict control over standards of construction that effectively required builder/tradespeople construction. Owner-building continued in outer suburbs and distant communities. | | Architect-designed houses |
| 1930s–1946 | Depression and World War II limited general building operations though owner-building during this period was active, but poorly recorded. | | |
| 1946–1955 | After World War II, Australia had a scarcity of both raw and processed building materials which caused all levels of society to do the best they could with what was available. Most of the pre-war tradespeople were now limited and new base levels of construction skills had to be developed. The people who filled these new positions commonly had their first taste of building in constructing their own home. | | |
| 1955–1965 | Growth of project home industry, decline in owner-building, because of low-cost land, low-cost housing and low interest rates. | | Architect-designed house revived. |
| 1965–1975 | Boom of project home industry. Highly refined product, no pressure on an OB. | | Architect-designed houses, but a growing move by the project home industry into 'up market' houses. |
| 1975–2020 | Movement into house/land package through pressures from lending industry, particularly in New South Wales. | Project homes geared into this sector of community. This sector, however, also rejecting the project home and making up the bulk of owner builders | Project home industry extensively established in this market. Architect-designed houses. Some owner OBs building the week ender, the holiday cottage, often through use of precut kit houses. |
| * This table follows NSW history, but is applicable in other states, if one varies the dates.<br>+ Projected trend. | | | |

# A HISTORY OF OWNER-BUILDING • 3

## OB 1909 … $100

The following article is an extract from a building publication of September 1909 entitled *Owning A Home Of Your Own*. It could almost have been written today.

'This is the time of year when the warm spring sunlight makes one long for the open air, and calls one away from the humdrum of city or suburban life.

'In the spring, man's first idea is "back to nature," and this "back to nature" cry is made a rule of health in many communities, not only in the spring, but the year round. In Australia particularly, this desire for the open life tingles the blood very keenly, hence it is most opportune, not only at this season, but also at this stage of these notes, that we should touch upon a simple construction to which home-lovers with the desire for the open air may have that desire gratified comfortably and inexpensively.

'A desire lies at the back of every achievement, and although the call of the wild may appeal to most of us, yet the usual horrors of camp life are often sufficient to stifle any desires for living "far from the madding crowd".

'The writer camped last Christmas at Narrabeen Lakes. Every possible comfort was provided for, with the exception of a solid building, yet although the days were pleasantly spent in fishing, prawning, bookreading and hill climbing, the evening convenience of civilisation, namely, a soft bed, was sadly missing. In the present instance, therefore, a few simple hints will be given on the construction of a primitive home for summer spending, and no doubt this first stage in your bush home construction will be but the first step in building up what may eventually become a substantial little out-of-city homestead.

'It is written by a homelover, who built. Our desire for a "house in the bush" was strong enough to make us plan for such a house even when there seemed no way of getting it.

'After a good deal of planning, estimating and reducing, we were able to build a little cabin for less than 50 pounds [$100], one that fills present requirements, and is capable of enlargement by and by.

'You may think that nothing worthwhile can be built for so little money; but many friends who saw our bungalow were pleased with it, and amazed when told what it cost.

| The cost of material | |
|---|---|
| 21 posts for foundations cut from site | |
| 3—4 x 6 hardwood 20 ft long x 120 ft | |
| 1—4 x 6 hardwood 8 ft long x 16 ft | |
| 2—2 x 6 oregon 20 ft long x 40 ft | |
| 24—2 x 6 oregon 16 ft long x 384 ft | |
| 11—21cxc6 oregon 15 ft long x 165 ft | |
| 4—2x4 oregon 20 ft long x 54 ft | £5 17 6 |
| 800 lineal ft 4 x 2 oregon dressed 534 ft | £4 8 0 |
| 2500 sq ft 6 x 1 baltic T & G | £23 15 0 |
| 600 sq ft Ruberoid 2 ply | £4 10 0 |
| 40 ft shelving for tables and shelves | £5 19 6 |
| Incidental expenses, carpenters etc | £5 0 0 |
| | £50 0 0 |

So it seemed worthwhile to write about it, in the hope that others who have the same desire, and who have not supposed that it could be gratified at so little cost, may hear of what we did.

'Of course, certain limits had to be fixed before we began at all. We were ready to be satisfied with a small and simple camp cottage. We wanted a comfortable bed, and decent cooking arrangements, and shelter against a storm. But we were willing to do without accommodation for servants, and for the present without rooms for guests; and since our scheme of living in the bush did not demand evening clothes or ball dresses, we could do without excessive space.

'But we wanted a house that could be enlarged easily when we were able to spend more, and we wanted it well built, to stand for some years.

'It had an asset of some value in the ability and willingness to do a considerable part of the work of construction myself, and when a delightful suite for such a house as we wanted was available, we set to work definitely to plan the bush home that we had so long dreamed of.

'The first plans, though modest, were impossibly expensive; and then began the process of cutting, which continued until we reached what we believe to be a minimum of present expenditure with a reasonable degree of present comfort and a maximum of future possibility.

'It is easy to do without things if you can persuade yourself that you are shortly to have them, and while we have enjoyed our little house just as it is, and expect to enjoy it for one more summer at least, yet it is a pleasure to think that it will not always be so small, and that it can be made habitable in winter as well as summer, when we can afford to do it.

'We wanted a good, large living room, with a fireplace. So we planned a living room 15 feet by 20. Chimneys are expensive, and we concluded that we could omit that for the present. We wanted a good verandah, and we decided to make that eight feet wide, and the length of the living room, 20 feet. With no servant, we knew that a small, compact kitchen would do very well, and we planned that seven feet by eight. Then came the question of bedrooms—how many, how large, how arranged with respect to the living room? We stopped to estimate the cost of what we already had, to see what was left for bedrooms. We found that we were already up to our limit. What to do now?

'Recovering from our display, we adopted the expedient of building a temporary partition, to separate a portion of the living room for one bedroom, just for ourselves, postponing entirely the building of bedrooms until more money accumulated. With the bedrooms we postponed, too, the possibility of entertaining a guest, but we found out later that a comfortable bed could be made out of the living room couch for a night or two during summer, and we did this several times during the summer.

'Our present floor plan gives us a living room 13 feet by 15; a verandah eight by 20; a bedroom seven by 15; and a kitchen seven by eight.

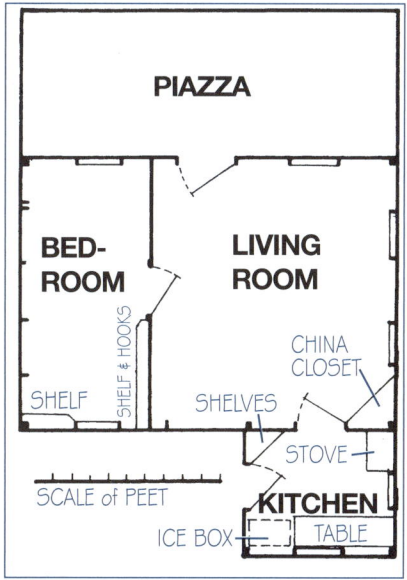

'The floor plan agreed upon, we began to plan the outside.

'As finally built, it is a one-storey building, with a low-pitched roof projecting outward in front to cover the verandah, and projecting the same way in the back over the kitchen. It is boarded with pine, to be covered with shingles later. The foundation is of posts. The roof is the same pine, covered with portable roofing. The pine also makes the floors and the partitions and the doors, except the front door of the verandah of which I shall tell later on.

'The interior is all unfinished, with the timbers of the frame all exposed. Having planned to have the interior this way, not ceiled or plastered, we hit upon a way of framing that makes a better appearance inside if unfinished than the ordinary method does. Instead of setting the upright timbers, or studs, 14 or 15 inches apart, and having on the inside a succession of narrow upright panels, we set the studs further apart, and 'bridged' between them with two rows of horizontal timbers. This method divides the inside walls into panels approximately square, less monotonous and more decorative.

'Our scheme for decoration, when we get that far, is to stain the timbers, and cover the boards with burlap or heavy paper, putting a small moulding around the edge of each panel to hide the tack heads. With this method of framing, the outside boards are put up and down, to save waste; but as they are to be covered with shingles, that does not matter.

'The casement windows are a distinctive feature of the house. They are hung on

# 3 • A HISTORY OF OWNER-BUILDING

hinges like doors, and open outward. We decided on this form of windows because they open their whole area to the breeze, and because they are cheaper, requiring neither frame nor trim. They are hinged on to the two-by-four studs.

'The front door I made with some care, and it excited no little favourable comment. It is a 'Dutch' door, upper and lower parts hinged separately. Each part is made of white pine six inches wide, in the form of a frame, and backed all the way across with narrow beaded yellow pine laid on diagonally. The panel thus formed on the front side is outlined with a moulding.

'If you want such a house as thus, you will probably ask for some assurance that the materials can really be bought, and the house built for the amount I have mentioned. Against such a demand, I have gone over my bills and compiled a list of all the materials needed, and I append the prices I paid last April. My house is located not 50 miles from Sydney and the prices could probably be duplicated anywhere without that radius. In other localities the prices would, perhaps, vary somewhat from my list.

'I had consulted a friend who is a practical builder as to strength of materials, and I followed his advice with satisfaction. I used four-by-six hardwood for joists under the floor beams, two-by-six hardwood for floor beams and oregon rafters, spaced two feet apart. The rest of the frame was two-by-four, doubled into four-by-four for corner posts and plate. With a carpenter's help anyone who is "handy" with tools can build this house, from the framing plans herewith. Indeed, modifications of the plan may suggest themselves to many. It is the general style of the building, rather than the exact details, that I want to offer to those who desire such a house. I merely reduced building to its lowest terms.

'As to furnishings, we did not buy much—a "wickless" oil cook-stove, and iron washstand and some wicker-bottom chairs. An old bureau and a crib we sent out from home, also a couch in the living room. We found a bedstead could be dispensed with; four legs fastened directly to the frame of the woven-wire spring did admirably. Tables I made.

'I have not given my plan for enlargement. We expect to build on the west end, and therefore we put no windows in that end. Anyone who builds such a house will be able to alter that to plan his own enlargement. If the narration of my experience induces anyone to attempt to reproduce "a summer home" exactly or with modifications, I am sure the home builder will be satisfied with the result.

'Hence, then, is a simple method of building a weekend home, that may, as your worldly prospects improve, develop into your "castle". It is worth trying, and if you are of limited means, who are poking in the crowded city and the adjoining suburbs, obey the "call of the wild" for, say, one season, you will be a convert to the Build Now Campaign.'

## The future of owner-building in Australia

What will be offered by the residential building market in the future? Trends appear to indicate:

1. The low-equity first home buyer will be offered house and land packages (including medium density and cluster developments) which are geared to whatever is the current lending policy of the money lenders and the government. These low-equity house purchasers will be unlikely to make up a large section of the OB activity, due to the constraints of low availability of time and underdeveloped money management skills.

2. The middle-equity second house and middle-income buyer will have a wide range of project homes to choose from, but these homes are very likely to be constructed from a limited range of mass-produced components, giving limited ability to the prospective purchaser to customise.

This limiting factor on customising may increase the activity of owner-controlled building in this sector. That is to say, in an effort to achieve home designs that suit them, OBs will extend their influence: from employing a builder under close supervision, through employing subtrades-people and ordering materials, to nearly complete hands-on building activity.

OBs are required to make a significant time contribution to the process. This often needs careful forward planning for, in the busy world of the 21st century, leisure time is often severely restricted.

OBs who can work from home rather than having to travel to an office, have professions that allow for flexible work hours or have jobs that provide adequate leisure time, are likely to turn to owner-building as a stimulating endeavour and as a way to achieve 'a hand' in their world that is otherwise extensively remote controlled.

3. The socioeconomic groups, with high debt servicing ability will tend to purchase; wherever possible, conservative existing houses in established locations and confine their new building activities to upgrading, particularly kitchens and bathrooms, alterations and additions and weekender/holiday houses.

The major hardware retail warehouses have increased the products and services suited to OBs; these have fuelled the number of upgrading projects that can be carried out by OBs within the confines of their credit card spending limits.

Houses in established locations are likely to be changed to accommodate improved insulation, solar power systems and contemporary lifestyle desirables including:
- Modern kitchens with an open presentation space and concealed preparation zone,
- Open plan dining/living space.
- Living spaces that open out through expansive glazed doors to decks and terraces
- Modern bathrooms are more desirable if ensuite with bedrooms
- A powder room close to the kitchen/dining areas
- A separate space devoted to video entertainment

OBs are undertaking extensive projects to modernise existing houses this requires them to increase their understanding and knowledge of building construction and management systems.

The impact of the Corona Virus Pandemic of 2020+ is likely to increase alterations and additions family homes, to provide *work from home spaces*. This can be challenging as *Capital Gain Tax* may become applicable under certain circumstances, particularly if the ATO considers working from home to have diminished the *Capital Gains Tax Free Status of the Family Home.*

The kit homes now coming on the market show the trend towards supplying the quality holiday house market.

The future of owner-building in Australia remains active, but the growth is likely be supported by families that are socioeconomic secure and the OB projects are likely to be a second house, alterations and additions or holiday home project.

For the low-income/low-equity first home buyers to achieve the obvious advantages in quality, quantity and flexibility that OB brings would require a significant change in government policy and a move towards deregulation of the building industry, which appears highly unlikely in the foreseeable future.

There have been some changes that were supposed to reduce the time and complexity required to gain an approval for a new family home or additions to an existing family home.

In New South Wales, it is possible to gain building approval (now called a Construction Certificate in that state) without ever going to the local council. Principal Certifying Authorities (PCAs), which can be a Local Government Authority (LGA) or registered building surveyor, can issue permission to build or add to a house under a provision of the Environmental Planning and Assessment Act (EPA). For this option to be available the local council has to have prepared the necessary Local Environmental Plans (LEPs) and Development Control Plans (DCPs) to allow for houses to be recognised as Complying Developments.

The ability of gaining recognition, acceptance and permission to carry out building work that the authorities have designated as complying or exempt varies from jurisdiction to jurisdiction throughout Australia. It is essential to contact the local authorities to make sure the OB project proposed meets all the requirements of laws, regulations and codes in the area where the project is to be constructed.

Multi-page checklists are likely required

# DOMESTIC ARCHITECTURE • 4

to be submitted for any building activity ranging from a fully Land Use Development Application to a minor backyard shed. They often provided for matters including, but not limited to:
- scale drawings showing the existing plans of any building and the extent that they are to be demolished or altered
- waste management, from excavation and demolition spoil to worm farms
- energy conservation
- statements of environmental effects
- heritage reports-some LGA require a professionally prepared heritage assessment of all buildings, regardless of age, when an application is requested for demolition
- shadow diagrams
- surveys of the applicant properties and the neighbours properties to Australian Height Datum (AHO).

An LGA in Western Sydney appears to have a fair attitude to Complying Development but on investigation, the requirements to comply effectively only apply to:

single-storey pitched roof, 2400 mm ceiling height, brick veneer houses with concrete slab floors, and a land fall across the house of less than one metre. This effectively eliminates custom-designed houses with elevated floors or ceiling heights over 2400-and OBs and architects-are excluded from the fast-track 10-day Complying Development system and are required to lodge a fully documented Development Application (DA) that often adds an extra 50 per cent to the cost of applying and means that approval in less than 40 days is unlikely.

In other states the systems used to assess development and building applications vary from place to place-always approach authorities as early as practicable to collect information and forms.

These hurdles may not have been placed to restrict or dissuade OBs but they certainly do not welcome them. OBs need to point out to their LGA officers that shelter is a basic human right and owner-building is one of Australia's great traditions. LGAs should have clear guidelines on what has to be done to ensure an approval-ask for them. Resist the drive by authorities and others with vested interests to limit the number of OBs.

There may appear to be a large number of house types in this country, but all of these are variations of six plan types (below). These types come in numerous variations: two, three storey, city and country versions, more recently (1960 onwards) split levels, courtyard, circular, pinwheel and pavilion. These types were built in a variety of styles or fashions by speculative builders who judged carefully the tastes and desires of the average family. Major developments in stylism were seen in the middle-class bracket, usually 3–4 years after they appeared in limited instances in well-to-do suburbs.

### Primitive 1788–1880s
Living, sleeping spaces, usually two rooms, with entry directly into living room. In South Australia, New South Wales and Victoria, a verandah usually ran across the front. In Western Australia and Queensland, the verandah ran around the house. Sometimes the verandah was infilled at the back or sides for a kitchen or additional sleeping quarters.

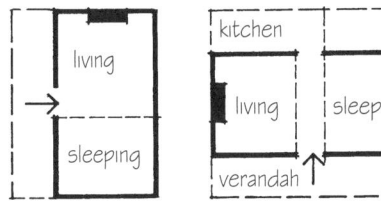

### Bungalow 1800s–1840s
Derived from the English cottage, the living and sleeping zones were separated by a central passage. Verandah usually ran around the house, and was infilled if additional accommodation was needed. Sometimes these plans were built in double-storey versions.

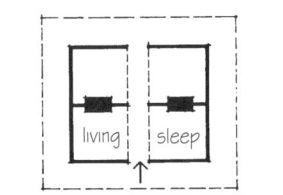

### Asymmetrical Front 1840s–1950s
Developed during the early Victorian era. Either the living or sleeping zone was projected forward. The side verandah was seldom incorporated and the front verandah shrunk until it was little more than an entry porch. The verandah at the back remained, was again infilled if additional accommodation was needed. This infilling led to setting down the concept of the 'extension' in society's mind.

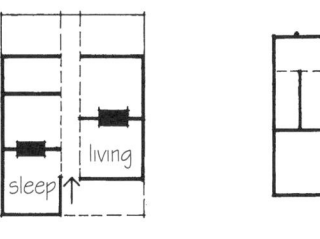

### L-Shape (and T-Shape) 1930s–Present
Originally developed in the 1880s as a servants' quarter wing and became vogue in the 1930s with the developing suburb. The bedroom wing was historically the projecting one, with living and kitchen at the back, the entry at the internal corner. This plan form lends itself to 'open planning'.

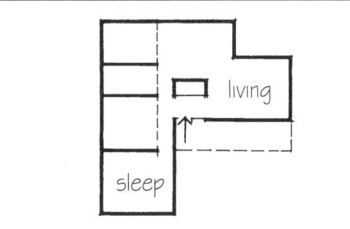

### Multiple Front 1945–Present
Originally developed for wider sites, evolved from turning the bungalow plan on its side and creating an entry through the bedroom zone. This form lends itself well to open planning. Progressively this plan form has been developed by the project home builders to suit modern sub 500 square metre allotments. The planning has become free-form and loose fit with living spaces often wandering through the house. Well suited to slab and truss construction, often with the garage under the main roof.

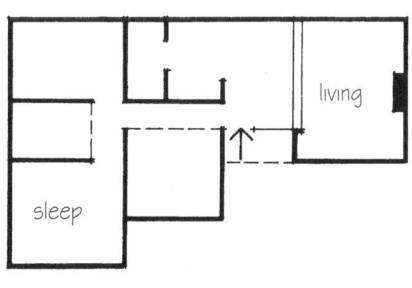

### Eclectic 1960s–Present
Evolved during the early 1960s by those who were anti 'project home'. Few examples until the early 1970s, when growing dissatisfaction with packaged houses made these houses 'socially acceptable'. Generally no rules, other than a dwelling tailored to the specific needs of a family and site. Can be both architect-designed or owner-designed. These house plans vary from neo-Classic styles to Minimalist design. Variations of eclectic designs have been developed to suit environmentally conscious designs.

# 4 • DOMESTIC ARCHITECTURE

## Architectural styles of cottages in Australia

| Style | Approximate period | Sketch | |
|---|---|---|---|
| Georgian Primitive | 1788–1815 | 1 2 3 | Constructed from available natural materials, few tools, initially rudimentary shelter but later became substantial dwelling. |
| Colonial Georgian (Regency 1811–1830) | 1810–1840 | 4 5 | Wattle and daub, both self-built and convict-built. Beginnings of echoing the prevailing style of the times. Arrival of tradespeople from England, better tools, more refined detailing, quality materials. |
| Gothic Revival | 1840–1880s | 6 | The break of Georgian symmetry. Gothic style competed with Classic. Gothic, the first conscious exotic or romantic style was a revival from the treatment of churches in Europe from the 14th to 15th century. Typical features high slated gables, carved barge boards, stucco and timber porches. |
| Italianate | 1850–1890s | 7 | Arrival of Italian craftsmen into the colony. This style became competitive with Gothic. It is typified by heavily ornamented facades, slate roofs. |
| Victorian | 1860–1900 | 8 9 10 | In later Victorian examples, cast-iron columns and lace work are typical. A front porch with metal roof is common. Toward the end of the century, ornamentation became heavier, slate roof often hidden behind a parapet, introduction of brick patterning with coloured glass feature panels around the entry door. |
| Federation (Queen Anne) (Art Nouveau 1902–1910) | 1890–1920 | 11 12 13 | Predominantly a brick finish, broken roof shapes with false gables, ornamented with terracotta work. Extensive timber work replaced the cast iron of the Victorian era. Early examples were geometrical but this gave way to free-flowing curves (Art Nouveau style). This spread from timber frieze work to lead light windows and fanlights. |
| Bungalow (Picturesque) | 1915–1940s | 14 | Californian import style, rugged homestead image, heavy timber work, rough textures. Lower pitched roofs, clad with terracotta tiles, metal in cheaper examples, dark brick was dominant in New South Wales; other states and NSW country areas used local 'red' brick or weatherboard. |
| Spanish Mission | 1925–1950s | 15 | Rendered brickwork, arches and colonnades, often with Baroque or candy-stick columns, Spanish-style tiles, boxed eaves and wrought-ironwork, extensive use of tiled flooring. |
| Suburban Moderne | 1935–1950s | 16 | Streamlined housing, also called Hollywood Jazz style—horizontal lines stressed, curved glass, combination render brickwork, steel windows, boxed eaves, circular windows. |
| Austerity | 1945–1954 | 17 | The exterior of houses lost their sense of flair, a result of rising costs and material shortages. Windows were of standard sizes, brickwork plain, the traditional house styling had come to an end. |
| Contemporary 50s | 1955–1965 | 18 | Renewed interest in the suburban house, development of open planning, houses changed shape—L-shape, flat roofs, use of picture windows. |
| Contemporary 60s | 1965–1975 | 19 | Project house era, brick veneer becomes the dominant building system, at least in the eastern states, and horizontal sliding aluminium windows tend to dictate exterior design. Family rooms are incorporated and the house size approaches 150 square metres. |
| Contemporary 70s | 1975–1985 | 20 | As the boom time of the early 1970s slowed down the project home mould was cast and the single-storey 'ranch'-style brick veneer cottages with concrete slab floor and tiled roof seems to be assured a long life. Planning has changed little over the years from the mid-1960s only the family room has been added and multiple bathrooms and fad kitchens. |
| Eclectic | 1960s–Present | 21 | This is OB territory—a bit of this and a bit of that. All in perfect harmony. Can be refreshing and exciting in the hands of an expert designer but is often a total mess when tackled by the inexperienced. |
| Federation Revival | 1980–2010 | 22 23 24 | A mixed style taking elements from the period around 1900s but using brick veneer, cement tile and aluminium windows. |
| Resort Style | 1990–Present | 23 | Anything goes but it must be open plan, colourful and have heavily modelled exteriors. |
| Environmental | 1990–Present | 24 | Considerate of a wide range of environmental issues. |

# DOMESTIC ARCHITECTURE • 4

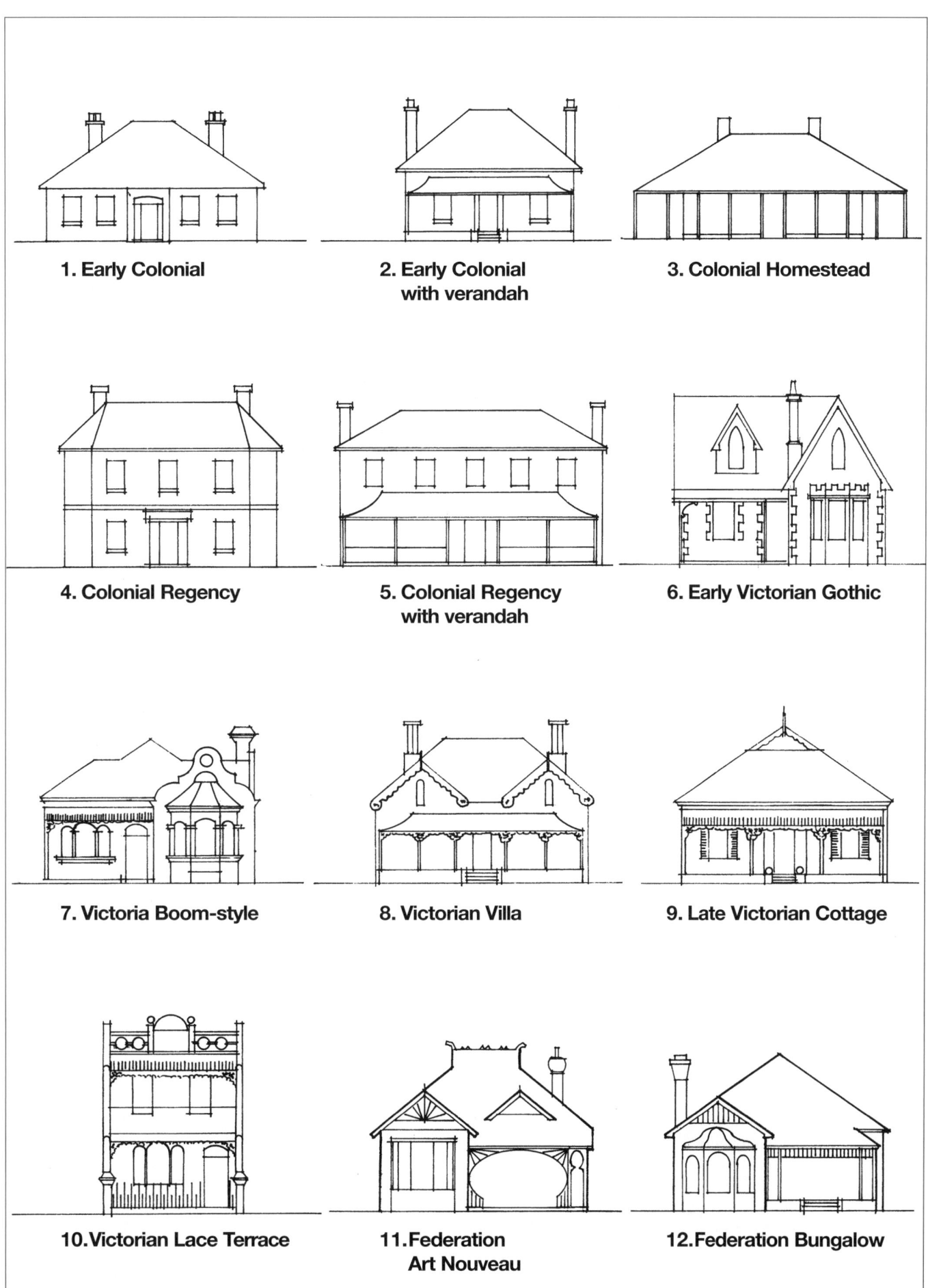

# 4 • DOMESTIC ARCHITECTURE

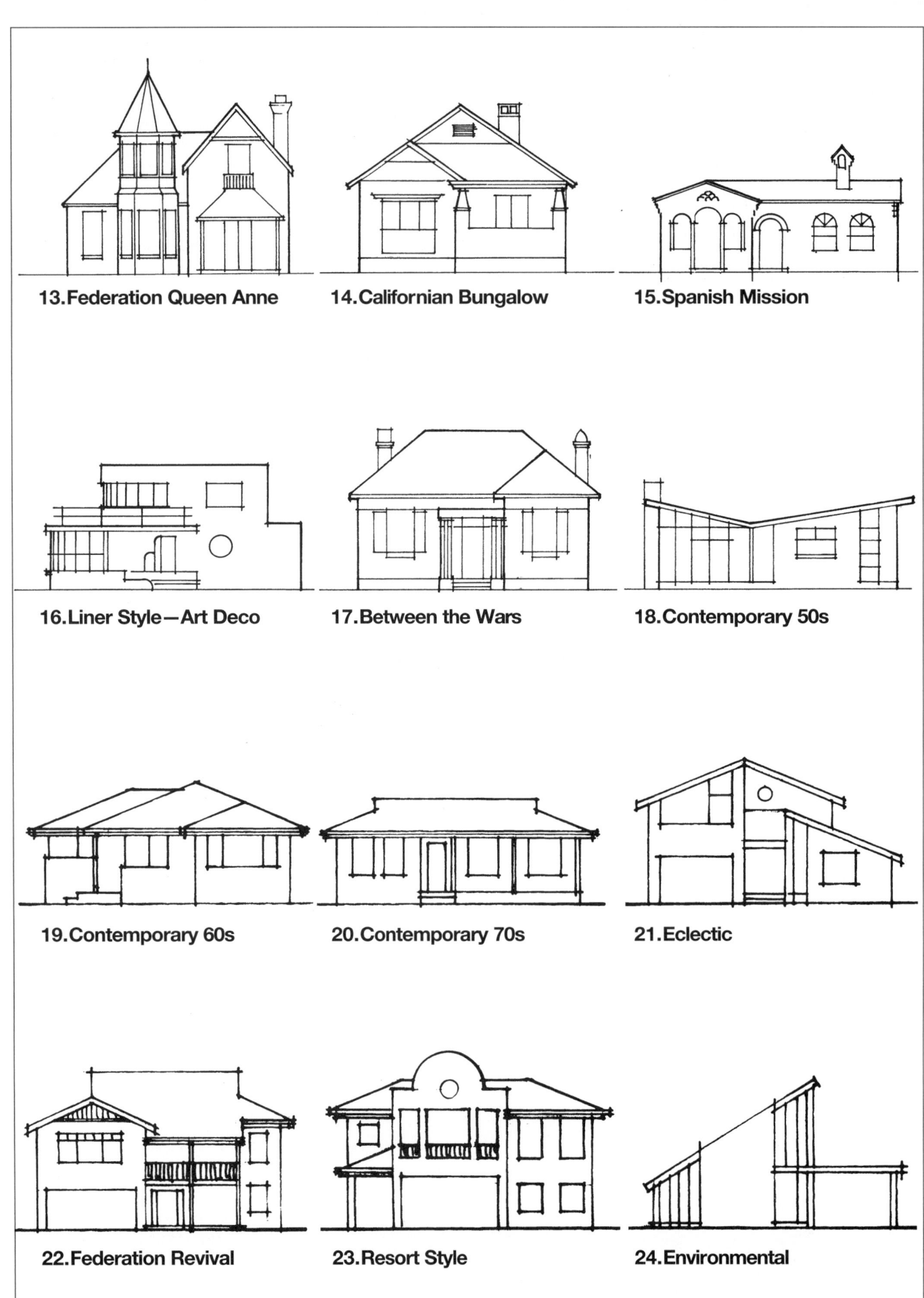

13. Federation Queen Anne
14. Californian Bungalow
15. Spanish Mission
16. Liner Style—Art Deco
17. Between the Wars
18. Contemporary 50s
19. Contemporary 60s
20. Contemporary 70s
21. Eclectic
22. Federation Revival
23. Resort Style
24. Environmental

# A HISTORY OF BUILDING MATERIALS • 5

The building industry is traditionally a conservative industry and this shows most markedly in the development of the materials used to build our houses. There have been few, if any, revolutions in the use of building materials. The process of the development and use of building materials has been a slow evolutionary one.

Building materials can be broken up into four main categories:

**1. Natural materials** – materials with little or no processing from the way they are found in nature. These would include the use of tree trunks and branches, the use of tree bark, the use of tree wattles, i.e., generally in the form of intertwined twigs, the use of mud and clays in either mud brick adobe or rammed earth pise construction, natural stones to build rubble walls and the like and the use of natural grasses for thatching.

These materials are used less today but are still found being used on the fringe of the OB movement particularly as alternatives to more expensive modern processed equivalents. There are, in Australia, movements that favour strongly built houses of mud bricks or houses on steeply sloping sites constructed on a forest of tree-trunk-like poles.

**2. Worked natural materials** – materials that occur naturally in nature but can be mechanically worked to make them more satisfactory. These are often have improved economical and flexible benefits: the worked natural materials; include sawn timber, dressed timber, slates, dressed stone, lime and gypsum. These worked natural building materials have been used in Australia almost since colonisation.

One of the very first industries set up in colonial times was that of timber sawyers and gangs of convicts were used to hand saw natural logs into square or rectangular building timbers. Another early Australian material production industry was quarrying. Sydney being a city built on a natural sandstone bedrock, made this material ideal for walls in our earliest buildings.

Sydney, as a beachside city, also had access to materials that were necessary to make the mortar on which to bed the stone and from the earliest days the naturally occurring lime in seashells was used in combination with the abundance of riverside sand in the area.

Early roofs were covered with split cedar or hardwood shingles but, as trade developed between Australia and Europe, empty ships returning to Australia to pick up cargoes of primary products were required to ballast their hulls, and one of the most convenient ballast cargoes that could be brought to the colony was split stone slate from England and Wales. For many years the only imported component other than furniture in an Australian colonial house would have been a slate roof.

**3. Processed natural materials** – the next stage in the working of natural materials and, as distinct from just taking the natural material and reforming it by mechanical means, processed materials generally need specific high-energy and often chemical processes.

Common processed natural materials used to in the Australian building industry include:

- reconstituted timber products, plywood, hardboards, chipboards, particleboards, fibreboards and laminated structural components
- Chipboard T&G flooring sheets have replaced timber strips as the most common timber flooring material, particularly where the floor is to be carpeted or tiled.
- medium (MDF) and the less common high-density fibreboard (HD3) are made from compressed timber fibres (generally *Pinus*) set in an adhesive matrix. They are used for casegoods, built-in cabinets and trimming mouldings.
- plywood was probably the first type of reconstituted timber product used in Australia. In the manufacture of plywood, the process is basically mechanical the continued development of improved adhesives has increased the range and adaptability of plywood and other glue laminated (glulam) structural products
- bricks manufactured from clay, concrete and silica lime
- terra cotta clay based roofing tiles
- cement and concrete based tiles
- cement and concrete combined to produce a range of pre-cast and insitu building components
- glass for windows
- metals, including ferrous and non-ferrous types, for structural, cladding, mullioning, tools and decorative elements.

From the 1880s until the 1950s, the commonly available bricks were kiln fired machine-pressed clay. Pressed bricks were slowly replaced by extruded wire-cut bricks. Although not necessarily an improvement in quality, this process allowed more controlled manufacturing process and a wider variety of colour and texture.

Concrete bricks and blocks have not gained universal acceptance in Australia, although they are a high quality manufactured product. Except in limited locations they are used almost exclusively in commercial work or as a secondary choice of product.

Silica-lime bricks have been made in Australia since at least the 1950s and were fashionable as *Colortone* bricks in the 1960s when they came in a standard brick-size unit in a wide range of colours including greens and blues. Silica-lime bricks are still available in their natural white colour, in sizes suitable for loadbearing face rendered internal wall skins, in some locations they are standard brick size in plan but twice a normal brick size in height. They are an economical alternative to clay bricks.

During the late 1980s autoclaved aerated concrete (AAC) was introduced into Australia from Germany, where it was developed. This is a cement-based product that has been foamed during manufacture by the addition of aluminium dust. The blocks and slabs are not cast to size but sawn from a large block of base material. AAC can be used to build ceilings, roofs, stairs and so on but it has its greatest acceptance in Australia as a wall and suspended floor material.

Roof tiles were introduced to Australia in quantity early in the 20th century. The first tiles that had mass availability were terra cotta clay tiles imported from France. These tiles are of a pattern that is still in common use and often referred to as the *Marseilles* pattern. Modern terracotta tiles come in a variety of different profiles and are often glazed to create other than the natural terracotta colour.

Cement roof tiles were introduced to Australia during the 1920s and for the next 30 or so years were of relatively unpredictable quality and had a tendency to crack, erode and lose the applied colour that was painted on them. Since the 1960s the development of the cement tile has moved forward significantly and modern cement tiles are now claimed to have at least the same useful life as the terracotta alternative and are generally offered in a wider range of colours. Some problems with colour fading are still apparent, however, particularly in the more gaudy colours.

Cement and concrete are two products that have undergone significant development in Australia. Cement was introduced in the latter part of the Nineteenth Century but the main impetus for the use of poured concrete in residential buildings came with the premixed concrete truck. This reduced the builder's on-site labour in mixing and pouring concrete and very rapidly non-concrete footings disappeared from most residences.

Another significant development in the use of cement for the residential building industry in Australia was the development of the fibre cement range of products. The development of fibre cement sheets allowed for very cheap, but sound, houses to be built very quickly. Although in recent times there has been less fibre cement used as a wall sheeting product, the basic concept has been developed to produce high tensile fibre cement flooring for wet areas, extremely durable fibre cement weatherboarding, a range of cladding profile sheets and fibre cement shingles as an alternative to the terracotta tile.

Originally, the fibre used in *fibrocement* sheets – fibro – was asbestos. This highly dangerous product was removed from fibro during the second half of the 20th century.

The manufacturers of modern fibre cement products indicate that all asbestos has been removed from their products and

# 5 • A HISTORY OF BUILDING MATERIALS

the new boards are no, longer a health risk if used to the manufacturers' published instructions. There are old asbestos cement sheets still built into many old houses; it is highly durable and does not rot. Take care if removing any old fibre cement sheets and **use only licenced operators who use approved methods for handling and disposal**.

The use of fibre cement product was. often considered a cheap alternative to, bricks and even weatherboard, and houses with the tell-tale battens over the sheet joints were less desirable. Then in the 1990s a new fibre cement product was released – *blue board*. This comes. in flat sheets that can be flush-jointed and coated with long-life, coloured, textured coatings. The blue board sheets have the added advantage of being strong enough to brace the building frame – a fibro revival. Currently the manufacture of blueboard sheets has been expanded and not all high tensile strength cladding sheets are blue and they are supplied in a wide range of sheet sizes and thicknesses. OBs considering using high tensile compressed fibre-cement products are advised to follow manufacturers recommendations as these are structural grade products and should not be treated lightly, they are heavy.

Most project homes in Australia have moved from the use of timber floor construction to the use of reinforced concrete floor construction, particularly with the development of the integral floor slab and footing system, commonly called the raft slab. Modern versions of this often use a waffle type construction, where blocks of rigid plastic foam are used to elevate the floor level of the house and to assist in accommodating ground slope.

Glass was introduced to Australia in about the middle of the 18th century and was initially brought out as ballast from England and other European ports. Today the glass that is used in Australian houses is either manufactured in Australia or imported from Europe and Asia, it is of very high quality and available in many forms to suit specific purposes. Window glass ranges from traditional drawn glass, float glass and cast glass. There has been a significant growth in the use of specialist environmental window glass products released in Australia during the C21; including products to reduce the impact of adverse acoustic conditions and for temperature control.

The technical development of windows, to ensure high efficiency, includes advances in reducing the heat transfer through framing materials by the use of insulation separators, long life gaskets, and durable sealants. OBs should consider carefully the orientation, local climatic conditions, privacy and energy conservation requirements, when designing their home.

Metal products have also provided important components and assemblies for the construction of houses. Wrought iron and wrought copper-based alloys have been used for fixings and fittings in Australian houses since the earliest European settlement. Hand wrought iron spikes, nails and other fixings and fasteners were precious commodities, they were used sparingly until the introduction of machine made nails.

Iron and steel fixings were prone to corrosion and were generally used where there was less exposure to moisture, this made them unsuitable for fixing roofing shingles and weather board cladding. Copper based alloy, brass and bronze, nails were used for these weather exposed fixings.

The shakes, shingles and slate roofs often relied on overlapping to achieve reasonable weathertightness. Only home constructed for people of influence and wealth could consider using lead, zinc and copper; flashing, spouting and tubing. These products had to be imported from Britain, this made them expensive and not easily accessible.

The use of lead flashing, gutters and lead based soldering of roof water collection systems introduced the danger of high lead content in drinking water. The high use of lead on roofs and in water reticulation systems remains a concern in many older buildings throughout Australia. OBs should take great care if they are working with older buildings to ensure all lead contamination is eliminated, *lead products were used in water supply systems for much of the C20.*

Corrugated iron was introduced to Australia in the 1850s, from that time zinc (galvanised) coated iron/steel metal products made a significant impact on the Australian building industry.

Australia has been a leader in developing profiled iron/steel based roofing products. The ubiquitous corrugated iron roof was used on houses from the tropical north, along the temperate east coast, on isolated homesteads throughout the outback and on Tasmania.

Corrugated galvanised (zinc coated) steel can be jointed using lead based solders and many existing galvanised steel water tanks still in use made still have lead leaching into stored drinking water.

Progressively the pure zinc galvanising has been replaced with more manufacturing tolerant alloy coatings and bonded paint finishes (Colorbond).

In the early 1960s the aluminium framed window began to make significant inroads into the timber-framed window market, by the 1980s the aluminium-framed window had all but superseded the timber-framed window in most cottage construction.

The quality of aluminium framed windows varied considerably over time. Many early frames were fabricated with basis grade mill finish extrusions, which discoloured with an unsightly corrosive coating. The quality of the joints, fixing and fittings was often so poor that it was the glass that provided the structure to the window as the framing was too flimsy. Window openings were often limited to horizontal sliding sashes.

Modern aluminium windows offer a wider variety of options, expect manufactures to offer:
- Technically efficient extrusions
- Efficient weather sealing
- Thermal barriers to improve thermal efficiency, reducing conductivity through the frame
- Thermal efficient and sound reducing glazing, including tinting and double glazing
- Welded frame connections and high quality durable hardware and fittings
- Openings are available in all sash types, including; sliding, awning, sash (double hung) and folding.

During the residential boom period of the late 1970s it became apparent that Australia was facing a shortage of good timber framing material and a number of manufacturers began producing alternative framing systems, using either steel or aluminium sections. These systems have become continually more refined and today are almost approaching the flexibility of timber as a wall framing material. However, their market penetration in most areas of Australia is still relatively small. Fluctuations in the housing market and the increased availability and use of plantation grown timbers, glue laminated sections, composite sections and *finger jointing,* has slowed the entry of metal framing in residential buildings and the pressure to replace timber as a framing material has somewhat diminished.

Timber engineered sections, framing and connections particularly in the project home industry has allowed for wider rooms due to lightweight trusses and fabricated long span joists. The total above floor mass of an Australian project home has been controlled, allowing larger houses without a significant increase in mass.

**4. Synthetic materials** – these materials have no semi-natural content and include those items loosely termed plastics. From the building industry point of view there has been relatively little impact from plastics, although the use of laminated plastics as surface coatings on kitchen and other joinery products has been significant, and high-quality plastic piping was introduced into drainage and plumbing tubing. And connections.

## WALLS
**Then**, walls were built on-site from rough-sawn local old-growth timber or imported conifer timbers. There were many interlocking timber joints and the nails were often handmade.

Timber weather board cladding was local durable native conifer timbers, including the highly prized Australian Cedar and native pines from Tasmania, though some were made from imported Red Conifer timbers or even some native hardwoods such as Jarrah in Western Australia.

**Now**, walls may be built on-site but the majority of stud wall frames are pre-fabricated

# A HISTORY OF BUILDING MATERIALS • 5

from plantation *Pinus* or Eucalypts and delivered to the site ready for erection.

Plantation-grown Cypress or *Pinus* (often treated to resist termite infestation) are the common real timber weatherboards. Alternatives are manufactured in traditional profiles from fibre cement or hardboard; there are weatherboard profiles in aluminium and plastic but these struggle for acceptance.

## BASE

**Then**, if a brick base was used the bricks were handmade and the mortar was a sand/lime mixture; interestingly, the bricks were much the same size as now (9 inches (230) x 4.5 inches (110) x 3 inches (76)). In many places, houses were supported on durable timber stumps – Red Gum, Jarrah, Iron Bark – from old-growth forests. The floor frame was universally constructed with a simple structure of bearers and joists from the most durable hardwood timber locally available. The flooring boards varied through a wide range of timbers from clear conifers through rich red Eucalypts to highly durable Tallowwood.

**Now**, the brick base and the timber floor frame have changed little; however, the growing shortage of durable hardwood timer has increased the use of galvanised steel floor frames. The flooring is commonly tongue and grooved particleboard sheets glued and screwed to the frame. If a polished timber floor is required, it is often fixed over these sheets. The available timbers are more limited as the best trees have long been cut out of our forests but beautiful recycled timbers can be purchased for a premium. There are plywood boards with a thin veneer of a real or manufactured identifiable timber.

---

### The development of the main building materials used in residential house construction in Australia

#### Bricks
The first bricks were made in Australia within a few short months of the First Fleet dropping anchor in Sydney Cove, and some of these original bricks have recently been uncovered during archaeological excavations on the site of the first Government House in Sydney.

The early bricks are often referred to as 'sandstock' not so much because they contained sand, but because sand was an essential part of the handmaking process that was used in their moulding.

By the turn of the 20th century bricks were being manufactured in all states of Australia. By this time most of the bricks were machine pressed moulded and fired in sophisticated coal-fired kilns, which gave a strong and almost infinitely durable product.

After the World War II brick manufacture in Australia embraced the extruded wire-cut brick moulding technology that allowed, in construction with modern oil or gas-fired tunnel kilns, the production of an almost continuous brick manufacturing process.

#### Timber
The first settlers were dismayed at the hardness of the local Australian timbers for, with the tools then available. They were exceedingly hard to work and required that tools be continuously sharpened.

The few easy-to-work species, including the great cedar forests of eastern Australia were soon cut out and a timber-importing industry was developed.

Currently a significant proportion of building timbers are imported, but this has been modified by the plantation program of Radiata Pine and specifically bred Eucalypts suitable for plantation agriculture. Only Queensland and Western Australia have use a of indigenous timbers in house building, the other states make wider use of plantation and imported species. The common imported species included Oregon from the United States, Canada and New Zealand, the latter being plantation grown. Other species have included Canada Pine (Hemlock) from Canada (and other sometimes suspect sources) and Meranti, and other rainforest species from the Pacific Islands and South-East Asia.

#### Iron and Steel
The use of wrought and cast iron dates back to the early days of Anglo-European settlement, when it was all imported from England and had to be used sparingly. With the development of the rural agricultural industries in Australia, particularly wheat and wool, ships began looking for high mass ballast to load their ships for the return journeys from Europe. Iron was a very suitable cargo and the growth of the use of iron in Australia can be traced to this period.

Cast Iron for decoration was among the first of these ballast cargoes but, as technology improved in both Australia and England, pig iron for local casting and sheet iron for profiling began to be available in Australia.

Eventually the iron was improved and steel became the base product for sheeting and emerged as a highly suitable structural material in beams and columns. Late in the 19th Century, iron was smelted in Australia for the first time at Lithgow in New South Wales. By the beginning of World War 1, the manufacture of iron and steel was well established in Australia.

#### Cement and concrete
It is hard to imagine building without the aid of cement and concrete, but these two important building materials, although known to and used by the Ancient Romans, were not rediscovered until 1756 in England. Modern Portland cement was not invented until 1824 and took until late in the 19th Century to come into common usage.

Although cement and concrete were used in the construction of Australian homes from the turn of the 20th Century, it did not attain the dominant position that it holds today until the 1950s when cement finally replaced lime as the main material in bricklaying mortar and continuous concrete footing replaced bonded brick footings.

#### Fibrous and board plaster
Finishing of masonry houses had traditionally been by the wet application of rendered plaster directly to the masonry and to this day this method is still applicable in some solid masonry homes.

During the Victorian era in Australia, heavily decorated ceilings and cornices were in fashion and tradespeople, in an effort to reduce the on-site problems associated with running elaborate plaster cornices in place, began to experiment with moulding the cornices at ground level and then lifting them into position. This process needed a material to reinforce the brittle plaster and a fibrous materials including sisal and horse-hair were cast into the wet plaster.

Sheets of plaster could also be cast at ground level if sisal was used to reinforce the sheets. Progressively the manufacture of fibrous plaster sheets became a factory process. Motorised trucks became available and allowed easy transport to building sites. So started the fibrous plaster revolution in Australia that was to see lath and plaster almost disappear by the outbreak of World War I.

Fibrous plaster dominated the wall and ceiling lining trade in Australia until late in the 1960s, when it was replaced by paper-reinforced plaster sheet, called plasterboard. Plasterboard can be produced by more mechanised methods than fibrous plaster and offers the advantage of a useable face on either side of a sheet.

# 6 • A HISTORY OF BUILDING CONSTRUCTION

Historically speaking there were two structural systems available to building designers:

Mass construction, in mass (masonry) construction the loads of roof and floors were carried to foundations by means of walls, which also provided the weather protection envelope to occupants within the building.

Frame construction. in frame construction, the loads of roof and floors were carried by a frame that concentrated the loads until they were redistributed by the foundation. The weather protective envelope was not load-bearing (but carried its own weight) and could be independent of the frame or combined with the frame by in-filling spaces between the members. Both of these methods and their developments are used in Australia, but we also devised local variations.

### EARLY ROOF FRAMING

16th Century King Post

17th Century King Post and Struts

18/19th Century Tie Beam and light collar tie roof

18/19th Century Tie Beam with angle struts

### English masonry construction

More than any other single component, the choice of wall material establishes the character of an architecture style. It was the wide range of masonry materials available in England that gave unique variety to vernacular construction. The problem with the mass walling or masonry, was the laying of walls from separate pieces of material (stones, blocks or bricks, while achieving a high degree of stability. Stability of foundation was not usually a problem for small, low buildings erected by masonry construction methods. For foundational stability, large boulders were common at the base of masonry walls, to evenly distribute the building mass onto the foundation and to provide a level plane. {what we now call footings).

Collapse of walls by outward buckling was avoided by using heavy floor beams that tied opposite walls together (a reason why such heavy beams were used). Outward pressure on the top of the wall was countered by the use of heavy tie beams, as part of the roof. These became members of a stable balanced triangulated roof truss arrangement, whereby loads were transmitted vertically down through the walls. To avoid the external walls of a building bucking under load or rotating through outward forces at the roof to wall junction, external walls were built with significant thickness, had regular buttressing piers or made use of internal cross walls where the building could be divided into cambers or rooms.

The strength of masonry walls was reduced by openings (windows and doors) and there became a limit to the number, size and proportions of these openings – in masonry walls, this contrasts to frame construction, where every panel could be theoretically transparent or open.

However, adequate the design proportions against collapse, the strength of a mass wall depended on the concerted action of the individual units. In England good quality stone, which splits naturally into block shapes, presented few problems whereas, in Australia, stone was not of this high quality – and not plentiful.

Although many colonial period buildings in Australia used stone masonry, the material that became the essence of masonry walling in this country was fired clay brickwork. Brickwork had advantages over other forms of masonry construction; it was cheap, could be built in a way to give decorative effects cheaply, and was durable. Australia is fortunate to have a good range of clays, so brick construction flourished. Progressively the size and quality of bricks became standardised and evolved from solid mass walling to the more climate proof cavity wall system

The roof loads of masonry buildings were carried onto the load bearing walls by assembling heavy timbers into a roof truss arrangement. This was common practice from the 14th to the 17th centuries. During the 18th century there was a tendency for roof trusses to pass out of use for domes tic buildings. Purlins carried by the gable and intermediate walls, usually some variation of the basic triangle of tie beam and principle rafters, were used, with a light collar tie or with angle struts (see illustrations). In all cases, a heavy ridge was used and side purlins were supported on the backs of the

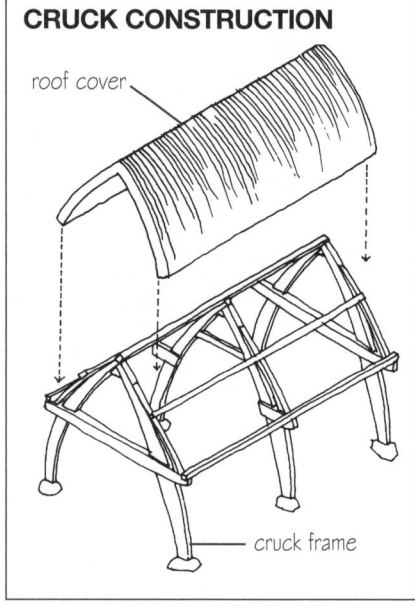

### CRUCK CONSTRUCTION

roof cover

cruck frame

### ENGLISH MASONRY CONSTRUCTION

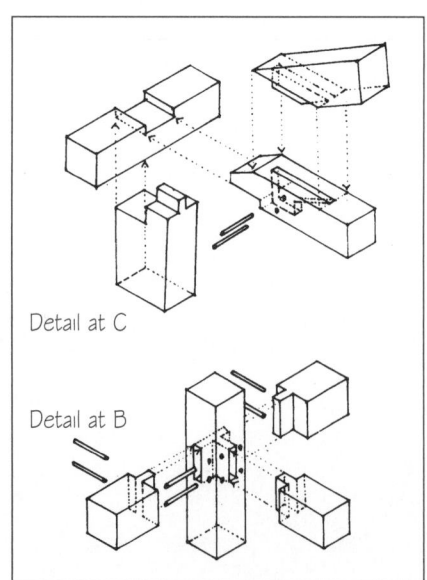

Detail at C

Detail at B

# A HISTORY OF BUILDING CONSTRUCTION • 6

principle rafters, not tenoned or passing through the rafter as was previous practice. Although roofing systems have been improved, the principles have not changed.

**Heavy timber construction**

Two types of frame were used – the cruck frame and the box frame. The former roof loads were transmitted by means of inclined crucks to the ground. In the latter they were taken on framed walls. It appears that the process of erection was radical in the two versions.

In cruck construction the A-frames were assembled on the ground then raised one by one into a vertical position so that ridge, purlin, side purlins and wall plates could be dropped into their sockets to tie the frames together. In box frame construction the posts were erected separately and propped into position. Next, the wall plates were placed onto the top of each post and finally the tie beam was dropped into position, simultaneously locking post and wall plates. The intermediate studs were placed piece by piece at the same time.

Box frame construction methods were taken to America during the colonisation period, although the framing system under went radical evolution on American shores. The Americans developed a lightweight timber framing system that they called balloon framing. This was composed of many small, closely spaced members that could be handled easily and assembled quickly by nailing instead of the more cumbersome box framed dwelling. The balloon frame is based, as was heavy timber construction, on post and lintel principles. Posts rest on a waterproof footing (usually masonry) on which a sill or base member is attached.

The upper storey is laid on cross beams that are supported in the exterior wall by horizontal members. Interior walls provide additional support. The light frame is then sheathed with vertical or horizontal boarding, which is over lapped for weather protection. This helps to brace as well as protect the frame, so the frame is not structurally independent.

**The American balloon frame**

There were originally three main types of timber-based wall construction:

1. horizontal log, in which the wall was composed of solid timbers laid one on top of the other and jointed at corners
2. post and plank, in which the wall consisted of a series of heavy planks slotted between even heavier posts.
3. timber frame:
   - heavy timber construction in the 10th to 18th centuries.
   - American balloon framing in the 18th to 19th Centuries.
   - platform framing in the 19th to 21st Centuries.

The characteristics of the balloon frame are:

1. studs (vertical members) that rest on the sill extending in one piece from it to a top plate, regardless of whether it is a single or double-storey dwelling.
2. studs that are usually 100 x 50 mm at intervals of 300, 450 or 600 mm.
3. horizontal plates (top plates) at the top end of the studs that supports the roof rafters.
4. ground floor joists that rest on the sill plates and upper floor joists resting on 150 x 25 mm ribbon boards placed horizontally between the inner faces of studs and corner posts.

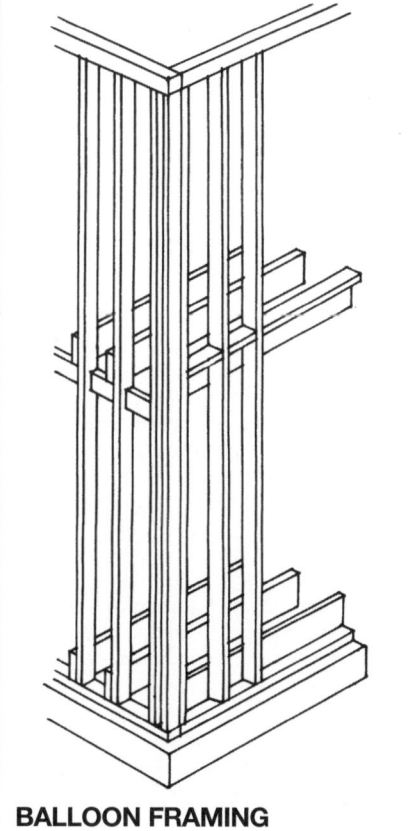

**BOX FRAME CONSTRUCTION**

5. topplates 100x100mm members and carry the weight of roof rafters.
6. a roof system essentially unchanged from that of 18th century construction in England.

The primary advantages of the balloon frame come from a saving in labour and materials. It has a construction advantage of a relatively small amount of vertical shrinkage. Balloon framing was developed into a more sophisticated framing system called platform framing. In this system the entire ground floor frame can be completed before work on an upper floor frame has to commence. This is achieved by placing the ground floor joists on the sill and sheeting them in timber or plywood. The ground floor studs are erected including the top and bot

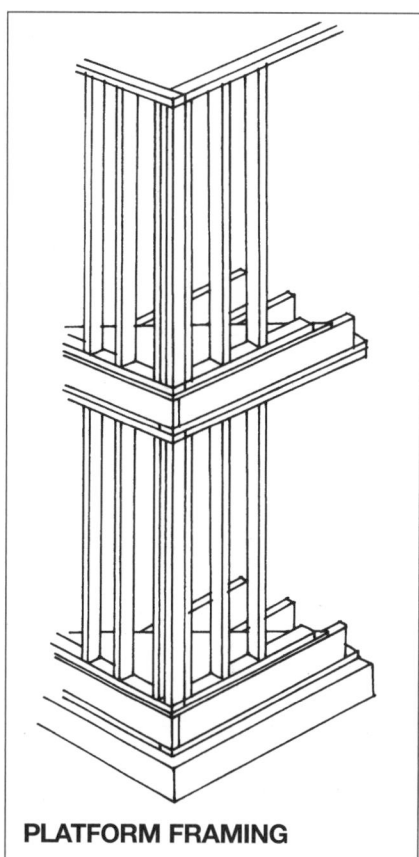

**PLATFORM FRAMING**

tom plates. The upper floor frame can then be constructed in a similar manner. The advantage of platform framing over balloon framing is that it provides a work platform from which to complete the construction.

**Brick veneer construction**

The method of cottage construction called brick veneer was developed and refined in Australia. It is the meeting of two historical methods of construction that were used in Australia up until World War II:

1. timber-framed construction clad with either weatherboards or fibre cement sheeting or similar cladding systems ;
2. the solid brick house generally constructed with brick external and internal walls from the footing to the

**BALLOON FRAMING**

# 6 • A HISTORY OF BUILDING CONSTRUCTION

underside of the roof structure.

In Australia after World War Two, the industry suffered a major shortage of building materials from 1945 through to the early 1950s this shortage of materials caused many architects and builders to look for hitherto unused and often untried methods of construction. The demand for housing at this time in Australia was very high, partly because of the baby boom immediately after the war years and also the mass immigration of European families looking for a new start after the devastation of the war.

No one seems to know where the first brick veneer cottage was built in Australia. Most likely it was in Melbourne, for it was in that city that this construction method was refined over a few short years and then exported to Sydney and slowly then to other states. Brick veneer construction takes the best of both timber framed and all-brick construction and combines them into a very flexible and economic system.

Brick veneer cottages are constructed with a timber frame, exactly the same as a timber frame would be constructed for cladding in weatherboard or fibre cement, but instead of this cladding a one-brick (110mm) thick skin is built around the outside of the timber frame, just clear of it (25 to 50mm), and tied back to the frame.

Therefore, the brick veneer method of construction allows the flexibility and adaptability of timber framing combined with the long life and low maintenance qualities of brick.

Over the years since the early 1950s, when brick veneer construction emerged as the dominant cottage construction method, the system has been continuously refined. One of the major benefits of the brick veneer system is that clear separation of sub trades has been achieved. This means that OBs, when using this method, can isolate the areas they require specific assistance, to let subcontracts on, and be reasonably assured that the tradespeople will understand the extent of the work required. Modern brick veneer construction has become so sophisticated that today individual trades subcontractors would not expect to attend the site on more than two clearly defined periods to carry out their works.

## Abba houses

The 'abba' house or the 'all bricks, balustrades and arches' style has developed in most Australian cities from the early 1970s through until the end of the 20C. This style is a major departure from the lightweight and flexible system that evolved with the development of brick veneer construction and is a reversion to heavyweight all brick and concrete construction more akin to residential flats than to residential housing in Australia.

Abba houses are generally solid brick buildings with concrete floors, double arched garages, projecting concrete balconies surmounted with classic revival balustrades and all topped off with a French-patterned terracotta roof.

The abba style construction method is favoured by people who want a solid-feeling house and are prepared to forego planning flexibility both initially and for all time, for it is virtually impossible to modify the interior layout of an abba-type house without confronting major structural engineering problems.

Generally, the abba-type construction method is not ideally suited to the OB, as it requires predominantly skilled tradespeople to carry out the majority of its construction and therefore denies the OB virtually any opportunity for hands-on construction.

## Faux-Federation and resort-condo styles

In the 1990s the predominant new house style in many parts of Australia was one that borrowed heavily from the styling elements of the Federation period (1901 to 1920. It mimics the fretted timberwork, the window glazing, colours and general proportions of Federation houses. These houses are brick veneer with steeply pitched, terracotta coloured, cement-tile roofs. Unlike their single-storey historical model, these two storey constructions have low ceiling heights rather than high ceilings. The simple project home design and construction techniques applied to the production of these houses allows them to be easily reproduced by OBs.

When climatic conditions make heavily decorated red brick houses unsuitable, a new set of clothes was used to dress up the same basic two-storey, brick veneer plan. A pastel-painted resort-condo emerged in the warmer latitudes. These are also winners for OBs: simple planning; lots of large diameter, tubular columns; over scale latticing; pergolas and half a dozen Cocos palms.

## Environmentally and ecologically sensitive houses

There has been a movement in Australia that grew to prominence during the second half of the twentieth century and it has been strongly supported by OBs-eco housing.

Houses in this group have no particular style, although they are often novel and sometimes radical in appearance and use of materials – they are always striving for a low impact and a minimum wastage design.

There have been many books written on the design of this type of house, some that are serious and provide excellent information.

In considering eco-housing, every location has different advantages and constraints, so all solutions should be unique. This means that the mass-produced project home is unlikely to provide an adequate solution for families who are eco-sensitive.

OBs who tackle their own eco-sensitive designs have not always been as environmentally and ecologically sensitive as they desired, and there is a scattering of their clumsy efforts on bush blocks at the edge of cities.

A partnership between an architect experienced in designing eco-sensitive houses and an OB with a desire for a responsible solution to their housing needs will often lead to highly suitable results.

Modern eco-sensitive houses do not have to be based on hippy commune concepts, nor should they become project homes with a three-and-a-half start energy rating that satisfies some bureaucratic formula. They can be worthwhile and still be modern and convenient.

Using sensible application of solar design principles, low wastage materials and techniques, recycling, plus human and environment safe products – should be everyone's goal.

The dangers facing OBs becoming involved in environmental and eco-sensitive residential design are manifold.

Three particularly confusing areas are:
1. Poor information sources, particularly books written in the Northern Hemisphere for places with climates much different to Australia, and with the sun in the southern sky.
2. There is confusion on what are the scientific facts, on environmental and ecological issues. Some of the New Age philosophies that creep into the eco environmental literature are based on less than perfect study. Concepts such as Feng Shui may assist some people in better understanding space/time issues but they do not change the facts on how physical and chemical actions work. Always right concepts, including all windows are better if they are double-glazed and thermal mass is critical to a well designed environmentally conscious residence. This is a dream come true for the project home builders. They can make extra money selling double-glazing and special internal brick walls all in the name of eco-environmental sensitivity. Everything has its place and it is possible to design a nature-sensitive residence without double glazing, thermal mass or even bulk insulation if the environment is fully understood. Take care of any one concept fits all situations philosophies.

## Conclusion

OBs would be well advised to consider timber-frame construction with board, sheet or brick veneer cladding, for these methods of construction will give them the most opportunity for flexibility both in design and construction and allow a wide range of hands-on trade classifications to be carried out by themselves.

# BUY, BUILD OR EXTEND?

The great householder's dilemma. what is best to do: buy a house and land package, build a new home or extend an existing home? The answer is yes. no or maybe.

There are advantages and disadvantages in each of the three basic routes to home ownership. These advantages and disadvantages vary significantly from house to house. from location to location and from family to family.

## Buy

This is not an OB's option. but there are many families that would much rather see the home they wish to purchase in its final built form before they purchase it. This market is satisfied by the existing house real estate market and by the project home builders and land packages – as well as the flat and unit market.

If a family purchases a house and land package, they are generally potential OBs, as the likelihood that they will extend or in some way modify it is high. However. the flat or unit purchaser is generally bound by the original layout and size of unit.

There is of course the family who will purchase a house in an especially desirable area (or for a very low price) with the intention of making alterations and additions to the house immediately or at least. in the short term.

## Build

In this category it is assumed that a new dwelling is built on a block of land. It could be a house on an infill block in the built-up urban fabric or a house on a newly subdivided block, farmlet or rural holding.

There are many ways a family can have a house built on their block of land: from employing an architect to prepare plans, call for tenders and administer the contract. to all processes being completely in the hands of the OB.

The family that chooses to become OBs should have weighed up the likely longer construction time that it will take as an OB. as well as the complete erosion of leisure time before being seduced by the potential savings.

The build category really has a further two sub-groups of factors underlying it. The first is the custom design category – which includes architect-designed and owner-designed houses. The second is the pre-designed house, which includes all project and kit homes.

OBs can either have a custom-designed house – designed by an architect or themselves – or a pre-designed house if they purchase a kit home as the basis of their OB activity

## Extend

The extend option generally requires a house to start with, although it is noted above that some families buy houses with the predetermined requirement to extend them.

Some houses are much easier to extend than others. and great care must be exercised when determining the suitability of a residence for extension. There must be a point where the cost to extend outweighs the benefits: demolition or relocation can become more desirable options.

The house that is to be extended should be structurally sound and free from any expensive defects. It should have an appearance acceptable to the family or further funds may have to be diverted from purely extension activities to cover the costs of visual improvements.

Most houses can be extended at ground level. assuming the land area and slope allow it. But it is more complex when upper floor additions or roof conversions are considered. Although most houses can have a second storey added. particularly timber-framed and masonry houses built between 1930 and 1970. Some houses need such extensive modifications that the cost outweigh the benefits.

Houses built using the Light Timber Framing Code – AS 1684 – and lightweight timber roof trusses. which have become common since the 1970s. are mostly too difficult and too expensive to support an upper floor conversion – due to the lack of internal load-bearing walls and because roof trusses generally require all members to remain intact for structural integrity.

OBs considering extensions should contact an architect, structural engineer or other building professional to survey the house to be extended before becoming committed to this option.

## Dual occupancy

Dual occupancy is the building of two dwellings on a block of land that was originally zoned for a single house. In some cases. the building of a new house alongside an existing house is accepted by the neighbours and the market. while in others there are difficulties.

For owners of houses on large blocks of land that have easy access to the rear of the property. It is tempting to consider a free-standing granny flat. villa or even a full-sized family home on the surplus land. There may be benefits including extra income from rent or sale, reduced cost of caring for elderly relatives (and they are close by if help is needed), capital gain and less garden to maintain.

Take care. when analysing the cost/benefits. The original market value of the property plus the cost of production should be less than the new marketable value of the property.

Developers buying a block of sufficient width can often get a better capital gain by building semi-detached dual housing rather than one house on the land. However. legislation in some states limits OBs to building houses for their own occupation and requires all other residential construction to be designed by registered designers/architects and that the construction is carried out by licensed/registered builders. Check the local rules

## Conclusion

The dilemma to buy, build or extend cannot be easily addressed. Each project has its own set of costs and benefits. These must be equated to the requirements of the family. All houses are compromises and a family should attempt to find. build or extend a house that has the least number of compromises.

OBs have the option. when they build or extend. of making adaptations as they go, to tune the compromises out of the project. Therefore, they are most likely to get a home closely tailored to their requirements. However. they should make sure that in achieving a house so heavily tailored to their requirements that they have not created a house that no other family would ever buy in the future.

Family homes remain an asset that should always be considered with an eye to the future.

Family requirements will change over time and with a family home often being the largest asset a family has it is wise to make sure it has the greatest flexibility to provide the family with future options; to Sell, Extend or Alter

# 7 • BUY, BUILD OR EXTEND?

**Cost or benefit?**

All families faced with the need for extra space. extra facilities. better kitchens and bathrooms, or even new curtains and floor coverings should analyse the cost of carrying out the work required at their present address. Compare this expense to that of picking up and moving to a new address. Moving often appears more radical than renovating, altering or extending, but all options should be considered if the most satisfactory financial outcome is to be achieved.

If the existing house is a small two bedroom with a living room. dining room. kitchen and bathroom. in a locality where the demand is for three and four bedroom houses with an ensuite bathroom off the main bedroom and a combined kitchen-family room. then additions seem logical. If the existing house has a low debt load then it may even be able to fund the mortgage value of the cost of the new work without stretching the family budget.

If the house can be made to fit the local market demand, at a cost that still leaves a surplus between the market value of the house before the additions plus the total cost of the additions, and the realisable market value of the improved property – less all costs of sale then there is sound economic sense in building the addition.

Economics do not always line up neatly with family finances. Unless the family can meet the repayments on any borrowed funds used to finance the additions. the additions are a poor choice. even if there is a potential profit when the house is eventually sold.

Spending hard-earned savings on extra space in a house has traditionally been favoured by Australian families over making quality improvements to a house without any increase in size. Many houses in middle socioeconomic localities have not had major improvements to their bathrooms or kitchens since they were built, but many of the same houses have had extra space added.

In the more expensive residential areas. well-presented houses with well-designed bathrooms and gleaming kitchens attract buyers there is a sense that most people would prefer higher quality living environments. as long as they do not have to be involved in the process of planning and constructing them.

In other parts of cities where there is closer settlement, terrace houses and apartments, there is an established demand for value-added houses and flats. Once unflatteringly called gentrified slums, many high-density localities have well-established premium prices. In these locations, space is at a premium and high-quality alterations and renovations are the proven way to tempt up the market value of residential property.

Alterations and additions will disrupt a family's daily routine throughout the construction period. even if the family can afford to leave the noise and dust far behind. Often alterations and additions are the best choice because no matter what the cost of disrupting a family's routine of schools, churches, social life and general attachment to their house, a more suitable home is assured. A family that has been through major alteration and addition activity at their home will generally have a heightened sense of achievement.

Answer the following questions. They will solve nothing, but hopefully they will help you avoid dreaded over-capitalisation. Altering, adding-to or moving-on are emotional decisions, but be careful that your emotions do not get the better of you.

- What did it cost to buy your home?
- For what price can you sell your home today?
- How much will it cost for another house with the features you want?
- How much will it cost to alter and/or add to your home to get what you want?
- For what price can you sell your home after the alterations and additions? One year on, three years on, five years on?

# PRE-DESIGN • 8

## Holding a family/stake holders conference

The idea of holding a family conference is to ensure that all the perceived needs of all people with a stake in the project are discussed.

This is a critical step in the process of achieving a satisfactory result when undertaking the preparation of the scope of the works and design brief for any building project. It is likely that the images in the minds of the adult members of the household will vary significantly from those in the minds of the teenage and child members.

Everyone's view is valid and all views should be discussed. Adults and those who control the funds should take care to not dominate the conference. The decisions will affect everyone living in the house, so it is critical there is a sensible degree of tolerance and consensus among all members of the household.

Family members may have hidden agendas when considering new or extended houses. Spouses may want a retreat area where they can be together, close to the rest of the family, but with privacy. Parents may want to create a house where their children will want to stay for an extended period of time rather than leave home and set up their own apartment. Children may just want their own bedroom because they are sick of sharing with a sibling.

What is happening in the above examples is that householders with a specific relationship problem or problems are trying to solve these interpersonal glitches with a physical solution. Take care; this will not always work.

Before the family conference, try to get everyone who will be involved to write down what they want from the new or extended house.

Can a spouse say that what they want is anything that will entice their partner to spend quality time with them? A retreat may just end up an expensive way to watch TV alone.

Can parents discuss with their teenagers: who is allowed to visit the family home; the amount of modern music acceptable before the threshold of pain is reached; who is allowed to stay over and do they have to have breakfast with mum and dad the morning after? Young adults may stay at home if they have freedom of association in the family home, but a bigger bedroom and a grand ensuite may not be as attractive as a live-in relationship with a special friend.

Can a child express the need for personal space in a way that is not translated into a three-metre square box with a built-in robe on the south side of the house? It may stop the midnight squabbles, but does it address the real problems – studying, TeeVeeing, music, junk, friends, smells? Is it a bedroom or an activity space? How many children really want a private sleeping space? Is this just the politically correct way of saying 'just let me be alone, to do my own thing – but don't close the kitchen'?

Deal with the issues as honestly as possible; not all issues are suitable for completely frank discussion and the parents' agenda is likely to prevail. Listen carefully to what the every member of the family is saying and try to provide an economically sound list of preferences.

After the participants in the family conference have prepared their submissions, the gathering itself can be free-for-all. Someone should take notes of the most important points and identify where there appears to be a reasonable level of consensus.

There will always be a lead time from concept to realisation. It is difficult to leap through to the completion of a new home or home extension in less than six months in any part of Australia, but it is not impossible. More likely the process will take 12 to 18 months.

## NEEDS CHART

The NEEDS chart is a way of allocating priority to needs. This chart is to assist families in seeing the BIG picture; it is not the definitive answer to everyone's problems.

| NEEDS | No. required | Size | Priority | | |
|---|---|---|---|---|---|
| | | | High | Medium | Low |
| **Living Spaces** | | | | | |
| *Sitting Room* | | | | | |
| large—30 sq.m+ | | | | | |
| medium—20–30 sq.m | | | | | |
| small—under 20 sq.m | | | | | |
| *Dining Room* | | | | | |
| large—20 sq.m+ | | | | | |
| medium—15–20 sq.m | | | | | |
| small—under 15 sq.m | | | | | |
| *Family Room* | | | | | |
| large—30 sq.m+ | | | | | |
| medium—20–30 sq.m | | | | | |
| small—under 20 sq.m | | | | | |
| **Sleeping Spaces** | | | | | |
| *Main Bedroom* | | | | | |
| large—20 sq.m+ | | | | | |
| medium—15–20 sq.m | | | | | |
| small—under 15 sq.m | | | | | |
| *Other Bedrooms* | | | | | |
| number required | | | | | |
| large—20 sq.m+ | | | | | |
| medium—15–20 sq.m | | | | | |
| small—under 15 sq.m | | | | | |
| **Kitchen** | | | | | |
| lavish | | | | | |
| sensible | | | | | |
| budget | | | | | |
| **Bathrooms** | | | | | |
| full | | | | | |
| medium | | | | | |
| small | | | | | |
| ensuite | | | | | |
| **Laundry** | | | | | |
| **Other Spaces** | | | | | |
| | | | | | |
| | | | | | |
| | | | | | |

# 9 • DESIGNING THE HOME

## How do OBs approach the task of designing their house?

1. Use an architect

This is the obvious way to go, but there are constraints, not the least of which is the architect's fees.

Architects are of particular value when OBs desire to build a home that is not conventional in plan and/or construction or they wish to build on a block of land with particular problems including, steep slope, unstable foundation, remote location, or in a restricted area.

OBs may find that an architect can assist them in reaching a design that either reduces the size or complexity of their home or scheduling the construction stages, to reduce subtrade attendance and/or recalls.

2. Building designers

Many people – from graduate but unregistered architects to the completely unqualified – advertise as building designers. Poorly qualified people may formulate adequate house or addition designs, but results can be unsatisfactory.

Many house designers/documenters use computer-drawing packages that generally ensure accurate working drawings and seductive perspective views. The lack of flexibility in computers drafting software may result in conservative unsatisfying designs. Some TAFE Construction Communication courses teach basic design theory and application.

The education, experience and regulations governing Building Designers may vary from location to location. It is important to check if a selected Designer is registered to provide design and documentation services in your State. Registration generally requires the designer to carry professional indemnity insurance.

3. Do it yourself

It is possible for competent people to design their own home, particularly if they are not attempting too complex a plan or using unconventional construction systems.

There are many sources of plans available in the community:

   a. project home brochures
   b. plan books from newsagents
   c. magazines.

These can be used by OBs to build up a stock of information and planning concepts. Note, most house design plans and associated documents are covered by copyright. Copyright is automatically applicable as soon as any creative work is published. The ideas and information contained in them are not allowed to be used without specific approval from the copyright owner, care must be taken by OBs not to infringe the copyrights.

Note: in Australia all plans remain the property of the designers or their assignee for their lifetime and for 50 years after their death. The copyright process is automatic and does not require any formal registration, therefore even if the © symbol is not on the plan it does not mean it is free from copyright restrictions. If OBs do produce an original design, it is automatically covered by copyright in their favour.

4. Use a plan service

In most states in Australia, building industry bodies, such as building information centres, offer home plan services.

These services generally operate by having a wide selection of home plans and alternate construction systems available for prospective home builders to peruse and choose from. The service then adapts the chosen plan to suit individual requirements and fits it to the proposed building block.

This is a good service, is normally value for money and is only limited by the range of available designs.

## Developing a brief

In all of the above options the OBs should scope out in detail their accommodation and other requirements. Scoping is the first stage in any design process, any mistakes or omissions made in determining the requirements of the occupier family has potential to reduce significantly the liveability of the completed house.

It is important when developing the brief to write down a full description of the family's make-up.

This requires the identifying of:
1. The family members family individually and collectively
2. How the family fits socially and economically into the fabric of society
3. The physical condition of family members, are there:
   - any or are there likely to be any young babies,
   - any school age children,
   - are there active teenagers,
   - are there any members of the family with physical disabilities; and
   - are there any elderly members of the family?

This assists the OB and the designer to become more aware of the requirements and constraints generated by the family on the house design.

When the members of the family have been identified, and categorised, it should then be possible to do a preliminary analysis of each member's need for space. Decisions have to be made on what special requirements are to be met in the proposed house design to meet the individual requirements of each member of the family. This is the time to decide on:

(a) how many bedrooms are required and how large they should be
(b) what relationship is needed between bedrooms. Should they be grouped at one end of the house or should they be separated to specific areas of the house or related to function? Do the parents of the house want to be as far from noisy teenagers as they can be, or vice versa?
(c) what type of living spaces are required, large, small, formal, informal and what are their functions?
(d) if bedrooms as an individual's private space, and any living areas as the family's collective space, how much consideration should be given to intermediate semi-private spaces such as studies, workshops, music rooms or libraries and similar functions
(e) how the kitchen should be defined. Not so much the colour of the laminated plastic on the benchtop, but rather whether the kitchen is to function as a mass-production family galley, or at the other end of the scale, a gourmet's pantry. The decision on the function of the kitchen, other than the simple fact of food preparation, is essential in determining its eventual size, the amount of bench area, storage, and the quantity and quality of major appliances to be installed. If the kitchen is open through to the dining/living spaces, consider a separate preparation area, often called a butler's pantry.
(f) the relationship of the kitchen to the living area of the house. Many families today automatically see the kitchen as an extension of the family room. This is a sensible location for a kitchen that is to feed a young and active family, but it may not be as sensible for a family of more diverse individuals, or for a family that sees the partaking of food as a distinctly different function from the watching of television or other activities that may take place in a large family room.

If the kitchen has to serve more than one eating area, and in most homes there is an informal and a formal eating area, care must be taken so that food can be easily conveyed from the kitchen to all eating areas.

(g) the requirement for bathrooms. It is not necessarily just a matter of providing an ensuite for the first bedroom and a main bathroom for the rest of the house; there may be other special family requirements to be catered for. If, for instance, there was a handicapped member of the household, then a well-located toilet and shower area may be specifically designed into the house. Maybe two teenagers need to shower at exactly the same time, regardless of the time of day. Therefore, there may be sense in providing two separate shower rooms rather than a shower and a bath in one room. There may even have an active sporting family that makes use of a backyard pool or tennis court and needs a shower and bathroom close to those activities.
(h) the features of a laundry. It is often the forgotten room in the house but major decisions have to be made about what is to be incorporated into it. Most laundries would include the tub and a washing machine, at the very least. Often there are also clothes dryer facilities, a hot water service, ironing

space, maybe even a corner for sewing (or maybe the sewing room should be one of our semi-private spaces).

Further, the laundry is often asked to double as the rear entry to the house. Choosing the room most likely to be a jumble of unwashed clothes as the rear entry to a house seems more traditional than practical.

(i) how the house is to be entered and exited. Consideration should be given to whether a formal or informal entry is required; local weather conditions will affect the design of the entries. Certainly, in cold climates, a separate entry to act as a wind and weather lock is almost essential – in warmer climates this requirement may diminish. Don't forget to consider access from car accommodation into the house. Not only from the car to the front door but from the car undercover in inclement weather and from the car with the heavy grocery packages.

(g) car accommodation. realise that the family car, when garaged, takes up about the same space as a medium living space. A double garage may be the largest space in a house. Consider how many cars are needed and if there are any boats or caravans or other trailered vehicles that need to be accommodated, and make sure that a position is allocated for them, not necessarily undercover.

(k) the storage space that you may require in bedrooms, linen cupboard, kitchens, laundries, living spaces and in the yard to accommodate gardening tools

How big a linen cupboard? Is a walk-in pantry required in the kitchen? Are there large amounts of collected paraphernalia to store? Where does the family lawnmower park? Should you include large built-in robes in teenagers' bedrooms or just make the floor larger where they normally hang their clothes anyway?

(l) other items. Make a list of all the furniture that is presently owned and will be taken to the new house, and a list of the furniture required to be purchased for the new house. Items like pianos, book shelves, large dressers and other similar items may need special attention in the planning. Allocate the major items of furniture to the rooms in which they are most likely to be installed.

This now-formidable list of spaces and functions should be discussed collectively with the family members and, when a close consensus has been established, the list should be written up clearly and provided to the designer to assist in the preparation of the plans.

## Appearance

Another family activity should then take place, to decide on the basic requirements for appearance and for materials to be used in the new house. An interesting way to handle this discussion is to have each member of the family bring along one picture or photograph of a house that particularly appeals to them to a family meeting.

It can be surprising, that even when members of a family believe they are talking about the same thing when there is no picture, often when each member of a family produces his or her photograph of how they see the proposed house, amazing differences arise. There is no formula for arriving at consensus, but some effort should be made to reach agreement on the style and materials that should be used in the new house.

When deciding on the external materials of the house remember that the windows are not only external but also internal elements and therefore they often set a significant impact on the character of the house. The roof shape is another influence on the exterior and interior. For instance, if a flat-roofed house is selected then it is fairly obvious that only flat ceilings can be used internally. However, a sloped roofed house may offer flexibility to use some of the roof space to produce vaulted or skylit rooms internally. If vaulted ceilings and skylights are part of the design scenario then they will further constrain the planning options.

Material choice at this stage can be relatively superficial but, it assists, a designer to know whether or not you have particular likes or dislikes. For instance, there may have strong objections to some materials: weatherboards that require painting or timber floors that may be attacked by white ants – particularly if you have experienced a problem in a particular area previously. There would also be certain budgetary constraints. Timber panelling can cost up to eight times as much as a wall lined with standard plasterboard.

When reasonable consensus has been reached on the style and materials to be used in the house, again, write these down together with a photograph of a house that most closely resembles your collective image.

It is important that when making this decision on style, particularly if an architect is required, that one is chosen who is experienced in the style desired.

It is not prudent to take along a photograph of a pseudo-classic Georgian mansion to an architect who has just won an international award for modern house design. The chances are that the OB and the architect will never see eye to eye. Much better to choose an architect who has an affinity with the design style that you prefer.

## Budgeting

Consideration of budgetary matters is then important. Much has been mentioned on the savings that are available to OBs, but there is little hard evidence available on the savings that they can achieve. Investigation shows that OBs are more likely to use more expensive finishes than project home builders. It is generally expected that in an average project home the cost of labour and material is split approximately 50–50. If an OB decides to build to exactly the same quality as a project home builder, and provides their own labour, then the maximum saving that can be achieved is about 50 per cent of the project home price. Even if the builder's profit component is added into the calculations, in most states and locations in Australia electrical, plumbing and sewer/drainage work must be done by licensed tradespeople, this factor is quickly eroded.

OBs who believe that they can build their dream home for half the price it would cost as a project home are unlikely to achieve this goal. If the cost of materials in a project home is 50 per cent of the total then an OB's house, which is more likely to use higher quality materials, is unlikely to achieve a 50 per cent saving. The saving is more likely to be in the 20–30 per cent range – and only when the OBs operate the project with the same cost control determination as project home builders.

Therefore, the recommendation on budgeting would be to take the price per square metre of an average project home and add 10–15 per cent for its site adaption. Then take this figure and reduce it by 30 per cent and you will have arrived at the most likely budget price per square metre for your new owner-built home.

This rough estimate should take into account the add extra cost associated with excessive land form, verandas, car accommodation, and all those items not normally included in project homes; adequate number of power points, roof and wall insulation, space heating or air-conditioning, external paving and pergolas, fencing, landscaping – and the inflation of building material over 12 months (the average owner-built building period) against the three to six months of the normal project home construction time.

This section on budgeting should not frighten away OBs. Even if the home ends up costing exactly the same as a project home of the same size, it will generally be built of higher quality materials and will be an individual home specifically for the family who designed it.

## Preliminary planning

When the stage of spatial organisation of all the data is reached. This is normally under taken by a designer by a method loosely called bubble diagrams. In bubble diagrams all the information collected is sketched around on paper under a loose arrangement of rooms and other factors, until what is achieved appears to satisfy most of the requirements and constraints imposed.

A bubble diagram has been included for interest but it must be realised that many diagrams will be drawn before a reasonable solution is achieved.

# 9 • DESIGNING THE HOME

When architects use bubble diagrams to plan buildings, they normally use semi transparent paper and slowly build up a jigsaw puzzle of parts by drawing, drawing over lays, redrawing, thinking, drawing again, studying, reflecting, sleeping, discussing, drawing, and so on until the solution begins to appear. It is generally not recommended that this initial planning stage be carried out with formal measurements or by mechanical drawing means. Further, the use of a rubber in this stage is actively discouraged. Often, somewhere along the process of sketching, keep all sketched ideas as they may be useful to refer to at a later stage in the development of the plan layout.

## Final plan

When the bubble diagram begins to take shape, the designer moves into the final stage of the design. The first step in this final stage is to rationalise room sizes and the building dimensions.

A sketch showing this is included and it can be seen that in this case the designer has used dimensions that are multiples of 300 mm to establish the basic dimension network. The use of 300 mm as a planning grid is a useful one, for most building elements are a reduction or addition of this dimension.

Once the designer has achieved a dimension sketch then doors, windows and other elements can be added to make sure the whole is a working network.

The final stage in planning a house is to prepare a scaled-up sketch drawing showing wall thicknesses, door swings, built-in furniture and fittings, as well as any special pieces of furniture that must be included in the design. All this is done in relation to the site data information that has been collected, so as to achieve an integrated house and land package.

## An example

Every stage in the design process explores a number of points so that there is usually a rational development of the design. It could change direction radically during this creative stage. However, in the design used as a reference in this section, a focused evolution is followed.

The design example we have included is deliberately fictitious.

Identifying features are:
- living room to the front of the house works best if house has a northerly aspect
- living rooms at one end of house and bedrooms at the other
- family room directly linked to kitchen
- entry nearly in the centre of the main wall
- terrace or deck with direct access through patio doors from the family room.

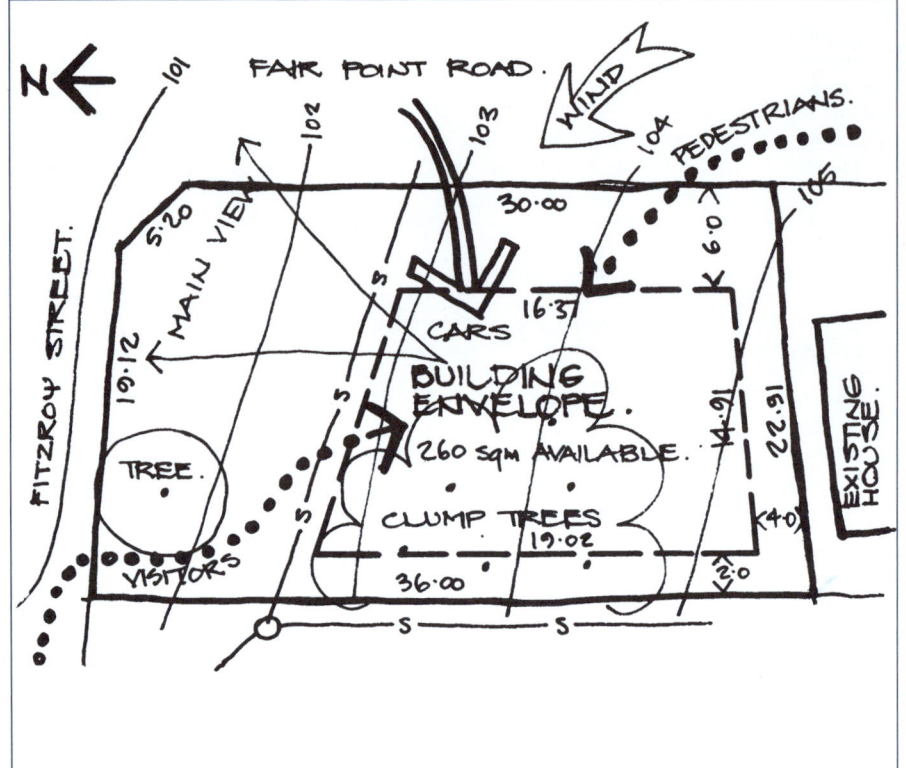

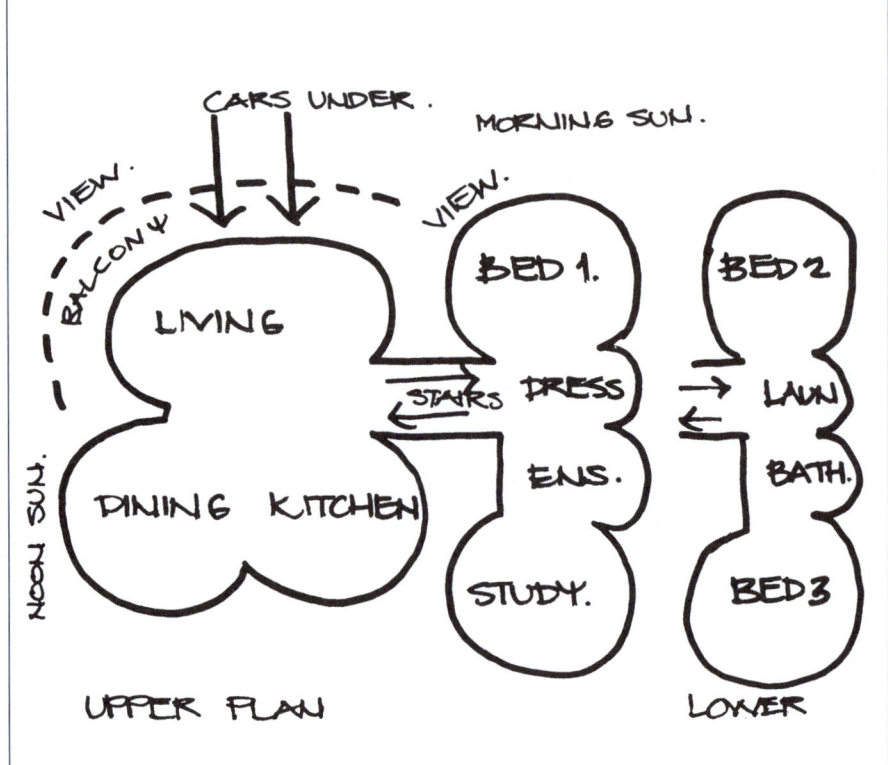

Special features of this design include:
- under-floor double garage
- split level floor.

These features take advantage of a sloping site. By stepping the floor, the living room end of the house has a higher ceiling height than the bedroom end of the house, but the ceiling actually remains on the same plane throughout the house – this means that it is relatively easy to construct the roof.

Note: many OBs produce highly developed plans only to find it is difficult to roof them suitably and economically.

# DESIGNING THE HOME • 9

## Where does the price blow out?
Higher than normal ceiling heights
Wider than normal room widths
Selected timber polished floors
Timber wall and ceiling linings
Panelled doors
Large section skirtings and cornices
Polished marble and granite
Multiple power points
Designer light fittings
Extensively tiled bathrooms
Designer ranges in sanitaryware and taps
Selected door furniture
Custom-designed kitchen cabinets
Super stoves and other appliances
Non-standard windows and
French doors
All the things you want?

Even premium quality paint sanded between coats and with a minimum of three coats is a quality cost-inflating requirement.

## PRICE BLOW-OUTS
If a bedroom has a normal minimum ceiling height of 2400, then an increase of height to 2700 would be a 12.5% increase. The walls would therefore cost over 10% more.

A basic double bedroom is about 3600 x 3600 or around 13 square metres.
A larger bedroom may be 4500 x 4500 or about 20 square metres.
A floor increase of 7 square metres or about 35%.

Comparing the basic bedroom 3.6 x 3.6 x 2.4 = 31 cubic metres to the larger bedroom 4.5 x 4.5 x 2.7 = 55 cubic metres or 77% bigger.

If a 3.6 x 3.6 x 2.4 bedroom cost about $40,000 then a 4.5 x 4.5 x 2.7 bedroom is likely to cost about 50% more or $80,000.

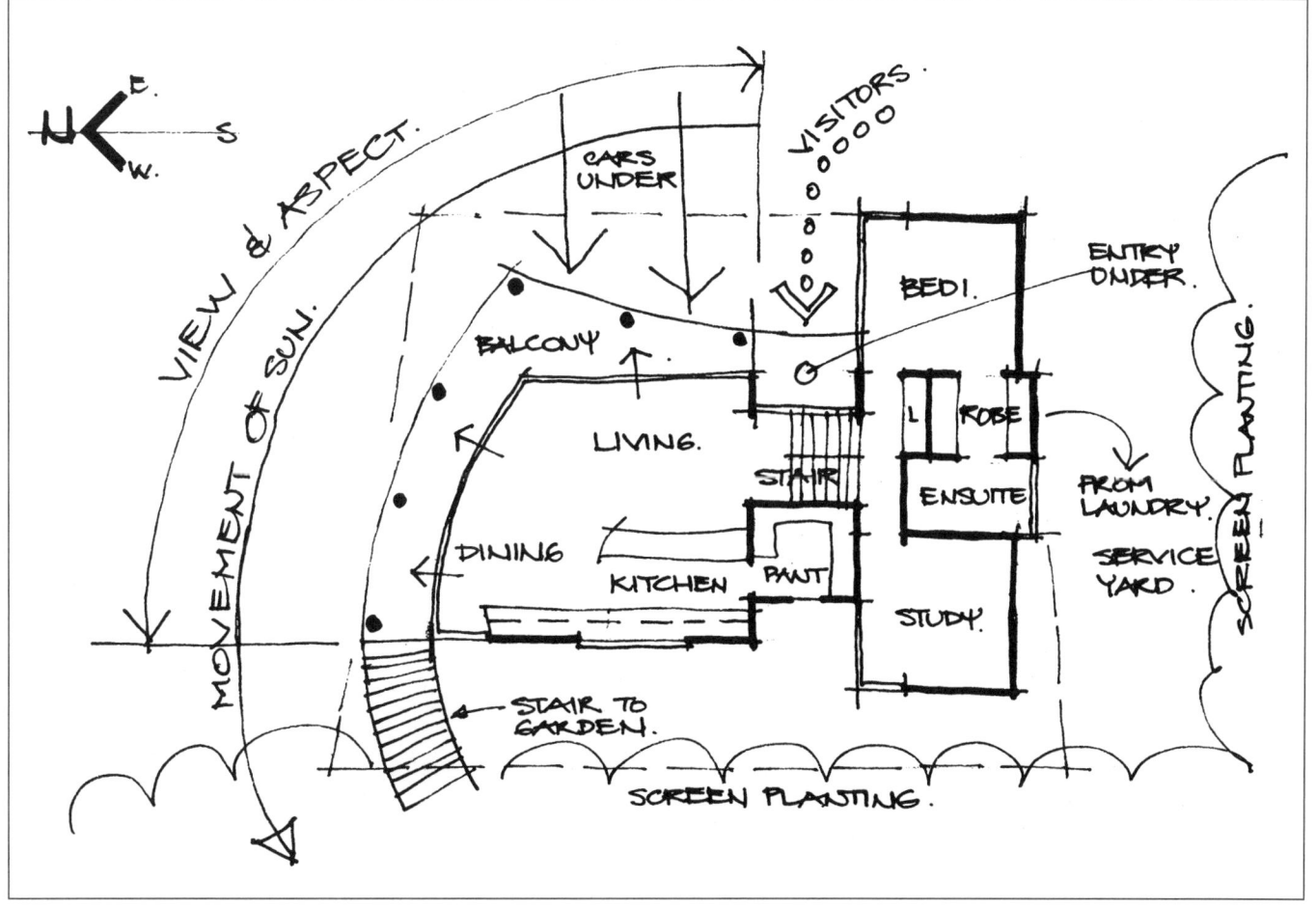

# 9 • DESIGNING THE HOME

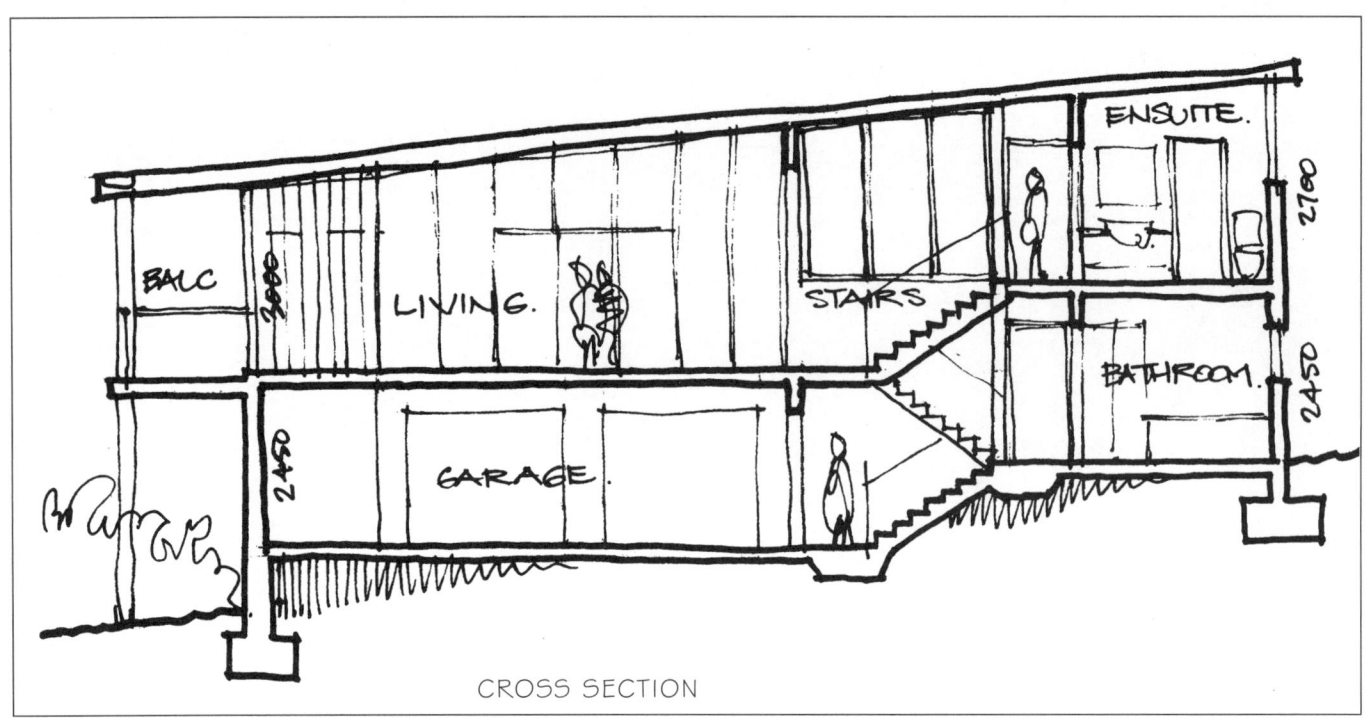

CROSS SECTION

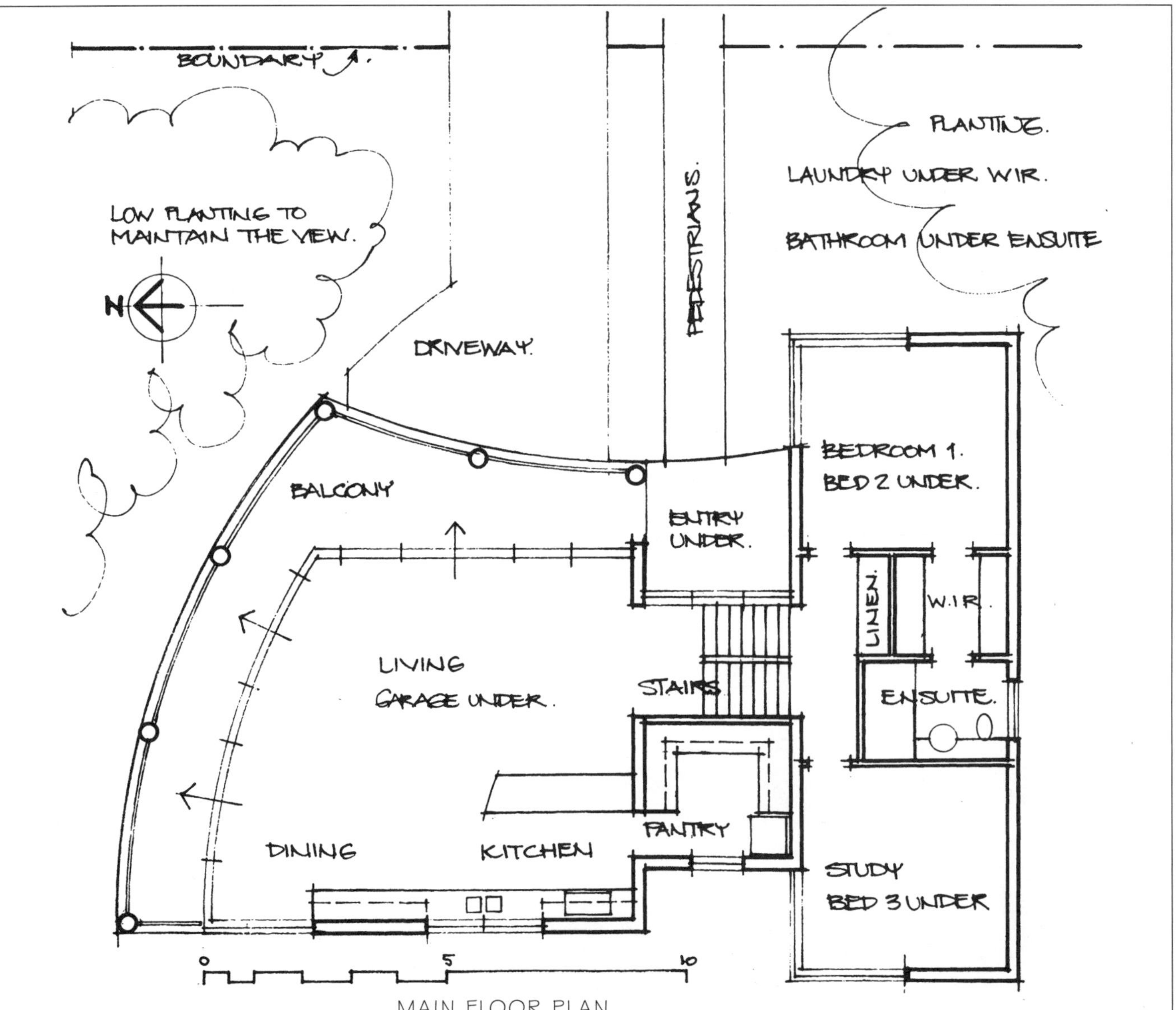

MAIN FLOOR PLAN

# CONSULTANTS • 10

## Consultants

Consultants provide professional advice, specialist knowledge and skills to the building process. It is possible to construct a dwelling without employing consultants, but in today's complex world, the chance that OBs will be required to employ consultants is increasing.

Consultants are often required by OBs by the requirements of regulations or by authority bodies. In particular, lending and approval providers often require the use of a land surveyor to make sure the improvements to be mortgaged are on the site to be mortgaged, and that the improvements are within all the statutory building lines and setbacks.

## Who are the consultants?

The consultants that OBs are likely to engage include the following:

Architect, building designer, drafting service, valuer, town planner, land surveyor, structural engineer, electrical engineer, mechanical engineer, hydraulic engineer, geo-technical engineer, heritage adviser and building surveyor.

They may also consider engaging:

Project manager, interior designer/decorator, landscape architect/designer.

In remote cases the services if traffic engineers may be required.

These professional consultants are experienced in providing specialised advice, designs, documents, reports and administration within the building industry. In some locations they will be registered professions, requiring current registration certificates to practice.

In the residential dwelling industry, OBs are likely to be required to employ; architects (or building designers), valuers, land surveyors, structural engineers and building surveyors in almost every case.

Architects fall into a special category and have been treated in detail in this section. All the others are used less frequently, often only when special constraints demand their special knowledge, except the interior designer who puts the final touches to the project.

## Architects

There has been reluctance in the community to engage architects to design, document and/or administer residential building projects. The feeling appears to be that architects are expensive. and provide few worthwhile benefits.

There are few arguments for the employment of an architect on a conventionally constructed house of unimaginative planning to be built by a small building company. These houses are generally built to conventional standards using standard components and assemblies that have evolved over many years, they are cost efficient and provide reasonable accommodation and appearance.

Owner builders are likely to benefit from the working with and architect when they require a carefully designed house that satisfies the full extent of their brief and is cognisant of the restraints of the site, the local climate the available resources and the available budget

Not all OBs want a two-third size palace replica using precast adobe panels with render, but some do, and many others at least want a dwelling that expresses individuality, lifestyle and personality. Architects can be of significant assistance in being able to organise individual concepts into understandable documentation, so that buildings can be approved and constructed.

## What is an architect?

In all states of Australia there are architect registration boards. It is the function of these boards to administer Architects' Registration Acts, that determine the minimum qualification and experience an architect must have to be able to practise as an architect.

To be considered for registration it is normal for architects to have studied at a university for five to eight years to obtain a degree in architecture, then to have completed a period of on-the-job experience followed by an examination by the board on professional practice, attitudes and knowledge. It takes about six to ten years of study and training before any person can be registered or chartered.

Many architects are also members of the Australian Institute of Architects (formerly the Royal Australian Institute of Architects) and can be identified by the use of either AAIA or FAIA after their names. The former denotes an associate of the institute and the latter a fellow of the institute. There are also life fellows but these are not common and unlikely to be in a residential-based practice.

Architects may specialise in certain areas of building with a distinctive design philosophy. Younger architects are most often practising in small partnerships specialising in house designs, whereas the more experienced architects are often involved with practices geared towards specialist commercial, industrial or civic buildings.

## How do you choose your architect?

From your list delete those architects that you sense do not have a first-class rating for your project. Attempt to have a list of three or four final candidates. Contact the architects on your final list and have a short discussion with them. Try to ascertain whether they are interested in a project of the size and type you are proposing, and whether they have sufficient time to devote to you. In other words, make sure the architect does not have an over-full calendar. By this process you should be able to isolate the most likely architect to commission.

When you have narrowed the field to the one most likely, arrange to meet with them in a mutually acceptable location, your house, the site of the proposed work or the architect's office. Try to make the location and time so that you are in comfortable surroundings and as far as practicable, free of interruptions.

To this first meeting with the architect bring all the people who are going to reside in the proposed dwelling. This makes your 'family' involved in the process of designing the house they are to live in and gives the architect a clear indication of the number and personalities of those to occupy the proposed home.

## What can your architect do for you as an OB?

Architects are trained to do many jobs in the building industry, but in particular their roles are:

1. to take instructions from a client and from these instructions develop a suitable, functional, aesthetic design solution
2. to develop the design into drawings and other documents that allow the proposed dwelling to be approved by all authorities having jurisdiction, and to allow the builder to determine the quantities and quality of the materials to be used, and assemble the components into a finished dwelling
3. to administer or act as the superintendent for any contract that may be entered into between the client and a builder, tradesperson or supplier, and to monitor and report on the progress and quality of any such contracts.

These three parts of an architect's service are clearly definable when a traditional client/architect/builder relationship is used. The difference between services can be less clear in an OB/architect relationship, but it remains the same in essence.

Architects can be engaged to carry out a full service or may only employed for part service, such as design, documentation and/or administration.

Most architects are prepared to assist OBs in the design process, prepare the minimum documentation required to gain building approval from the responsible authority, and then offer other documentation or administration services on an hourly, or fee for service, basis.

Architects can be less enthusiastic about documenting or administering projects when they were not involved in the initial design. This is due to the many problems that can arise through not having worked the project through from its inception. So, even if you have worked out your own plan and external appearance, you may find it worthwhile to consult with an architect early in the process.

# 10 • CONSULTANTS

## How much do architectural services cost?

Architects' fees have traditionally been based on a system that linked the cost of the project to the fee charged. This fee structure is normally expressed as a percentage of the estimated or actual contract price. The method is straightforward and easy to administer in agent/architect/builder contracts but is less applicable in OB/architect projects. It is possible for an OB to save up to 50 percent on normal builder-built projects.

Therefore, if the architect's fee were based on the actual cost of construction to the OB, there is a good chance that the architect would earn less than if they were working in the traditional mode.

This suggests that a special system is required for OB/architect projects. This can be either:
1. percentage fees based on an estimated builders contract figure
2. percentage fee multiplied by the estimated difference between OB and a builder contract
3. a negotiated fee, generally paid in stages
4. an hourly rate.

All four methods have a place but in OB/architect projects a combination of design and documentation based on a negotiated fee and a set rate schedule for visits, consultations and reports during the construction period would probably work satisfactorily.

For a project that is estimated to cost $400,000 if built by a builder, architects' fees are likely to be in the range of 5 to 12 per cent of the estimated contract sum, i.e., $20,000 to $48,000 for full services. Individual services would be approximately as follows:
- Design, 1.5 to 3%-$6,000 to $12,000
- Documentation, 2 to 4%-$8,000 to $16,000
- Superintendence or Administration, –1.5 to 5%-$6,000 to $20,000

## What do you tell your architect?

Everything! The architect needs to know how the family that is going to occupy the house lives. They need to know the individual and collective likes and dislikes of every member of the household, including the dog, cat, tropical fish and birds. Architects can only do their best work when all the preconceived images of the ideal new house have been brought out into the open, discussed, analysed and incorporated where applicable.

Clippings from the popular monthly home and decorating magazines on the perfect fireplace, kitchen, bathroom, street elevation, and so on, are generally disliked by architects, who often find these items restrictive, particularly if presented in a scrapbook at the first meeting. Therefore, use the clippings carefully; they are a good tool to illustrate a particular point but should be used judicious, as no one would really like to have a house looking like a scrapbook collection.

It is often better to spend an extra two or three hours talking to your architect and aiming for design consensus rather than telling the architect to come back with two or three alternatives.

## What is design?

Design is that process whereby a designer (in this case an architect) works to convert the ideas, briefings and instructions of the client into an aesthetic and functional solution within the financial resources available. Good design occurs when the functional and economic constraints are met, and the aesthetic presentation of the design is accepted by enough other people to achieve the status of acceptable good design.

OBs often have a greater opportunity for a new dwelling, addition or alteration to be designed to suit their particular requirements than families who take the traditional track and employ a builder. This is because architects often avoid labour intensive details in a dwelling to be contract-built as they add excessively to the final cost, whereas in an OB dwelling labour is only restricted by the owner's skills and availability of time.

## What is documentation?

When the client family accepts an architect's design proposal, the next move is to prepare the drawings and specifications required by the authorities having jurisdiction over the works, and to communicate information on the quantity and quality of the materials, layout, assembly and components that go to make up the building.

These documents are generally made up of two distinct components:
1. Working drawings including; layouts, assemblies and details.
2. Specifications of the work to be done and the materials to be used, including specific schedules of fixtures, fittings and finishes

OBs should discuss the format of the working drawings and specification with their architect who can tailor the documents to make them easier to read and understand.

This two-part documentation assumes a simple application under the National Construction Code (NCC) (formerly Building Code of Australia (BCA). In some locations, an approved qualified building surveyor and appraiser can issue certificates of likely compliance for the NCC. Check this with the local council or the state department responsible for local government.

The NCC has generally reduced the requirements imposed on home builders but, the complexity of town planning, environmental planning, social planning, neighbours' rights, heritage provisions and townscape provisions have increased the difficulty to reach the stage where an application for building permit is lodged and approved.

In many parts of Australia, all applications to build a new house or to add to an existing one, require the lodging of a development application. This will contain all the information the authorities insist on to determine approval for the use of the property.

The applicant has to provide documentation that could include:
- three dimensional representational drawings or models of the proposed buildings
- statements of environmental effects or environmental impact statements

This is required not just for major residences in a highly sensitive areas. It is possible and probable for a municipal council to require most, if not all of the above, to be lodged before processing any kind of application. The majority of this information is made available to your neighbours, in case they wish to make representation about your application.

## What is superintendence or administration?

There is sometimes confusion by OBs and others associated with building work about the role of a superintendent. It should not be confused with supervising; it is effectively ensuring that any agreement (contract) between two parties is carried out to the terms and conditions of the agreement. It can be a very powerful role and the parties to a contract are bound to adhere to the decisions of the superintendent as defined in the contract conditions.

In many cases, in OB projects there are no contracts between the OB and any building contractors. In these cases, the architect can be engaged to visit the site on a regular basis to determine that work is progressing at a sensible pace, is of correct quantities, of acceptable quality and to answer any questions relating to the drawings and specification the OB may have. Here the architect is simply reporting facts to the OB, to aid in the systematic and efficient progress of the project.

The OB may, with the architect's approval, nominate the architect in any subtrade contracts as the superintendent of that contract. When the architect is nominated as the superintendent in a contract then they are required to work equitably for the contract and for the client, they are reimbursed by the OB.

This process of contract superintendence/administration should be explored by the OB who is tackling a large dwelling where individual subtrade contracts may exceed $100,000. The architect can also check all accounts where applicable, particularly where provisional sums, prime costs, rise and fall contracts or cost plus contracts are involved.

Many modern contracts use the term

# CONSULTANTS • 10

superintendent and require that they are an individual (not a corporation). Some older contracts used the term administrator and, although this is basically the same role, the term is now generally considered obsolete.

There have been contracts issued by building contractors' associations that uses the term agent; here this appears only to nominate a person to act for the owner but not to superintend the contract. Agents may be useful in some cases but do not appear to have the same powers as a superintendent.

If there is any doubt, ask a lawyer to give an opinion, particularly if a registered architect is not involved.

## Heritage adviser

State and local councils list buildings that are deemed have historic or cultural heritage. This information often originally intended to assist in identification of historic buildings of a particular style or appearance, is used by government authorities when determining approvals for building, alterations and additions. and has become lists of 'must keep, do not touch' items.

Although the lists contain most of the important buildings in localities, they are not definitive. Many of the buildings are basic family homes, and owners may not be aware of this until they seek council approval for further work on their property.

The impact of heritage does not end at the title boundaries of listed properties; neighbouring, adjacent and properties within a heritage precinct may also be affected.

Council regulations place the onus on the owner of a heritage listed property to prepare a report proving that the work they are applying for will not affect the heritage value of the property or the precinct. The owner of the property can be required to find, brief and pay all the costs of employing a special heritage adviser (or architect or consultant) to write a report which may or may not support the owner's case to be allowed to renovate or extend the house.

The Office of Heritage or Heritage Council in your state may provide lists and details of accepted heritage advisers.

## Project manager

These consultants can be engaged by OBs to manage the construction of a building, to achieve on-time completion and on-budget expenditure.

By employing a project manager, the OB becomes simply the owner, as the manager takes on the administrative role of the builder and absorbs most of the savings available to the OB. They can make the job of owner-building much easier. When you negotiate the contract with a project manager, establish whether you or the project manager sets the deadlines and budget.

Some project managers may hold a building license; this could be useful in avoiding the need for the OB to gain an Owner Builders Permit (check the requirements in the state where the work is to be carried out). Asking a licensed builder to be a project manager could raise some conflict of interest issues and an implied building contract may exist – always seek professional advice before entering any such agreement.

## Certifier/Building Surveyor

Private certifiers and building surveyors do some jobs previously handled by local government town planners and building inspectors. These consultants may charge more, but they often have more flexible contact times and can give advice, unlike some government officials. Ensure the consultant you choose is fully accredited. The local council may assist in finding a suitable local certifier or building surveyor.

## Interior designer/Interior architect

These consultants are specialists in designing the interior spaces of buildings, whether unbuilt or existing. They can be of great assistance to the OB when the brief for the interior of a house is complex. The location of doors, windows and furniture, for example, can have more to do with the efficient use of the room than with its dimensions.

Interior designers are often confused with interior decorators. An interior designer should have completed a course of three or more years at a university or institute of technology and have appropriate professional experience.

Note: in some States universities issue degrees in Interior Architecture. This qualification does not make the person both an interior designer and an architect, so take care. In many alteration and addition projects, a well-qualified and experienced interior designer/architect may be an asset to the OB.

## Valuer/Land economist

Valuers are a special branch of the real property industry and in essence constitute the link between it and the building industry, and a further link between the building and the lending industries.

Valuers are not normally employed directly by OBs, but often it is an indirect requirement of a mortgagor that progress payments during the construction of an OB project are assessed and approved by a qualified valuer.

Valuers are required to be registered by a Valuers' Registration Board in all states and may be employed directly by OBs who wish to independently value residential unimproved or improved property when considering a purchase.

The Institute of Valuers in all Australian States will assist OBs with lists of valuers who are skilled in the particular area of valuation required.

## Town planner

Until quite recently the consulting town planner was almost unknown and those who were in practice tended only to hold briefs from governments, semi-government authorities and large companies.

That was when houses were constrained only by health and building regulations and ordinances. Progressively, more restrictions were placed on residential development by government planners. Today, in many areas, even single-family freestanding residences require specific town planning development approval.

This phenomenon has caused many prospective house building families to be obliged to consult a qualified town planner to determine what development, if any, is allowed on their chosen allotment and to ensure the applicant documentation covers all requirements of the Local Environmental or Planning Scheme, and the often plethora of attached development control management codes.

## Land surveyor

Surveyors may be engaged by owner builders on up to three occasions during the development on any site:
1. Boundary survey –on purchase of the property, to ascertain the location of site boundaries, possible encroachments. This survey may be widened to include:
   - a tree location survey
   - a contour (ground levels) survey
   - utility services survey.
2. building set out survey –to position the building in relation to building alignments and title boundaries. This survey can be of benefit to OBs on steep or awkward sites in locating critical corners.
3. Check survey – this is normally only required to ensure side boundary and frontage setbacks have not been accidentally encroached upon during construction.

## Structural engineer

OBs may need to employ a structural engineer to design, document and possibly supervise particularly critical structural components. The most common areas of service to owner builders are:
1. The design of foundations and footings
2. The design of steel and sometimes concrete beams, particularly over large spans or openings
3. The design of suspended reinforced concrete floor and roof slabs, particularly those on upper floors and with critical design details such as cantilevered balconies and upper-floor spa tubs.

It should be noted that in some localities, authorities require engineering details and computations for structural work other than that listed above.

# 10 • CONSULTANTS

## Electrical engineer/ Mechanical engineer

Consulting electrical/mechanical engineers are normally used on commercial and industrial projects and seldom engaged in single-family housing. However, when complex electrical and/or mechanical installations are contemplated by OBs, then consideration should be given to the employment of an engineer.

The employment of these engineers can be very useful to develop and install energy management systems, particularly when central air-conditioning systems are required.

## Hydraulic engineer

Hydraulic engineers are involved with the design of fluid movement systems and, although not often required on simple residential projects, their advice can be important if particularly complex or restricted water/sewerage/drainage systems are involved.

## Geo-technical engineer

Geo-technical engineers make test holes into the substrata of land to be developed. They are able to ascertain from these test samples significant information about bearing capacities, stability and sub-surface water tables, this is required before design of structures on difficult sites can commence.

In some localities, they are required to provide an absorption index to assist in the design of septic tanks and other ground water systems.

Note: it would be normal to employ a geo-technical engineer under instruction from the structural engineer or architect.

### How can you find an architect?

The Australian Institute of Architects (AIA) represents many architects and will supply lists of architects practising in your locality, who specialise in assisting OBs. However, not all architects are members of the AIA and you may discover smaller community-based architectural firms by accessing the list of State Registered Architects.

Architects can advertise in local news papers and other magazines and journals. Many will indicate a specialisation including: interest in heritage, solar design, alterations and additions.

Some architects will indicate a willingness to provide building services or project management-OBs should think carefully before commissioning these services and, if they feel this need. should consider contracting a licensed builder on a fixed-price, lump-sum contract instead.

Searching the Internet can provide lists and information on architectural practices

You could add to your list by asking friends who have used an architect's services to give comments. Some site signs provide the architect's name particularly if the design under construction is appealing.

You could ring the AIA in your state and ask for a list of architects practising in your locality who have experience in new dwellings, additions to existing dwellings or alterations to dwellings.

### Australian Institute of Architects

**NATIONAL OFFICE**
Level 1, 41 Exhibition Street
Melbourne, Victoria 3000
p: +61 (03) 8620 3877
Toll number: 1800 770 617
Enquiries: national@architecture.com.au

**STATE AND TERRITORY OFFICES**

ACT CHAPTER
2a Mugga Way
Red Hill, ACT 2603
PO Box 3373, Manuka, ACT 2603
p: +61 (02) 6208 2100
e: act@architecture.com.au
w: architecture.com.au/act

NSW CHAPTER,
NEWCASTLE DIVISION
& NSW COUNTRY DIVISION
Tusculum
3 Manning Street
Potts Point, NSW 2011
p: +61 (02) 9246 4055
e: nsw@architecture.com.au
w: architecture.com.au/nsw

NORTHERN TERRITORY CHAPTER
Level 16, Regus Centre, Charles Darwin Centre
19 Smith Street Mall, Darwin, NT 0800
PO Box 1017
Darwin, Northern Territory 0800
p: +61 (08) 7969 6000
e: nt@architecture.com.au
w: architecture.com.au/nt

QUEENSLAND CHAPTER
2/270 Montague Road
West End, Queensland 4101
p: +61 (07) 3828 4100
e: qld@architecture.com.au
w: architecture.com.au/qld

SOUTH AUSTRALIAN CHAPTER
L2, 15 Leigh Street
Adelaide, South Australia 5000
p: +61 (08) 8402 5900
e: sa@architecture.com.au
w: architecture.com.au/sa

TASMANIAN CHAPTER
Level 1, 19a Hunter Street
Hobart, Tasmania 7000
GPO Box 1139
Hobart, Tasmania 7000
p: +61 (03) 6214 1500
e: tas@architecture.com.au
w: architecture.com.au/tas

VICTORIAN CHAPTER
Level 1, 41 Exhibition Street
Melbourne, Victoria 3000
p: +61 (03) 8620 3866
e: vic@architecture.com.au
w: architecture.com.au/vic

WESTERN AUSTRALIAN CHAPTER
33 Broadway
Nedlands, Western Australia 6009
p: +61 (08) 6324 3100
e: wa@architecture.com.au
w: architecture.com.au/wa

# AUTHORITIES • 11

There are many authorities that have jurisdiction over OBs. These authorities may be departments or commissions of Common wealth, State and Local Government, or quasi-government boards as well as private sector corporations and organisations.

The type and names of authorities will vary from location to location and State to State throughout Australia, but the general mechanism and process of control will remain generally constant.

Some of the common areas of authority control are:
1. Local Municipal Council
   - town planning departments
   - health and building departments
   - engineer's department
2. Water supply
3. Sewerage drains
4. Stormwater drains
5. Electricity supply
6. Gas supply
7. Scaffolding regulations
8. Telephone connection
9. Builders' Licensing Boards

The following are notes to assist in understanding the wiles of controlling authorities. Many factors will vary from place to place including the names of the authorities, the names of the divisions and departments within the authorities and the names given to specific officers – don't worry – there will be plenty of forms to fill in, many dollars to spend in fees, charges and deposits. At least make sure that you check off all the following steps:

**1.** In most areas of Australia the local shire, municipality, town, or city council has control of the main building approval requirements. Most local councils have at least five main divisions;
a. general manager/chief executive manage administration
b. infrastructure engineer's division
c. health and/or building division
d. town planning division
e. parks and gardens division

Some large municipalities may have a greater number of divisions and departments; some small shires may have only a general manager and a few staff, and some municipalities may arrange the department structure differently, but all in all this basic formula remains.

Most OBs will encounter the planning and building divisions at the local council.

In some localities the building of a single family residence on a suburban block of land is an 'of right' use and may only requires a local council building permit or a permit issued by a registered building surveyor under the NCC.

Other OBs may live in areas where town planning approval as well as building department approval is required. This requirement was once restricted to areas with special environmental impact risks, such as water frontages. Some municipalities, however, do require all buildings to have town planning department approval either separate from, or in conjunction with, the building applications.

Many town planning departments employ heritage. officers or engage the services of a heritage adviser to assist with applications that affect listed heritage buildings, places and precincts. It is more likely that an OB will be required to deal with a heritage adviser in older, more established locations than in recent subdivisions.

Generally, the building department will also encompass the local health department. OBs are unlikely to have any contact with this department unless they intend to install a septic tank or similar on-site effluent disposal system.

Some LGAs will refer all development applications to their development engineers, they will assess a list of matters including, property access, vehicle parking, tree and vegetation removal, and the adequacy of water supply, sewage disposal, stormwater disposal utility services.

**2.** The water supply authorities vary significantly between locations.

Where there is no mains water reticulation; rainwater collection and sub-terranean bore water is generally administered by the LGA. Where there is mains water reticulation the supply authority it may be a statewide authority, a regional authority or a local government division.

Most land released for housing in urban areas must be provided with water reticulation and therefore all the OB must do is tap into the main. Note, however, all water supply authorities require this tapping to be carried out by an approved/licensed plumber and often require a water meter to be installed.

Water reticulation in dwellings connected to main water supplies must be carried out by properly qualified plumbers, and OBs should make sure what constitutes a properly qualified plumber in their area. Incorrect plumbing is dangerous to health and expensive to alter.

**3.** The wastewater, sewage and sewer drainage are generally administered by a special department of local government or a board responsible for a statewide, regional or inter-municipality sewage collection and disposal system.

OBs should make themselves aware of the regulations of their local sewerage authority, particularly with respect to the need for any applications to be lodged in conjunction with building applications, and what requirements are placed on drainage contractors and OBs by way of registered or licensed trades.

Inspections of sub-surface sewer drains may be required by the sewer authority before they are backfilled or built over.

Most plumbing and sewerage pipes and fittings are stamped by a water or sewerage board to make sure that the fittings you purchase are allowable in your area.

At the completion of plumbing and sewerage installation OBs should make sure they obtain the certificate of compliance from the plumber/drainer and/or control authority.

**4.** In most municipalities, the main stormwater and inner allotment drainage systems are controlled by the engineer's division of the LGA although in some locations centrally controlled stormwater drains apply.

OBs should check carefully on the requirements that apply to stormwater drainage and disposal in their locality.

**5.** Throughout Australia the methods of electrical power distribution vary significantly. In some states, the electrical supply is the exclusive domain of a central generating and distribution corporation, but in other states various generating and distribution authorities operate.

From the OB's point of view, it is generally the applicable distribution authority or corporation that controls the requirements for approvals and inspections.

As all electrical wiring is to be installed by trained electrical tradespeople, care should be taken by the OB to ensure that the electrician employed is licensed to work in the applicable distribution area. Some electrical authorities require their inspectors to test completed electrical installations and OBs should make sure that their installation is inspected and approved.

**6.** The gas supply companies vary from large government corporations to private companies operating in small towns or cities. Most supply companies have a list of approved/ licensed gas fitter plumbers who must be used for the installation of gas piping in their area of operation. The companies may require inspections by their employees but in many instances the control of quality is maintained through a tough licensing system.

Private liquid petroleum gas systems are controlled by state acts that frequently refer to appropriate Australian Standards and OBs should check with their local council to see if there are any specific codes applying.

**7.** The scaffolding regulations that apply to OBs vary significantly from state to state and location to location. OBs should check with their local council and/or work safe authority to ascertain what, if any, regulations are applicable. Note: roof tilers and other trades are required to use safety harnesses and roof edge barriers attached to the eaves of the roofs they are working on.

Some authorities may not appear interested in scaffolding safety regulations, this does not lessen the responsibility of the OB to use and maintain fit for purpose, sound and safe scaffolding.

In some locations a fee must be paid before erecting any scaffolding and a safety inspection must be carried out before the scaffolding is used by tradespeople.

# 11 • AUTHORITIES

8. National Broadband Network (NBN) connections are privatised and there are a wide range of carriers. These service providers work to the regulations issued by Austel. OBs should, where possible, determine what carrier they desire and then contact them for connection to the system. Once the cables have been brought into the property, any licensed, Austel-approved installer can carry out the cabling within the property. Note: Telstra, Optus, Foxtel and other similar corporations are also satellite and cable television carriers and have interests in the Internet.

OBs should check if the NBN that is available to their property is a satellite, microwave, fibreoptic or copper wire system. This information will assist in determining the quality of the signal for the electronic communication equipment installed on the property.

9. In some states, governments have legislated to license residential house builders and building tradespeople. These laws were enacted to control the quality of the product being offered by builders and to introduce an insurance protection system for house purchasers against poor workmanship, sub-standard materials and/or structural failure.

It is hard to protect OBs from themselves, but in some States, OBs are required to take out an Owner Builder's Permit from the local builder's licensing board in their state. OBs may also take out special insurance to protect any person who buys the house from them against their poor workmanship and structural insufficiency.

OBs should check carefully their particular state's legislation on builder licensing or registration, tradesperson licensing or registration and any requirements of building tradespeople's long service leave and superannuation portable contribution schemes

This last item may require OBs to send returns and contributions to the authority controlling the long service leave fund, the tradesperson's superannuation fund or reach an agreement with the tradesperson on non-contribution by the OB. It is wise to seek specialist advice on these matters as there can be significant penalties for non-compliance.

## Summary of authorities

OBs should consult with their Local Government Authority (LGA) building department at the earliest possible opportunity before purchase of the proposed property, either as; vacant land, a building to be demolished or an alteration and addition project, to ascertain the authority's requirements and regulations that are likely to apply to a building application.

Local architects are often a good source of local knowledge on the requirements of authorities with jurisdiction in a particular locality. They can add interpretation to the often esoteric regulations and save the OB from the problems and costs that can arise during the building period due to lack of information or even misinformation.

## Questions to ask at the Local Council

There continue to be changes to the methods of gaining approval to build a new home.

There was a time when local councils had a series of enquiry counters: City Engineers Department, Town Planning, Health & Building Department, and the Town Clerk's Rates Collection counter. There was a bell on the counter, and when you rang for assistance on a building matter, a building inspector came out and spoke to you. These counters are now being increasingly grouped into one 'Help' counter dealing with all council matters, including collecting the multitude of fees. Here is a checklist that may help in obtaining useful information from a local council. If you are contemplating building your own home, ask these questions of the council:

1. Do you have an easy-to-read document fully explaining the process of lodging an application to gain permission to build a new house or to carry out alterations and additions to an existing house? Do not be surprised if you receive a startled look.

2. Can I speak with an official who can explain to me in simple terms what I have to do to lodge an application to gain permission to build a new house or to carry out alterations and additions to an existing house?

If you are not at the counter between 8:30 am and 9:30 am, or between 3:30 pm and closing time, you will probably find that there is nobody who can help you. Even at these times officials may be at a seminar or a meeting, doing inspections, on a rostered day off, on a flexi-day or sick. Attempt to make an appointment to discuss your proposal with a suitability experience officer.

3. Do I need to read any Planning Documents, Environmental Plans, Development Plans, Control Plans, Heritage Lists, Codes or Guidelines? If so, can I get them here and how much do they cost?

Many of these documents were never meant to be read by the lay person, while others are easy to read and are very helpful try to peruse them before you buy., they can be expensive. Attempt to gain advice on how to access and utilise the electronic information available online at the Council's web site.

4. How much work can be carried out without requiring any permission from local authorities?

In some localities, small sheds, patios, decks, fences, kitchen alterations and bathroom upgrades may be exempt from requiring formal approval. The responsibility remains with the owner and contractors employed to ensure all laws and regulations are respected.

5. How much work can be approved under complying development rules? Is there a checklist I can use?

In some areas, relatively fast approval may be granted to applicants who can design their new home or alterations and additions within a prescribed set of conditions and requirements.

6. If the proposed building work cannot be approved as a complying development, what is the extent of the drawings and other documentation required?

These extra documents can include statements of environmental impact, shadow diagrams, detailed identification surveys, Australian Height Datum (AHO) levels, waste disposal and spoil control proposals, energy efficiency calculations, wind pressure modelling, material and colour schemes, perspective drawings and scale models – the list grows continually and varies alarming between jurisdictions.

The distinction between simple houses that comply and others that require highly detailed and expensive applications to the authorities tends to favour mass-produced house designs, restricting exciting individual designs.

7. Is the land to be built on or the house to be altered or extended affected by any heritage controls? Sometimes this is obvious – few high profile houses up to and including the Federation period, Inter-war Bungalows and Art-Deco villas, escape heritage listing. In other locations whole precincts are listed as having heritage value; this can mean that even vacant land or a fibro wreck is affected.

8. How many steps are required to gain permission to commence construction? In some jurisdictions, development-planning approval is required before documents can even be lodged for building permission. Other areas will allow a joint development, planning and construction applications, or may require only a single application.

9. Is it possible to carry out any building work without involving local government authorities at all?

In some states there are rules that allow private building certifiers/building surveyors to approve and certify certain designated building projects. OBs may find that commissioning private certifiers, where local laws permit, could allow wider consultation and flexibility than the services provided by local government – although this is likely to cost a little more and should be assessed carefully.

10. How much credence will the local council give to objections and representations

from neighbours or the wider community?

Many councils will advertise proposed building projects for the benefit of neighbours. If objections or representations are raised, an extended period of deliberation may follow or the matter may be referred to a third party, a tribunal or court for adjudication or decision. Sometimes meeting with neighbours before lodging an application can be an advantage – but not always.

11. Does the council have performance based codes? If so, is there a schedule detailing what is deemed to comply?

Older building regulations often stated the minimum distance the wall of a house could be built from a side boundary fence (900 mm in New South Wales). Houses can now be built closer than this, even right on the boundary. This requires the applicant to argue that the ventilation, solar access, fire separation, and so on meet all performance requirements. If this sounds like hard work, 900 mm is generally the deemed-to-comply distance.

Significant changes like this have been instrumental in causing the State based Building Regulations to grow from a paperback-sized book, costing a few dollars, to a multi-volume, ring-bound, loose-leaf code, the National Construction Code holding hundreds of sheets and costing hundreds of dollars. There is even a significant cost to gain access to the on-line version.

## Costs associated with satisfying authorities

Progressively the cost of satisfying the requirements of the authorities that have jurisdiction over building permission has increased. These costs are presented as being in the public interest, and so they may be, but they certainly add cost to building a home.

What follows is a representative list of some of the most common impositions. The names given to the services/requirements may vary from place to place.

Two are not spelled out in detail. These are:

1. the Goods and Services Tax imposition by the Commonwealth Government – not included as there is technically no cost to the OB in processing this tax system. However, significant time is expended in filling in returns to the Australian Taxation Office (ATO). If the returns to the ATO are inaccurate, deliberately misleading or incorrect, the OB could face fines and other penalties. Always check with the ATO, what are the most current requirements and returns applicable.?

2. impositions on land developers by state and local governments and authorities for contributions towards the provision of infrastructure projects sometime after the subdivision is approved, sold and houses built. Most of these contributions are used for the purposes defined at the time of subdivision, but often the cost of these contributions is not communicated to land purchasers. Landowners may wish to check with their local authorities to determine how much they have contributed and whether the infrastructure paid for has been provided.

## Prepare a development application

This planning mechanism was introduced to improve the quality of planning. The application is not always required but, if so, can introduce a package of documents including environmental planning policies, zoning plans, local plans, neighbourhood plans, control plans, townscape guidelines and acceptable standards for colours, shapes, styles, trees, fences, heights and so on. Even the same family, staying in the same house but wishing to build a new garage, will be asked to submit a new development application (DA) in some localities.

## Providing environmental and heritage statements

If a DA is required for a house, the LGA may require a statement of environmental effects. In some cases, a specialist consultant may need to be engaged to ensure the adequacy of this statement. If the house is considered by the authorities to have heritage impact, there could be a request for a heritage report.

A detailed heritage report could mean that the home-owners are required to employ an expert to research the property and its social and cultural, as well as physical, history. This is submitted to the authorities to assist them in deciding whether the development applied for will be detrimental to the heritage value of the locality. There are families who have discovered that the house built by their grandparents is considered to be of sufficient heritage value so that no alterations or additions could be undertaken.

## Notifying the neighbours

The authorities can charge a fee to cover the cost of notifying neighbours and the general public by mail and newspaper advertisement that a family has applied to build a house. If the neighbours make representation or object, the authority may not approve the application, even if all the other requirements of the building codes have been adhered to. In New South Wales, the only method of appeal is the Land and Environment Court; there is no small-matters tribunal. In other states there are various methods of dealing with refusals from approval authorities. Some have reasonably simple procedures to assess appeals – but will cause delay and charge a fee (win or lose).

## Gaining a development approval

Some states still believe if a block of land is zoned for single-family, residential housing, then no development approval is required. All matters are dealt with at the building application stage – check with local authorities for details and requirements.

## Preparing working drawings and specifications

For a long time, working drawings and specifications have been required to gain building approval. In many locations there are now a number of new components to the traditional building approval – energy ratings, environmental impacts, light and shadow rights, waste management, landscape plans, vehicle access and parking and so on.

## Pay certain prescribed insurance policies/ home-owner warranty

Laws have been passed in some jurisdictions requiring home-owners to take out insurances to protect themselves and others, including future purchasers. Careful checking of these requirements is essential, as some are mandatory and there are steep penalties for not complying.

## Gaining an OB's permit

The rights of OBs have been progressively restricted and controlled by Laws and Regulations. Once anyone could be an OB so long as they obeyed the basic minimum habitable and construction codes.

There still are a few places in Australia where an OB can simply build what they want, without any reference to any authorities but these are few and are generally in very remote areas.

Most jurisdictions now require any habitable building to gain approval from the LGA, obey the National Construction Code, that references a host of Australian Standards, use structural engineers for work once considered capable of being understood by builders, carry a range of insurances and gain certificates of compliance and occupancy from a designated authority.

Most states require Owner Builders to apply for and pay a fee to gain an Owner Builders Licence/Permit. This is often just an administrative impost and the fees are generally small.

In New South Wales effective from July 2002, OBs for building work costing above a relatively modest sum are required to attend approved courses that deal with:
- Information on the (NSW) Home Building Act;
- Home Warranty Insurance Requirements;
- Building Approval Procedures;
- Taxation Issues;

# 11 • AUTHORITIES

- WorkCover Issues.

Not all jurisdictions have moved to this extreme level of control but all potential OBs should thoroughly check with their LGAs and any State Government Department administering OB matters – in New South Wales in 2002 this is the Department of Fair Trading.

## Courses for Owner-Builders

There have been courses for OBs for many years; in fact, this book grew out of just such a course. It is always sensible for any potential OB to review their building industry and construction knowledge before commencing any construction.

In some places, it is essential for OBs to provide evidence of construction knowledge before they can be issued with an Owner Builder's Permit. The NSW Department of Fair Trading states: 'All owner builders supervising building and renovation projects above a specified cost must complete an approved course'.

Do not assume that any other qualification is equivalent; always check with the authorities to ensure compliance. Book early to ensure gaining a place as they could be limited, causing, therefore, a delay in commencing the course. Expect to commit 8 – 15 hours' attendance.

A course that satisfies the requirements of the regulations is only part of an OB's education and a review of courses offered by local technical colleges, community colleges and building information centres is worthwhile. Choose extra courses to fill gaps in knowledge or skills.

## Employing a structural engineer to calculate sizes of all concrete and steel

More and more items are being included on the list of structural components and systems requiring engineering verification. Computers, tabulated information and better training of builders means that many steel and concrete building components and assemblies are easy to determine. Even timber roof trusses are required to come with an engineering certificate.

## Having a check survey prepared of house location

Even where there is already a survey existing, some authorities require a set out and check survey to ensure the OB obeys the rules. They often will not accept the on-site measurements of a builder – thus, another cost.

## Gaining building approval or construction certificate

The modern National Construction Code (NCC) is a performance-based document with limited prescribed clauses. Since the implementation of the NCC, some authorities have taken to writing their restrictive interpretation of the performance requirements as conditions of the Building Approval. There are, in fact, very few areas where the NCC requires authorities to take building applications, but the authorities keep collecting their fees and issuing often detailed and confusing conditions of approval. Most building approvals will require certain inspections to be carried out by building surveyors/inspectors/certifiers conforming to these inspections is obligatory.

## Registering with the local workers' compensation authority

In most areas, OBs will be required to take out a basic workers' compensation policy; even if all on-site workers are employed by contract or contractors, there are circumstances where the law can determine that they are simply employees and able to make compensation claims against the OB.

## Registering with the Australian Taxation Office (ATO)

Governments change the rules regarding taxation and the OB. Currently OBs are required to register with the ATO and have a reference number to allow them to:

- make returns to the ATO regarding the GST component of all labour and material purchases; and
- ensure that any person deemed to be an employee of the OB pays the requisite income taxation.

Always correspond with the ATO and find out what the rules are at the time your house is being constructed.

## Notifying the water and sewerage authority

This is normally simple, as these are generally well-organised bodies with the responsibility of supplying water and piping away sewerage. They still charge a fee for their service.

In some regions there are drainage boards that control stormwater as well as sewerage – always check.

## Insurances

Every OB family should spend time with an experienced insurance professional. It is too easy to overlook an insurance requirement or one of importance to the family's protection.

The following is a general introduction to the insurances that should be considered by OBs. It is not definitive, nor does it attempt to cover all local requirements.

OBs who are building alterations and additions should be aware of the requirements for joint insurance policies between them and any or all contractors

## Statutory insurances

These cover a wide range of matters and vary from locality to locality. They must be checked carefully, as non-compliance can lead to penalties as well as lack of coverage.

In some jurisdictions home-owner's warranty insurance is required for all works above a specified value. This must be checked in each state, as in some cases failure to gain this cover is unlawful and could invoke a penalty or even limit a home-owner's ability to sell a property. It could mean that most of the trades contractors on an OB project have to give the OB insurance certificates.

## Public liability

OBs must take out a public liability policy for a reasonable cover – the amount of coverage and the liabilities of other contractors engaged should be discussed with an insurance provider.

The basic rules of insurance are:
- Are all the people who reside at the site of the works protected?
- Are all the people who work on the site of the works protected?
- Are all the people who could enter the site of the works protected?
- Are all the people who could pass the site of the works protected?
- Are the existing property improvements and contents covered?
- Are the works in progress covered?
- Are the completed improvements covered?
- Are the on-site plant, materials and equipment covered?
- Are the neighbours covered?
- Is the wider environment covered?
- Is the next purchaser of the property covered – by some form of home-owner's warranty insurance?

# DESIGN DRAWINGS • 12

Dictionaries state that architecture is 'the art or science of building', but it can be more satisfactorily defined as a creative activity that uses science to good purpose. Architecture, in other words, is the art of building wisely, of providing what is a practical and aesthetic or pleasing answer to the question of how to best enclose space for human dwelling. Much of the early history of efforts to solve this problem are now lost, but the path of progress is marked by the emergence of a succession of construction procedures, for example, different roofing techniques, post and lintel construction, the arch and the truss. Early construction was by trial and error; those designs that stood up and withstood the elements were used again. Those that collapsed were dismissed.

Now ideas and construction methods for producing houses are set down in drawing form so that others can follow what work is to be done. All architectural drawings are prepared using special, but quite simple, language. When a new house or the remodelling of an existing house is being planned, various types of architectural drawing are required.

There are several reasons why. After architects or designers have discussed project scope with their clients, they have a fairly complete mental picture of the required house and they know pretty well how it should look, how many rooms it is to have and what materials are to be used in its construction.

However, a mental picture is not completely accurate, there are too many items to be resolved and too many details that are impossible to design and correlate by means of mental pictures alone. It would be impossible for an architect to design and plan any building without the mental picture being converted into drawing where it can be further developed. In the development of these architectural and building ideas two types of drawings are used:
1. Design or sketch drawings that set down ideas on room relationships, room sizes, the orientation of the house, the stylistic treatment of the house, colour schemes, possible furniture layout and landscaping ideas.
2. Working drawings or construction drawings that show the finalised design. They indicate the exact arrangement of spaces, their sizes, the materials of construction and specific details, to enable the builder to produce the house.

## Design drawings: how are they produced?

The architect draws some freehand sketches that help to develop the mental picture and give the owner an opportunity to see the design in picture-like form. Discussions between architects and owners continue until final agreement is reached. Once the owner is satisfied with the concept sketches, the architect or draughtsperson converts them into accurately scaled and visually complete design drawings.

Sketches A and B are the type of sketch drawings architects may use during design development.

The drawings have to be produced in a size that can be easily handled, so they are miniature as well as picture-like reproductions of the house they represent. They are called scale drawings. Because architectural drawings are so small compared with the actual size of the house they represent, the architect must employ a great many abbreviations and symbolic representations of the many materials and details necessary. A special kind of language is employed to indicate the hundreds of items that cannot be actually pictured on such small drawings. Symbols are used to represent much of the information about materials, windows, doors, bathroom fixtures, walls, footings and floors.

Several different types of architectural drawing are required in order to show the required information concerning a new or remodelled house. The drawings are prepared in such a way that each of them shows this information in a standardised manner.

## CAD

Many architects and designers use computer-aided design/documentation (CAD). The use of computers has not changed the design process or improved designs, they have simply replaced pens and pencils with a keyboard. The operator is able to develop and modify a design electronically and then produce high-quality documents from printers. A specific benefit of CAD is that many programs can produce three-dimensional perspective views of the design with a speed and accuracy not so easily attained by hand-drawing techniques.

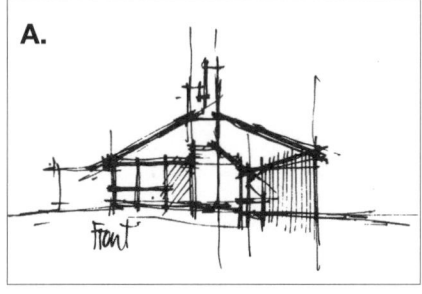
A.

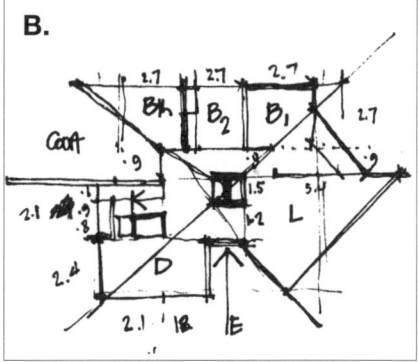
B.

## Architectural conventions

Because architectural drawings are the graphic language of building, and because this language must be standardised so it can be easily understood by people from different backgrounds (and even in different places), all drawing types comprise one or more specific architectural conventions or rules. These conventions are a series of related drawings that allow us to view the entire house to be constructed. By looking at these drawings in conjunction with each other we are able to understand the overall three-dimensional form of the design and how its parts relate one to each other.

### 1. Plans
The term plan is generally thought of in two ways. All of the drawings for a proposed house are conventionally known as 'the plans'. More accurately, a plan is a view that shows the layout or arrangement of rooms and other parts of the interior of a house. They are referred to as either plan views or floor plans. In plan views, annotation, symbols and other graphic conventions are the means of showing the required information. So far as plan views are concerned, these symbols constitute the language by which the plan is to be understood.

### 2. Elevation views
These are imagines accurately representing the vertical faces of buildings. Each side is called an elevation view. In architectural drawing, architects refer to the sides of a house as elevations. Elevation views are the representation of how the exterior sides of a house will appear after all the structural work has been completed. In order to name elevation views so they can be referred to without confusion, they are annotated to as north, south, east and west elevations. The elevation of the house that faces north is the north elevation, and so on. Not all building elevation drawings will be accurately square on to the exact compass reference annotated but are the most appropriate representation.

### 3. Section views
In most cases elevation and plan views cannot show sufficient information to enable a builder or tradesperson to see exactly how the various structural parts of the house are to be built or assembled. The section is a view of the house after it has been cut vertically at some point. Like the floor plan, it shows the size and thickness of structural members and relationships between spaces. For example, it shows the height from the ground to the floor, the height between.

## Comparing hand drawn sketches to CAD images

The hand drawn sketches by architects are a direct product of the architect's creative thoughts and are a component of

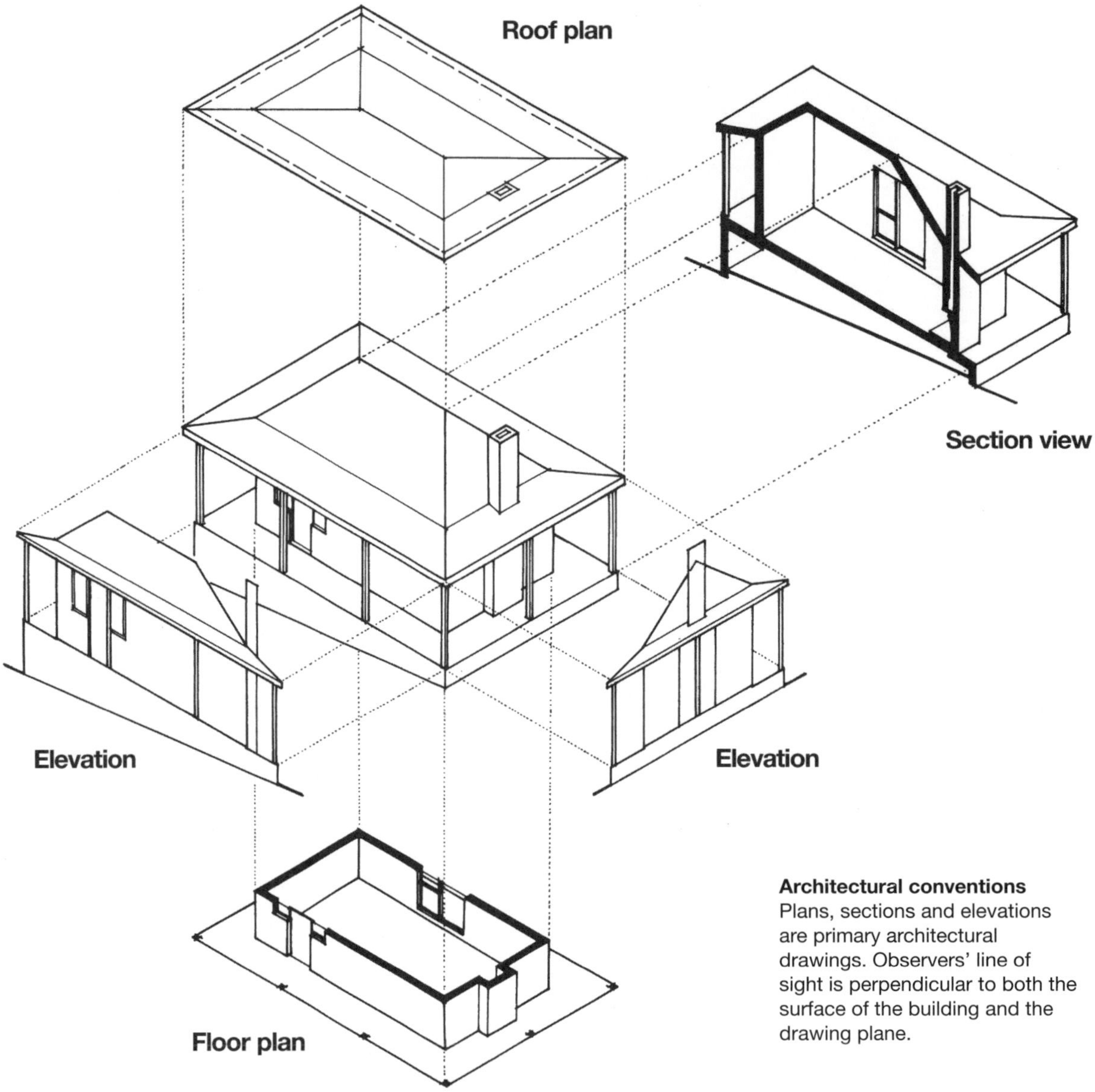

# 12 • DESIGN DRAWINGS

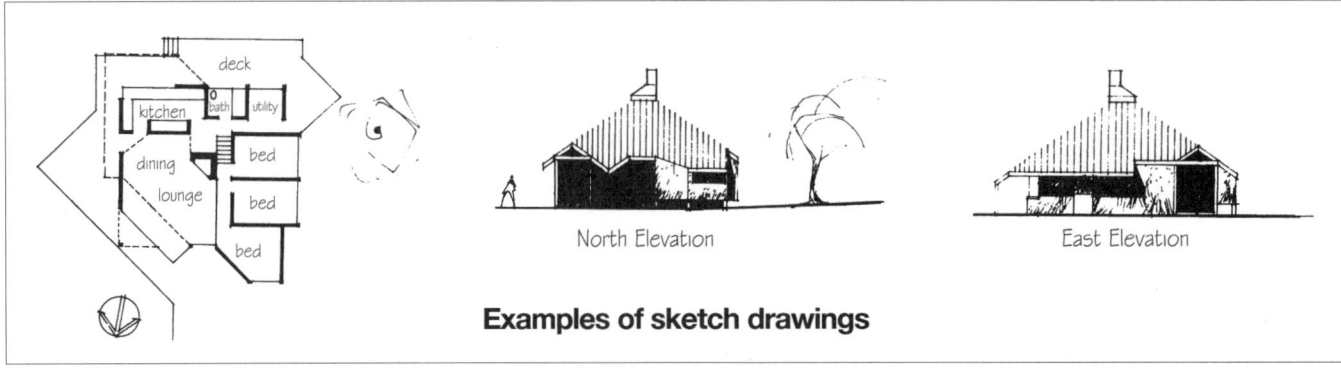

**Roof plan**

**Section view**

**Elevation**

**Elevation**

**Floor plan**

**Architectural conventions**
Plans, sections and elevations are primary architectural drawings. Observers' line of sight is perpendicular to both the surface of the building and the drawing plane.

North Elevation

East Elevation

**Examples of sketch drawings**

the preliminary development of the design and foster direct communication with the client. Hand drawn sketches can easily be preproduced on paper and allow anyone with access to the sketch to; draw over the sketches, write notes on to them and trace over them. They can be easily copied, coloured, discussed, changed, rolled, posted, scanned, folded, rolled up and filed.

Computer Aided Designs require highly specialised equipment, software and input skills plus precise information on the likely construction materials and methods to be employed in the design. This can have the effect of forcing decisions on these matters too early in the design process. CAD can compromise flexibility and creativity and limit design concepts due to the restrictions in the software. The design data is stored electronically and can be transferred electronically to other computers if they have suitable software installed, they can be viewed on computer screens, pad computers and smart phones.

Both methods can produce equally satisfactory design concepts, whatever method is used to provide the visual representation of the design they are both processes to aid communication between architects and their clients.

# BUILDING DRAWINGS • 13

Building drawings or construction drawings set out the finalised design of the house and all of the information necessary for a builder and tradesman to construct it. These drawings should be neat and perfectly clear.

## Scales

The process of drawing large objects such as the parts of a house to a proportionate size that can be contained on handy sheets of paper is called drawing to scale. The small drawings must be in exact proportion to the actual size of the house they represent. On most drawings the scale used makes the drawings 1/50th or 1/100th the original size. This means that instead of drawing something one metre long it is drawn 1/50th or 1/100th of a metre long. The finished drawing looks exactly like the full-sized object. The scales in most common use are as follows:

| | |
|---|---|
| Location drawings | 1:2500, 1:1000 |
| Site plans | 1:500, 1:200 |
| Plan views | 1:200, 1:100, 1:50, 1:20 |
| Elevations | 1:200, 1:100, 1:50, 1:20 |
| Sections | 1:200, 1:100, 1:50, 1:20 |
| Detail drawings | 1:10, 1:5, 1:2, 1:1 & full size |

## Value of drawings

Buildings consists of many elements: the structural frame, the walls, the partitions, the roof, heating, lighting, plumbing and so on. Without any one of these elements the building is incomplete and therefore it follows that the set of drawings that tells the person on-site how the building is to be constructed is not complete if any one of these elements is not shown or catered for. All these elements affect the architect's design. It must be remembered that the only value of drawings to a scale of 1:50 or smaller (1:100, 1:200) is to show the principal outline and dimensions of a building. Details of construction can seldom be designed or shown to a smaller scale than 1:20.

In the construction industry it is normal to produce drawings for houses at a scale of 1:100, possibly 1:50. This has come about because domestic construction is relatively standardised and there is no need to show details that are common knowledge throughout the building industry – excepting builders or tradespeople in special situations. The OB can therefore be presented with a dilemma. If drawings are produced at a scale of 1:100, sufficient though they may be for a builder or tradesman, the OB may have difficulty understanding the details, components and assemblies.

This is not of great consequence if the OB is planning to subcontract work. However, if they are going to construct part of the house themselves more detailed drawings may be required. It is wise to tell the architect, or draughting service you employ, of your intentions, so working or construction drawings can be tailored to your particular situation and needs.

## Content of drawings

1. **Survey plan**
Usually produced by a land surveyor. Contains:
   a. existing site and surroundings
   b. position of major natural features, trees, ponds, rock outcrops
   c. sufficient spot levels and contour lines related to a specified datum (this can be a local site datum that is allocated a convenient numerical level however it is recommended that building should be set out using the Global Positioning System (GPS) coordinates and the Australian Height Datum (AHD).
   d. dimensions, bearings and GPS coordinates of title boundaries
   e. position of roadways, easements, existing stormwater and sewerage drains, and utility service supply mains
   f. location of the dividing boundaries, relative to all existing improvements on adjacent properties. The shadows cast by adjacent developments should be marked. If there are no local code requirements, calculate the shadows at; 0900, 1200 and 1500 local time at the equinox, summer solstice and winter solstice. This will assist in determining the height of proposed buildings and the effect of solar access.

2. **Site plan**
   a. outline of site boundaries, showing location of proposed building
   b. position of boundary setbacks
   c. significant changes in site levels and where they may occur
   d. new roads and pathways
   e. soil and surface-water drains, complete with pipe sizes
   f. service runs from the house to mains
   g. location of utility services (sewer, water, gas, electricity, NBN)
   h. point of connection of those services to the house itself
   i. indication of banking and cutting and areas for depositing and spreading surplus soil
   j. new levels on the site in connection with the new house
   k. landscaping. Note: if the site is undulating or steeply sloping, sections should be added to show principal areas of cutting and filling
   l. location of any existing improvements, their plan shape, floor levels and heights where applicable.

3. **Floor plans**
   a. overall dimensions
   b. dimensions of openings
   c. internal dimensions necessary to establish positions of internal walls or fittings
   d. thickness of walls
   e. door swings
   f. windows
   g. location of fittings and fixtures
   h. names on all rooms
   i. floor finishes
   j. position of stairs and number of stair treads.

4. **Other plans that are used either where requested or for more complex houses.**
   (a) **Footing plan**
   (i) width and depth of all footings to wall, piers, stanchions
   (ii) location of footing system
   (iii) position and levels of drains, gullies close to footings
   (iv) walls above footings with thickness noted.
   (b) **Roof plan**
   (i) shape of roof
   (ii) slopes of levels
   (iii) types of coverings
   (iv) falls to gutters and gutters
   (v) roof lights
   (vi) possible type of construction.
   Note: on simple houses the roof plan is superimposed over the floor plan.
   (c) **Services plan**
   (i) electrical layout
   (ii) plumbing and internal drainage
   (iii) air-conditioning or other mechanical.
   Note: on simpler houses the electrical layout and plumbing layouts are superimposed on the floor plan.

5. **Sections and elevations**
   a. elevations of all faces of the building
   b. size and shape of openings
   c. external finishes
   d. new and old ground levels showing cut and fill
   e. positions of floor level and ceiling level
   f. elevations and should show the relationship of existing and proposed levels (reduced levels/RLs) to the new buildings. The sections should clearly indicate the floor and ceiling levels. Floor levels should be shown as finished floor levels (FFL), i.e., the top of the carpet, tiles, polished floors, etc
   g. sections of buildings with staircases clearly showing the height of the flights. This requires a calculation to determine the floor-to-floor separation dimension. Note: always identify stairs by the number of risers; if these are specified as equal then the calculation of the individual risers can be left to the stair-maker.

6. **Construction details**
   a. sections through external walls, footings and roof
   b. plans, sections and elevations of stair cases
   c. any room or part of the building, where the setting out is difficult or involves extensive installation of fittings or fixtures such including:
   (i) kitchens
   (ii) bathrooms
   (iii) utility rooms
   (iv) special purpose rooms
   d. windows and doors
   e. part elevations of any part of the building containing special features such as:
   (i) entrances
   (ii) special forms of construction
   (iii) balconies
   (iv) ornamental work.

# 13 • BUILDING DRAWINGS

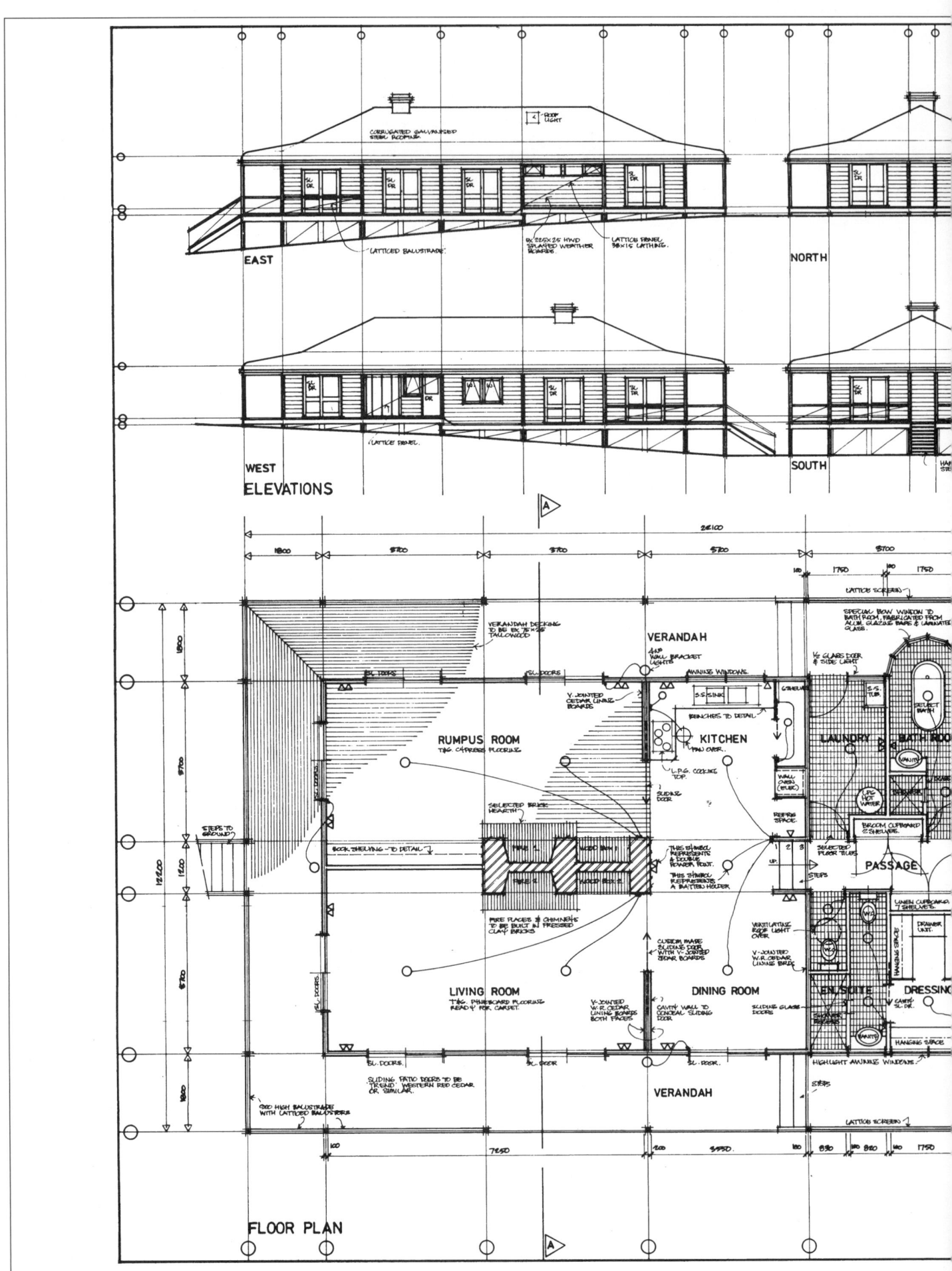

# BUILDING DRAWINGS • 13

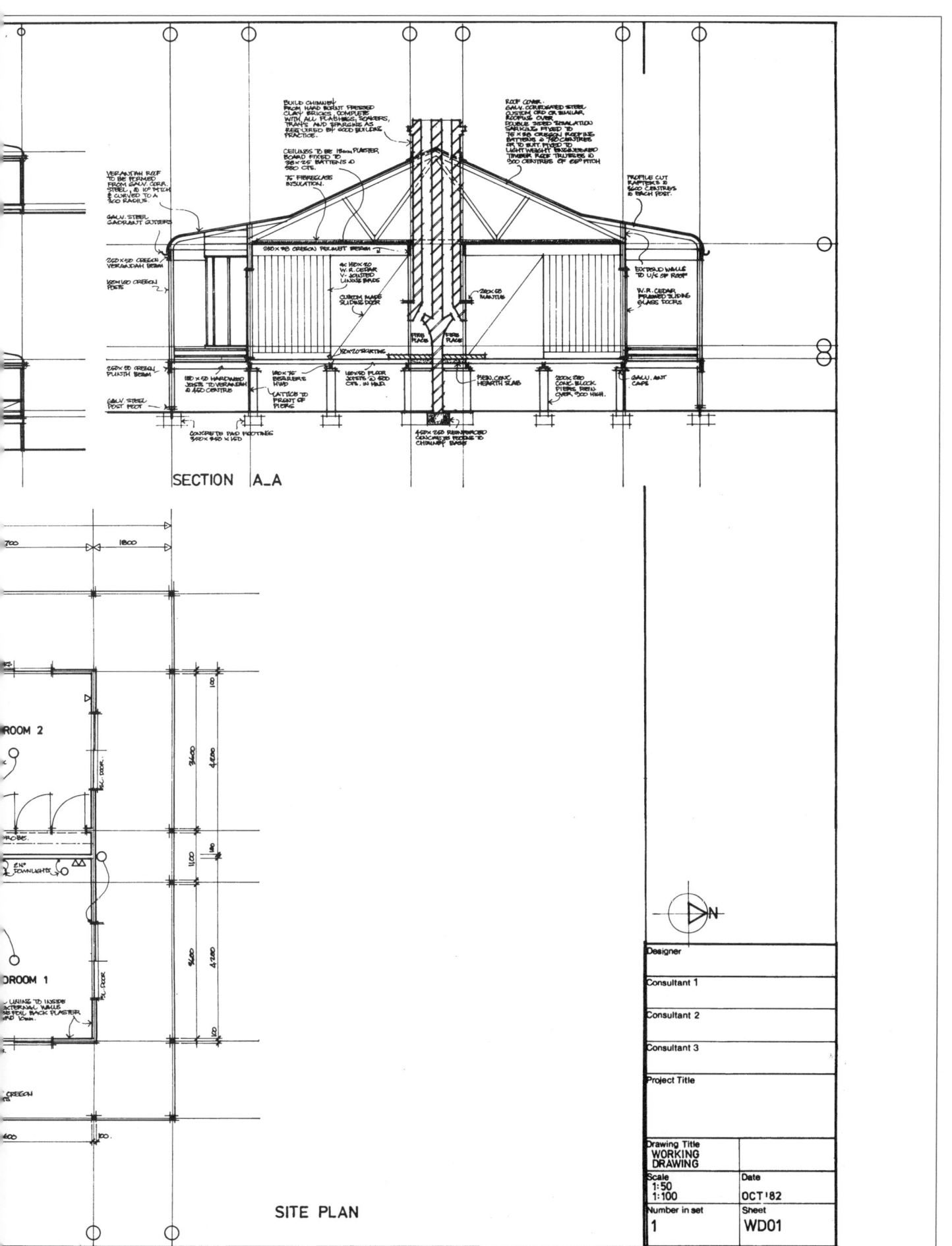

# 13 • BUILDING DRAWINGS

## Checklist

The following is a checklist of information that can be included on basic working drawings. This checklist does not include specific detail drawings (component and assembly), which will be discussed separately.

This priority guide shows which items should be included or specifically excluded:
- \*\*\* must include
- \*\* wise to include
- \* include if considered helpful or if demanded by authority.

Note that all drawings should at least show the property-owner's name, the address of the site of the works, drawing titles, annotation, scales, date first issued, any amendments and date amended, and the copyright owner (architect, designer, proprietor, builder).

## The Site Plan

- \*\* title reference information including Lot number
- \* street number
- \*\*\* north point
- \*\* street location, width and cross-street reference
- \*\*\* boundary drawing
- \* location of buildings, overlooking windows and trees on adjacent properties
- \*\* location of all easements
- \* location of utility points of attachment:
  sewer
  stormwater
  water meter
  electricity
  gas
  telephone
  cable TV, etc.
- \*\* on-site septic or other sewerage treatment systems
- \*\* on-site location of stormwater absorption systems
- \*\*\* location of verandahs, terraces, decks, pergolas, etc
- \*\*\* location of outbuildings—sheds, carports, garages, pools, etc
- \*\* location of driveways, parking areas and paths
- \*\* reduced levels of important points on the site (RL)
- \* Finished floor level of floors (FFL)
- \*\* site contours
- \*\* location of retaining walls, battens, steep grades, etc
- \* location of trees as defined by preservation orders:
  to be retained
  to be removed
  proposed planting
- \* garbage bin area
- \*\*\* land surveyor reference if applicable

*Extra information specific to alterations and additions*
- \*\*\* outline of existing building
- \* proposed demolitions (if extensive, a separate demolition plan is recommended)
- \*\*\* hatched or coloured drawing of proposed additions
- \*\*\* dimensions from boundaries to new work

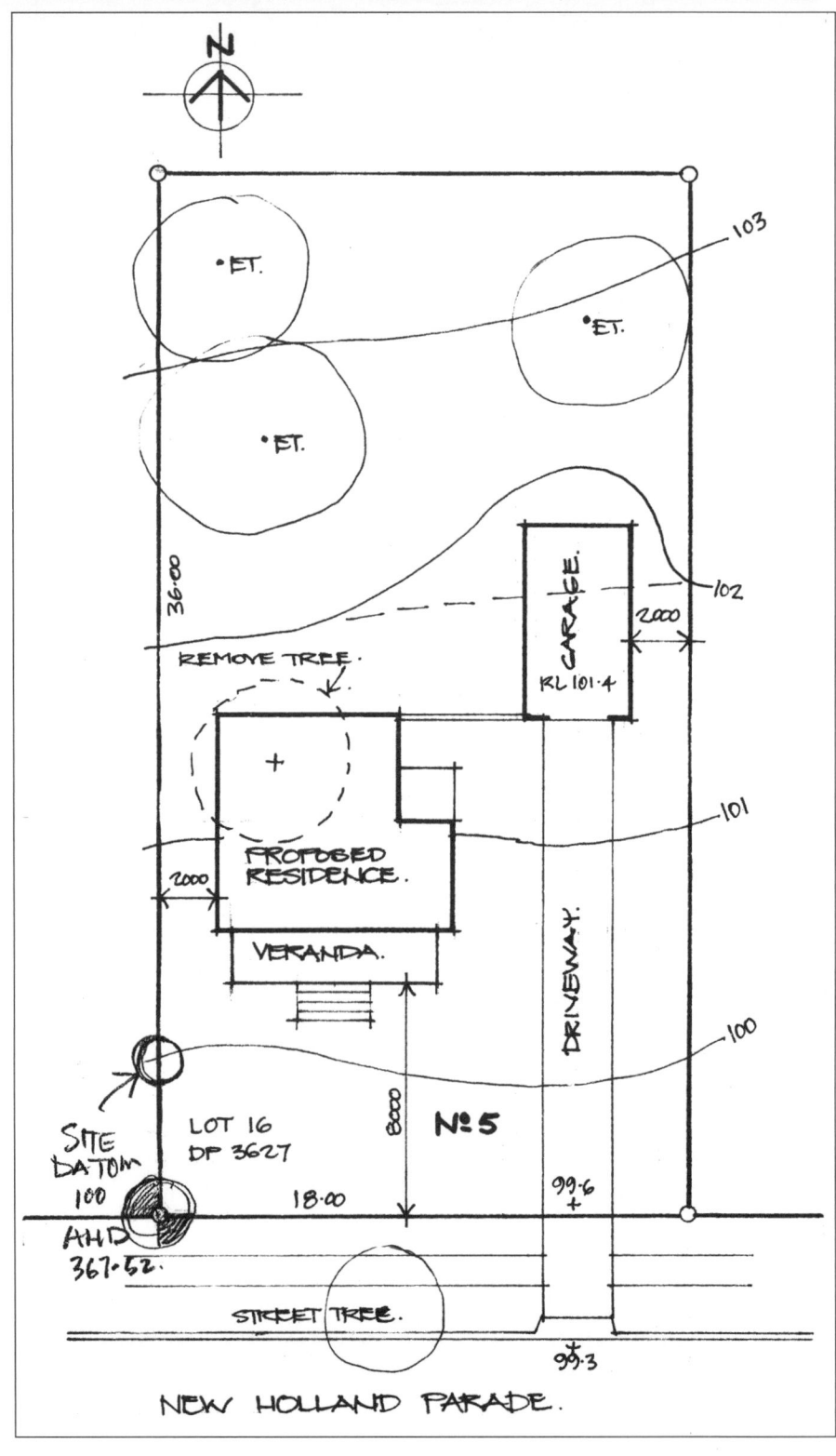

# BUILDING DRAWINGS • 13

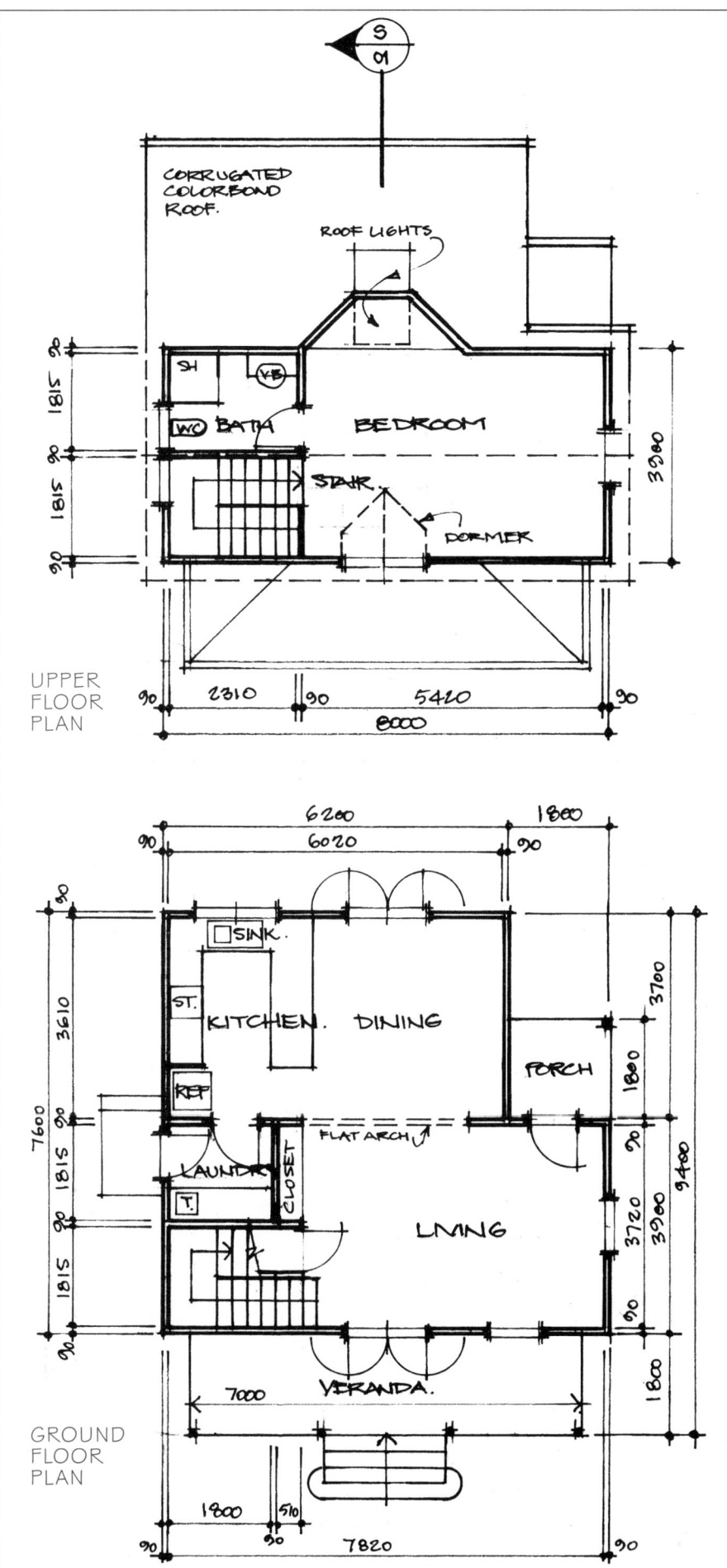

## Floor plans

** subfloor plan
* cut and fill under floor
* extent of subfloor structure, base walls, piers, stumps, frame
** extent and sizes of footings
* extent, sizes and locations of all sub-floor structures
* subfloor drainage system
***all ground- and upper-floor plans
***extent of all walls, doors, windows, room names
***all dimensions to locate walls, windows, doors, wall thicknesses
* all finished floor levels (FFL)
* all floor finishes
** location of all sanitaryware
***all steps or staircases

*Extra information specific to alterations and additions*

** extent of existing footings adjacent to new work
** existing building showing all walls, doors, windows
* extent of all demolition
***extent of all alterations to existing plan

43

# 13 • BUILDING DRAWINGS

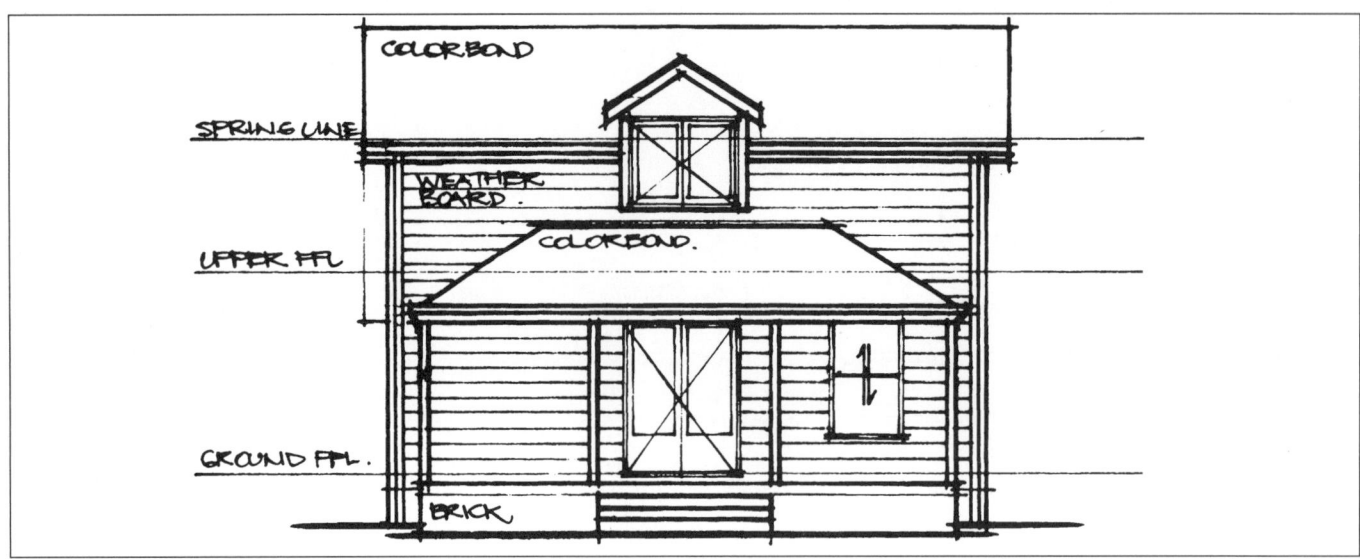

## Exterior elevations

\*\*\*all faces of the building
\*\*\*existing and altered ground line at elevations
\*\* window material dimensions and sash information
\*\* base, wall and roof materials
\*\* downpipe locations
\*\* chimneys and flues

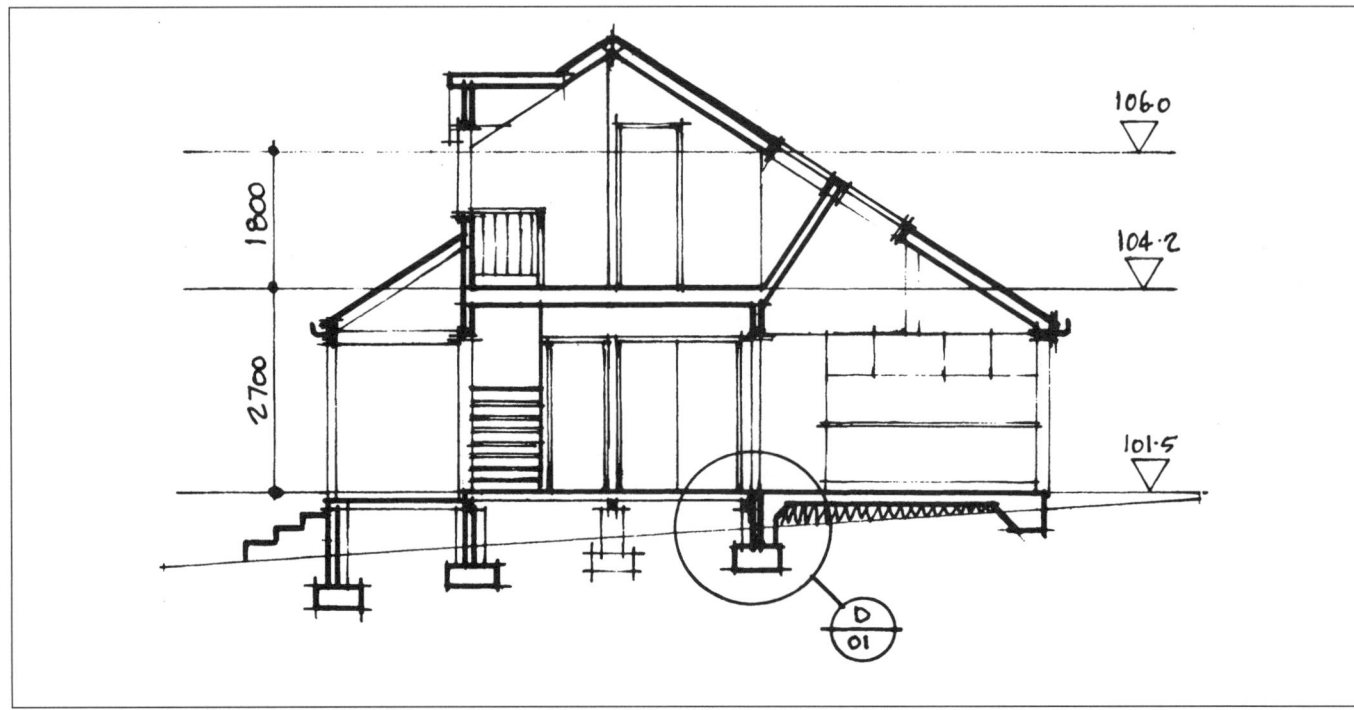

## Section drawings

\*\*\*existing and altered ground line at sections
\*\* location of any agricultural or other sub-floor drains
\*\* section at stairways
\*\* sections showing ground to structure clearances, FFLs and ceiling heights
\*\*\*floor construction system, materials and member sizes

\*\*\*wall construction system, materials and member sizes
\*\*\*roof construction system, materials, member sizes and pitch
\*\* ceiling heights
\* window head and/or sill heights
\*\* internal materials annotation
\*\* wall and roof cladding materials
\* flues, cappings, flashings, barges, gutters and associated works

*Extra information specific to alterations and additions*

\*\*\*section showing connection between existing and new work

# BUILDING DRAWINGS • 13

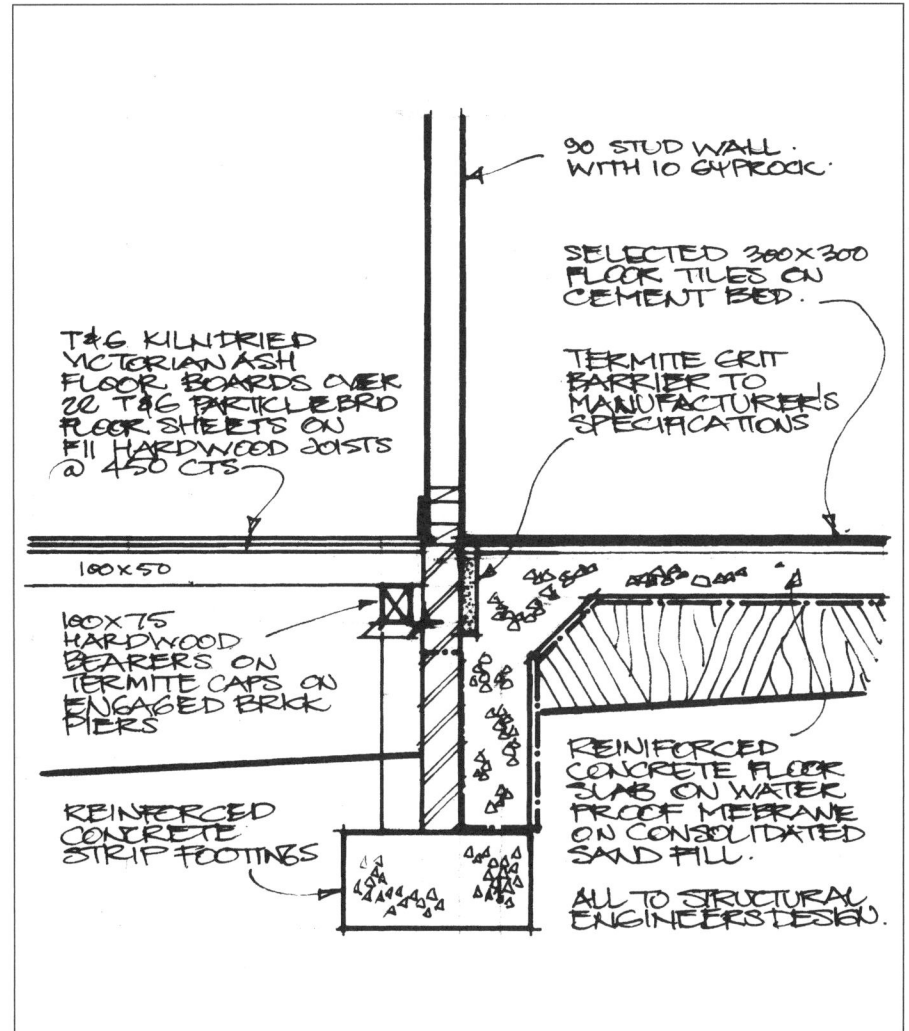

## Construction details

There is no specific checklist for construction details but the basic rules are:
- If it is a construction technique that is likely to be unknown to the builder, then detailed drawings will assist.
- If there are sections of the building that are critical to the final appearance of the design, then detailed drawings are essential.

Detail drawings should be to a scale that provides the easiest information transfer; they can be two-dimensional or three-dimensional and they can be ruled or freehand—the key is to communicate.

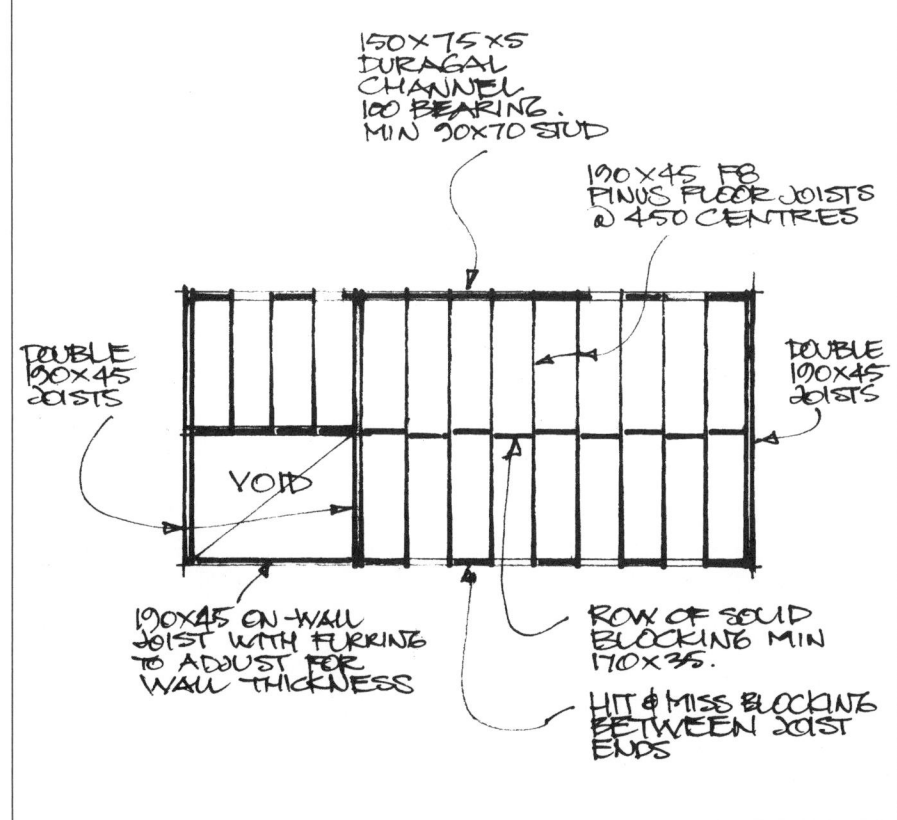

## Structural engineer's drawings

These are drawn by engineering draftspeople to conventions that are quite different from the conventions used by architects and building draftspeople.

Reinforced concrete drawings concentrate on the locations of steel fabric and bars; structural steel drawings are often single-line diagrams. It is important that OBs are sure that the people who are using these drawings on the site are competent to do so—take time to have the engineer explain them.

# 13 • BUILDING DRAWINGS

## Landscaping plan

* plan with all building outlines, paths, drives and other hard features
* plan showing all existing major plantings, proposed plantings and removals
* plans of specific beds or areas
* contour plans and sections showing existing and proposed garden form
* drawings of garden features, arbours, seats, ponds, paths, etc
* drainage and irrigation system
* Structural systems, retaining walls and banks drawn by an engineer
* all plants annotated with botanical and common names
* alterations to existing fences and new fences
* erosion and sedimentation control systems, temporary and permanent

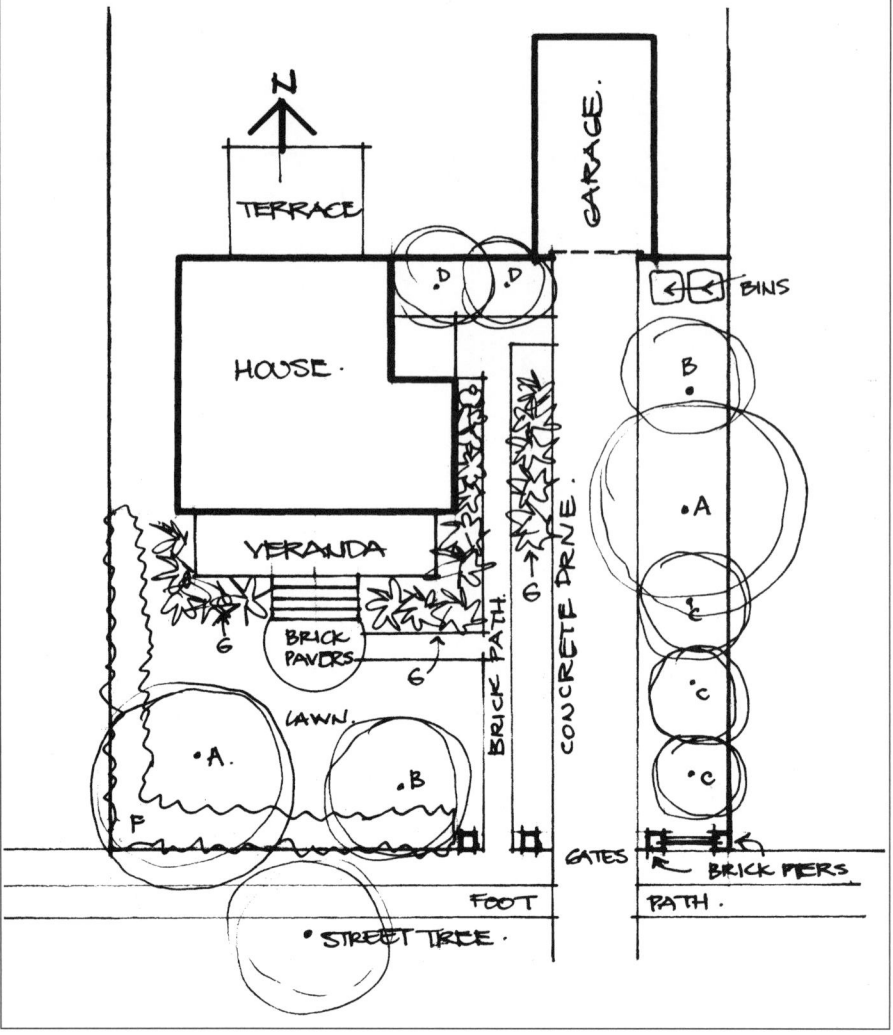

## Reflected ceiling plan (incorporating the electrical and lighting plan)

This is a plan of the ceilings of the house as if they were being reflected in the floor—see sketch.
* form of ceilings, bulkheads, cornices, roses, applique
* all ceiling-mounted light fittings
* all wall-mounted lights
* power outlets, to keep electrical info on one sheet

This drawing is not required on simple alterations but is highly recommended on more complex projects.

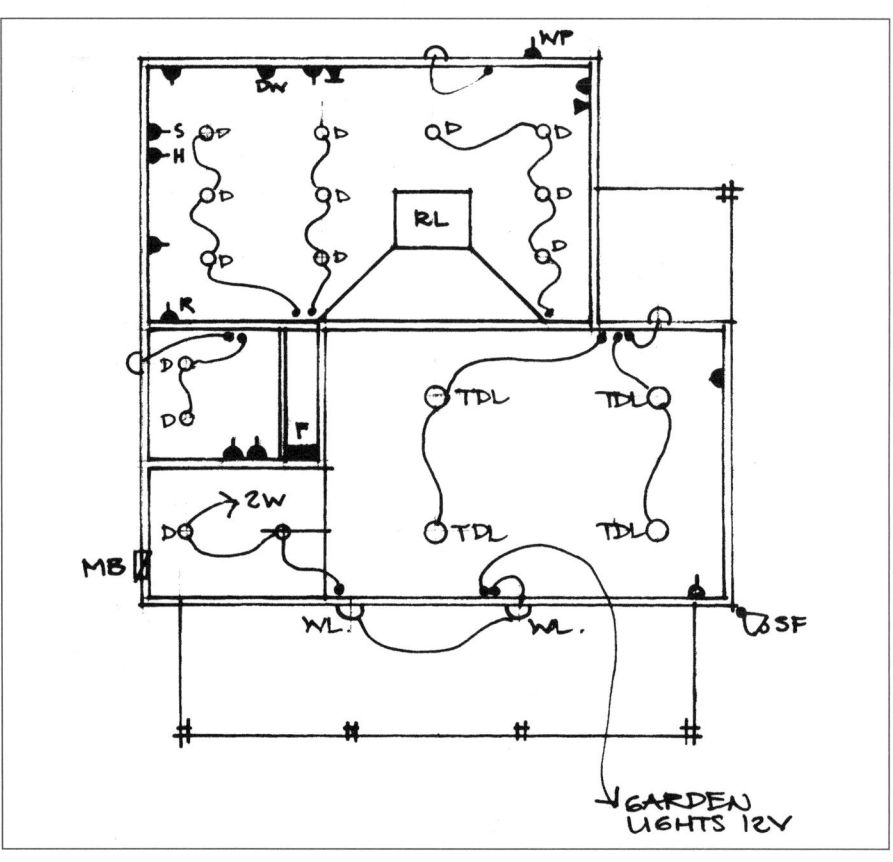

# BUILDING DRAWINGS • 13

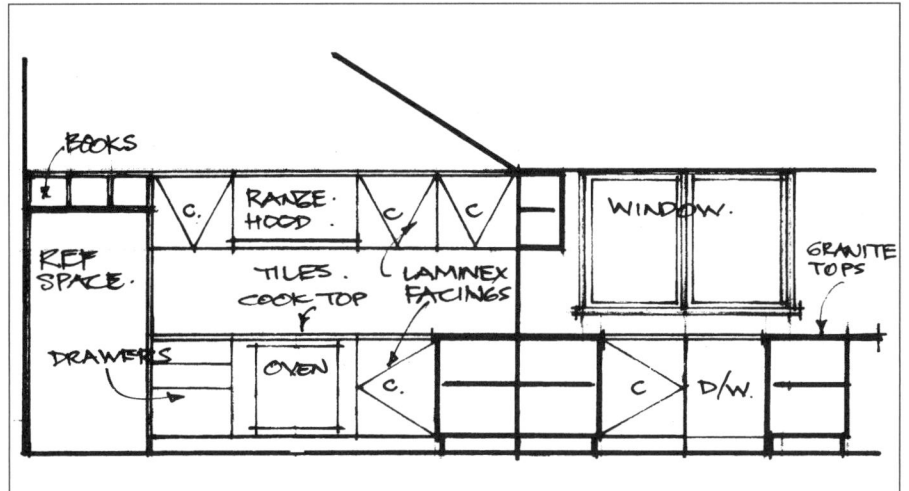

## Kitchen details

** Kitchen plan at 1:20 showing all bench-tops and overhead units
** kitchen cabinet elevations at 1:20 showing all doors, drawers and appliances
* detailed construction sections if non-standard construction is required
* appearance details for edges, panels, etc, if required

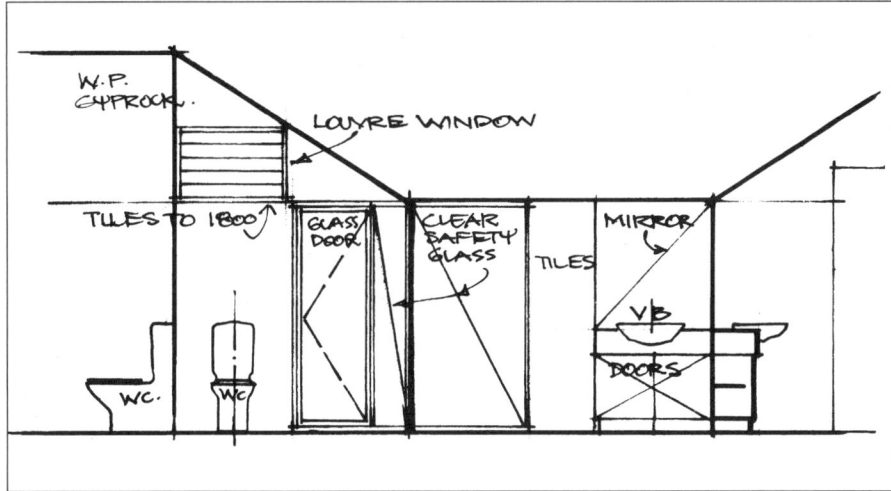

## Bathroom details

** plan at 1:20 showing all sanitaryware, floor patterns, waste, etc.
** Elevations at 1:20 including sanitaryware, extent of tiling, vanities and mirrors
* details of any special cabinets

## Cabinet Details

** elevations at 1:20 showing doors, drawers etc.
* sections at 1:10 showing basic construction
* Details at 1:5 showing specific connections

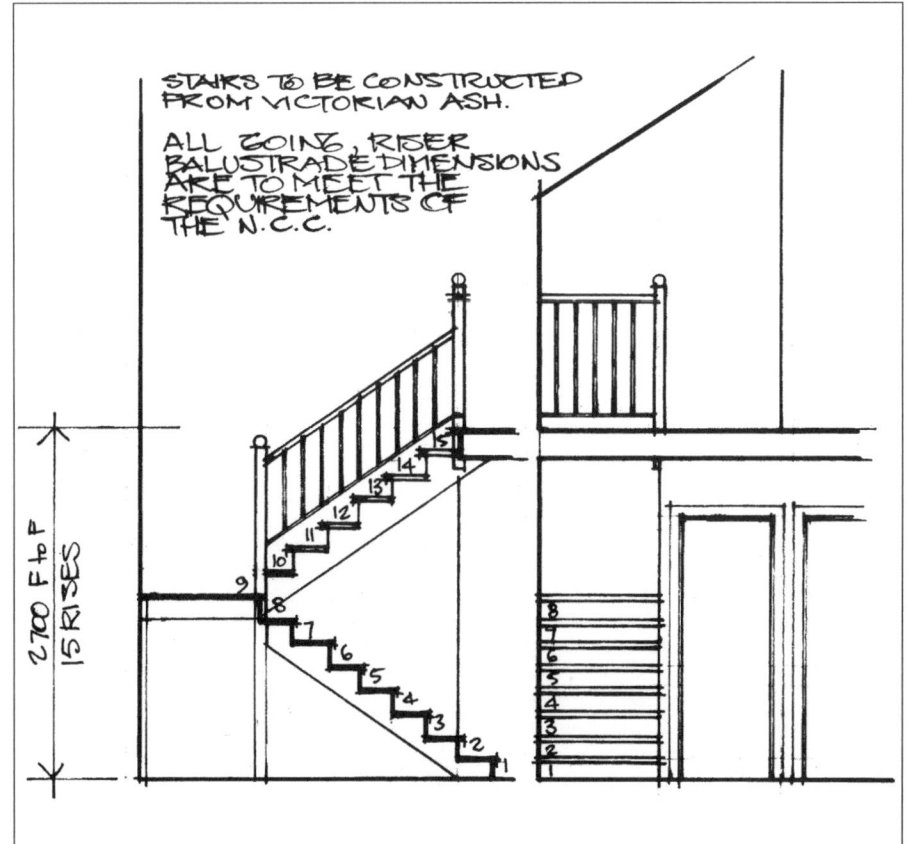

## Stair details

*** plan showing lower- and upper-floor key dimensions, treads, newels and balustrades
*** section at 1:20 or 1:10 to show construction and heights
*** elevations of balustrades and stringer appearance
* details of construction at 1:5

Although a stair can be ordered without drawings and good results are achievable, unless the stair to be used is virtually identical to the one on display, it is wise to prepare detail drawings or at least have the stair manufacturer provide detailed workshop drawings.

# 14 • SPECIFICATIONS

## What is a specification?

The specifications are one of the documents that make up a normal building contract. The specification is predominantly concerned with the quality of the materials and labour to be used in a particular project.

Specifications generally include a preliminary section giving some basic information on site conditions, access and other administrative matters. Care should be taken not to paraphrase sections of the conditions of agreement, as there is a danger that the contract, as a whole, can be weakened by doing this.

## Do OBs require a specification?

Any building application in Australia requires a set of working drawings and a specification to be lodged with the building surveyor as part of the approval system. The specification is required so that it can be determined whether or not the proposed building will meet the quality Standards required by laws and other regulations covering building construction.

OBs require specifications for at least this purpose, but the specification is also particularly valuable to any person constructing a building as a guide to the minimum standards of quality that are required.

If an OB is borrowing money from a lending authority then it is often a condition of the lending authority that the building to be constructed and used as collateral for the mortgage should be to a quality deemed desirable by them.

Some lending authorities can provide OBs with standard forms of building specifications that have been written to suit various types of construction and to meet the local authority's minimum quality standards.

In most cases an OB will sub-let certain parts of the construction to specialised tradespeople. For example, electricians, plumbers, drainers and bricklayers. When this occurs, the specification becomes an invaluable guide to both the OB and the tradespeople concerned, to the extent of the work and the quality of the materials and labour required. Without a specification the OB could be at the mercy of the tradesperson in matters of quality. This is obviously not desirable, and OBs should make sure a specification, particularly relevant to their project is available to them for the purposes outlined.

The main sections of the specifications are listed under trades headings, and these are listed in roughly their chronological order. There are other ways to write specifications, from a few notes on the margins of drawings, to a custom-written set by an architect or other suitably qualified person.

Specifications notes on drawings are not recommended, except for on small projects using standardised methods. Custom-written specifications are justified unless there are complex trade descriptions, very high standards of quality required, or the building system to be employed is not standard practice.

## Where do OBs obtain specifications?

Specifications are available from lending authorities, but this is not the only source and in certain instances not always the most desirable. Other sources of specifications are:

### 1. Architects
Architects are trained to prepare specifications for specific projects. Architects working on projects for a client has the capacity to write a specification that is particularly tailored for the project. Where OBs are employing architects to design and document their projects, they should give careful consideration to allowing the architect to prepare a specification directly applicable to their project.

This can be handled in one of two ways.

The architect writes the specification from scratch. This is a fairly lengthy exercise but is particularly relevant where an OB is undertaking a project that uses highly technical building systems or methods of construction that are different from the generally accepted norms, for example, a house constructed predominantly of fabricated steel, a mud brick house or similarly uncommon construction methods.

The architects can use a standard form of specification and modify this document by the addition of appendix, addendum, schedules or the like. This approach is particularly useful where the OB is undertaking a project that is relatively conventional in construction but where special particular information is required for the project.

If an architect prepares a specification, it is worthwhile considering having the architect prepare the specification in a format that allows it to be used in a manner directly related to the way the OB intends to carry out the building program.

### 2. Natspec
A range of specification are available from Natspec Construction Information, these are revised regularly to ensure compliance with the National Construction Code (NCC) and the relevant publications of Standards Australia they are of great assistance to all OBs.

(The main body of the text to give sound protection of the principal to a contract without placing excessive or difficult to achieve conditions on the contractor.

The Natspec specifications list the Standards affecting residential building in Australia. This is a useful reference if an OB wants to make sure work is up to the correct standard. Not all local libraries will hold copies of all standards so check with Standards Australia to obtain information on the closest reference library to you. Many TAFE colleges and universities will hold reference sets.

Natspec also contains information to help the OB bridge the jargon gap.

### 3. Lending authorities
In some states the lending authorities have combined to produce a range of standard specifications that outline the minimum standards of quality against which they'll lend money.

Those standard specifications are often of sufficient standard for use by OBs, although care must be taken to upgrade any sections that do not fit the method of construction or level of quality required by the OB.

It is normal to adapt these specifications by deleting irrelevant clauses and by inserting additional clauses by way of addenda. Also, provision is generally made to allow OBs to insert particular member sizes, especially in the carpentry section, to suit individual projects.

The number of specifications from lending authorities in circulation have been reduced in recent times with less demand for custom-built homes and more project homes being built.

### 4. Manufacturers' specifications
Manufacturers of many specialised building products, components and/or systems often provide standard specification sections to suit their products. These can often be useful to an OB in briefing a sub-tradesperson, as they generally set out clearly the correct methods of installation.

It is a good practice to convey information for items such as window and door hardware by means of schedules. These are an itemised or tabulated means of conveying the architect's or owner's wishes to other people concerned with producing, providing or fabricating activities during the construction process. The builder may find that the information contained in the manufacturer's specifications is easier to understand than the working drawings and specifications.

### 5. Information to Trades and Builders
Sometimes a short document, often labelled Information to Tenderers, is distributed with the working drawings, specifications and the conditions of agreement to the builder or other contractors who have been asked to tender quotations for all or part of the works.

This document should contain a disclaimer noting that its contents are to assist tenderers in providing a quotation, but that anything contained therein will not automatically be included in the final contract agreement unless the tenderer indicates the desire to do so in their tender. The tenderer can propose amendments to the details included within the Information to Tenderers, or even add further conditions attached to the official tender form.

# SPECIFICATIONS • 14

These should be the only documents that are amended during the tender process. The working drawings, specifications and the conditions of agreement should remain intact at least until a tenderer has been issued with the final contract documents.

## Some other notes on the value of specifications

Houses to be constructed in mud brick, using poles and beams, structural steel framing, extensive use of reinforced concrete, heavy precast concrete elements or to elaborate energy conservation principles may benefit from the use of custom-written specifications. However, if additional effort was used in the drawings, particularly through the extensive use of, compenent and assembly details, an addendum to a standard specification may be suitable.

Many proforma conditions of agreement documents require additional information and choices to be provided by the tenderer. The explanation of this information should be included in the Information to Tenderers document. OBs who are not familiar with building contracts are advised to gain the assistance from an architect, consultant project manager or solicitor when completing this information, and discussion with an insurance company or broker is advised when choosing the insurance options.

The biggest gap between the working drawings and a standard specification of materials to be used, and labour to be employed, is the definition of particular materials and fittings.

The drawing's role is to provide all important dimensions and the necessary information to allow the quantities of materials to be measured and then placed in the correct location.

The specification's role is to nominate the standard of quality to be used when purchasing and placing materials and fittings.

Particular materials can be annotated on the drawings, but if too many notes are used, the drawing ceases to be a quantity document and broaches the area of quality. Ideally, the particular materials and fittings should be included in the specifications with only pointer references on the drawings, but this can be cumbersome and contractors can spend inordinate time scanning through the specification list.

If the Information to Tenderers document is used to bridge the gap between the drawing annotation and the specification's general quality standards, a fair working document can be attained.

A simple method is to write down all the trades that are to be employed on the project:
Demolisher, Concreter, Bricklayer, Carpenter, Plasterboard fixer, Plumber and Drainer Electrician, Steel worker, Tiler, Painter.

This list may not be as extensive as in the specifications; it should be kept simple and only deal with trades where particular materials and fittings need to be expanded. In the above list it may be decided that the Demolisher and the Concreter are adequately described in the standard specification, and the Plasterboard fixer may not require any further explanation.

Consider then what may be listed under bricklayers:

A  The drawings may provide pointer annotation which indicates the general extent of face brickwork, common brickwork and brickwork to be rendered.
B  The specification will indicate the basic quality of bricks, mortar, coursing and laying tolerances.
C  The Informantion to Tenderers can expand the annotation by defining face bricks by manufacturer, colour, texture, size, course style and bond type.

Example: All face bricks will be Bloggs, Heritage Blue, clinkers, Slim lines, laid in a one third, overlap stretcher bond with natural ochre tinted recessed compo mortar, as determined by on-site samples (allow for five sample mixings). Joints and general appearance are too closely match adjacent existing brickwork.

This information could be annotated on the drawings, but it is becoming a fairly wordy statement, and if it was included in the specification, the information could be spread over three or four clauses. This method is simple and easy for the tenderers to understand.

Tenderers need to be provided in clear terms what they are tendering on, and owners need to be sure they have maximum protection should there be any disputes on site. A building contract often ends up being a conglomerate of simple information encased in a set of hard-edged legalese documents which, in many cases, neither side really understands or wants to worry about – until one or the other defaults.

## Understanding quantity and quality

The concept of defining quantity and quality is a constant concern to all people associated with building projects. At least a family who is altering or adding to their home can often point to an existing feature and simply say: 'Match it'.

To try to define the difference between quantity and quality, consider a standard Australian metric brick – it is approximately 230 x 110 x 76 mm.

If this is expressing the size of an individual brick, this is a statement of quantity, but if this is the size of all the bricks to be used on the project, then this is a statement of quality. To say that any brick that is not this size is substandard and should be rejected could be unreasonable because the manufacturing techniques used to make bricks cannot deal with close tolerance critical dimensions. So quality standards have been devised; the brick is allowed to be slightly bigger or slightly smaller than the perfect brick. There are set limits, though, and a face brick may have to meet higher standards of dimensional tolerance than an exposed common brick, or a brick to be built into a wall that will be covered with render.

It is much easier in the real world if consumers do not have to worry about whether a brick is manufactured to all relevant standards or not. Although we could check a brick's dimensions, it is too hard to check its composition, durability, absorption properties and the like.

To ensure that the bricks are of a controlled quality, most manufacturers have introduced quality assurance. This means the manufacturer has initiated systems that guarantee every brick sold will conform to certain published qualities. This is not a money-back guarantee, but a guarantee that there is no need to have a money-back guarantee. The product will not be supplied if its quality cannot be assured.

There are international standards for quality assurance systems, and Standards Australia has been instrumental in preparing a number of local reference codes.

Some builders apply quality assurance (QA) principles to their projects and there are certain subcontract trades and installers who are qualified to issue certificates of quality. QA means labour, materials, fittings and construction are checked systematically against accepted standards. If all the QA checks are conducted on a building process, no order should be late, no quantity should be wrong, no tradesperson should be underqualified. If the standards chosen are comprehensive, and the quality levels are achievable, then a product of consistent quality should be achieved with no weak links.

OBs can apply QA principles to their projects and, although it takes some effort, there are real benefits to be achieved. Read a book or two on the subject and check with your local TAFE system to see if they offer a convenient course.

# 15 • SCHEDULES

Where words and pictures fail, try a table. This is good for colours and finishes, fixtures, fittings, windows, doors and door furniture, lockware, sanitaryware and bathroom ware.

## Advantages of Schedules

1. Checking for errors in omission or duplication of items is simplified.
2. The counting of similar items for obtaining estimates or placing orders.
3. The omission of information relating to an item becomes redundant as the appropriate space in the schedule would be then blank.
4. If the information is set down systematically, prolonged searching through a specification or trade literature is avoided.
5. Provides a means of checking orders on site.

The recording of information in schedules makes the task of reference more foolproof than that of searching through drawings for itemised information on assemblies (windows, doors). Schedules make the building administration process efficient and tabulates the information about assemblies in three areas:

1. Item, is the thing described, e.g., door, window, finishes.
2. Size, includes all dimensions relevant to the particular item, e.g., the door schedule will tabulate thickness, width and height of each door, but a finishes schedule would not record room dimensions.
3. Characteristics, includes all the information necessary to provide a complete description of the item such as, quality of material, method of construction and finish.

Some information may be more clearly conveyed in a schedule than by a dimensioned drawing. The schedules are not intended to supplant the drawings, but to extend the drawings into checklists.

Examples of types of schedules are:

1. Window schedule, including frames.
2. Door schedule, including frames and thresholds.
3. Hardware schedule (unless incorporated in the door and window schedule).
4. Fittings schedule, including small standardised fittings such as towel rails.
5. Decoration schedule.
6. Sanitaryware fixtures and fittings schedule.

This is not a comprehensive list, there are many opportunities for using schedules.

The column headings under sizes and characteristics selected for each schedule cannot be standardised as they vary by the following criteria-

1. Who is going to use the schedule (tradespeople, suppliers)?
2. What information do these people need?

If these rules are borne in mind, the schedules created will be comprehensive without being cluttered with irrelevant information.

### Door and door frame schedule

| Door no. | Hand | Leaf Size | Door Type | Frame type | Hardware Furniture |
|---|---|---|---|---|---|
| 1 | L | 890 x 2040 40 thick | WRC Traditional Four Panel | WRC Solid Rebated | 3 Stainless Steel Butt Hinges Double Cylinder Dead Lock Lever Latch |
| 2 | L | 820 x 2040 40 thick | WRC Half Glass | WRC Double Rebated | 3 Stainless Steel Butt Hinges Double Cylinder Dead Lock Lever Latch |
| 3 | R | 820 x 2040 32 thick | WRC Insect Wire | WRC Double Rebated | 3 Stainless Steel Butt Hinges Door Closer Lever Latch |
| 4, 6, 7 | L | 820 x 2040 40 thick | Flush Panel Solid Core Vic Ash Veneer | Vic Ash Solid rebated | 2 Stainless Steel Butt Hinges Lever Latch |
| 5, 8, 9 | R | 820 x 2040 40 thick | Flush Panel Solid Core Vic Ash Veneer | Vic Ash Solid rebated | 2 Stainless Steel Butt Hinges Lever Latch Privacy Snib |
| 10 | FS (Face of Wall Sliding) | 870 x 2040 | Flush Panel Solid Core Vic Ash Veneer | Vic Ash | Head Track Flush Pulls Vic Ash Pelmet |
| 11 | CS (Cavity Sliding) | 870 x 2040 | Flush Panel Solid Core Vic Ash Veneer | Vic Ash | Cavity Head Track Flush Pulls Vic Ash Pelmet |
| 7 | Pair | 2 x 620 x 2040 | Vic Ash Full Glass Rebated Stiles | Vic Ash Solid Rebated | 2 x 2 SS Butt Hinges 2 x Recessed Bolts Rebated Lever Latch |

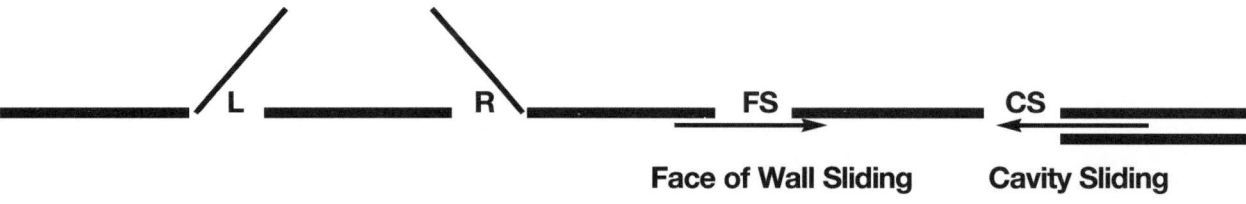

Face of Wall Sliding     Cavity Sliding

# SCHEDULES • 15

## Window schedule

| Ref. no. | Window Size<br>Width x Height (mm) | Opening Size<br>Width x Height (mm) | Material | Detail | Supplier and Code |
|---|---|---|---|---|---|
| 1, 2, 3<br>3 required | 3125 x 2100 | 3150 x 2125 | WRC<br>Safety Glass | Patio Doors | TimbaWindos<br>TFS-3121 |
| 4, 10, 11, 20<br>4 required | 1706 x 2100 | 1731 x 2125 | WRC<br>Float Glass | 4 Light Awning | TimbaWindos<br>TFA-1721 |
| 5, 16, 24<br>3 required | 873 x 2100 | 898 x 2125 | WRC<br>Float Glass | 2 Light Awning | TimbaWindos<br>TA-0921 |
| 6, 18<br>2 required | 1090 x 2100 | 1115 x 2125 | WRC<br>Float Glass | 2 Light Awning<br>As 5 | TimbaWindos<br>TA-1121 |
| 7 | 2140 x 675 | 2165 x 700 | WRC<br>Reeded Glass | Fixed Hi-Lite | TimbaWindos<br>TF-2107-RG |
| 8, 14, 15<br>3 required | 645 x 2100 | 670 x 2125 | WRC<br>Float Glass | 2 Light Casement | TimbaWindos<br>TC-0721 |
| 9, 25<br>2 required | 910 x 600 | 960 x 650 | Aluminium<br>Powder Coat<br>Float Glass | 2 Light Sliding | AlumeWindos<br>ALFS-0906 |
| 12 | — | 800 x 800<br>Nominal | Acrylic in<br>Aluminium | Roof light | RoofLiteCo<br>RL800SQ |
| 13 | 2756 x 1190 | 2781 x 1205 | WRC<br>Float Glass | 3 Light Awning | TimbaWindos<br>TFAF-2812 |
| 17 | 2410 x 600 | 2460 x 650 | Aluminium<br>Powder Coat<br>Obscure Glass | 3 Light Sliding | AlumeWindos<br>ALFSF-2406-OB |
| 19 | 946 x 1836 | 970 x 1870 | WRC<br>To Match Existing<br>Float Glass | Double hung | CustomsashCo |

## Finishes schedule

| Room/Space<br>Name | Floor | Skirting | Walls<br>North | East | South | West | Ceiling | Cornice | Code |
|---|---|---|---|---|---|---|---|---|---|
| Entry | A | E | J | J | I | I | I | E | |
| Bedroom 1 | B | E | I | I | I | I | I | E | |
| Bedroom 2 | B | F | H | H | H | H | I | G | |
| Bedroom 3 | B | F | H | H | H | H | I | G | |
| Study | A | E | J | I | J | I | I | E | |
| Lounge | B | E | I | J | J | I | I | E | |
| Family | A | F | H | H | H | H | J | F | |
| Kitchen | C | F | J | J | H | J | J | F | |
| Bathroom 1 | D | F | K | K | K | K | H | G | |
| Bathroom 2 | D | D | K | K | H | H | J | E | |
| Dressing Room | C | E | I | I | I | I | I | E | |
| Laundry | D | D | H | H | H | H | I | G | |

A: T & G brush box strip flooring, stained and lacquered
B: Carpet—wall-to-wall
C: Vinyl tiles
D: Ceramic tiles
E: Finger-jointed *pinus*, period style
F: *Pinus*, plain splayed
G: Plaster cornice, standard scotia moulding
H: Satin paint finish
I: Matt paint finish (for paint colour see Schedule 6)
J: *Pinus*, lining boards, varnished
K: Ceramic tiles to 1500 mm, then painted in satin above

# 16 • COSTING

The greatest problem facing many OBs is determining how much their proposed house is going to cost. In many cases OBs' homes appear to cost more than the original budget, and there are many reasons for this. However, this need not be the case and in this chapter a general method to control cost blow-out is outlined. All OB should apply a similar method to their cost determination and control.

It is often stated that OBs can save up to 50 per cent of the price of a contract-built home by doing the work themselves. This figure, though not impossible to attain, is highly unlikely because the cost of a contract-built house that matches exactly the design and quality of the OB's home is often impossible to determine.

To achieve a saving near 50 per cent requires OBs to do all the labour themselves.

If the OB finds a project home that almost meets their requirements, they must be careful to add the base price of the project home to the 'on-site' costs:

| | |
|---|---|
| Base price of project home | $300,000 |
| @ 150sqm ($2,000/sqm) | |
| Cost to add double garage | $40,000 |
| @ 40sqm ($1,000/sqm) | |
| Cost to adjust land fall | $30,000 |
| Cost of services connection | $20,000 |
| Cost of fencing/paving | $15,000 |
| Cost of variations* (extras) | $15,000 |
| (nominally 5% of base cost includes | |
| * Relocation of windows, extra electrical points, and other minor variations). | |
| Estimated cost of base house | $420,000 |
| (say $2,800/sqm) | |

Is the quality of the materials in the project home up to the standard of finish required by the OBs building for themselves? This is unlikely, and the following extra/over material allowances are likely:

| | |
|---|---|
| Higher quality roofing | $5,000 |
| Higher quality wall facings | $5,000 |
| Double glazed insulated windows | $7,500 |
| Premium kitchen cabinets | $7,500 |
| Premium sanitaryware | $3,500 |
| Premium appliance | $3,500 |
| Higher quality internal doors | $2,000 |
| High ceilings | $20,000 |
| Quality additions to a project home | $54,000 |

(nominally a 12.5% cost escalation)

The base price of the project home converted to OB quality is now likely to be $474,000 or about $2,500 per square metre for a 150sqm house including an attached 40sqm garage.

These calculations are simple guidelines that require information that is current and location defined to provide a reasonable approximation of the cost of building a house.

Where, then, are the savings in being OBs, when it appears that if an OB house is completed for anything less than the advertised base price of a project home of similar size, good cost control has been exercised?

The savings are, that careful OBs can greatly improve the quality of construction, materials and inclusions of a house and still build the house for a cost under that of the accepted market minimum indicator, i.e., a project home of similar size and materials.

So, for rule-of-thumb pricing of a project home, choose a project home of similar size and layout and use the base price of that house as an indicator of the likely cost of the completed owner-built home.

Or consider, the cost of an architect-designed quality home is likely to be twice the cost of a project home of similar size. Therefore, if an OB can save 50 per cent of an architect quality home, then the savings are obvious.

Comparative estimates may assist for determining the general building budget for the owner-built home but more exact cost control is needed if the OB is to meet the allowable budget. To achieve this, the OB should determine in advance the likely cost of each section of the building work. To assist this process, a cost control sheet is included here for that purpose.

The cost control sheet is a guide document only and OBs will have to adapt and add to it to suit their particular project.

With GST, most items of labour and material associated with the construction of a house attract 10 per cent tax. The law requires most prices be inclusive of GST, but always check every quote requested. Some individual traders may not include GST.

---

### Reality check

**Can Owner Builders build a home?**
Realistically OBs cannot do all the work themselves, regulations require that electrical, plumbing and sewer drainage are licensed trades in most locations in Australia, and few OBs are skilled at laying brickwork or tiling roofs.

Therefore, it may be necessary to adjust the reference budget for the electrical, plumbing and draining works plus the cost to lay the bricks and roof tiles.

Consider also what costs may be associated with:

| | |
|---|---|
| **Design, documentation and project administration** | Architects |
| | Engineers |
| | Land Surveyor |
| | Building Surveyor |
| | Estimator/Quantity Surveyor |
| **Authority's fees and charges** | Owner Builders Licence/Permit |
| | Planning and Building Application Fees |
| | Compliance Deposits (Refunded after occupancy certificate) |
| | Utility charges, fees, rates, taxes levied during construction |
| **Insurances** | Worker's compensation |
| | Public risk |
| | All risk (materials and building) |
| **Materials** | Allow for rise in cost of labour and materials, reference current building industry CPI |
| **Equipment** | Purchase of tools and equipment |
| | Hire of tools and equipment |
| **Labour** | Day labour allowance |
| | Unpaid labour (OB's family) |

---

When estimating your own labour content do not forget to include a sizeable amount of time for running around organising labour and materials. and hours of phone calls before dawn and after dark.

A normal OB will find that the hours that can be devoted to the building on-site will average out at no more than 30 hours per week assuming that a full-time job is pursued at the at the same time.

OBs must be careful that they do not allow more of their time dedicated to the project to exceed the time that is possible and sensible to allocate.

# COSTING • 16

| Cost Control Sheet For Owner Builders | Materials | | Labour | | OB's time | |
|---|---|---|---|---|---|---|
| *Est. = estimated cost<br>Act. = actual cost | On site cost | | Subtrade Cost | | | |
| ITEM | Est. | Act. | Est. | Act. | Est. | Act. |
| **1. Pre site costs**<br>a. Architects or draftsman<br>b. Engineer<br>c. Other consultants<br>d. Insurances<br>e. Surveyor<br>f. Fees         **Sub total** | | | | | | |
| **2. Preliminary on site costs**<br>a. Temporary services<br>b. Workmen's services<br>c. Storage shed(s)<br>d. Setting out         **Sub total** | | | | | | |
| **3. Excavator**<br>a. Site clearance<br>b. Topsoil storage and/or removal<br>c. Excavations, bulk<br>d. Excavation, footings<br>e. Excavations, drainage<br>f. Excavations, other services<br>g. Grading and filling<br>h. Granular fill<br>i. Road opening and restoration         **Sub total** | | | | | | |
| **4. Concreter**<br>a. Readymix concrete<br>b. Steel reinforcement<br>c. Formwork and centring<br>d. Concrete, placing<br>e. Concrete, finishing<br>f. Concrete, stripping<br>g. Concrete, curing<br>h. Concrete, coring<br>i. Membranes<br>j. External steps and landings<br>k. Driveways and paths<br>l. Mower strips, etc.<br>m. Vehicular footpath crossing<br>n. Terrazzo thresholds         **Sub total** | | | | | | |
| **5. Bricklayer**<br>a. Bricks<br>b. Cement and lime<br>c. Sand<br>d. Colour and additives<br>e. Mortar, ready mixed<br>f. Damp-proof coursing<br>g. Termite strips and caps<br>h. Flashings<br>i. Switch board cupboard<br>j. Hoop iron and ties<br>k. Brick laying<br>l. Brick paving<br>m. Vents         **Sub total** | | | | | | |

# 16 • COSTING

|  | Materials | Labour | OB's time |
|---|---|---|---|
| **6. Metalworker** <br> a. Steel lintels <br> b. Steel beams and columns <br> c. Balustrades and handrails <br> d. Gates <br> e. Galvanising <br> f. Aluminium windows and doors <br> g. Aluminium screens <br> h. Aluminium awnings <br> i. Aluminium shower screens <br> **Sub total** |  |  |  |
| **7. Carpenter** <br> a. Floor framing <br> b. Wall framing <br> c. Roof framing <br> d. Roof trusses <br> e. Eaves and fascias <br> f. Insulation <br> g. Flooring <br> h. Fibrous cement cladding <br> i. Soffit linings <br> j. Hardboard cladding <br> k. Timber weatherboard <br> l. External vents <br> m. Clothes hoist <br> n. Bath riser <br> o. Compressed asbestos flooring <br> p. External steps and decks <br> q. Verandahs <br> r. Sheds/Garage <br> **Sub total** |  |  |  |
| **8. Joiner** <br> a. Door frames <br> b. Doors <br> c. Window frames and sashes <br> d. Kitchen cabinets <br> e. Linen/broom cupboard(s) <br> f. Laundry cabinet <br> g. Bathroom cabinet(s) <br> h. Vanity unit(s) <br> i. Skirting boards <br> j. Timber stairs <br> k. Timber panelling <br> **Sub total** |  |  |  |
| **9. Fencer** <br> a. Side and rear <br> b. Front <br> c. Internal <br> d. Gates <br> e. Letter box <br> **Sub total** |  |  |  |
| **10. Roofer** <br> a. Sarking <br> b. Battens <br> c. Tiles <br> d. Sheets <br> e. Built up roof system <br> f. Patent roof system <br> g. Roof lights <br> h. Flashings <br> **Sub total** |  |  |  |

# COSTING • 16

| | Materials | Labour | OB's time |
|---|---|---|---|
| **11. Internal linings** <br> a. Wall and ceiling straightening <br> b. Ceiling battens <br> c. Wall linings <br> d. Ceiling lining <br> e. Cornices <br> f. Angles and trims <br> g. Vents <br> **Sub total** | | | |
| **12. Renderer** <br> a. Sand/sponge finish cement render <br> b. Set finish render <br> c. Waterproof render <br> d. Stair topping <br> e. Cornices <br> **Sub total** | | | |
| **13. Plumber** <br> a. Road opening <br> b. Main to metre water service <br> c. Cold water service <br> d. Hot water service <br> e. Taps and fittings <br> f. Sanitary plumbing <br> g. Roof plumbing <br> h. Flashings <br> **Sub total** | | | |
| **14. Drainer** <br> a. Pipes and fittings <br> b. Pipe laying sewer <br> c. Pipe laying stormwater <br> d. Sewer connection <br> e. Septic tank and disposal <br> f. Sullage disposal (inc. grease trap) <br> g. Drains under pavements <br> h. Agricultural drains <br> i. Pits and other works <br> **Sub total** | | | |
| **15. Gas fitter** <br> a. Mains to meter service <br> b. Consumer service <br> c. Installation of appliances <br> d. LPG installation <br> **Sub total** | | | |
| **16. Electrician** <br> a. Mains to meter connection <br> b. Meter to switchboard <br> c. Electrical rough-in <br> d. Power points <br> e. Light points <br> f. Appliances and fans <br> g. Heating and air conditioning <br> h. Telephone wiring <br> i. Television antennae and wiring <br> **Sub total** | | | |

# 16 • COSTING

| | Materials | Labour | OB's time |
|---|---|---|---|
| **17. Glazier**<br>a. Glass<br>b. Installation<br>c. Special glazing<br>           Sub total | | | |
| **18. Painter**<br>a. Paints<br>b. Stains<br>c. Preparation and surfaces<br>d. Application<br>e. Non paint materials<br>           Sub total | | | |
| **19. Tiler**<br>a. Wall tiles<br>b. Floor tiles<br>c. Trims<br>           Sub total | | | |
| **20. Appliances and equipment**<br>a. Hot water service<br>b. Stove<br>c. Cooktop<br>d. Wall oven<br>e. Microwave oven<br>f. Dishwashing machine<br>g. Garbage disposal<br>h. Garbage compactor<br>i. Range hood<br>j. Exhaust fans(s)<br>k. Bath(s)<br>l. Basin(s)<br>m. Vanity unit(s)<br>n. Shower base(s)<br>o. Shower screens<br>p. Laundry sink(s)<br>q. Laundry tub(s)<br>r. Toilet pan(s)<br>s. Toilet cistern(s)<br>t. Toilet seat(s)<br>u. Soap and paper holder(s)<br>v. Towel rail(s)<br>w. Shaving cabinet(s)<br>x. Spa bath(s)<br>y. Space heaters<br>z. Air conditioner(s)<br>aa. Light fittings<br>bb. Security systems<br>cc. Music/communication system<br>dd. Door chimes<br>ee. Taps and outlets<br>ff. Door furniture<br>gg. Clothes hoist<br>hh. Insect screens<br>ii. House numerals and name<br>           Sub total | | | |
| Total estimated | | | |
| Total actual | | | |
| Difference | + or − | + or − | + or − |
| Total cost of project | $ | | |
| Total owner builder hours | | hours | |

# TENDERING, CONTRACTING AND LEGAL • 17

The building industry and the laws that control and regulate it vary significantly from state to state, so this chapter can only act as an introduction to the general concepts applying to legal relationships and the law.

Owner Builders should consider that, although a contract is an important tool in laying out the extent of service offered by one person to another, and the payment that is to be paid for the service, it will not make for a smooth personal relationship between the signatories if they are not prepared to work together.

When negotiating any contract bear in mind that although the tendered price is of primary importance, it does not make sense to give a contract to anyone that you do not feel at ease with. It is generally better to keep looking until you find a contractor who has a sensible price and a compatible personality.

All workers who contract out parts of a building project are called sub-contractors, particularly on projects where a builder is organising the labour and materials. OB projects, however, are quite different because, rather than a head contract between a builder and the proprietor and a series of supporting sub-contracts, every sub-trade has an individual contract.

## Tendering

One of the hardest tasks for some OBs is to find reliable sub-tradespeople from which to get tenders. Tendering means two or more people, or firms, are approached and requested to give a price for the completion of a section of work.

When preparing the information to tenders for any section of work, it is recommended that OBs provide the same clear statement of the work to be undertaken to all tenderers. The tender is generally limited as far as practicable to being required to provide only the price in the tender, avoiding variables such as time, alternative materials or methods, or reduction in quality of materials to be used. Complexities such as hourly rates, rise and fall clauses, pre-payments, cost plus and discount for cash alternatives should be treated with care. If there are too many variables in tenders received comparison can become impossible and the OB would have been as wise to have employed the first person talked to.

Considering some of the variables and alternatives, individually.

### Time
Some tenderers may request a limited period of time in which they are prepared to carry out the work. Check such limits carefully and make sure that, if a time-limited contract is accepted that it is realistic to accommodate the contractor on-site by the due date. Avoid allowing a contractor to work outside the time nominated in the contract, as this may increase liability for extra cost.

### Alternative materials
In many cases alternative materials do not cause any particular problem, but if one tenderer requests a material variation and you approve it, it is good practice to alert the other tenderers, this increases the opportunity of receiving the best possible price. Before approving alternate materials request a samples of the materials so that they can be compared with the originally specified materials.

### Alternative methods
As with alternate materials, alternative methods may not cause any particular problems, but it is wise to check carefully before granting approval, as in some instances alternative cheaper methods may reduce life expectancies. than more accepted materials and traditional practices.

### Reduction of quality
There are few circumstances when an OB should approve a reduction in quality for a cheaper price. An obvious exception is where a section of work is tendered at a price much higher than the budget can allow and quality reduction is inevitable. But take care not to reduce quality too far and exchange low cost now for high maintenance costs later

### Hourly rates
OB are going to come across some trades where hourly rates seem to be the accepted method of charging, but take care, for if it is not known what is the sensible time to complete the task then there is a chance that the tradesperson may take advantage of your ignorance. Much better to get a price you can compare with another tenderer. Few hourly rates work out cheaper over a job when compared with a competitively tendered price. Consider: tradespeople aim to collect an amount of money each week to meet normal business and living expenses, therefore a tradesperson will normally set an income goal per day to reach the weekly target income. It is possible, but not necessarily likely, that a tradesperson who could finish a job in 7 hours and then go home will extend the job out to 8 hours to maximise the day's pay.

It is important that an efficient project management system is maintained on projects that have hourly day labour or equipment hire rates

### Rise and fall contracts
These are contracts that provide a tender price that can be varied to accommodate any rises or falls in the material and labour price indices. OBs should attempt to avoid this form of pricing where possible, if only because the information on relevant price movements is often hard to obtain, and disputes over whether work was done before or after the effect of a price rise are common.

### Pre-payments
There is almost never a time when an OB should give tradespeople money in advance. If there is a large material component to a section of the work in question then the OB may see an advantage in paying for the material direct to the supplier, but only if the goods or materials are delivered into the control of the OB. The general rule is only pay for what you have received and where possible retain a proportion of the payment, between 5 per cent and 10 per cent is reasonable, to ensure quality and completion.

If there is no other choice but to give a supplier or contractor money in advance, make sure there is a real contract in place. Try to have clear descriptions of the quantity and quality of materials and services to be provided at a lump sum value; avoid hourly rates or materials at cost conditions. Include in the description of the work, when the materials and services are to be provided; a delivery or commencement date and a completion date is a good start. Pay the smallest possible advances and pay them as close to delivery or service as possible. You may want to check the offices, factories, warehouses and yards of contractors who want payment in advance to see if it appears a viable business.

A simple method of ensuring a contractor starts on a project and gets on with the job is to agree to pay an advance on the first invoice claim after the work has commenced. For example, a tradesperson may be working on the site over a few weeks. At the end of the first week, pay the value of the work completed plus a percentage in advance (perhaps the equivalent of half a week's pay). All other claims are paid at net value of work completed except the last claim, which is adjusted to take into account the advance as well as any agreed retention sum (this is kept until the end of the defects liability period).

### Cost/Plus
This is where the tenderer offers to do the work for its real cost plus an overseeing charge. It generally does not work in small projects and even in large projects experienced project managers/contract administrators are necessary to ensure everything is continually efficiently documented .

It is too easy for material and labour costs to be inflated to the disadvantage of the OB, undeclared discounts, commissions and overcharging for labour are devices that have been used in cost plus contracts in the past, and there is no reason to believe that there are still not disreputable operators. A good tradesperson should be able to estimate the quantities of labour and material in most jobs; cost plus may be covering a less capable contractor's inability.

### Discounts for cash
It can be tempting to accept an offer that is based on a sizeable discount for cash. Before taking up any cash offer think about why it can be offered. The usual reason that will come to mind is that the tenderer is probably avoiding income tax. When people do not pay their share of income tax, all others have to pay a larger share. It is against the law to assist anyone to avoid paying income tax so be sure you have a clear conscience before taking the 'for cash' discount. A further problem with 'for cash' deals is that there is seldom, if ever, a written contract, which leaves the OB open to a host of liabilities and little or no protection. Legislation is now covering the cash payment question through the GST and Homeowners Warranty schemes.

OBs building a whole house or a major

# 17 • TENDERING, CONTRACTING AND LEGAL

extension will be required to manage the GST for all labour used on their projects. They may be required to fill in reports that are sent regularly to the ATO. Be aware that as soon as a permit is issued from a regulatory authority that the ATO is advised of the particulars; they may contact applicants for building permits but, if not, the applicant is responsible.

## Contracting

After the OB receives a price for a section of the work, a contract should be executed between the two parties to describe the extent of the work involved, the materials to be supplied, the plant to be used and the price to be paid for the work done. If possible, a schedule of the times and dates that the contractor should be available to carry out the works contracted should be part of the contract agreement. A schedule of critical activities and material deliveries may be required to ensure the contractor is aware of their time responsibilities, the critical date information should be updated continuously and all contractors providing critical time activities kept informed.

Some states mandate that a contract must be executed for work above specific values. OBs must check their local requirements every time they consider a project, as the requirements are seldom stable and require constant monitoring. Many OBs (and experienced builders for that matter) do not give sufficient attention to the agreements they have with tradespeople and suppliers. Although a contract cannot make the two parties perform to their best ability if they do not want to, contracts can protect the OB (and the contractor) from misinterpretation…even in a contract where both parties have a good commercial relationship.

Generally, even a simple contract is better than none at all; in fact, the simpler the contract is the better it is, so long as no important areas of protection are omitted. The majority of trades contracts should be able to stated clearly the requirements of the parties in between one and two printed pages.

There are many pre-written contracts that are available for the OB to use if they can gain access to them. However, they are often prepared by professional, trade or corporate bodies that do not want their copyright infringed by outsiders. In NSW the Master Builders' Association produces a special contract for the use of the OB, designated OBI (Jan '68), which is very useful.

An internet search will bring up a host of proforma for building construction contracts, choose carefully.

If an OB cannot gain access to a pre-printed contract then the following is a guide to the parts of a normal building sub contract that should be considered. Specific legal requirements should be checked by solicitors.

1. The agreement document should show clearly the date that the contract was signed, the name of the OB (the Principal) and the name of the contractor (the Contactor) .
2. The work to be done should be clearly defined by:
   a. special instructions or drawings relevant to that part of the work being contracted
   b. reference to the drawings and specifications that make up the overall building documents.
3. The method of paying the contractor and the amount to be paid should be clearly set out. It should be clear if the contract is:
   Fixed price, lump sum.
   Rise and fall, lump sum.
   Cost (of materials) plus (labour).
   Cost (of materials and labour) plus (administration).
   And it should include a schedule of rates.
4. The agreement should indicate the earliest and latest dates that the contractor can commence the work, the maximum period that is allowed for the work and any liquidated and ascertained damages provision provided for late completion.
5. The method of payment to the contractor should be set out, the OB should attempt to pay for work progressively (weekly or fortnightly or monthly) as it is complete, avoid in-advance payments, and where possible retain an amount of money to protect themselves against the builder's default. The retention sum is normally 10 per cent of the value of the work completed up to 50 per cent of the total work, then 5 per cent of the total contract sum.
6. A statement should be included in the agreement to the effect that the contractor shall be responsible for the payment of all wages (including superannuation) of the workers before any claim is made.
7. Variations to the contract are inevitable and OBs should provide a mechanism for their approval in the agreement. It is generally best to say that no variation shall be approved until the contractor has furnished a written quotation to the OB and the OB has signed the variation quote. It is also worthwhile stating in the agreement how much allowance over the net cost of the variation is to be allowed to the contractor to cover administration (attendance) and profit.
8. Many building permits issued by local councils restrict hours of work. OBs should take careful note of these restrictions, for work can be restricted to normal weekdays (between 7 am and 5 pm is common). Restrictions of this type are a restriction on the rights of the OB and should be challenged on a political level. However, where they exist it is the OB's responsibility to make sure that the contractors that work for them know of the restrictions and, therefore, a clause should be inserted in the agreement alerting the contractor to the hours during which work is allowed.
9. Some of the sub-trade sections of the owner-built house will require the payment of fees to statutory and other authorities and the giving and receiving of notices to the same. Make sure that the agreement with the contractor makes them responsible for payment these fees and handling any notices.
10. Insurances are a major cost to building trade contractors. Make sure that the contractor you employ has all the necessary insurances for their and your protection. Most important is insurance to cover homeowners' warranty, workers' compensation and public liability plus any other insurance to satisfy local requirements. The agreement should provide for the contractor to take out the necessary insurances and provide certificates of their currency.
11. The agreement should have a clause to protect the OB from default on the part of the contractor and to protect the contractor from default on behalf of the OB.
12. In most states of Australia there are specific regulations covering the erection and use of scaffolding and other safe working provisions. There should be a clause in the agreement clearly laying out the responsibilities of the OB and the contractor with respect to this matter.
13. From time to time in any contract there may be need for formal notices to be given by the OB to the contractor (or vice versa) and a method of sending and receiving such notices should be outlined in the agreement.
14. A clause is necessary in the agreement to require the contractor to gain the approval of the OB before subletting any part of the job. This is important protection for the OB, as it may be necessary to check the competence of the proposed sub-contractor or to check insurances, registrations, licences, etc.
15. The OB should have a clause in the agreement that allows for the removal of incompetent or misbehaving workers from the site.
16. Most workers in Australia are covered by some relevant award or industrial agreement and the contractor should be required to meet all conditions of any such award or agreement.
17. Contractors should be required to clean up after themselves, and the agreement should state that the OB can take any money necessary from that owing to any contractor who has not cleaned up at the completion of the day or activity.
18. Sometimes one contractor will damage another contractor's work, so the agreement should require the contractor causing the damage to be responsible for its repair or replacement.
19. A statement of quality should be included in the agreement. The best form is to nominate the Australian Standards that are to apply (see specification section).
20. A defects liability period should be stated in the agreement. This is the period during which the contractor will be responsible for the making good of any defects in the work.
21. If there is a requirement for any authority to have access to the contractor's work (lending or statutory), then a clause should be inserted in the

# TENDERING, CONTRACTING AND LEGAL • 17

agreement requiring the contractor to take direction from the OB with regard to any such inspections.

22. A clause should be inserted in the agreement as to what is to be done in the case of an unresolved dispute. It is normal to nominate the president of the local Institute of Arbiters to arbitrate.

23. Any other special conditions can be added to suit the specific requirement of the contract, but care must be taken to stay within the law.

   Remember that it is not possible to contract around the law and any attempt to do so could render part, or all, of the contract void and leave the OB open to penalties.

24. Finally (or firstly, depending on how you look at it) the contract must be signed by both parties, dated and witnessed. The contract documents should consist of all clauses of the agreement, all other written instructions and specifications and any sketches or drawings. Both parties should have a bound full set of all the documents to the contract and each separate page should be signed by both parties and all handwritten sections or changes should be initialled. If possible, have a solicitor or architect act as witness to the contract.

The OB (Principal) should assess whether all of the above clauses and recommendations are included in every contract that is executed for a project. Some contractors shy away from formal contracts. It is recommended OBs be careful if they allow work to be carried out without a properly prepared contract agreement. Tradespeople usually fulfil the requirements of the work that they are employed to carry out, but, if the unexpected happens, the OB, will need adequate protection against delays, costs, charges and other disasters.

## AS 4905-4906

The AS 4905 (Superintendent Administered) and 4906 (Principal Administered) are standardised Minor Works contract documents for use in the building industry. They are suitable for agreements between two parties (the principal and the contractor), use:

AS 4905 when there is a third party acting as superintendent, i.e., the person who wants the work done, the person who will carry out the work, plus the person who ensures that all agreements are met.

AS 4906 if an OB decides to be less of an OB and more of a principal, consider using this contract, as it is generally accepted to be free of excessive influences from either the owners or the builders.

It is possible to use these contracts as an agreement between an OB and a tradesperson but it is not recommended. However, it is recommended that OBs consult an architect or project manager with sound experience in superintending building contracts to determine they should tackle this difficult dual role.

## Legal obligations

The legal obligations of an OB are far too wide and varied to be covered in this chapter. It is, however, important to understand that there are many requirements of law that affect OBs.

Careful planning of your OB home includes a detailed study of the requirements of all laws and regulations to be met and understood. Information is available through a number of channels including:
- solicitors
- architects
- valuers
- lending sources
- local government
- government publications
- building information centres courses.

The professional advisers (solicitors, architects, valuers) have usually done extensive courses on the law (generally, in the case of solicitors and particularly in the case of architects and valuers) and normally charge for advice, they are bound by professional ethics to give considered advice and to be responsible for their actions.

Other sources may be more general in their advice but in many cases the advice is free, or only a nominal charge is made.

## Goods and Services Tax GST

The general tax rate for GST is 10 percent and it applies to all goods and services used in the construction of houses. It is important to review all tenders and quotations carefully to ensure that they are inclusive of GST, and that any tax credits have been deducted. For example, assume a quotation is received from a roof tiling contractor: ideally, the calculations in the table below should be included in the quotation.

The quotation may only show the inclusive GST price, i.e., $3190. Then the GST payable by contractors is one-eleventh of this price: $290.

If the quotation is not set out in this fashion, but instead shows a GST-inclusive price, there is little that can be done, as that satisfies the requirements of the GST laws. The quote to be concerned about is the one that shows a price plus 10% GST, as in this case tax credits may not have been deducted.

Most OBs will be building a house for their own occupancy, and so will not be looking to gain tax credits to offset a selling price. But it is still important that every amount of money paid for goods and services has a tax invoice that shows the total price for taxable supply and the value of GST included. This, if nothing else, provides confirmation that the supplier has an ABN (Australian Business Number) and has taken the responsibility to pay GST to the ATO.

It is recommended that OBs who intend to carry on a business at their home, to rent all or part of the home after completion, or are rural producers, see their taxation consultant before proceeding with any OB project.

If a supplier of goods or service (labour) does not provide a tax invoice and an ABN, they may not be registered with the ATO for GST purposes. In some cases, the receiver of the goods may be required to withhold a percentage of the value of the claim and to forward this to the ATO. If tax is withheld from a supplier, then the OB will have to register with the ATO as a withholder.

Discounts for cash may save some money but could be illegal if the receiver is actively avoiding GST or other taxes, in which case the payer (the OB) may be liable. Transactions that do not have a taxation receipt could also void any warranty or insurance cover associated with the work.

For more information on GST, contact a taxation accountant or visit the ATO website.

## Taking care

Many trades require registration with a state government authority and are required to carry identification to show their validity. OBs are advised to view these licence/registration cards and to ask for copies of the current insurance policies/renewals required by local regulations.

## White Card

The white card (or general construction induction card) is required for all workers who carry out construction work. People who need a white card include: architects, engineers, project managers, supervisors, surveyors, labourers and tradespeople. All workers whose employment causes them to routinely enter operational construction zones, including an Owner Builder who carries out any onsite activity associated with the construction of a building.

## Private certifiers

Private Building Surveyors and Principal Certifying Authorities can be used in some states instead of the local council building inspectors. This can be useful to OBs who desire out-of-business-hours contact with the authority in order to certify that the work on their building project meets the required laws, codes and standards. An OB who chooses to use a private certifier or building surveyor should ensure that any trade contracts they issue note which certificates will be required to be issued by the certifier and any limitations this may impose on the contractor-particularly time in advance required for booking on-site inspections.

| | |
|---|---|
| Tiles purchased by contractor for $4000 plus $400 GST (10%) | $4400 |
| Battens purchased by contractor for $1000 plus $100 GST (10%) | $1100 |
| Labour wages estimated by contractor including income taxes, etc | $2000 |
| Net outgoing estimated by contractor including GST | $7500 |
| Plus contractor's profit and administration (say about 20%) | $1500 |
| Total cost to contractor is | $9000 |
| Less GST credits | - $500 |
| Net cost to contractor is | $8500 |
| GST@ 10% is | $ 850 |
| The GST-inclusive quotation then should be | **$9350** |

# 18 • SUBCONTRACTORS

## Administering an owner built home

The one thing generally common to all OBs is the responsibility to administer the integration of the sub-trades and suppliers who make up the building team. No matter whether the OBs are building the home totally with their own hands or totally with subcontract labour, the administrative content remains. In most traditional owner/builder contracts the builder takes on the responsibility of administering the contract. Sometimes, when an architect is employed by the owner, the owner puts certain requirements on the architect to administer or superintend certain parts of the contract. However, it is generally the builder who bears the brunt of the day-to-day site administration.

In administering their own building work OBs must develop a clear understanding of the requirements and conventions that apply. Not only do OBs have to realise that they have become the administrator of labour and materials, but also that they have become the odd-job person. What is meant by this is that on normal building sites it is the builder who either by his own labour does the tidying up of the bits and pieces of work around the site or, if the job is big enough, employs a labourer to do that.

This means that an OB must be prepared to spend a significant part of their leisure time at the site, looking at and being aware of what has happened and what is about to happen.

An extremely important part of the administration of the building work is to determine when the sub-trades should be approached, how far in advance they need to be notified and how to get them onto the site on the day agreed. The answer to the first two parts is easy-as early as possible. The earlier you talk to the sub-trades people the more chance you have of get ting them to give you a good price and of being able to predict reasonably accurately when they will be able to arrive on your site. The answer to the last part is less easy to answer. Subtrades people, by their very nature, are working on numerous jobs and it is often very hard for them to predict exactly when they will be available to attend yours. The best thing to do is to keep them well informed on the progress of your job. Sometimes you can even bring them in earlier than you originally thought if it fits in with their program. The best answer of all, however, is when you go out looking for the sub-contractors that you require, employ the ones that you like, employ a person that you can talk to. Don't employ any person you feel will take advantage of you or that you just don't like.

The first and most important thing to do in developing your administrative role is to list every sub-trade that you are likely to employ on the job. Do this regardless of whether that person will be a subcontractor employed by you or whether it is a sub-contract area that you envisage carrying out by yourself. Secondly, write down all the different types of material that you are likely to require and a list of the possible suppliers of those materials.

After you have put together your list of subcontract labour, write down a list of all of the subcontractors that you currently know. It is ideal to have at least two subcontractors for each trade grouping. This allows you to get comparative prices that will keep both tenderers honest. To fill in the gaps on your list of subcontractors, ring up all the friends and acquaintances that you know who have had building work done recently in your area and from that you may pick up a number of other contractors.

Another source of subcontractors can be the local newspaper. Many subcontractors run a continuous advertisement in the local newspaper, advertising their trade. Thirdly, drive around the suburb or area where you intend to build and stop at all building sites where building work is going on. Stop and talk to the subcontractors on those jobs and find out whether they are looking for any work in the future. Note: Never enter a construction site without approval of the person responsible for the site of the works. Always carry a full set of PPE (personal protection equipment] in the boot of your car, you will need it to enter any construction site, including your own projects. The minimum requirements are, high visibility jacket (HiVis), work boots with sole and toe protection, a safety helmet, safety eyeware, all of this equipment should be labelled as complying with the appropriate Australian Standards.

Fourthly, which is the most enjoyable part of all, go to all the clubs and pubs in your area and ask loudly in the bar who in this bar is a building subcontractor looking for work. Or use more subtle means, like looking around the bar for a guy obviously covered in paint or plaster or with a trowel or builder's measuring tape on their belt. These are reasonable give-aways that the people you are looking at are building subbies.

Once you have established your list of building subcontractors, then commence establishing prices for the work required. It is absolutely essential that if you are getting prices from more than one subcontractor you give the same instructions to all subcontractors so that the prices you received are comparable. Instructions to the subbies must be clear and you should make sure that you cover all the areas that you need that subcontractor to work on. A good question to ask the subbies after they have looked at your plans 'is there anything else on the plan that you can see that you would like to give me a price for?'.

Once a prices are established for the work with the subcontractors, tie them up with some form of contract that will afford you reasonable protection from default and/or insurance claims. This is not as easy as it sounds because many subcontractors have never signed a contract for work in their lives, or maybe they don't realise they have, so the subtle approach is worth considering.

Instead of dangling a pink-coloured 35 clause contract in front of the subcontractor's eyes when they is only going to do a few hundred dollars of work, it may be better to consider writing a clear letter of confirmation to the subcontractor stating all of the work that has to be done and the conditions that you expect to be applied should he default, or if there is an insurance claim made. Send the original and a copy of the letter to the subcontractor, or better still take it to them personally, and have them sign the copy and return it to you so that you at least have some record of agreement.

Most major building companies issue orders to their subcontractors that on one side are just a series of lines in which they fill in the information that they wish the contractor to know about for that particular job. On the back there is a 20-clause (or so) set of conditions that will apply to that particular works order. Understanding the content of documents of this type may assist in avoiding particular problems during a building project and establish methods to mitigate subcontractors' faults/errors or damage.

The most appropriate advice when a subcontractor's payment is due is don't pay anything unless the work has been done'. If possible, work out in advance with the subcontractor how much money will be paid for what stage in the process of the work. Sometimes it is 100 per cent of the money for 100 per cent of the work, but in other cases it may be that the electrical contractor is doing work in two stages, the first stage being the rough-in stage and the second stage being the finishing stage.

It may be that you agree to pay that electrical contractor 50 per cent of the money at the end of the roughing stage and the balance at the completion stage. The second part of the money in this case would probably not be paid until the electrical contractor was able to give you certificates from the local supply authority showing that the system had been tested and was proven to be adequate, safe and so on.

With some subcontractors, particularly those working in larger areas or for larger amounts of money, it may be possible to withhold some of the money in a retention fund. In this case, a retention of between 5 percent and 10 percent of the contract value for a period not exceeding 90 days seems to be generally accepted.

However, there are two things to be remembered about retention funds. Firstly, make sure that if you are levying a retention fund that the subcontractors are aware. It is not valid after the contract has been signed, and when the contractor comes for their first payment, to say 'we are retaining 5 or 10 per cent' unless that has been previously agreed.

The employment of contractors varies from place to place but OBs must check the legal requirements applicable in the jurisdiction in which they are building. They may be required to contribute long service leave or superannuation to contractors.

Authorities responsible for the licensing and registration of building contractors and building trades contractors should be able to provide information on current obligations. Some government regulations may determine that an OB contracting to a trade provider is considered to be an employer,

# SUBCONTRACTORS • 18

and the trade provider an employee. Check the regulations.

## Security of Payments Act

The Building and Construction Industry Security of Payment Act 2002 (known as the SOP Act) helps ensure that any person who carries out construction work or supplies related goods and services under a construction contract gets paid. It does not nominate adjudicators or take part in payment disputes.

All people who engage construction contracts are bound by the Security of Payments Acts in all States, if a person meets their conditions of engagement and completes the tasks in the agreement between the parties, they must be paid promptly what has been agreed or what is a fair price for the work done.

All OBs should consult with the local authority having jurisdiction of the Act and ensure that they understand its requirements and apply it diligently.

## The subcontracting process

1. Select two to four tenderers.
2. Issue clear tendering instructions.
3. Order any materials required by the contractors.
4. Assess tenders.
5. Select contractor.
6. Confirm material deliveries.
7. Exchange contracts and ratify insurances.
8. Confirm date for contractor to commence (and complete) on-site.
9. Check that work is completed before paying any claims.

## Golden rules

1. Only use contractors that you have met and who you believe you can work with.
2. Avoid making advances or overpayments.
3. Make sure all workers on-site are covered by workers' compensation insurance.
4. Only pay claims to people with whom you have a contract. Do not pay site workers directly whatever the circumstances.
5. Wherever possible, have a signed agreement with the contractor that clearly defines the extent of the work to be done.

## The taxing work of an OB

GST means that every monetary transaction for goods and services incurs a tax at the nominal rate of 10 per cent. OBs should receive a tax invoice from every tradesperson or material supplier prior to payment being made. Ideally, the tax invoice should show:

- Date of issue
- Name and address of supplier
- ABN of supplier
- Your name, address and ABN, if applicable
- Place of delivery of activity
- Description and quantity of goods and/or services supplied
- Total value of taxable supply
- Value of GST included
- Total amount payable-GST included.

A fully detailed tax invoice is required for all sales over $1000. Sales under $1000 can use a slightly simpler format. The ATO has published a pamphlet, The Building and Construction Industry & The New Tax System; write to ATO, PO Box 9935 in your capital city to obtain a copy.

Some providers-very small suppliers or tradespeople-will not be required to collect GST. Always check that they have exemption from GST, as in certain circumstances OBs may have to withhold 48.5 per cent of the value of the supply.

From the first day of the Owner-Building project, keep accurate records, carefully file all tax invoices, and lodge on time any returns required by the ATO-particularly if any withholding taxes are to be paid.

OBs who are tempted to employ discount-for-cash workers may not be able to explain to the ATO how a whole trade section of work happened as if by a miracle. In some states, prescribed insurances are required for contracts for even quite small sums, and there is always a risk that discount-for-cash workers are not protected by workers' compensation insurance.

## Outline of Trades and Labour

There are many trades that make up the home building industry; some, like electricians and plumbers, are restricted trades. In most locations in Australia where a house is connected to the electric and sewer mains, only licensed electricians can carry out any work in 240-volt wiring and only plumbers can make connections to and reticulate the water supply and carry out all sanitary and sewer drainage.

Most other trades are licensed or registered to some extent; the authorities believe that this way the public is protected. In some areas there is sufficient restriction that OBs may find it difficult to use unlicensed tradespeople, even if they are friends and relatives. It is important to check with the authorities to determine what the limits are, and to check all insurances , particularly compulsory ones required by the authorities.

Most of the trades used in house residential projects are covered in detail in this book, but there are less obvious members of the building team that deserve some coverage, as understanding their role is crucial to a well-run, on-budget project. They are the labourers, truck drivers, cleaners, and they are the people that link the trades together and provide crucial support.

Builder's labourers will be used for fetch and-carry jobs around the site; they will dig any holes, knock down anything to be demolished, pick up the rubbish, sweep the floors and tell the longest stories at smoko. Many OBs try to make do without a builder's labourer and do the work themselves; this is a good idea if there is sufficient time available to the job and the OB is fit enough to carry out some of the heavier tasks.

Other trades may require labourers:

The excavator mainly digs the trenches by machine, but sometimes someone has to dig the trenches out with a shovel and place reinforcing steel, lay sand beds, and similar tasks.

Drainers may dig their own trenches, but often prefer if the OB to organise and excavate the trenches.

Concreter may on small projects fix the steel and pour the concrete themselves, but it may be an economical option to provide labour to carry and tie steel reinforcement, and to move heavy barrow-loads of concrete. Warning to all OBs: if you have never wheeled a full barrow load of concrete, try to ensure that you are capable before volunteering to shift the six or seven cubic metres of concrete spewing out the back of a big concrete truck. It is recommended that most OBs should consider employing the labourer option.

Bricklayers and other masonry trades generally work in small teams; on residential projects two trade bricklayers are often supported by a brickie's labourer. The task of this labourer is to build scaffolding, mix mortar, move bricks, tidy up around the work, and clean the brickwork down when it is complete. This is a specialist job and it would be a very fit OB who believed that they could do this work.

Steelwork labourers and riggers carry out the heavy sitework of structural steel erection. They support the specialist trade welders and fabricators. Modern steel, used in residential construction, is often relatively light and strong and can be cut, fastened and erected by relatively untrained yet intelligent people. If the use of heavy sections of structural steel are limited to simply supported spans, and site welding is eliminated, much steel work, particularly where rectangular hollow sections (RHS) are used, can be carried out by OBs. Some equipment may be required to aid in cutting, drilling and for safety; this is available at most builders' hire services.

Carpenters are often central to many on-site activities and carpenters working on residential projects will generally work without the assistance of skilled or unskilled labourers. OBs should take a rational position when they carry out carpentry work and assess carefully when it is more sensible and economical to employ skilled tradespeople and labourers than undertake tasks beyond their skill or physical capacity. Some carpenters will happily work alongside the OB at an hourly rate; this assists the OB by bridging the experience gap and lending a hand to lift the heavier components.

Roofers either specialise in fixing profiled metal roof systems or in tiling roofs with, tiles, slates or shingles. Preparing the roof framing for a simple skillion roof to support metal roofing is often within the capacity of

# 18 • SUBCONTRACTORS

an OB; even fixing profiled steel sheeting, like corrugated steel or steel decking, can be handled with satisfactory results-just follow the suppliers' published guidelines. Metal sheet roofing on complex hip, valley and gable roof structures will require careful measuring, cutting, cappings, flashings and rainwater collection, this work should be carried out by skilled fixers and roof plumbers. Tiled, shingled and slate-type roofs require specialist tradespeople and labourers. In many locations it is nearly impossible to buy tiled roofing materials unless it is a supply and fix contract-even if the local tile company will provide the materials, it is often at a price that is uneconomical.

Plasterboard fixers are a specialised trade that can efficiently sheet out the interior walls and ceilings of residential projects with the minimum of waste. Modern plasterboard and fibrous cement lining products can be fixed and placed by confident OBs but seldom will this be a value option except on smaller alterations and additions projects. Large 1200 x 4800 mm sheets of plasterboard can be beyond the capacity of inexperience people, particularly if they are to be fixed to the ceiling. The jointing and flush finishing of plaster board is critical to the appearance of the interior of a home this work is highly specialised and best left to experience finishers.

Renderers and wet plasterers have for millennia been a specialist trades that require extensive training and experience to achieve high quality durability and finish. It is general recommended that OBs avoid wet trade rendering wherever possible in a residential project. It is possible in many projects to fix plasterboard sheets directly onto the internal face brickwork so as to avoid the messy, wet and unpredictable trade of rendering. If rendering is unavoidable only use experience tradespeople.

Electricians are the only people allowed to carry out any work on mains supply electrical installations. OBS can assist in some sections of the electrical work, under the direct supervision of the licenced electrician, this is seldom wise or economical. In all locations where solar photovoltage panels are used to provide electrical energy, particularly when the panels are supplying power into the main grid all work is required to be carried out by a specifically registered electrician.

Electronic cablers are a recent addition to the home construction team, they specialise in suppling and installing the cabling required to support telecommunications, video systems, sound systems, security systems, and computer networks and an ever growing range smart homes systems. Whereas, in mains power electrical systems simple insulated copper wire is installed in the wall framing or chased into masonry where connections are made with simple clamp connectors, this is seldom the case with electronic cabling. Most of the cabling used for electronic applications requires highly specialised cables and connections. If a house requires specialist electronic cabling a specialist installer should be engaged to assist in determining the most suitable locations to the cabling, associated conduiting and points of connection. Some of the cabling may utilise fibre-optics, these demand accurate sweep bends and specialised connections.

Plumbers must be used, it is their skill that, ensures clean potable water supply, correct temperature hot water, safe and the hygienic disposal of sewage and stormwater. If the home requires a split liquid waste system that separates black water (sewerage) from grey water (from laundries, bathrooms and kitchens) and rainwater a specialist plumber or a hydraulic engineer may be required to design the systems, particularly if there is to be a treatment installation, irrigation system, grease extraction, and/or on-site disposal system.

Tilers are a trade that requires knowledge on achieving the technical and visual qualities required for individual installations, including plumb vertical surfaces, correct floor grades, appropriate wet sealing behind the tiles, while providing level grouted joints. Modern adhesives have made tiling easier carryout tiling work, than it was when all tiles were bedded onto wet mortar beds. With care, many OBs will be capable of tiling simple projects. If there are large areas, curves or complicated cuts to be performed, then it is wise to stick to a trade tiler. The labour component of tiling is roughly the same cost per unit area as the materials, so if the OB gets it wrong and has to start again, all advantage is lost.

Cabinetmakers provide either on-site or prefabricated cabinets, many offer design, fabrication and installation services, these companies vary from providing basic products to extreme high quality designer installations, complete with expensive appliances. Modern, pre-coated sheet product, medium-density fibreboard and particleboard have revolutionised general cabinetry; therefore, OBs may be able to fabricate the carcasses of kitchen, bathroom and other built-in cabinets. It is more difficult to hang doors, install drawers and make benchtops, but it is possible. There are companies in most large cities that will take an OB's cutting list and supply all the pieces of board ready for assembly. Failing this, there is always the flat-packed, pre fabricated cabinet packs, these are most suited to OBs who speak Swedish and played with Meccano when a child.

Painters once a highly specialist field when highly toxic lead oxides and volatile solvent ingredients were mixed from scratch by painters. Now, this is the OB's territory, because anyone can paint? But, consider this information, an average house with 200 square metres of floor area has 600 square metres of paint on the ceiling alone, plus over 1000 square metres of wall paint-assuming three coats. That's over 1500 square metres, or a strip one metre wide stretching 1.5 km. Then there are the doors, the windows, the skirting boards, the architraves and metre after metre of cutting in. Get a price from a painter and contemplate the agony.

## Contracting labour

Hired help, from labourers to plumbers, know their place in the pecking order; they have well-practised rules of survival. If you have opted for the OB option you need to master this game without delay.

Who will give a quote and who wants to work on an hourly rate?

Let's tackle the simple ones first. Most plumbers and electricians will provide detailed quotations based on rates per point (light points cost $X-water points cost $Y). These trades stick to their quotes most of the time, and only charge for extras that you have specifically asked for. It may be because they do not want the other trades to know their hourly rate that they are so eager to provide reliable quotes.

Company structured labour and material suppliers will also readily issue lump- sum quotes; the concreters, plasterboard fixers, the roofers and painters all fit into this category.

Trades that work in assembling units, such as bricklayers and floor and wall tilers, work on rates $Z per metre or $W per square metre {take care who supplies all the materials, scaffolding, and specially hired tools and equipment, and who cleans up after them). Some measuring techniques are inventive, like counting the bricks that would have been used if a window was made of bricks, because this offsets the extra work needed to install the sills. Another method is that the area of the tiles charged is the area that could be covered by the number of tiles taken from the boxes, including all the offcuts.

Observe that many bricklayers and tilers only make cuts from full bricks and tiles, even when the ground around them is strewn with bits of brick and tile?

Carpenters will work for rates from project home builders because these jobs have a known time allocated. OB projects, particularly alterations and additions, are seldom as precise, so carpenters will argue for an hourly rate. They know that many OBs will need their site knowledge to complete the project, and that they will end up doing many other tasks than first discussed.

An experienced, leading-hand carpenter with a ute full of tools, is a worthy discovery. Do not get carried away, negotiate a fair rate, and pay a small, but worthwhile, bonus. Good quality coffee is often a winner, and pastries at morning tea are seldom left to the flies.

Labourers can be difficult to find for a small projects – the experienced ones are working on the big projects, and the ones left are not always the pick of the bunch. Try to work beside any labourers to determine their productivity.

The Australian house building trades have a high reputation for honesty when working around people's houses; they guard this reputation jealously and summary justice has at times been meted out to a petty thief. Even so, keep a wary eye on where people are on the job-if anyone is prone to wandering into parts of the site or into existing buildings where there is no work to be done, it could be wise to keep

# SUBCONTRACTORS • 18

## The Subcontractors

**Who are the subcontractors?**
**What work do they carry out?**

| | | | • | ? | X | SF | LO | PM |
|---|---|---|---|---|---|---|---|---|
| a. | Site clearing | | • | | | | LO | PM |
| b. | Excavation | – bulk | • | | | | LO | PM |
| | | – particular | • | | | | LO | PM |
| c. | Concreting | – footings | • | | | | LO | PM |
| | | – floors | | ? | | SF | | PM |
| | | – structure | | ? | | SF | | PM |
| | | – paving | | ? | | | LO | PM |
| d. | Bricklaying | – base | | ? | | | LO | PM |
| | | – top | | ? | | | LO | PM |
| e. | Carpentry | – framing | | ? | | | LO | PM |
| | | – fixing | | ? | | | LO | |
| f. | Drainage | – stormwater | | ? | | SF | | PM |
| | | – sewerage | | | X | SF | | PM |
| g. | Plumbing | – cold water | | | X | SF | | |
| | | – hot water | | | X | SF | | |
| | | – roof | | ? | | SF | | |
| | | – gas | | | X | SF | | |
| h. | Electrical | | | | X | SF | | |
| i. | Roofing | – tiles | | | X | SF | | |
| | | – sheets | | ? | | | LO | |
| j. | Lining | – plasterboards | | ? | | SF | | |
| | | – timber boarding | • | | | | LO | |
| | | – other | | | X | SF | | |
| k. | Rendering | – external | | ? | | SF | | PM |
| | | – internal | | ? | | SF | | PM |
| l. | Tiling | – floor | | ? | | | LO | |
| | | – wall | | ? | | | LO | |
| m. | Metal working | – steel | | ? | | SF | | |
| | | – aluminium | | ? | | SF | | |
| n. | Painting | – external | • | | | SF | | |
| | | – internal | • | | | SF | | |
| o. | Cleaning | | • | | | | LO | |
| p. | Sanding | | | ? | | SF | | PM |

**Code:**
- • Can be done by OB with little or no training
- ? Can be done by OB with some training
- X Cannot be done by OB in most cases
- SF Subcontractor normally supplies materials and labour
- LO Subcontractor normally supplies labour only
- PM Plant and machinery required

their period on the site to a minimum.

All construction sites are required to be a smoke and alcohol free areas, you are right to inform the trade workers of this restriction, Random checks for compliance can be carried out, there are organisations that offer this service.

Check how much cover the labour employed has, from insurances they or their employer holds, and have this checked by your insurance provider to ensure you minimise your risk. OBs are required in most areas to take out workers' compensation insurance and to provide a safe working environment-never take these requirements lightly. Read the literature from the safe work authorities and keep a watchful eye.

## In your state the rules are ...

Most Australian states have a government department called the Department of Fair Trading. This is the department that attempts to protect consumers from unfair suppliers and contractors.

These departments have different ideas on the most efficient methods of protecting the public from unscrupulous building suppliers and tradespeople.

In New South Wales it is mandatory to have a signed contract for any building work exceeding a published value. The Department of Fair Trading has issued 'plain English' standard form contracts for repairs and maintenance work. They also have a renovations contract for significant work and a major works contract for work that are be indexed to CPI.

Although these contracts are relatively easy to fill in, they have some problems. For instance, they have fill-in boxes where the normal rate applying to a specific contract clause is also shown. In the case of Prime Cost (PC) allowances the contract indicates that it would be normal for a contractor to be allowed to claim up to 20 per cent extra on any additional cost of the item purchased by the contractor above the allowance. This is good income and is more than most architects would allow in purpose-written contracts.

The option remains to use a contract prepared by any competent person, as long as it includes some specific clauses to do with insurances and government regulations.

This option is well worth considering. For example, if the PC allowance in the contract for a stove is $1000 and the actual cost is $2000, the contractor could be entitled to claim 20 per cent of the difference, i.e., $200 extra for the same time as it would have taken to order a $1000 stove.

In the Australian Capital Territory, no formal contracts were required at the time of writing. In Victoria and South Australia contracts are only required for work over a published minimum.

In Queensland no formal contract is required but the local building services authority has two Standard Contracts available-one for work to $25,000 and one for work over $25,000. The authority recommends their use.

In Western Australia a contract is required for work over $10,000.

In the Northern Territory and Tasmania there is no legislation requiring formal contracts. These appear to be the only states where contractors are not required to be registered or licensed.

Before commencing any building work, it is important to check with the local authorities to determine the current requirements, as these can change.

# 19 • SUPPLIERS

There was a time when only established builders could get discounts and long-term credit facilities from building trade suppliers, but many suppliers now offering discounts to all. The hardware and building supply retail warehouses advertise they will match or better the prices of their competitors and their system of drive a ute to the store select the goods and materials required, then pay by credit card before driving the purchases to where they are required.

This retail based system is very convenient and has eliminated wait time on most carry out products. However, when bulk purchases of framing timber, steel sections, bricks, blocks, pavers, sand, crushed rock, specialist suppliers may provide the best quality materials, delivery systems and convenience.

OBs should shop around and ask for discounts, particularly when paying cash. Remember, it costs a supplier at least $1 per $100 of inventory per month to hold stock and give credit.

To gain the best advantage of material supply discounts OBs are advised to gain access to a large four-wheel trailer or a 1- to-2 tonne small truck. Often savings can be made by avoiding delivery charges, and most suppliers deliver to sites during the week, when often OB sites are unattended, therefore increasing the risk of theft.

The important factor in purchasing building materials is determination to achieve the best deals available. Request discount, ring around to get prices, ask the question 'is this trade price?' The money saved can be used to purchase higher quality materials, or simply to achieve on budget completion.

Not all materials are easy for OBs to purchase and some do not have discounts, so ask around and explore the disposal and second-hand building material classified advertisements in the weekend newspaper and on the internet.

Who then are the suppliers and how do they operate? Suppliers will vary from location to location but the following list will assist most OBs to identify sources and expected savings.

### 1. Demolisher
It is good to identify and quantify reusable materials at this stage. Be conservative here; a 50 per cent reusable factor, even for bricks (only those laid in lime-rich mortar) and Oregon wood, is a fair estimate. The main supply item for a demolition is the delivery and removal of dump bins, so check prices from local suppliers.

### 2. Excavator
For simple excavations for footings and utility services on relatively level sites where the excavation averages about 600mm then this work may be conveniently carried out by one or two labourers.

For more complex excavations, in hard clay, shale or rock, excavator machines will be required. The excavation contracts can supply; the floatage (trucking the machines to and from the site of the works), the correct machine for the excavation required, the tip truck to remove excess spoil and any specialised labour required.

Excavation works can be unpredictable and the costs can escalate when there are unforeseen circumstances or latent site conditions. Excavators may provide quotes by the hour for the time they are at the work site, it is reasonable to expect that they work diligently during this time but the cost time clock keeps running if the machine is delayed because on site factors limit their ability to operate. Due to the complexity of floating equipment to another project and the time taken for establishment many excavators may require to bill at a minimum number of hours per day.

If there are any deep excavations, shoring materials may be required; this information is best obtained from the chosen contractor. If fill is required, the volume needs to be calculated. This is supplied as loose material, and a compaction factor is required. If the loose fill is about 70 per cent compacted and needs to be 95 per cent compacted on site, at least 20 per cent extra volume should be considered. Some material will be sold by the cubic metre and some by the tonne, so check the volume of the material per tonne to assist calculations.

Modern excavation machines are available in a wide range of capacities, always attempt to ensure the machines to be used are the most appropriate for the work required. If excavation is required close to underground utility services consider using vacuum exaction, this method sucks the spoil from the excavation into a tanker and reduces significantly the risk of damage to the utility services.

### 3. Equipment hire
Most OBs will need to hire or buy equipment. Check time/rates offered by two or three hire companies and if it's cheaper to rent, do so. In some instances purchase may be desirable, particularly if the tool has a long-term usefulness to the purchaser.

Check special late-Friday to late-Sunday rates offered by some hire companies, and take care if you have equipment delivered and picked up by the hire company, the charge for this service can sometimes exceed the hire charge.

### 4. Ready-mix concrete
Ring two or three companies and check their current rates including minimum charges and waiting time fees. Include a mini-crete company if you only require a small load. Order the largest load you can handle, but make sure you have enough labour and equipment to handle the order.

If you are not in a hurry to complete driveways or paths it is sometimes possible to prepare the formwork and wait until the ready-mix company has a truck with surplus concrete that you may be able to purchase at a reduced rate.

### 5. Reinforcing steel
This is a tightly controlled industry but savings can be made by ordering from the supplier/fabricator in the largest possible amount.

### 6. Formwork
It is generally uneconomic to purchase or hire formwork for other than small jobs. It is often more economic to employ a company to supply the formwork and the labour to erect it. It is worth making the comparison. Considering permanent-formed steel formwork in some applications.

### 7. Bricks and blocks
Few brick or block companies offer significant discounts and most require cash in advance of delivery.

### 8. Hardware
Most OBs would find it an advantage to form an association with two hardware shops for the duration of their project. Many of these shops will allow monthly accounts with a range of discounts related to the terms and the quantities purchased. Credit cards can be an advantage but remember the shopkeeper is paying a fee on all purchases made on these cards and you can not expect to get high discounts as well. The large chain operated hardware warehouse will carry a large range of hardware, door furnisher, locks, and other small fittings, choose carefully as the items on their shelves can vary significantly in quality.

### 9. Scaffolding
The regulations relating to scaffolding vary significantly from state to state, but in generally less expensive to hire scaffolding than to purchase it. Program its use so that the hire time is kept to an absolute minimum. Consider using specialist scaffold erection teams, they will ensure the work they do satisfies the regulations and the work safe requirements.

### 10. Scantling timber
Here it is wise to make one large list of all the scantling required for the project and get prices of a number of timber yards. Remember to add in delivery charges where applicable. Be prepared to vary the timber species or sections if this will not reduce quality but will decrease cost. Check the yard giving the lowest price to ascertain that the quality is OK. If it is, proceed. If not check the second lowest price and so on until the best quality/cost balance is achieved.

### 11. Flooring timber
Buy only the quality required. Chipboard flooring is ideal under carpet. It is wasteful to use expensive timber boards where they are to be covered with carpet or vinyl. It generally doesn't matter mixing flooring materials between areas of exposed boards and areas of carpet, vinyls or other

# SUPPLIERS • 19

covering-and the savings can be significant.

### 12. Fixing timber
When timber is ordered for fixing out the house including skirting boards, architraves, trims and the like, try to order the full requirements at the one time and get two or three comparative prices. The larger timber merchants generally give better prices than small 'do it yourself' stores but local comparison is recommended.

### 13. Cabinets
Much of the cabinetry used in modern houses is factory-built modular units that can be purchased through a sales consultant, or often direct from the manufacturer in either built-up or flat-pack boxes. Check the best price you can achieve from these mass manufacturers with a couple of local custom joinery companies and there is a good chance you will be pleasantly surprised. Mass production does not always equate with a reduction in selling price and seldom offer the range of finishes and quality of the custom fabricator.

### 14. Plumbing and drainer suppliers
Purchase of these items is generally best left up to the plumber-as any unused pipes are generally refunded if the purchaser is an established customer-a service not always offered to a one of OB.

### 15. Roof tiles
In most cases roof tiles are purchased from the manufacturer and the price includes the tiler's labour. The tile companies prefer to use tradespeople that they know as they generally give a weather tightness guarantee with the roof and this guarantee is void if the tile laying is not carried out by their nominated tiler. Check that the roofing quote includes all necessary, pointing, sarking, battens, flashings, scaffolding, safety barriers

### 16. Roof sheeting
Roof sheeting comes in three main materials:
a. Steel-based-either galvanised (zincalume) or Colorbond, in many profiles allowing either pierced fixing (nails or screws) or clip fixing. All these can be fixed by an OB but a roof plumber may be required to measure, supply and fix flashings and gutters and down pipes. Supply is generally from the manufacturer or manufacturer's agent and discounts on small orders are seldom available.
b. Aluminium-based-as either natural or coloured material, in either pierced or clip fixing. May be fixed by OBs but flashings are often complex. Supply is generally direct from manufacturer and discounts may be obtained if an order is large enough.
c. Fibrocement-corrugated fibrocement sheets are no longer extensively used on houses but this product can be fixed by OBs. The flashing, capping and gutter systems are complex, however, and require skilled installation. Supply is through the sole manufacturer's agent and discounts can sometimes be obtained.

### 17. Electrical cable
The electrical contractor normally buys cable in bulk and it is seldom sensible for OBs to purchase this material.

### 18. Sanitary fittings
The range of styles and colours available in sanitary fittings has been extended and good buys can often be made through building material discounters. However, care must be taken to purchase fittings that are compatible with the local sewerage authority's requirements.

### 19. Appliances
Cash purchases of major appliances from electrical discount houses generally result in the lowest price, but make sure the appliances purchased are compatible with the electrical installation.

### 20. Floor and wall tiles
Special buys are often available through specialist tile suppliers but also check other discount stores, as there are wide price variations in these items.

### 21. Plasterboard
Plasterboard is available to the owner builder but discounts are hard to find and often the on-wall price offered by a plaster board fixer is not significantly different from the delivered price to an owner builder. It is essential that plasterboard is put under cover as soon as it is delivered as a short shower of rain can render it useless.

### 22. Fibrecement products
These products include flat wall and ceiling sheets as well as moulded wall sheets and weather boards. They are normally supplied through the sole manufacturer's agents and discounts are often available.

### 23. Windows
Most windows are made in either aluminium or timber and are factory produced in a fixed range of sizes and styles. Aluminium windows vary extensively in quality and price. Check the quality thoroughly, as some of the cheapest brands tend to use poor quality corner fixings and often have unbeaded glass, which means that broken windows may not be able to be reglazed without removing the window frame from the wall. Timber windows have fewer variations in quality and are less likely to be sold through discount building material centres, checks of quality and prices are recommended.

Custom-made aluminium windows are almost always a premium-price product and are seldom an economic alternative to the mass-produced product. However, aluminium windows to suit a particular architectural requirement can be sourced in most major cities. Windows requiring; large undivided glass areas, insulated framing, multi sheet glazing for thermal comfort and noise reduction, will generally require custom fabrication. Large areas of double glazing are very heavy and specialist installers are recommended.

Joiner shops, however, do still make timber windows to suit purchasers' requirements and these are not always more expensive than the factory-made timber window.

### 24. Doors
Most internal doors used in houses today are mass-produced flush-panel hollow-core units that are so much cheaper than joinery-built alternatives that few people take the alternatives. Flush-panel doors are available at discount prices from building material suppliers, and OBs should give serious consideration to pre-hung door units from these sources.

### 25. Prefabricated wall and roof frames
There are many companies specialising in systems-based prefabrication of timber and metal wall frames and roof trusses They will usually use the building drawings to give you a quote on fabricating and delivering the frames. It is critical to check the fabricator's workshop drawings before approving manufacture, to ensure accuracy. These systems work best for conventional homes; care must be taken if satisfactory results are to be achieved in homes with split levels or other unconventional design features.

### 26. Autoclaved Aerated Concrete (AAC)
A relatively underutilised product in Australia that can suit OBs more than conventional masonry products. Manufacturers are limited but provide good back up detailed information on use and ordering. Available in blocks and panels, AAC is sold through building supply retailers and prices vary.

These materials cannot be supplied over the counter, so check delivery delays and allow for these delays plus contingencies like strikes and simple delays. It is not uncommon for incorrect items to be delivered even after two months' delay, so make sure you make a precise clear order-if the error is yours, so is the cost.

### 27. Insulation
There are many types of insulation suitable for use in modern houses. Most insulation products can be purchased and installed by OBs. There are sheet, blanket, batt, foam and loose-fill insulation products, manufactured for use under floors, in walls, ceilings and roofs. Many can be bought at hardware stores but others may require ordering from a specialist supplier. Remember to buy or hire protective clothing to use when handling mineral wool or fibreglass-type products.

# 19 • SUPPLIERS

## Kitchen Quotations Checklist

1. Who is to prepare the design and is there a cost?
2. Is there a salesperson gaining a percentage commission that may induce them to sell more-expensive materials, finishes and appliances? Always check appliance prices with major retail houses before agreeing to kitchen manufacturer deals.
3. Who is responsible for taking on-site measurements?
4. Who is responsible for the accuracy of on-site measurements, including levels and angles?
5. Who is responsible for the provision and accurate location of utility services?
6. Who is responsible for making on-site connections to utility services?
7. Are accurate workshop detail drawings prepared by the kitchen fabricator?
8. Who is responsible for checking and accepting the workshop details?
9. Is the quotation a lump sum fixed price?
10. Does the quote include GST?
11. How are any on-site or after-acceptance-of-the-quote variations dealt with?
12. Is there a published schedule of labour and material rates for extras?
13. Is a deposit required, is it within the maximum allowed by applicable laws?
14. When are payments due? Remember that until a supplied item is actually attached to the building it may not be considered by law to belong to the purchaser.
15. Is there any real protection provided for a defects liability period?
16. Is the supplier required to provide a Homeowner Warranty or any other guarantee demanded by regulation?
17. What procedures are in place if the quality of the materials, fabrication or labour is not up to the standard offered by the quotation?
18. Is the supplier a quality-assured company, certified under ISO 9000 or a similar standard or code?
19. What recourse is allowed if the work is delivered or installed later than agreed?
20. Does the supplier provide for liquidated damages in the offer?

### 28. Personal safety equipment (PPE)

Laws require that OBs protect all workers and others associated with construction, to the current standards of the government safety regulations. To achieve this, certain occupational health and safety equipment may have to be purchased or hired. Always check with local authorities (Work Safe) before commencing any work.

### 29. The cabinet maker. kitchens

Some specialist kitchen makers use totally modular systems, and others will custom fabricate to a specific design. If an OB uses a modular system, the supplier will generally give assistance with designing the kitchen to fit the available space. When the design is approved by the OB, the cabinetmaker will provide a price to manufacture, deliver and fit the kitchen. Consider the risks; read the contracts carefully, as cabinetmakers often require a significant amount of the contract price to be paid up-front. Be cautious if there is any doubt about the supplier's stability or ability to complete the work.

If the kitchen has been specifically designed, or has uncommon features or materials, it may be more appropriate to use an experienced custom kitchen fabricator. It is not necessarily more expensive to use a good cabinetmaker, as they do not have the overheads of showrooms and sales/design consultants.

If there are special materials involved in the fabrication, such as polished stone, stainless steel and special finishes, these may need direct involvement of the OB, and even direct purchase.

### 30. Light fittings

The electrician, if requested, may supply some light fittings, such as down lights and fluorescent fittings, but the OB will have to purchase other fittings.

### 31. Prime cost (PC) items, including sanitaryware

Prime cost does not particularly apply to OB projects, but it has been used generally to refer to all items not selected before building work commences and that can be given an indicative price for normal budgeting purposes.

**PC items normally purchased by OBs**

- stoves (cooktops and ovens)
- rangehoods
- kitchen sinks
- dishwashers
- mirrors
- toilet suites (including pans, cisterns and seats)
- basins (including fixing kits, waste kits and brackets)
- baths (including waste kits and spa kits)
- shower bases (pre-made, not tiled in place)
- shower screens (for shower bases and baths with showerheads above)
- basin tapware (check the number of holes required)
- bath tapware (check if a shower diverter is required)
- bidet tapware (special reflux protection fittings may be required)
- shower tapware (including a shower rose)
- toilet-roll holders
- towel rails
- stop cocks (ask the tapware supplier if they are required for the fittings selected; do not forget one for the cistern-it is often supplied by the plumber, but check)
- vanity cabinets (often supplied with a basin in a mass-manufactured item but cabinet-makers can make them to specific designs)
- kitchen cabinets (see elsewhere in this chapter)

Note: refrigerators and microwaves are not normally considered to be PC items, except when they are to be built into the cabinets.

# SUPPLIERS • 19

## Prime Cost Items Checklist*

| Item | @ Purchase | Subtrades |
|---|---|---|
| Stoves | Gas supply required | Gas plumber |
| Cooktops | Electricity supply required | Electrician |
| Ovens | Water supply (Combi-ovens) | Plumber |
| | Cutouts and cabinet requirements | Cabinet maker |
| Rangehoods | Electrical supply required | Electrician |
| | Ducting and cowl required | Carpenter/Roof plumber |
| | Cutouts and cabinet requirements | Cabinet maker |
| Kitchen sinks | Wastes and plugs | Plumber |
| | Tap hole(s)—filter & boiling water units | Plumber |
| | Cutouts and cabinet requirements | Cabinet maker |
| Dishwashers | Water supply (hot and/or cold) | Plumber |
| | Drainage system | Plumber |
| | Cutouts and cabinet requirements | Cabinet maker |
| | Facing panels | Cabinet maker |
| Toilet suites | S or P and waste location | Plumber |
| | Water supply entry point | Plumber |
| | Special brackets (cantilever pans) | Carpenter/Plumber |
| | Duct space required—concealed cistern | Carpenter/Bricklayer/Plumber |
| Basins | 1, 2 or 3 tap holes | Plumber |
| | S or P trap and waste location | Plumber |
| | Water supply entry point and stop cocks | Plumber |
| | Standard or ceramic washer taps | Plumber |
| | Special brackets (cantilever basins) | Carpenter/Plumber |
| | Vanity Cabinet requirements | Cabinet maker |
| Baths | Trap and waste location | Plumber |
| Spas | Electrical point for spa motor | Electrician |
| | Building-in details | Carpenter/Bricklayer |
| Bidet | 1 or 3 tap holes | Plumber |
| | S or P trap and waster location | Plumber |
| | Water supply entry point and stop cocks | Plumber |
| | Standard or ceramic washer taps | Plumber |
| | Special brackets (cantilever basins) | Carpenter/Plumber |
| | Any special water supply requirements | Plumber |
| Shower bases | Waste size and location | Plumber |
| | Sub-base requirements | Concreter/Carpenter |
| | Wet seal requirements | Wet seal contractor |
| Basin Tapware | Three hole mixer set/One hole lever mixer/Two separate taps | |
| Bath Tapware | H&C taps with spout or Single lever mixer—wall or hob location | |
| Shower Tapware | Diverter bath tapware/H&C taps/Single lever mixer/fixed or movable rose | |
| Bidet Tapware | Three hole mixer set/One hole lever mixer/Above rim/Bottom of bidet spray | |
| Toilet Tapware | External stop cock or concealed stop cock | |
| Refrigerators | Electrical supply required | Electrician |
| | Water supply required (ice maker type) | Plumber |
| | Building-in and cabinetry door if required | Cabinet maker |
| Microwaves | Electrical supply required | Electrician |
| | Building-in information if required | Cabinet maker |

* This checklist is not exhaustive but it provides a useful starting reference for PC purchasing.

# 20 • ADMINISTRATION

There are two main roles that OBs take on while building their own home. The first is the obvious role of hands-on labourer, but the second may be less obvious, that of the administrator.

The administration role is one by which most OBs will make or break the potential saving of being an OB. Under administration, the OB must be capable of overseeing the project at the highest level of efficiency and perform all the functions that would normally be carried out by a builder and all the functions normally attributed to the owner.

The OB must become a three-in-one person:
- owner
- builder
- labourer

and it is the division of effort between these three roles that demands a high level of administrative skills from an OB.

The key to good administration is good planning and the OB should devote as much time as is necessary at the beginning of the project to make sure an appropriate plan of action and control is produced.

## Time control

The OB should take the time to produce a thoroughly developed critical path program for the job. (A sample is included for reference.)

This critical path chart should concern itself with producing a flexible time program that allows for flexibility in the work schedule wherever possible but avoids unnecessary delays.

The most important concept that must be understood in preparing a critical path program is that the project cannot be built faster than the sum of the time required for the individual processes on the critical path. The critical path is the lineal progression of processes that must follow one another.

Once the processes required on the critical path are established and minimum times allocated to them individually, then the minimum possible building time can be established.

The processes along the critical path can be given a numerical start and finish number, commencing from day 0 (zero) and adding through to the completion. These numbers can be marked on a calendar or diary and weekends, holidays, and other time-out periods allowed for.

As the job progresses these dates may be altered to suit delays, improved times or other influences and the target completion date is always available.

Parallel to the critical path, a series of non-critical processes are plotted and the earliest and latest start days established.

The OB should keep the critical path program and attendant calendar up to date on a daily basis. Only by using a program of this type can the building program be kept in control.

## Labour control

The OB should keep a chart of all labour expended on the project, whether by way of paid contractors or unpaid labour provided by self, family or friends. This should be charted on a sheet similar to the 'Cost Control Sheet' included in the chapter on Costing.

Comparison of estimated expenditure of money and hours to actual expenditure will assist the owner builder to reassess the labour cost budget and allow modifications to be made as necessary.

If the owner builder's family are not able to produce the hours estimated then an administrative decision must be made to allow the project to extend over a longer period, or for paid labour to be employed.

Or, if the paid labour has exceeded its budget a decision may be required to increase the unpaid family labour involvement-even if the project must be extended over a longer period.

Also, in labour control, the owner builder should make sure that sufficient lead-time is built into the program to allow labour to be contracted. Contractors must be immediately informed of any changes that involve them in the building program, so that they can advise the owner builder administrator that they can still do the work at the new programmed time-or advise what adjustments may be required.

## Materials control

The control of materials is twofold. First, the OB is required to order the materials direct from the supplier and arrange for the delivery to the site. Second, the contractor is required to supply the materials as part of a supply-and-fix contract.

When supply-and-fix contractors are used, the ordering and delivery of the materials is the responsibility of the contractor and the OB does not have to be concerned with the quantity of the material ordered, except if the contract is on a cost plus basis. When a cost plus contract is negotiated with a contractor, OBs should be careful that any material purchased under the contract is purchased for the best price from the supplier and passed on to the OB at the best price by the contractor.

To gain control over the materials used on the project OBs should:
1. Separate the materials to be used on the project into two categories:
   a. those supplied by the OB,
   b. those supplied by supply and fix contractors.
2. Carefully estimate the quantities of materials that the OB is responsible for purchasing; check with suppliers to make sure that the method used is compatible with the method used by the supplier, i.e., timber is only sold in multiples of 300 mm. In most areas, it could cost a lot more money if this fact is not understood.
3. Group all materials that can be purchased from the same supply source, i.e., scantling timber, fixing timber, hardware and the like.
4. Get at least two prices for each group of materials from selected suppliers so that you have already eliminated quality variables and can choose the lowest price. It is very difficult to compare two prices where there are both quality and quantity variables to take into account.
5. Tabulate all material quantities and material groups and show preliminary estimates and quotations. As the project proceeds, show a running total of the positive or negative state of the materials purchase account.
6. Note the advance warning suppliers need to assemble and dispatch orders and make sure this advance warning is given. OBs do not want either late or early deliveries.
7. Note the best discount and credit terms that are available from the suppliers and attempt to gain the most advantageous position.
8. Compare all delivery dockets with the load delivered to check against shortfalls in delivery, and then check the delivery docket carefully against the invoice to avoid overcharging.
9. Take care when asked for payment in advance. Check whether this is an industry norm for that type of supplier-brick companies commonly want 100 per cent before delivery-then pay only the minimum that is required and as close to the final delivery date as possible.
10. Check that all GST is paid and any credits gained.

## Money control

If an OB controls time, labour and materials then the money should just about look after itself. OBs should keep a continually updated set of account books, and any apparent upward movement in the cost of the project should be identified and dealt with immediately.

If money is being borrowed for the project, then it is wise to inform the lending body of any increase in the cost of the project as soon as it becomes apparent, so that a decision on whether extra funds can be advanced or not can be made. It is no good to anyone, either the OB or the lending body, to have an 80 percent completed house.

When borrowing money, attempt only to pay interest on money that has actually been paid to suppliers and labourers. If the lending body insists on lending a lump sum, make sure the unused portion of the funds is kept in interest-bearing accounts.

# ADMINISTRATION • 20

## Introduction to critical path planning

Time is the essence of all work; there is never enough of it and every job needs more of it than was planned. But it is possible to plan a building project and to achieve sensible results.

The basic rule is that it will take at least a fortnight to get any project rolling, even after everyone has been planning for months. Also, it is nigh on impossible to build any residential project in under 60 working days.

Add 12 weeks (60 days divided by 5 working days = 12 weeks) to the 2 weeks to start, and the realistic minimum completion time for a project, using a full complement of trades, is 14 weeks. With wet weather losses, wrong and late deliveries and the sometimes hung-over trade workers, add another 2 weeks-that's 16 weeks.

A well-organised project can be completed in 16 weeks with reasonable confidence, but any slip will cause a delay.

The simple rules for a smooth-flowing project are:
- list all trades-then select a short list, check their rates or get quotes and make an agreement for when they will be needed to carry out their section of the work.
- list all material suppliers-check their credit terms, prices, discounts, delivery conditions, stock limitations, lead times, and place preliminary orders.
- revise and upgrade dates regularly and keep all labour and material suppliers informed.
- use a simple computer spreadsheet to see the big picture, and then purchase a simple planning program to keep control. If computers are not for you, then at least draw up a bar chart to keep track of progress.

The biggest part of this planning, after you have mastered the skill of getting the labourers to the site on or near the correct day and keeping them there until they have finished their work so the next trade can start, is to find places for the deliveries to be dropped. Conversely, you also have to find a place for the dump bin.

The deliveries will need careful planning, particularly if you are working on a confined site. Plan for the 30-tonne concrete truck, which turns the lawn into a quagmire, through to the brick delivery truck that will always plan a delivery when you are called away. Be prepared for occurrences such as eight-packs of bricks delivered in the gutter only a day after the local ranger has given you your last warning about deliveries left in the public road.

Make the order and labour lists while you clean up the site at the end of the day's work and be prepared to spend hours ringing tradespeople during the evening-all trades' spouses remove the phone from the hook and turn their partners' mobiles off during the evening meal and until after the best TV is over. Try their email, hop in the car and check if there are lights on, then ring from your mobile phone.

If you use a critical path plan, at least there is specific allowance for unplanned events.

### Conclusion
A well-controlled and administered OB home building project is one where the OB knows at any point in time how much has been spent on the project, how much is still to be spent on the project, has placed all necessary material and trades orders and is closely monitoring the time control

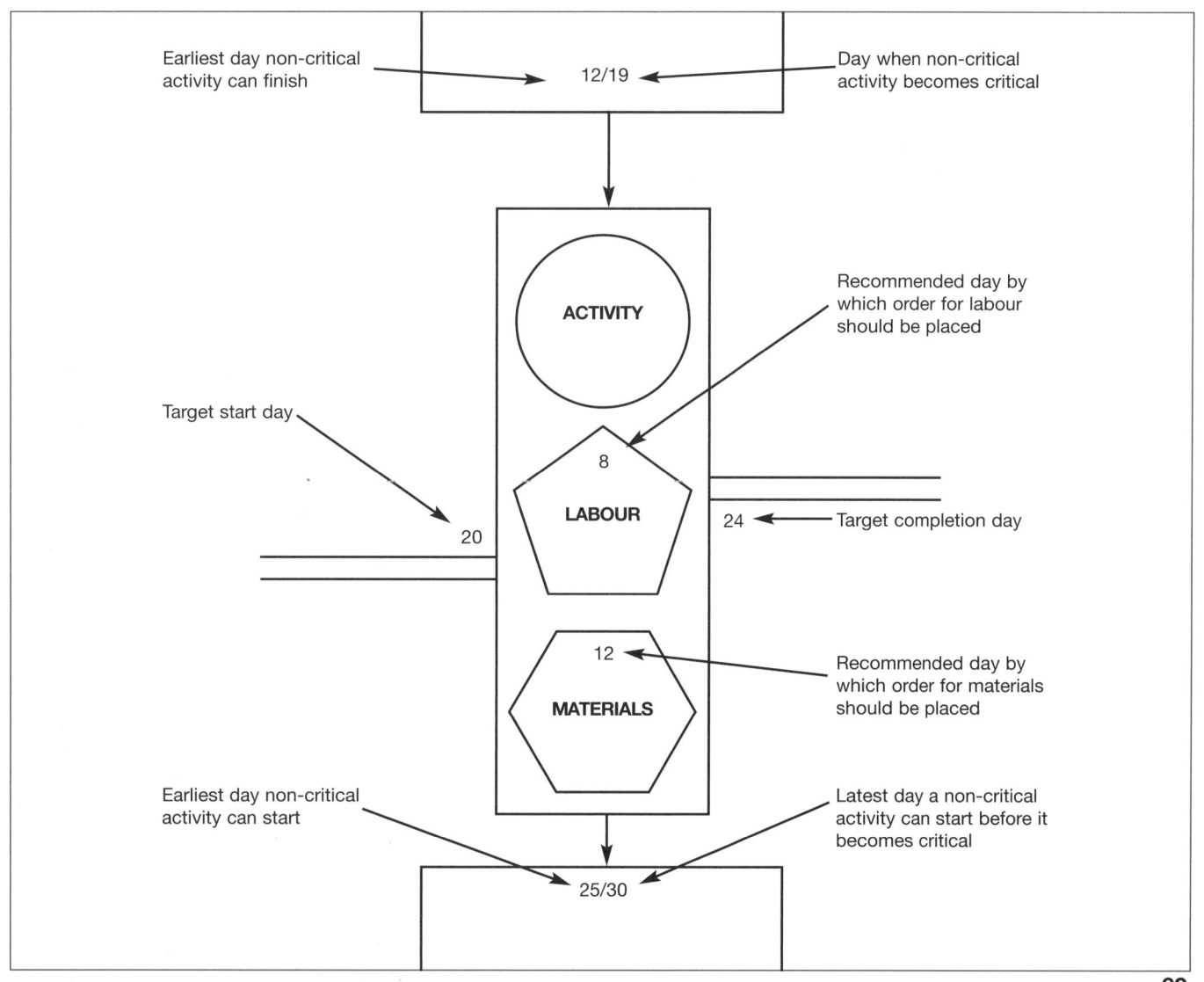

# 20 • ADMINISTRATION

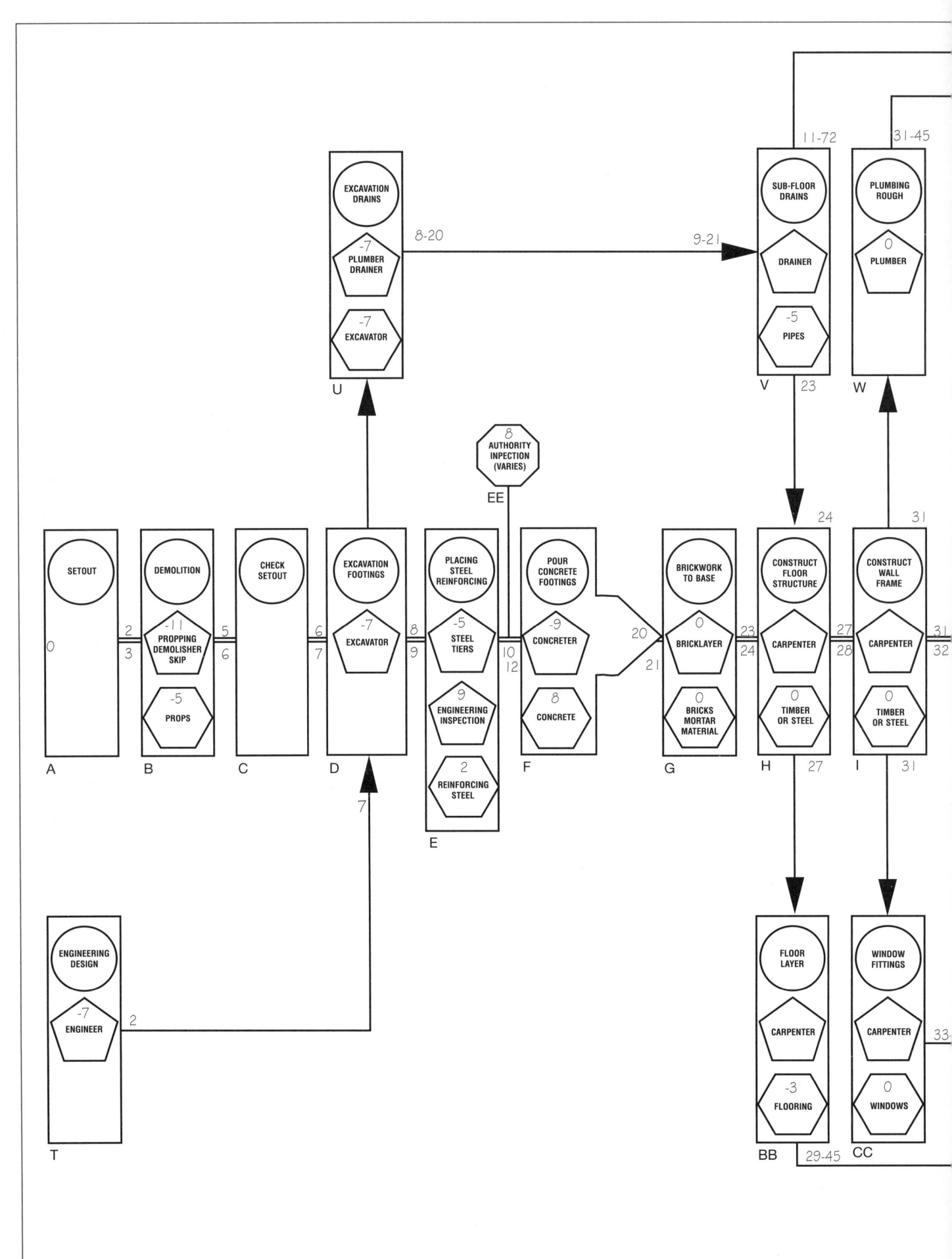

# ADMINISTRATION • 20

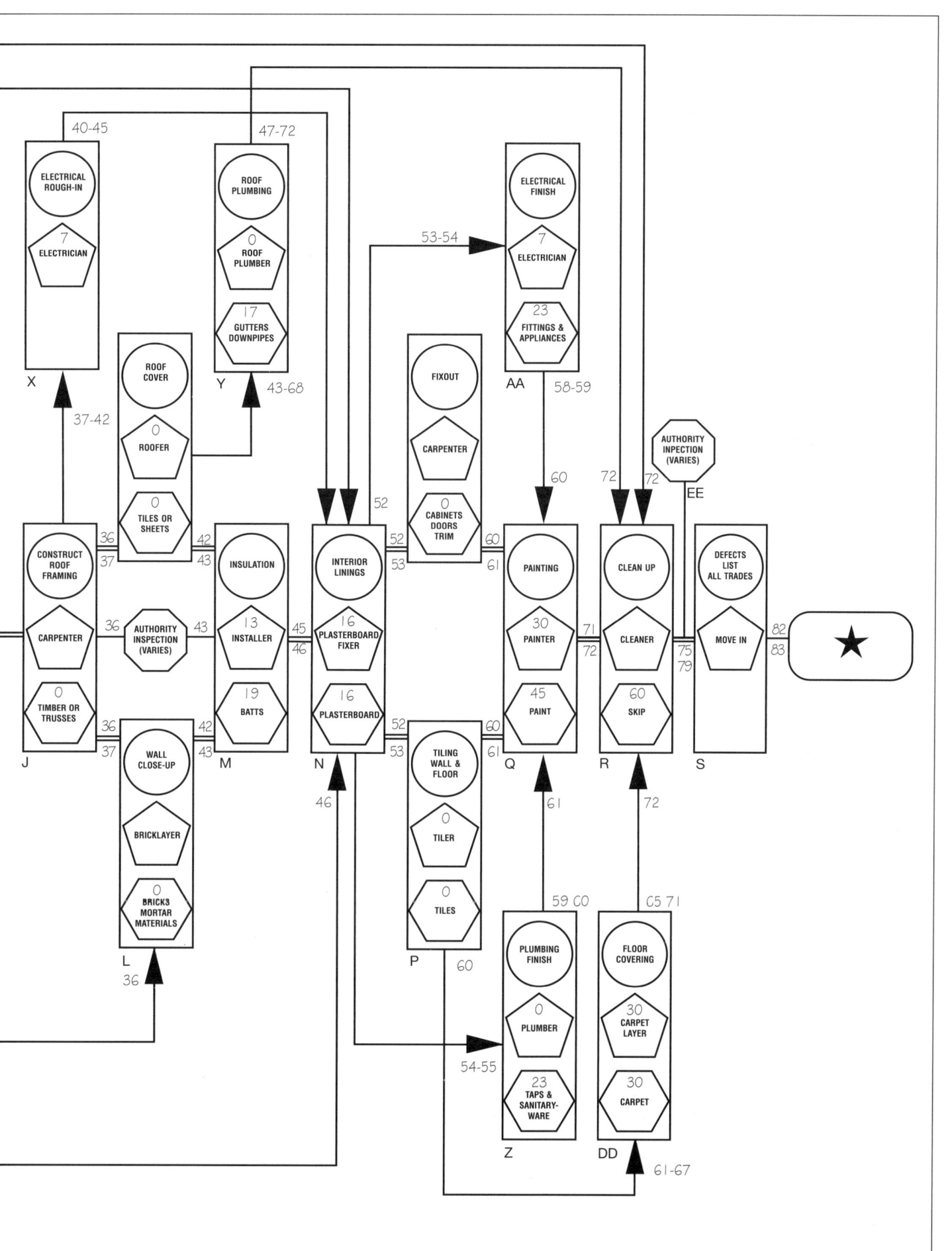

71

# 21 • ACHIEVING COMPLETION

Achieving completion of a building project is often the hardest job of all. It is difficult enough when there is a single building contract but for the OB this great achievement of actually finishing the home is almost unobtainable.

An appropriate method to obtain a completion is to commence the project with a detailed plan of progress and then stick to the plan all the way through the project; list all trades, material deliveries and constraints.

Whether the OB uses a well-planned job approach or the muddles through as time and funds allow, when the project nears completion it is essential that a final list be drawn up of all materials and labour necessary to achieve completion. This list should include all the items still requiring completion, including the names of the tradespeople or suppliers who are responsible to action the tasks, always show dates when the tasks should be complete.

The final list of actions should not extend over four weeks period and, as actions are completed, they should be deleted from the list. Rewrite the list for each week and only when that list will fill one A4 page of text will projects have reached the stage, that in the building industry is called, the State of Practical Completion. At this time most of the final inspections by authorities can be booked and the target completion of all remaining work be set at one week.

Some OBs may not want to reach completion of the whole project at the one time, that is, they want to stage the project over an extended period of time. If this is the case, make sure the lending institution holding a mortgage or lean over the property, and the local council that has control over building compliance and occupancy, are also aware of the intention to build the project in stages and that they give approval for this method to be followed. Many finance institutions and local councils do not like unfinished projects. Financial institutions may deny release of money and the local council could deny occupancy.

## Effects of weather

| Phenomenon | Effect on construction |
|---|---|
| Rain | 1. Affects site access and movement<br>2. Spoils newly finished surfaces<br>3. Delays drying out of buildings<br>4. Damages excavations<br>5. Delays concreting, bricklaying and all external trades<br>6. Damages unprotected materials<br>7. Causes discomfort to personnel<br>8. Increases site hazards |
| Low temperatures | 1. Damages mortar, concrete<br>2. Slows or stops development of concrete strength<br>3. Delays painting, plastering, etc<br>4. Causes delay or failure in starting mechanical plant<br>5. Freezes unlagged water pipes and may affect other services<br>6. Disrupts supplies of materials<br>7. Increases transportation hazards<br>8. Creates discomfort and danger for personnel<br>9. Deposits frost on formwork, steel reinforcement and partially completed structures |
| Snow | 1. Impedes movement of labour, plant and material<br>2. Blankets externally stored materials<br>3. Increases hazards and discomfort for personnel<br>4. Impedes all external operations<br>5. Creates additional weight on horizontal surfaces |
| High wind | 1. Makes steel erection, roofing, wall sheeting, scaffolding and similar operations hazardous<br>2. Limits or prevents operation of cranes and cradles, etc<br>3. Damages untied walls, partially fixed cladding and incomplete structures<br>4. Scatters loose materials and components<br>5. Endangers temporary enclosures |

# WHERE TO NOW? • 22

This drawing outlines the following sections of this book. They have been segmented into clear division in the construction process.

**Sites** covers selection, physical and legal considerations in choosing a block of land.

**The base** covers that part of the structure between the foundation (site) and the top of the ground floor.

**The walls** covers the construction from ground floor through to roof (second-storey work is covered under Options).

**The roof** examines roof structure, covering materials, ceilings and special roof options.

**Services** looks at domestic services, drainage work, plumbing and electrical.

**Fixing out** covers that work done after the external envelope of the building has been made secure and weathertight. This involves joinery work, kitchen and bathroom completion, fireplaces and stairway construction.

**Options** looks at a range of miscellaneous things that do not readily fit into the above sections. These include double-storey construction, garage accommodation and surface finish considerations.

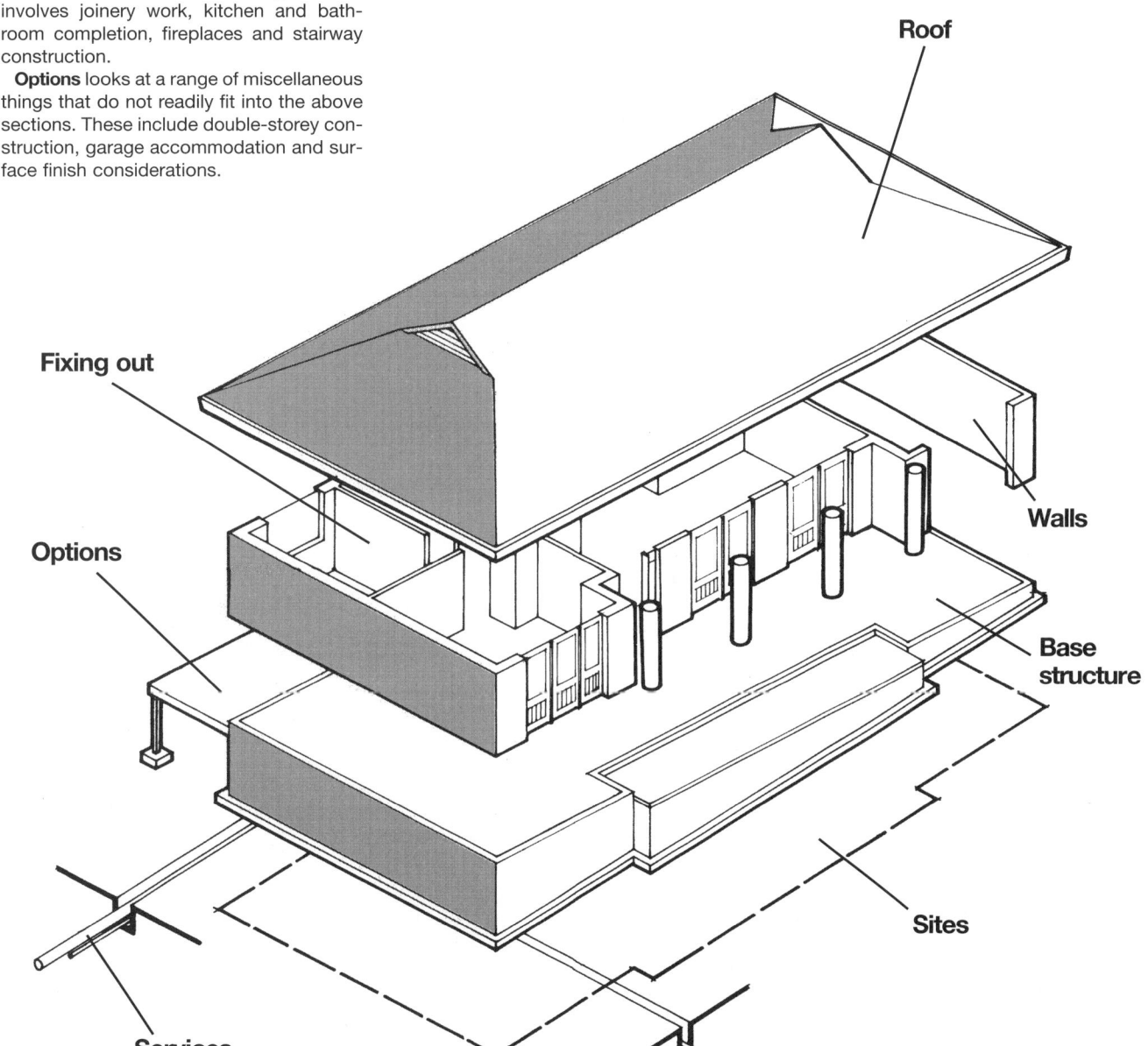

73

# 23 • SITES

Moving into the more practical site analysis area, OBs should spend as much time as practicable at their chosen site determining where the sun is at any point in the day, which way the winds blow, where the views are (a stepladder can often help in determining that), where the neighbours' windows are, and how much the land really slopes. Can you drive your car onto the site without it becoming bogged when you try to leave?

Site analysis is an extremely important part of assembling information from which to design a house, and should not be taken lightly. The following data should be collected for your site and, again, this information should be tabulated to be used later in the planning process.

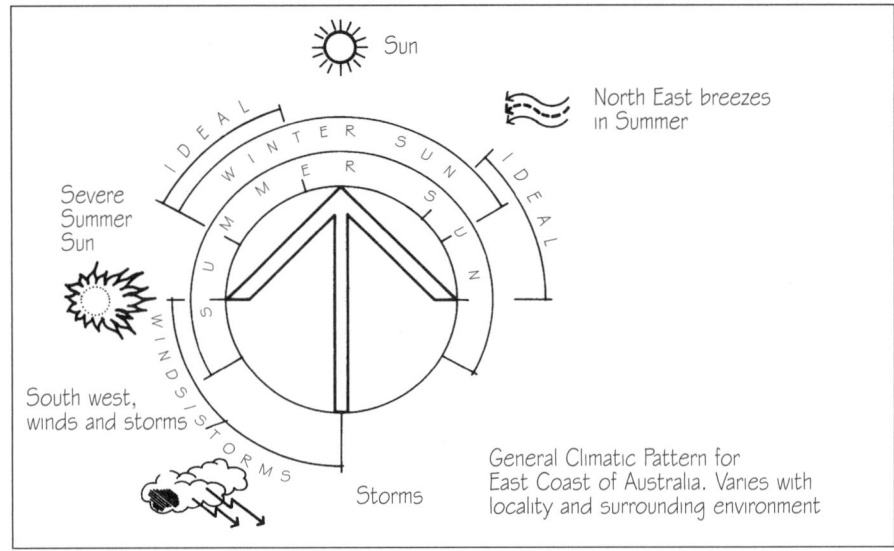

General Climatic Pattern for East Coast of Australia. Varies with locality and surrounding environment

1. The site should be accurately surveyed and all corners pegged so that you can see clearly the extent of the land you own.
2. It is also a good idea to measure and peg the main building line constraints. It's surprising how easy it is to imagine that the front of a house could be somewhere closer to the street than the council code allows.
3. The slope of the site should be accurately determined. This can be done by a surveyor or by OBs, and instructions to assist you with this are included elsewhere in this book. All sites have some slope and all slopes have some determining influence on the final planning and appearance of a house.
4. Determine where the sun rises and sets on your site. As it is often inconvenient to wait a full cycle of seasons to watch the sun rise and set, the information you gather from site observations should be combined with known sun angles available from local meteorological authorities. Meteorological tables cannot tell you accurately everything you need: what effect shadows cast by trees will have; how early or late on a winter's morning the sun warms the frosty ground; or that the best view from the block looks directly into the setting afternoon sun. The latter may be pleasant on a winter's evening, or on a site with a long view and glorious sunsets, but normally the western sun will be a severe constraint in the long summer months experienced in most of Australia.
5. The site should also be surveyed as to wind direction. It is important that this survey should be done on as many windy days as possible throughout the cycle of seasons. Again, this can be impracticable if the site is to be built on in less than a year and information may have to be used from secondary sources. Talk to the neighbours in the area, particularly those who have been living in their houses for more than a year. Information on wind direction is also available from local bureaux of meteorology.

The problem with information gathered from bureaux of meteorology is that it may not take into account local factors, and sometimes a city may expect southerly breezes on summer afternoons whereas, because of ridges and valleys, your site might experience northerly winds at the same time. There is no better way to collect information on breezes than to experience them first hand.

6. Other aspects of weather should be investigated—is there a large rainfall in the area and if so is there adequate drainage available to dispose of it? Is there higher than normal likelihood of thunderstorms and hail? If so, this may affect your choice of roofing material, for instance. Is there any chance of local flooding? Most states now require flood-liable land to be clearly indicated at the time of title transfer; however, there are still areas of Australia where good accurate flood information is not available. Again, check locally.
7. The substrata of the site that is to be used as the foundation for your new home should be checked. Sometimes this can be done easily by just inspecting trenches dug on adjacent properties, especially if you are in a new area and there is a lot of building activity. However, there is no better way to determine the likely foundation material on your site than to go out and spend a weekend digging holes. If you suspect that there is any chance of landslip, or similar foundation problems, then in some instances it is worth the investment to have a geo-technical survey done of the property.
8. The location of major trees and other vegetation on the site should be clearly recorded so that the position of any vegetation you wish to maintain can be located and the proposed residence built around them.
9. A survey of available services should be made as early as possible. The services that OBs should be most interested in are:
   a. electricity supply. Is it available? If not, when will it be available? Is it underground or above ground? This can have an effect on the cost of connection.
   b. the main sewer line. Does it exist? Will it

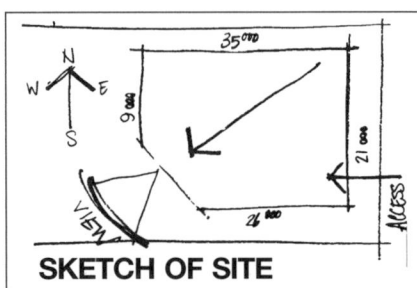

**SKETCH OF SITE**

ever exist, and is it in a position that allows easy connection? (This means a sewer on the low end of the site allowing for normal gravity fall.) Is it somewhere else on the site, maybe requiring a pump system or some other equally expensive method of disposal? If sewer is not available in the foreseeable future, discussions should be had as soon as possible with the local authority controlling septic tank disposal systems. In some areas quite extensive areas of land must be set aside for the disposal of sewer effluent.

   c. street water. Is there any past the property? Is the water on your side of the street or the other? It can be relatively expensive to bring a water supply from one side of the road to the other for, in some council areas, road opening is actively discouraged and conduits may have to be driven under the road pavement. If you are in a bushfire area, then the position of the neighbourhood fire hydrant may also be of some interest to you.
   d. stormwater drainage. Is there any in the area? In some states and cities there is an extensive network of good stormwater drains, but in other cities and towns stormwater drainage is almost non-existent and often relies extensively on street curbs and gutters to dispose of residential runoff. More and more councils are now insisting that stormwater drainage be either taken by gravity to a trunk stormwater main or, if this is not practicable, are requiring stormwater sys-

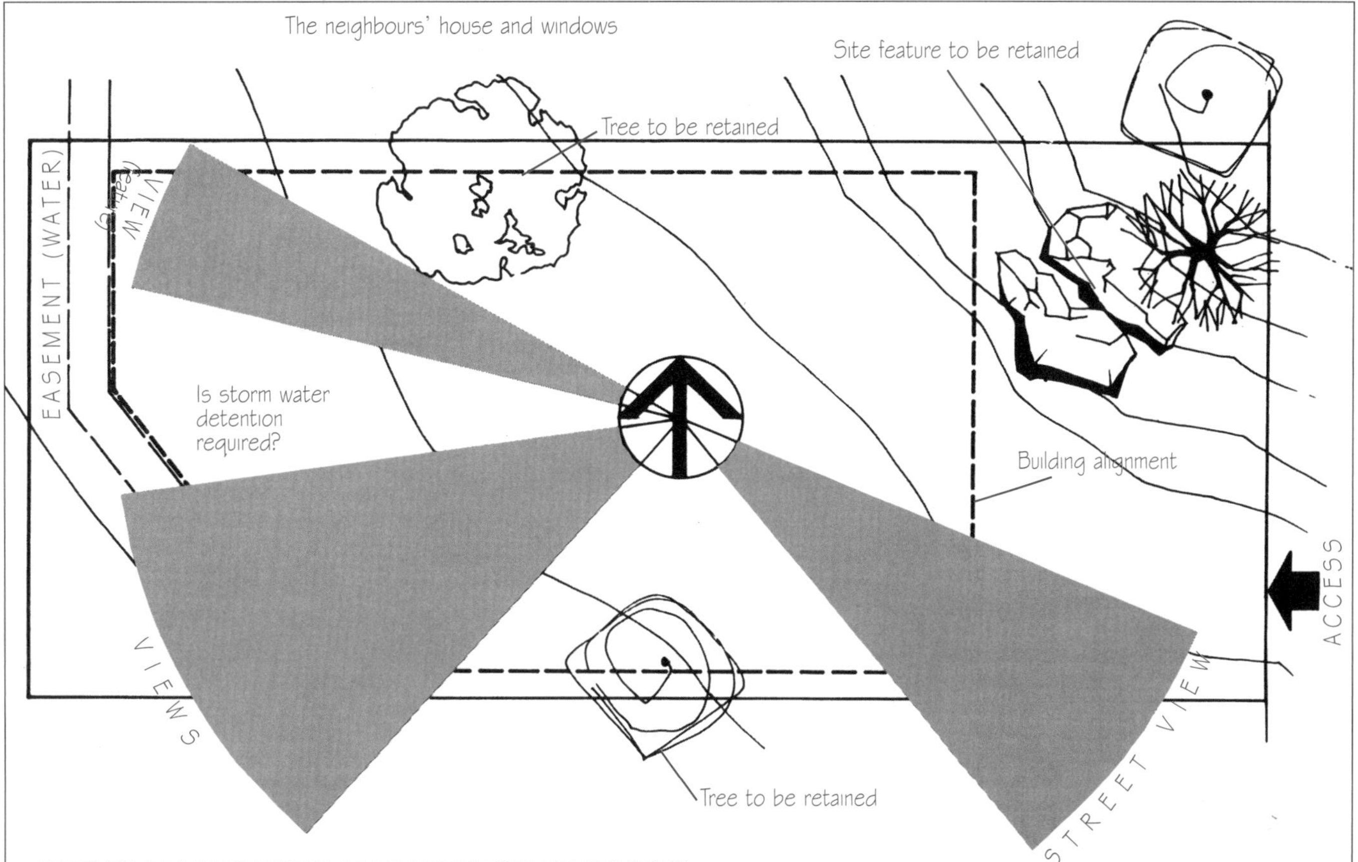

# SITES • 23

**SITE PLAN SHOWING MAIN DESIGN FEATURES**

tems that pump the water from the property out into the street curb and gutter. A discussion with the engineering department at the local council can often assist in giving information on the method of stormwater disposal required.

e. gas supply. With the ever-increasing cost of electrical energy in some states, this has become an important consideration in residential areas. Check whether gas is available in your locality. Sometimes you will find that local gas companies are prepared to run gas mains a reasonable distance from existing systems even if initially they are only to service one household. Check with the local gas company, it may be worth the effort.

f. telephone lines. have they been brought into your area? These days service providers are generally well up to supplying telephones in just about any area of Australia but, from time to time, there are delays caused by assorted reasons often too complex to understand. It is worth making an application on the day that you purchase the land for the installation of your telephone.

Consideration must also be given to site access. The location of the entry for vehicles onto the building block will have a significant effect on the placement of the car accommodation on the site and indirectly on the design of the house that can be built.

Once the OB has collected all of this site information, it should be placed on a sketch of the site so that, during the planning and design stage, the available site data is easily obtained by the designer. This will assist in achieving the most functional design for the particular block of land.

## 1. Locations

You should feel good about the place you choose to live. The most pleasant house will be diminished if its residents hate the neighbourhood. Before selecting a building site, let us examine some characteristics of localities in which we might choose to live.

a. Where do you want to live? An inner city environment, a rural setting, a suburban community?
b. What type of people like to live in the community—majority retired, young families? What are the children like?
c. Is there an active political and community consciousness in the area?
d. Are there adequate utility services available—water, electricity, gas, sewer, fire control?
e. Is there an adequate shopping area?
f. Is transport available—to work, to shops, to other areas, or will you have to depend on a private vehicle?
g. Are adequate social utilities available—churches, libraries, parks, clubs, recreation facilities?
h. Are adequate educational facilities present—preschools, kindergartens, primary schools, high schools, tertiary colleges?
i. How high are land rates, water rates?
j. Are the other houses in the immediate

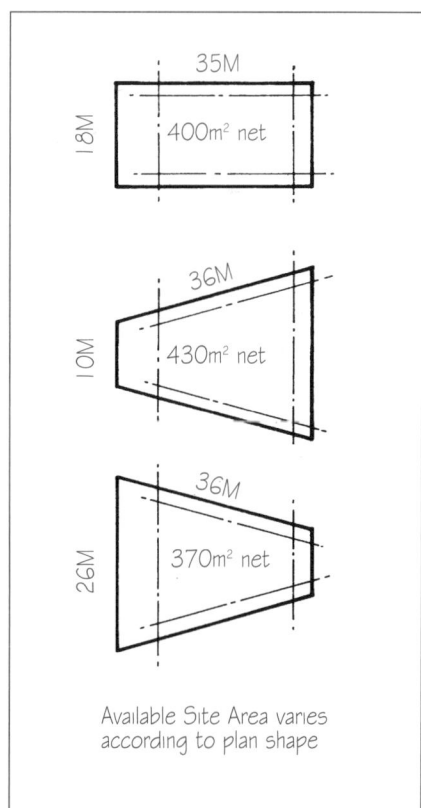

Available Site Area varies according to plan shape

vicinity well maintained? This can affect the future investment potential of your block, eg. are the houses neglected because a proposed highway is planned to pass nearby?

# 23 • SITES

k. What development plans, if any, are proposed by federal, state or local government bodies for the area—airport, highway, garbage dump?

## 2. Access
What access is there to the site? Will you have to build a private road?

Do obstructions exist in the path of available access, eg. telephone electrical poles, side entry pits, trees, junction boxes, rock outcrops, water courses, vents. Is it possible to circumvent these, or do they have to be moved—how difficult is it going to be to liaise with authorities over this?

Differential heights between street and site may lead to excavation, filling or ramping. Unformed council road or verge/gutters may need to be bridged.

## 3. Site amenities
What natural features does the site have?
a. trees, vegetation—are these worth retaining? Can they be maintained. Is building work (recontouring of site) going to give problems. eg, kill trees?
b. views, both local and distant—can they be utilised, or will they be built out by others?
c. dam, creek, watercourses—will they give flooding problems? What are your rights and responsibilities over these?
d. rock outcrops—can they be utilised, or will they have to be excavated? Are they part of the bedrock or are they floating boulders?

## 4. Site services
What utility services are available, and what will it cost to connect into them?

The cost to connect to utility services can be expensive, especially if the service has to travel some distance to reach your house or site. Problems such as cliffs, extensive rock or neighbouring properties may hinder connection.

If some services are absent, how much will it cost to set up alternatives? The cost of alternatives can be expensive, in some cases offset by long-term running costs, but also aggravated by running/maintenance costs.

**Substitutes:**
a. Sewer—septic tank, grease traps, absorption trenches, higher initial cost, long term maintenance.
b. Water—tank water, higher installation costs, no water rates.
c. Electricity—photo voltage, only gas generator.
d. Gas—town or bottled.
e. Telephone—mobile phones and other digital communication systems.

## 5. Site geology and substrata
Is there a good foundation on which to build?

It is wise to avoid building on organic matter (tree roots, leaf mould). These should be removed to a distance of three metres from the building area; excavation down to an undisturbed soil stratum is desirable.

Do you want to keep the topsoil for later landscaping? If so pile it up to one side. Do not build over or on holes/hollows filled with rubbish or organic material—fill these with clean filling material.

What type of soils are present on the site? Are they amenable to water retention, drainage. Clay soils (reactive soils) absorb water and swell—this can give footing problems in some instances. Sandy soils are porous and surface water is absorbed down to the ground water level.

Are there any soil erosion or possible flooding problems? If these possibilities exist, can they be overcome? How much will it cost? Advice from a geotechnical engineer may be in order.

Will the site require recontouring? If so how difficult will it be? Is bulk excavation or fill going to be needed? How much will it cost? Is excavation in rock necessary? Is this expensive? Are specialised tools (eg. jackhammers, compressors) needed? Will the neighbours complain? Remember the noise nuisance laws for your area.

## 6. Site orientation
What direction is north? This is a desirable direction in which to face living areas. Will the location of the site affect the amount of sunshine and prevailing winds for your locality.

Does the site disposition allow penetration of winter sun and avoid some summer sun?

| | |
|---|---|
| Winter sunrise | 60° east of north |
| Winter sunset | 60° west of north |
| Summer sunrise | 120° east of north |
| Summer sunset | 120° west of north |

These vary somewhat depending on latitude. Will the sunshine be blocked off by nearby buildings or hills, trees?

Is the site exposed to wind? Winter wind is in the south-south west—west sector. Summer breezes are in the north-west, north-east sectors.

Check to see if desirable sun times or breezes will be blocked by existing buildings or trees. This, however, can work to your advantage in harsh climatic zones, or orientations. Check to see your house will not obstruct sunlight to adjoining properties. This situation could be legally set down in your locality.

## The 'legal' site

### 1. Site boundaries
**(see A in diagram on next page)**
Has the site been surveyed, and have the boundaries and boundary corners been located?

Even if you are absolutely sure that the boundaries of your building block are clear, it would be wise to get a boundary survey. This is a certified drawing from a registered surveyor that notes the location and exact lengths of boundaries. The surveyor usually pegs the corners of the block to indicate a change in direction.

If your survey identifies encroachment from adjoining properties then give this information your immediate attention and obtain legal advice as to what your next step should be. In some instances encroachments across titled boundaries can jeopardise your ability to mortgage the property and obtain building finance.

Note: subdivision blocks are numbered from No. 1 consecutively at the time a subdivision is registered with appropriate authorities. Therefore it is possible to have two Lot No. 1s in a street, if separate subdivisions have taken place. Make sure the block you have seen is the one you have purchased, and make sure your block is clearly identified for builders, tradespeople and suppliers, so that work in fact happens on your block, not somebody else's.

### 2. Site zoning
What is the site zoned for? Is it prime residential? How are the surrounding sites zoned?

The provisions of land-use zoning vary significantly from locality to locality. It is therefore imperative that prospective purchasers glean from a responsible authority whatever local knowledge is necessary so that a clear picture is built up about what current and future development may be expected.

### 3. Covenants
Does the site have any covenants governing the type of construction and use of materials?

Covenants are specially framed private conditions that are applied to particular blocks of land. Normally those provisions are made applicable by the original subdivider/developer of the land over which they apply. Covenants are often used by developers in an attempt to control the quality of development in the subdivision and most often are concerned with items such as brick, exterior walls, tiled roofs, minimum floor areas, maximum heights and the like. In some localities government agencies may use covenants as development control devices.

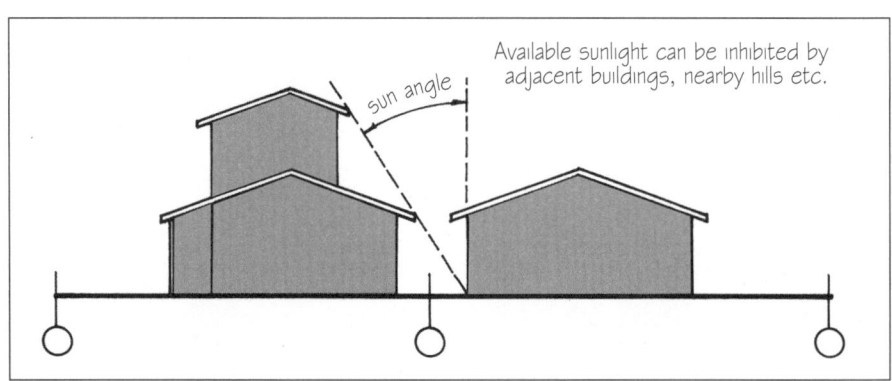
Available sunlight can be inhibited by adjacent buildings, nearby hills etc.

# SITES • 23

### 4. Building alignments
All sites have one or more building alignments. These are (letters refer to diagram):
A. Site boundaries (refer 1 above)
B. Front building alignment—runs parallel to the front site boundary; if the front boundary is curved or changes direction, so does the front building alignment. The distance back from the front boundary varies with the local council, the type of road your block fronts, eg. the setback on arterial roads is normally greater than on domestic roads. The common distance ranges from zero to over 10 metres.
C. Side building alignments—vary for each locality, but can range from zero from the boundary to the outside surface of a sidewall, in excess of 3 metres. Where a block has a side street, a special site setback provision for that side street may apply.
D. Rear building alignments—again vary for each locality. They can range from zero to over 3 metres.
E. Vehicular access to backyard—some localities require that car access be provided from the street alignment to the rear of a block, past any building obstruction. To achieve this, a minimum setback from the side boundary of approximately 2.5 metres is normally stipulated.

Sometimes this can be taken to mean access through a garage or carport.

### 5. Site area
Do you have enough room on the site to build what you need? Is there room for planned future expansion—a garage, swimming pool, granny flat?

Most residential subdivisions today provide a gross site area of between 400–1000 square metres. At the bottom end of this range, development is generally restricted to modest cottage construction due to the net available building area (after deducting all setbacks) of approximately two thirds of the gross area, ie. 300–400 square metres.

It is important to understand that in most localities the setbacks are set by the imposition of a numeric figure that does not take into account specific site variations, ie. wide frontages. Small gross area sites can be disadvantaged.

Although most state building codes allow up to two-thirds site cover many local authorities restrict this further to half gross site cover or less. This has the effect of at least applying an equitable restriction.

Examples (see diagram on p.73).

Consider a hypothetical local regulatory body that requires the following site development conditions:
1. minimum gross site area of 630 sq.m
2. minimum front setback of 7 m
3. minimum side setback of 1 m
4. minimum rear setback of 3 m
5. minimum frontage of 18 m for rectangular sites and 10 m for fan shaped sites.

The setbacks do not take these variations of site shape into account.

Within the coverage provision for the net useable area of your site, in most localities you have to put your planned development, ie. house, garage, swimming pool, future granny flat, extension, garden shed or workshop. Note that these provisions relate to ground area covered; this does not affect double-storey developments and except in most localities, numerically larger side setbacks apply.

If on our hypothetical block constant conditions were applied to the following site, then the net useable site area would be 720 square metres or 72 per cent of gross site area (percentage of useable land increases significantly—see example A above). This means smaller blocks of land tend to be penalised by the regulations on two levels:
(i) there is obviously less gross area.
(ii) there is proportionally less available area leading to a reduction in flexibility of site.

### 6. Easements
Do any easements for utility services pass through the site?

The authorities that are charged with the provision of utility services are normally also given extensive rights of access and easements.

Normally, gas, electricity and telephone services are provided within the roadway zone. However, stormwater drainage and sewer drainage may run through properties because of the need of these drains to follow natural falls and contours. Most sewer and drainage authorities attempt wherever possible to position their main drainage lines adjacent to property boundaries, with little or no detriment to the available site building area, but sometimes these mains are found away from the boundaries, crossing the site detrimentally (as illustrated). Sewer and drainage are seldom less than 1 metre in width but are unlikely to exceed 3 metres in width. As the authority responsible for the service, located in the easement has right of access to the service then normally landowners are not permitted to build over or into the easement area. Some authorities do permit buildings to be constructed over easements if certain conditions can be met.

### 7 Neighbours' rights
Many modern planning and environmental laws require that any new buildings or alterations to existing buildings be advertised. Some authorities may do this in the local paper, others may send letters to adjoining landholders or refer the application to a citizens precinct committee. Check your local authority and provide them with the information they require.

Although neighbours can lodge objections in most areas, it is the local authority that will make the final decision. If you are refused a building permit there will almost certainly be an appeal procedure.

These processes have added time and cost to the application procedure and it is wise for owner builders to discuss the impact of their proposed developments with the local authorities at an early stage in design.

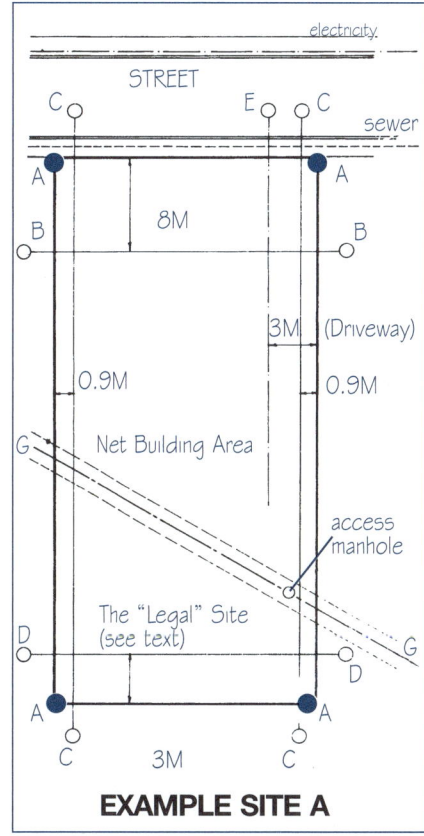

**EXAMPLE SITE A**

# 23 • SITES

| | SLOPE TYPE 1 | SLOPE TYPE 2 |
|---|---|---|
| The site can be viewed from two different but complementary standpoints. **The physical site**—its characteristics, limitations. |  |  |
| 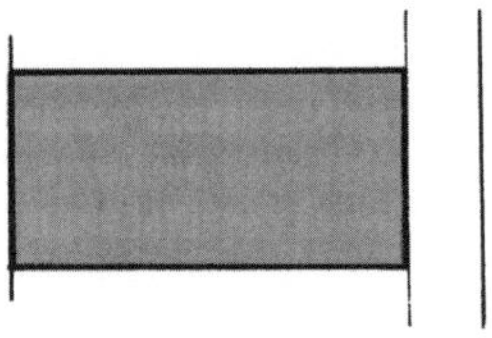 **PLAN TYPE A** | **A1**<br>• Vehicular access, whilst building and when house is designed.<br>• Position of drainage systems can be a problem—from house, sub-soil drainage.<br>• Suitable for split levels, elevated houses, but avoid looking down from street effect.<br>• Problems—retaining walls, excavations.<br>• Connections to utility services, especially if they are at road level.<br>• Slab on a cut and fill bench can cause very steep entry for cars and added danger of water entering the house. | **A2**<br>• Vehicular access can be restricted, may require excavated driveway and car accommodation.<br>• Drainage generally good.<br>• Avoid excessively high base walls to house.<br>• Roofs may not be visible from street.<br>• Services generally not a problem.<br>• Location of swimming pool can give problems. |
| 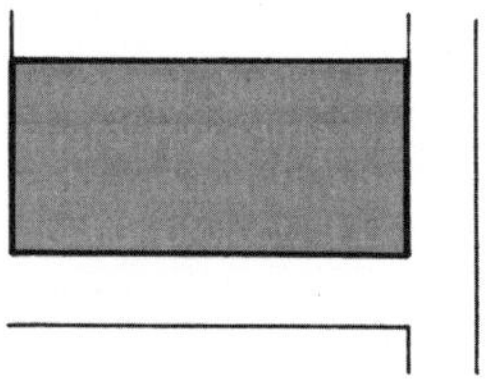 **PLAN TYPE B** | **B1**<br>• As for A1.<br>• May get subfloor garage from side street. | **B2**<br>• As for A2.<br>• Car accommodation to rear of site. |
| 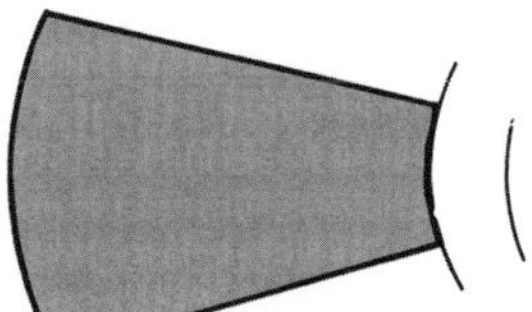 **PLAN TYPE C** | **C1**<br>• As for A1.<br>• Take care to consider house being obscured by car accommodation.<br>• A good block if views to rear of site. | **C2**<br>• As for A2.<br>• Vehicular access is limited. |
| 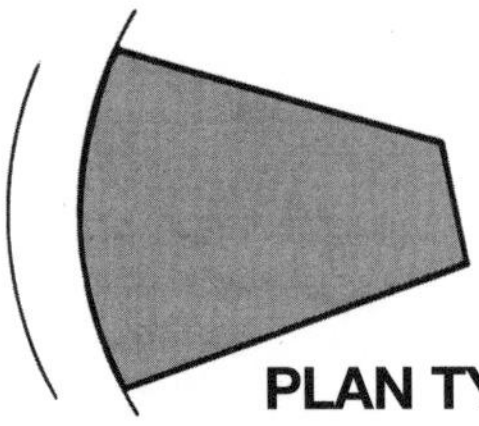 **PLAN TYPE D** | **D1**<br>• As for A1.<br>• Wide frontage may force the available building area to be confined (front building line works against development). | **D2**<br>• As for A2.<br>• Provision for swimming a major problem.<br>• Building area is confined. |
| 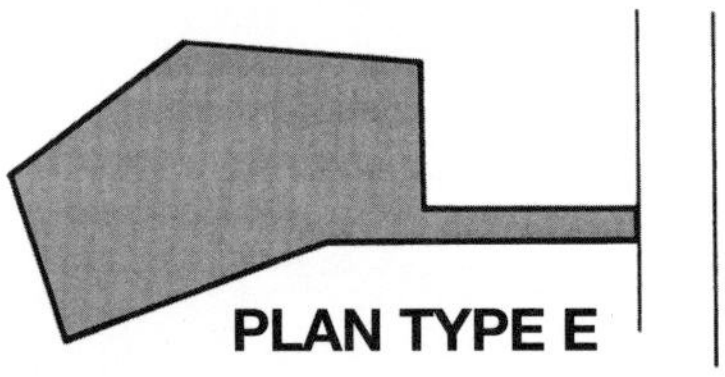 **PLAN TYPE E** | **E1**<br>• As for A1.<br>• Danger in leaving site to enter road—view obscured.<br>• The 'handle' of block must be capable of allowing access to trucks during building.<br>• Increased water runoff down driveway can cause drainage problems. | **E2**<br>• As for A2.<br>• Truck access a problem.<br>• Driveway design can become a problem.<br>• Water runoff into neighbour's property. |

# SITES • 23

| SLOPE TYPE 3 | SLOPE TYPE 4 | SLOPE TYPE 5 |
|---|---|---|
| **A3**<br>• Some slight fall is desirable (1:40).<br>• Access good, consider boat access and tidal range.<br>• Check 'king tide' and flood levels.<br>• Suitable for elevated houses.<br>• Check height restrictions for rights of views from neighbouring properties.<br>• Swimming pools may need special design because of ground water fluctuation.<br>• Check combination of street and foreshore building lines.<br>• Services may be at high level.<br>• Drainage (sewerage) could be a problem (especially septic). | **A4**<br>• Some slight fall is desirable (1:40 is ideal) otherwise drainage can be a problem.<br>• Access good.<br>• Suitable to most building types.<br>• Could be in low lying zone—may be prone to flooding, fogs, inversions.<br>• Views can be easily built out. | **A5**<br>• Don't trust your eye—get a survey of land levels.<br>• Check that vehicle accommodation can be incorporated within budget.<br>• Access to rear of site a problem.<br>• Normally rock geology, drainage systems tend to be complex and expensive—large surface runoff.<br>• Lack of sewerage to sites of this type can delay development as septic systems are often not permitted.<br>• Septic systems difficult and expensive.<br>• Could have substrata problems in supporting house.<br>• Generally requires specifically designed house—few project homes are applicable.<br>• Swimming pool a problem. |
| **B3**<br>• Not normally applicable, but as for A3. | **B4**<br>• As for A4.<br>• Vehicular accommodation from side street.<br>• Flexibility in orienting front of house.<br>• Not applicable to split level. | **B5**<br>• As for A5. |
| **C3**<br>• As for A3.<br>• Highly desirable block.<br>• Care not to have excessive exposure to winds, or provide for these in design of house. | **C4**<br>• As for A4 and other C categories.<br>• No special problems. | **C5**<br>• As for A5.<br>• Can give aggravated access problems. |
| **D3**<br>• As for A3.<br>• As for D category problems.<br>• Narrow water frontage may present problems if boat accommodation is desired. | **D4**<br>• As for A4.<br>• No special problems. | **D5**<br>• As for A5.<br>• Building area can be restricted. |
| **E3**<br>• As for A3.<br>• Possible right of access to waterfront problems. | **E4**<br>• As for A4.<br>• No special problems. | **E5**<br>• Aggravated access problems for trucks.<br>• Runoff to neighbour's property. |

# 24 • THE BASE

The base of the building is defined as that part of the building between the foundation and the top of the ground floor.

The base work for a cottage includes, but is not necessarily limited to:
- the footings
- the base walls
- the piers or stumps
- the floor frame
- the floor sheeting
- the concrete floor slab
- the external access steps
- the retaining walls
- the subfloor garage, and so on.

The base of a house has many functions but the main reason that it exists is to provide support for the floor platform, so that the slope of the land can be converted to a horizontal level floor.

The base structure of the house can vary from a simple raft slab on level ground to a complex double garage and basement on a steeply sloping site.

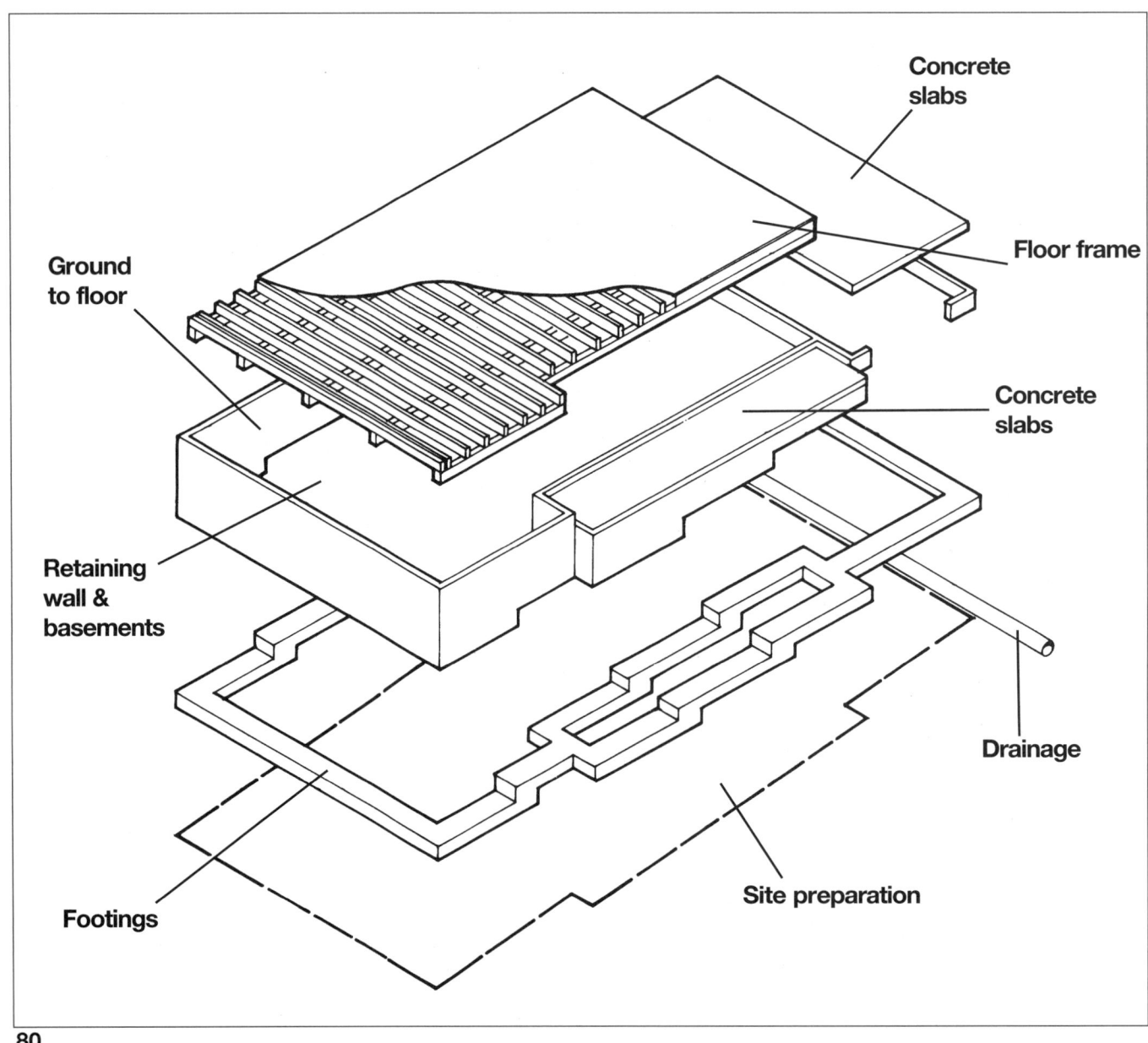

# SITE PREPARATION • 25

Before any work on the building proper can commence, the site has to be prepared to enable work on and around the building. Site preparation comprises clearing the site, setting out the building and excavation for the footings.

## 1. Clearing the site

Clear the site of underbrush and trees over the area designated for the building proper. Clear away any roots, stumps and tree limbs. Remove any underground wood and rocks to a distance of approximately three metres from the building zone. Check with local authorities for any old or existing water, gas or sewer lines that may pass through the site. If the topsoil is of good quality, remove it with a bulldozer for later use in landscaping.

When the site is sloping, create a level platform at the approximate location of the house. If the slope is steep you may need a bulldozer to level the estimated building area.

There are two types of excavation. The first is called bulk excavation. This is excavation that clears a level area for the building. Usually a bulldozer or tracked loader type vehicle is employed.

The second type of excavation is called specific or footing excavation. The same type of vehicle with a back-hoe or shovel can be used, but for footing excavation to take place, the building must be set out in its proper place with respect to site building alignments so that trench excavation for footings can proceed.

## 2. Site setout

The following refers to the adjacent diagrams in a simplified procedure of setting out or locating the building. The setout lines are the outside surface of the base walling materials. Allowance will have to be made for the width of the footing. Operators of excavation equipment prefer either the centre line of the trench marked or the outside edge of the trench marked. Marking usually takes place by running along the ground with a bag of lime, marking out in a straight line the wall outline.

a. Check to see if survey pegs have been placed at the exact corners of the site. Run a string line between the nails in the top of the stakes across the front boundary, using the boundary line as a guide.

The front building line can be found by measuring back into the property at 90 degrees to the string line. Note: the front building line runs parallel to the front boundary line. If the front boundary of the property curves or has an angle in it, then so does the front building line.

A layout square is used to make measurements at right angles to the string line.

Measurements are taken at two different points along the front boundary line.

Make sure the stakes are carefully pointed otherwise they will twist when driven in and an accurate indication of where the corner is will not be obtained (see Diagram 1).

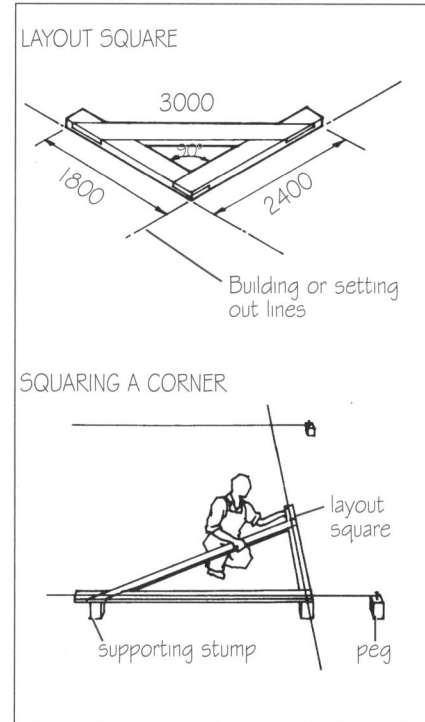

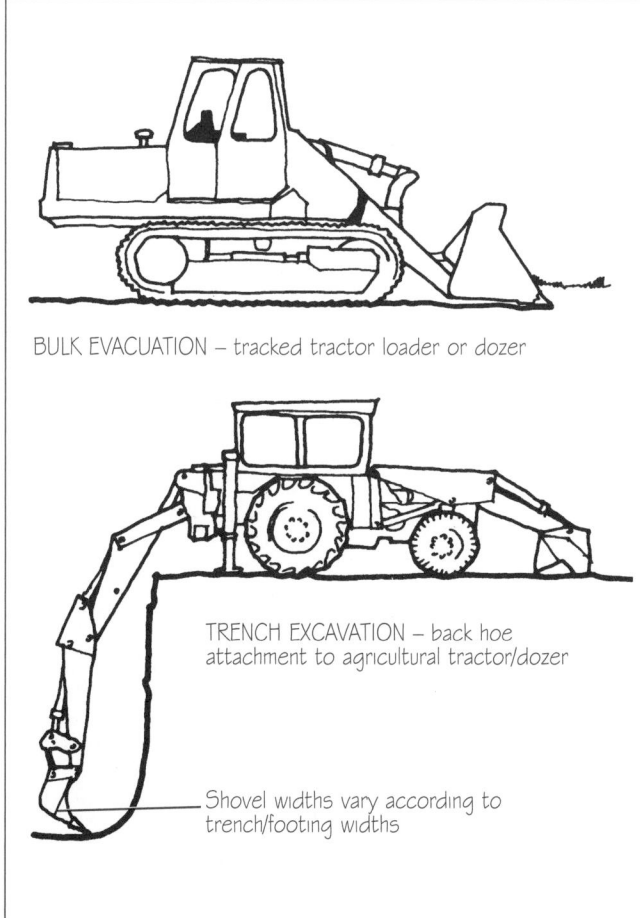

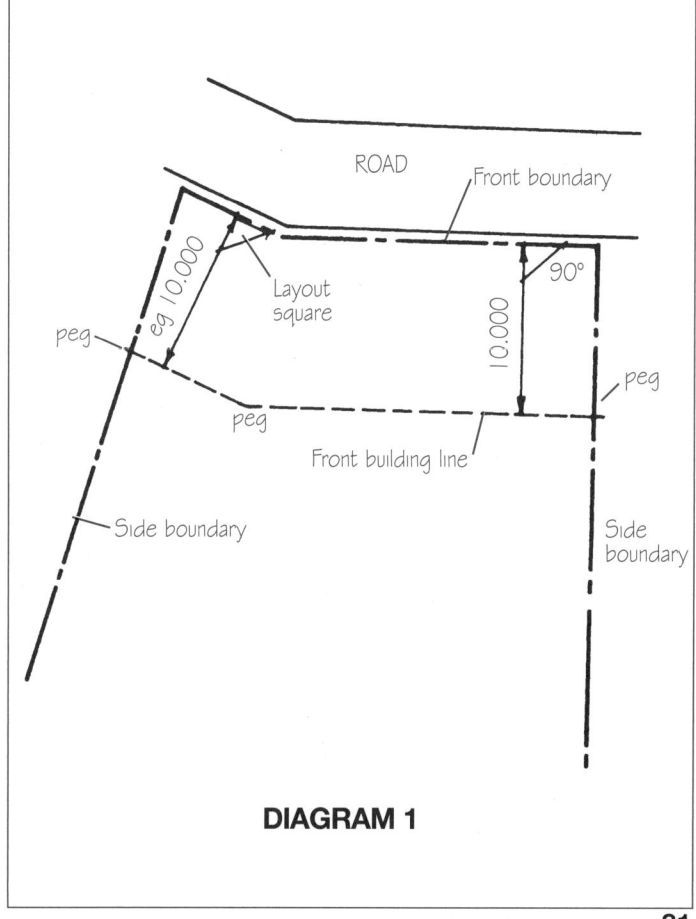

**DIAGRAM 1**

# 25 • SITE PREPARATION

b. Establish the position of the first corner. Do this by measuring along the front building line the appropriate distance from the side boundary line. This locates corner one. Drive in a stake (see Diagram 2).

c. Starting at corner one, measure along the front building line the length of the front of the house and locate corner number two. Check by measuring along the remainder of the front building line and see if the summation matches the front building line total (see Diagram 3).

d. Once the front two corners are found use a layout square to find corners three and four. Place one leg of the square directly beneath line one/two; the adjacent leg will indicate the perpendicular. Measure along this line approximately two to three metres longer than the dimension of the side of the house. Drive a stake at point 3A. Run a string line between 2 and 3A. Measure the side dimension and drive a stake at corner 3 (see Diagrams 4, 5).

e. Repeat this procedure to find corner number four (see Diagrams 6, 7).

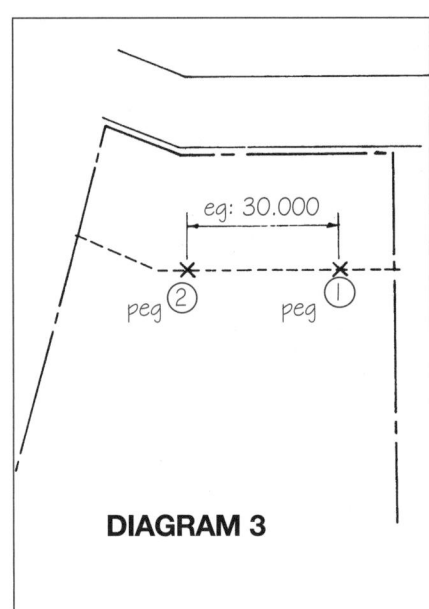

**DIAGRAM 3**

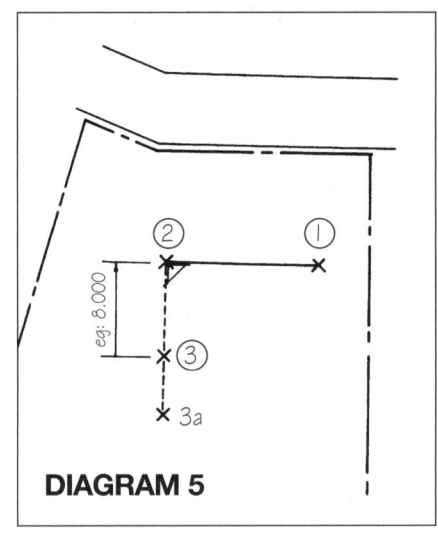

**DIAGRAM 5**

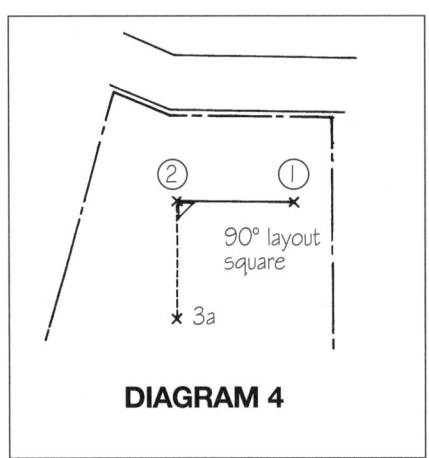

**DIAGRAM 4**

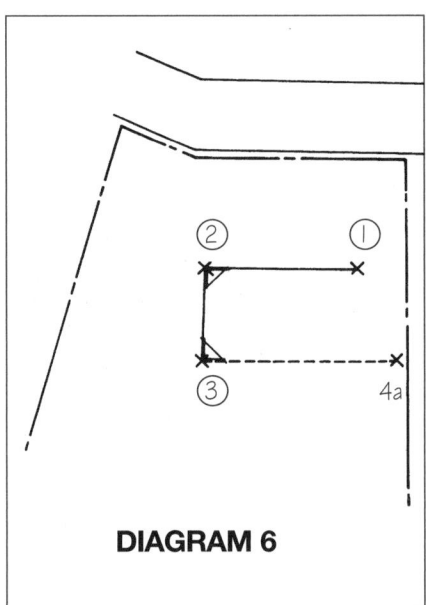

**DIAGRAM 6**

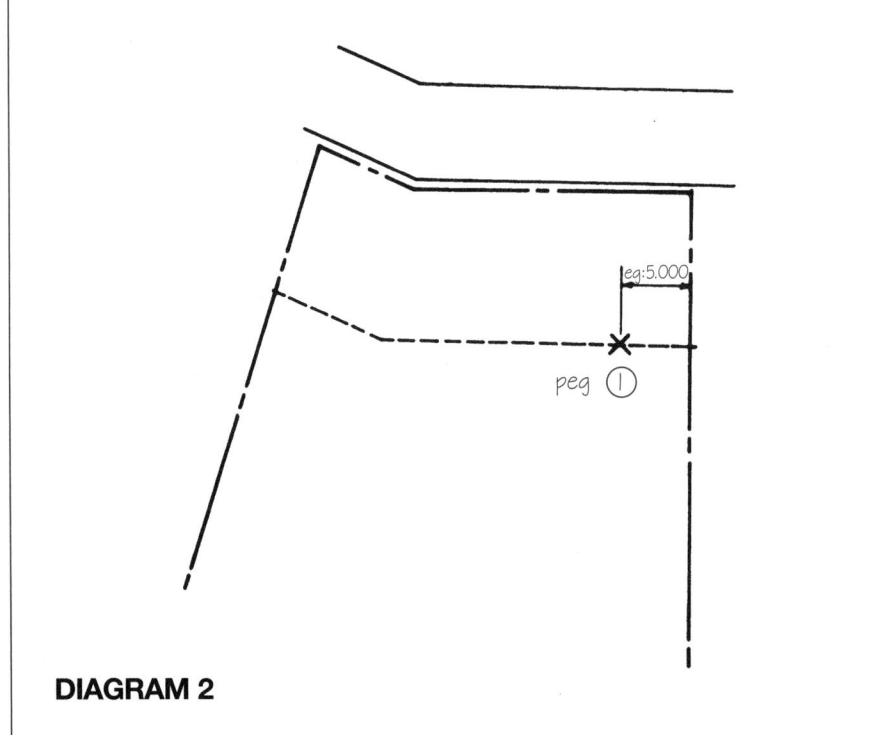

**DIAGRAM 2**

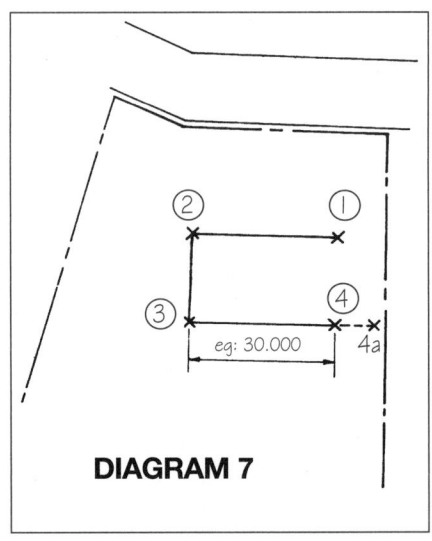

**DIAGRAM 7**

# SITE PREPARATION • 25

f. The accuracy of the corners should be checked by measuring the lengths of the diagonals one/four and two/three. The diagonals of a perfect rectangle are always equal. If there is a discrepancy, adjust the lines accordingly by repeating the above procedure. Double check al dimensions. Please note: all other construction is based on these first steps.

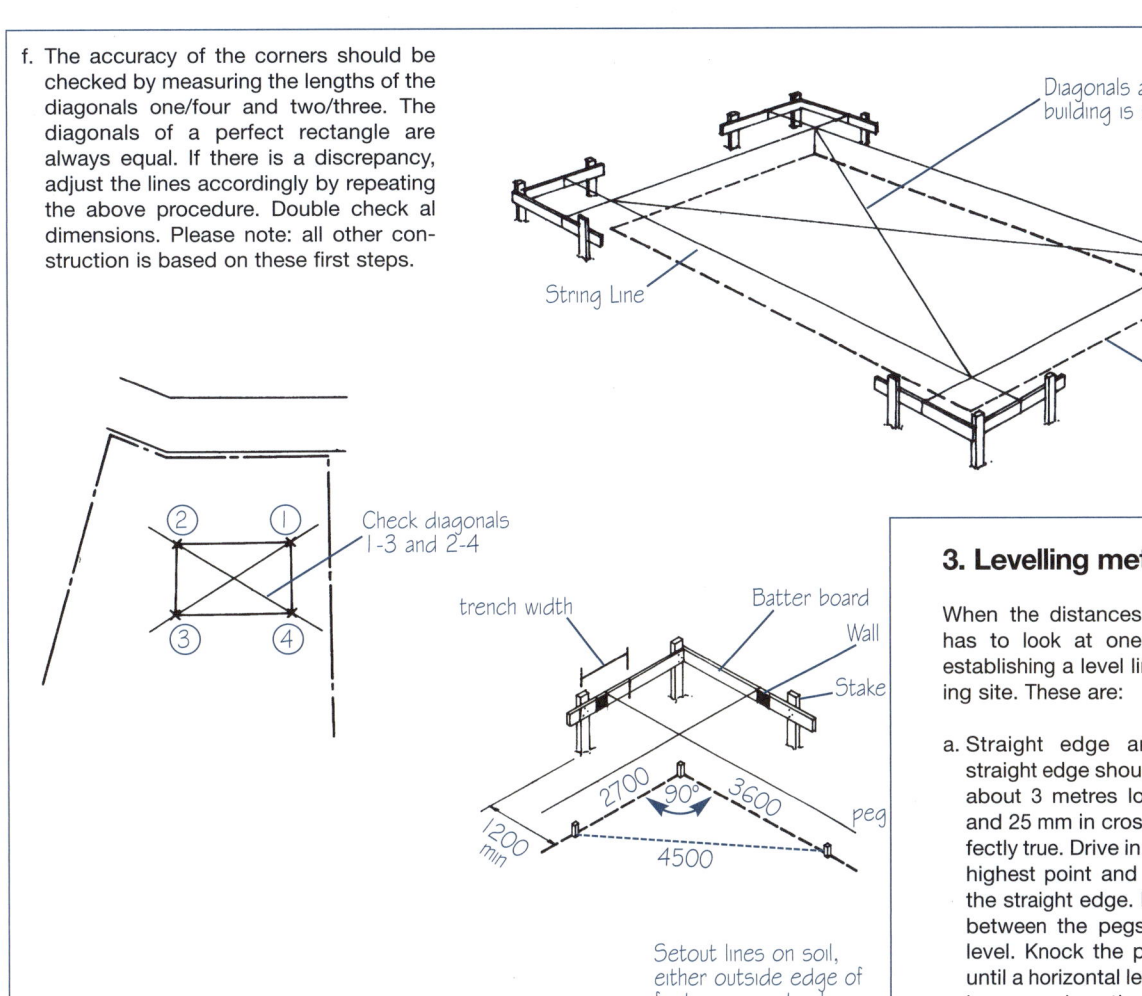

## 3. Levelling methods

When the distances become longer, one has to look at one of four methods of establishing a level line or plane on a sloping site. These are:

a. Straight edge and spirit level. The straight edge should be a piece of timber about 3 metres long by 150 mm deep and 25 mm in cross-section, planed perfectly true. Drive in a peg at point A at the highest point and point B the length of the straight edge. Rest the straight edge between the pegs and apply the spirit level. Knock the peg B into the ground until a horizontal level is achieved. Check by reversing the spirit level on the straight edge. Drive in another peg at C and repeat the operation.

If the setout is on sloping ground, measuring along the surface of the sloping ground will not give the correct horizontal distance between two points. Over a short length of slope measure the distance in stages. Always keep the tape horizontal and the ranging rods vertical.

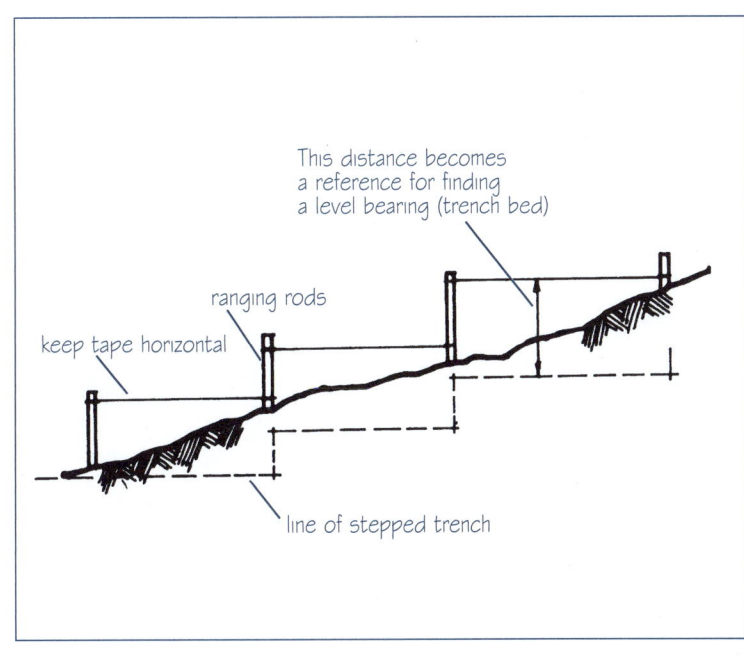

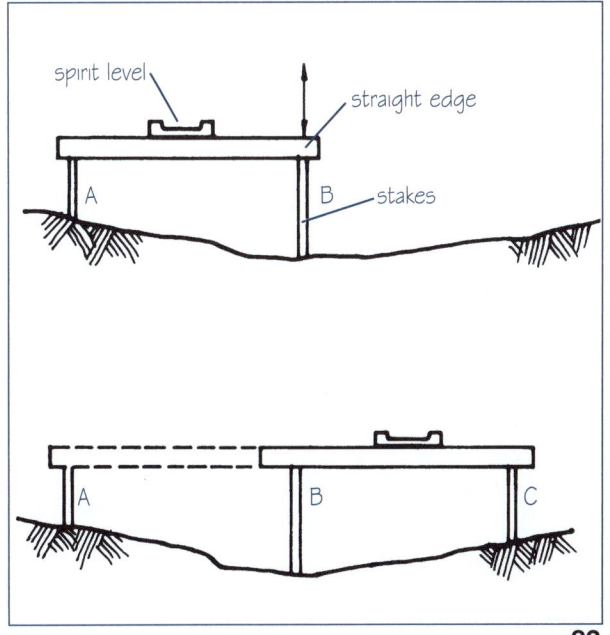

# 26 • FOOTINGS

b. Water level. This consists of two glass or plastic tubes connected by a garden hose or alternatively a long run of clear plastic tubing. The water surfaces are always equal and give two equal levels. This is a good method for transferring levels around corners. It has an accuracy to +5 mm independent of the distance covered.

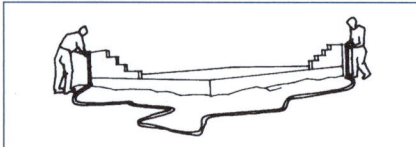

c. A line level (bricklayers line). A string line and a small spirit level hung by two hooks onto the line at its middle. One end is attached to the reference peg, the other pulled tight. This level can be accurate but (i) the level should be at the mid point of the line, (ii) no part of the line is in contact with anything except where secured at the ends, and (iii) the line is pulled tight.

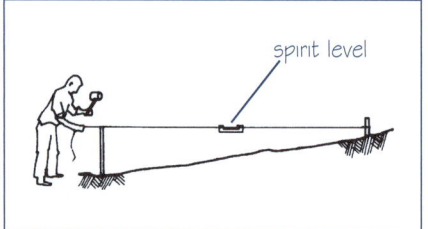

d. Surveyor's tripod and dumpy level. These require a small amount of training although they can be hired. They provide the most accurate method of site setout.

e. Laser level. This is the really easy way to level with maximum accuracy. They can be hired; just follow the simple instructions—it's all done with lights and beeps.

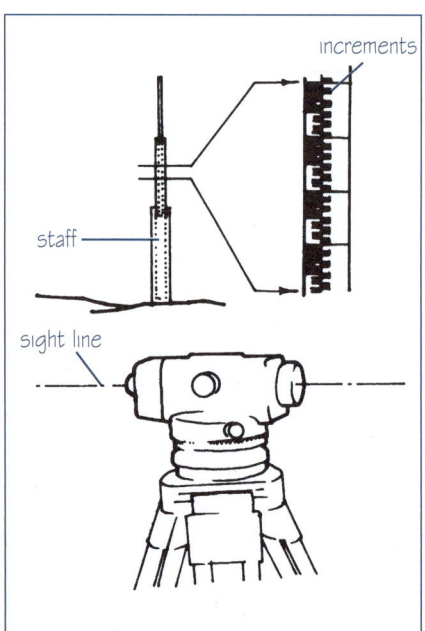

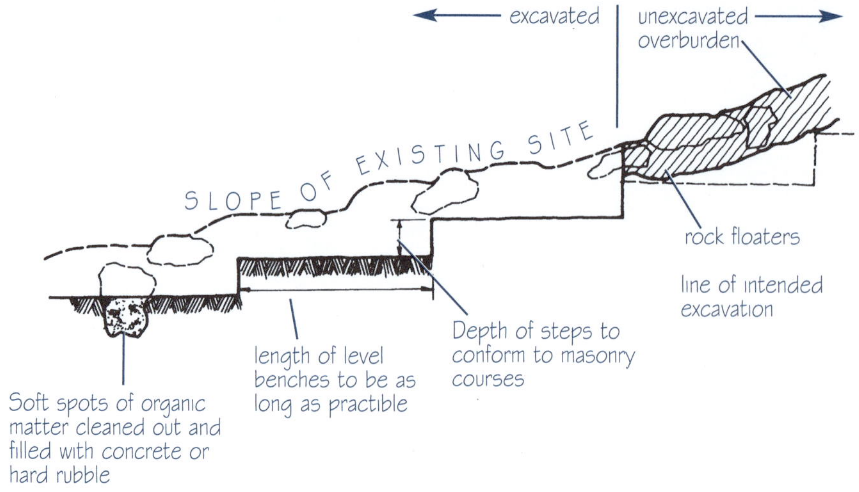

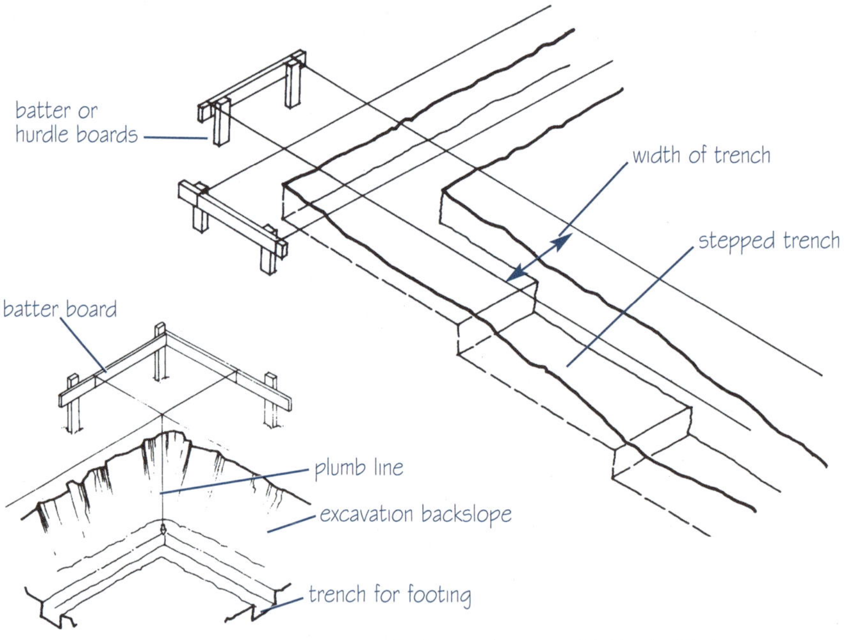

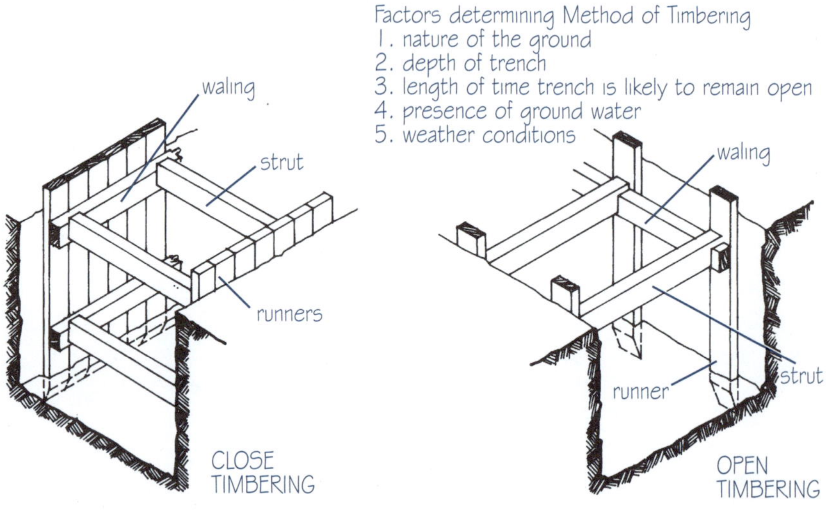

Factors determining Method of Timbering
1. nature of the ground
2. depth of trench
3. length of time trench is likely to remain open
4. presence of ground water
5. weather conditions

CLOSE TIMBERING

OPEN TIMBERING

84

# FOOTINGS • 26

The footings of a house are that part of the structure that makes contact with the ground (foundation), and distributes the total load of the building onto it.

Footings for houses are normally of three basic types:
1. isolated—including pads, piers, piles.
2. continuous—including strip, beam.
3. integrated—including raft slabs, pier and beam.

The size of footing systems for houses has always been arrived at by a combination of a lot of local knowledge and a bit of engineering know-how. The forces that can work on a footing over the life of the building are complex and not always predictable or understandable.

To ensure a reasonable degree of quality and sufficiency, local government building departments have often issued minimum requirement tables for the cross-sectional size of footings. This method is a departure from the normal role of local government building departments, which is to require the builder to have anything of an engineering matter designed by a consultant engineer and only checked by the council.

A judgement of the Western Australian Supreme Court may have changed the traditional practice of footing design, at least in that state, but with a likely flow on to other states if upheld under challenge. In the case* brought before the court an owner alleged that the builder did not construct an adequate footing and that the house cracked. The builder contended that he had constructed the footing system to the guidelines of the local council and therefore had done all that could be reasonably expected of him to provide an adequate footing system. The judge did not think so and found that 'It was not a sufficient compliance with the contract merely to comply with the local government by-laws and the conditions of the building approval, without producing the result the contract required to be produced. Accordingly, the liability was not in this case discharged by complying with the local government requirements'. That is to say, the judge believed that the building contract called for a crack-free building regardless of any other extraneous condition.

Rulings like this can upset traditional relationships and accepted standards of the building industry. If the WA ruling was to be applied in the reactive clay belt suburbs of Sydney and Melbourne then the acceptable standards for building footings would be changed extensively, away from a current attitude that accepts some wall cracking as almost inevitable to a position that would find any footing movement that induced wall cracks completely unacceptable.

Note: many authorities require all footings to be designed by a qualified engineer.

*Mansard Homes and Mr and Mrs H.W. Brockenshire 1982 Supreme Court WA.

## Isolated footings

These are the footings that are separated one from the other to form a series of support points for a building. In house construction, isolated footings are nominally used under isolated brick piers and stumps. They are also used under timber posts or poles, steel columns and reinforced concrete columns. These footings can take the form of pads, bored piers or driven piles.

When isolated footing systems are used, consideration must be given to the fact that the isolated footings rely on the foundation material around them to keep their position one to the other. If the material between isolated footings, particularly pads, is not sufficiently stable or restraining enough then the footings may move around, causing material failure that will manifest itself as cracks in walls at the least, or building collapse at worst.

It is absolutely essential that all isolated footings supporting the same building be taken down to the same foundation strata. If the foundation material varies then generally so does the bearing capacity of the foundation. Variation in bearing capacity can (and often does) lead to differential footing/foundation settlement, resulting in structural failure and often unsightly appearance.

Although most isolated footings are formed in reinforced concrete there are some occasions when masonry units may be used (brick, stone, concrete block), particularly in New South Wales where isolated brick piers supporting timber floors and walls are normally built on brick footings.

It is necessary to tie the piers and stumps to the footings. OBs should check local conditions and requirements thoroughly before deciding how to tie the piers or stumps into the footings.

When posts, columns or poles are used they must be firmly connected to the footings by:
- steel brackets and bolts
- steel reinforcing starter bars cast into the concrete footing, or
- casting the pole, post or column directly into the concrete footing.

All these methods have their place in the design of isolated footing systems, and each should be used to suit particular local conditions.

### STUMP DIMENSIONS
**Assumes stable even bearing**

Reinforced concrete stump

Normal bearing area required approx 0.05 sq m or 225 x 225, 250 diameter

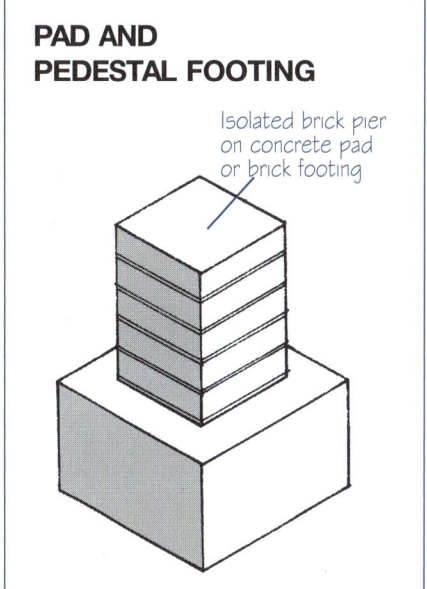

### PAD AND PEDESTAL FOOTING

Isolated brick pier on concrete pad or brick footing

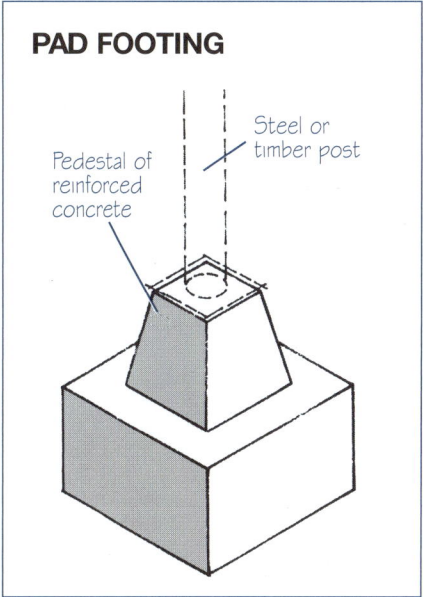

### PAD FOOTING

Steel or timber post

Pedestal of reinforced concrete

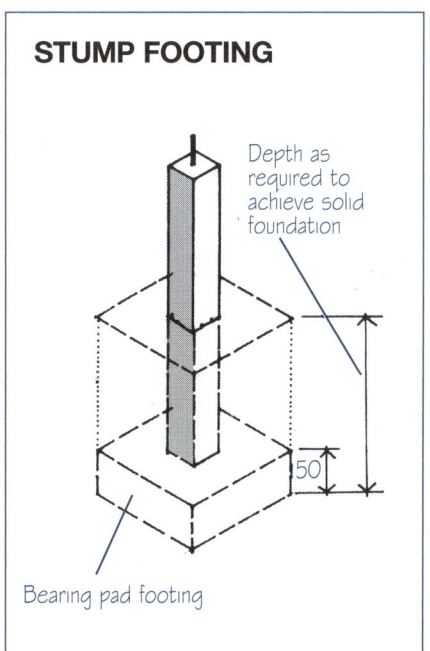

### STUMP FOOTING

Depth as required to achieve solid foundation

50

Bearing pad footing

# 26 • FOOTINGS

## CLAY REACTIVITY CYCLE

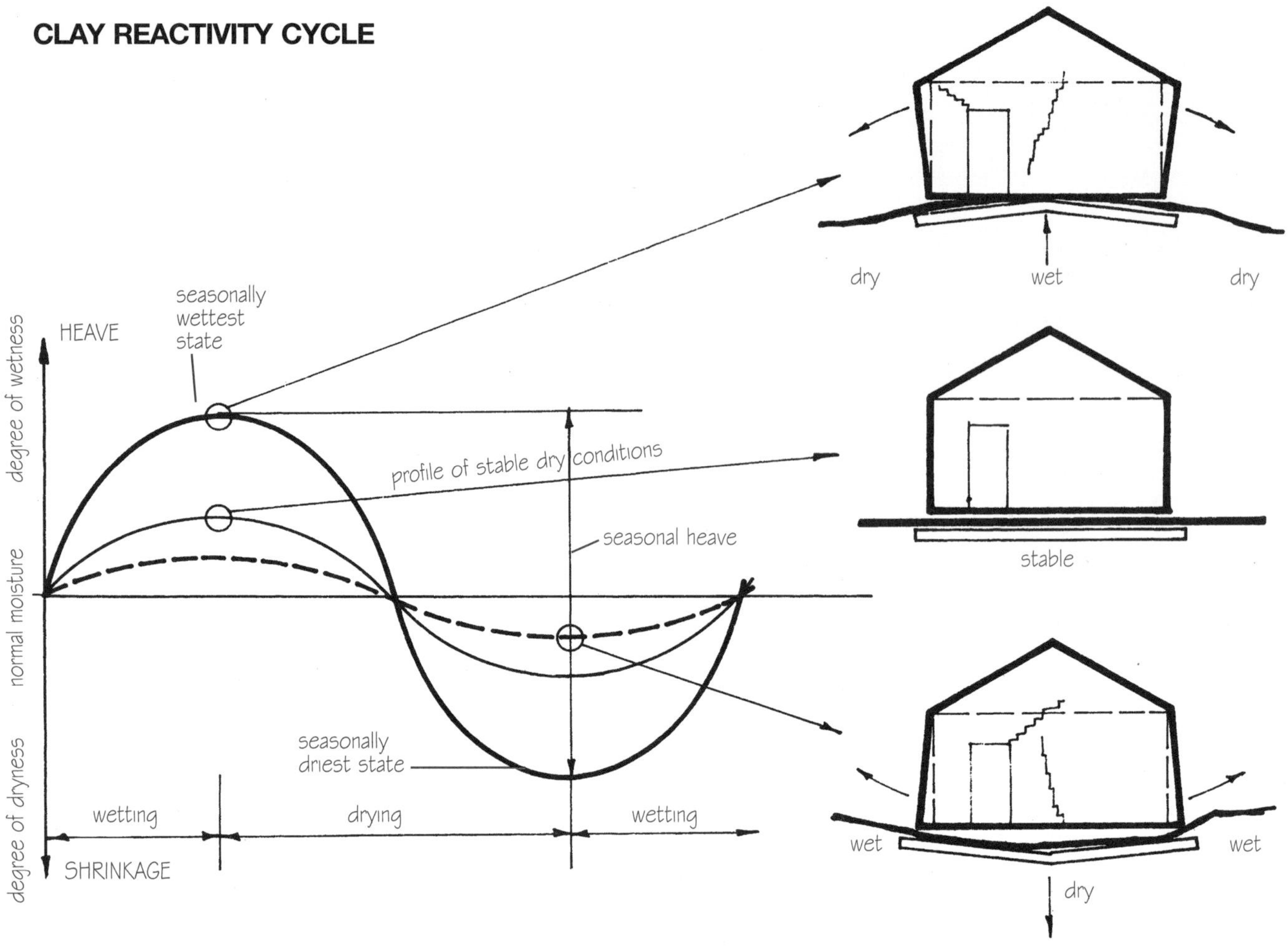

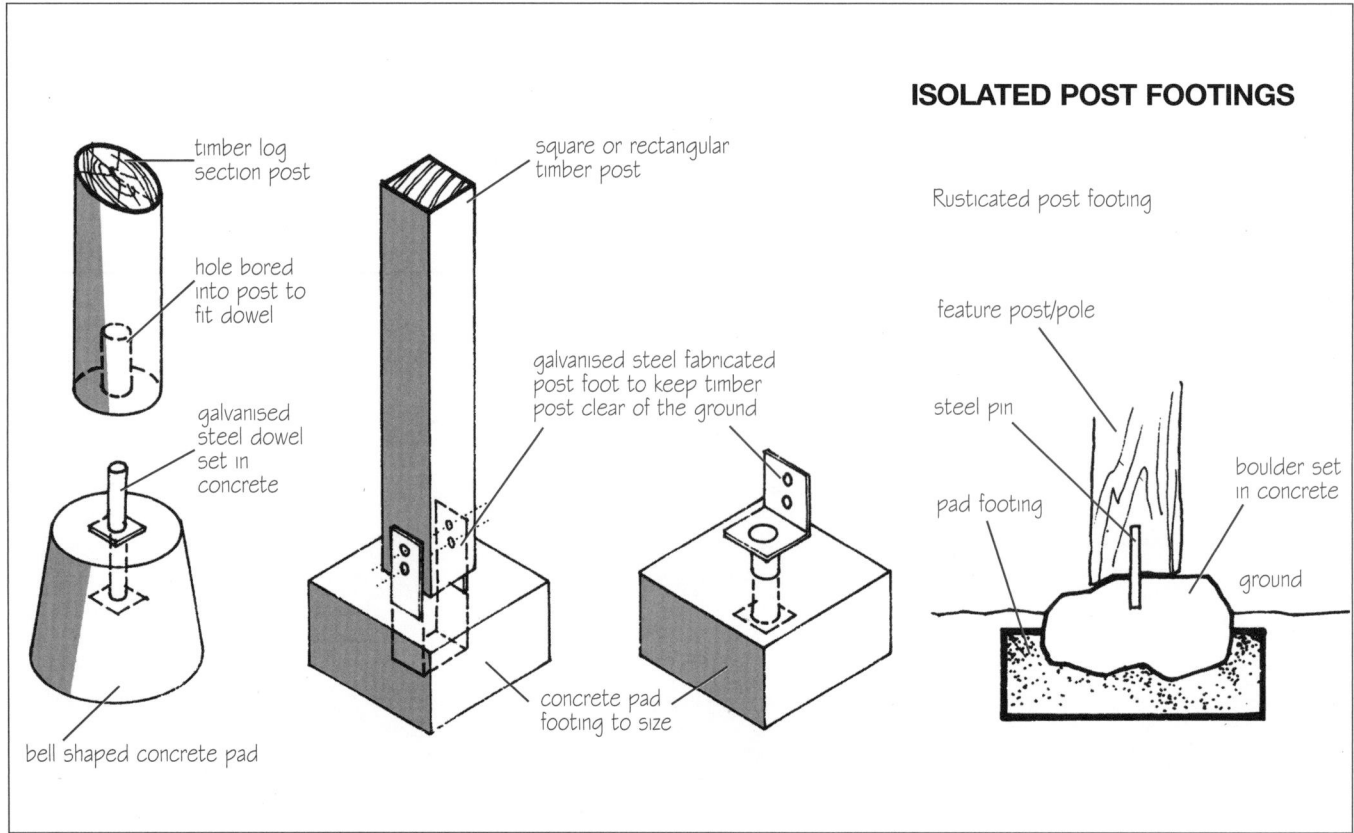

## ISOLATED POST FOOTINGS

# FOOTINGS • 26

## Continuous footings

Continuous footings are generally used in conjunction with masonry walls. They come in two basic forms.

1. Strip footings. These are basically continuous reinforced concrete footings that are generally wider than they are deep. The reinforcement counters any tendency for the footing to move due to local settlement. This is a very basic form of footing that is used extensively throughout Australia.

2. Beam footings. Although the normal strip footing is, to some extent, a beam footing it is not generally purpose-designed to resist bending forces, as a designed beam footing is.

A beam footing system is normally designed to a grid pattern and is capable of continuing to give stable support to a building structure even when the foundation strata may be unstable. A beam (or, as they are sometimes called, deep beam) footing is normally deeper than it is wide. Beam footings are sometimes combined with bored piers to make beam and pier footing systems.

There is a third option, which is to build a strip footing using masonry units, bricks or stone. But this is seldom used in modern construction and in many areas is banned. The masonry unit footing is, however, generally acceptable in stable foundation areas.

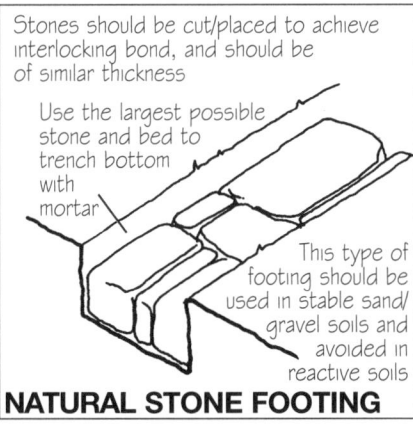
**NATURAL STONE FOOTING**

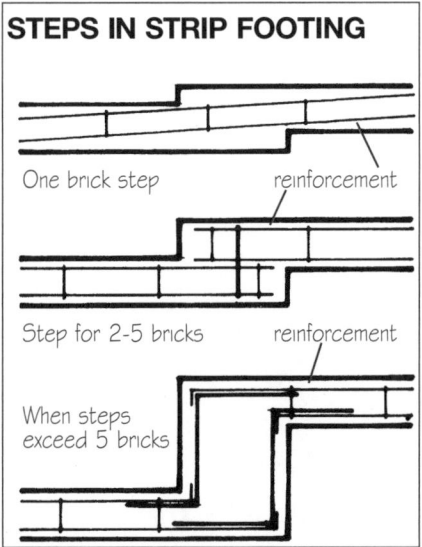
**STEPS IN STRIP FOOTING**

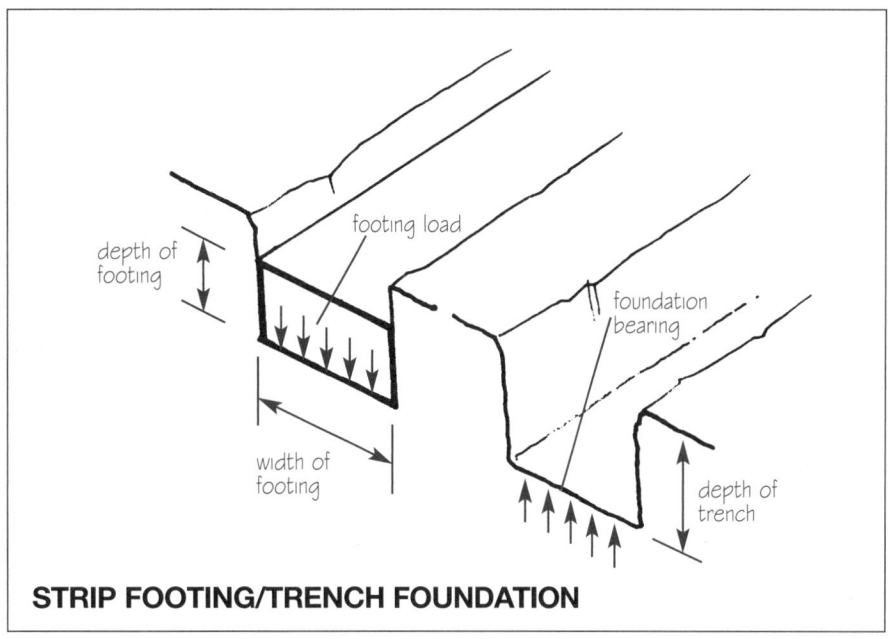
**STRIP FOOTING/TRENCH FOUNDATION**

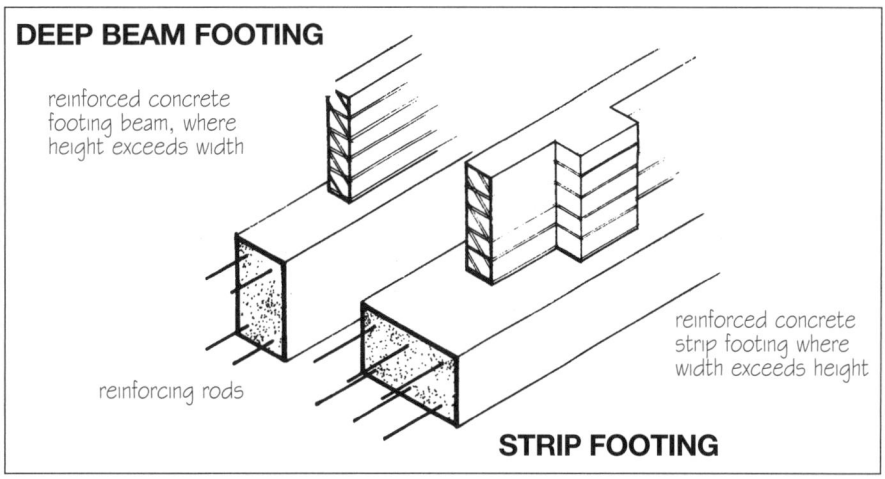
**DEEP BEAM FOOTING** / **STRIP FOOTING**

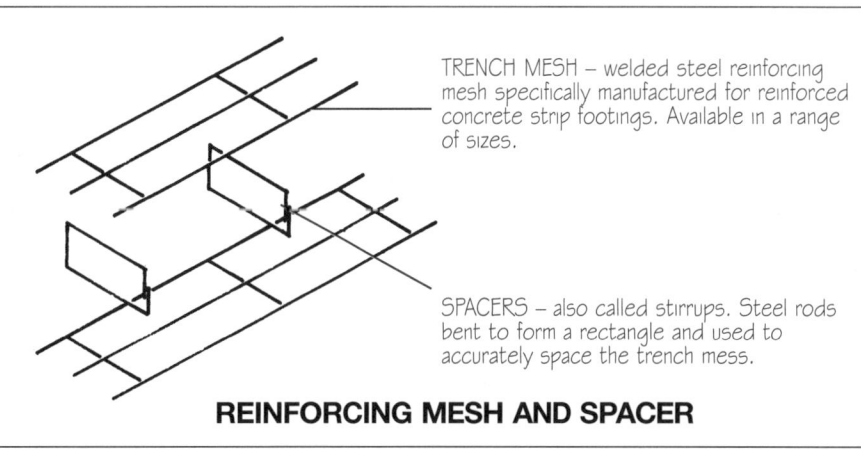
**REINFORCING MESH AND SPACER**

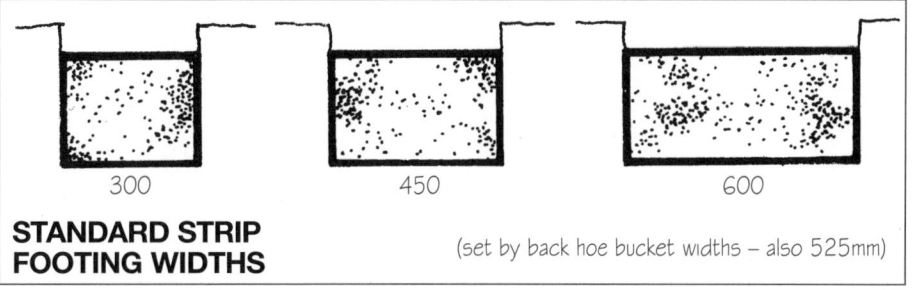
**STANDARD STRIP FOOTING WIDTHS** (set by back hoe bucket widths – also 525mm)

# 26 • FOOTINGS

## Integrated footings

The most common form of integrated footing used in domestic construction is the slab-on-ground system with an integral edge beam. This, combined floor footing system, is often referred to as a raft slab. Some engineers challenge the term raft slab being used with the common rudimentary cottage slab/beam combination. They say that these slab/beam systems are not true raft slabs because they lack the complete rigidity that is associated with a raft slab.

Very few houses are built using the refined integrated slab structure that engineers prefer. The term is generally used to denote all concrete floor slab systems where the footing beams are poured integrally with the floor.

The domestic raft slab is almost the most common footing (and floor) system used in Australia. It is a safe, problem-free system when used in well-drained stable foundation areas, but needs special care when used in unstable foundation areas. Particular problems can be experienced in areas with reactive clay soils and careful design of any cuts and fills must be undertaken along with a sub-surface drainage system to attempt to stabilise the sub-surface moisture content.

The second type of integrated footing used in domestic construction is the pier and beam system, which we touched on in the section on continuous footings. Pier and beam footings are used when there are not adequate foundation-bearing strata near the ground surface. The system requires that pier holes are bored (or in some cases hand dug) along the lines of the proposed footing beam trench until solid bearing is struck. In residential construction the piers would normally be spaced out along the trench at about 2000–3000 mm centres. When a reinforced concrete beam is cast on top of the piers, this provides a system where continuous masonry walls can be built onto a continuous beam—which is in turn supported on a deep stable footing by the use of the bored reinforced concrete piers.

Pier and beam footings are very useful when houses have to be built on filled, low bearing or unstable sites.

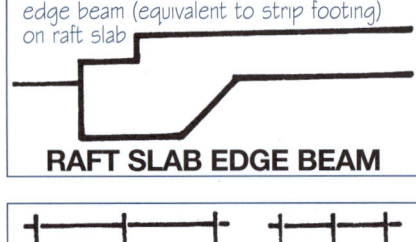
**RAFT SLAB EDGE BEAM**

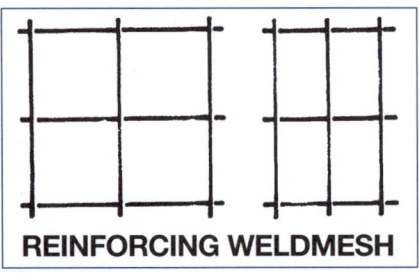
**REINFORCING WELDMESH**

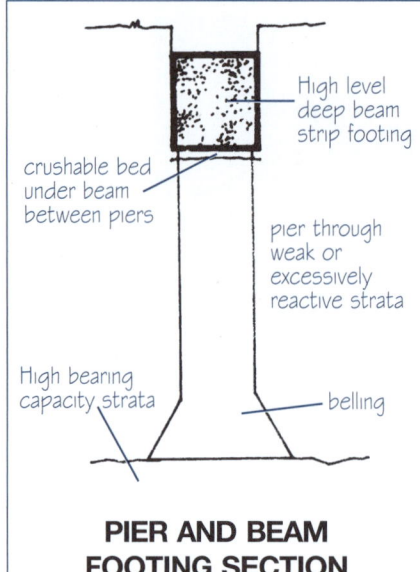
**PIER AND BEAM FOOTING SECTION**

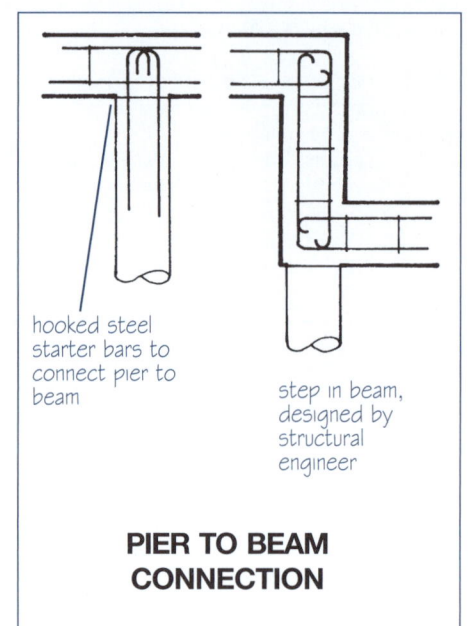
**PIER TO BEAM CONNECTION**

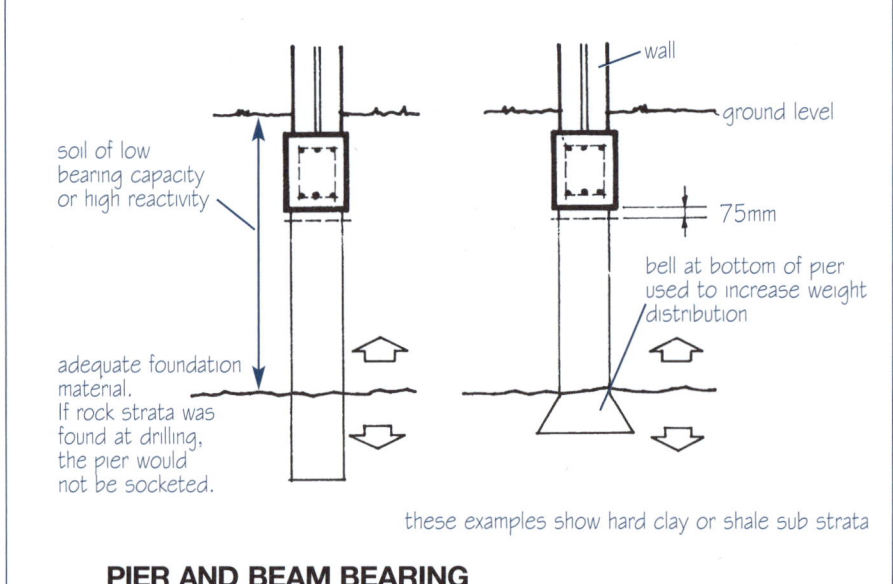
**PIER AND BEAM BEARING**

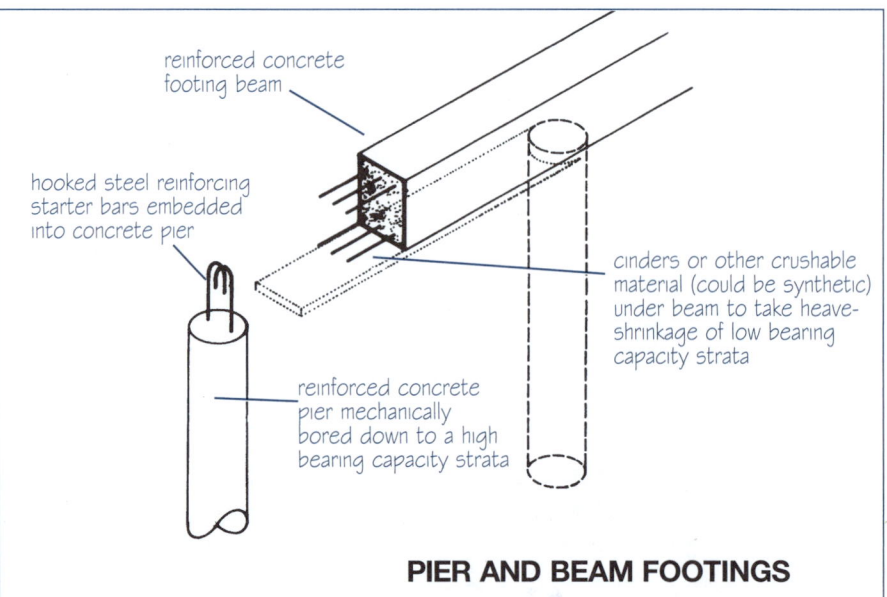
**PIER AND BEAM FOOTINGS**

# FOOTINGS • 26

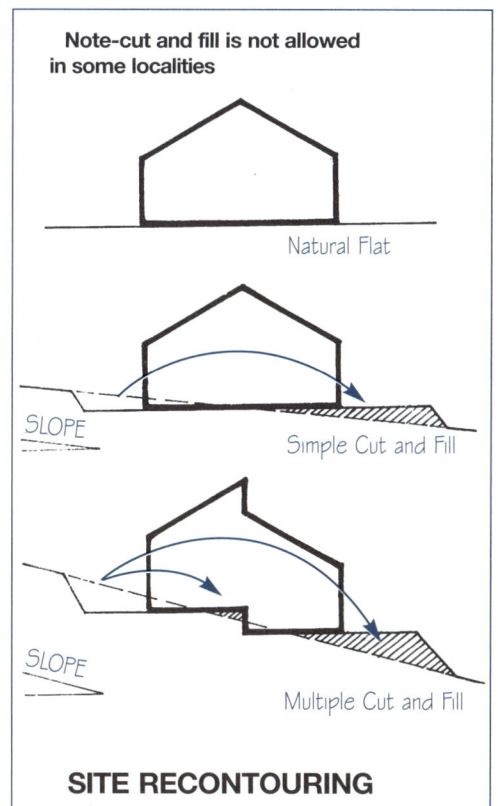

Note-cut and fill is not allowed in some localities

**SITE RECONTOURING**

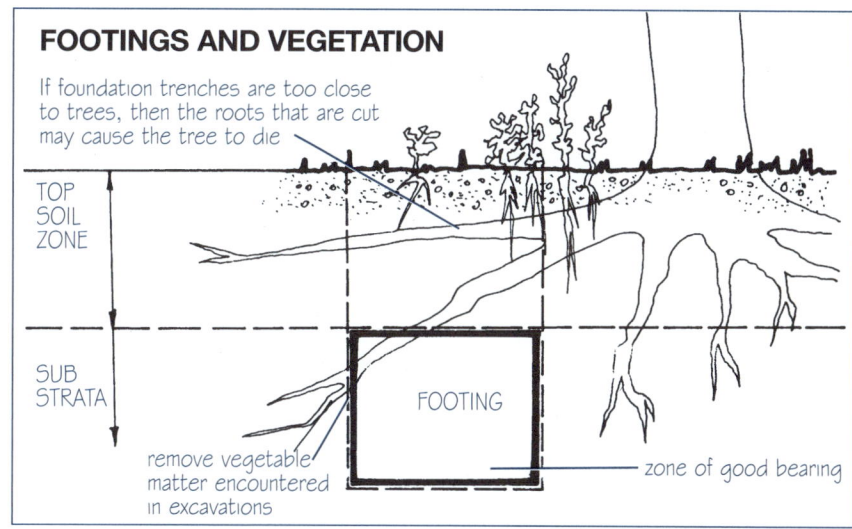

## FOOTINGS AND VEGETATION

If foundation trenches are too close to trees, then the roots that are cut may cause the tree to die

TOP SOIL ZONE

SUB STRATA

remove vegetable matter encountered in excavations

FOOTING — zone of good bearing

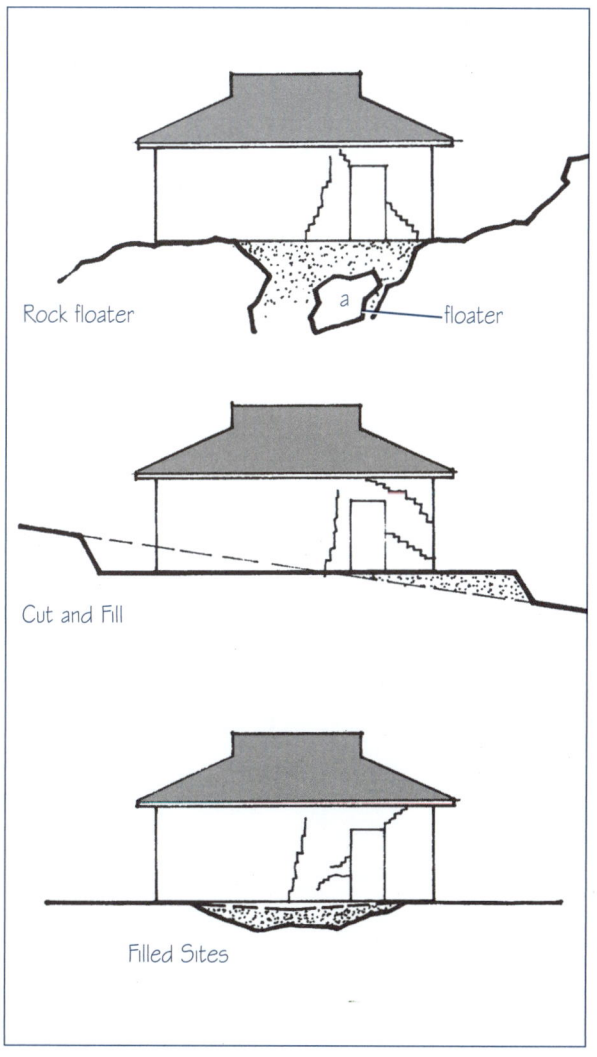

## Structural failures due to buildings straddling various foundation conditions

### Rock floater
A common problem encountered in areas with sandstone foundations is the condition of the floating boulder. In the illustration the house bridges from a solid rock foundation to another zone of solid rock, but the space in between is filled with non-rock material containing a large boulder.

Builders can easily identify the difference between rock and non-rock; it is much harder, however, to identify a floater rock. The common method is to hit the rock with a crowbar to check for a dull thud—any ringing echo could indicate a floater and allow the necessary mitigation to be carried out. Sometimes, however, the floater may be so large that the normal crowbar-striking technique may not work and so a building is built partly supported on bedrock and partly on a floater. This condition is shown in the illustration.

Problems occur if the floater moves and causes the footings that it supports to move independently. This, in turn, reduces the support for the wall structure above, causing at least superficial cracking or even structural collapse.

### Cut and fill
With the greater use of reinforced concrete raft slabs as house footing systems, there is a tendency to cut and fill sloping sites to form a horizontal area capable of taking a raft slab. This method can be successful when the soils are sands/gravels and the fill is laid in consolidated layers. Supervision by a skilled engineer is recommended. Structural cracking will result if part of the house's footings rest on one foundation bearing condition and the rest on another foundation bearing condition.

If cut and fill is used to level sites it is generally recommended that pier and beam footings be used in the fill area to take the building weight, through the filling to the same foundation stratum or condition as the cut section. When in doubt, have a structural engineer give specific advice or design a footing system to suit the particular condition.

### Filled sites
Often residential building blocks will include an area that has been filled. It is a major disaster if a site is purchased and found to be a filled-in brick pit, but this is uncommon. More common is when a rural area is subdivided for residential lots and stock watering holes and dams are filled with soil moved from another part of the subdivision.

This and other filled sites can cause trouble for the house builder, as the filling is not always fully consolidated and over a period of time can compact and therefore sink. This is a problem if a house has been built partly on the filled ground and partly on the natural ground. Structural cracks are likely in all cases. In cases where a major subsidence is experienced the house may lean over, which is expensive to rectify.

# 27 • SUB-SURFACE OR AGRICULTURAL DRAINAGE

Many houses require a system of sub-surface drains to be constructed to reduce the adverse effects caused by excess water below the ground surface. Nearly all houses built on reinforced concrete slab or ground footing/floor structures need sub-surface drainage. Houses on sloping sites normally require an up-hill drain to cut off the flow of sub-surface water that could adversely affect the stability of the house.

Agricultural drains are normally constructed by making a strip excavation across the path of the sub-surface water flow. The excavation is taken down to the lowest invert to which the water flow can be traced. The excavation is graded to give an adequate flow of water, then a bed of crushed insoluble stone or brick waste is placed. On top of this bed a perforated plastic pipe or short length of terracotta pipe is laid, and the trench is then filled to the topsoil line with crushed rock/brick. A cheaper method is to leave out the pipe and substitute a layer of large rocks (about the size of a quarter brick) then cover this with a plastic sheet membrane or corrugated iron. This method works in many circumstances but is not as predictable in the longer term.

Where possible, make the agricultural drains grade to either end and pick up both ends into the main stormwater disposal system. This means that, should part of the system silt up and cease to function, then the whole system is less likely to fail. Connect the ends of the agricultural drains via silt traps to the main stormwater system. These silt traps should have removable tops grated, if possible, so that checks can be made to see if the agricultural drains are functioning and to clear out the silt traps. A silt trap is simply a pit that has an open water flow and a bottom lower than the invert of both the input and take-out pipes. Sometimes baffles are added to increase the silt trapping efficiency, but these are seldom needed in residential applications.

**EXAMPLE LOCATION PLANS OF AGRICULTURAL DRAINS**

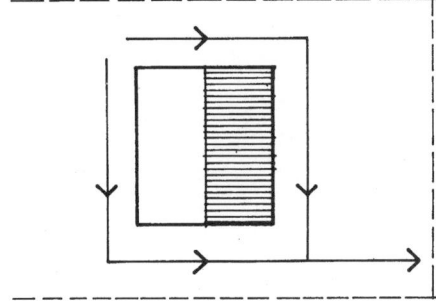

'Moat' pattern, applicable to wet sites or sloping sites

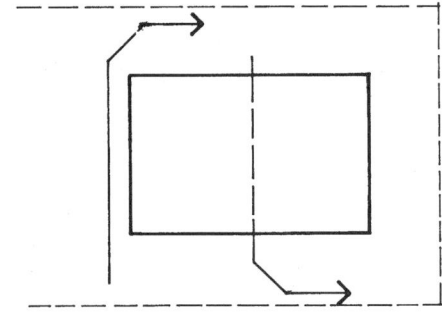

Plan of agricultural drains for retaining walls of split level houses or houses on cut and fill sites

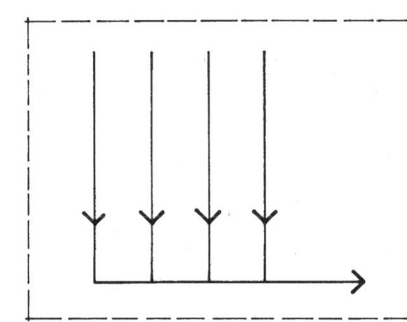

'Grid Iron' pattern providing subsoil drainage to flat areas (see bottom diagram, facing page)

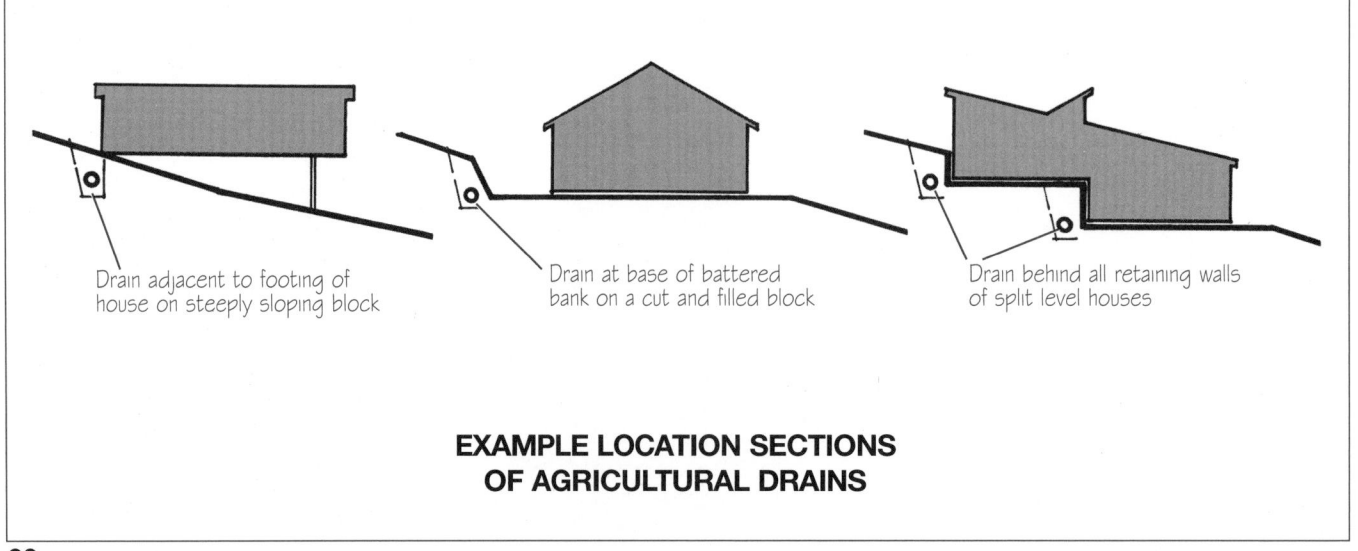

Drain adjacent to footing of house on steeply sloping block

Drain at base of battered bank on a cut and filled block

Drain behind all retaining walls of split level houses

**EXAMPLE LOCATION SECTIONS OF AGRICULTURAL DRAINS**

# SUB-SURFACE OR AGRICULTURAL DRAINAGE • 27

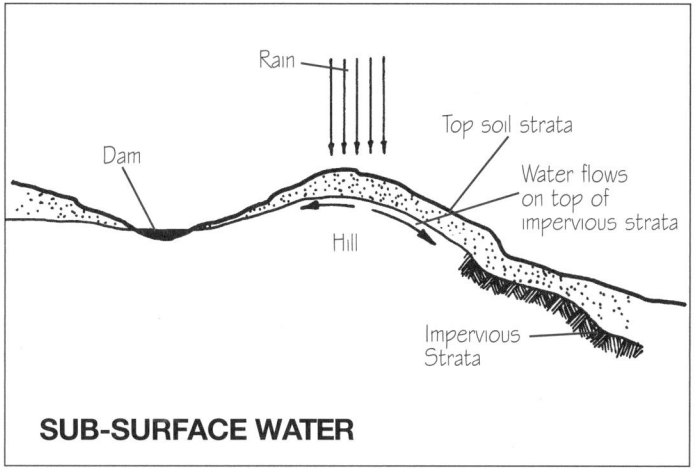

**SUB-SURFACE WATER**

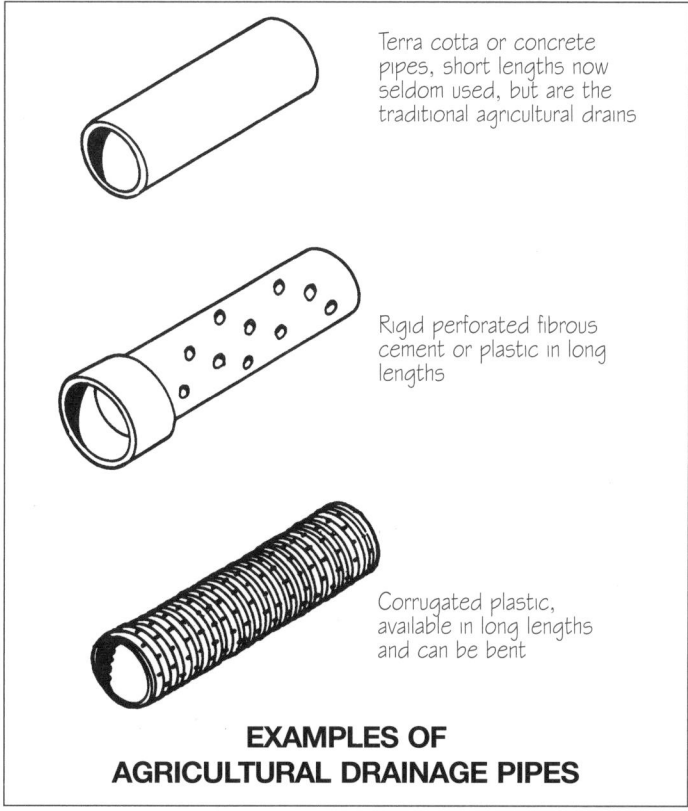

**EXAMPLES OF AGRICULTURAL DRAINAGE PIPES**

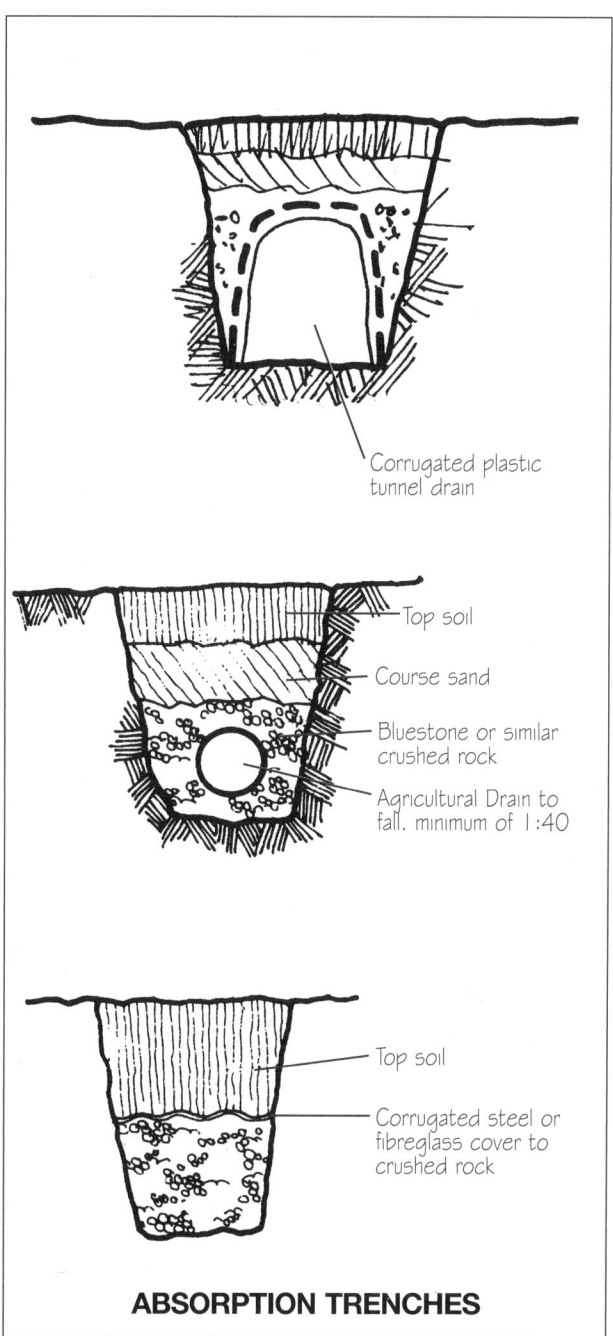

**ABSORPTION TRENCHES**

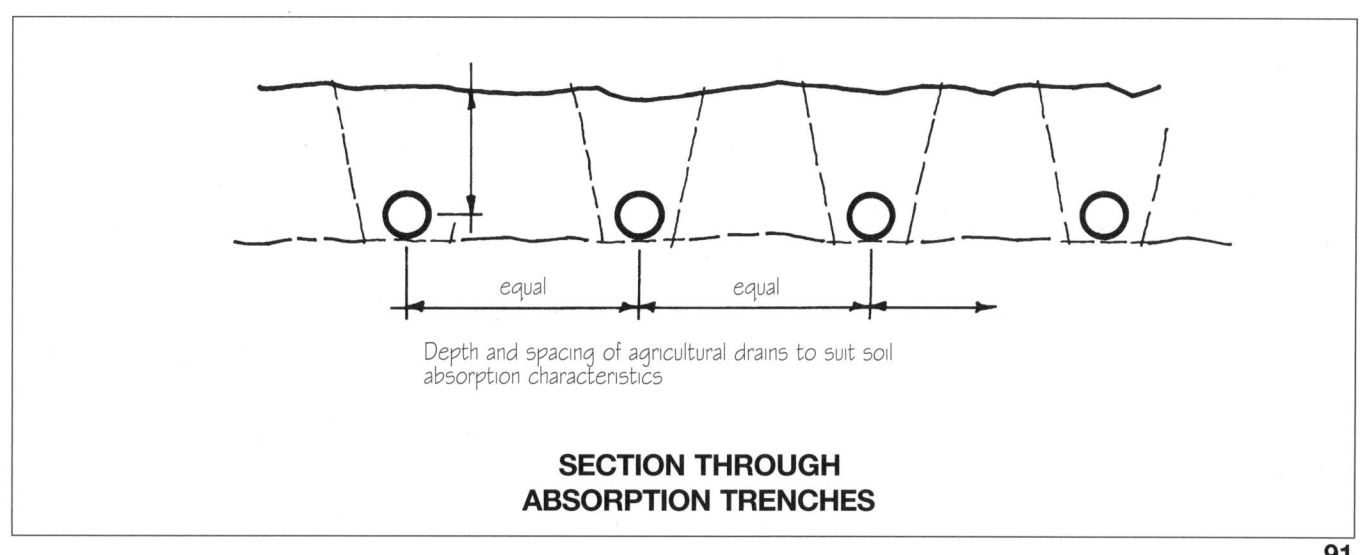

Depth and spacing of agricultural drains to suit soil absorption characteristics

**SECTION THROUGH ABSORPTION TRENCHES**

91

# 28 • RETAINING WALLS

## Types of retaining walls

Three distinct types of retaining wall can be identified:
1. inclined or stabilised banking.
2. gravity retaining wall.
3. cantilever retaining wall.

### 1. Inclined or stabilised banking

This is a type of gravity retaining wall. Usually constructed from masonry materials, eg, brick, stone, concrete block. it can be constructed from interlocking units either pre-cast concrete or hardwood. Here the backfill is deposited inside the interlocking crib structure and acts integrally with the crib structure to form the gravity retaining wall.

Construction materials:
a. pre-cast concrete units
b. hardwood units
c. concrete blocks
d. bricks
e. stone.

### 2. Gravity retaining wall

This is the commonest type of retaining wall used in domestic construction. They depend on their own mass to resist the overturning forces exerted on them.

Gravity retaining walls can be constructed out of almost any masonry units or poured concrete. In most cases, gravity-type walls are generally well bonded and bedded on mortar courses, but it is possible to construct them using dry laying techniques—particularly when natural stone is used.

Bonded and mortar-bedded walls using bricks, concrete blocks or natural stone are generally restricted in height to approximately three times their width measured at the base. If the wall is tapered or stepped in section then, at any point in the wall's height, the 3:1 ratio should be maintained.

Although it is possible to build a mass concrete gravity-type retaining wall, it is generally more economical and structurally better to use reinforced concrete and design the wall as a cantilever system (see the next subheading).

Unmortared rubble stone retaining walls require significant skill to build to a standard of reasonable safety and, as a general rule, the height of these walls should not exceed twice the thickness of the wall.

### 3. Cantilever retaining wall

The cantilever retaining wall makes use of the extra strength that can be obtained by using reinforced concrete—either poured in place or purchased as precast units—or by reinforcing specially manufactured concrete masonry blocks. The base to height ratio may remain in the 3:1 range, but the wall thickness above the base can be significantly reduced.

Cantilever retaining walls need individual engineering design and, as they are normally used in more critical locations than either the inclined or gravity types, skilled tradespeople are generally required for their construction.

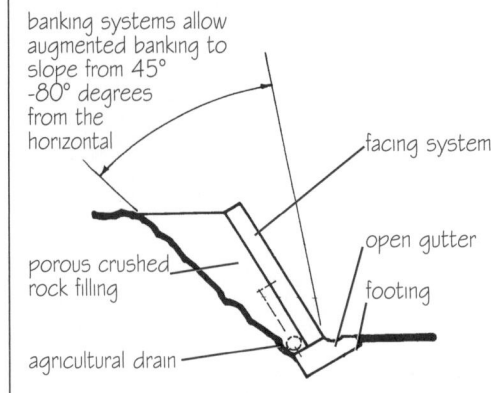

## IMPROVED ANGLE OF SLIP–SLOPING TYPE

Many retaining systems simply retain the natural bank by augmenting its natural angle of slip. These systems include: pre-cast concrete planks, timber sleepers, interlocking timber or concrete patented systems and natural stone.

## GRAVITY TYPE – MASS RETAINING WALLS

These walls are generally built up from masonry units, including bricks, concrete blocks and natural stone.

Type A:
Parallel and vertical faces

Type B:
One vertical face and one sloping face.

Type C:
One vertical face and one stepped face.

## CANTILEVER TYPE – REINFORCED CONCRETE RETAINING WALLS

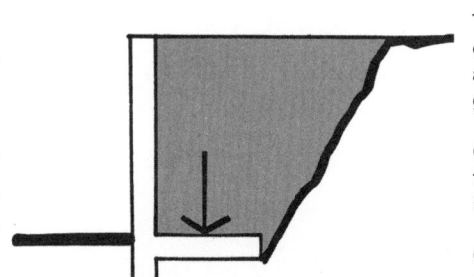

These walls make more economic use of materials than masonry walls but are more expensive to construct and generally less pleasing in appearance.

Crushed rock fill is used to reduce the pressure on the wall and to assist in restraining the overturning effect.

Cantilever reinforced concrete retaining walls require a structural or civil engineer's design.

# RETAINING WALLS • 28

## HOW DOES A RETAINING WALL FAIL?

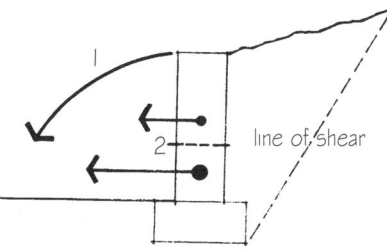

There are two main causes of retaining wall failure.
1. The pressure acting on the wall is so great that the wall is pushed over, that is the wall fails by overturning.
2. The pressure acting on the wall is not evenly distributed or restrained and one part of the wall fails by shearing. This is common in unreinforced walls and there are many examples of walls that have moved horizontally forward but have not yet overturned.

## THE OVERTURNING FORCES ON A

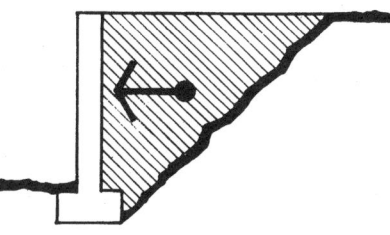

The filling used to bridge between the natural bank and the retaining wall must be selected to minimise the pressure on the wall – usually crushed rock filling gives the best results.

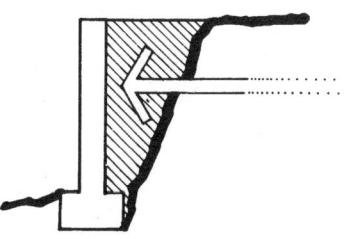

Water that is likely to flow from the natural bank into the filling must not be allowed to build up pressure on the wall and therefore must be drained away from the area.

## Brick retaining walls

**Design & quantities table**
**Based on height to thickness ratio not exceeding 3:1**

| Thickness | Maximum height | Nearest brick course | Number of bricks in parallel wall per metre run | Number of bricks in stepped wall per metre run |
|---|---|---|---|---|
| 230 | 690 | 8 | 64 | 64 |
| 350 | 1050 | 12 | 192 | 120 |
| 470 | 1410 | 16 | 384 | 288 |
| 590 | 1770 | 20 | 640 | 448 |

NB—Walls over 1500 high should be checked by a structural engineer

| | | | | |
|---|---|---|---|---|
| 710 | 2130 | 24 | 960 | 640 |

## Soil types and problems

| Site types | Soil types | Characteristics |
|---|---|---|
| Stable | Rock, sandstone, shale, sand, sandy clay | No movement in adjacent construction, small amount of movement in foundation material, less than 10 mm |
| Intermediate | Sandy clay, dry clay, medium clay | Some movement in adjacent construction surface cracks develop in clay soils during dry season and close in wet season. Movement in foundation material approximately 10–25 mm |
| Unstable | Heavy plastic expansive clays | Deep surface cracking in dry season. Evidence of movement in adjacent buildings. Foundation material has considerable movement greater than 25 mm |

In the design of retaining walls we can classify the soil types into two major groups. Cohesionless soils or cohesive soils.

### Cohesionless soils

These are soils composed of grains large enough to be perceived and which can be readily separated, eg. sand, sandy clays. If a trench is dug into a soil of this type the sides of the trench collapse, ie. the soil will fall under gravity to its own angle of repose. If one were to build a retaining wall in this material, the material to be held back between the back of the wall and the angle of repose of the soil type would exert a force on the wall. It is the weight of this material that determines the thickness of the soil. These soils are generally classified as those found on stable sites.

### Cohesive soils

These are soils that consist of extremely fine particles and require distinct force to break pieces apart. The size of this force is the measure of cohesion of the material. If a trench is dug in a material of this type, the sides of the trench will stand vertical without collapsing under gravity.

From this, you may think that retaining walls in this type of material can be thinner than those supporting cohesionless soils. However, when these soils become wet they swell and exert a pressure on the retaining wall from the amount of water retained in the soil. This pressure is many times greater than the weight of the backfill.

# 28 • RETAINING WALLS

**4. Reinforced concrete block retaining wall.**
Concrete blocks for reinforced concrete block retaining wall construction are a special large cavity block specifically manufactured for this use. Concrete is poured into the block cavities and is vibrated to remove air pockets and ensure contact between steel reinforcing and concrete.
Reinforcing steel vertical bars are spliced to starter bars.
Reinforcing steel horizontal bars are wired to vertical bars and laid in the groove of special concrete blocks.
Reinforcing steel starter bar is cast into footings. This is part of the vertical and sheer reinforcing of the wall.

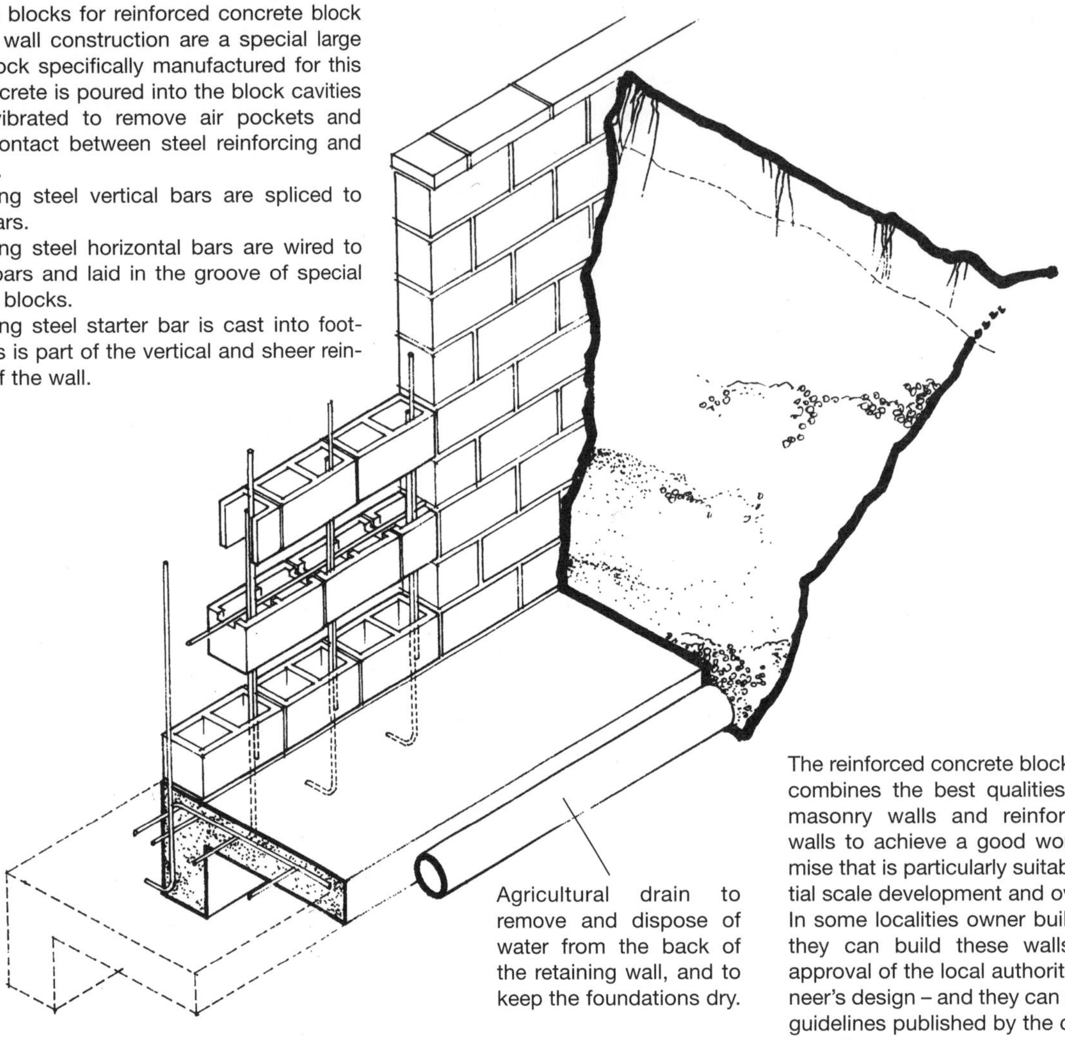

Agricultural drain to remove and dispose of water from the back of the retaining wall, and to keep the foundations dry.

The reinforced concrete block retaining wall combines the best qualities of reinforced masonry walls and reinforced concrete walls to achieve a good working compromise that is particularly suitable for residential scale development and owner builders. In some localities owner builders may find they can build these walls without the approval of the local authorities of an engineer's design – and they can simply use the guidelines published by the concrete block manufacturers.

## DIAGRAMS OF COMMON SYSTEMS USED TO REDUCE WATER PRESSURE BEHIND RETAINING WALLS

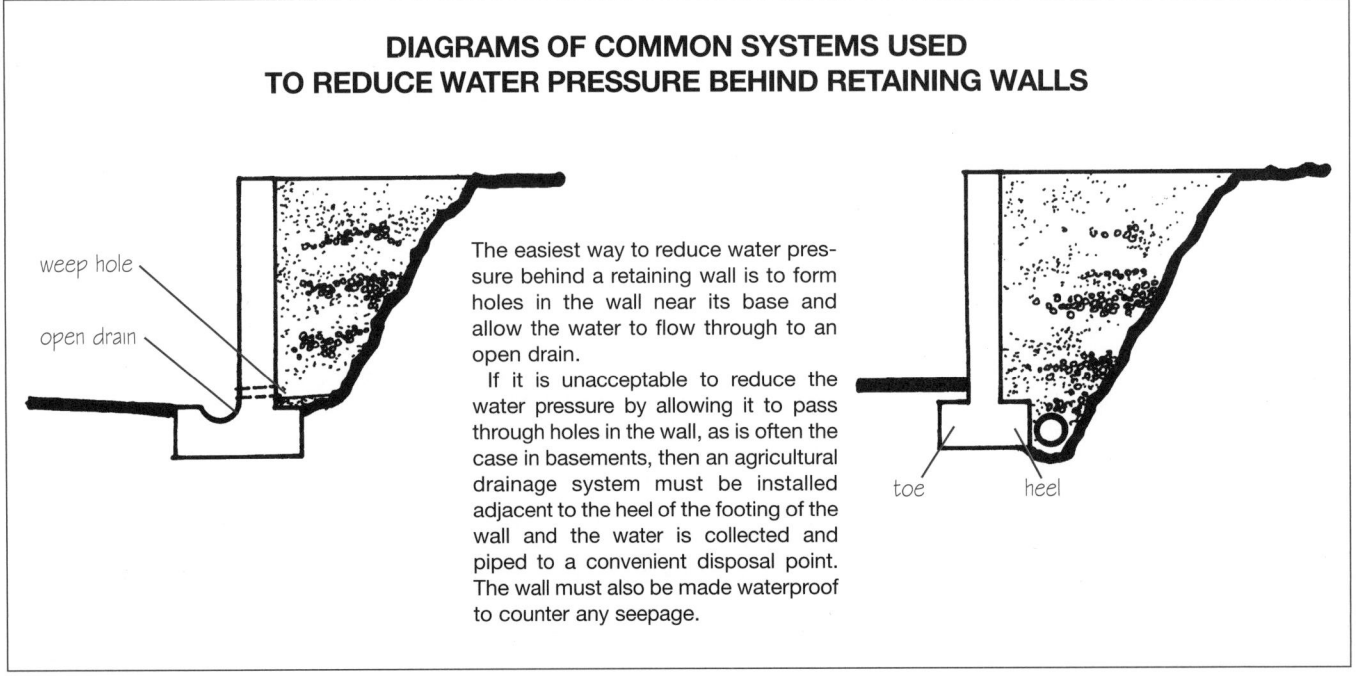

The easiest way to reduce water pressure behind a retaining wall is to form holes in the wall near its base and allow the water to flow through to an open drain.
  If it is unacceptable to reduce the water pressure by allowing it to pass through holes in the wall, as is often the case in basements, then an agricultural drainage system must be installed adjacent to the heel of the footing of the wall and the water is collected and piped to a convenient disposal point. The wall must also be made waterproof to counter any seepage.

# RETAINING WALLS • 28

Adequate drainage is essential, with outlets to conduct water away from the wall. Drains must be constructed so that they will remain clear and unblocked under all conditions and, so water will not accumulate behind the wall, outlets must be provided. It is desirable to include a facility for inspection and cleaning. The complexity and general efficiency of the drainage system depends on the quality of water to be disposed of and the moisture conditions in the soil type. Water that enters the backfill area must be removed from the base of the retaining wall by one of two methods.

1. Weep holes—holes that pass through the wall. These allow water to seep out as it enters the backfill area. The holes should be approximately 50–75 mm in diameter and equally spaced along the wall (approximately 1.5 metres). The weep hole may be formed by leaving a vertical joint (perpend in masonry) completely open and free of mortar or by building a pipe into the space left by a cut in the masonry work. The water is taken through the hole and collected in a drain on the front side of the wall.

2. Agricultural pipes—this method is more complex than weep holes but is far more effective. It comprises an agricultural pipe located at the base of the retaining wall footing, covered with the granular backfill. The outlets of the agricultural pipe must be taken beyond the ends of the wall (see Subsoil Drainage). This type of drainage system allows the front face of the wall to be kept dry, eg. in a split level house or basement floor.

## Backfill

It is essential that there be a suitable filler material between the back of the wall and the natural ground to be retained. The backfill should be a granular material placed in thin layers, eg. gravels, sandy soils or slag. This is of no consequence in cohesionless soils, but in cohesive or clay base soils the function of the backfill is to reduce the ground water pressure exerted by the soil alone. The backfill also acts as a water collector for sub-surface drainage, so water can be diverted away from the base of the wall.

The choice of materials for backfill will greatly depend on those available, the site conditions and loads to be placed on the backfill. Clays and organic soils should not be used as a backfill material, since they are subject to seasonal variation, swelling and in the case of organic soils, deterioration. These factors will cause increases in the ground water pressures on the wall.

## Drainage

Water that enters the backfill area must be removed from the base of the retaining walls by one of two methods. The first of these is by weep holes of 50–75 mm in diameter equally spaced along the wall (approximately 1.5 metres). Here the water is taken through the wall and collected in a drain on the front side. The second method is used where the front of the wall must be kept dry, eg. in a split-level house or basement floor), water must be taken away from behind the wall by agricultural pipes.

## Tanking

Where retaining walls must resist moisture penetration it is often necessary to place a waterproof membrane in the wall.

These membranes can be as simple as common and readily available sheet plastic vapour-proof membranes or can be multi-layered bitumen/felt membranes that require highly specialised tradesperson for installation.

The multi-layer bitumen/felt membranes are built up on site by making a sandwich of special high-durability felt sheets bonded with hot bitumen.

All membranes, whether plastic sheet or multi-layer bitumen/felt, should be placed so that the pressure of the retained material holds them against the back of the retaining wall—never attempt to use the membrane on the visible face of the retaining wall.

To avoid punctures in the waterproofing membrane a protection wall or covering should be constructed between the retained material and the membrane itself.

### HOW TO KEEP WATER OUT OF A HOUSE ON AN EXCAVATED SITE

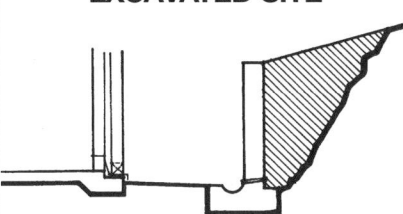

Separate the retaining wall from the house wherever possible.

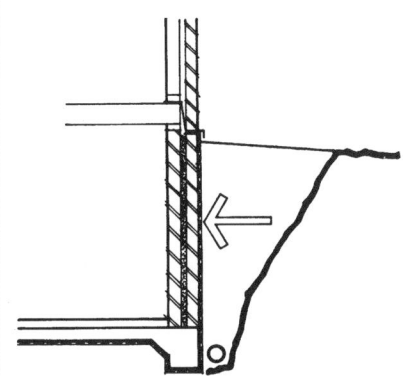

Use an external marking membrane plastic sheeting or bituminous felt. Great care must be taken to make sure there are no holes in the membrane for even a pin hole will allow the water through.

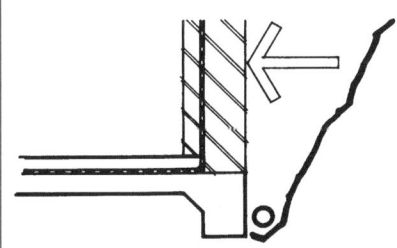

Using a tanking membrane sandwiched between two wall skins (and two floor slabs). The membrane may be plastic or bituminous sheet. Care must be taken to avoid water pressure building up through the outer skin, then pushing against the membrane, causing it to put excessive loads onto the inner skin. It may cause the failure of this skin.

Where the external walls of a house are required to retain excavations the wall must be engineered as a retaining wall.

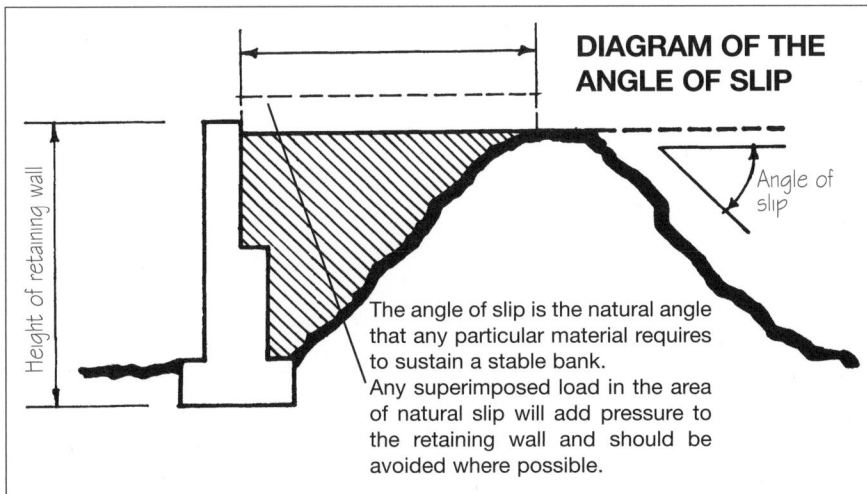

**DIAGRAM OF THE ANGLE OF SLIP**

The angle of slip is the natural angle that any particular material requires to sustain a stable bank.
Any superimposed load in the area of natural slip will add pressure to the retaining wall and should be avoided where possible.

# 29 • GROUND-TO-FLOOR SYSTEMS

Ground-to-floor systems are divided into two basic types that can either be used separately or in combination.
1. **Isolated support system**—includes all isolated piers, stumps, posts, columns and poles.
2. **Continuous support system**—includes all continuous wall-like base structures.

## Isolated piers

Isolated brick piers are the common floor support system used in New South Wales and to a lesser extent in other states. Brick piers are generally 230 mm square, often built on a brick footing 350 mm square and two bricks thick (or a concrete pad of same size). Brick piers are generally not connected to the floor frame above by any tie members, and the building they support simply stays there because of gravity. In some areas where there are high winds tied down rods may have to be built into the piers.

Brick piers are seldom braced but the maximum recommended height for a 230 mm square pier is 1500 mm and for any piers that require to be higher than this that part of the brick pier further than 1500 mm below the floor structure should be increased in section to 350 mm square. At the same time, the footing should be increased to three bricks thick and 470 mm square.

## Stumps

Supporting stumps were originally round tree logs standing on end. Later they became square durable timber sections resting on timber sole plate footings. Today, most stumps are precast in reinforced concrete. The reinforced concrete stumps have a steel rod projecting from one end that is used to connect the timber floor frame to the stump.

The footing for the stump is normally a poured concrete pad in a hole bored or dug down to stable bearing. The stumps are cast with holes through their faces so that bracing can be attached. It is normal to brace stumps that extend out of the ground more than 900 mm. The standard section of a stump is 100 x 100 mm, but 125 x 125 mm and 150 x 150 mm are also made to allow for taller stump heights than that which can be achieved with the standard section.

## Columns:

Reinforced concrete columns can be used for the base structure of a house either in its raw form (like stumps) or as an infill within a masonry pier. Reinforced concrete columns can be poured in place as distinct from being precast, as stumps normally are. This is called in situ concrete and needs the use of formwork for its placement. Column formwork can either be fabricated from timber or metal boards and battens or factory-manufactured cardboard tubes can be used. Reinforced concrete columns allow for much longer and stronger supports than can be provided by stumps or brick piers.

Reinforced concrete columns are very useful on steeply sloping sites and, when combined with a reinforced concrete floor, create an almost indestructible frame.

The problem of the use of in situ reinforced concrete piers is that the formwork adds a significant item of expense without adding anything to the final building. Alas, reinforced concrete is considered by many to be a less-than-attractive finished building material.

The use of hollow concrete blocks to build piers that can be filled with reinforced concrete allows the two problems of unproductive formwork and poor appearance to be solved at the same time. Reinforced concrete block piers can be tied down to the footing with reinforcing steel and the timber or concrete floor they support can also be tied to the piers. This ability to tie the structure together means that reinforced concrete block piers or columns can be used for elevated houses in high wind (even cyclonic) areas.

## Posts

Posts are nominally combined with beams when they are used in a base support system. The posts (and the beams) can be either of timber, or steel, or a combination of the two. Posts are normally connected to the footings by the use of steel post feet, either purpose made or off the shelf. The post feet may be either cast into the footing or sometimes they are bolted to it by holding down bolts or expanding nut fixtures.

Posts are generally used in houses that are raised off the ground or are on sloping sites. This means the posts are often over 2500 mm long, which, because they are fairly slim and flexible units, means that extensive and well-designed bracing is essential.

Post construction requires careful design and election of materials, as the materials often have a much closer tolerance to safety than brick or concrete support systems. Timbers must be selected that will combine long life with high strength and durability, while steel posts must be carefully treated to avoid rust weakening them in the future.

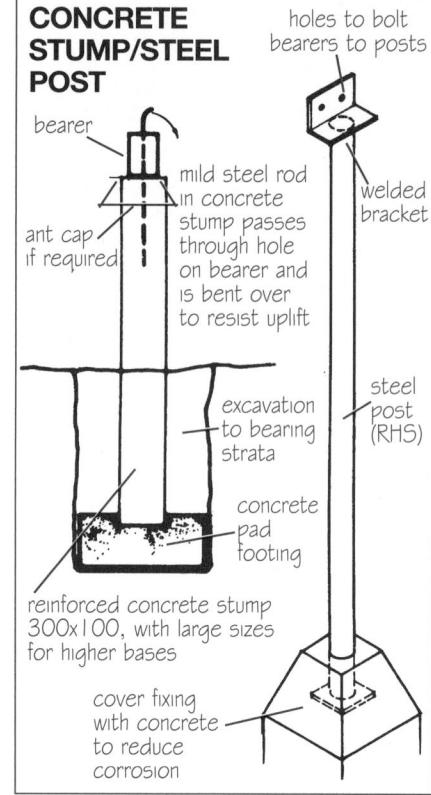

**CONCRETE STUMP/STEEL POST**

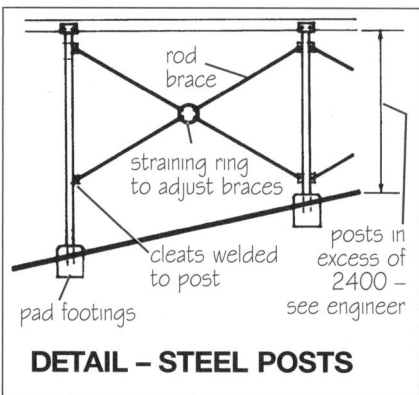

**DETAIL – STEEL POSTS**

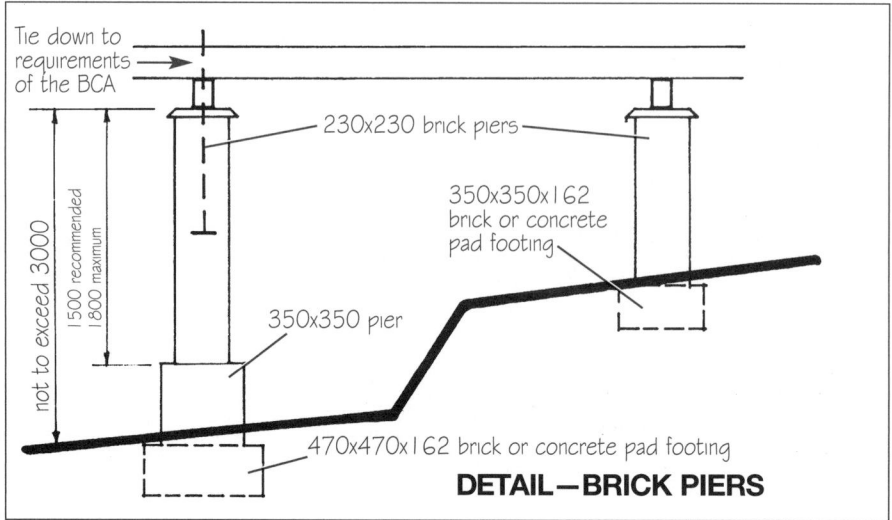

**DETAIL — BRICK PIERS**

# GROUND-TO-FLOOR SYSTEMS • 29

## Poles

Recently a new type of post construction has become popular in Australia, termed pole construction. It is a highly integrated post and beam system.

## Continuous base walls

Continuous base walls are used under most masonry walls in house construction and often around the perimeter base walls of weatherboard- or fibro-clad houses.

Continuous base walls are most commonly built in brickwork, but a significant number are constructed from concrete blocks and natural stone. Where a base brick wall supports an upper brick wall the base wall should have the same thickness as the wall it supports—that is 110 mm for single skin, 230 mm for double skin and 270–280 mm for cavity brick work.

For brick veneer construction, and for weather board/fibro construction, only the perimeter base walls are constructed in continuous brickwork—110 mm thick with engaged piers at approximately 1800 mm centres. The internal base work is generally isolated brick piers or stumps.

For cavity brick construction, the perimeter base walls are 270–280 mm plus an engaged pier. Internal walls are supported on continuous base brick walls 110 mm thick plus engaged piers to either side, while the timber floors within rooms are supported on isolated brick piers or stumps.

When brick base walls are used with brick veneer or timber-clad houses the single 110 mm skin wall should not exceed 1500 mm in height. If the wall has to be higher than this then the wall from the ground level to within 1500 mm of the underside of the floor structure should be a double brick wall of 230 mm.

Engaged brick piers are normally 230 mm along the wall and 110 mm out from the wall. They should be bonded to the wall every fourth course with brick or wire bonding.

The base brick wall extends from ground level to the underside of the timber floor frame. In most states the minimum distance from the ground to the underside of the bearers is 200 mm. The damp-proof course and the termite strips are installed

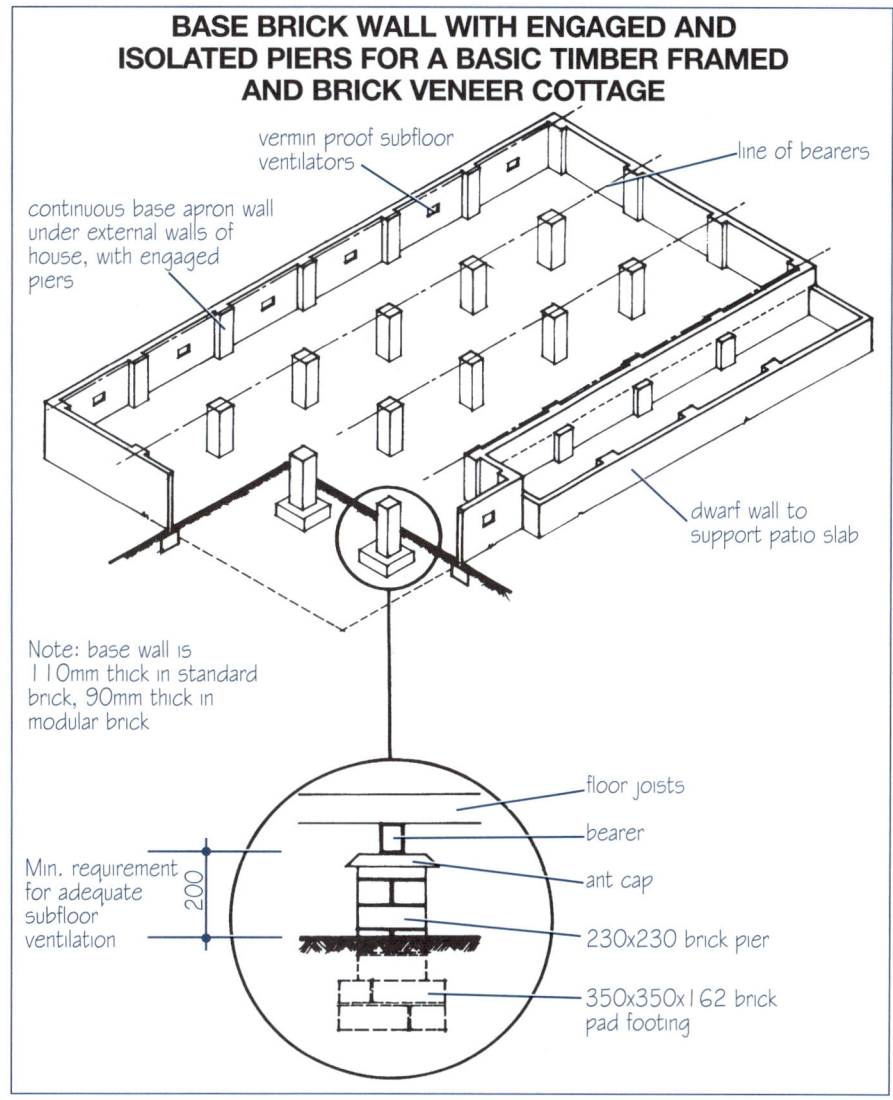

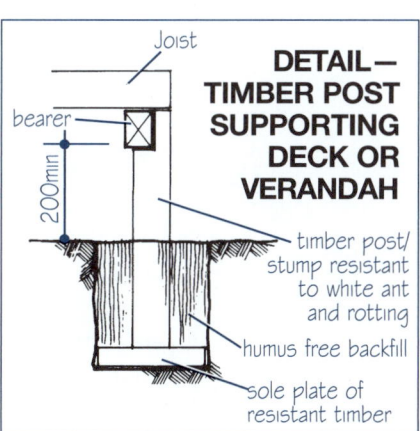

at this level and any ventilators should be installed so that the top of the ventilator is in line with the top of the base wall.

Base walls built in concrete blocks are built in much the same way as brick base walls. Some localities require solid blocks to be used in all base work.

Natural stone base walls are generally 300 mm thick but this varies from locality to locality and with the type of stone available for construction. With some stone it may be easier to use brick engaged piers, and in other instances a complete base brick wall behind the stone work may be required.

## DETAIL—TERMITE AND DAMP-PROOFING OF BASE BRICK WORK

Ant capping to piers and ant stripping to base walling is an essential measure required in all areas where there is white ant (termite) activity. The capping must be thoroughly soldered or bonded and riveted to the stripping, any small hole could let the insects through to attack the structural timbers of the house.

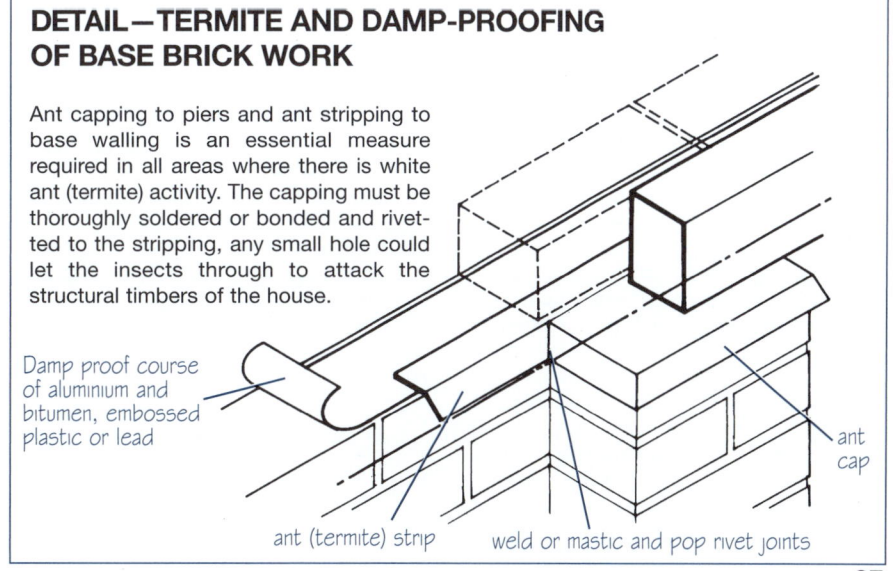

# 30 • TIMBER FLOOR FRAMES

The timber floor frame has developed over many years, and there is currently a major transition happening. The movement from what has been the traditional method of floor construction since the 1950s, towards a modern material efficient system, has been developed and promoted by the timber development associations and is called the platform floor.

## Traditional

This system is part of the traditional method of timber framing that tied the floors, walls and roof of a house together and made them totally interdependent. The floor frame was arranged in such a way that all walls were supported by either a bearer line or a double joist line, which were, in turn, supported by brick piers/stumps that followed the line of the walls and provided a pier/stump at every corner.

The procedure, therefore, to build a traditionally framed timber floor for a timber wall framed house (whether brick veneer or sheet clad) is to first set out the subfloor piers/stumps so that there is support under all walls and subdivide the area under the floors so that there is a pier or stump on a grid not exceeding 1800 x 1800 mm. This grid can be varied by using none of the standard timber sizes, but this would be a break from the traditional.

When the base structure is completed, the ant (termite) caps are put in place, along with the damp-proof course if required. On top of the base go the bearers, traditionally 100 x 75 mm, generally running parallel to the longest side of the building, but this is not essential. The bearers should be set to be level in all directions and fixed to the piers/stumps if applicable.

On top of the bearers are the floor joists. These are traditionally 100 x 50 mm set at 450 mm centres, under all walls, running in the direction of the joists. There are double joists to allow for the stable seating of the wall bottom plate and to provide a ledge on to which to seat the flooring boards.

## Platform

The main difference between the traditional framed floor and the platform type is that the internal walls are not supported on piers/ stumps directly, as it is important to use platform floors in conjunction with timber roof trusses, which put no load on internal walls.

Piers/stumps are still set out on a 1800 x 1800 mm grid but, as no reference has to be taken of internal wall locations, there is generally a saving in the number of piers/stumps that have to be provided.

Bearers are laid parallel to the longest side of the house, and joists are laid at 450 to 600 mm centres, depending on the floor sheeting thickness, with only double joisting at the ends.

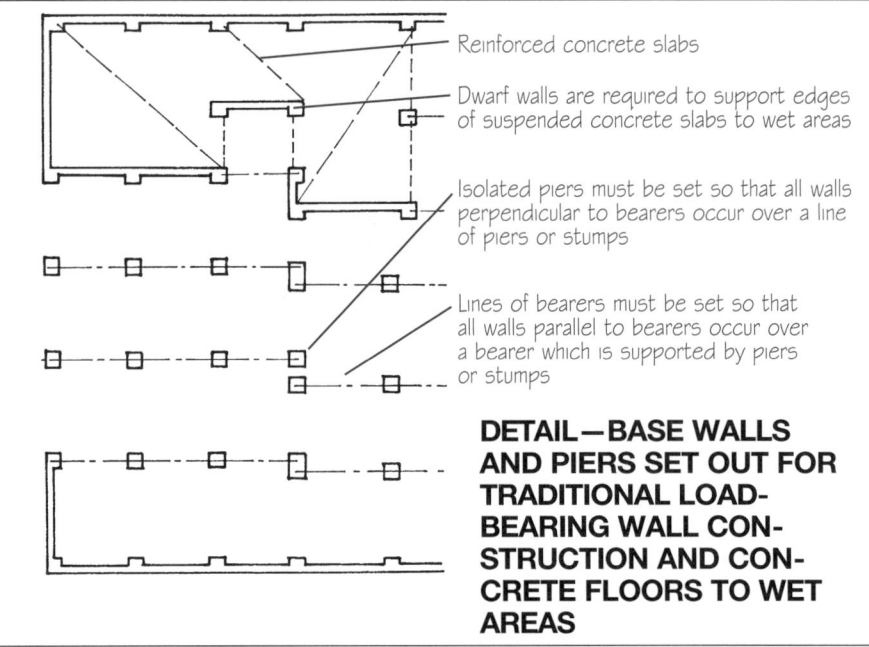

**DETAIL—BASE WALLS AND PIERS SET OUT FOR TRADITIONAL LOAD-BEARING WALL CONSTRUCTION AND CONCRETE FLOORS TO WET AREAS**

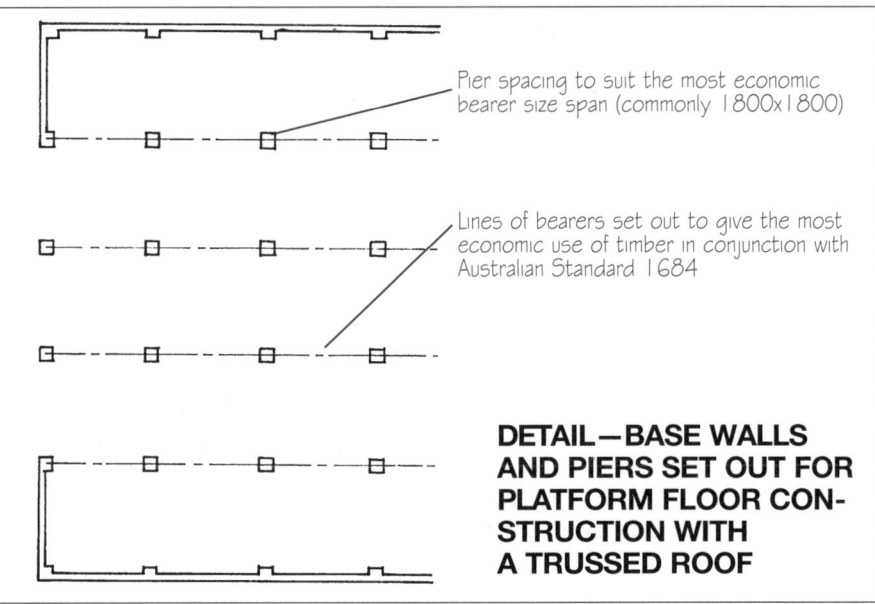

**DETAIL—BASE WALLS AND PIERS SET OUT FOR PLATFORM FLOOR CONSTRUCTION WITH A TRUSSED ROOF**

**FLOORING BOARD SECTIONS**

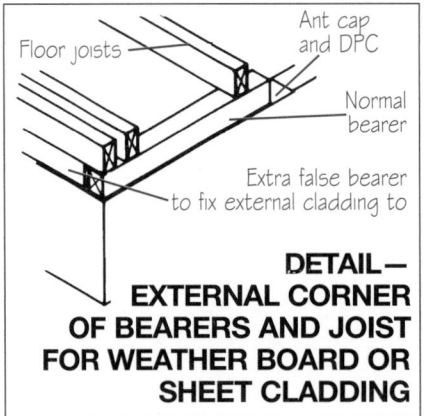

**DETAIL— EXTERNAL CORNER OF BEARERS AND JOIST FOR WEATHER BOARD OR SHEET CLADDING**

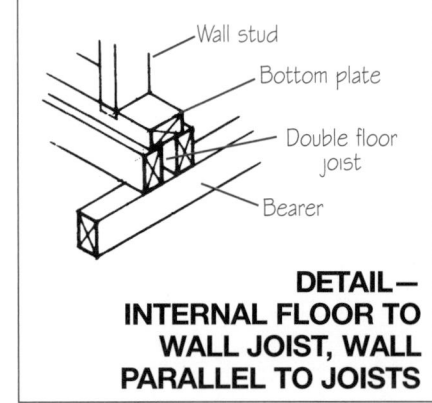

**DETAIL— INTERNAL FLOOR TO WALL JOIST, WALL PARALLEL TO JOISTS**

# TIMBER FLOOR FRAMES • 30

## Flooring materials

Traditional flooring ('cut in' flooring) was laid on the joists, strip by strip, room by room. If platform construction is used, then sheet flooring is generally used and laid before the walls are erected, giving a safe working platform for the workers as well as simplifying the construction.

Traditional strip flooring boards were joined one to the other with simple tongue-and-groove joints moulded from the material of the strip. The tongues are inserted in to the groove and the whole is clamped together to give a draught-free floor. The strips are nailed to the joists, then the nails are punched below the surface of the strips, which are then sanded to give a level surface ready for carpeting or other floor covering. If the floor is to be stained and sealed then a second fine sanding is required and the building should be dust-free before any sealing is carried out. If the nail holes would concern you it is possible to get strip flooring with secret nailing.

Many timbers are used in strip flooring and these vary significantly from location to location. In New South Wales the most common strip flooring is Cypress pine, in Victoria it is kiln-dried hardwood and in Western Australia Jarrah is the most common. Other strip flooring timbers include rainforest species that are becoming scarce and there are environmental reasons for limiting the use of these timbers.

Platform flooring structures can be covered with strip flooring boards, but are more commonly covered with weather-proofed plywood, or chipboard sheeting. This sheeting has been specifically developed as a flooring medium and it is capable of withstanding up to three months or so of wet weather, which sometimes occurs during a construction period. The edges of the sheets have tongue-and-groove edges, sometimes these are cut into the material, but more often the sheets are grooved and the tongue is made of plastic. Sheet flooring should be used strictly to the manufacturer's instructions, as this is critical in avoiding bulges.

## Wet areas

In some states it is a requirement of the building regulations that floors to wet areas, particularly bathrooms, toilets and laundries, must have an impervious subfloor. In the past this has been interpreted to mean poured-in-place concrete, but in recent times this restriction has been broken down; first by the use of compressed fibre cement sheets, then more recently by the use of waterproof plywoods and chipboards. The compressed fibre cement floor was the greatest breakthrough in providing a sensible floor for wet areas. It avoids the inconvenience of pouring concrete in small lots while providing a material that is rot proof and, if used correctly, virtually indestructible.

By using fibre cement sheets, plywood or chipboard platform floor construction methods greater versatility can be achieved, something that was nearly impossible if a poured-in-place concrete floor was used.

If poured-in-place reinforced concrete floors are used in wet areas then dwarf brick base walls must be built to support the edges of the concrete slab, and the sewer drains must be accurately placed in position before the concrete is poured.

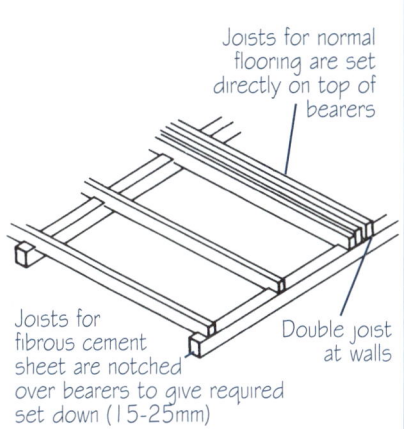

**DETAIL—SET DOWN FLOOR IN WET AREAS USING FIBRE CEMENT FLOOR SHEETS**

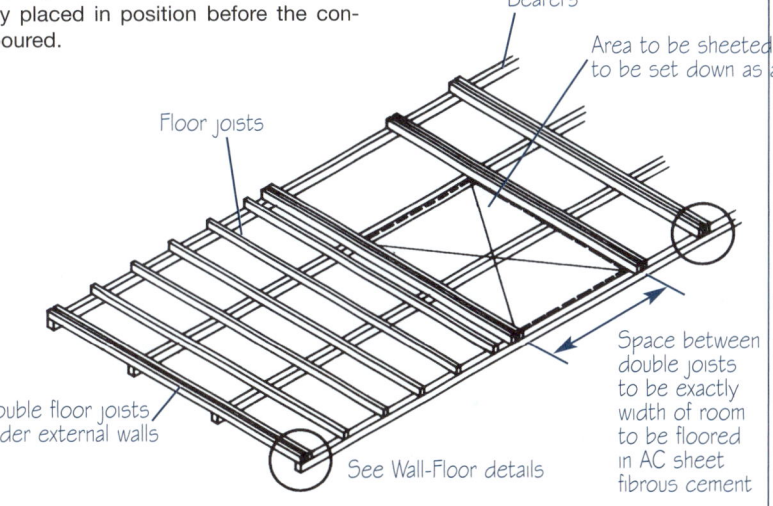

**GENERAL ILLUSTRATION — LOCATING SET DOWN FIBRE CEMENT SHEETS**

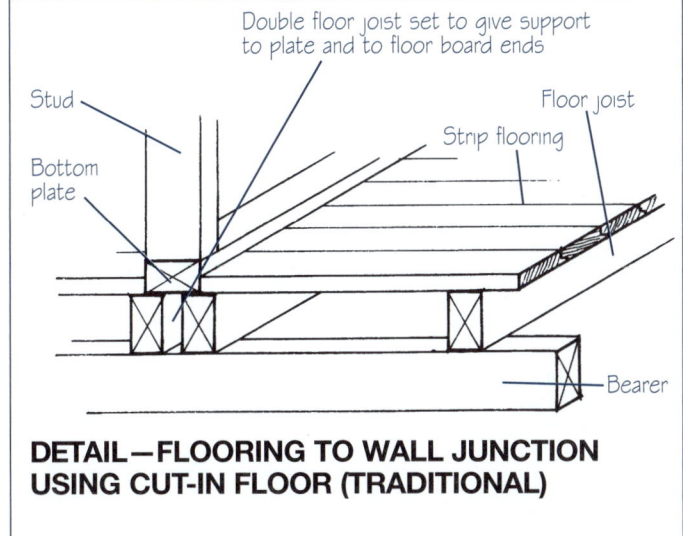

**DETAIL—FLOORING TO WALL JUNCTION USING CUT-IN FLOOR (TRADITIONAL)**

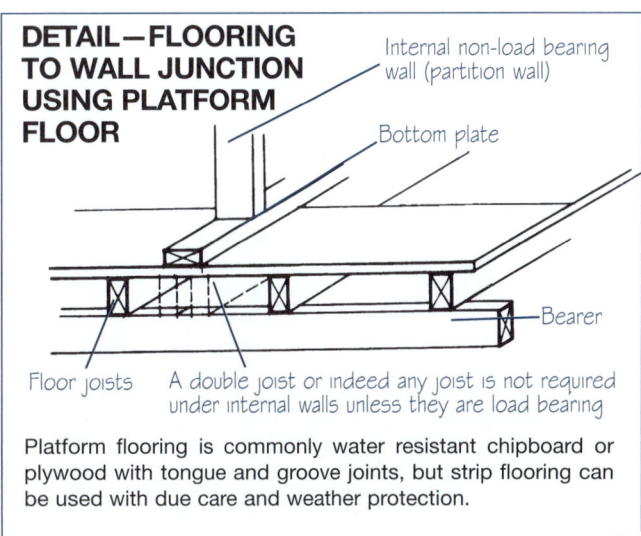

**DETAIL—FLOORING TO WALL JUNCTION USING PLATFORM FLOOR**

Platform flooring is commonly water resistant chipboard or plywood with tongue and groove joints, but strip flooring can be used with due care and weather protection.

# 31 • STEEL FLOORING SYSTEMS

As the availability and quality of timber flooring decreases and its price increases, there is increased opportunity for steel-based products to gain acceptance. There are a number of galvanised steel systems available on the market that offer a floor frame for a conventional house—including steel posts, bearers and joists—at prices competitive with traditional hardwood, Cypress pine or treated *Pinus* frames.

If used in conjunction with masonry veneer construction, the steel posts replace the isolated piers or stumps, but if used with a lightweight cladding system like weatherboard, fibrocement sheet or even AAC panel, all the subfloor support can be steel posts. The posts are commonly pre-galvanised, square hollow sections (SHS). The square section is generally easier to fix than the alternative circular section and is available in a wider range of wall thicknesses. The pre-galvanised SHS can also be welded and, although the galvanised area around the weld is damaged by heat, this can be reduced by using modern MIG welding techniques and high-quality, cold galvanising products.

This is a major advantage. In the past, if a steel frame was to be galvanised, it had to be taken to a major industrial plant to be hot dipped in a tank of molten zinc.

The posts are either cast into concrete pad footings or bolted to pre-poured concrete pads. If the slabs are pre-poured, then either holding-down bolts have to be cast into the concrete, or the concrete is drilled to take holding-down bolts, which are fixed by expanding friction anchors or bonded to the concrete with special chemset adhesives.

A base plate and bearer bracket can be welded to the bottom and top of the post and then the post is bolted down to the concrete and stabilised by packing under the base plate with a non-shrinking grout. Alternatively, proprietary bases can be fixed to the concrete and then the posts are fixed to them using self-tapping screw fixings. In this system the bearer bracket is often fitted with an adjusting jack to assist in levelling the bearers, so you get a perfectly level floor—a good idea for an OB.

The simplest steel floor frame can be built by simply using techniques similar to timber framing. Generally this involves using pre-galvanised rectangular hollow section (RHS) steel tubing as a direct substitute for traditional timber bearers and joists; even if the same section size as timber is used, the system will be sound. If the bearers are welded to the bearer brackets that are attached to the posts and the joists are welded to the top of the bearers at the correct spacing for the flooring material to be used, the whole will be stable and rattle-free.

Alternatively, if a welder is not available or a proprietary system is used, self-tapping screws and brackets may be applicable. These are satisfactory but are not as practical as a fully welded frame.

Some proprietary steel floor framing systems use cold-formed, folded, galvanised steel sections in lieu of the SHS and RHS. These sections are sometimes specifically designed for the task, but often they are existing standard sections—roof purlins and wall girts are favoured.

There is a small difference between steel and timber that can be felt when walking on the floor—steel is more elastic and the floor is likely to feel springier. However, the steel will increase in housing when the cost is comparable to alternative methods and where there is danger of attack by subterranean termites.

There is a further problem. Many local building inspectors treat timber and steel differently so that, although architects and builders can select timber section sizes, an engineer is required to determine the section size if there is any steel involved in a building.

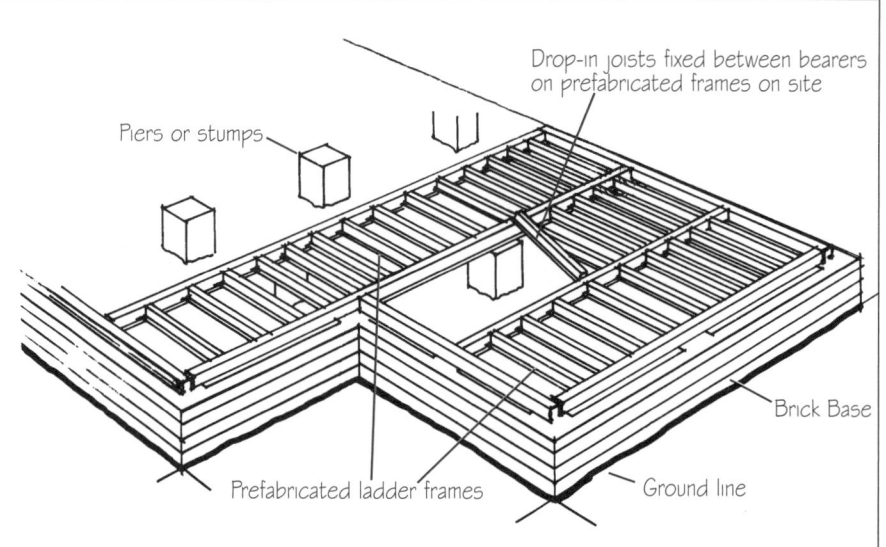

### SEMI PRE-FABRICATED LADDER FRAMED SYSTEM

Site assembled ladder floor systems. Special bearer sections are installed on piers with drop-in joists fixed between them using self-drilling screws. These are typically used in lower floor applications.

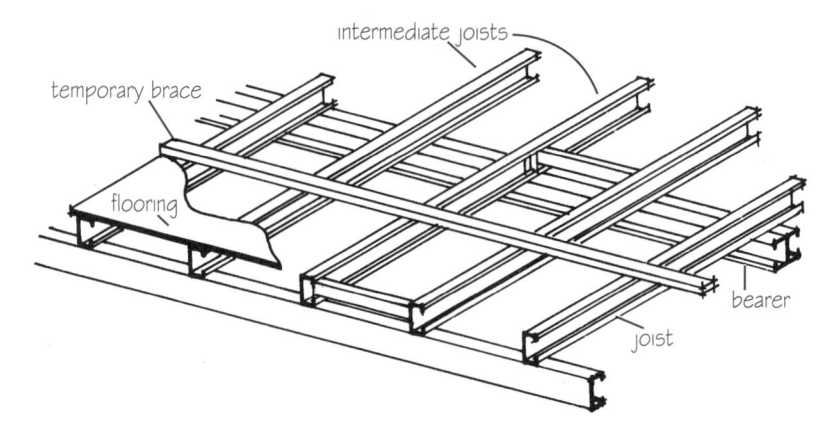

### LONG SPAN SYSTEMS

These are usually installed in a similar manner to timber systems. Rolled steel sections (typically C sections) are installed on bearers or on lower floor wall frames. Typical applications are upper or intermediate floors or longer span applications.

All systems can be installed on steel stumps, concrete or brick piers. Sheet flooring (particle board, plywood or fibre cement sheeting) can be installed on top of the floor frame.

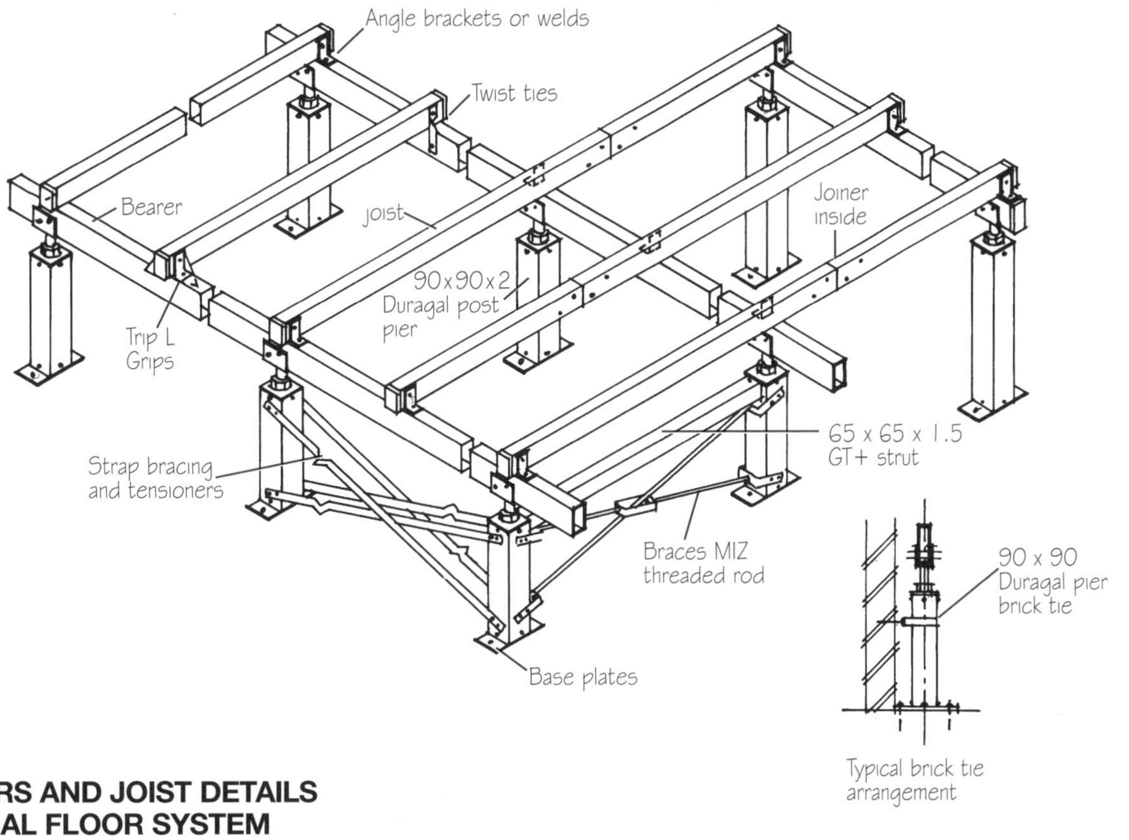

# STEEL FLOORING SYSTEMS • 31

**BEARERS AND JOIST DETAILS
DURAGAL FLOOR SYSTEM**

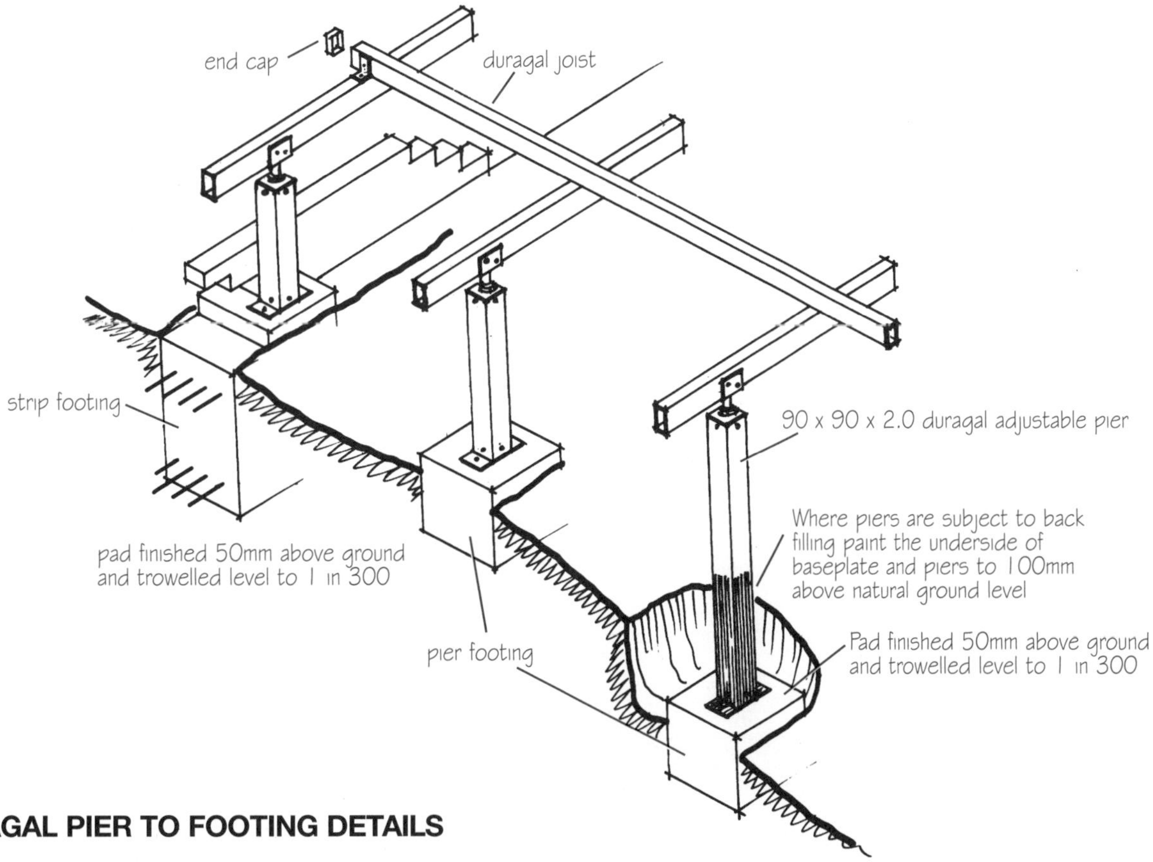

Labels: Angle brackets or welds; Twist ties; Bearer; Joist; Joiner inside; 90 x 90 x 2 Duragal post pier; Trip L Grips; Strap bracing and tensioners; 65 x 65 x 1.5 GT+ strut; Braces M12 threaded rod; Base plates; 90 x 90 Duragal pier brick tie; Typical brick tie arrangement

**DURAGAL PIER TO FOOTING DETAILS**

Labels: end cap; duragal joist; strip footing; pad finished 50mm above ground and trowelled level to 1 in 300; pier footing; 90 x 90 x 2.0 duragal adjustable pier; Where piers are subject to back filling paint the underside of baseplate and piers to 100mm above natural ground level; Pad finished 50mm above ground and trowelled level to 1 in 300

# 32 • CONCRETE SLABS

Concrete slabs used in house construction are almost without exception reinforced concrete. Concrete floor slabs can be used in two basic ways:
a. supported by the ground
b. suspended above the ground.

The first type used in house construction in two basic forms:
(i) raft slab,
(ii) floating slab.

The second type is used in house construction also in two forms:
(i) suspended slab cast on permanent filling
(ii) suspended slab cast onto removable or permanent framework.

In all states there is a requirement to have reinforced concrete slab design done by qualified structural engineers, except in some specific instances where standard designs can be used.

Reinforced concrete is an unforgiving material, and many OBs have found out to their dismay that if a mistake is made when placing this material it is often very expensive to rectify. Concrete sets very hard, very quickly. The golden rule for all concrete work is get it done right the first time. Plan all concrete pours carefully, making sure that when the first cubic metre of concrete begins to be placed you have allowed sufficient labour to complete the work—and that there is sufficient time left in the day of the pour to finish the slab in the manner required.

A critical part of all concrete work is the adequacy of formwork. Concrete is a very heavy fluid, and many a disaster has happened when a section of formwork has given way during a concrete pour. At worst this can be fatal, but even at best a very expensive jackhammering job may be required.

## Slabs on the ground

### Raft slabs

The raft slab has been discussed in part under Ground-to-Floor Systems. The most important part of the pouring of a raft slab is the preparation of the sub-slab foundation area.

The area must be made level by excavation, and all vegetable-bearing soil should be removed. The area should be protected by agricultural drains so that the moisture content of the sub-slab area is kept as stable as possible; this is particularly important in reactive clay soils.

The edge beams and any stiffening beams are excavated to adequate bearing foundation. Some areas may require the excavation of extra deep edge beams to assist in vermin control. The area under the slab is filled with fine aggregate or sand, levelled and consolidated. Remember to place any sewer drains that pass under the floor, being particularly careful to get them in their correct locations—there is no second chance.

On top of the sand is placed a moisture-proof membrane normally made of a specially developed reinforced plastic. The membrane is also placed under the edge beam about 300–450 mm overlapping out of the trench.

The sheets of membrane must be sealed at the joints with special moisture-proof tape, and all penetrations through the slab—sewer pipes—are sealed to the membrane with tape.

After the membrane is in place the reinforcing steel is placed in its correct position in the trenches and the slab. Take care to overlap the reinforcing steel as required by the engineer, and tie it adequately with wire. The reinforcing steel should be held to its design height above the membrane with purpose-made bar chairs of a design that will not penetrate the membrane.

The edge of the slab may require special formwork to achieve an edge profile that will allow the building of timber frames and masonry work as required to achieve water tightness.

Some areas of the slab may require different surface finishes from other areas. Make sure profile panels are made up before any concrete is poured so that special finish areas will be finished correctly.

Around the extreme edge of the slab a board is set to the finished level of the slab.

After the local government building inspector has checked the steel and formwork, the concrete can be ordered. Try to calculate as close as possible to the required quantity. If the order is short the small amount required to finish the job can be very expensive, as many concrete companies charge a base hiring rate for a truck regardless of the load. If you have excessively over-ordered you cannot return the concrete left in the truck for a credit and, if the driver is in a bad mood on the day, you may have surplus setting concrete to dispose of.

Check where the concrete truck can park, and make sure that the concrete labourers can barrow the concrete from that spot to the area to be poured. If the barrowing distance is too long or too steep then a concrete pump may have to be hired—a marvellous piece of machinery, but like all machinery prone to break down at the most inopportune moments. Try to find a truck-parking spot close to and slightly uphill from the area to be poured.

Order the concrete for delivery in advance, but do not confirm delivery until the last possible moment on the morning of the pour, just in case the concrete labourers do not arrive or the morning dawns with a heavy downpour of rain. Do not pour concrete in the rain—better to be sure than sorry.

When the concrete arrives on the site, make sure that it is of the quality you ordered. This can be confirmed by a simple slump test that the driver should be able to do for you. Concrete should be placed carefully into the prepared area and worked around all the steel reinforcing with a vibrator if possible. All care must be taken to make sure that none of the tools used by the labourers penetrates the membranes, for these cannot be repaired and damage may cause damp spots in the concrete floor, or worse, in the future.

When the concrete is placed and the surface finish complete, then the concrete must be cured in a manner that will prevent it drying out too rapidly. A simple method that is generally adequate for all slabs on ground is to cover the slab with polythene membrane for a period of up to 30 days.

### Floating slabs

The floating slab varies from the raft slab in that it is generally built without edge beams and within the perimeter of a masonry base wall. In floating slab construction, the floor slab rests directly on a stabilised base, generally of consolidated sand, with a moisture-proof membrane directly under the slab.

The loads of the building are transmitted to the foundation through the walls to strip/beam footings. No loads are transmitted to the slabs other than the weight of furniture and occupants.

Floating concrete slabs are often favoured by architects, as they eliminate many of the problems associated with raft slabs. However, they are inevitably more expensive. Floating slabs allow more flexibility, to break up floors in the future to replace or alter drainage pipes or other sub-floor services, than raft slabs.

Care must be taken when preparing the sub-slab area. Pressure can be exerted on the inside of the base masonry walls and can cause their failure if they are not adequately designed or if mechanical consolidation has not been carried out.

---

### TYPICAL GROUND FLOOR REINFORCED CONCRETE FOOTINGS AND FLOOR SYSTEMS

Reinforced concrete strip footings, cavity brick wall, reinforced concrete floor slab on compacted sub-base and a sand bed.

Reinforced concrete strip footings, cavity brick base wall, brick veneer top, reinforced concrete floor designed as a suspended slab but poured on non-expanded filling, e.g. cinders.

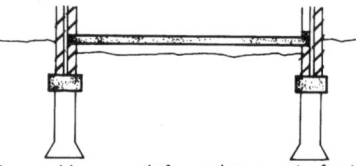

Pier and beam reinforced concrete footings, cavity brick base wall, cavity or brick veneer top, reinforced concrete floor slab poured on temporary or permanent formwork as a suspended slab.

Edge beam and floor slab integral reinforced concrete, brick cavity or brick veneer walls.
Sand bed and moisture/vapour membrane required under slab.

# CONCRETE SLABS • 32

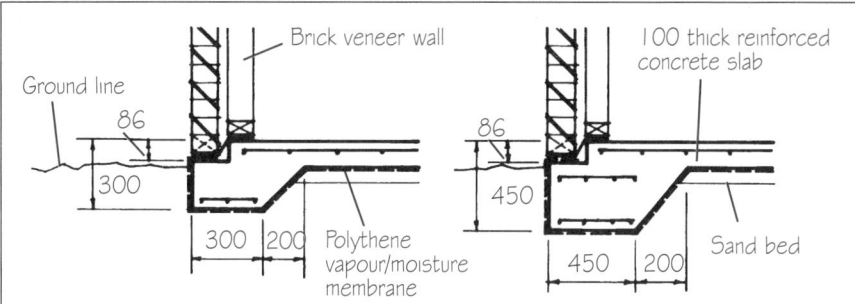
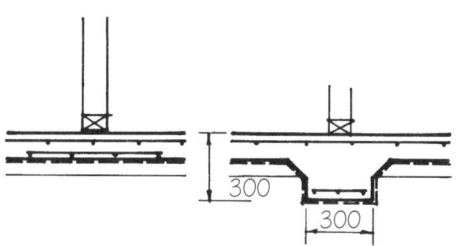

**SINGLE STOREY BRICK VENEER**—edge beam.

**DOUBLE STOREY BRICK VENEER**—edge beam.

**SINGLE STOREY TIMBER FRAME**—internal wall slab stiffening.

**DOUBLE STOREY TIMBER FRAME**—internal wall beam.

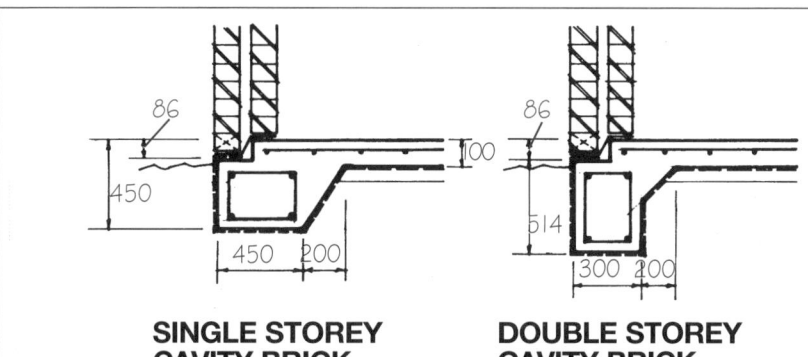
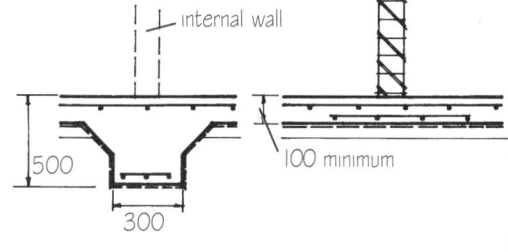

**SINGLE STOREY CAVITY BRICK**—edge beam.

**DOUBLE STOREY CAVITY BRICK**—edge beam.

**SINGLE STOREY BRICK**—internal wall slab stiffening.

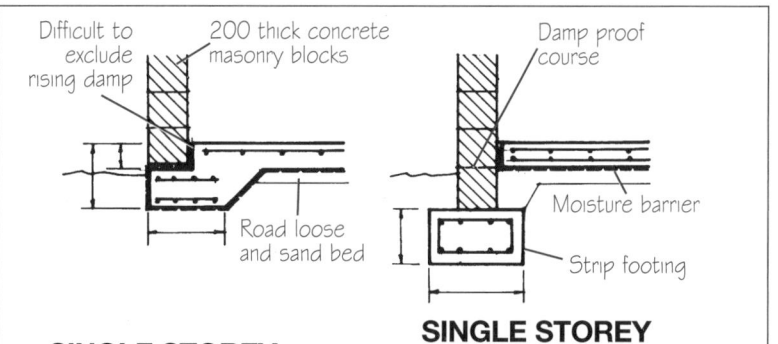
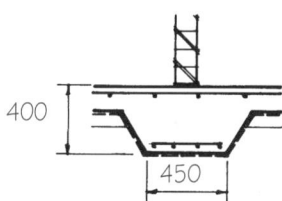

**SINGLE STOREY CONCRETE BLOCK**—edge beam

**SINGLE STOREY CONCRETE BLOCK**—strip footing, slab on ground

**DOUBLE STOREY BRICK**—internal wall beam

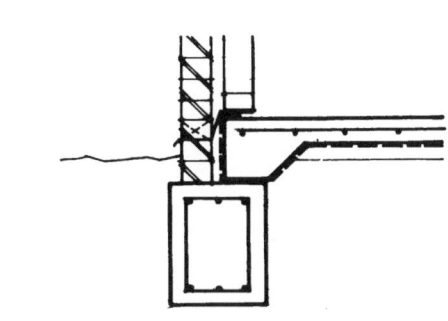

**SINGLE OR DOUBLE STOREY BRICK VENEER**—deep beam footing, suspended slab with edge beam

Dimensions indicative only—final dimensions and reinforcing details should be provided by a structural engineer.

Example of a complex footing and concrete slab system designed by an engineer to satisfy a particular set of site requirements and foundation constraints. Well designed footings and slabs are likely to reduce maintenance.

All dimensions are for guide purposes only, final dimensions should be designed by a structural engineer to suit local soil conditions and the requirements of local authorities. Raft slabs are generally most suited to well drained non-cohesive foundation soils, e.g. sand and gravel.

# 32 • CONCRETE SLABS

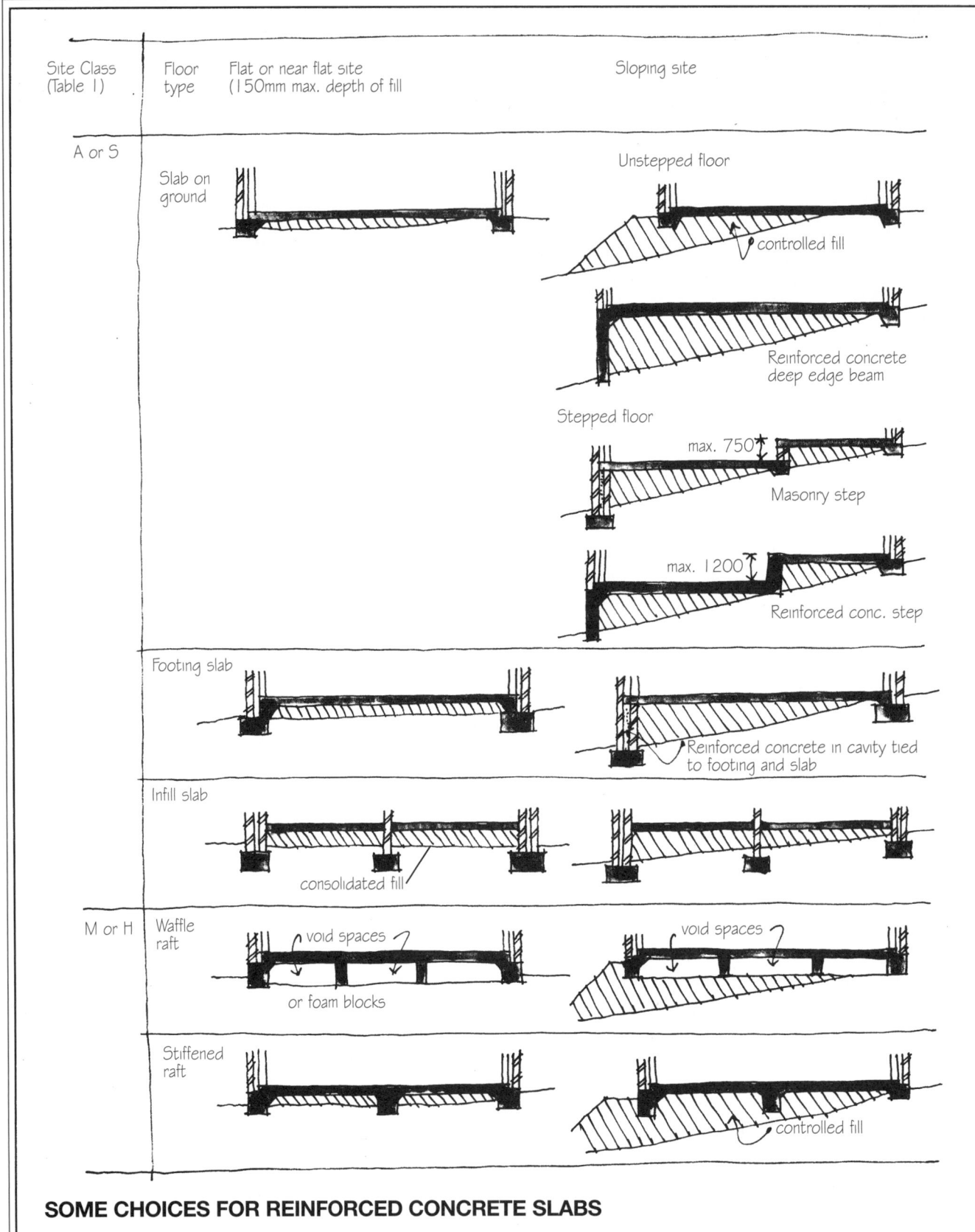

**SOME CHOICES FOR REINFORCED CONCRETE SLABS**

Every site is different and needs to be checked by a structural engineer.
These sketches are to assist at the building design stage only

# CONCRETE SLABS • 32

## Suspended concrete slabs

### Suspended slabs on fillings

This method of constructing concrete slabs has all the simplicity of the floating slab method but the slab sits on top of the base walls rather than between them. The slab is reinforced to span the distance between supports. The filling is generally of poor-quality unconsolidated fill, which will settle away from the underside of the slab with time. However, as the slab is designed to be suspended, the result is a simple-to-construct concrete floor in which all the building loads are transferred to the footings.

Suspended concrete slabs on fill are only used on ground-floor slabs where the height from the ground to the underside of the slab is 600 mm or less. If the clearance under the slab exceeds 600 mm then removable formwork is generally more economical.

### Suspended slabs on formwork

Reinforced concrete slabs cast on formwork are generally used where access to the underside of the slab is desirable—as in basement garages or first floors. Without doubt, suspended concrete floors on temporary or permanent formwork are expensive, and, if they are to be used by homebuilders, good design and planning are essential to gain the best value for the dollars outlaid.

The design of suspended reinforced concrete slabs is very definitely the province of a structural engineer, no one should attempt to design their own suspended slabs unless they are thoroughly convinced of their ability.

Although it is possible for OBs to build their own formwork, it is generally not economical, except if permanent formwork is being used. The formwork shuttering boards and the joists and props can be used on multiple projects; therefore it is nearly always cheaper for the OB to employ a formworking gang that supplies its own material, than it is to buy or hire (although this is not always possible) the materials for the formwork.

If OBs decide to place and tie the reinforcing steel for their own suspended floor, they should ask themselves this simple question: Am I convinced that the job I will do is going to be good enough for me to sleep peacefully at night, on or under the concrete slab in which I placed and tied the reinforcement? If the answer is not a resounding YES, then employ someone who knows what they are doing and reduce the risk.

Remember to cast knock-out holes in the slab for any penetrations needed for drainage pipes or other services. Electrical conduits should be cast into the slab.

The underside of a suspended concrete slab is the ceiling of the floor below, so if a quality strip finish is required then quality formwork will be required first.

When the concrete is placed, adequate vibration is required to make sure that the undersurface of the slab is free from defects such as 'honeycombing', which not only looks unsightly, but in extreme cases can result in structural failure.

## COMMON USES OF CONCRETE IN HOUSES

**Single storey cottage**

Floors to wet areas, e.g. bathrooms, laundries, toilets. (This is required by regulation in some states)
Floors to patios, verandah and steps.

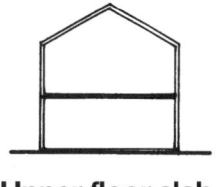

**Upper floor slab**

Upper floors should be designed by a structural engineer.
Ground floor walls generally required to be masonry.

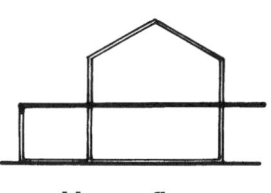

**Upper floor terrace and balcony**

Weatherproofing of slabs over rooms needs special consideration.
Cantilevered slabs require engineering design.

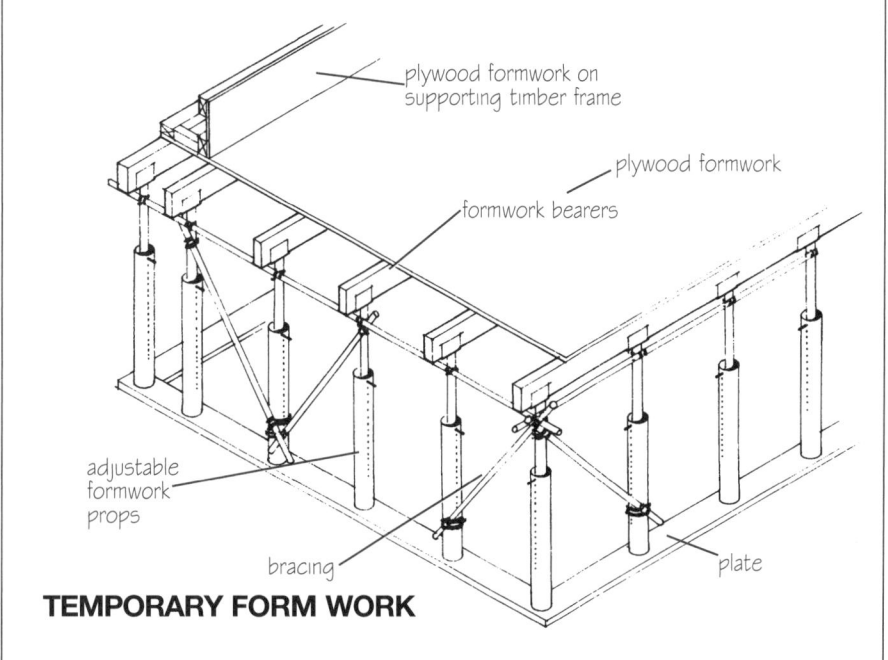

**TEMPORARY FORM WORK**

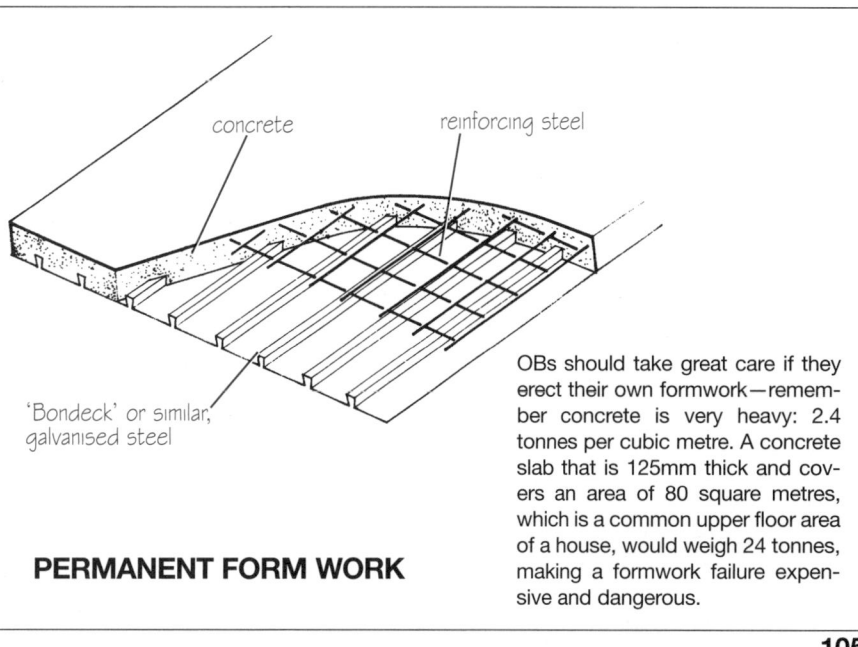

**PERMANENT FORM WORK**

OBs should take great care if they erect their own formwork—remember concrete is very heavy: 2.4 tonnes per cubic metre. A concrete slab that is 125mm thick and covers an area of 80 square metres, which is a common upper floor area of a house, would weigh 24 tonnes, making a formwork failure expensive and dangerous.

# 32 • CONCRETE SLABS

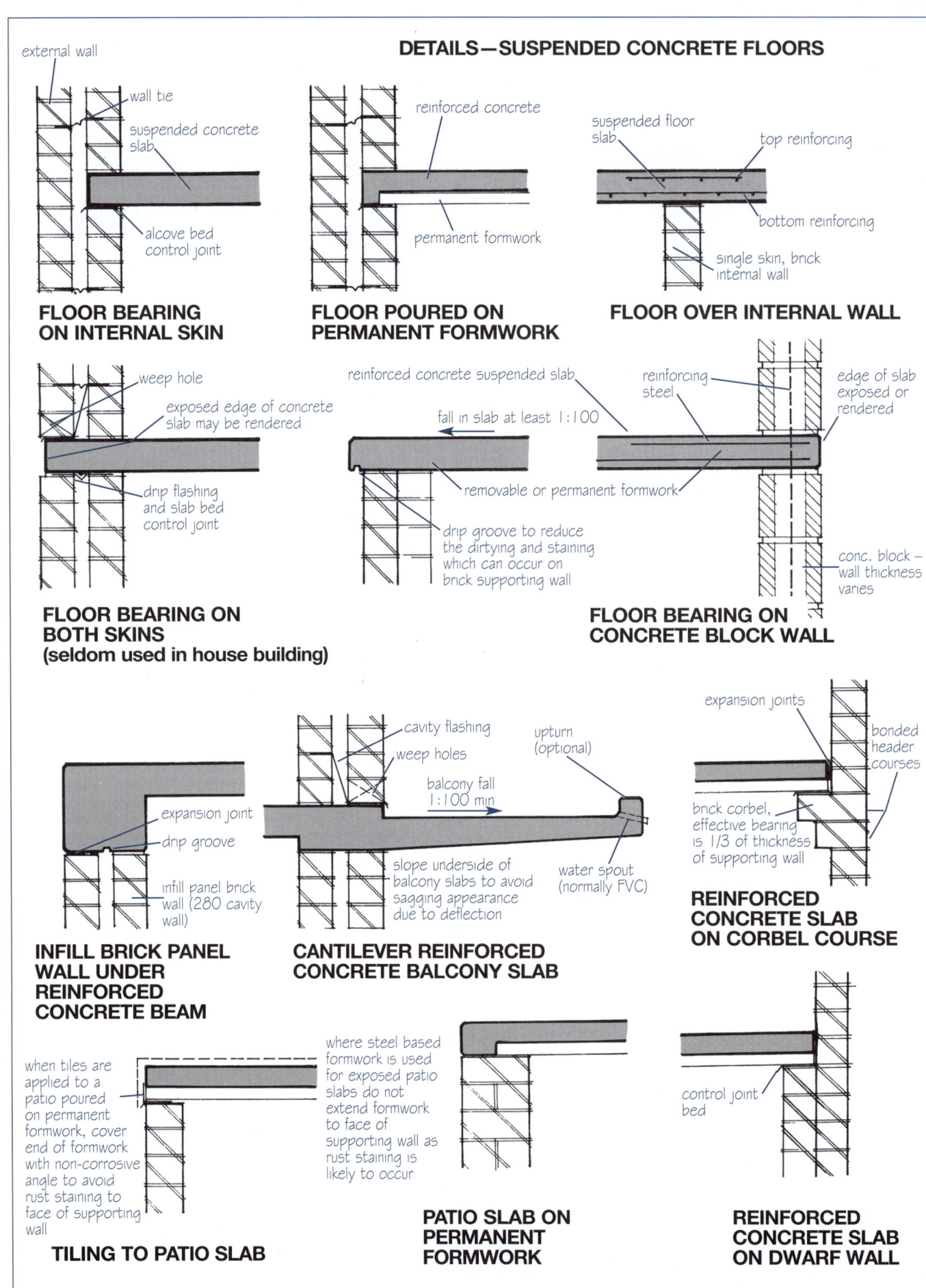

DETAILS—SUSPENDED CONCRETE FLOORS

# CONCRETE SLABS • 32

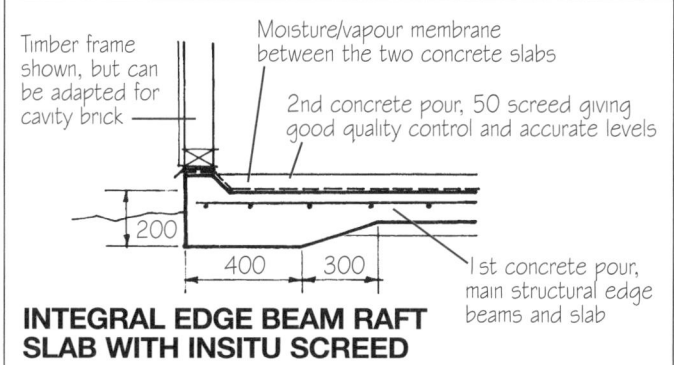

**INTEGRAL EDGE BEAM RAFT SLAB WITH INSITU SCREED**

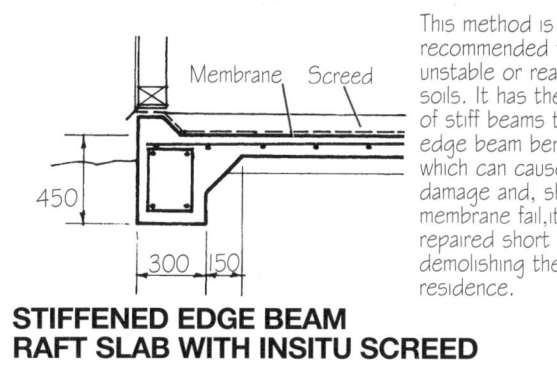

**STIFFENED EDGE BEAM RAFT SLAB WITH INSITU SCREED**

## Termite protection for timber above concrete slabs

Where concrete slabs are in contact with the ground or are elevated by inaccessible sub-slab support structures, termite protection is required. Some jurisdictions may still allow chemical treatment of the soil to kill and maintain a barrier against termite infestation. However, this method is not considered desirable in most areas, and in many places is now illegal. There is a risk of poisoning humans, and the active life of the chemicals may be short, meaning they need regular re-application.

Barrier protection is now the recommended method and there are many new products coming onto the market. Traditional galvanised steel termite barriers (traditionally but incorrectly named 'ant caps and strips') are still highly successful if used correctly. Stainless steel mesh and fine grit barrier systems give buildings lifetime protection. These products have a guaranteed effectiveness if supplied and installed by an Australian Standards certified contractor. It is worth considering the use of chemically treated timber or metal framing members in areas where a high incidence of termite infestation can be expected.

Termite proof membranes for use with concrete slabs are available as are some marine grade aluminium barrier systems. There are continuous developments being made in termite resistant products. A trip to a building information centre or a large home building materials centre before building will expose OBs to new products, and a surf through the building information web sites will unearth many novel approaches to the termite problem.

Take care there have been many changes to anti-termite methods over the decades, some like simple physical barriers have remained in use for many decades and are still applicable, others including most chemical treatment systems have been found either wanting or dangerous to health. It is recommended to use only proven systems or systems supported by a standards approved quality assured certificate.

**TYPICAL DETAILS OF REINFORCED CONCRETE SLAB ON GROUND WITH INTEGRAL EDGE BEAM**

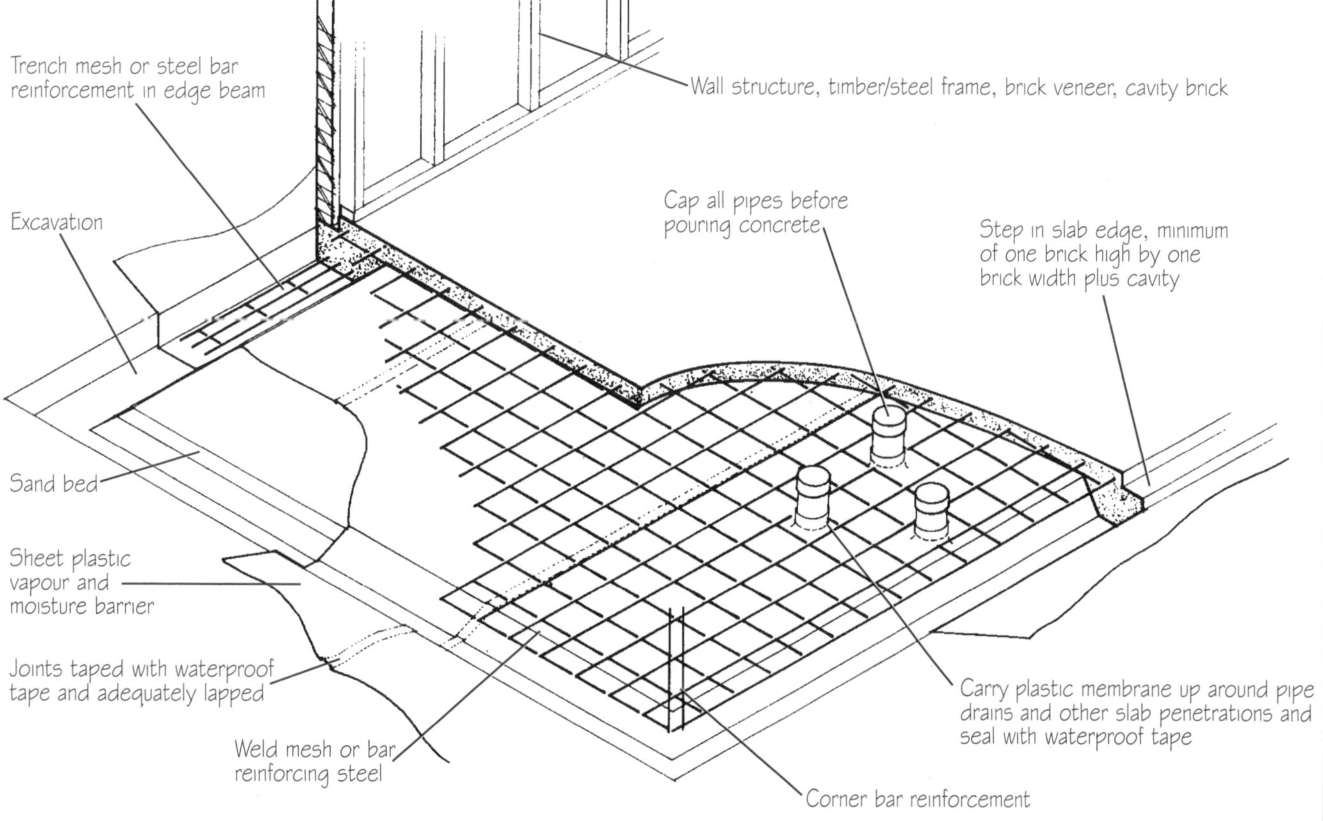

# 32 • CONCRETE SLABS

## BAR CHAIRS

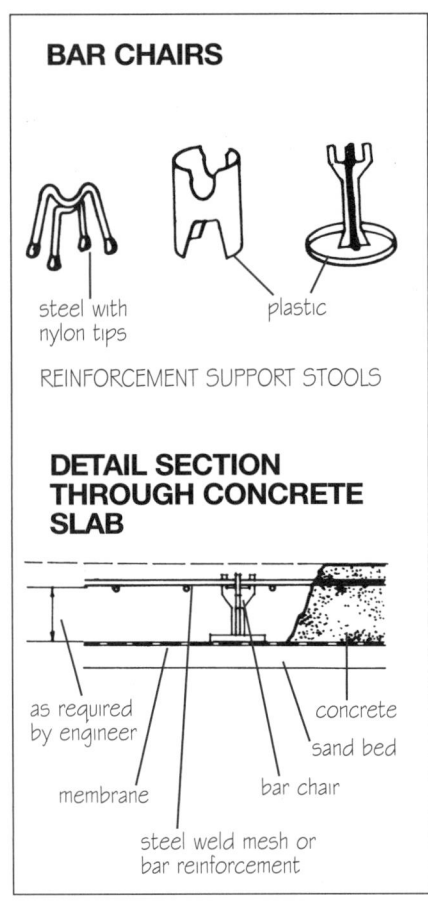

steel with nylon tips

plastic

REINFORCEMENT SUPPORT STOOLS

## DETAIL SECTION THROUGH CONCRETE SLAB

as required by engineer
membrane
concrete
sand bed
bar chair
steel weld mesh or bar reinforcement

## SLAB ON GROUND MOISTURE BARRIERS

This method gives good results but membrane failure cannot be repaired

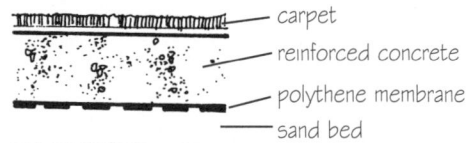

- carpet
- reinforced concrete
- polythene membrane
- sand bed

### Traditional location of polythene membrane

This method is expensive but allows membrane failure to be repaired by chopping out the tapping screed replacing the membrane and repouring

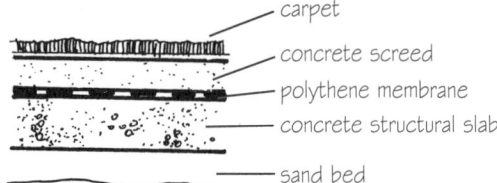

- carpet
- concrete screed
- polythene membrane
- concrete structural slab
- sand bed

### Screeded slab location of polythene membrane

This method has a high risk of failure but at least the membrane can be replaced easily. Generally usatisfactory

- surface membrane
- reinforced concrete slab
- sand bed

### Top membrane location of polythene membrane

## Location drawing—section through split level house floor system

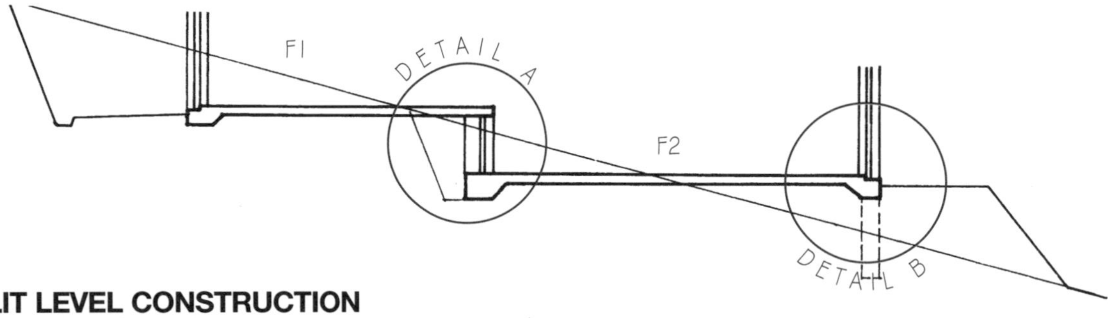

## SPLIT LEVEL CONSTRUCTION WITH RAFT SLABS

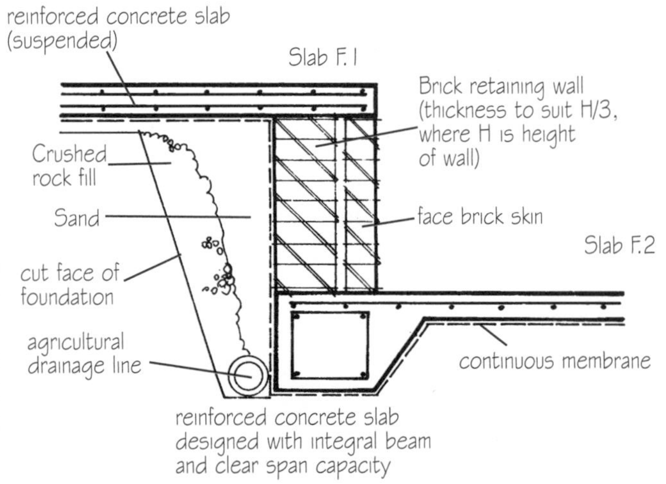

reinforced concrete slab (suspended)
Slab F.1
Brick retaining wall (thickness to suit H/3, where H is height of wall)
Crushed rock fill
Sand
face brick skin
cut face of foundation
Slab F.2
agricultural drainage line
continuous membrane
reinforced concrete slab designed with integral beam and clear span capacity

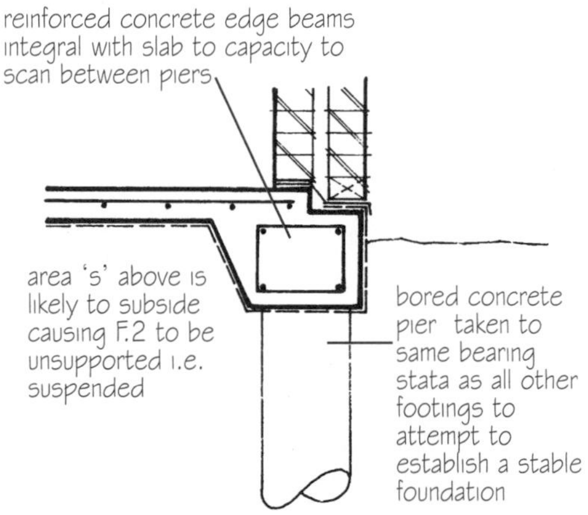

reinforced concrete edge beams integral with slab to capacity to scan between piers
area 's' above is likely to subside causing F.2 to be unsupported i.e. suspended
bored concrete pier taken to same bearing stata as all other footings to attempt to establish a stable foundation

# WALLS—INTRODUCTION • 33

The walls both support the roof and enclose the dwelling against the weather. They comprise a structural frame or solid mass that transfers the load of the roof structure and roof cover to the base structure. They also provide support for various types of wall penetrations (windows and doors).

Wall structures are usually clad in either brick (brick veneer) or sheet and board materials. This cladding serves to protect against weather and provide aesthetic amenity. On the inside the wall structure is lined with either sheet or board material or rendered/left with natural masonry.

Window and door assemblies can be either purpose-made or stock items with now-limited variation in type, but with wide application.

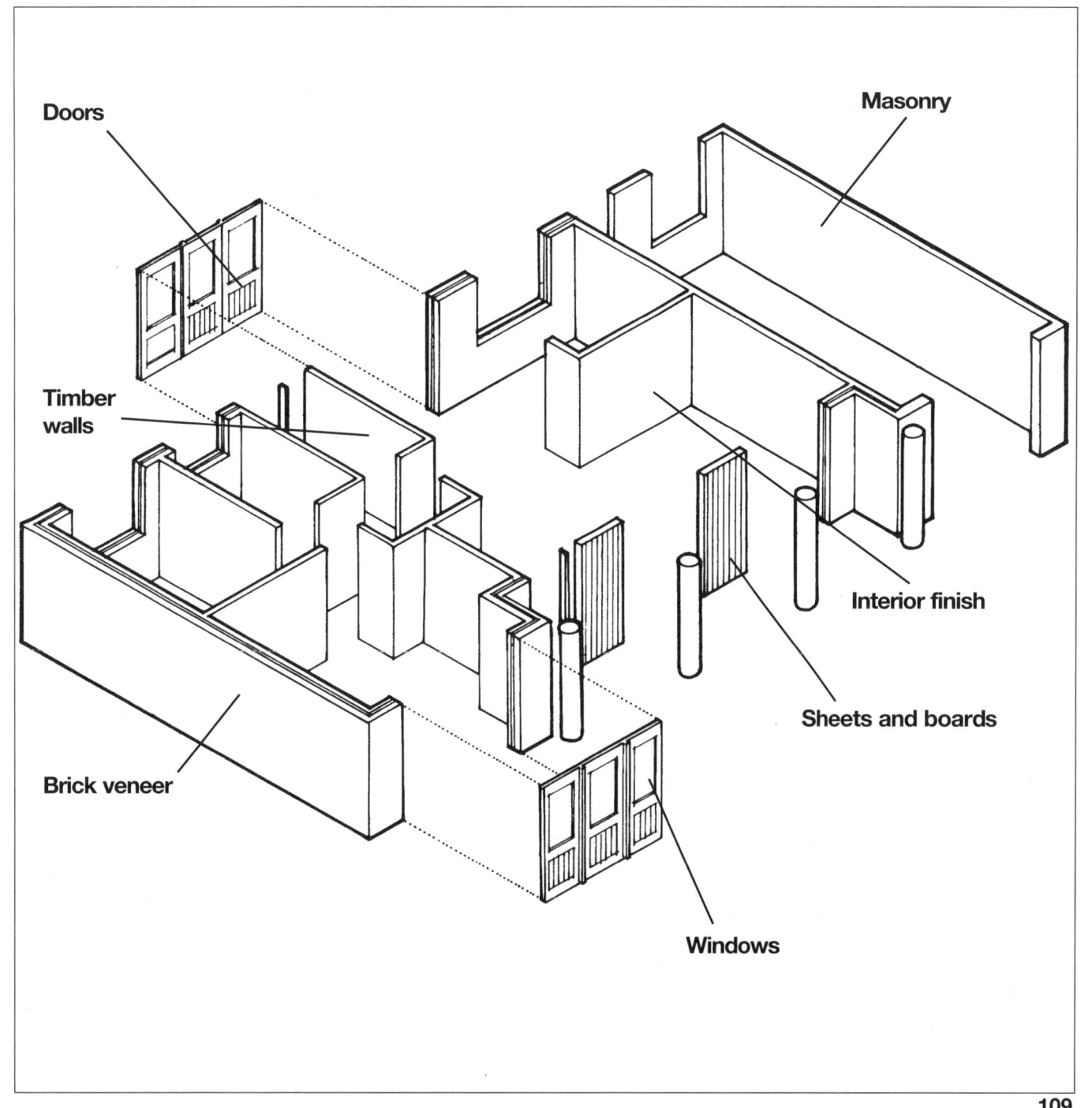

# 34 • TIMBER WALL FRAMING

Timber wall framing developed in North America and has slowly evolved until it has become, in the Australian context, one of the most efficient and economic building systems used in the house building industry. It is also forgiving—make a mistake when working in timber and it is likely that it can be fixed with the minimum of inconvenience. This makes timber framing the almost ideal OB system.

Timber wall framing aims to make use of relatively small section timber members, called scantling, to build strong, rigid wall frames. The timber sections used range from 75 x 38 mm to 100 x 50 mm, with larger sections than this only being used to span window, door and other openings.

Wall frames are fabricated lying down, either on-site, or, as is becoming popular—particularly in the project home industry—in a prefabrication factory.

The frame consists of horizontal members at the bottom and at the top of the wall, called the bottom and top plates respectively. The evenly spaced vertical members between the top plate and the bottom plate are called studs. Between the studs are one, two, three or more rows of horizontal members called noggings.

The plates, studs and noggings are the basic components of a timber-framed wall, but a wall framed of these members only will rack. To stop this, diagonal members are added to brace the frame.

Wherever walls intersect, the plates should be adequately fixed one to the other by means of halving joints, or purpose-made metal plates.

Openings in framed walls require beams over the head of the opening. Traditionally these beams have been in timber, but there are also many engineer-designed, mass-manufactured steel and aluminium head beams available. The metal heads have gained wide popularity because they are strong and light, and their lightness reduces the weight of the wall that has to be lifted from the ground or floor where it was assembled, to its final vertical position.

## Assembly methods

It was traditional in timber frame construction to make trenches in the top and bottom plates to receive the studs. This trenching was primarily to make the distance from the bottom of the bottom plate to the stud the same all the way along the plate. So that studs of the same length could be used, the distance from the top of the top plate to the top of the studs was also trenched to an even thickness. The need to trench plates to achieve even thickness is still there when off-saw timber is used, as there is a distinct possibility that the thickness of plates will vary significantly along their length and from plate to plate. In many locations it is now possible to purchase sized (dressed) scantling for much the same price as off-saw material. This sized timber seldom needs to be trenched when used as top and bottom plates. This is a benefit to OBs for it means that the frame is much simpler to assemble, less skill is required and expensive and dangerous trenching saws are eliminated from the job.

Plates are normally joined to studs by driving 100 mm nails through the plate into the stud; a stronger joint is achieved if these nails are driven so that they enter the plate at an angle to it and when two nails cross, locking the joint.

When nailing the noggings into place, it is not necessary to keep them in a straight line. If the wall is to be lined with plasterboard, it is much easier to nail the noggings if they are staggered. If the noggings are to be used to nail-fix vertical or diagonal boarding then it is advisable to keep the noggings in line so that all nail heads can be lined up. It is still advisable, even if secret nailed boards are used.

Bracing was traditionally a timber batten let into the face of studs and plates and nailed to them. This method has just about been replaced by using a bracing member made out of a perforated metal angle. When fixing the metal angle bracing it is simply a matter of making a saw cut into the face of the studs and plates, pushing the leg of the angle in and nailing off when the frame has been squared and plumbed. A tip is to use two pieces of angle bracing, each about 300 mm longer than half the length of the required bracing member. This allows the bracing to be fixed to the top and bottom plate but still be racked to achieve squareness. When the frame is square and plumb the bracing can then be easily nailed to the studs.

Window and door heads must be given adequate support at the sides of the opening. This can be done by increasing the size of the opening studs, ie. the studs at each side of the opening, or by doubling the studs at these locations.

If a raked or vaulted ceiling is to be used remember to make up the walls that follow the sloping ceiling with sloping top plates.

## Scantling materials

As noted above, the most common material used in framed walls is timber. This can be either local hardwoods or pines, or imported timbers, which generally come from North America, New Zealand or the Pacific Islands.

The timbers used vary from state to state. Cypress pine is common in Queensland, Northern or Southern Hardwood and *Pinus* in New South Wales, ordinary builder's hardwood and oregon in Victoria, *Pinus* in South Australia and Jarrah in Western Australia.

Each timber has its own special properties and problems. Some are more durable (Cypress and Jarrah), some are more stable (Oregon) and some are less expensive (*Pinus*), but they all are capable when used correctly of giving a strong stable frame.

In many locations *Pinus* and some other softwood scantling is available in sized sections. Some is available chemically treated to avoid termite attack, and some, mostly softwoods, is seasoned to give greater stability and avoid excessive shrinkage.

A number of metal wall frame systems are available in either formed steel or aluminium sections. These can be very good systems if their use is preplanned. However, on-site amendments or errors may not be as easily fixed as they would in timber. If you are contemplating using a metal framing system, check carefully as an OB that you will benefit from its use. Go and have a careful look at a house under construction that is using the system. Check that the cost is comparative to timber framing and take careful note of the connection methods, making sure you have access to any special tools that may be required, as some systems require welding. Discuss the system with your plumber and electrician to make sure that they will not charge you a premium because it is more complex to install services.

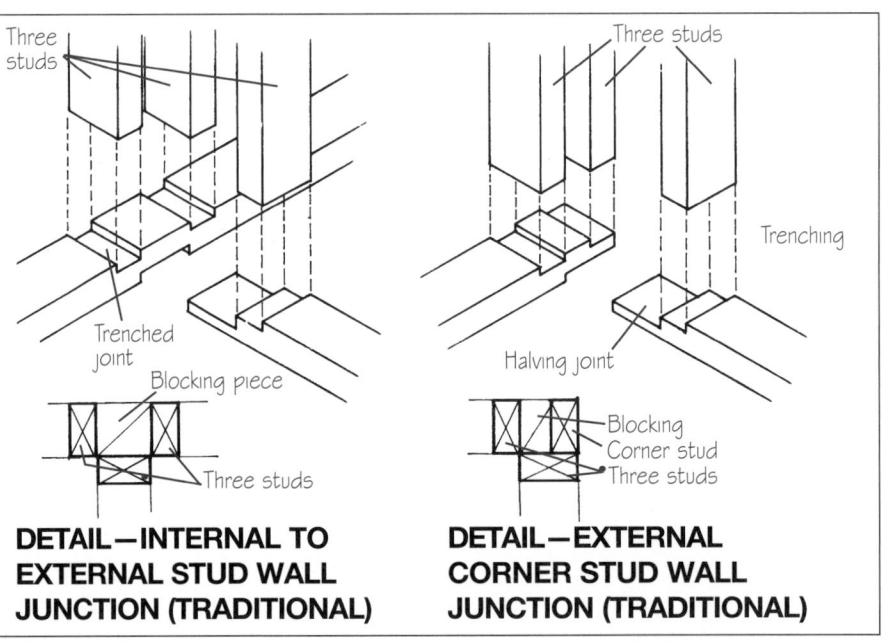

**DETAIL—INTERNAL TO EXTERNAL STUD WALL JUNCTION (TRADITIONAL)**

**DETAIL—EXTERNAL CORNER STUD WALL JUNCTION (TRADITIONAL)**

# TIMBER WALL FRAMING • 34

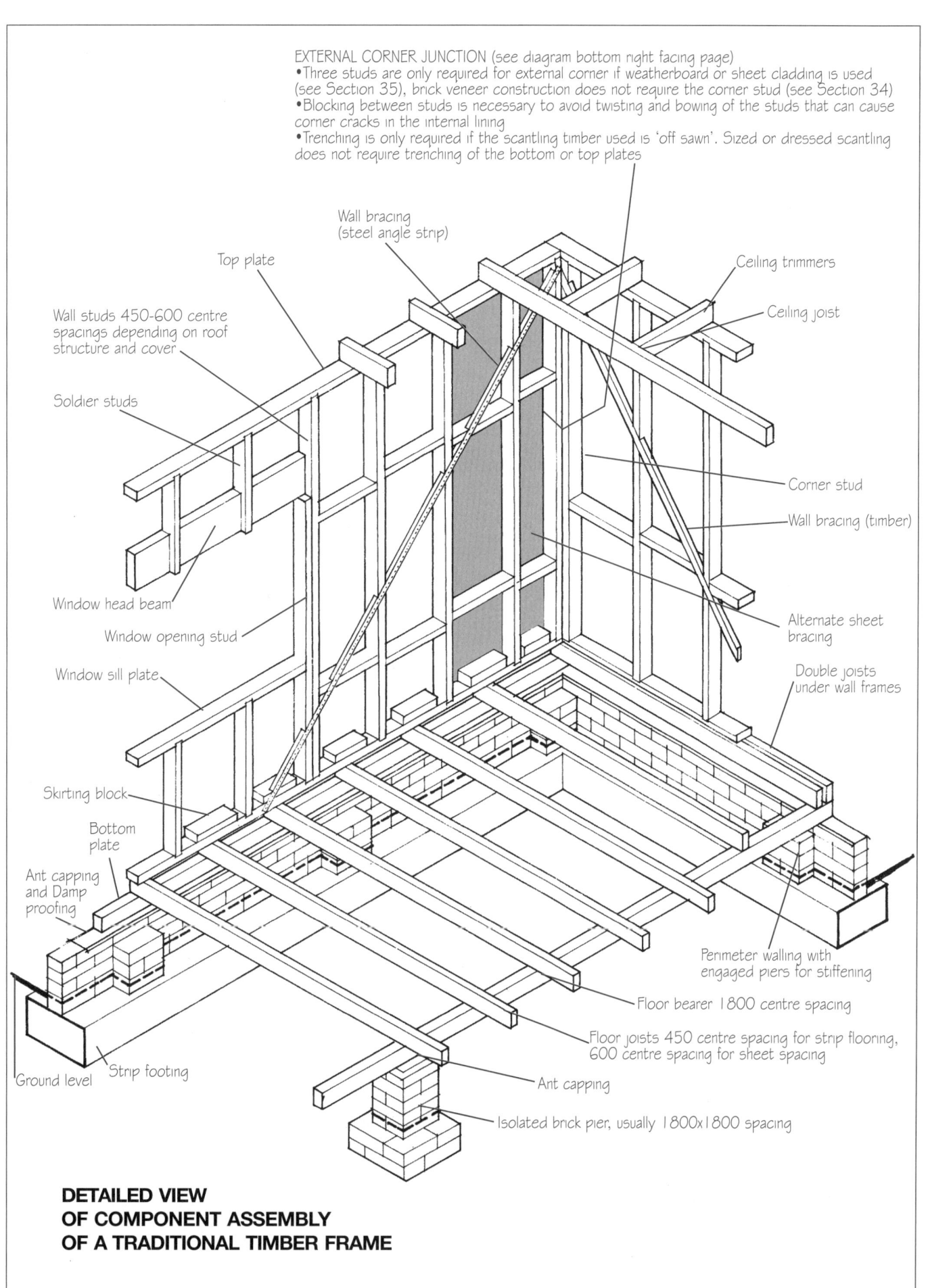

EXTERNAL CORNER JUNCTION (see diagram bottom right facing page)
• Three studs are only required for external corner if weatherboard or sheet cladding is used (see Section 35), brick veneer construction does not require the corner stud (see Section 34)
• Blocking between studs is necessary to avoid twisting and bowing of the studs that can cause corner cracks in the internal lining
• Trenching is only required if the scantling timber used is 'off sawn'. Sized or dressed scantling does not require trenching of the bottom or top plates

**DETAILED VIEW OF COMPONENT ASSEMBLY OF A TRADITIONAL TIMBER FRAME**

111

# 34 • TIMBER WALL FRAMING

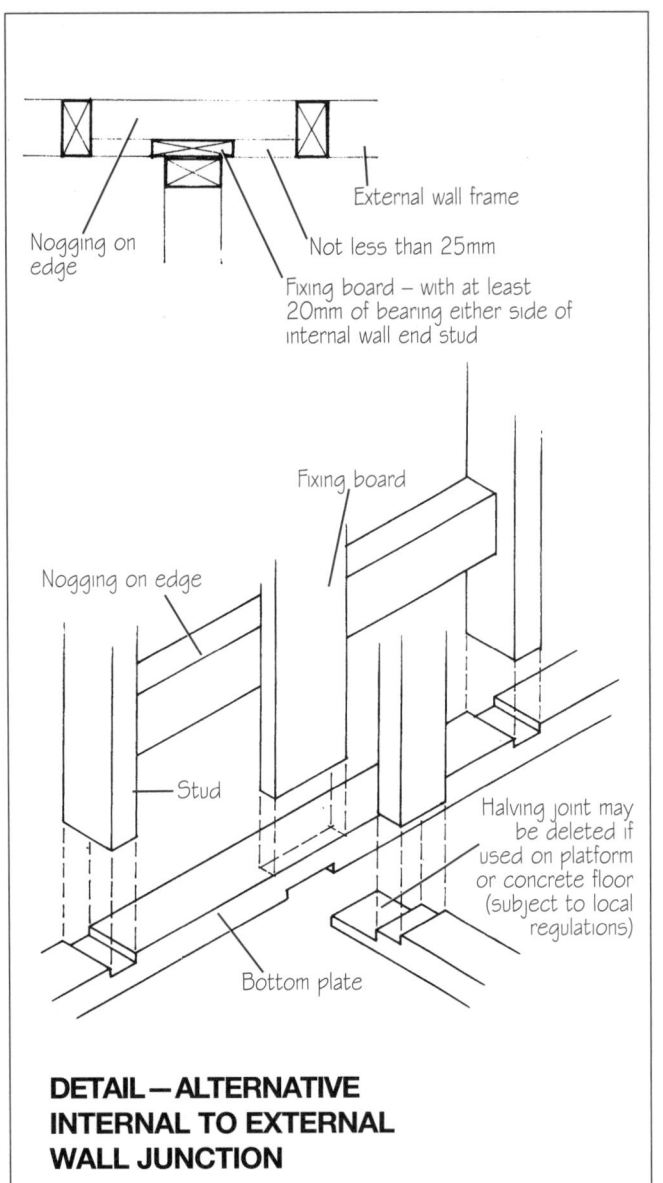

**DETAIL — ALTERNATIVE INTERNAL TO EXTERNAL WALL JUNCTION**

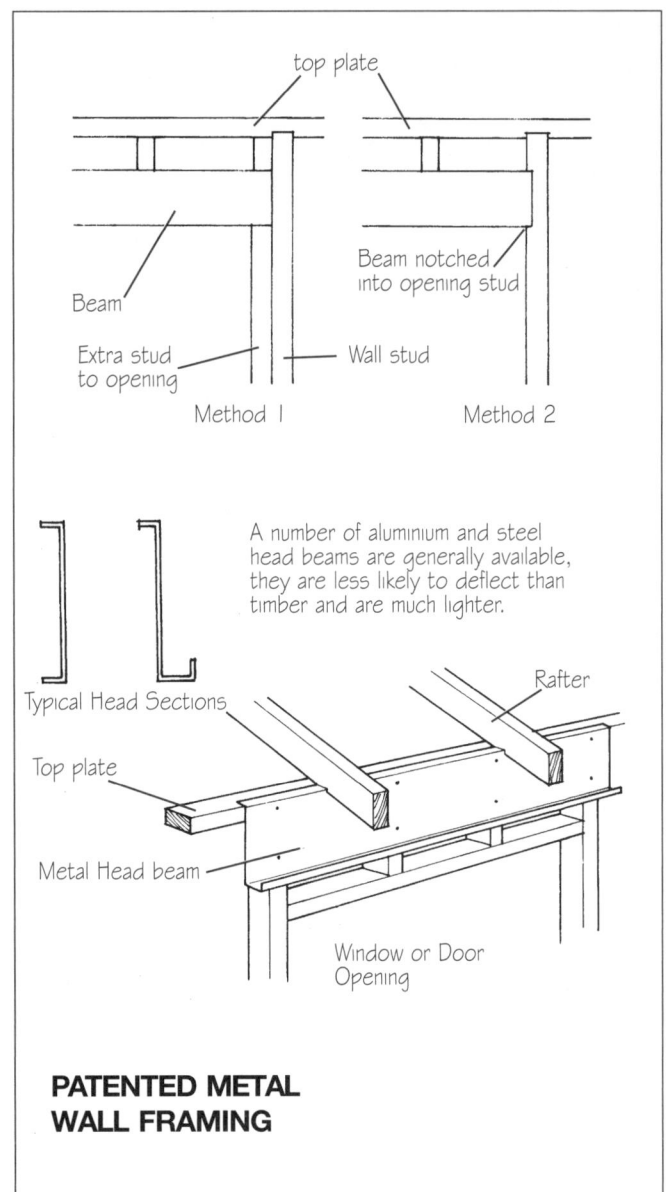

**PATENTED METAL WALL FRAMING**

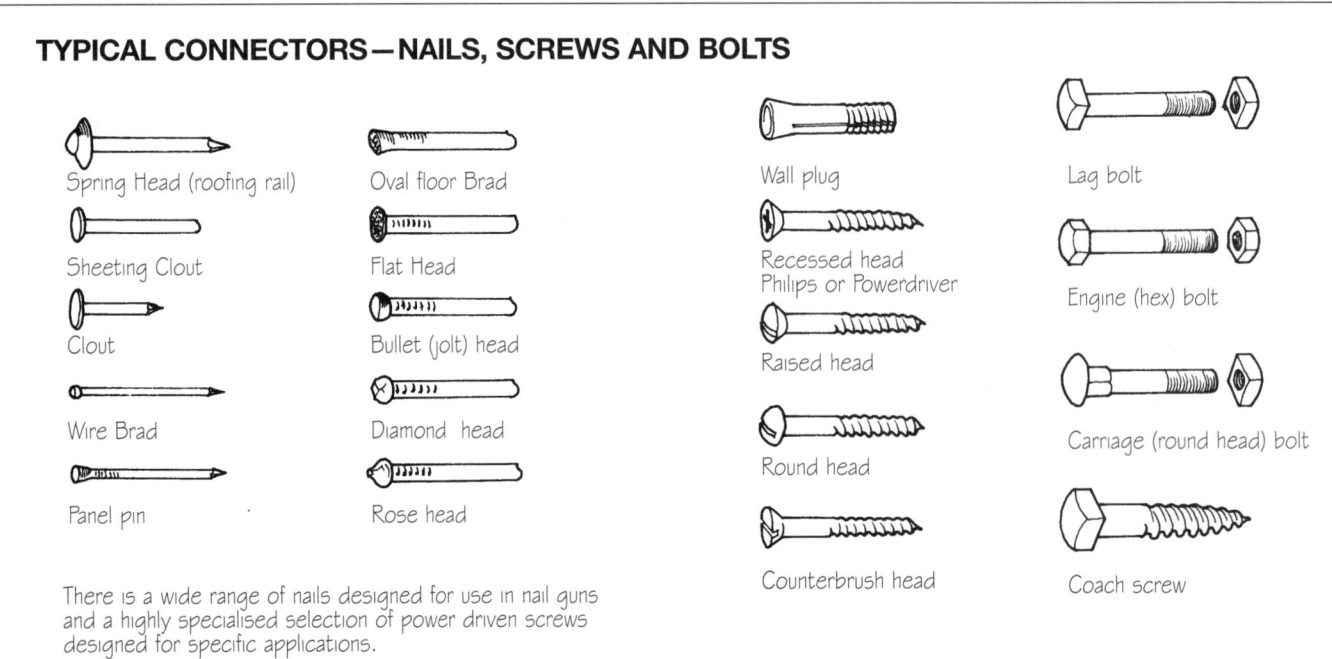

## TYPICAL CONNECTORS — NAILS, SCREWS AND BOLTS

There is a wide range of nails designed for use in nail guns and a highly specialised selection of power driven screws designed for specific applications.

# TIMBER WALL FRAMING • 34

## TIMBER CONNECTORS

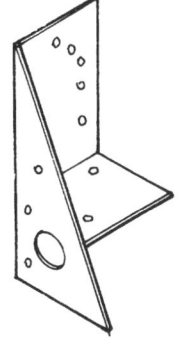
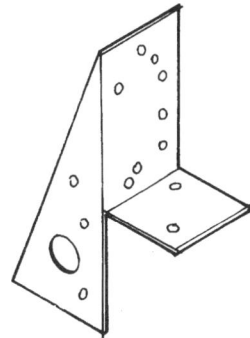

**Three-way connectors**
This connector is a general purpose connector for use in timber framing and allows connections between members to achieve high resistance to tensile loads, shear loads, and allows for stiffer corners.
Easy, simple and efficient but time consuming to drive the multiple nails.

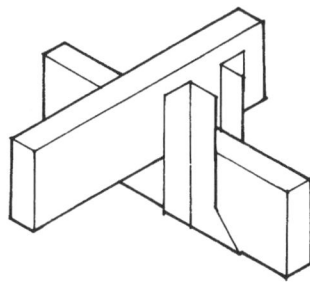

**Joist connector**
Allows easy and strong connections in upper floor joist construction, fixed by nailing.

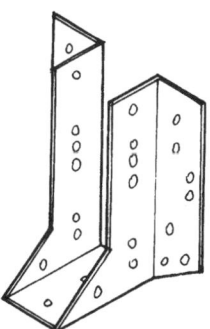

Cold formed Galvanised connector

Holes to suit engineered nails

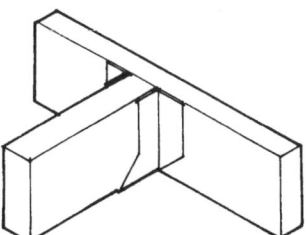

## TYPICAL APPLICATIONS OF THREE-WAY STEEL NAILED CONNECTORS

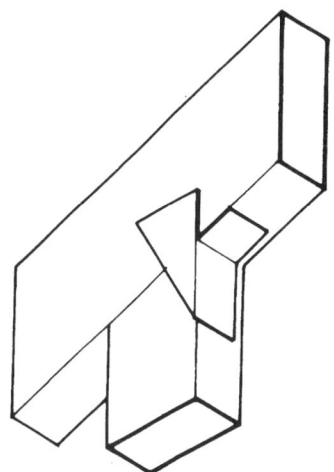

Rafter to stud

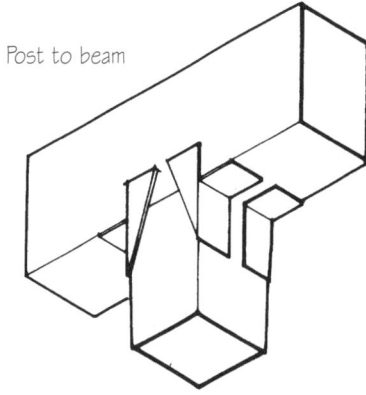

Post to beam

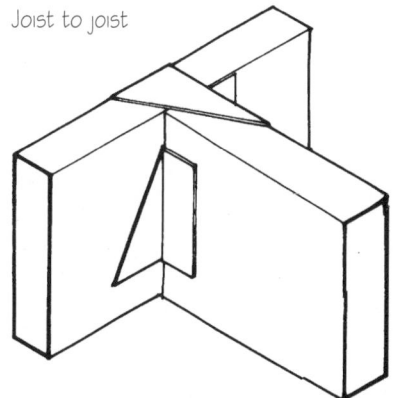

Joist to joist

# 34 • TIMBER WALL FRAMING

## Post and beam construction

The majority of discussion has dealt with the use of traditional framing methods. There is, however, an alternative to the methods that is generally known as the post and beam method.

Post and beam construction is based on a system of supporting posts and walls under a grid of primary and secondary beams that make up the roof system. There is nothing very complicated about this system, but great care must be taken to achieve frame stability through bracing or, in some instances, rigid joints.

OBs contemplating using post and beam construction are well advised to seek out a local architect who has had experience with this building method, as it is important that special care be taken in designing all the joints in the system so that minor problems are not built into the structure.

Post and beam construction, which on the surface may look simpler than traditional framing, requires much finer tolerances in assembly. Because of the large engineering-quality timber sections used, it is nearly always more expensive.

In pole frame construction it is generally considered that dual floor beams bolted with split ring shear connectors to notched joints at the poles is the most satisfactory building method.

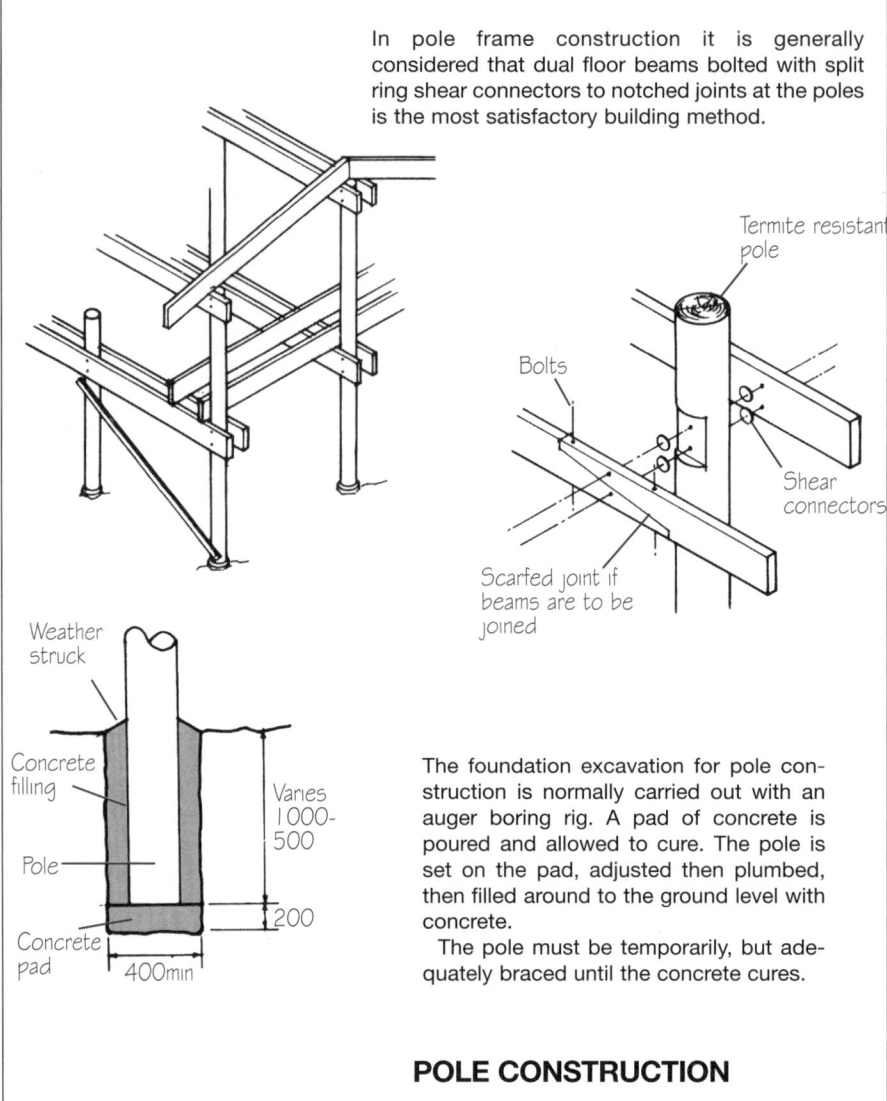

The foundation excavation for pole construction is normally carried out with an auger boring rig. A pad of concrete is poured and allowed to cure. The pole is set on the pad, adjusted then plumbed, then filled around to the ground level with concrete.

The pole must be temporarily, but adequately braced until the concrete cures.

**POLE CONSTRUCTION**

**OPEN WEB JOIST**

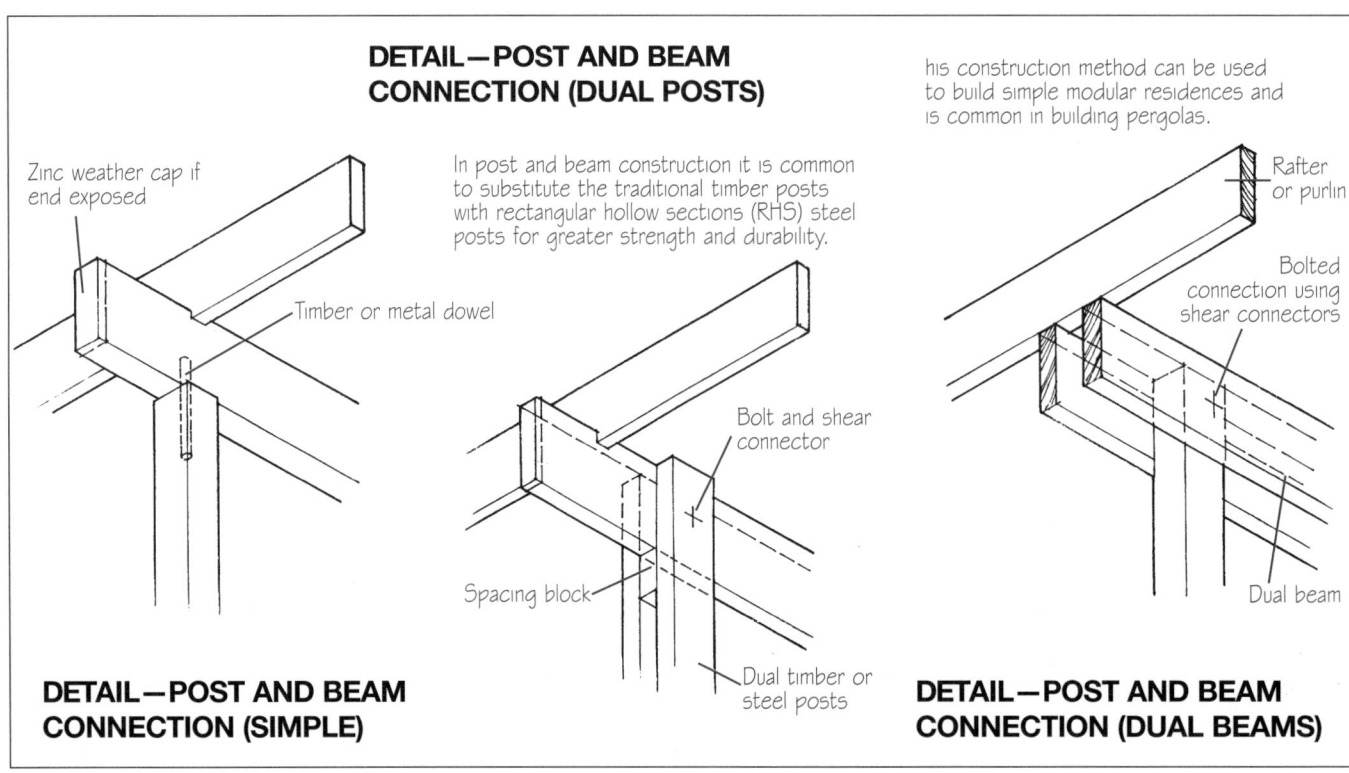

**DETAIL—POST AND BEAM CONNECTION (SIMPLE)**

**DETAIL—POST AND BEAM CONNECTION (DUAL POSTS)**

In post and beam construction it is common to substitute the traditional timber posts with rectangular hollow sections (RHS) steel posts for greater strength and durability.

This construction method can be used to build simple modular residences and is common in building pergolas.

**DETAIL—POST AND BEAM CONNECTION (DUAL BEAMS)**

# METAL WALL FRAMING • 35

Metal stud wall systems have become progressively popular—they are light, strong and termite resistant.

Modern metal wall framing systems generally use zinc-coated steel members fixed together with pop-rivets, power-driven screws, tabs and slots, welding, gluing or by mechanical interlinking.

The systems are often of the same wall thickness as equivalent timber walls, 70 or 90 mm but there are systems that use less conventional thicknesses.

The sections are mostly of plain channel section, cold-formed from galvanised steel. These sections are designed to fit inside each other, eg. the studs fit into the top and bottom plates—this makes the plates thicker than the studs by the thickness of two sheets of metal (about 2–3 mm). The nogging may then be thinner by a similar amount to fit inside the studs.

It is not generally difficult to purchase the steeling framing members and, with the right cutting and jointing equipment, steel wall framing is no harder than traditional timber framing to use on-site. Many OBs will, however, opt to have a fabricator supply pre-made frames to the building site. This is a reasonable way to go and eliminates the waste that can occur from having to order minimum length or bundles from a steel merchant.

When using prefabrication methods, the OB must request that the fabricator supply accurate workshop drawings for checking. This task has to be carried out with great care. Check:
- openings for windows and doors
- wall heights
- accumulated wall lengths
- window heights
- against-site setout dimensions
- alignments of lower to upper walls
- openings.

If the workshop drawings are approved by the OB then there should be little danger of incorrect fabrication and, if there is, and the workshop drawings show fabricator's errors, then getting a replacement or amended work from the fabricator should be simple. But if the workshop drawings are not checked or the check is not done carefully enough, then OB beware—it is difficult to blame the fabricator.

When the wall frames arrive on the site check them as soon as possible, while the truck is still on-site is the best time but if that is not possible then ASAP.

Check all the things that were checked on the workshop drawings, and look for dents, bends, kinks, broken joints, missed fixings, and so on. If there are major problems and they are clearly in contravention of the supply contract, attempt to reject the delivery and have the delivery truck remove any substandard material or fabrication. If the truck refuses to take a back load then at least delete the items in contention from the delivery docket before signing to accept delivery.

Note: the check-before-you-accept-delivery policy should apply to all items delivered to the site. OBs should get to know the acceptable tolerance benchmarks for all materials and fabrications, if the items are below the benchmarks then they should not be accepted, but take care not to be excessive in demands for unattainable quality. Also check there is not compensation in a load, eg. the fabricator may have included lengths of steel framing in the load to allow for replacement of any in-transit damage.

One of the reasons for using steel framing instead of traditional timber framing is that it does not shrink like timber. This is of course true, steel is much more dimensionally stable than timber and, unless it rusts, it remains the same strength for life. There is one action that can make it seem like the steel frame has in fact shrunk—this is when during fabrication the steel studs are not installed so that the will bear any superimposed loads directly from plate to plate. The stud is then bearing on the radius of the channel section of the plates and not hard against the horizontal part of the plate. If this stud is loaded beyond the capacity of the radius or any other fixing to resist the load, then the wall will be reduced in height just like it has shrunk. Always check load bearing wall sections to ensure they will support their design loads.

All fixing to steel wall frames will be by either adhesives or power-driver type screws. This is OK for plasterboard and similar products but does not work well when traditional nail-on linings are required. In these cases timber battens can be fixed to the studs (either vertically or horizontally), allowing the use of nail fixing. Take care, as this will often thicken the wall by as much as 45 mm and this may impact on other alignments.

Plumbers and electricians may complain that it is difficult to run pipes and cables through a steel wall frame system. Generally these difficulties are as much to do with lack of familiarity as any real problem. Do ensure that the electrician uses safety insulation grommets, where required, at places where electrical cable passes through holes in the steel framing. Plumbers should consider using flexible plastic piping in steel wall frames rather than the much less flexible copper alternative.

A consultation with the electrician and plumber before finally approving the workshop drawings for the steel frames may alleviate particularly difficulty cabling or piping problems.

# 35 • METAL WALL FRAMING

**COMMON SECTIONS IN COLD FORMED METAL**

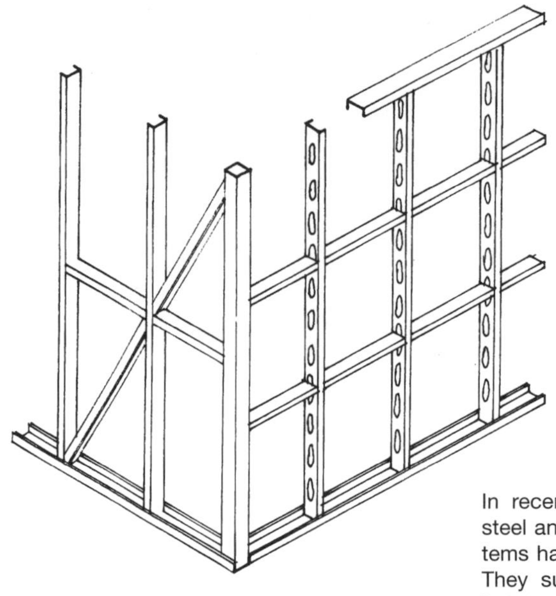

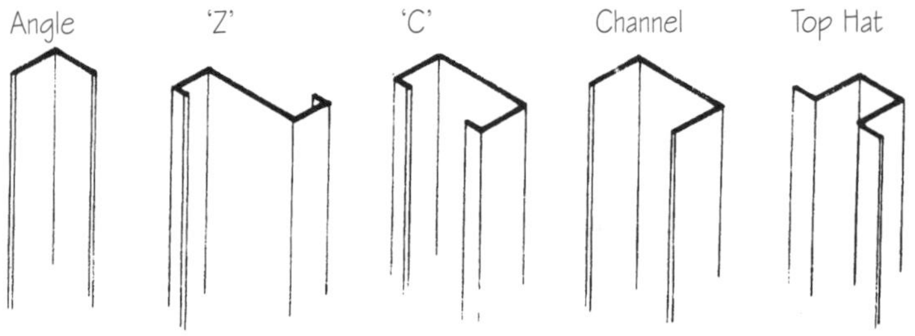

Angle   'Z'   'C'   Channel   Top Hat

In recent years a number of ingenious steel and aluminium cottage framing systems have been developed and released. They suit mass manufacture of houses but tend to be inflexible and unforgiving when used by owner builder. White ants cannot eat them!

Assembly is similar to timber but fixings are self tapping screws, pop rivets, welding or special lugs.

## TIMBER CONNECTORS

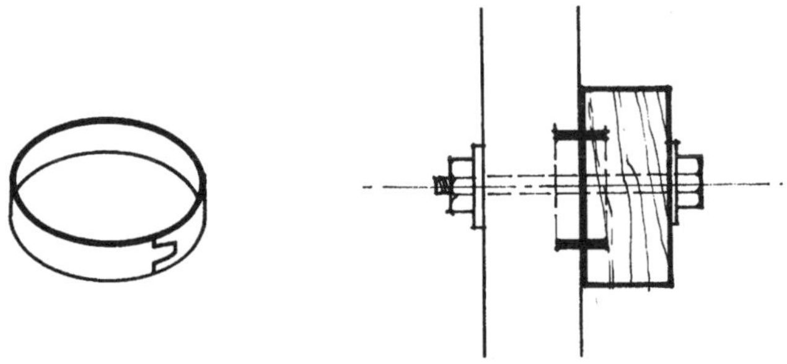

**Shear ring**
Timber to timber, used to increase effective bearing area of bolted connections, used in hardwood or softwood construction. Particularly heavy trusses and post and beam buildings.

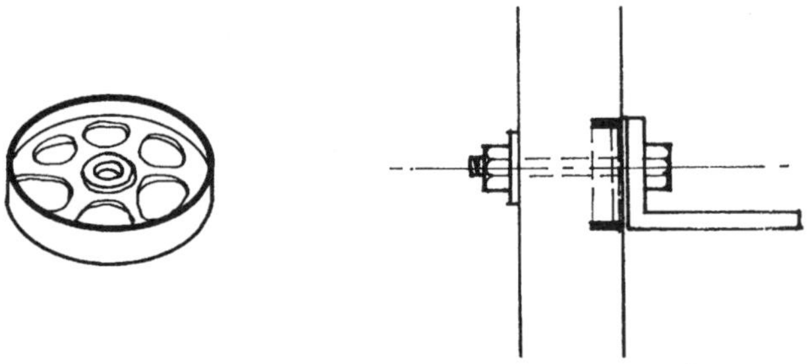

**Shear ring**
Metal to timber, used to increase the effective bearing area of timber structural components bolted to steel members.
Used in connecting steel gusset plates in heavy trusses and timber beams to steel posts.

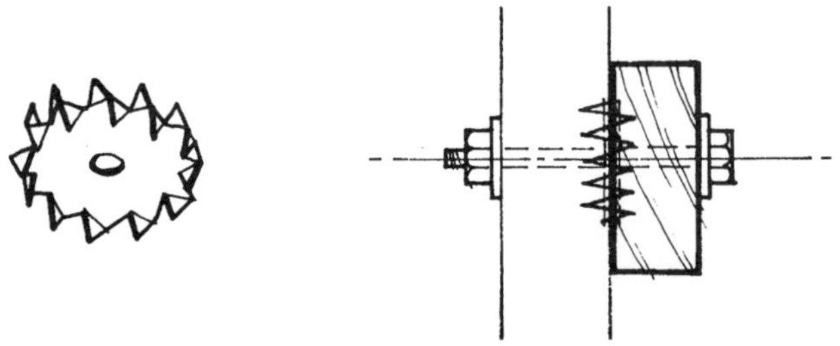

**Star connector**
Timber to timber, used when connecting softwood members to increase effective bearing and reduce rotation.
Used when constructing post and beam buildings particularly in Oregon.

# BRICK VENEER CONSTRUCTION • 36

Brick veneer construction is an Australian development. Although used in some other countries around the world it is at its most refined in the eastern states of Australia. It could be said Sydney and Melbourne produce the 'state of the art' in brick veneer construction and development.

In simple terms a brick veneer house is a traditional timber-framed house that has been clad with a brick skin in lieu of the board or sheeted cladding that would be found on this type of frame in the western United States.

There are good reasons why brick veneer houses have developed in Australia. Partly it is because of the merger of the English building methods of our pioneers with the imported Californian building methods of the mass-market house builder. It is interesting to note that in Texas a significant number of new houses have been built by Australian builders using the brick veneer tile roof model from Australia.

However, the most important reason for its development in Australia is that it produces low maintenance houses, and is a cheap and quick method of construction ideally suited to our subcontract method of trades and to the 600 square metre building block.

## The brick veneer building method

The method used to build a brick veneer skin on the timber frame of the house requires that a base brick wall or a concrete raft slab be constructed to extend beyond the perimeter of the concrete frame by the width of the brick skin. This distance is normally 110 mm, plus an allowance for a cavity, 30–50 mm, giving a total distance from the face of the outside of the timber frame to the outside of the brick skin of 140–160 mm.

This brick veneer skin is not a primary structural component of the house. That is, no superimposed structural load is carried on the brick veneer skin.

In single-storey construction the brick veneer skin is carried up to the underside of the eaves soffit lining, which in the case of a house with flat eaves, is approximately 2100 mm above the finished floor level of the house. If flat eaves are used in conjunction with windows set at a standard height of 2100 mm above the finished floor level, then no brickwork need be carried over the heads of openings, giving a significant saving in steel window lintels, arch bars or other beaming methods.

Not all house designs can make use of this saving and, where lintels are required over windows, these are commonly steel angles. It is advisable to have these galvanised to reduce rusting, which can cause unsightly staining of the building and eventually lead to the failure of the lintel.

In brick veneer construction, it is good practice to fix all external windows and doors into the timber frame before commencing the laying of the brick veneer skin. This allows the bricklayer to build the brickwork up to the sides of the door and window frames and reduce the chance of having too big or too small an opening for the windows and doors.

Flashings are necessary on the sides of all door and window frames, and a sill flashing is required under windows. Doors and windows that line up with the eaves soffit are not normally flashed but doors and windows with brickwork over them require a head flashing.

The bricklayer will bring the brickwork up to one or two courses under openings to allow for the later laying of brick or tile sills. The brick or tile sills should not be laid until the timber frame and the brick veneer skin have had time to settle; then they should be set so there is a gap of at least 15 mm between the underside of the window frame sill and the top of the brick or tile sill. This allows for further settlement of the building without the window frame coming into contact with the brick or tile sill, which in extreme cases can cause the bricks/tiles of the sill to be dislodged—which is very dangerous in windows high off the ground. If the sill bricks/tiles do not dislodge, the window frame may distort, causing sashes to jam or glass to break.

The brick veneer skin should be tied to the timber frame with galvanised steel veneer cavity ties. These restrain the brick skin from overturning, give added stiffness to the building frame and reduce the danger of uplift due to high winds. In Sydney, where the timber floor frame was not normally tied to the supporting piers, it is only the veneer cavity ties and gravity that hold the house down.

A standard metric brick is 230 mm long and 110 mm wide. Therefore, when the allowance is made for mortar between the brick it is 240 mm in effective length. Reasonable effort should be made to set out the dimensions of brick walls and openings to suit brick sizes. This is more critical the shorter the length of wall or the narrower the opening. Longer walls and wider openings can be adjusted by the bricklayers who can widen or reduce the vertical mortar joint, called a perpend, to suit full bricks.

For walls the formula for standard metric brickwork length is:

240 mm + (120 mm if a half brick is required) x number of bricks − 10

For openings the formula for standard metric brickwork openings is:

240 mm + (120 mm if a half brick is required) x number of bricks + 10

These dimensions can be obtained from brick industry publications; it also helps to know that for every 600 mm length of wall there are two and a half bricks and for every 600 mm of height there are seven bricks.

In some locations there are also metric modular bricks on the market. These are normally 290 mm long, which makes them 300 mm long with the perpend joint. Note, with these bricks, the joints are staggered 1/3 to 2/3, not 1/2 to 1/2 as in standard metric brickwork. This requires special care in working out dimensions, as this type of brick does not allow the flexibility of the standard brick.

- Standard metric brick—230 x 110 x 76 mm
- Metric modular brick—290 x 90 x 90 mm
- Metric modular (Norman) brick—290 x 90 x 65 mm.
- American brick—190 x 90 x 65 mm.

Concrete blocks are also available in a wide range of sizes and are fully modularised:
- Lengths—90, 190, 290, 390 mm
- Heights—40, 90, 190 mm
- Widths—90, 140, 190, 240, 290, 390 mm.

Other sizes may be made in certain locations, or as special orders. For concrete block veneer, generally only the 90 mm thick blocks would be used, with a standard block length of 390 mm, and a height of 40, 90 and occasionally 190 mm. The 190 mm blocks are often referred to as full heights and 90 mm blocks as half heights.

Natural stone is seldom used as a veneer skin over a timber frame but, when it is the stone used is most likely limestone or sandstone, in sawn blocks approximately 600 to 900 mm long, 200 to 300 mm high and 150 mm thick.

Bricks are made by pressing or extruding clay, or a clay/shale mixture. Pressed bricks are the ones with the hollow in the top (or is it the bottom?) called the frog. These are made in a brick-sized mould. Extruded bricks are the one many people call wirecut, because as the clay is extruded it is cut to size using a wire-cutter. Extruded bricks are those that have holes in them.

Pressed or extruded bricks are satisfactory for use in brick veneer walls—it is a matter of personal choice. Clay and clay/shale bricks are fired in a kiln to vitrify them and give them outstanding lasting qualities.

Bricks and blocks made of concrete are moulded, then carefully cured to give accurately dimensioned units. There are also bricks made from a combination of silica (sand) and lime, which combine chemically to produce sparkling white bricks.

All masonry units in brick veneer walls are set in mortar beds with similarly filled perpend joints. Mortars are a wet mixture of sand and a binder, either cement or lime or a cement–lime combination. The mixture recipe varies to suit the units to be bonded, but the general rule is for the set mortar to be of a strength slightly weaker than the masonry units being bonded. This means any forces acting on the wall that exceed the wall's ability to absorb the forces will cause failure in the joints, where simple repointing can be carried out, rather than cracking the masonry units, which are unrepairable.

Mortars with a high percentage of cement in the mix are much stronger (in controlled proportions) than mortars using a lime-only binder. Cement/lime compositions (compo mortar) have a strength factor in the middle. Lime makes the mortar mix easier to work with, keeping the mortar more plastic than cement. If lime is not added to a mortar a chemical wetting agent plasticiser may be considered. Many of these products are of very high quality, but must be used strictly to manufacturer instructions.

There is a saying in the building industry that could be applicable to 49 per cent of OBs—'...when all else fails, read the instructions...'

# 36 • BRICK VENEER CONSTRUCTION

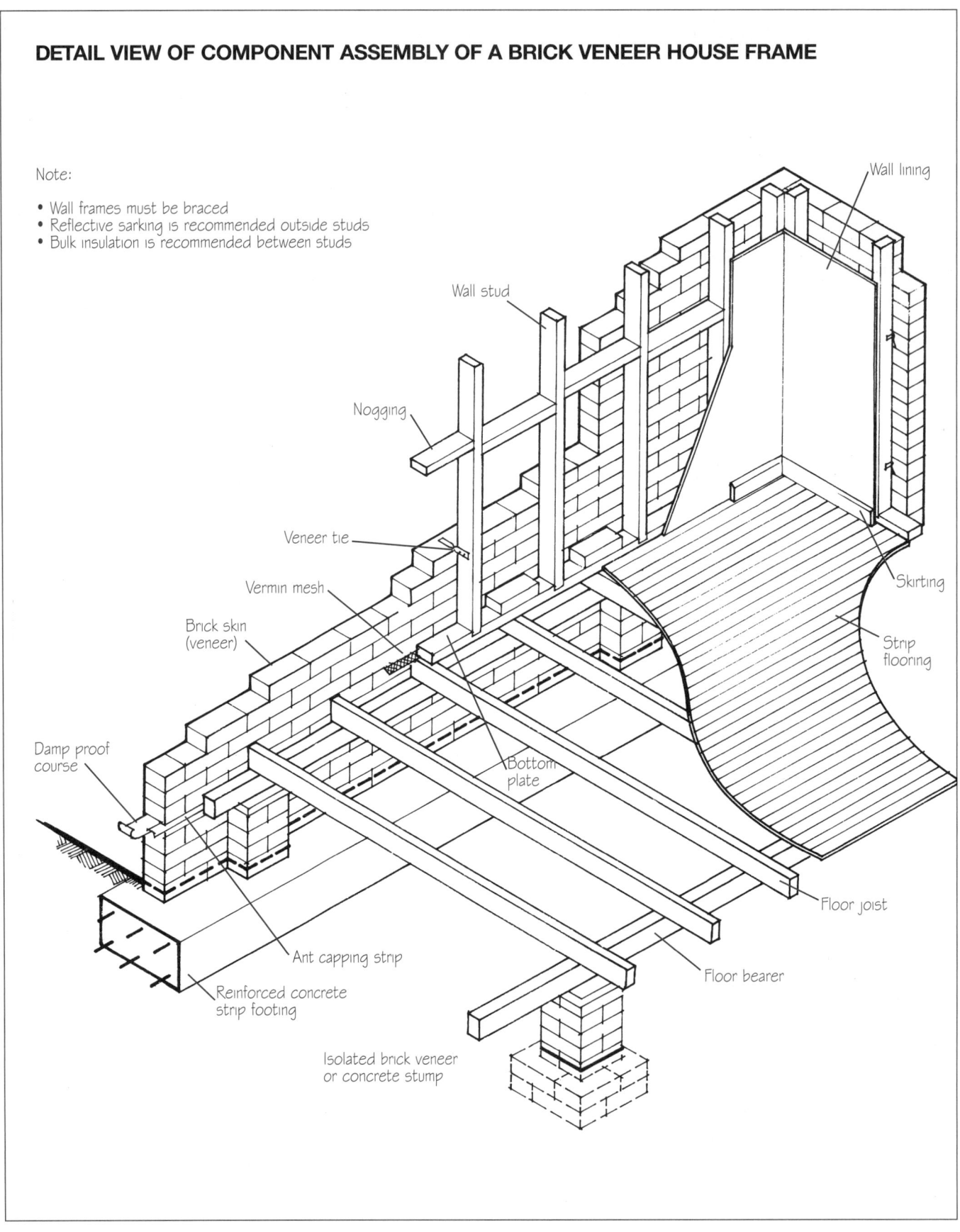

**DETAIL VIEW OF COMPONENT ASSEMBLY OF A BRICK VENEER HOUSE FRAME**

Note:
- Wall frames must be braced
- Reflective sarking is recommended outside studs
- Bulk insulation is recommended between studs

## DETAIL—BRICK VENEER: BASE, WALL AND FLOOR

Galvanised steel veneer ties are built into the mortar courses of the brickwork and nailed to the studs
Vermin barrier of galvanised steel wire mesh is fastened to bottom plate and built into mortar course, therefore bridging the cavity and stopping vermin entry
Zincalume steel or aluminium alloy ant strip to the inside of base brickwork to prevent termites from reaching timber frame
Bitumenous aluminium sheet or embossed plastic sheet damp proof course topre-vent rising damp. Chemical admix com-pounds to mortar may be permitted in some localities.

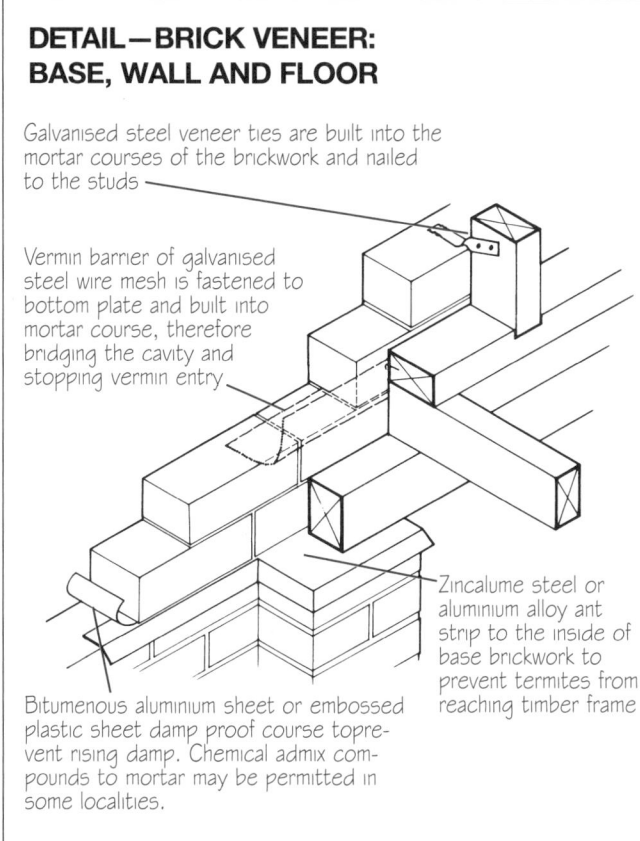

## SHEET BRACING

Top plate
plywood or tempered sheet bracing instead of angle bracing
corner blocking
floor joists
bearer

blocking

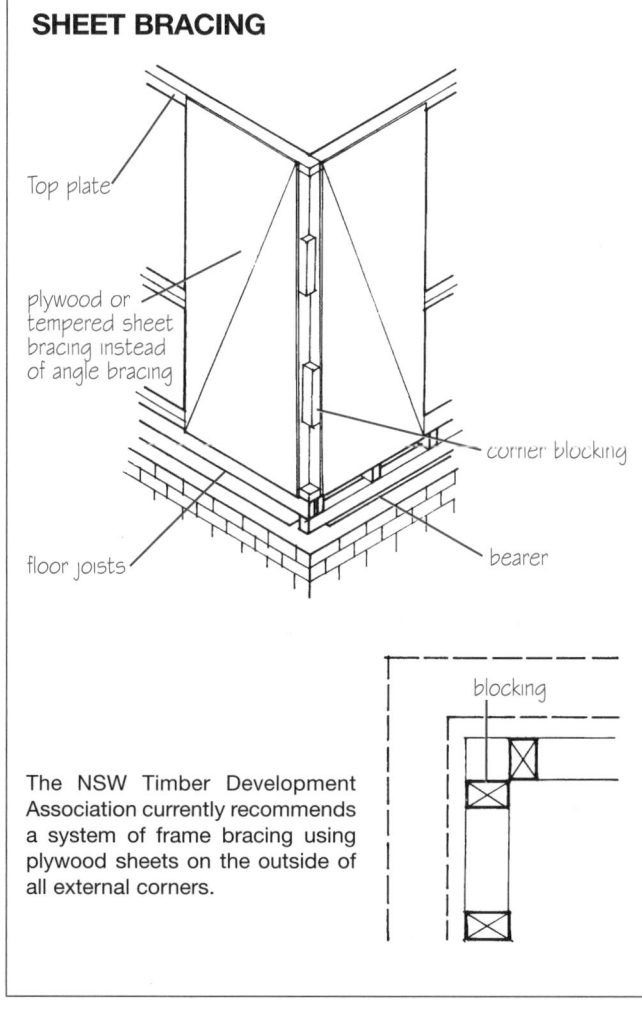

The NSW Timber Development Association currently recommends a system of frame bracing using plywood sheets on the outside of all external corners.

Rafter
Ceiling joist
Top plate
Soffit hanger
Fascia
Ceiling lining
Wall lining
Window head
15 minimum gap between head of window frame and underside of window head beam (to allow for deflection)
Soffit bearer
Soffit lining
Soffit trim. min 10mm gap between top of brick skin and soffit to allow for brick growth
Window sill lining
Architrave
Window sill plate
Skirting block
Skirting
Aluminium window frame
Packing
Brick on edge sill
Window sill flashing
Bottom plate
Vermin wire barrier
Flooring
Joists
Bearer
Ant cap/strip
Brick veneer skin
Subfloor vent
Engaged pier
Base walling
Ground level
Reinforced concrete strip footing

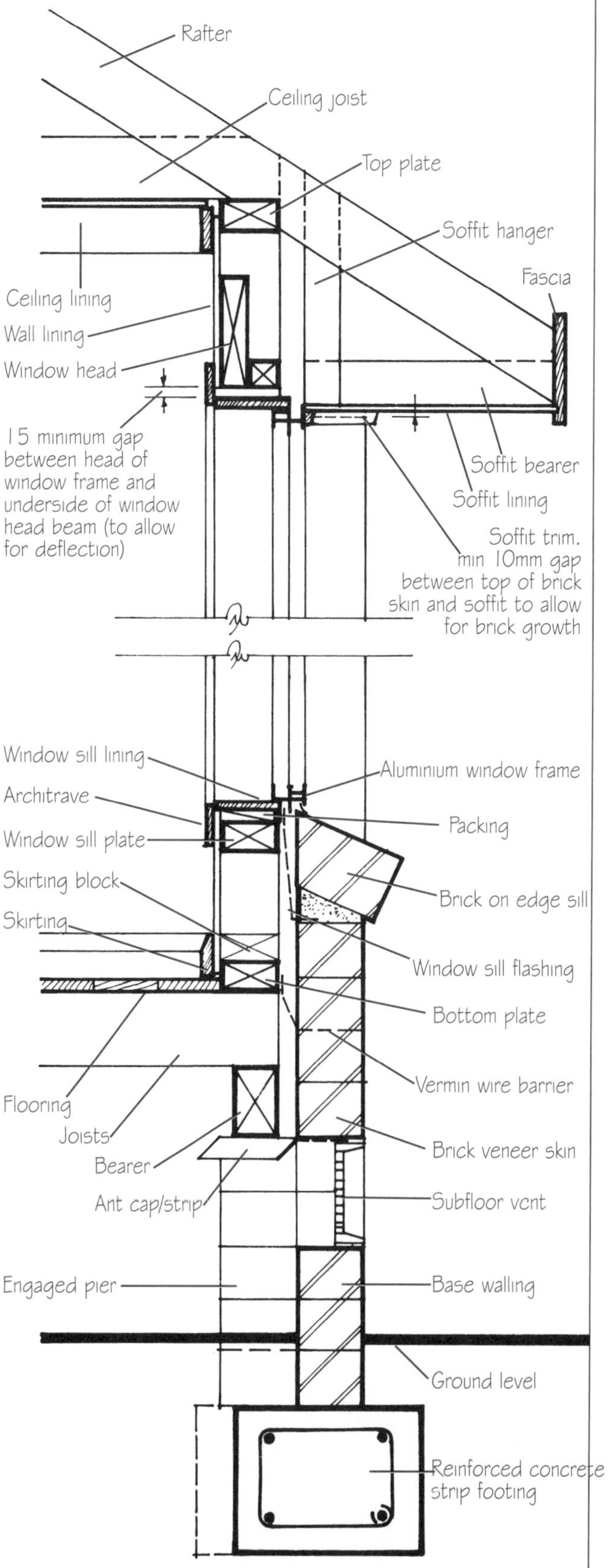

## DETAIL SECTION—THROUGH TRADITIONAL BRICK VENEER WALL

# 37 • AUTOCLAVED AERATED CONCRETE (AAC)

To some OBs, autoclaved aerated concrete (AAC) is the stuff that dreams are made of: a masonry product that does not require mortar, can be cut with a handsaw and is so light it will float on water. AAC blocks can replace other masonry products in most applications.

The blocks are very accurately dimensioned, which makes them easy to lay—if they start from a horizontal bed. With most masonry products—bricks, concrete blocks and stone—it is possible to adjust the mortar courses to correct irregularities in the starting bed. This is not possible with AAC blocks because they use a thin bed adhesive between courses in all except the starting bed.

If a house is going to have a series of interconnecting bonded walls meeting at one floor or roof springing level, then the top of all AAC blocks of the first course should be dead level to the same reduced level (RL). This is achieved either by correcting the levels in a thick adjustable bed of mortar under the first course or by laying all the blocks off a perfectly level concrete slab. Once the first course is level, OBs should be able to lay the blocks without excessive problems if they have access to the correct tools and follow the manufacturer's illustrated manuals.

Wall thicknesses will vary in relationship to the loads and stresses, and the manufacturer's recommendations should be carefully analysed. Bonding is essential in all masonry construction, including AAC, and all openings and wall intersections must be laid with care to avoid straight joints that may move and cause unsightly cracks.

ACC blocks are water-resistant to a significant degree but not waterproof. They can be used as single-skin external walls, which is often more convenient for OBs than the problems associated with cavity construction methods. The exterior has to be kept dry by using wide eaves or verandas and waterproof membrane coatings.

In general, AAC is not suitable for cement render because the blocks have very low water absorbency and the render will not key to the surface. External coatings will generally be synthetic compounds, eg. acrylic, with colour and texture additives.

Internal facings are possible using skim coat plaster finishes but many builders now favour the use of plasterboard sheets fixed with adhesive directly to the face of the AAC walls.

Fixings into AAC require careful evaluation, as the material is not suited to most expansion fasteners. The manufacturers have a wide range of approved fasteners for specific applications and their recommendations must be followed. This is particularly necessary when attaching a roof, and uplift considerations may be critical when dealing with low-mass walls in high-wind areas.

ACC has excellent noise and thermal insulating properties and it is effectively fireproof. It has, however, limited thermal mass and will not retain heat. This may limit its application in some climatic conditions and in designs on solar principles.

## BUILDING BLOCKS

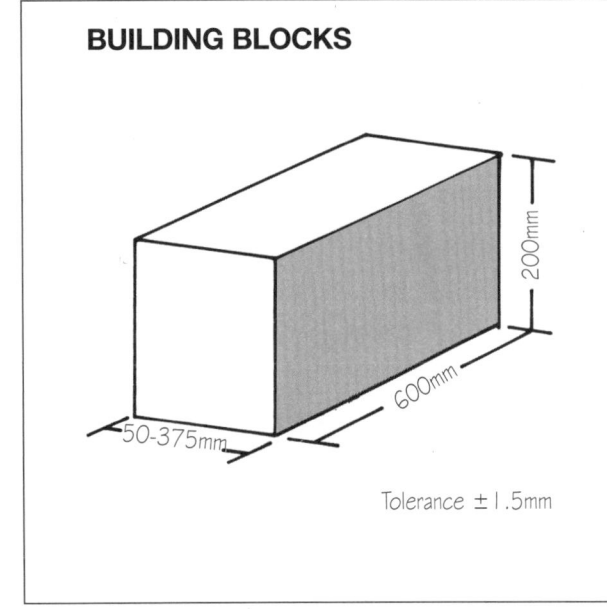

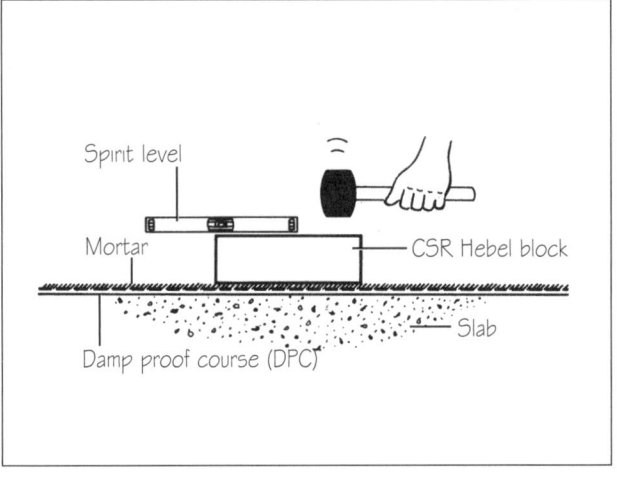

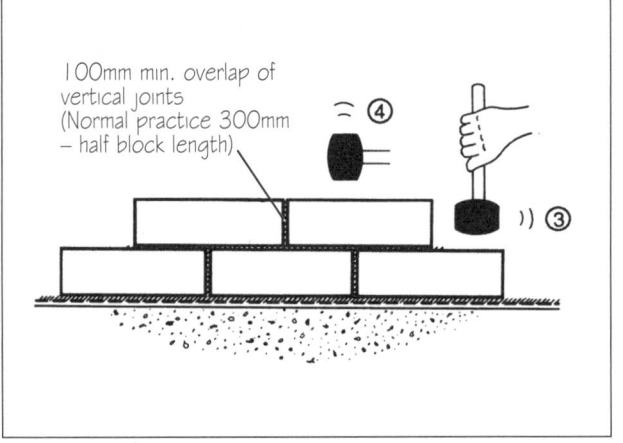

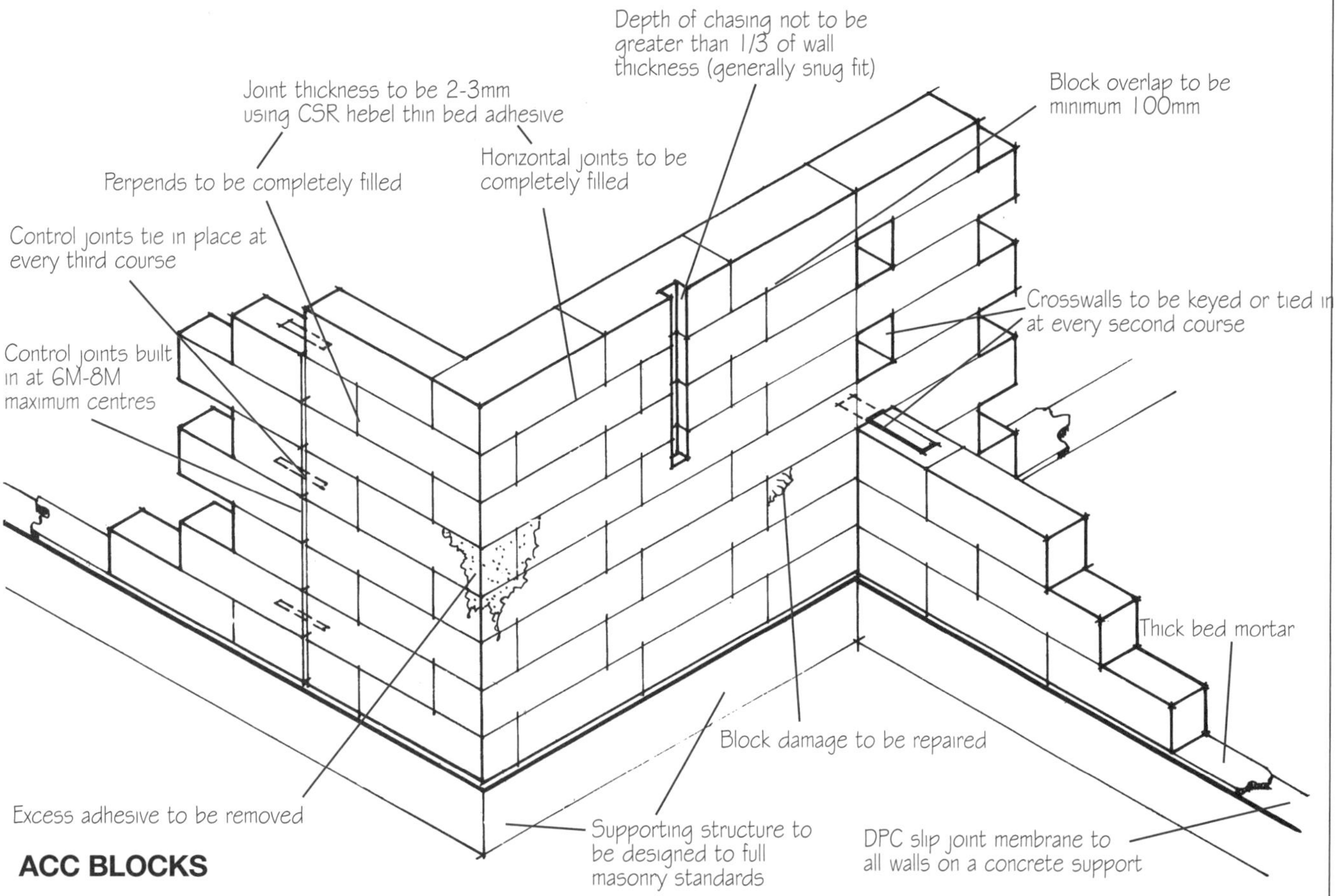

# AUTOCLAVED AERATED CONCRETE (AAC) • 37

## ACC BLOCKS

- Joint thickness to be 2-3mm using CSR hebel thin bed adhesive
- Perpends to be completely filled
- Control joints tie in place at every third course
- Control joints built in at 6M-8M maximum centres
- Excess adhesive to be removed
- Depth of chasing not to be greater than 1/3 of wall thickness (generally snug fit)
- Horizontal joints to be completely filled
- Block overlap to be minimum 100mm
- Crosswalls to be keyed or tied in at every second course
- Thick bed mortar
- Block damage to be repaired
- Supporting structure to be designed to full masonry standards
- DPC slip joint membrane to all walls on a concrete support

## ACC PANELS

- Tophat furring channel max. 150 from top of panel
- 600 wide panels
- line of corner panel beyond
- lines of 'tophat' section behind panels
- opening
- AAC power panels on end
- panels butt joined & glued with hebel thin bed adhesive
- top hat furring channel min. 100 above concrete door
- ground line
- edge of concrete slab
- Tophat section
- Timber stud or steel stud frame
- Typical panels 2400 or 2700 high
- Tophat section
- Edge of concrete slab
- Ground line

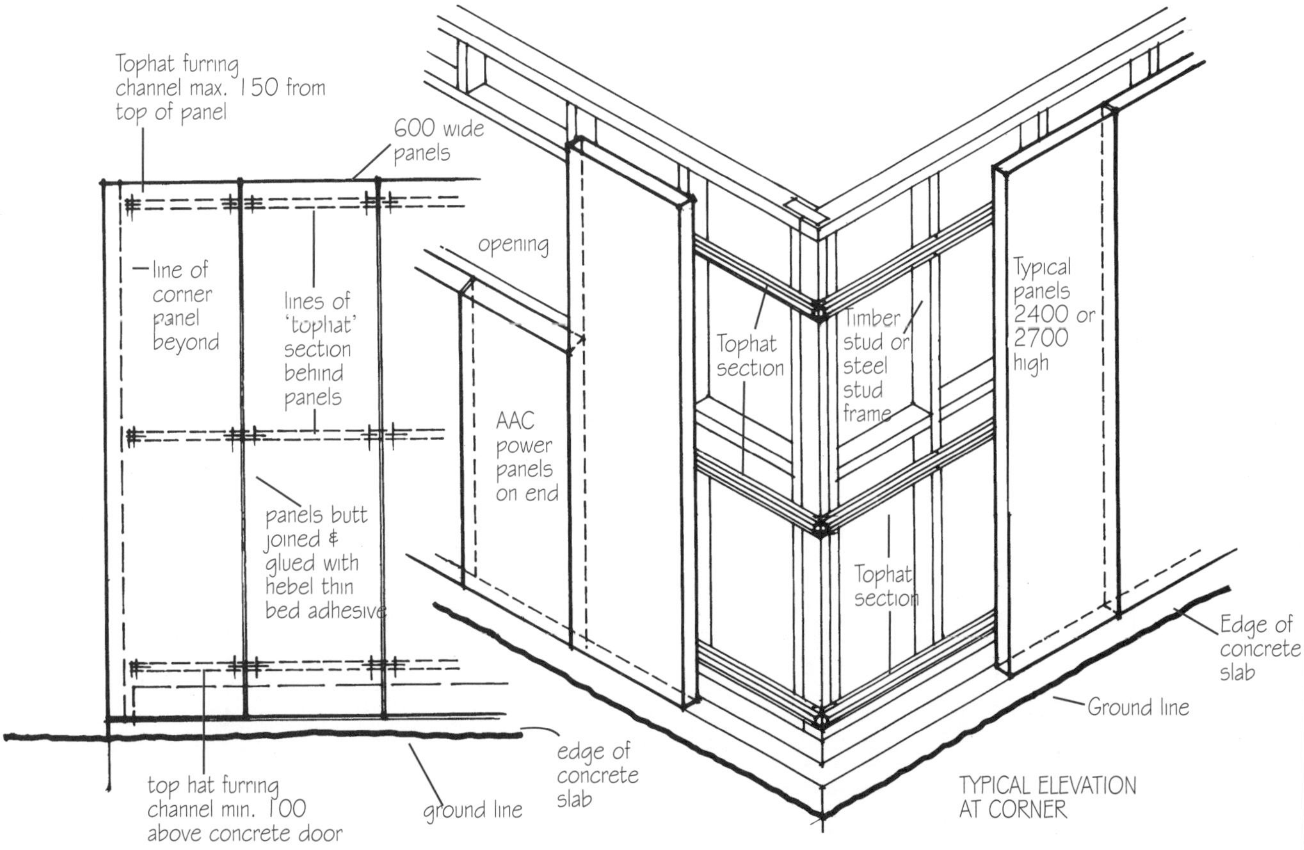

TYPICAL ELEVATION AT CORNER

# 38 • POWERPANEL®

## Powerpanel®

This product has been developed from the European technology of ultra-lightweight AAC blocks and panels. It is manufactured in Australia by the CSR company

AAC is used as a structural load-bearing material in most European applications. This Australian adaptation—Powerpanel®—is designed specifically to replace the brick skin in traditional Australian brick veneer homes. It is a cladding material, nominally 75 mm thick, that is attached to the outside of a timber or metal stud-framed wall to provide a weatherproof enclosure but not to carry any superimposed building loads.

A basic Powerpanel is 600 mm wide and 75 mm thick and comes in various heights (2400 mm or 2700 mm are the most common). It stands on its end outside a conventional stud wall frame with the weight of the panel supported on the edge of the floor slab by a subfloor brick wall or on a galvanised steel angle fixed to the base structure.

Folded steel top-hat sections are fixed horizontally to the external face of the studwork. The Powerpanels are then stood vertically against these top-hat sections and fixed through the section into the internal face of the panel.

The panels are 5–10 mm apart; these joints are fitted with packing rods and then filled to the manufacturer's recommendations to achieve a Vee groove or flush joint. Panels are cut on-site to provide under- and over-window panels.

The panels are coated with sealers, textures and skins to provide various colours and textures but specifically to provide the panels with weather protection. The building industry is always looking for a material that costs no more than a clay brick skin to replace bricks in brick veneer homes. Powerpanel will do this as long as it is coated with an economic waterproof coating. However, if a high quality coating is used, Powerpanel may cost more than conventional brick veneer.

Powerpanel may succeed with project home builders, as it is at least as flexible as bricks and can be used with existing house designs with minimum changes to the drawings. OBs should be able to use modestly skilled people and the Powerpanels provide a reasonably sound and thermal barrier skin.

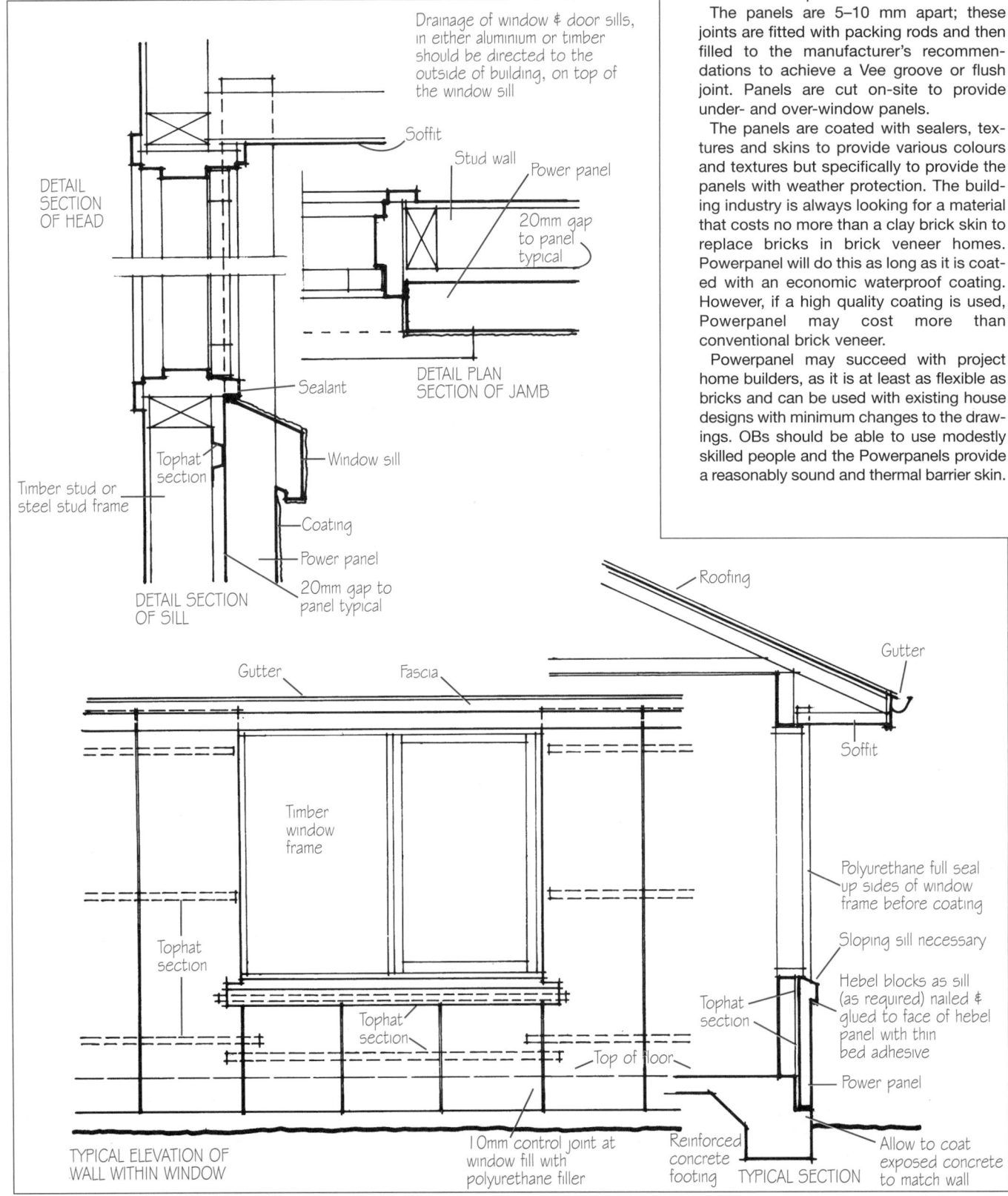

# SHEET AND BOARD CLADDINGS • 39

Traditionally timber frames were clad with weatherboards, that is, natural timber boards wrought to a section, generally tapered, that could be fixed in a weathertight lapped method. Weatherboards of this type are still used in some houses, but rising costs and diminishing resources of suitable timbers have caused the traditional weatherboard to be displaced by manufactured reconstituted timber and composite material sheets and boards.

There are many people who now believe that it is easier to insulate a timber framed board or sheet clad wall than it is to insulate a brick veneer wall. In fact, some modern passive energy house designs suggest that a house with an external insulated and sheet clad skin and an internal skin of masonry may give the best compromise on thermally efficient planning.

Other Queensland and the Northern Territory, no Australian State has used board or sheet clad houses to any great extent since WW2. The myth of brick is better—nay, best—has developed, along with the notion that boards and sheets must be painted every few years and maintenance is the curse of the homeowner.

Low maintenance boards and sheets are available including natural cedar through to factory finished boards of aluminium. Paints and stains have also improved markedly in the last 20 years. Painting is now much easier than ever before and the paint life is significantly longer—10 year paint frequencies are now not only possible but common.

## The product range:

### Fibrocement (Fibro)

Fibro, as this product was dubbed by the Australian community, started out as a grey-coloured, brittle, thin asbestos fibre reinforced cement sheets. It was the product that, after WWII allowed many State Housing Commissions to build cheap mass housing. Its association with that type of housing gave the product a bad name. Then there was the asbestos scare; asbestos fibres were linked with some deadly cancers.

Today fibrocement has changed. The asbestos has been removed from the wide range of products traditionally reinforced with this material and the replacement fibrocement products are believed to pose much lower health risks (a risk that was greatest if the asbestos cement sheets were cut and the asbestos fibres were allowed to become airborne), and are produced in a range of products that have gained wider acceptability than in the past, particularly the wide board planking ranges.

Fibrocement products are sold as a complete system, backed by a wide range of accessories and detailed instruction pamphlets.

### Hardboard

A product with a similar chequered past to Fibro, Masonite was considered an indoor only product—and only then when there was no better alternative.

Modern hardboards are well-developed quality products that include a range of external plankings. These compete with fibrocement planking for a share of the cladding market.

### Natural timber

Natural timber weatherboards are now considered by many people, including an increasing number of architects, to be a premium cladding product. There is a certain amount of pride in saying, 'I live in a natural Western Red Cedar home'.

Many timbers can be used to make weatherboards, from native species hardwoods through local pines to the imported Western Red Cedar and other more exotic species, including Palm Wood.

Timber weatherboards can be used in the traditional horizontal planked method, in a vertical mode or on any other diagonal angle. They can be painted, stained or left natural (certain timber species only).

There are also natural timber shingles and shakes that can be used on walls and are particularly suited to walls that are sloped, as may be found on a mansard roof design.

OBs should take some care when using weatherboards, as some are still manufactured from scarce rainforest and diminishing native timbers. Where possible use only weatherboards sourced from plantation or environmentally managed forests.

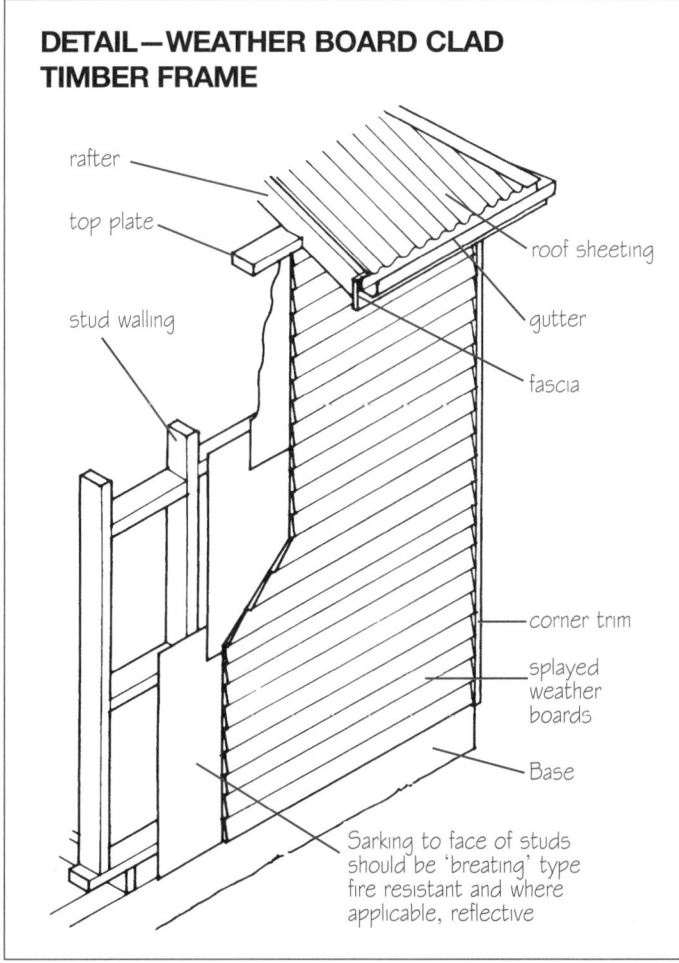

**DETAIL—WEATHER BOARD CLAD TIMBER FRAME**

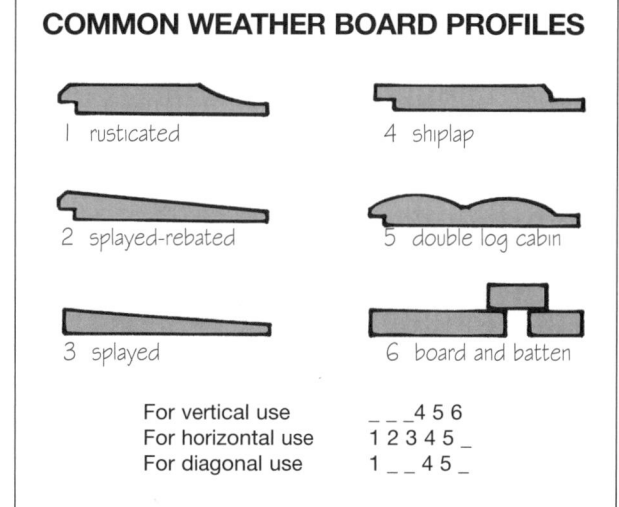

**COMMON WEATHER BOARD PROFILES**

1 rusticated
2 splayed-rebated
3 splayed
4 shiplap
5 double log cabin
6 board and batten

For vertical use _ _ _ 4 5 6
For horizontal use 1 2 3 4 5 _
For diagonal use 1 _ _ 4 5 _

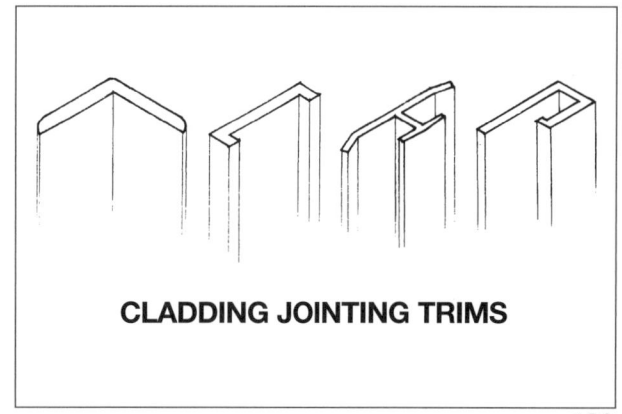

**CLADDING JOINTING TRIMS**

# 39 • SHEET AND BOARD CLADDINGS

**Aluminium**

Aluminium is used to manufacture a range of profiled sheets and boards that generally mimic weatherboards, with wood-grain texture on the board surface (fibrocement and hardboard planks also offer wood graining; only natural timber comes free of a deep embossed 'natural timber' look).

Aluminium is a very durable material, but the aluminium used in some aluminium claddings is of a light gauge and therefore likely to be damaged by stones thrown up by lawn mowers and the like.

The coloured finish is applied under factory-controlled conditions and this gives the finish a very long life, but it is not infinite. At some point in the future an aluminium-clad house is going to need repainting; the techniques that are needed to prepare and apply the new finish are beyond the abilities of most OBs.

**Plastic**

Plastic-moulded, wood-grained, self-coloured weatherboards are used extensively in the United States. Some of the products are marketed in Australia but few have had extensive solar testing in the harsh Australian sun. As sun, or more particularly ultraviolet light, is a destroyer of many plastic compounds, take care if you choose a plastic cladding material that you are sure its life expectancy is sufficient under Australian conditions to justify its use.

## Fixing methods

Most board or sheet products are fixed to the timber frame by the use of nails; non-corrosive nails are generally the most acceptable. Be careful that copper nails are not used with Western Red Cedar, for the natural esters in this timber attack the copper.

Modern power-driven, self-drilling screws can be used and are recommended for use with fibrocement products, as they cut down the danger of cracking, which is greater if driven nails are used.

Turning corners, whether internal or external, is the area that needs the most care when fixing weatherboards or planks. In natural timber, either a stop mould running vertically up the corner can be used, or the boards can be mitred. The latter is a skilled job and care must be taken to seal the end grain of the boards carefully, or there is always a risk that water will get into the joint and into the timber, causing fungal rotting.

Almost all cladding systems have specially developed corner mouldings that simplify the process significantly.

Some of the profiled horizontal sheeting products in fibrocement—and to some extent in aluminium—will require vertical joints along long wall runs. Take care if the products are not used sympathetically, as these joints can look terrible.

Building paper is normally fixed to the outside of the timber frame before the sheets or boards are fixed; this is considered an integral part of most modern cladding systems. The building paper can be reflective to improve the thermal efficiency of the house, or it can be brown kraft paper. It should always be fire-resistant and it is recommended that the paper be perforated to allow air circulation and to reduce the tendency for condensation between the paper and the cladding.

Flashings for all windows and doors are important and the choice of the moulds used to finish cladding to opening junctions can have a major bearing on the final appearance of the house.

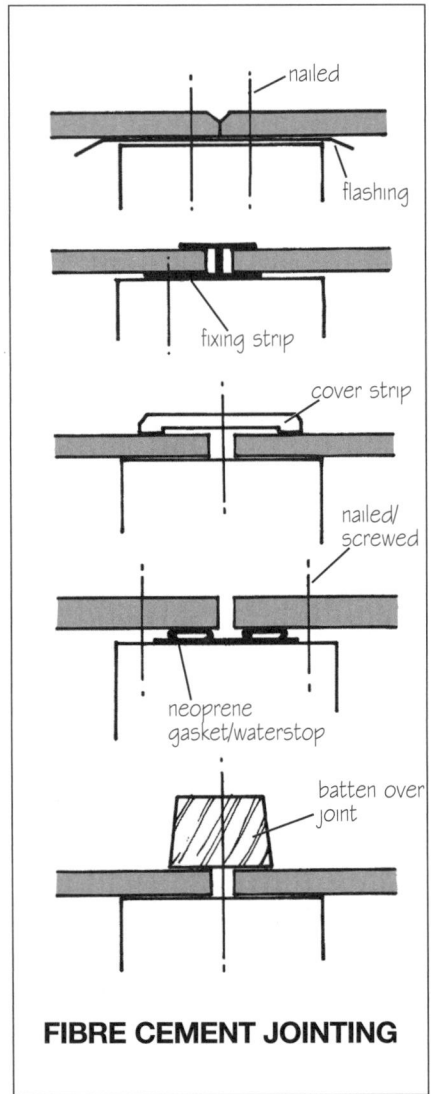

**FIBRE CEMENT JOINTING**

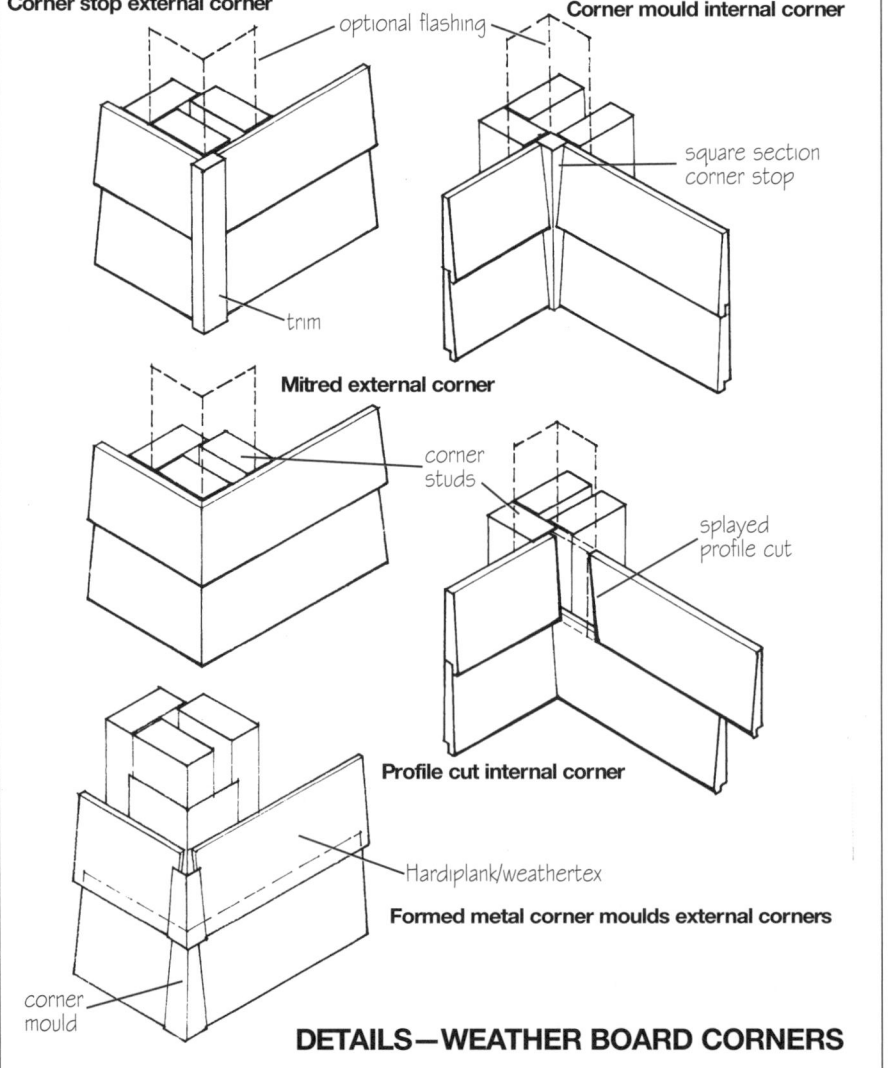

**DETAILS—WEATHER BOARD CORNERS**

# SHEET AND BOARD CLADDINGS • 39

## COMMON SHEET CLADDING SYSTEMS

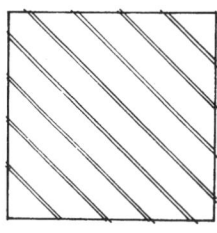

### Diagonal boarding
Diagonal boarding is generally built using rebated softwood weather boards, other profiles and materials can be used. Wall sarking is recommended and 450mm maximum stud spacing.

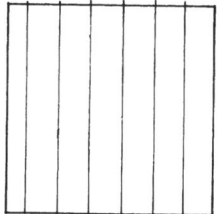

### Vertical boarding
Vertical boarding is generally built using rebated boards or boards and battens. Wall sarking is recommended and multiple rows of noggings are essential (keep nail holes in straight lines).

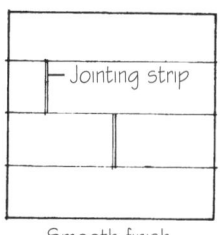

Smooth finish — Textured finish

### Composite weather boards
Composite weather boards are those manufactured from wood fibre (hardboard) or cement and fibre cement. These boards are normally used horizontally. Boards are available in smooth or textured surfaces.

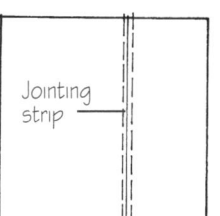

### Composite flat sheets
Composite flat sheets are generally manufactured from cement and fibre cement. They can be used as finished cladding or they can be covered with expanded metal lathe and cement rendered.

Horizontal — Vertical

### Profile moulded composite sheets
Profile moulded composite sheets are generally manufactured from cement and fibre (asbestos cement). There are a number of profiles to choose from for vertical or horizontal application.

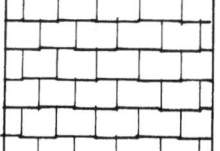

NB. Shingle must be twice as long as visible length

### Shingles
Wall shingles can be split or sawn hardwood or softwood or a manufactured material. Wall sarking is recommended or sub-wall cladding in marine plywood. Battens are generally required over the studs to fix the shingles.
NB Shingle must be twice as long as visible length.

# 40 • SHEET AND BOARD PRODUCTS

## Blue Board

For many years the building industry had an unfulfilled demand for a long-life of coloured coating on a simple sheeting product. Any options were either compromises or expensive.

In the 1950s, the Housing Commission in Victoria used two experimental building systems: one was a system of manufactured reinforced concrete walls often referred to as the Holmesglen system, the other was known as conite.

The Holmesglen system of cast reinforced concrete panels, produced multicoloured houses of modest dimensions with a traditional plan and a conventional tile or corrugated iron roof. The system was also used to produced horrid high-rise flats, much like the Stalinist flats of the USSR.

The conite system was simpler and allowed building of solid-looking houses with tiled roofs after WWII, when building materials were in short supply. Conite used a conventional timber stud frame covered with kraft building paper, wrapped with chicken wire, cement rendered and colour coated. It was a good system but it never found general acceptance—probably due to its Housing Commission associations.

From time to time, lightweight framed houses have been sheeted with fibrocement sheets or marine plywoods, covered with chicken wire or expanded metal lathing, and then cement-rendered to produce any number of stuccoed finishes. This had little impact on mainstream home building until the condominium developments of Port Douglas and Sanctuary Cove: neo Mediterranean classicism in coloured renders.

Designers and builders began to experiment with internal grade, recessed edge, fibrocement products with flush joints covered with thick, textured paint finishes. Initially the results were poor: the joints cracked, the coatings slipped on the sheet surface, colours faded rapidly and the details used were often fussy.

Then modern fibrocement sheets were specifically developed for outside use under textured coatings. They form a complete system, from the high bracing resistance of the sheets themselves through the specially developed board sealers, jointing tapes and fillers, to the various textured coatings and protective coloured skins.

It is now possible to have a light timber or steel-framed house with a coloured textured stucco finish that allows greater modelling of the building's exterior appearance than brick or weatherboard could ever achieve. Even better, the original paint should remain in good condition for over ten years and last more than 15 years. On many of these coatings a repaint with a single coat of exterior acrylic paint is all that will ever be needed every dozen or so years. If the coating is not kept spruce with repainting, it will simply become weathered and mellow with the passage of time.

A few points should be considered by those setting out to use this system for the first time:

- Try to keep the sheet joints vertical, as horizontal joints are more likely to show.
- Make sure the base wall frame is flat.
- Take care that there are expansion joints in large wall lengths, particularly near external corners.
- In two-storey buildings a horizontal joint at the shrinkage/flexure line is required.
- Remember that most of the textured coatings are rolled on and some are trowelled over, so try to avoid vertical planes greater than 3 metres high between ribbon projections.
- The details between the texture coating and windows, doors and other openings are areas of critical detail and need careful design.
- It is much easier and quicker if the whole exterior of the building is scaffolded prior to the commencement of the sealing, jointing and finishing.
- The average house may need more than a tonne of coatings.

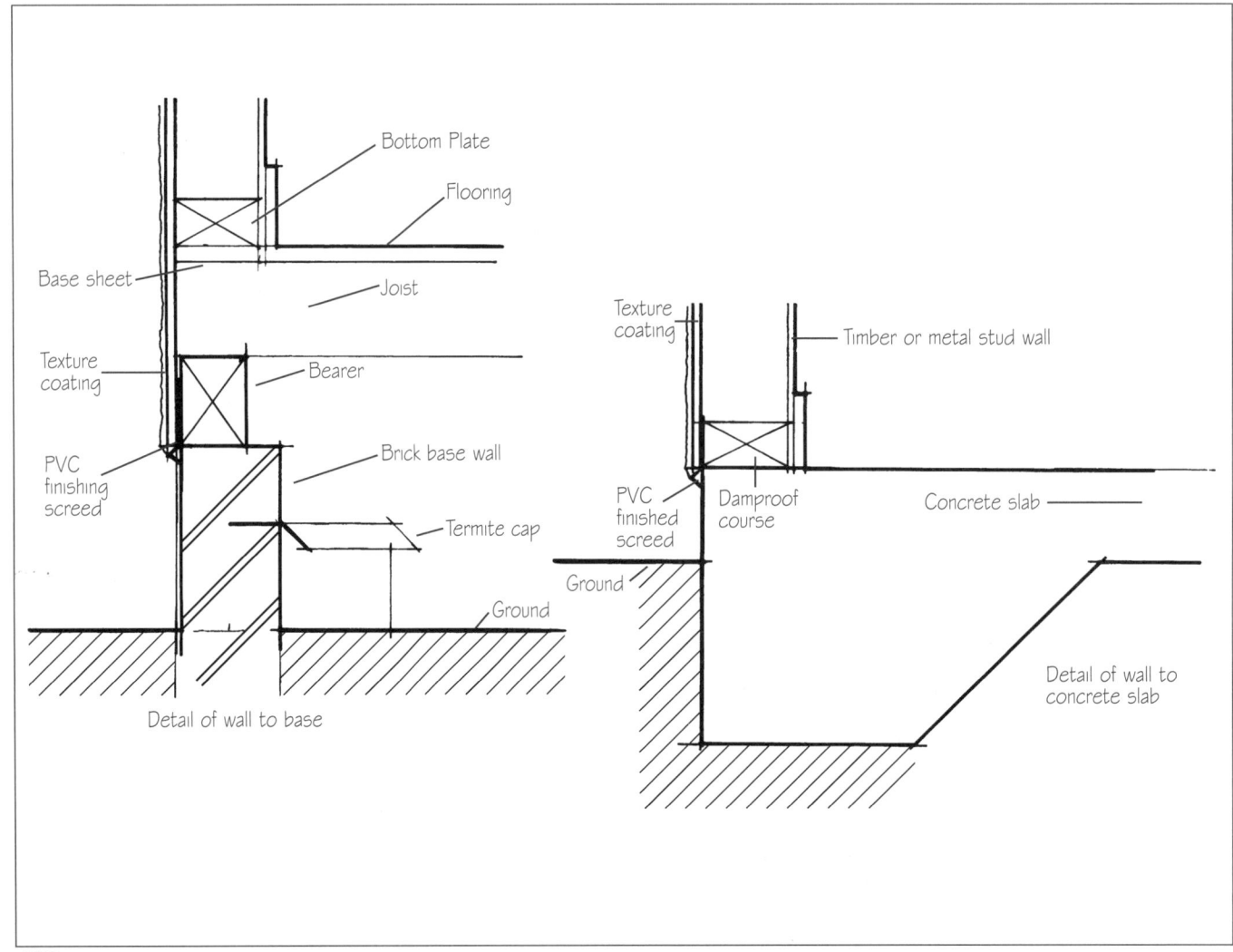

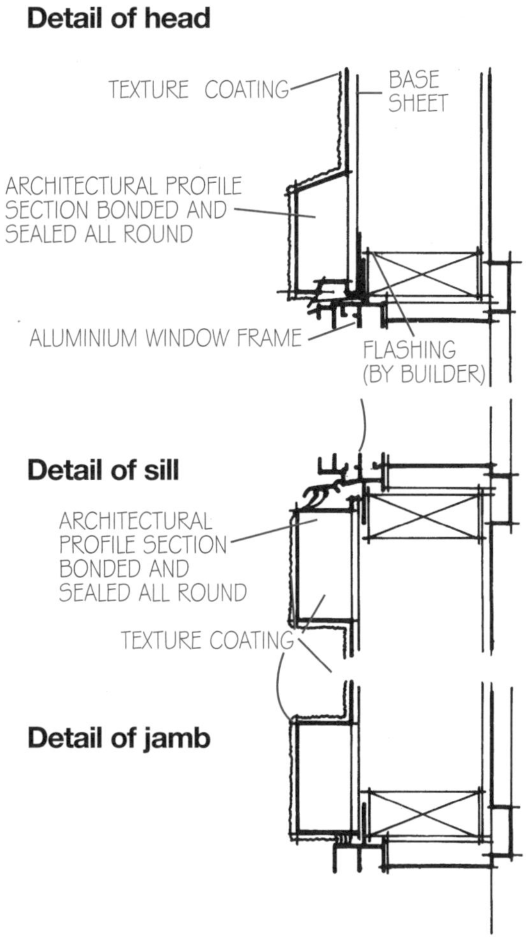

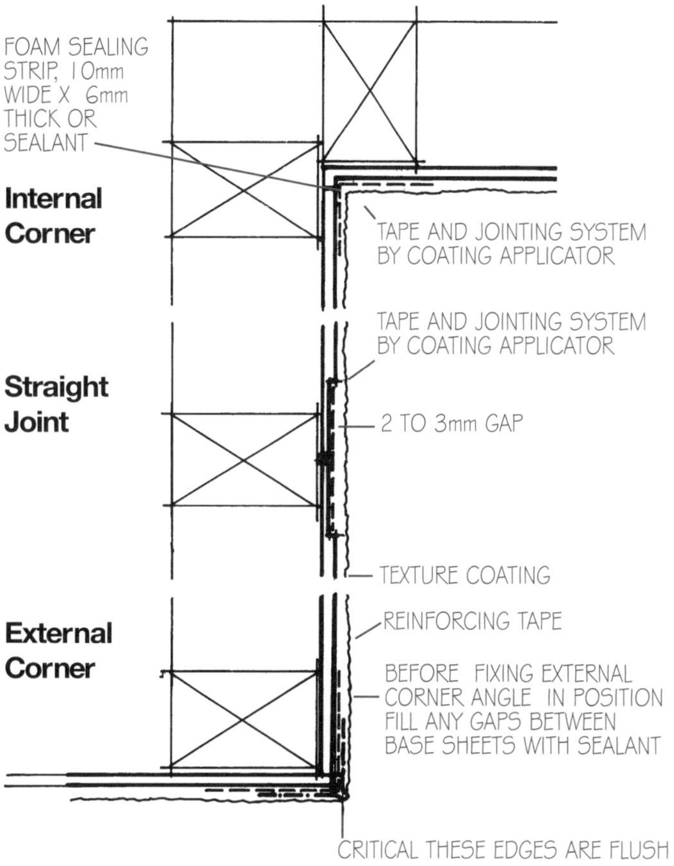

# SHEET AND BOARD PRODUCTS • 40

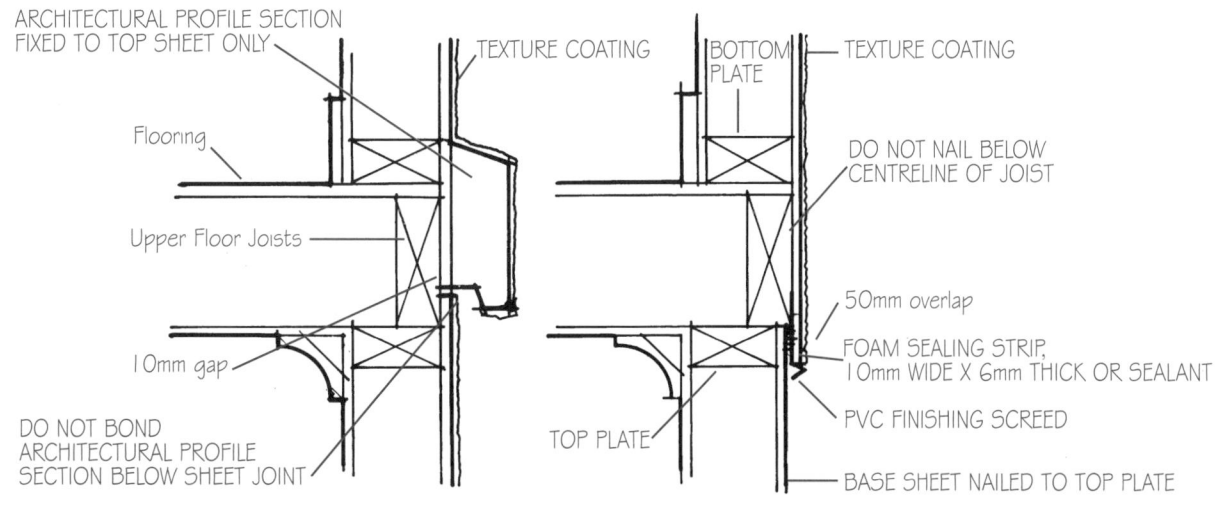

**SHEET MOVEMENT JOINT
2 STOREY CONSTRUCTION**

**ALTERNATIVE MOVEMENT JOINT
2 STOREY CONSTRUCTION**

# 41 • SOLID MASONRY CONSTRUCTION

Solid masonry construction includes the methods of construction used to erect buildings that use brick, block and natural stone as primary load bearing components in the form of walls. Mud brick (adobe), rammed earth (pise) and concrete could also be considered to fit this description, but this book will not attempt to deal with these relatively specialised low volume building methods, which are well catered for in specialised text books.

The use of solid masonry construction in Australia dates back to the early days of European settlement, when the good quality Sydney sandstone and locally made sand stock bricks were used to build buildings closely modelled on those being built in Europe, particularly England.

Bricks are one of the most durable building materials used by man and are sufficiently flexible to meet the demands of strength and aesthetics required for house construction. Over the generations in Australia, however, the cost to build an all brick house has escalated until timber frame, or more especially brick veneer houses are so much cheaper except in a couple of localities notably Western Australia and to a slightly lesser extent South Australia and Tasmania.

There are many people who strongly believe there is no alternative to a solid masonry house, as they feel the mass of the structure gives a greater sense of security and well being, less noise transmission and white ant proof. All this is true, but they are generally more expensive to build than alternative timber framed houses, and are often less flexible in layout, harder to insulate and prone to cracking in reactive foundation areas.

## Masonry materials

Bricks, blocks and natural stone have been covered under brick veneer construction. Traditionally, bricks used for structural walls have been of the pressed type as they are generally stronger and easier to cut than extruded-wire cut bricks. In residential construction, the ultimate strength of the bricks is seldom tested and many extruded bricks are used in load bearing applications, as are silica-lime bricks and concrete blocks.

Few houses are built of solid natural stone today, but there are still a considerable number of homes which combine a load bearing brick substructure with a natural brick facing to the outside of the house.

Limestone and sandstone are easily cut and dressed to form well finished blocks, and good deposits of these stones are still being worked in NSW and South Australia. Blue stone, a form of basalt, is used in Victoria, and is a very hard stone that is difficult to dress and mould.

In the country, bush or field rocks have been used to build houses. Although the rocks are cheap the mortar to fill the irregular joints between them can be excessive both in cost and in spoiling the appearance of the walls. Bush rock is generally suitable as a facing only and the substructure should be of brick or post and beam, so that there is no imposed load on the bush rock wall.

## Building with masonry

The most important part of any masonry wall is to establish a stable and adequate foundation/footing system, because masonry walls are only as strong as the foundation/footing system that supports them.

Once an adequate foundation/footing is achieved then the next step is to bring the masonry up in a uniform manner. It is generally poor practice to build sections of a building higher than other sections. When laying masonry, bond the individual units solidly one to the other by the prescribed method of overlapping the bricks.

Most masonry work used in Australian houses consists of either a single thickness brick wall or walls that are a combination of two single brick thick skins, the latter commonly referred to as a cavity wall. This use of mainly single brick skins means the common or stretcher bond is the most suitable.

Special complex bonding methods have been developed for brick walls that are two or more bricks thick. These are seldom used in residential construction—even when thicker walls are required, simplified methods of skin bonding are used, such as brick course reinforcing or galvanised steel ties.

Windows and doors need lintel beams over them, except in the case of the outside skin of a cavity wall used in a brick single storey cottage. In most cavity walls, only the inside skin is structural, the external skin acts as a cladding to keep the weather out.

Lintel beams can be steel or concrete, or structural arches can be used to bridge openings. Steel lintels are commonly cut from rolled steel bar, angle or channel, and in most instances galvanising is applied to the lintel to inhibit corrosion and eventual failure. Concrete lintels were very popular for a time before WW2 but are little used today except in special applications.

The correct installation of damp proof courses in solid masonry buildings is essential. It is the protection against rising damp and other moisture induced building deterioration.

## For something a little different

Modern brickwork is identified with bland, single colour, dead flat stretcher bond walls. For those of you who think this is boring, take yourself off to one of the brickworks, there is at least one in most of Australia's capitals—excluding Canberra—that still produce squint, bull nose, cant, plinth and other assorted shaped bricks. You may be able to design into the walls of your solid masonry house polychrome bricks, engaged pilasters, projecting plinths, natural stone quoins or anything your imagination, and a sneak look at some history of architecture texts, can conjure up.

**Polychrome**   A number of colours of brick generally in a pattern.
**Pilaster**   An engaged pier, often decorated.
**Plinth**   The base of the building.
**Quoin**   The brick, block or stone that shows its face and head at the outside corner of a masonry wall.

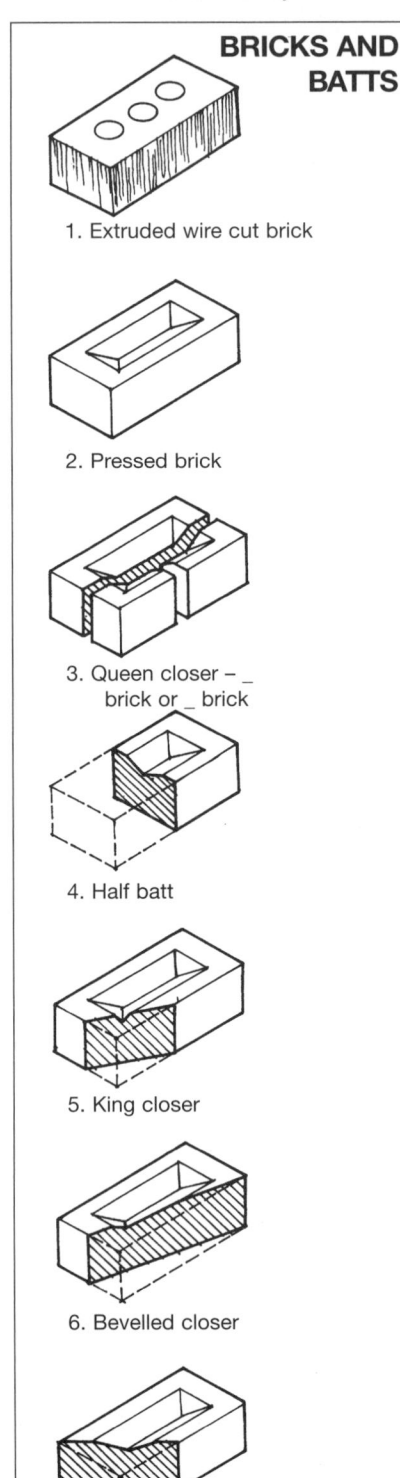

**BRICKS AND BATTS**

1. Extruded wire cut brick
2. Pressed brick
3. Queen closer – _ brick or _ brick
4. Half batt
5. King closer
6. Bevelled closer
7. Mitred closer

# SOLID MASONRY CONSTRUCTION • 41

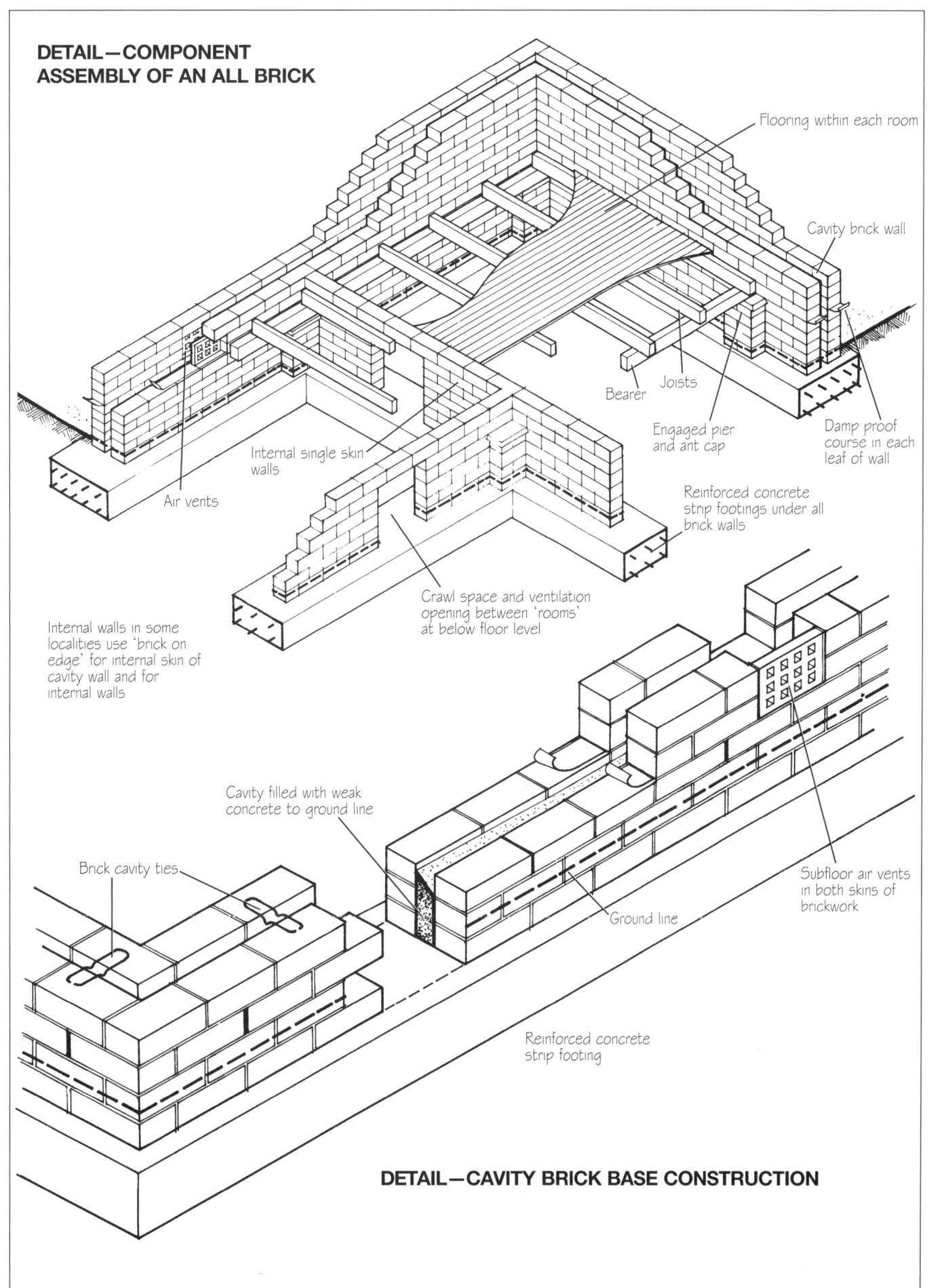

DETAIL—COMPONENT ASSEMBLY OF AN ALL BRICK

DETAIL—CAVITY BRICK BASE CONSTRUCTION

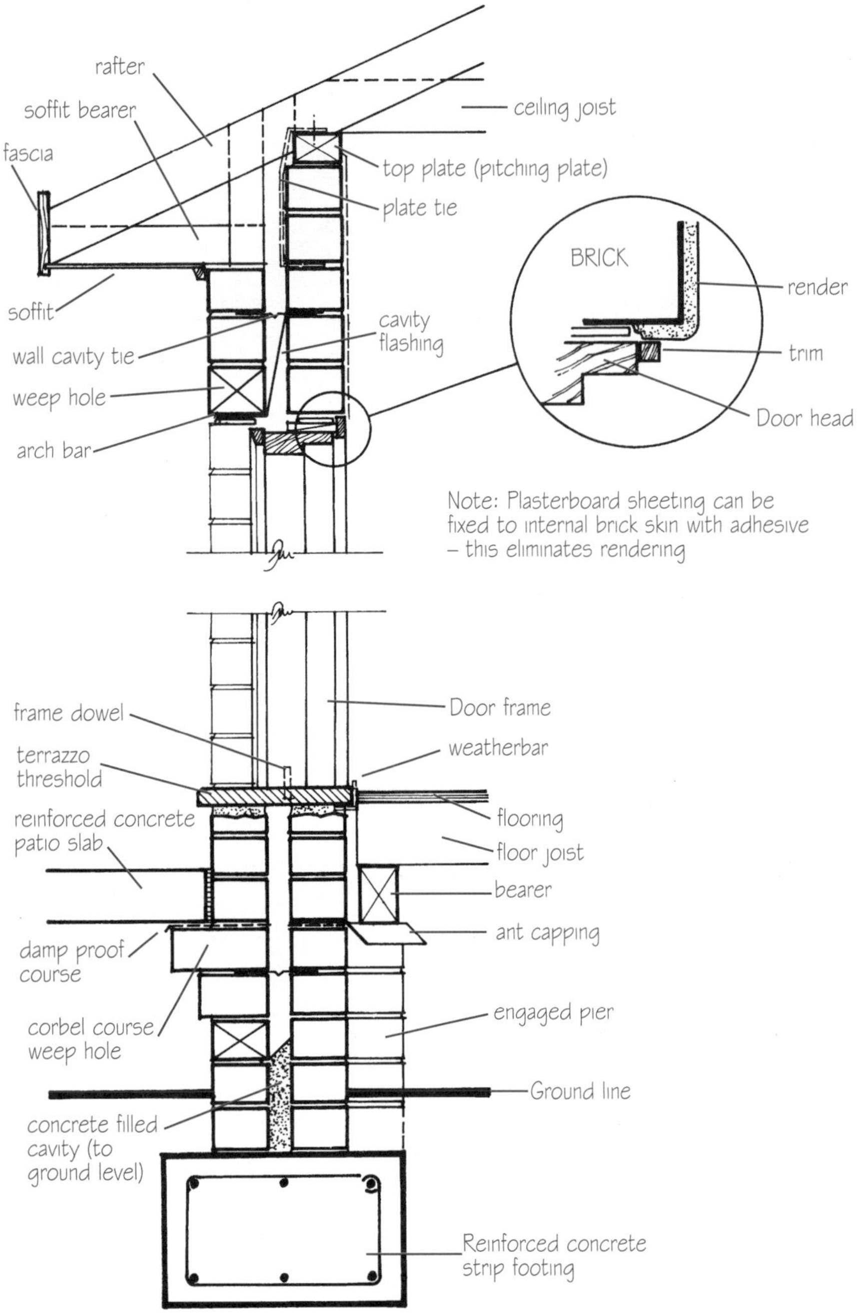

# 41 • SOLID MASONRY CONSTRUCTION

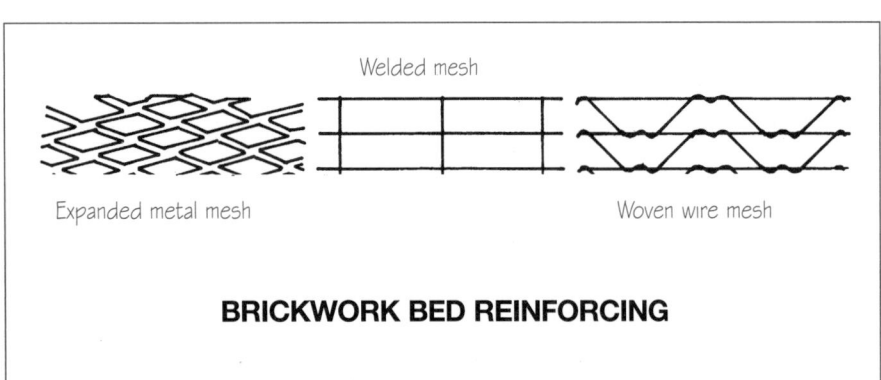

**DETAIL SECTION—TRADITIONAL CAVITY BRICK WALL**

**BRICKWORK BED REINFORCING**

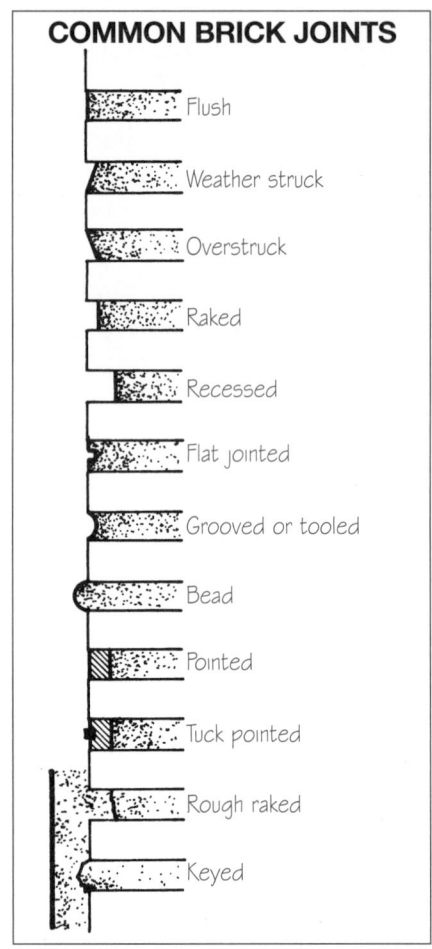

**COMMON BRICK JOINTS**

- Flush
- Weather struck
- Overstruck
- Raked
- Recessed
- Flat jointed
- Grooved or tooled
- Bead
- Pointed
- Tuck pointed
- Rough raked
- Keyed

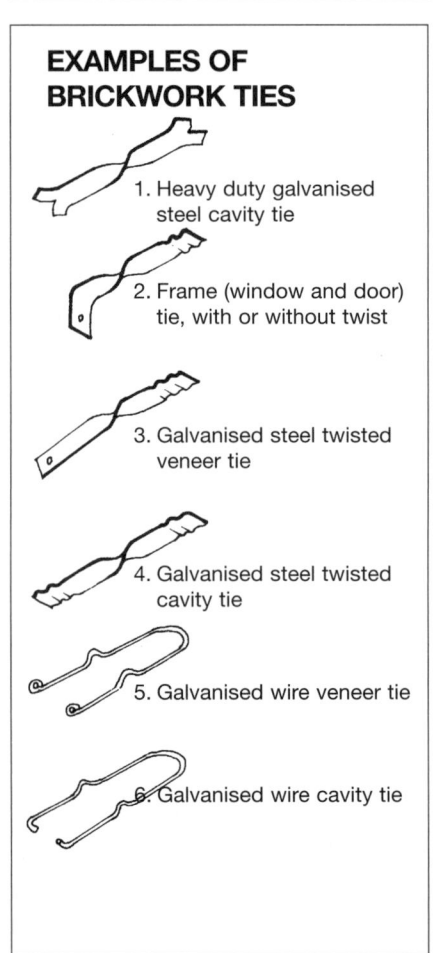

**EXAMPLES OF BRICKWORK TIES**

1. Heavy duty galvanised steel cavity tie
2. Frame (window and door) tie, with or without twist
3. Galvanised steel twisted veneer tie
4. Galvanised steel twisted cavity tie
5. Galvanised wire veneer tie
6. Galvanised wire cavity tie

# SOLID MASONRY CONSTRUCTION • 41

## SPECIAL BRICKS—LIMITED AVAILABILITY

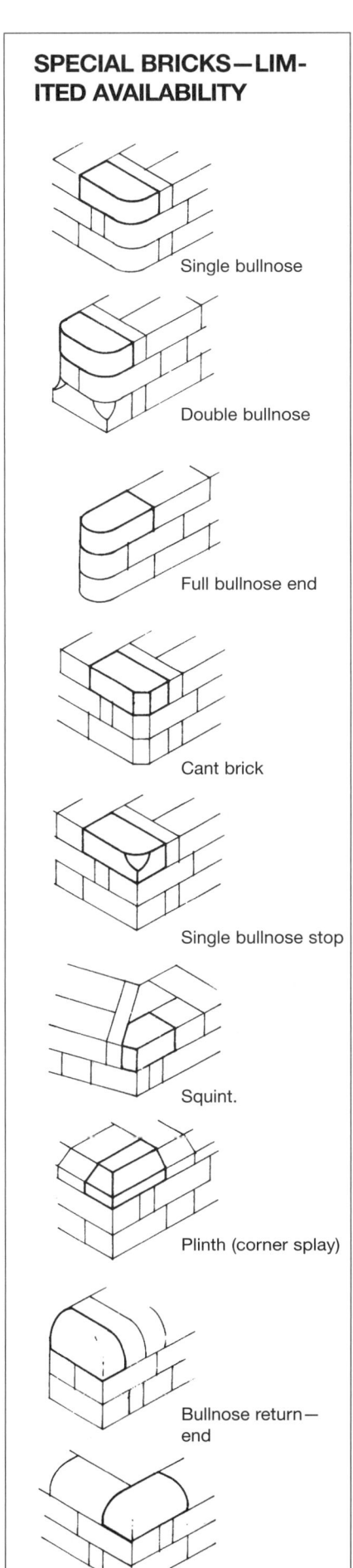

Single bullnose

Double bullnose

Full bullnose end

Cant brick

Single bullnose stop

Squint.

Plinth (corner splay)

Bullnose return—end

Stretcher Corner

## DETAIL—ELEVATION OF STEEL ANGLE LINTEL

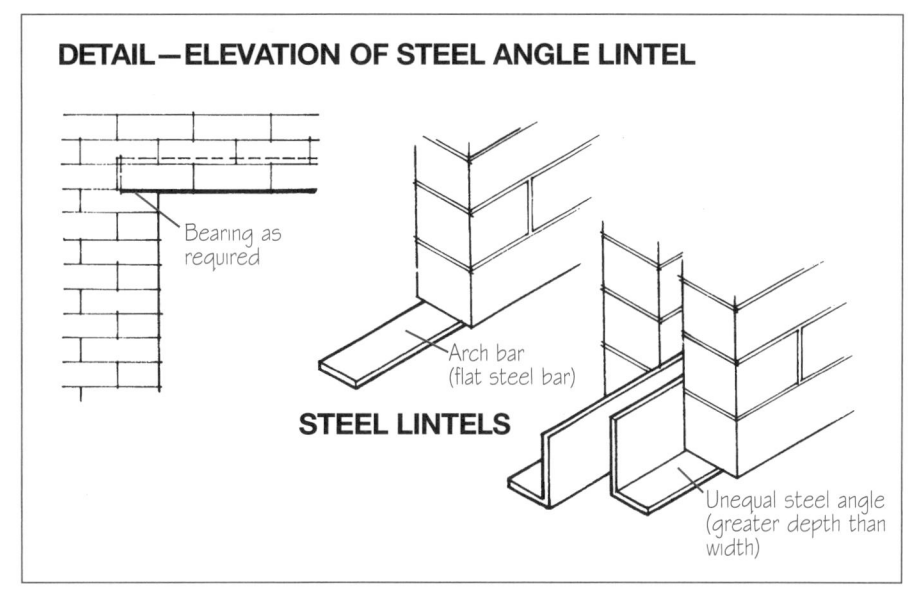

Bearing as required

Arch bar (flat steel bar)

### STEEL LINTELS

Unequal steel angle (greater depth than width)

## DETAIL—OPENING HEAD

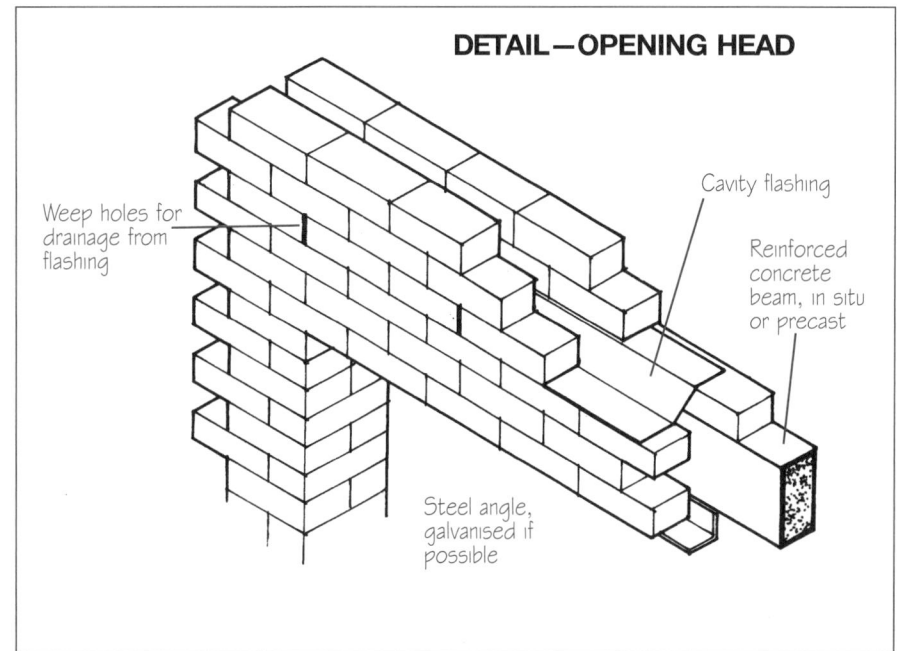

Weep holes for drainage from flashing

Cavity flashing

Reinforced concrete beam, in situ or precast

Steel angle, galvanised if possible

## DETAIL—LINTEL FOR BRICKWORK

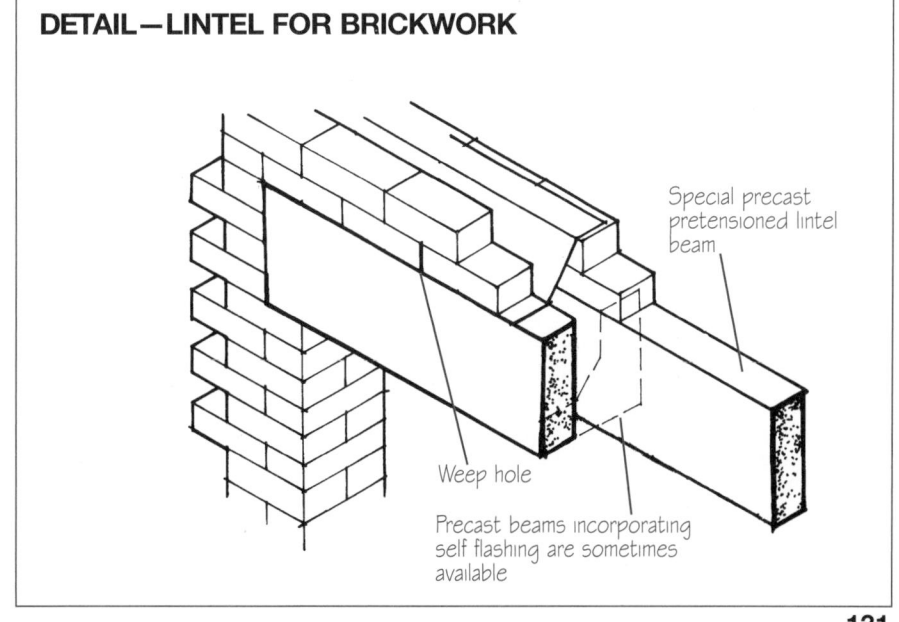

Special precast pretensioned lintel beam

Weep hole

Precast beams incorporating self flashing are sometimes available

# 41 • SOLID MASONRY CONSTRUCTION

## Arches

Arches have been used by masons to span openings from prehistoric times and are thought of by many people as the applicable head to openings in masonry walls today. Arches use gravity, the force that acts downwards to lock together the units of the arch. The arch will remain supporting until the ability of the walls on each side of the arch to withstand the outward pressure, exerted by the force of gravity trying to flatten the arch, is exceeded.

Therefore the rules of thumb which allow arches to remain structurally sound are: the greater the curve the stronger the arch; the more solid the opening of the supporting wall the more stable the arch.

Arches are built on a removable form or 'centre'. That is, the bricklayer or mason builds the arch onto a supporting structure the same shape as the arch will be when completed. Brick and block arches have one, two or three rings generally of the standard masonry unit. Sometimes specially-made tapered brick or block units are manufactured, but this practice is seldom applied these days. Stone arches, because of the larger size of unit being used, generally only have one ring and each unit is cut to a tapered voussoir.

In brick and concrete block walls it is advisable to build brick course reinforcing into the courses immediately above the arch.

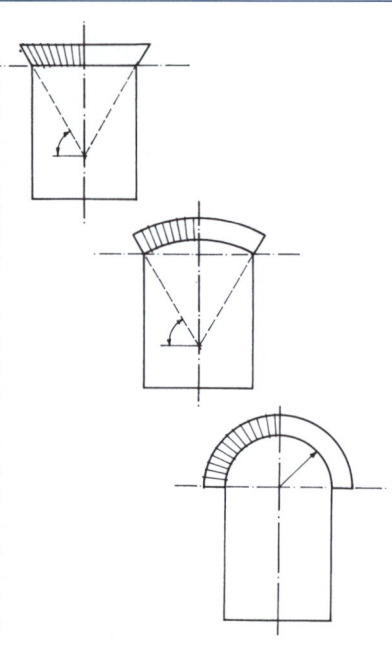

**COMMON ARCH TYPES**
Flat arch
Segmental arch
Semi-circular or Roman arch

## COMMON BRICK BONDS

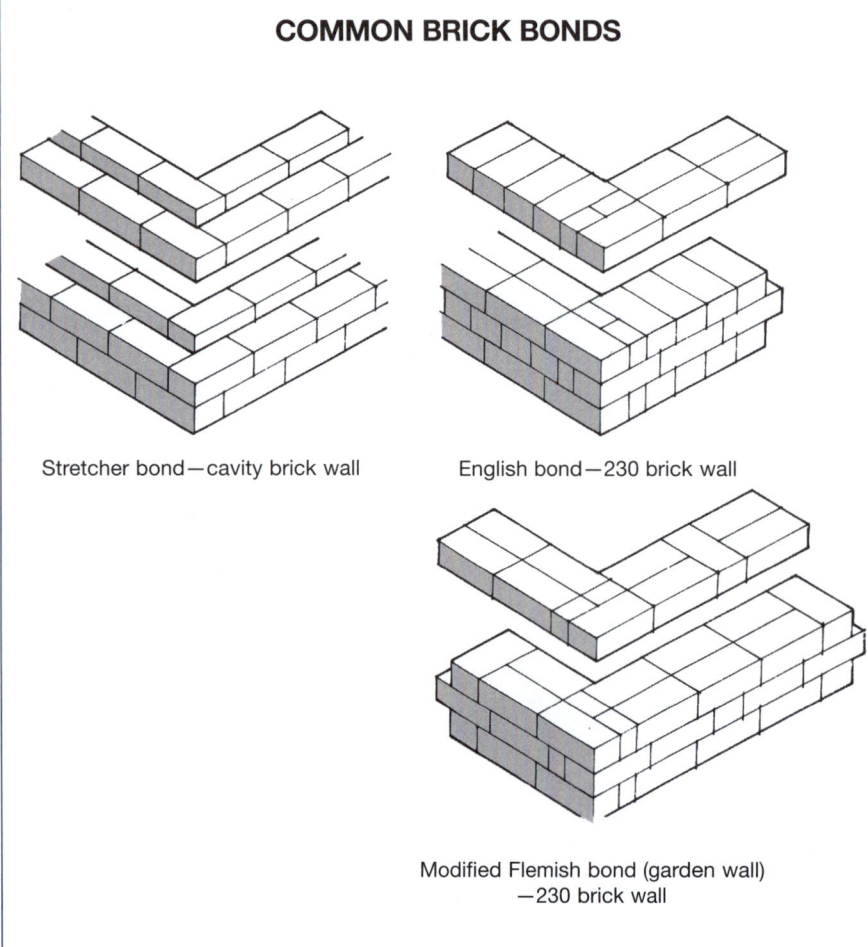

Stretcher bond—cavity brick wall

English bond—230 brick wall

Modified Flemish bond (garden wall)—230 brick wall

## COMMON CONCRETE BLOCK BOUNDS

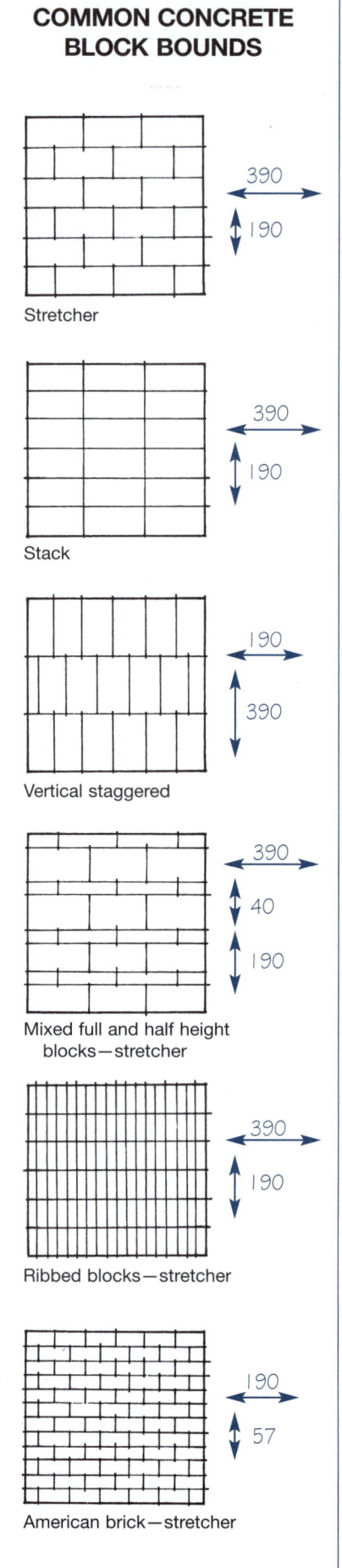

Stretcher — 390 × 190

Stack — 390 × 190

Vertical staggered — 190 × 390

Mixed full and half height blocks—stretcher — 390 × 40 × 190

Ribbed blocks—stretcher — 390 × 190

American brick—stretcher — 190 × 57

# SOLID MASONRY CONSTRUCTION • 41

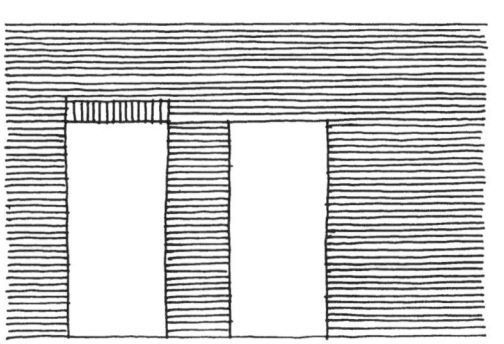

Steel lintels.
Lintel may be articulated with a soldier course or concealed

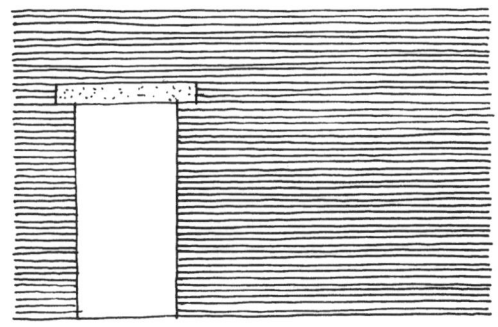

Concrete lintels.
Lintel exposed

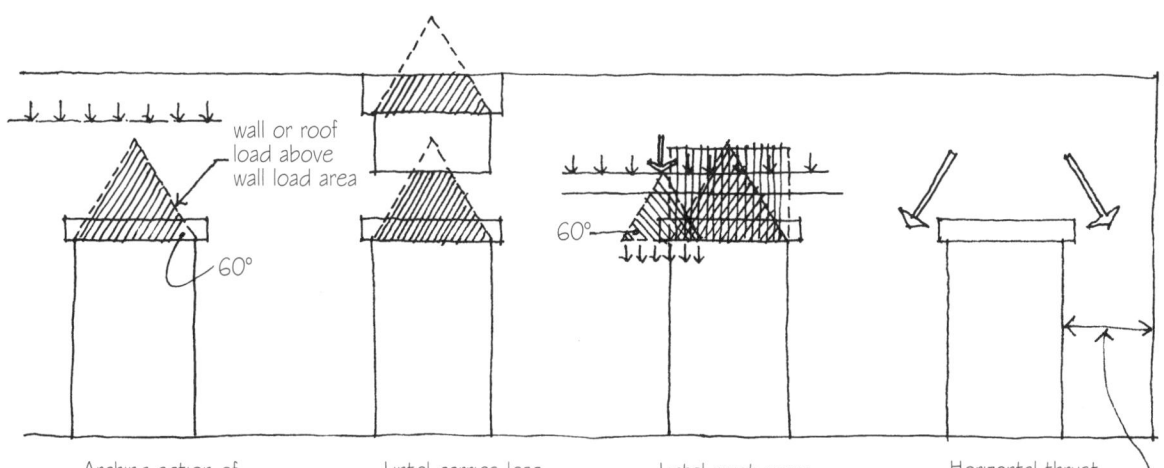

| Arching action of masonry above opening supports weight of wall outside of wall load triangle | Lintel carries less wall load than normal wall load triangle | Lintel must carry additional loads if concentrated loads and floor or roof loads fall within normal wall load triangle | Horizontal thrust resulting from any arching action must be resisted by the wall mass on either side of the opening |

**LOADED BRICKWORK**

**UNDERSTANDING BRICKWORK LINTELS**

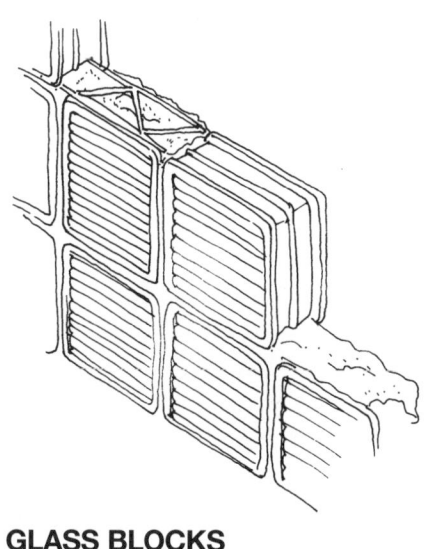

**GLASS BLOCKS**

## DETAIL SECTION—CONCRETE BLOCK COTTAGE WALL

pitching plate bolted to bond beam
metal cleat fixed to rafter/plate for high wind conditions
fascia
Bond beam, special hollow concrete blocks for top course, have reinforcing steel and concrete fill to bond wall and act as a lintel
Soffit bearer
Soffit lining on battens
This single skin form is not approved in some regions, particularly where there are wet winters
plywood flooring
floor joist
bearer
ant cap and strip
engaged pier
'monsoon' tie (optional) for high wind areas
reinforced concrete strip footing

c
b
a

| a | b | c |
|---|---|---|
| 90 | 90 | 40 |
| 140 | 190 | 90 |
| 190 | 290 | 190 |
| 240 | 390 | |
| 290 | | |
| 390 | | |

Plus 10mm for mortar joint

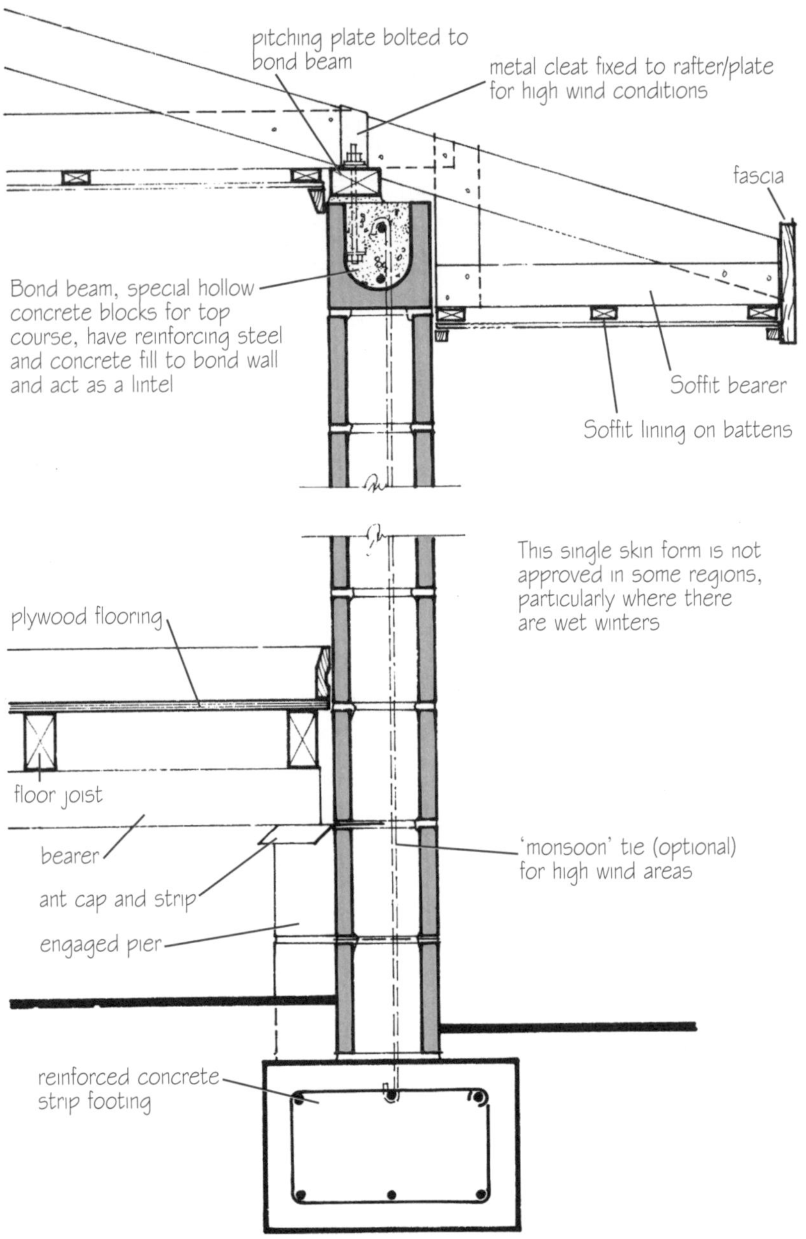

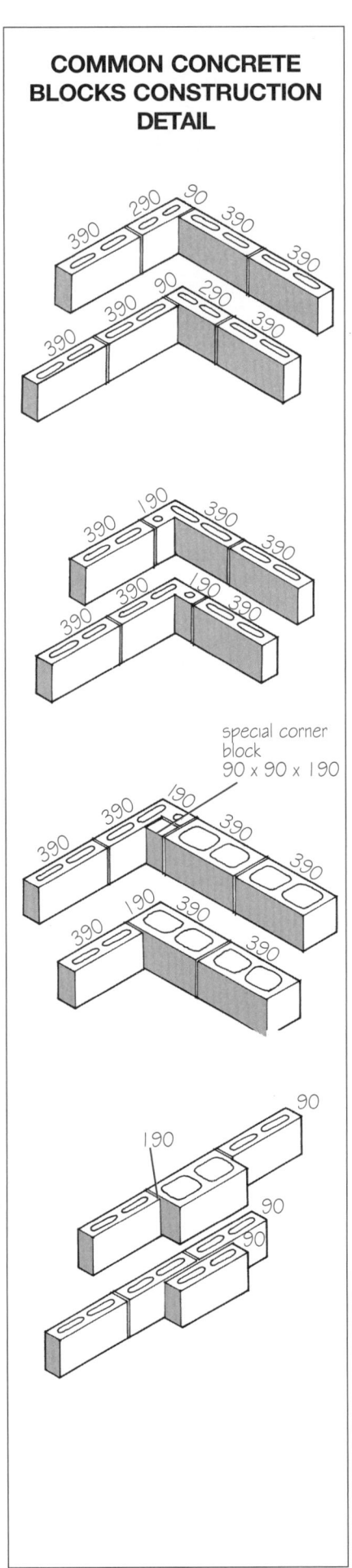

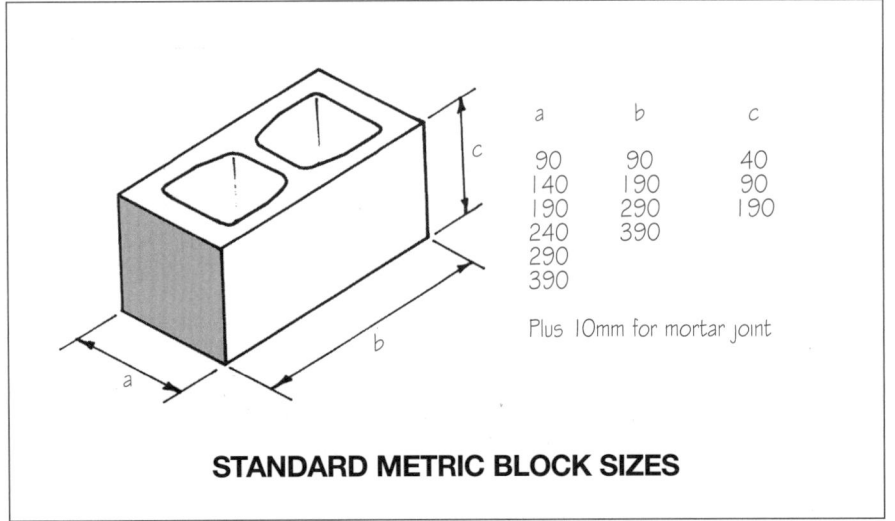

### STANDARD METRIC BLOCK SIZES

## COMMON CONCRETE BLOCKS CONSTRUCTION DETAIL

390
290
90
390
390
390
390
90
290
390
390
190
390
390
390
390
190
390
special corner block 90 x 90 x 190
390
390
190
390
390
390
190
390
390
90
190
90
90

# SOLID MASONRY CONSTRUCTION • 41

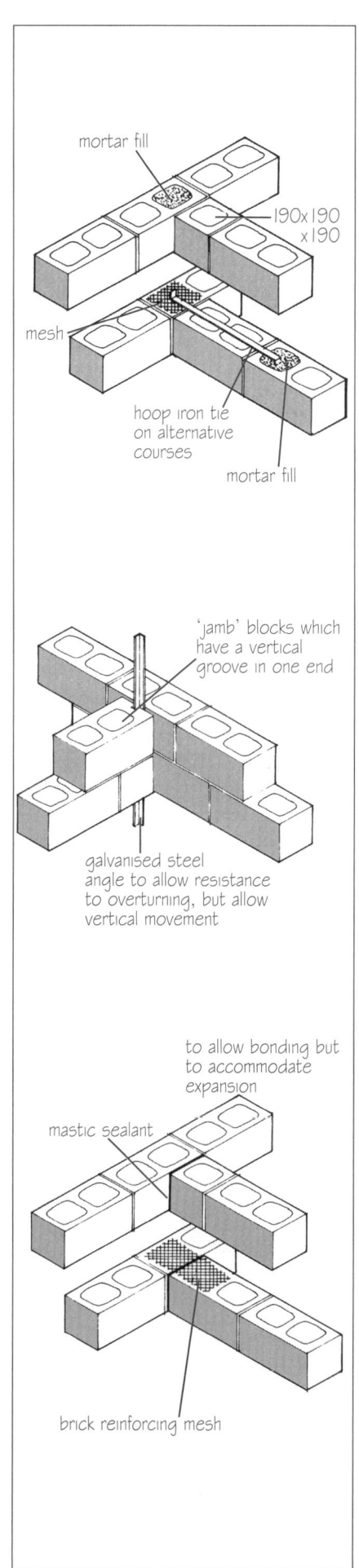

## COMMON STONE WALLS

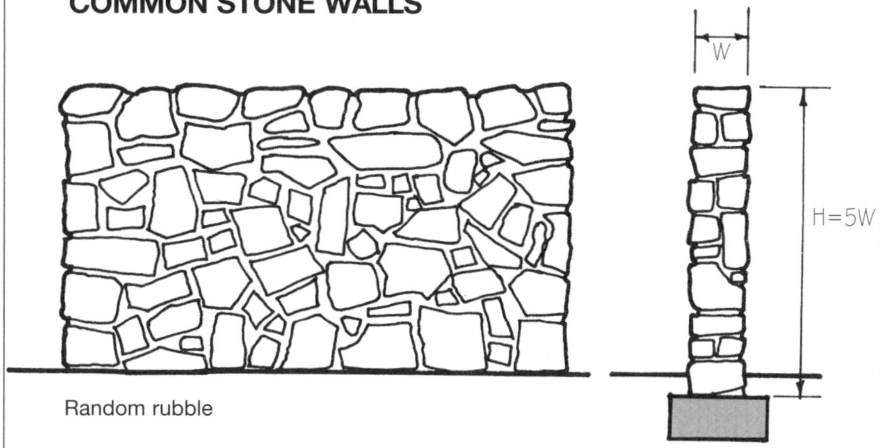

Random rubble

Coursed rubble

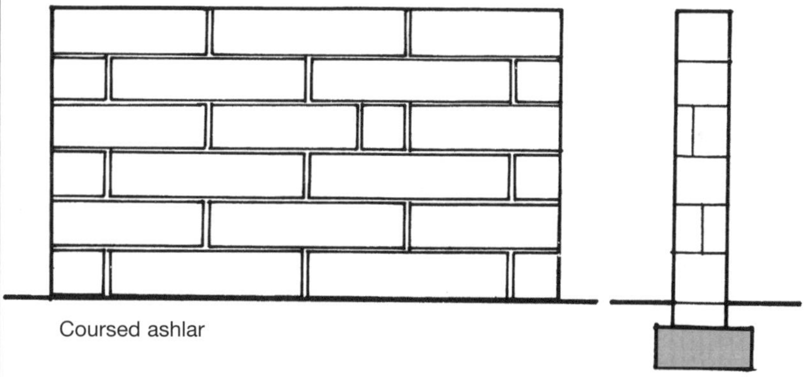

Coursed ashlar

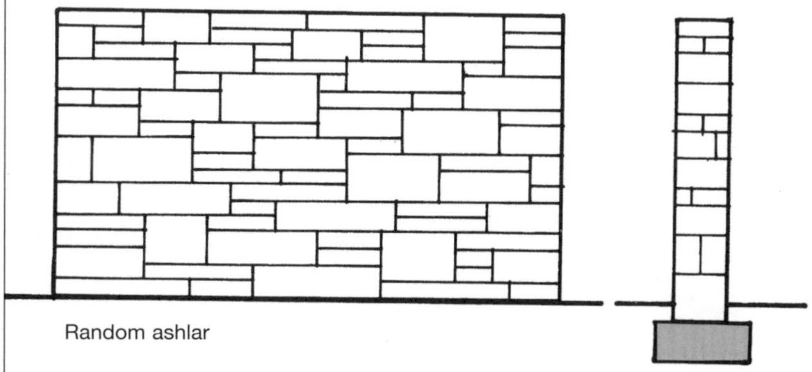

Random ashlar

H = 5W

# 41 • SOLID MASONRY CONSTRUCTION

## Stonework

Many OBs want to have solid stone houses. To follow this course, extra preparation is required, as it is not a simple matter to design, plan and produce a solid stone building.

Considers the following factors:

### What stone can be sourced locally or trucked in?

There is obviously lots of stone. Dig anywhere and eventually stone will be encountered, but how much can be used for building purposes? The common building stones include Hawkesbury sandstone from Sydney, Melbourne bluestone (basalt), Adelaide bluestone (shale), Mount Gambier stone (limestone), Blue Mountains drywall (ironstone) and granite from all over the continent west of the Great Dividing Range. In some rural locations, there are rocks scattered across the paddocks: basalt, granite, quartz, ironstone and even marble.

All of these have been used for building at some time in Australia's history and there is even evidence of drystone windbreaks constructed before the Europeans arrived.

The settlers who arrived at Sydney Cove soon found that the local sandstone was an easy-to-work freestone and it was eagerly quarried once the tools and masons arrived in the colony. There are many old sandstone buildings in Sydney. Today there are at least two quarries, one on the Central Coast and the other in the lower Blue Mountains supplying newly quarried stone and significant recycled stone from demolition sites.

Melbourne was blessed by its very durable but extremely hard-to-work Bluestone. This very heavy stone made good kerbs, gutters and cobbles and was used on a fair number of residences and public buildings. A very dark stone it is generally considered very sombre, when compared to the bright red of Melbourne's common bricks.

Adelaide bluestone remains a feature of many houses in the city and the surrounding hills. It is a very soft stone of low durability and is generally combined with burnt clay brick quoins and jambs to extend its lifespan.

The limestone from Mount Gambier is almost pure white, is easy to work and reasonably durable, but discolours readily and needs to be used carefully. It is still readily available and can be transported great distances due to its relatively light weight.

Paddock rubble can be used to construct interesting walls but is often difficult to lay because the stones are often rounded and are difficult to bed onto each other. The ironstone, which is more likely to be tile-shaped, is easier to build with but has a strata-like structure that provides homes for much animal and plant life.

Take care with all stone when building fireplaces (or even barbecues); many of these stones will explode into fragments when heated.

### What footings are required for stone walls?

Stone walls require highly supportive, stable foundations of even bearing. If the foundation is not suitable and the footing not well constructed, irreparable cracked and leaning walls will result.

It is tempting to use the same stone for the walls and the footings and this can often be successfully done. However, reinforced concrete footings will almost always provide a better result and precious stone is not wasted out of sight in the ground.

### How thick is a stone wall?

A high-quality ashlar (square-cut stone) sandstone wall 300 mm thick will stand to over 2500 mm high as an external wall and over 3000 mm as a bonded interior wall; external walls 3000 mm high need to be closer to 400 mm thick.

At the other end of the scale a paddock rubble wall of granite boulders may need to have a ratio of height to thickness of no more than 5:1, ie. a 3000 mm high wall needs to be in the order of 600 mm thick.

Using brickwork as the inner skin and for the internal walls can reduce the thickness significantly—the stone skin could then be reduced to as little as 150 mm or even thinner if bonded directly to the brick structure.

### How is stone cut to size?

Some is used as found but most stone needs to be worked to allow for sound construction. Limestone, sandstone and granite are often sawn. Basalt, sandstone and shale can be split and, traditionally, some of the very best stonework is hand-split and hand-worked to gain the best visual and structural properties.

### How is stone handled?

Stone is heavy and does not have handles. Most stone needs to be handled with mechanical devices—these can be traditional block and tackle devices hanging from scaffolding, but a tractor with a HIAB-type jib is generally easier to use and is safer. A forklift-type machine helps to move stone around building sites as just lifting a stone onto a stonemason's barrow is back-breaking and beyond the physical capacity of many people.

### What bedding mortars can be used?

Traditionally most stone is bedded on a secret compound mortar containing a high proportion of slaked lime putty. Track down local knowledge about the mortars recommended for your selected stone and try to find old books that give technical information. If there is a mason available, give serious consideration to making them an offer.

The basic rule with stonework bed joints is to keep them as thin as practical; mortar will seldom give as good appearance as the stone. Lime-based mortars are flexible and allow some movement in the stonework over time—if cement is introduced into the mortar it becomes brittle and, if it is smeared onto the stone face, it is extremely difficult to remove—take care.

### How is the stonework damp- and weather-proofed?

Most stones are porous and will take up a significant volume of water during rainy periods. This need not be a problem if the walls are thick enough or incorporate a cavity or other moisture barrier. Good ventilation around the house and direct sunshine will also help in maintaining low moisture content in the stonework. Dry stonework is also less likely to support organic growth; moss, lichen, ivy and the like will cause walls to remain damp and their roots can damage mortar joints and even cause cracking in the stones themselves.

Eliminating rising damp is a major consideration when building stone walls. Lead and other metal barriers have been used for centuries but modern plastic membranes work as well, are very flexible and, if used sensibly, will give a century or more protection. Watch out that easy access-ways are not provided for termites—building in termi-mesh barriers is worthy of consideration.

### How are openings spanned?

The most suitable method is to use an arch. This means the construction of a centre, which supports the arch stones while it is being built. Every stone must be carefully selected or shaped so that they key together to form a sound and stable arch. Arches also take up significant height: a semicircular arch spanning 1500 mm will be 750 mm above the springing. If the springing is 2000 mm them the rise of the arch plus the arch stones and the required wall over it will require in the order of 3500 mm.

Flatter arches are possible but the shallower the arch the more care required to ensure stability. Flat lintels can be used in steel or concrete if detailed carefully—a poorly executed arch soffit can reduce the visual quality of the arch significantly.

In some designs solidly constructed timber or moulded stone window frames can be used to support the stonework over an opening, as was common in vernacular English farmhouses from the 15th century onwards.

### Is the stone to be visible outside and inside?

Building a stone wall with exposed stonework to both faces adds a factor of difficulty of three or four. Either every stone must be carefully sized and dressed on both faces, and then end-matched, or each wall face has to have independent stones that are rough-cut at the back and an interior filled with mortar and stone chips—some bonding stones will go right through to stabilise the wall.

### Is the stone structural?

Stone is quite capable of supporting a building, but often it is either a facing to a brick, concrete block or reinforced concrete wall or it is used as infill between structural columns of timber, steel or concrete.

# SOLID MASONRY CONSTRUCTION • 41

**How are door and window frames made and built-in?**

Ensure there is a sound head over all doors and windows, as any sag will not only give a poor appearance to the building, but doors and sashes will jam. Take care to flash all around the door or window frame to eliminate drafts and weather penetration.

**What about rendering?**

Rendering the interior of a stone building is the normal way of finishing stonework. Again the general rule is—do not use cement-based renders—but use a traditional slaked lime putty based render. It is much less likely to crack or cause damage to the stonework. Furring over the stonework with top-hat section channels and lining with plasterboard is another method well suited to finishing the interior of a stone wall—it allows for the wiring to be concealed without messy chasing.

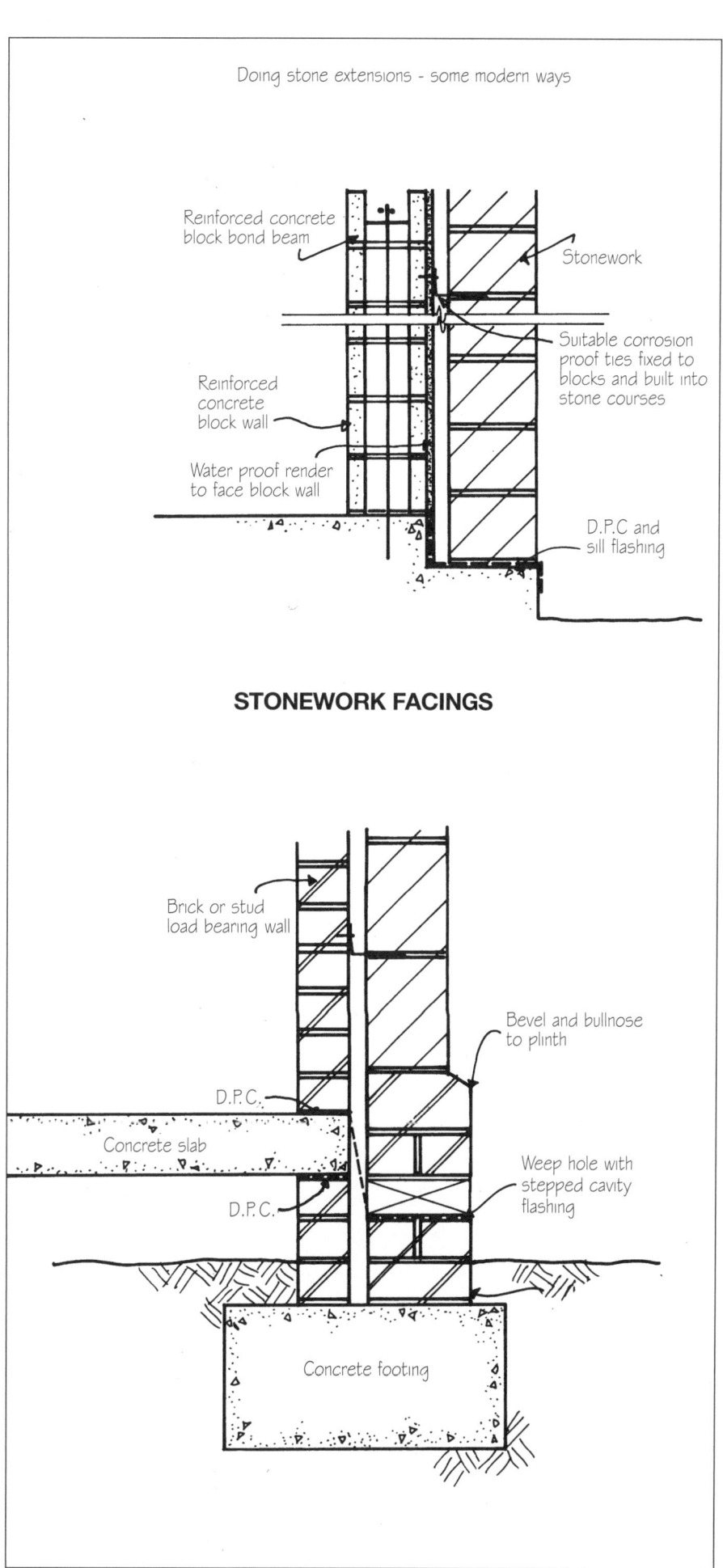

**STONEWORK FACINGS**

# 42 • INTERIOR LININGS

Interior linings are the exclusive domain of framed walls and ceilings and are therefore inescapably connected. The lining of framed walls has always posed a problem for Australian house builders; the frame was cheap and easy to erect but the linings were not.

Early lining options consisted of hessian bagging stretched over the frame then covered with newspaper and painted. This was cheap, but less than satisfactory. Another method was to nail lathing to the wall and ceiling frame, and then render this base with wet plaster. This gave a satisfactory result but required highly skilled tradespeople and, anyway, it was generally cheaper to build a brick wall and render it. As woodworking techniques improved, a supply of moulded timber lining boards became available. These made good wall linings and ceilings, but were not cheap, and were a fire hazard as well—eventually opening up at the joints to let the draughts and dust through.

Lathe and plaster ceilings remained the norm up until the 1920s and lining boards were used on timber-framed houses as internal linings. But these were generally considered an inferior finish to the smooth plaster finish that could be achieved in a rendered brick house. Also, lining boards were unsatisfactory in houses in an era when wallpaper was the fashion.

Technology did not seem to have an answer to the cheap, smooth, flush-jointed wall material needed for timber-framed walls either locally or overseas. The Europeans built only in masonry, and the western states of the United States appeared quite happy to use lathe and plaster—they could not build in masonry anyway because of the earthquake risk.

Wunderlich, a great old Australian building material company, did provide a partial answer with a pressed metal sheet product suitable for walls and ceilings. It became a success on ceilings but was less appealing on walls.

Finally, the breakthrough came with the invention of fibrous plaster, sheets of cast gypsum plaster bonded and reinforced with sisal fibres. Fibrous plaster sheets could be cast off-site at a factory then transported and nailed in position. The gaps were filled with gypsum putty that, when dry, could be sanded back to give a smooth flush finish to walls and ceilings, which sometimes featured cast decorative designs.

Fibrous plaster made the timber frame acceptable at reasonable cost, and cheaper than solid masonry. A house could be constructed where the interior walls appeared to be rendered-and-set plaster, unless knocked. The Australian expression to 'knock' something may have had its roots in the habit many Australians have of 'knocking' walls to check whether the house they are in is solid masonry or 'just' fibrous plaster.

It was the invention of fibrous plaster that paved the way for the brick veneer house, the house that has the appearance of a solid masonry house for lower cost.

Plasterboard, which is a laminated board consisting of outer stressed skins of special paper and a core of gypsum plaster, has in recent years taken over from fibrous plaster as the main flush wall lining available for timber-framed walls and ceilings.

Timber lining boards continued to be used; variously being in and out of fashion. Other wall linings: plywood, hardboard, asbestos and fibrecement have enjoyed only limited favour and have never seriously challenged fibrous plaster/plasterboard as the most favoured product.

## Plasterboard fixing

There are a few rules for OBs, who wish to fix their own plasterboard.

## COMMON TYPES AND APPLICATIONS OF INTERIOR LININGS

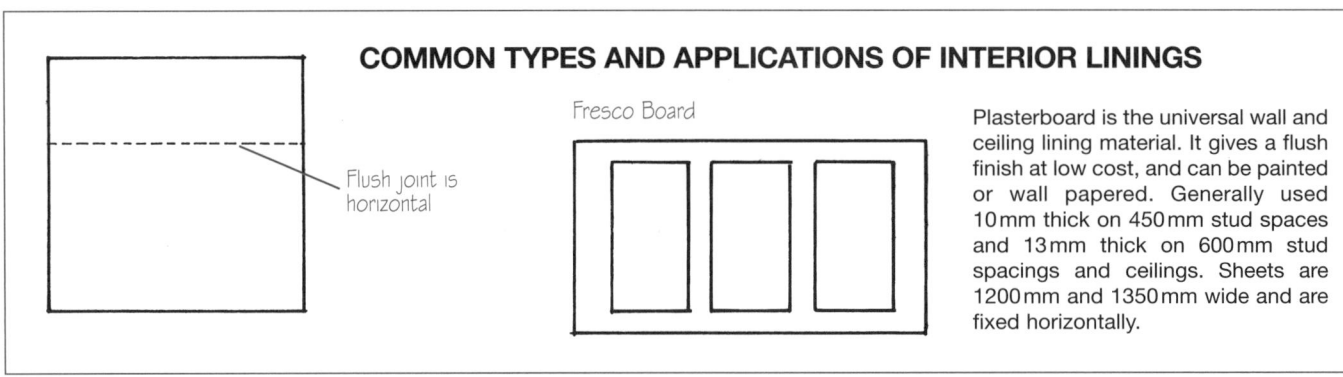

Plasterboard is the universal wall and ceiling lining material. It gives a flush finish at low cost, and can be painted or wall papered. Generally used 10mm thick on 450mm stud spaces and 13mm thick on 600mm stud spacings and ceilings. Sheets are 1200mm and 1350mm wide and are fixed horizontally.

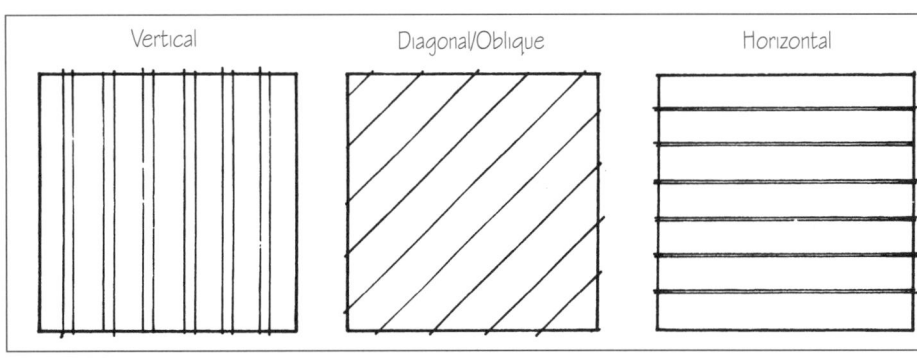

**Timber lining boards**
Timber lining boards come in many timbers from local hardwoods (Tasmanian oak), local pines (Pinus) through the USA grown species (Western Red Cedar) and European hardwoods (Oak). They can be moulded to many thicknesses, given many profiles and used horizontally, vertically and obliquely. Lining boards are not cheap, but every owner builder can fix them.

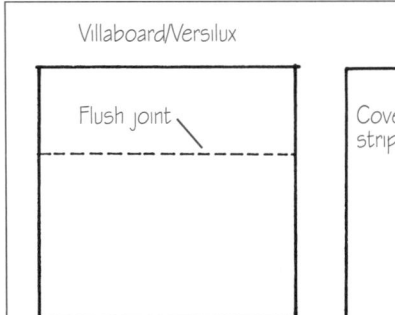
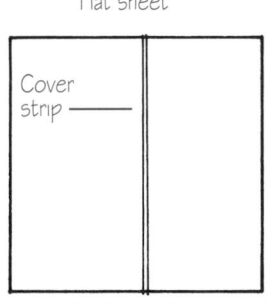
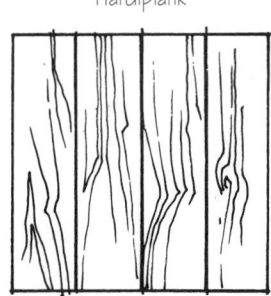

**Fibre cement**
Fibre cement does not contain asbestos any more, and can be used with flush joint (like plaster board), with covered or exposed joints, or with feature patterns from tiles to wood grain. Fibre cement is little affected by water, cannot burn and is good value. It makes a very good base for tiled walls.

# INTERIOR LININGS • 42

1. Check that you can really buy plasterboard sheets cheaper separately, than supplied and fixed. Strange trade this one.
2. If you must try to fix plasterboard yourself practise in the third bedroom. That way if you are not as good as you thought you can call in a tradesperson—nurseries are wallpapered, aren't they?
3. The sheets are fixed horizontally to walls with adhesive. Get instruction pamphlets from the Australian manufacturer. Do not use a book from the United States setting out how to fix plasterboard, as a number of these books advocate fixing the sheets vertically and this is not suitable to Australian methods.

## Plasterboard on masonry

Plasterboard sheets can be used over masonry walls, eliminating the need for the wet trade of rendering. If the walls are dry, straight, square and plumb, the sheets can be bonded directly to the masonry face and finished as normal. Less perfect walls may require timber or metal top-hat furring to correct irregularities, to provide a sound base for the plasterboard or recessed-edge fibrecement boards.

## MDF—medium density fibreboard

This material is somewhere between particleboard and hardboard in its structure.

Timber fibres are pressed into a rigid smooth-faced board that is highly suitable for cabinetwork.

MDF boards can be moulded to give the appearance of timber panelling by the machining in of joints of panels. Highly suited for painted wainscots, as MDF has a very smooth surface, cuts to smooth recesses, is tough and comes in fairly large sheet sizes.

## Timber lining boards

Cedar and local *Pinus radiata* are the most common lining board materials sold in Australia. These are both softwood timbers that are supported by kiln-dried Australian hardwoods (sold as Tasmanian oak in some localities) and other local species, such as Jarrah in Western Australia and Cypress pine in Queensland.

The decision on which board to use is a matter of picking a moulding. The range is often limited to two:

1. Vee-jointed
2. Shiplapped.

One smart manufacturer has a board that is vee-jointed on one side and shiplap on the other, which is good for people who have trouble making up their minds. Other mouldings, including the traditional reeded board, are available for those who want something a little different.

The timber selection is generally a matter of picking a grain and colour you like:

| | |
|---|---|
| Western red cedar | Reddy brown |
| Red pine | Pinky red with dark stripes |
| *Pinus* | Yellow cream |
| Tasmanian oak | Yellow to grey |
| Jarrah | Red |
| Cypress | Honey yellow |

Always go to the timber merchant and inspect the lining boards being offered before making a commitment to purchase.

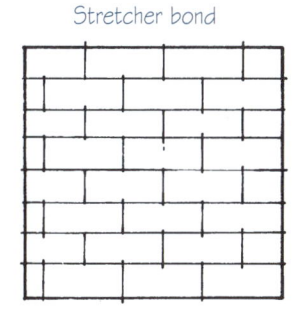
Stretcher bond

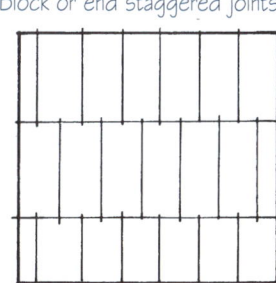
Block or end staggered joints

**Masonry**
Bricks, concrete blocks or natural stone are not technically interior linings, but many owner builders expose these materials in the interiors of their homes. Masonry units may be left in their manufactured colours, be built into particular patterns and they may be painted, bagged and rendered.

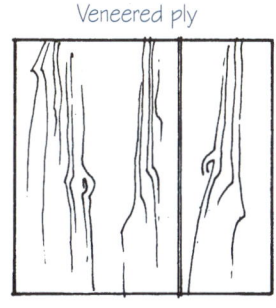
Veneered ply

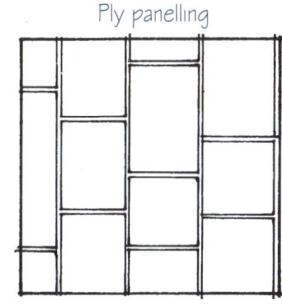
Ply panelling

**Plywood**
Plywood was a very popular lining in the 1920s and 1930s. It has not however enjoyed wide appeal in recent times. Plywood is available in either slice cut or rotary cut veneer faces and in timbers from the mundane to the exotic. It is a unique material in that it can be used to clad and line all the surfaces of the home including the floor and it is a material worth serious consideration by any owner builder.

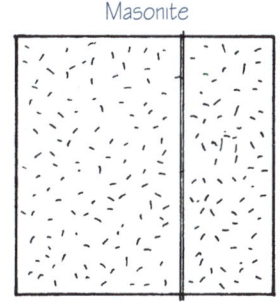
Masonite

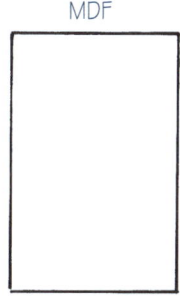
MDF

**Hardboard**
Hardboard wall boards are manufactured from compressed wood fibres, and although available in natural plain brown sheets, more commonly encountered in a home as the facing to flush panel doors and as pre-finished linings on feature walls. Melamine face hardboard can be used to give water resistant wall linings in bathrooms and laundries. Careful fixing is required to avoid buckling and it requires the use of exposed cup head screws.

# 43 • WINDOWS

Timber windows have been used since the beginning of housing construction in Australia. Very early windows were likely to be simple casement or vertical sliding types. Old windows that are handmade from Australian red cedar are very durable and may well be in excellent condition when over a hundred years old. By Victorian times less local cedar was available and window frames were made from endemic hardwoods and imported or local conifers. If they have been well maintained, Victorian era windows, which are mainly box-framed double-hung windows with counterbalance weights, can still give good service. If they have original drawn glass or leadlight they deserve special care. One of the reasons these old windows have lasted so long is that they were primed with high-lead-content undercoats and finished with leaded enamels. Few modern paints will protect so well, but take care when removing the old paint—lead is seriously toxic.

During the Federation period and between the wars, windows were well constructed; they used many variations of elaborate leadlight casement windows, often with bay or bow configuration. Double-hung windows remained in common use, but were generally relegated to the sides and rear of the house.

Since the 1960s, when Melbourne-based architect Robin Boyd designed the first mass-market, factory-produced windows made of western red cedar, this material has dominated timber window frames throughout Australia. Western red cedar (WRC) could be stained or painted and could have double-hung, awning, casement or hopper sashes fitted into its repertoire. It is durable, resisting attack by fungi and insects, but is very soft and quite weak. This was solved by gluing the glass to the frame. If WRC windows are in good repair, they can be used and reused for an extended period of time, but if their joints are damaged or they have many cracked panes, then it is worth considering their replacement.

Steel windows arrived in the 1930s, and through the 1960s were used extensively in modern houses. These are often large and elegantly proportioned but, sadly, many were not galvanised and have succumbed to the ravages of rust. There are companies that can recondition these windows if they are not too rusted, and they can even supply matching windows. During the 1990s there was a modest revival of the use of steel-framed windows by architects, but it is unlikely that steel frames will ever be a significant force again. Their sashes were universally catchment and there is a continuing difficulty in fixing suitable insect screens.

Aluminium windows have progressively replaced steel windows and have made significant inroads into the use of timber. The debate over whether to use timber or aluminium frames has not been resolved in Australia. Some families will replace their old timber windows with aluminium windows, hoping to gain the benefit of low maintenance in the process. Other families will tear out the aluminium windows to install high quality timber windows.

There is logic in both camps, but in the end you get what you pay for. Aluminium windows were often sold as a cheaper and more durable alternative to timber windows, and the market has been flooded over the years since the 1950s with some diabolically bad aluminium windows. Flimsy, narrow frames, poor proportions, permanent glazing, no security locking, highly visible corrosion and non-existent connections have often been the lot of aluminium windows. If the windows in your existing house are poor grade, do not try to save them. The alternative need not be timber, as there are very high-quality aluminium windows available that can more than compete with many of the mass-manufactured timber windows in price and quality.

Glazing in windows has not changed significantly over time, but glass has improved in its optical quality—although modern float glass does not have the often picturesque distortions common in old drawn glass. Look through the windows in your existing house and see if the glass has character before deciding to replace a drawn-glass window with modern float

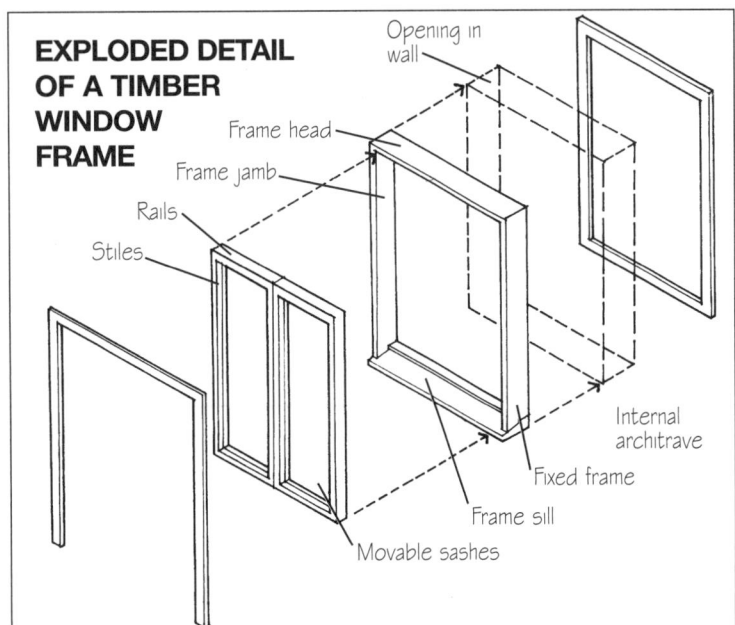

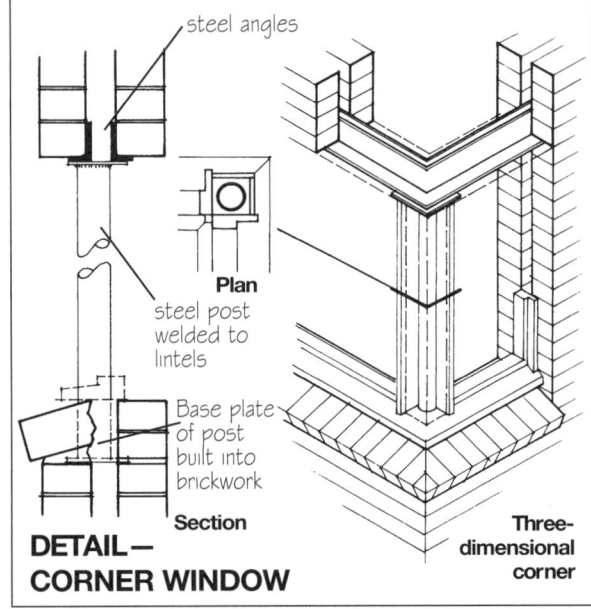

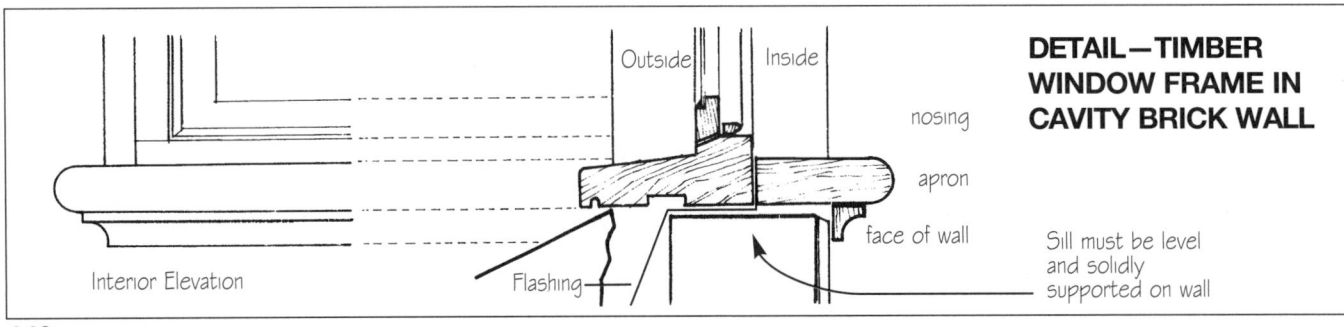

# WINDOWS • 43

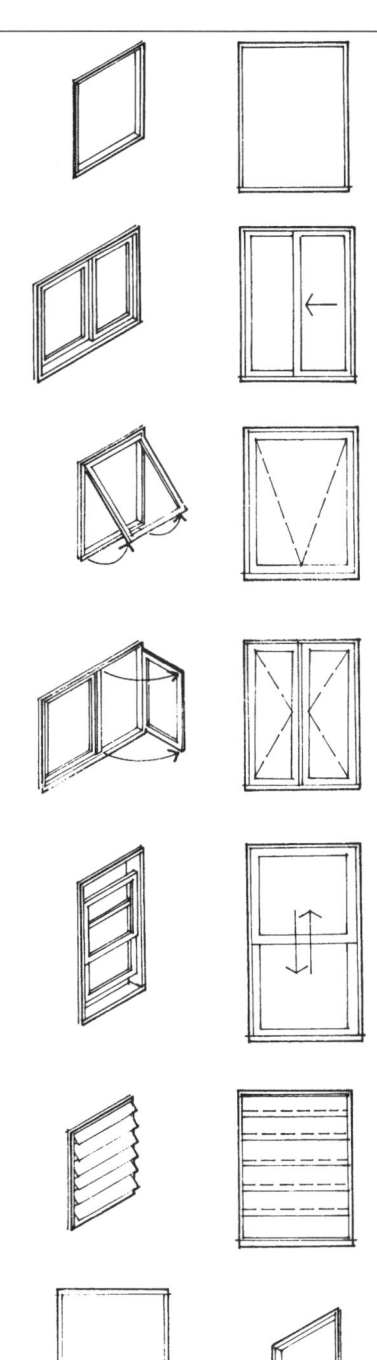

## COMMON WINDOW SASH TYPES

**Fixed light**
- Frame may be in timber or aluminium.

**Horizontal sliding**
- Frame generally in aluminium.
- Commonly produced item in standard sizes.
- Insect screen normally fitted on the outside.
- Wide range of prices and quality, not all are weather tight or secure.
- Range of natural or coloured finishes in anodising or baked acrylic.

**Top hung awning**
- Frame generally in timber.
- Commonly mass produced item in standard sizes.
- Insect screens normally fitted on the inside.
- Generally fabricated in Western Red Cedar.
- Some have clips and friction stays, others use positive action window openers.
- May be painted or stained.

**Side hung casement**
- Frame generally in timber.
- Generally purpose made, some standard ranges.
- Insect screens normally fitted on the inside.
- Generally fabricated in Western Red Cedar or (kiln dried or seasoned) hardwood.
- Stay-openers on winders may be used.

**Double hung**
- Frame traditionally in timber, but now often in aluminium.
- Generally purpose made, some standard ranges.
- Insect screen traditionally fitted on the outside, some patented roll-up screens now used internally.
- Fabricated from Western Red Cedar, hardwood or aluminium, using many counter-balancing systems including traditional lead weights and spiral friction balances.

**Louvre**
- Frame in aluminium, steel or timber.
- Generally purpose made.
- Insect screens normally fitted on the outside, but can be fitted internally.
- Generally fixed to timber sub-frame.
- Poor sealing makes their use in cold climates inadvisable, but suited to tropical areas.

**Bottom hung hopper**
- Frame generally in timber.
- Generally purpose made.
- Insect screen must be fitted on the outside.
- Seldom used these days as it is very hard to achieve 100% water-tightness and open sashes interfere with curtains.

glass. The older types of obscure glass have more defined patterns than most of the modern equivalents; consider keeping the old glass panels and reusing them in the new work. Leadlight and coloured glass from Federation and bungalow houses are not easily replaced; consider carefully before removing, reglazing or replacing any of these windows—modern coloured glass does not have quite the quality of the old. (This could have something to do with the highly toxic materials that were once used to obtain the colour that are totally banned today.)

There was a time when windows were solid frames of timber glazed with glass. Today many windows are becoming panels of glass to hold a frame of lightweight timber or aluminium.

The above statement may be an oversimplification of the facts but it is not far from the truth. There are windows that are sold as quality products that are so designed that a broken window cannot be replaced without removing the frame from the wall and dismantling it.

More and more window frames are being made of aluminium today. Aluminium windows are generally considered to offer longer life expectancy than timber, and normally cost less than a timber-framed window of similar area. Whereas it is probably true that a well-constructed aluminium window will last longer than an equivalent timber window, there are a number of issues that should be considered by all OBs who propose to use aluminium windows.

1. Is the sash type and general window appearance you want suitable for economic reproduction in aluminium? Horizontal sliding windows suit aluminium frames extremely well. Double-hung sashes also are well suited, but can lose appeal if fitted with clip-on plastic 'colonial' glazing bars. Awning and casement windows

# 43 • WINDOWS

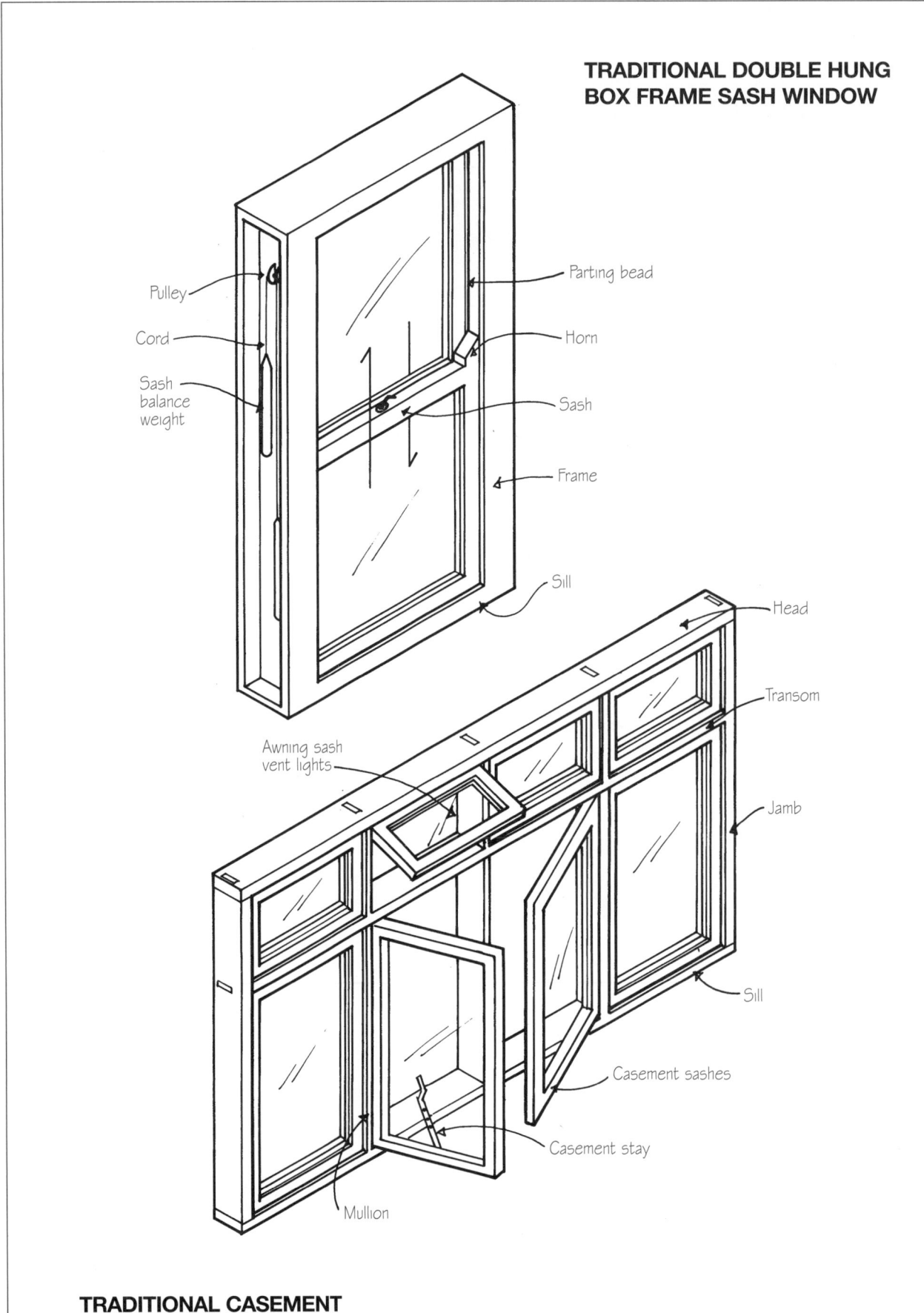

**TRADITIONAL DOUBLE HUNG BOX FRAME SASH WINDOW**

**TRADITIONAL CASEMENT FRAME WINDOW**

# WINDOWS • 43

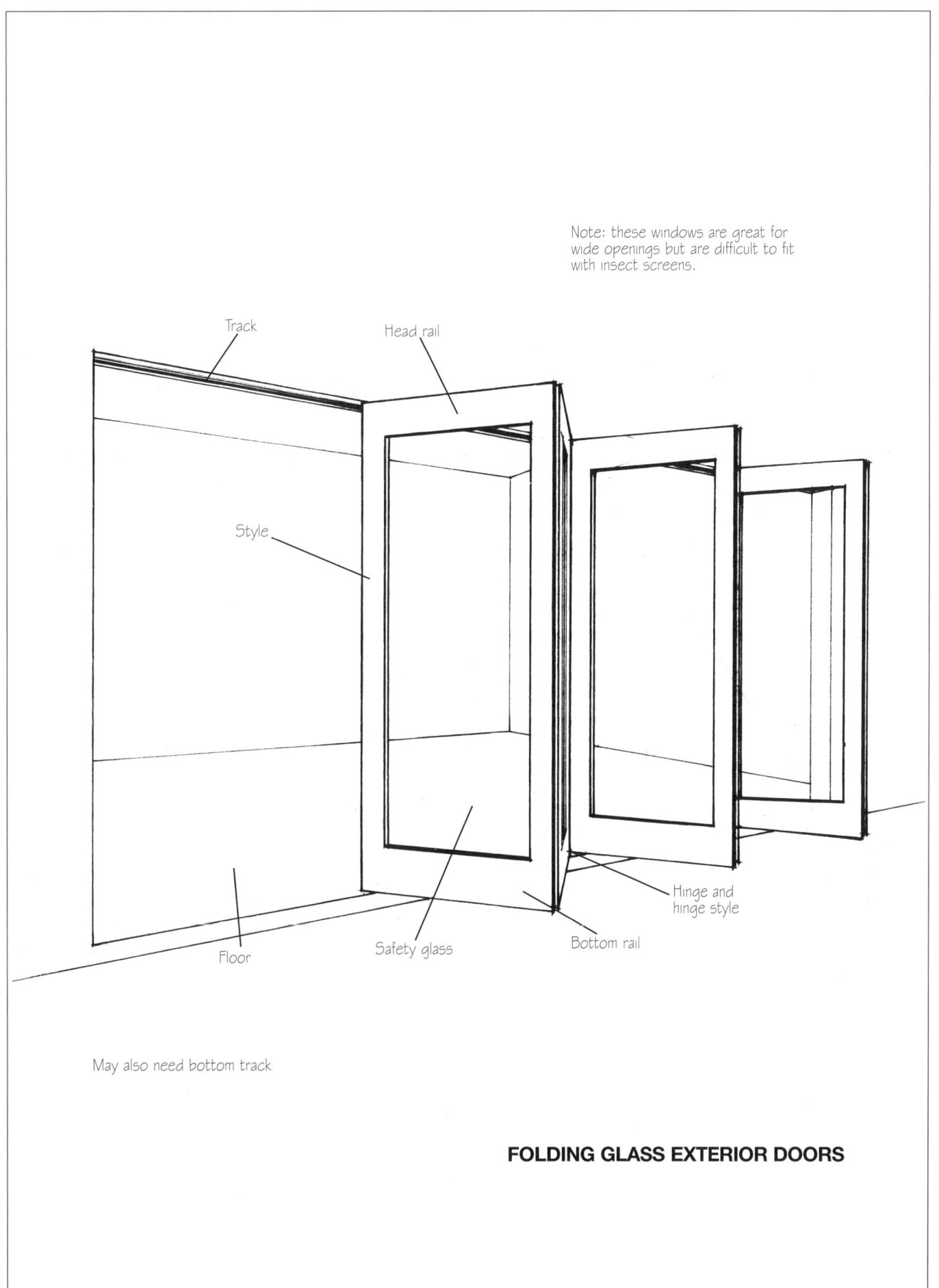

Note: these windows are great for wide openings but are difficult to fit with insect screens.

May also need bottom track

**FOLDING GLASS EXTERIOR DOORS**

# 43 • WINDOWS

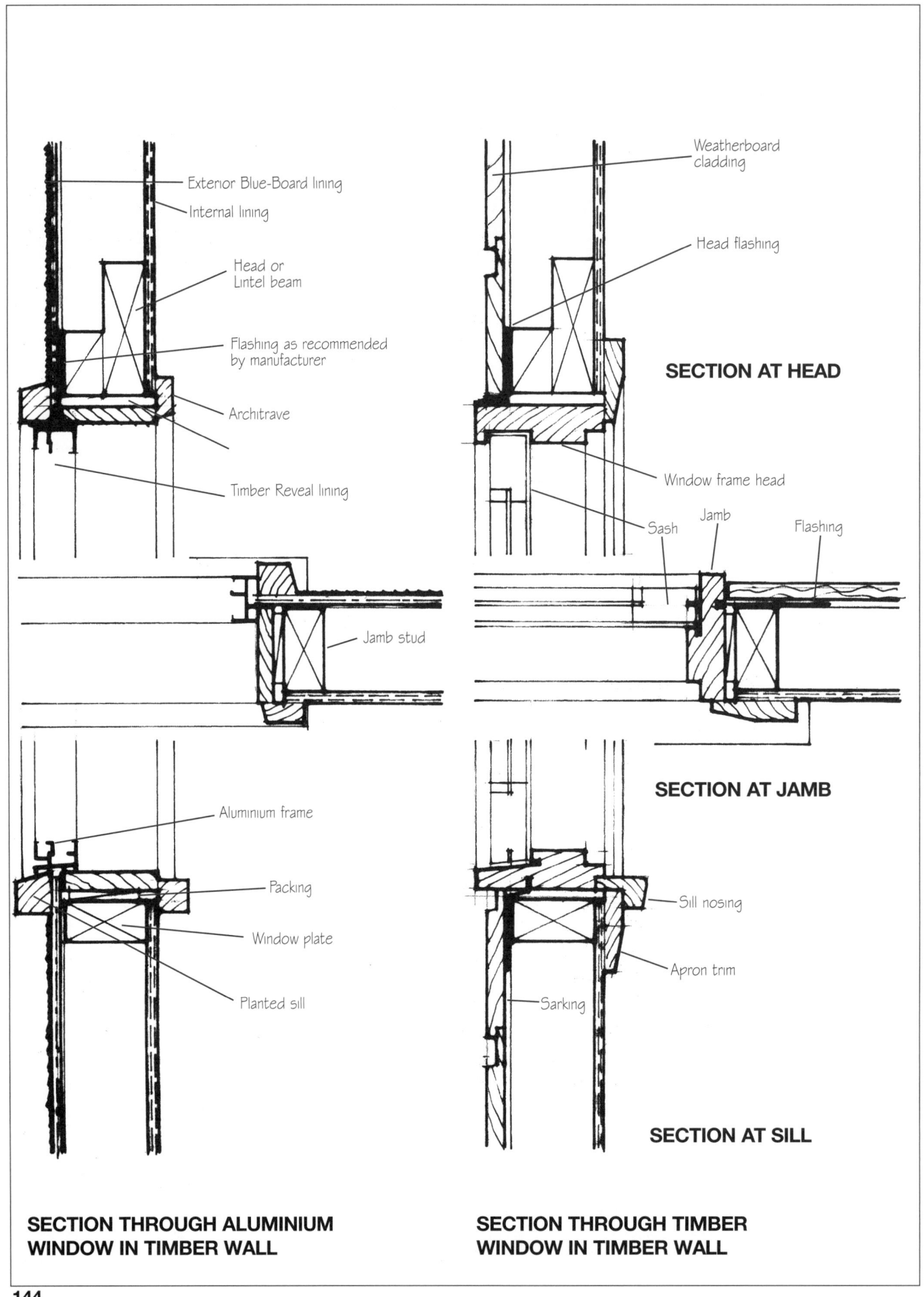

# WINDOWS • 43

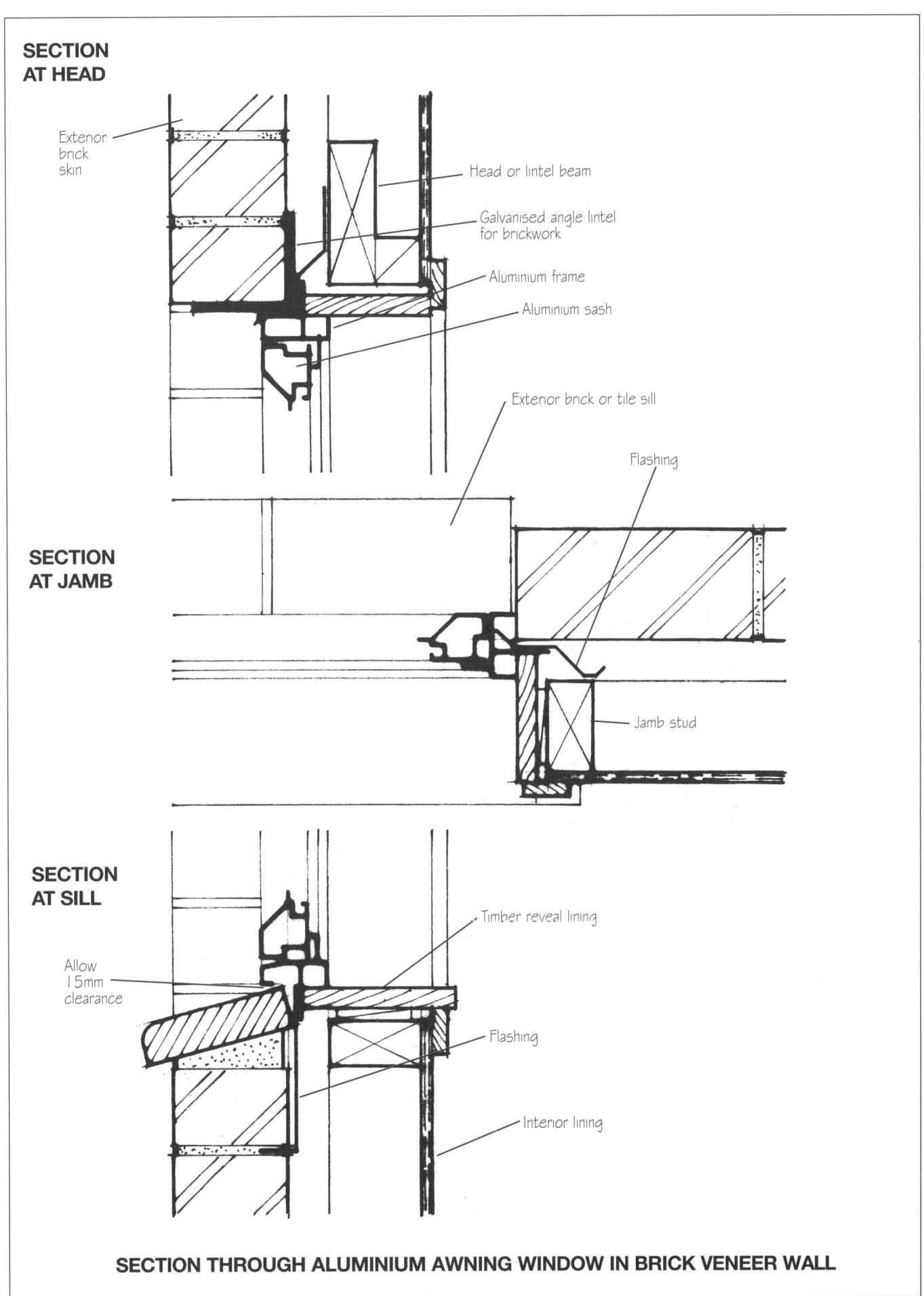

**SECTION THROUGH ALUMINIUM AWNING WINDOW IN BRICK VENEER WALL**

# 43 • WINDOWS

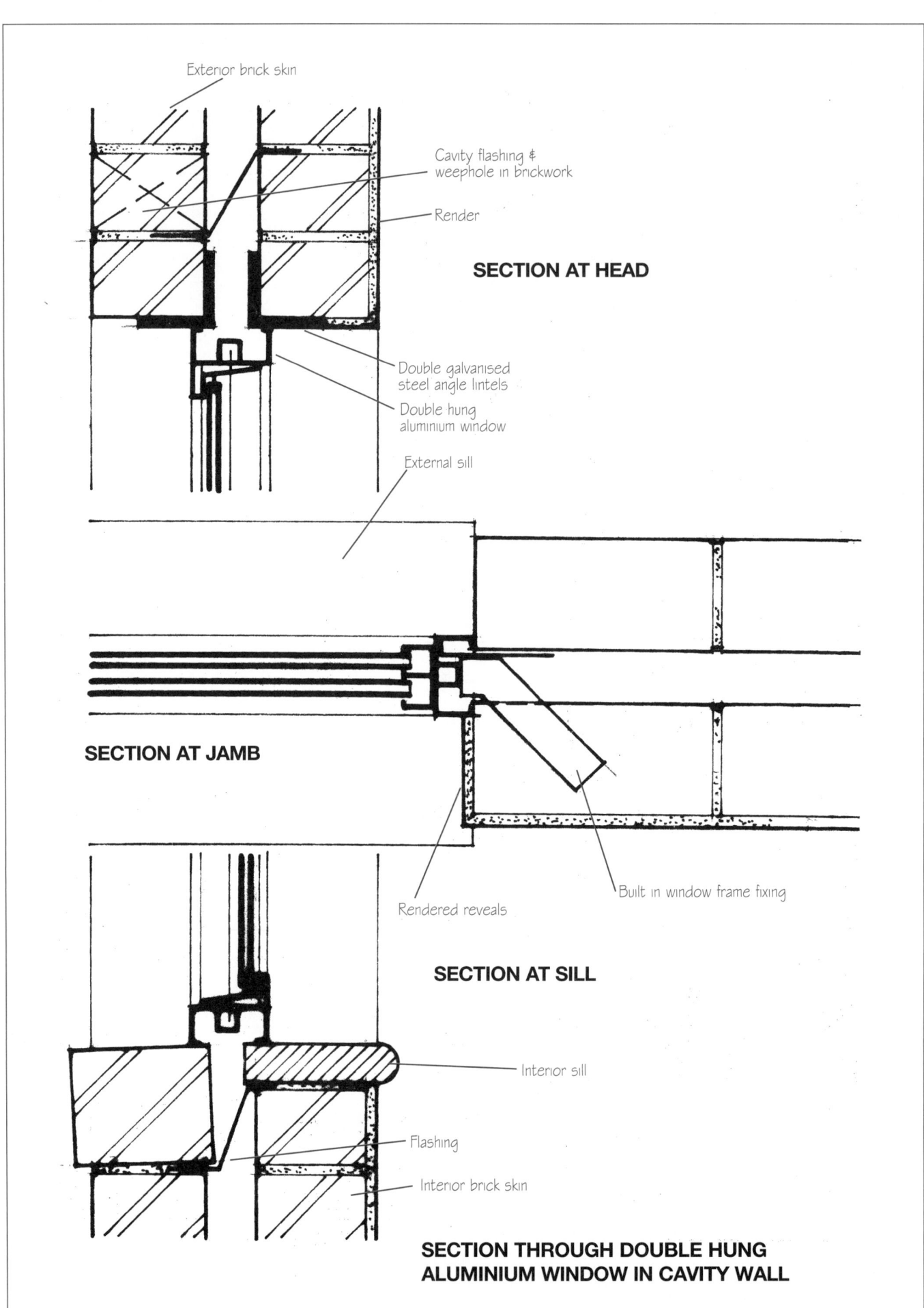

# WINDOWS • 43

are generally considered to be more suited to timber frames than aluminium.
2. Aluminium itself is an extremely durable material which will last almost infinitely, but some manufacturers of aluminium-framed windows have been known to skimp on the corner fixings they have used. OBs should check the fixings on the windows they are likely to purchase for flimsy or incompatible materials. Timber windows should also be checked to make sure that suitable joints are used.
3. Many aluminium windows sold today have a durable colour coating that is an obvious improvement in the appearance of aluminium for many residential applications. What OBs must establish is how long the colour coating on the aluminium windows they buy will last; no applied finish has an infinite life. Timber windows must be painted regularly, every 5 years or so. If this is done, the window frame has a very long life expectancy—over a hundred years is possible. But with an aluminium window, the factory-applied colour coat may last 10 or more years. The question that OBs need answered is how they will repaint aluminium window frames when they eventually require it?

The windows manufactured to suit single-storey suburban cottages may not be suitable for windows in more exposed locations, as they are not designed to withstand driven rain, snow, dust storms or similar unconventional weather conditions.

If an OB wants a window of a special design to operate successfully in abnormal conditions, to look different, to be thermally or acoustically more efficient, then the OB must contact a specialist window manufacturer. Specialist window manufacturers are generally separated into timber and aluminium window fabricators, and it is often worthwhile getting a price from these custom manufacturers. Even if the windows being priced are close to standard.

When a number of prices are tendered for windows, make sure the glass that is included or not included in the quotes is the same. Some manufacturers will quote including glass and others will quote frames only. Also, the quality and thickness of the glass can make a significant difference to the cost of the windows.

When the window is built into the wall, remember to fit all necessary flashing and storm moulds to the sill, jambs and head of the window. Incorrectly fitted sill flashings often cause dampness to appear on the walls at the skirting level 500 mm or so on either side (or sometimes only one side) of the window. This is often confused with rising damp, which seems to send the majority of home owners into panic, and thousands of dollars are spent trying to stop rising water that is really coming down from a poorly flashed window above.

## Window glass

There is no perfect glass for every application; selection is made by weighting up many factors.

Modern clear window glass is made by a process know as float glass and although not optically perfect, it is generally clear of distortion.

There is a library of Australian Standards dedicated to the manufacture and use of window glass; most of these standards have been incorporated in to building regulations and codes. This means that OBs will be responsible for complying with all the standards and codes.

The Building Code of Australia—since July 1999—has moved to recognise a number of standards associated with the manufacture and glazing of windows. There are many good reasons for the codification of these standards, but it is not all good news

## SCORE CARD FOR WINDOW ENERGY EFFICIENCY

| Summer Score | Winter Score | Average | |
|---|---|---|---|
| **Orientation of Living Areas** | | | |
| Living areas over 75% north facing | 10 | 10 | 10 |
| Living areas 50% to 75% north facing | 7 | 7 | 7 |
| Living areas 25% to 50% north facing | 3 | 3 | 3 |
| **Orientation of House Living Areas** | | | |
| Long direction of living areas faces north with windows | 5 | 5 | 5 |
| Long direction of living areas faces 15∞ of north | 4 | 4 | 4 |
| Long direction of living areas faces 30∞ of north | 1 | 1 | 1 |
| **Window Shading Devices - eaves, verandas, pergolas, etc.** | | | |
| North side width   450 to 1000 | -10 | 0 | -5 |
|                    1000 to 1500 | 20 | -8 | 6 |
|                    over 1500 | 22 | -14 | 4 |
| add if fixed east shading or no window | 9 | -3 | 3 |
| add if fixed west shading or no window | 9 | 3 | 6 |
| add if east/west shading is operable | 0 | ? | 1 |
| No eaves or shading on any face | -20 | 10 | -5 |
| **Glazing** | | | |
| 75% of glass on north side | 3 | 4 | 3.5 |
| 50% of glass on north side | 2 | 3 | 2.5 |
| add if double glazing/appropriate high efficiency glass | 14 | 14 | 14 |
| **Insulation - Windows** | | | |
| Aluminium frame - clear glass | 0 | 0 | 0 |
| Aluminium frame - toned glass (shading coefficient 0.66) | 15 | -12 | 1.5 |
| Timber framed or thermal break aluminium | 4 | 24 | 10 |
| add for 3/10/3 double glazing to south only | 2 | 4 | 3 |
| add for 3/10/3 double glazing all round | 18 | 16 | 17 |
| **Timber or thermal break aluminium** | | | |
| + 5/10/3 double glazing, toned glass | 60 | -28 | 16 |
| **Timber or thermal break aluminium** | | | |
| + 4/12/4 double glazing, argon filled | 32 | 54 | 43 |
| **Maximum possible score** | | | 100.5 |

# 43 • WINDOWS

for OBs. The main problems are:
- recycled windows may not be considered acceptable by building inspectors
- made on site windows will be difficult to certificate
- short guarantee periods of less than 10 years are offered and promoted by the window manufacturers as a valuable provision for the property owner. The problem with this type of guarantee and the associated quality certification is that it seems to miss the point that most families actually expect their homes to last a lifetime.

> Most windows are pre-glazed and, if the manufacturer and supplier have been briefed correctly at the time of ordering, then the window should display a sticker providing the following information:
> - the manufacturer's certification that the window has been designed in accordance with AS2047 (or whatever other standard or amended standard is applicable)
> - that the performance of the window has been verified at a NATA (National Association of Testing Authorities, of Australia) accredited testing laboratory
> - the manufacturer's name
> - the manufacturer's accreditation number
> - the structural performance of the window
> - the water performance of the window.

## Safety glazing

Safety glazing is required by regulation in a number of specific locations, the most common of these are:
- full-height glass window walls
- glass doors
- shower screens.

There are two types of safety glass commonly sold on the Australian market.

**Toughened glass**—heat-treated so that if it is broken there are no shards of glass, only relatively harmless crystals.

**Laminated glass**—consists of two sheets of glass sandwiching an inter-layer of a plastic material. If it is broken any shards will remain stuck to the plastic and are less likely to stab a person.

These two safety glass types are safer than plain float glass; they are stronger and resist breakage, but they are significantly more expensive than plain glass. They can still cause serious injury to human if broken, not the least because when people run into them they do not break.

Other safety glazing products existing include traditional Georgian wired glass, plastic sheets (Lexan, Perspex, acrylic, resin and fibreglass) and glass blocks.

## Obscured glazing

Traditionally, Australians have used obscured glass in their bathroom windows. Many beautiful glasses were made for this market and families gave serious consideration to their choices. Now most of the few readily available obscure glasses sold in Australia are effective but ugly.

Although more and more families have moved to clear glass in their bathroom windows with curtains to maintain their modesty, the use of obscure glass may still be growing. In many locations around Australia, authorities have decided to impose rules that restrict neighbours overlooking each other's properties through windows. They can still go outside and look over the fence—and in many places require windows facing side or rear boundaries to have their sills above eye level or have obscured glazing. Please note: there does not seem to be concern that people can open windows and by-pass the obscured glazing.

Check what is required by the local authority and try to get a definition of what constitutes obscured glass—stained glass, ozone glass, colour-back glass, recycled cast glass.

## Environmentally sensible use of glass

Check the local energy star-rating guide as the starting point for sensible application of windows and glazing in residential projects.

The table shows the guidelines relating to windows used by one council in New South Wales—extracted from *Energy Efficiency Scorecard*, Blue Mountains City Council 1999. This is for general information only; every location will have different criteria to suit the local climatic conditions. There are a number of considerations effecting energy scorecards: a score of over 40 per cent is required to achieve a 3.5 star rating and over 70 per cent to gain the maximum 5 stars. This table demonstrates the impact window design can have on the rating.*

It is possible to lose direction when dealing with the use of windows and glazing when designing a house. Much of the data published suggests that the more money spent on the purchase of windows and glass, the better the internal environment of a house.

Glass and glazing selections can be as much influenced by ventilation, shading devices, drapery, views, privacy, and personal values as by the charting of average human factors with perfect environmental factors.

All factors must be considered carefully; some people may enjoy full winter sun availability from a northerly aspect through a clear glass window and be prepared to use adjustable awnings and heavy drapes to control the excess heat load in summer and the low overnight temperatures in winter. If they lived in the Blue Mountains, the Council would consider them to score:
15 for orientation
−10 for shading—adjustable awnings and drapes do not score
3 for glazing to north
8 in total

If you spent many thousands of dollars for argon-filled double-glazing, their score would have jumped to over 43 for that factor alone. However, they would not feel the warmth of the sun in winter …

Just careful selection of windows and glazing will give a full 100 per cent score and 5 stars.

*Note: these scores are part of an attempt to ensure all house designers make an effort to consider environmental factors in their designs; they are neither absolute nor definitive.

## Double glazing

Double glazing is not a cure-all and must be used with great care. Always seek professional guidance and think carefully about what is to be gained from using this expensive product.

There is no doubt if a window must face a source of noise pollution, that impacts on the reasonable enjoyment of a residence - double glazing as likely to be successful. It must be used in conjunction with walls of at least an equal sound transmission coefficient, adequately sealed window frames and fixed windows - if the windows are opened then all sound insulation is lost.

Double glazed windows can be used to contain energy within buildings and reduce the transmission of heat from the outside - both from the latent heat held in humidity and the impact of the direct rays of the Sun. So can heavy drapes and sun-shading devises.

This indicates the down-side of some double glazing systems - in many instances the window is restricted to being a light and view device, losing its role as a ventilator and as a source of warmth.

Houses with double glazing will often require; complex air conditioning, fresh air and climate control systems to work efficiently. This can be a high up-front capital cost and a long term; maintenance and energy purchase impost.

Double glazing systems have their place but must be designed into the whole residence in conjunction with all other environmental factors to be efficient - used in an ad hoc manner and they are likely to be inefficient and expensive.

# DOORS • 44

Doors are used on the outside and the inside of houses; this means that there are those that have to keep out the elements and give security, and those that simply function to close openings between internal spaces. Special doors may be used to control the spread of fire or to reduce noise transmission.

## External doors

Doors used on outside walls will generally need to be resistant to environmental conditions likely to cause deterioration of the door leaf. The door frame must be draught-proof and, as far as possible, the whole assembly should be free of rattles and have a secure lock.

External doors can be broken into two distinct groups, although they are not always clearly separated by function in all houses.

**Service doors**—including the front and rear doors of a residence
**Informal doors**—including French and patio doors.

The front door of many houses is both ornate and of heavy-duty construction. This is obviously desirable, for it is the front door that is generally the first line of defence against burglars. Solid core or semi-solid core doors are acceptable if faced with waterproof plywood/hardboard and glued with waterproof adhesives. Framed doors with solid timber panels are also satisfactory. If glass panels are used in front doors, consideration should be given to the use of laminated glass for security and safety; even if a decorative glass is desired, a double glass panel should be used with one of the sheets in laminated glass.

The rear doors of a house from laundries, kitchens and the like are not always given as much attention as is sensible. It is recommended that a waterproof door leaf is used and that any glass used should be laminated or at least toughened to improve security and safety.

French doors are glazed doors, generally in pairs, that lead to the outside of houses from rooms such as living rooms, dining rooms, family rooms and bedrooms. French doors are nominally hinged doors, whereas patio doors are normally sliding. French doors are almost exclusively made with timber frames, while patio doors commonly have aluminium frames—although a number of manufacturers do make them with timber frames.

In most states of Australia, regulations require that safety glass be used in the glazing of French or patio doors. The mass-manufactured doors generally use toughened glass that forms safe non-jagged crystals if broken. Purpose-made glazed doors are more likely to use laminated glass, which is considered safe because of the high impacts necessary to break it, and because it has a tough perspex interlayer that reduces the risk of parts of the body being cut on any slivers of glass.

Most external door frames are rebated out of solid stock that reduces the likelihood of weather penetration, and weather strips should be fitted to all doors with flush-through thresholds.

## Internal doors

Internal doors are generally hinged or sliding; both methods of opening the doors have their uses.

Swinging doors have been used for centuries and if correctly hung in the first place will last for up to 100 years before requiring new hinges. Sliding doors may have a similarly long history but it is not until recent times that long-lasting, trouble-free sliding door hardware has been perfected. Even now, a sliding door is likely to give more trouble than a hinged door.

The rule, then, is to use hinged swinging doors wherever possible and use sliding doors only where their special features are critical.

Sliding doors are sold as pre-hung units designed to be built into timber-framed walls so that the door leaf slides out of sight into the wall. These are ideal for many applications, but a word of warning ... remember to leave an opening a bit over twice the width of the finished door in the wall frame to accommodate the unit. Also, pipes, conduits or some types of electrical wiring will not fit into the same part of the wall as the open door and this can give trouble in bathrooms, for instance.

There is a of an OB who went to a lot of trouble to keep the pipes out of the cavity of the concealed sliding door into the bathroom but later, when it came to the time to fix the towel rail in the bathroom, forgot where the door cavity was. The bathroom door is now permanently open—he screwed the towel rail right through the wall lining and into the open door behind.

Internal doors are often hollow core. This means that the door consists of a lightweight perimeter frame of timber, infilled with paper 'honeycomb' or similar ultra-lightweight core material and faced with thin 3 mm plywood or hardboard. These hollow core doors are amazingly strong, very cheap and perform their function admirably, but there is one point to remember: be careful when fitting the door lock/latch, as there is likely to be a restricted area available for morticing.

Many internal hinged doors are now purchased in a ready-hung form. This method of buying doors is ideal for the OB, for it means that the time-consuming and exacting job of hanging a door, mitring and fixing the architraves is reduced to a very simple task. Internal glazed doors should have safety glass.

### Folding glass exterior doors

The use of very large openings in houses has led to the development of multi-panel folding door systems. Special tracks, hinges and associated hardware systems are available to suit openings over 6000 mm wide.

These doors are suited to living rooms and family rooms that have at grade access to decks, terraces or even lawns—they allow the inside and outside to merge. They

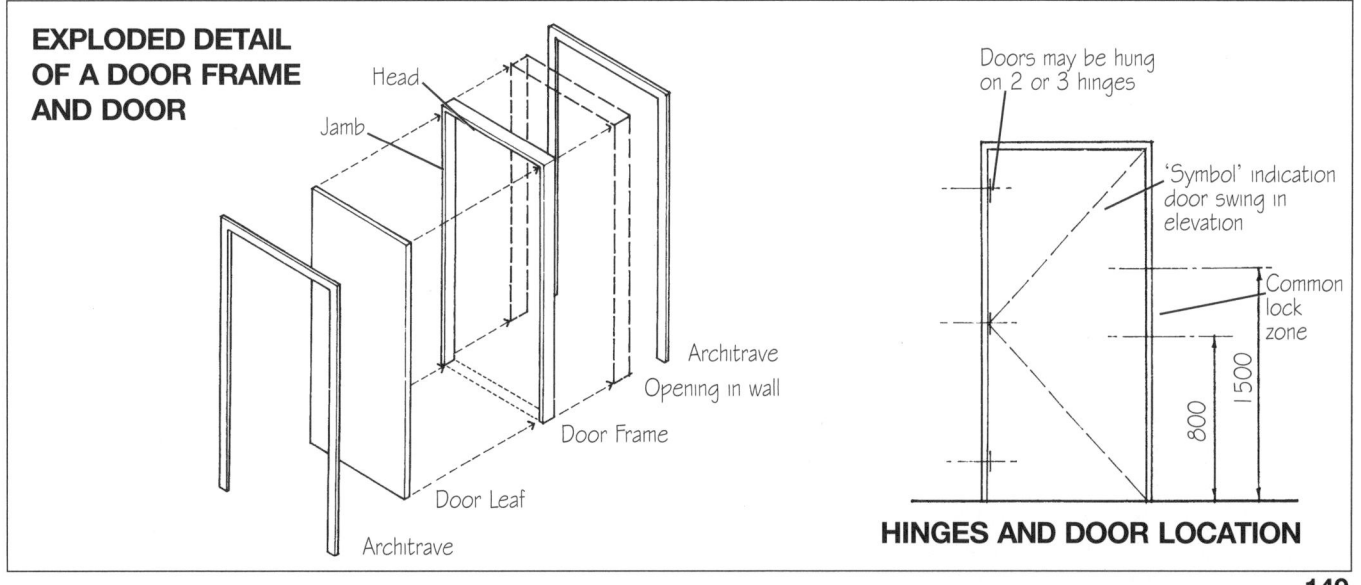

**EXPLODED DETAIL OF A DOOR FRAME AND DOOR**

**HINGES AND DOOR LOCATION**

# 44 • DOORS

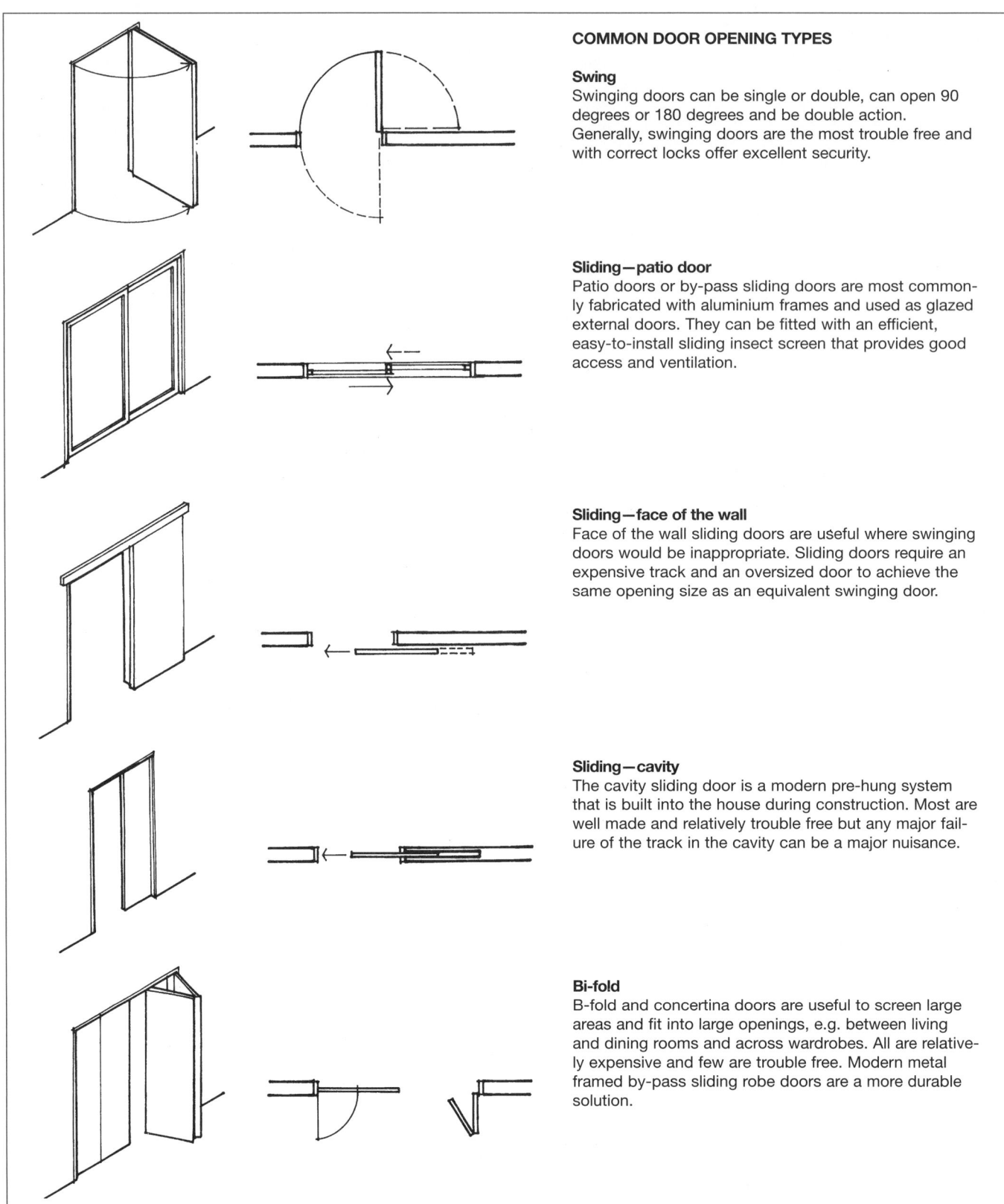

**COMMON DOOR OPENING TYPES**

### Swing
Swinging doors can be single or double, can open 90 degrees or 180 degrees and be double action. Generally, swinging doors are the most trouble free and with correct locks offer excellent security.

### Sliding—patio door
Patio doors or by-pass sliding doors are most commonly fabricated with aluminium frames and used as glazed external doors. They can be fitted with an efficient, easy-to-install sliding insect screen that provides good access and ventilation.

### Sliding—face of the wall
Face of the wall sliding doors are useful where swinging doors would be inappropriate. Sliding doors require an expensive track and an oversized door to achieve the same opening size as an equivalent swinging door.

### Sliding—cavity
The cavity sliding door is a modern pre-hung system that is built into the house during construction. Most are well made and relatively trouble free but any major failure of the track in the cavity can be a major nuisance.

### Bi-fold
B-fold and concertina doors are useful to screen large areas and fit into large openings, e.g. between living and dining rooms and across wardrobes. All are relatively expensive and few are trouble free. Modern metal framed by-pass sliding robe doors are a more durable solution.

are generally fully glazed and can be timber-framed, steel-reinforced timber-framed aluminium-framed or frameless.

Special care is required to ensure that the door head is well constructed, as the top of the opening is subject to the weight of the door, because most systems require a top track to support the doors. If the head sags when loaded, the doors are likely to jam and bind. Care must also be taken to ensure that the head will resist any wind loadings transferred from the doors.

If the door is to be used in an area where there are many flying insects, be aware that there is no easy or neat way to provide insect screening.

Generally, the last door in the chain will act independently, so that easy entry is available when the main doors are shut.

Ventilation is all or nothing, as these doors are generally fully open or fully closed. In some cases frameless double-hung windows can be built into the door frames, this also allows insect screening (see illustration in WINDOWS).

# DOORS • 44

## COMMON TYPES OF DOORS

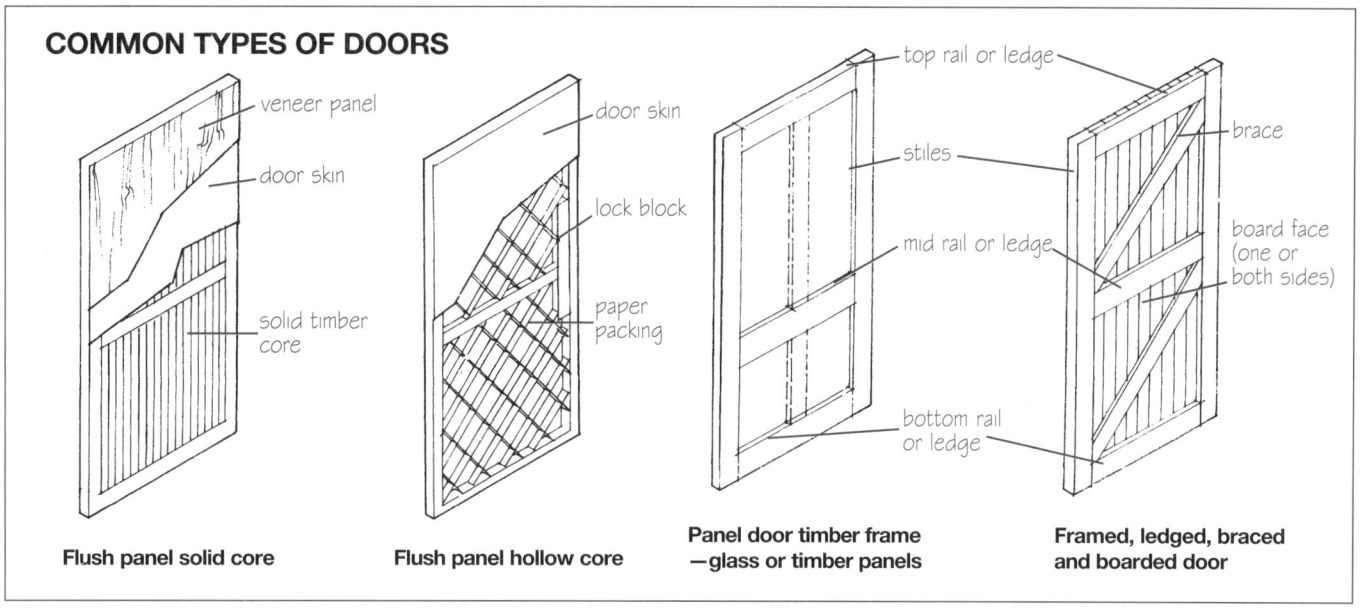

Flush panel solid core | Flush panel hollow core | Panel door timber frame—glass or timber panels | Framed, ledged, braced and boarded door

## COMMON DOOR TYPES

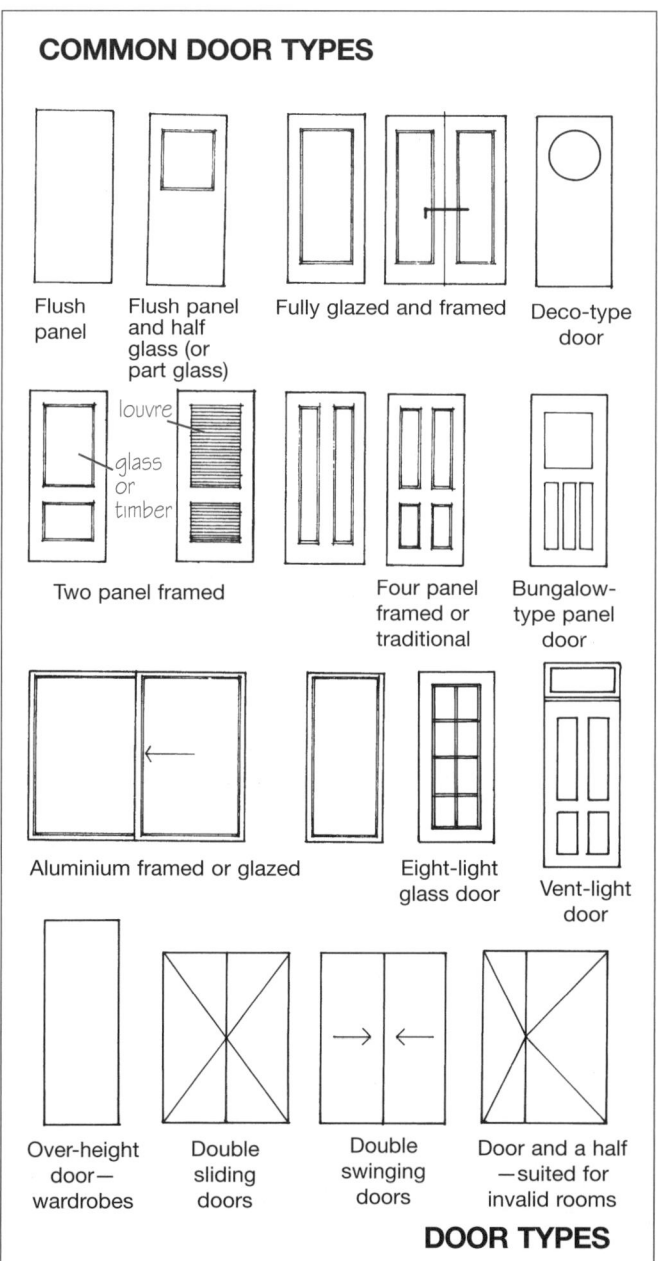

**DOOR TYPES**

## DETAILS— DOOR JAMBS

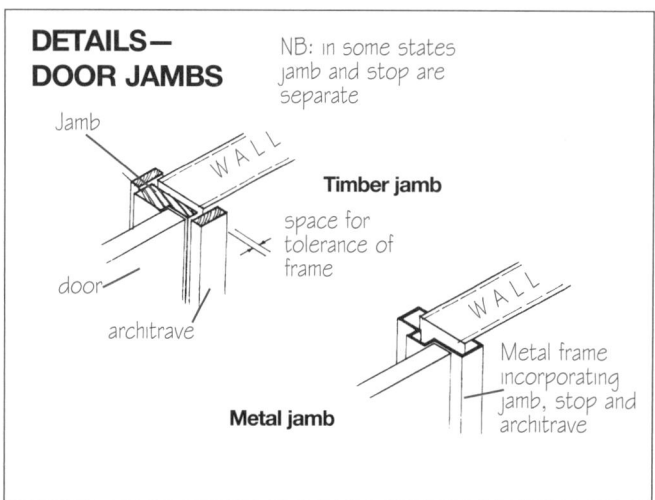

NB: in some states jamb and stop are separate

## MEETING STYLES

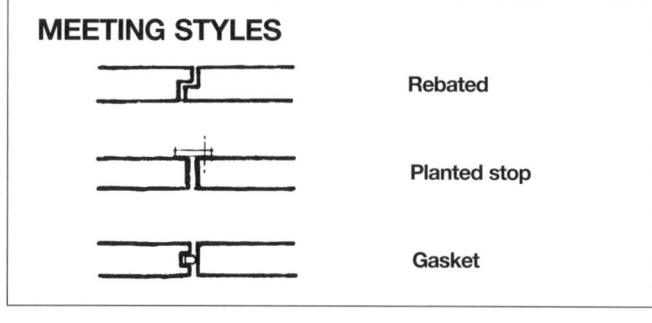

- Rebated
- Planted stop
- Gasket

## WEATHER PROOF THRESHOLDS

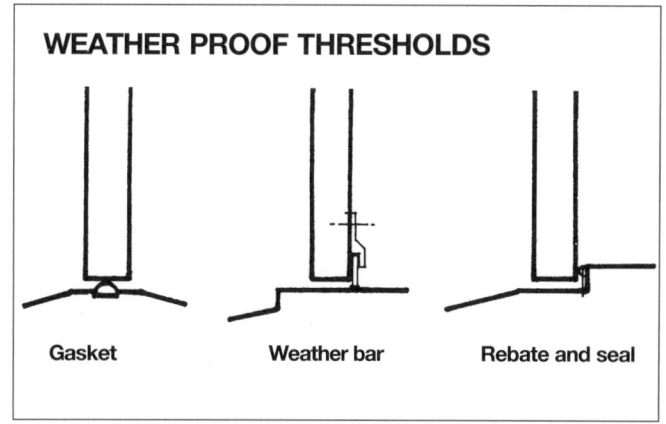

Gasket | Weather bar | Rebate and seal

151

# 45 • ROOFS

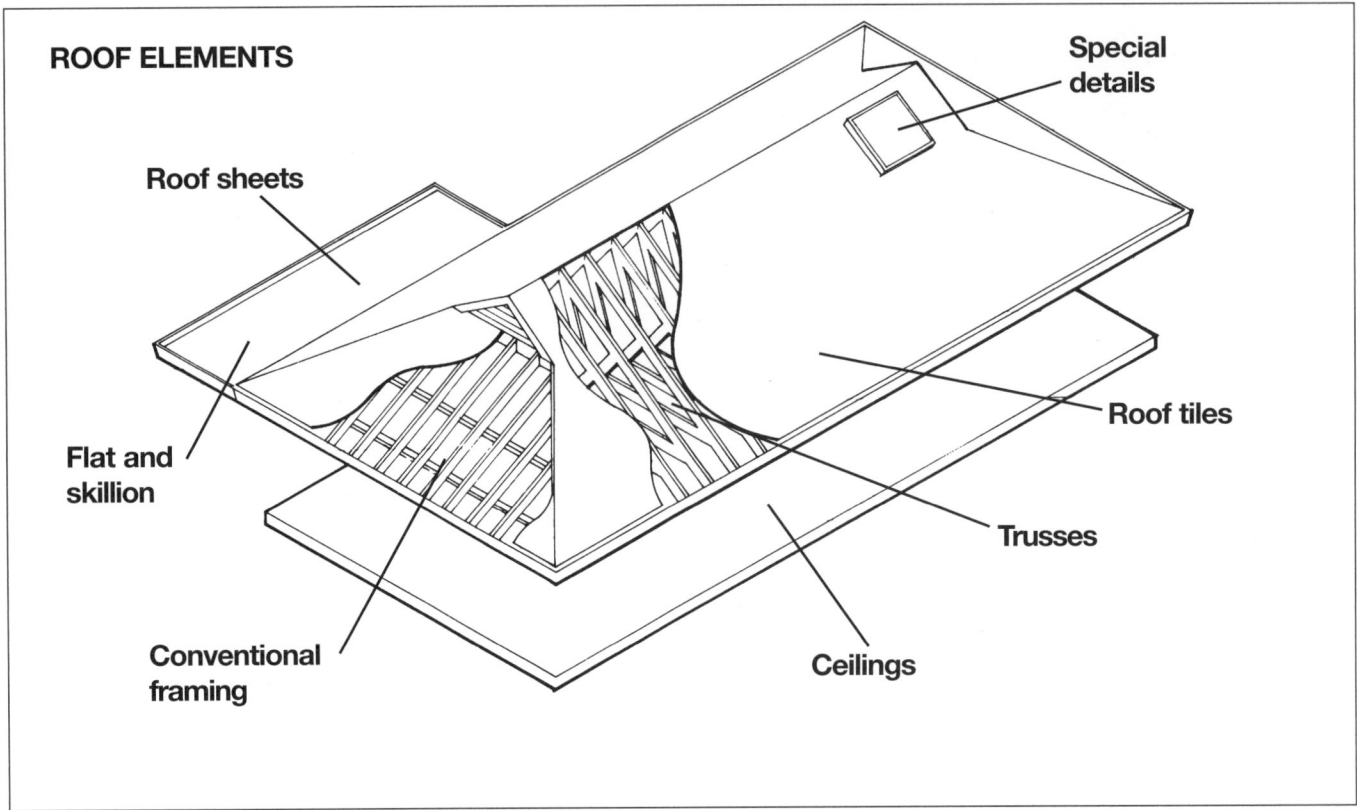

**ROOF ELEMENTS**
- Roof sheets
- Flat and skillion
- Conventional framing
- Special details
- Roof tiles
- Trusses
- Ceilings

## Structural frame

The frame of a roof in residential construction is generally built from timber elements. It is common to break roof frames into two basic types:
1. Stick-built—those constructed from assembling the same lengths of timber by the use of saws, hammers and nails, on-site. Stick-built roofs can be either pitched or flat.
2. Trussed—those constructed from prefabricated, triangulated assemblies that are hoisted onto walls as a complete two-dimensional frame. Trussed roofs can be of many shapes, but are most suited to gable and near-gable applications.

## Roof cover

The roof cover is the surface of the roof. Its main function is the weatherproofing of the house. Roof covering in Australia is normally split into two categories, although two further categories are possible.
1. tile roof—including interlocking terracotta or cement tiles, slate, fibrocement or composite material shingles, timber shingles or shakes and pressed metal tiles, together with some other more exotic systems that are not significantly applied in Australia.
2. sheeted roof—including steel and aluminium-based sheet products, from traditional corrugated zinc-alloy coated steel through many different profiles in steel and aluminium sheets, each with a particular purpose, and numerous coatings to satisfy aesthetic and durability requirements.
3. membrane roofs or built up roofs—roofs where a solid base of plywood or concrete is built and multi-layers of a waterproof membrane applied. These roofs are used extensively in Europe and North America but, because they generally require bitumen as a sealant, they have proved less satisfactory in Australia, where high summer temperatures have caused high maintenance cost and short lives.
4. Concrete roofs—used occasionally in Australia, particularly where 'walk on' roofs are required. But, as with built-up roofs, they are expensive to build and maintain.

## Ceilings

The ceiling of cottages can be constructed in many different ways to satisfy their functions of giving a top enclosure to internal spaces and stopping the intrusion of drafts and dust into the living spaces.

In Australia, ceilings have for many years, since the invention of fibrous plaster and its successor plasterboard, been predominantly made from this material. In the 1920s and 1930s these ceilings were decoratively moulded, but in recent years they have generally become smooth and flush jointed, sometimes raked to the roof slope and with exposed rafters or purlins.

The text on ceilings deals with preparing the structure to take particular ceiling forms and material applications, and explores the structural link between the roof, walls and ceiling system.

## Roof lights

In the past there were major problems in installing windows in roofs. Today most of the old problems have been, for the most part, overcome, although the installation of roof lights requires extensive planning to achieve satisfactory results. Roof lights come in four basic forms:
1. the dome light—a mass-manufactured unit designed to be installed in a flat or sloping roof of either tile or sheet covering.
2. the southlight—so named because these vertical or near-vertical glass windows on roofs traditionally faced south to avoid the build up of summer heat or excessive glare inside the house. Although care should always be exercised when installing roof lights, south-facing may not always be applicable. In fact, some north-facing roof lights are used in passive solar heating systems.
3. roof slope windows—generally mass-manufactured, complex imported units, which are most suitable for room-in-roof applications.
4. dormer windows—built into the roof frame where attic rooms are part of the design. They tend to be expensive and complex to construct, and OBs should plan carefully before deciding on their use.

# ROOFS • 45

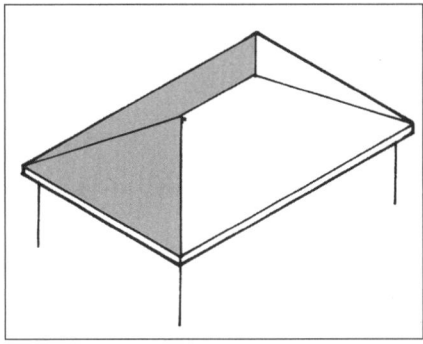

**1. Hip:** The hip roof is a traditional form of roof in brick veneer houses. It is extremely economic to construct by skilled tradesmen. It has the advantage of continuous gutter and one height walls.

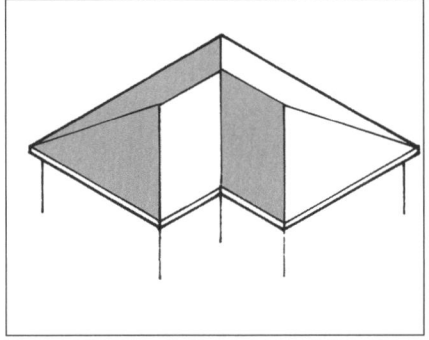

**2. Hip and valley:** Has the same general characteristics of the hip roof but by the addition of the valley(s) may be used to cover multi-sided plans.

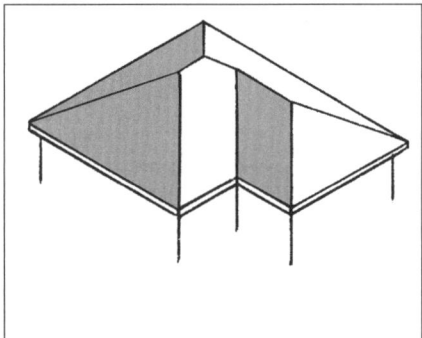

**3. Broken hip and valley:** Has the same general characteristics of the hip and valley roof but allows for variation in the width of the wings.

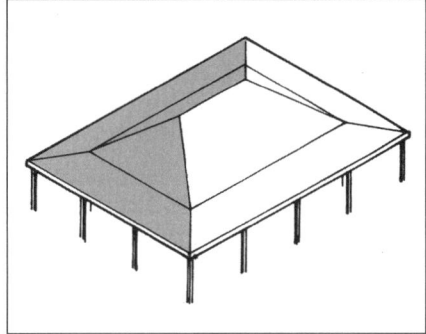

**4. Bellcote:** This is a hip roof with a continuous verandah/lean-to around its perimeter. Generally the roof pitch of the verandah is shallower than the main roof. The verandah roof may be a simple plane or can be curved galvanised metal sheets.

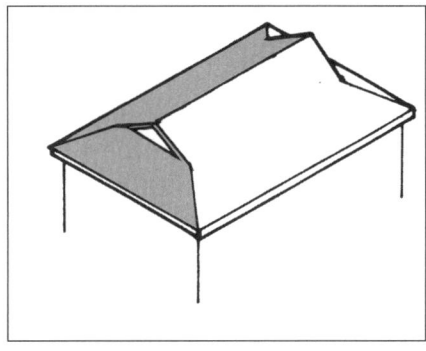

**5. Gambrel:** This is a roof combining the simplicity and economy of a hip roof with the addition of mini gables to provide ventilation to roof space, and to allow the roof to make economic use of roof trusses by reducing the number of truncated trusses required – or just to add character to the roof.

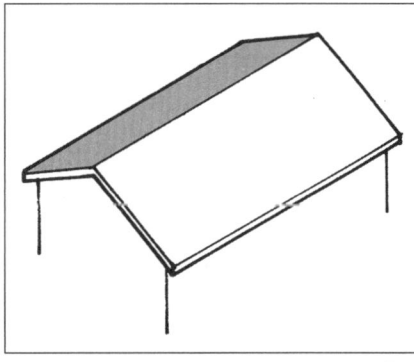

**6. Gable:** This roof is a traditional form of roof used in many historic period styles. It is probably easier to construct by an owner builder than the hip roof and is ideally suited to designs using light-weight roof trusses or vaulted ceilings. It has the economic disadvantage of not only requiring as much roof covering material as a hip roof, but also having exposed gables that have to be clad.

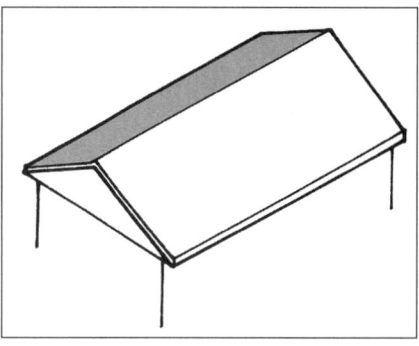

**7. Boxed gable:** A gable roof built so that the gable end of the roof is extended out beyond the building wall and the roof and walls become clearly separated components.

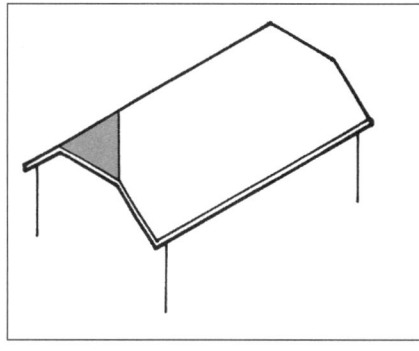

**8. Jerkin head or Dutch gable:** A roof which is a combination of gable construction with a hip added at the apex of the gable. It is most suited to houses where an attic room(s) is required – and therefore steep roof pitch – as the mass of the bale is reduced visually and the roof line is softened.

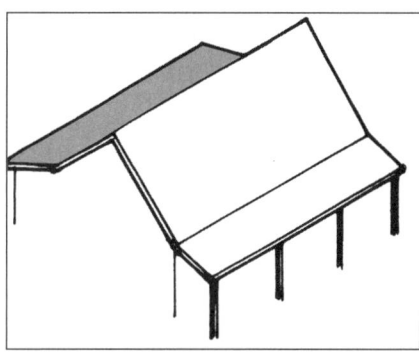

**9. Early Australian lean-to/verandah.:** A gable roof is constructed over the main part of the house and the service area to the rear is covered with a lean-to (skillion) roof and a verandah is attached across the front. This has been a common style in rudimentary cottage construction in Australia from the First Fleet until today. The roof form is much more Australian vernacular, particularly in cities and towns, than the bellcote type that has been recently romanticised. For the best effect, the main roof should be pitched at 3 in 4 (utilising the 3/4/5 triangle) and a ceiling height of 2700mm or more if necessary.

# 45 • ROOFS

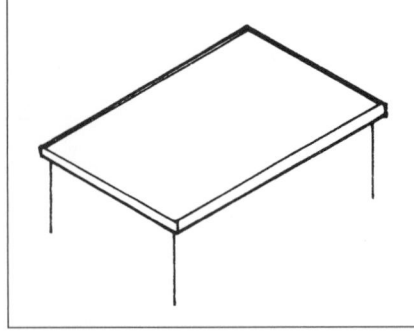

**10. Flat roof:** The flat roof is used in Australia mainly in conjunction with clip-fixed steel decking and allows complex plans to be covered simply. Owner-builders may find the flat roof useful as it is generally easier to construct than hip or gable types. A flat roof is normally taken to mean a roof with less than 3 degrees fall.

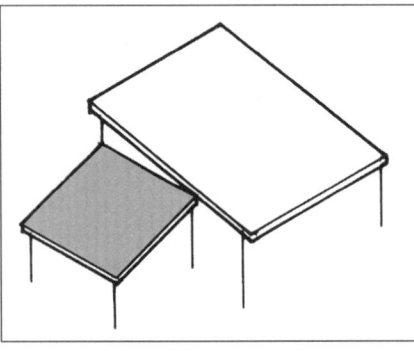

**11. Skillion/lean-to:** This roof is sometimes referred to as shed roof and, unless care is taken in the planning and massing of a house design using the skillion roof, a visual disaster may emerge. If used by skilled designers (architects) then very pleasing roof forms incorporating highlight windows can be created. Construction is generally straightforward and future additions can be easily accommodated.

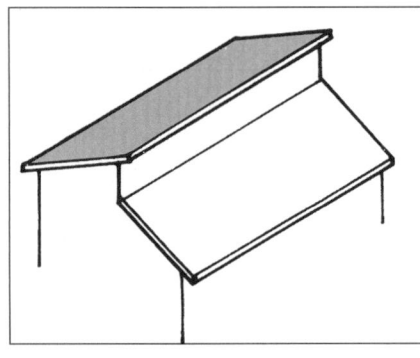

**12. Clerestory:** This is a combination of gable and skillion roofs to achieve a window along the roof ridge. It is a sensible roof where natural light and ventilation are required into spaces away from outside walls or in restricted sites.

## SUITABILITY OF ROOF TYPES FOR OWNER BUILDERS:

### Framing Systems

| Roof Type | Conventional frame | Trussed roof |
|---|---|---|
| 1. Hip | 1 | 2 |
| 2. Hip and valley | 1 | 3 |
| 3. Broken hip/valley | 1 | 3 |
| 4. Belcote | 1 | 2 |
| 5. Gambrel | 1 | 1 |
| 6. Gable | 1 | 1 |
| 7. Boxed gable | 2 | 2 |
| 8. Jerkin head | 2 | 2 |
| 9. Early Australian | 1 | 1 |
| 10. Flat | 1 | 4 |
| 11. Skillion | 3 | 3 |
| 12. Clerestory | 3 | 4 |

**SUITABILITY OF ROOF TYPES KEY:**
1. Most suitable
2. Suitable
3. Reasonable
4. Unsuitable
5. Not viable

V—Ventilating
F—Fixed closed

### WATERSHEDDING

| Roof Type | Roof Tiles | Roof Sheets |
|---|---|---|
| 1 | 1 | 3 material wastage |
| 2 | 1 | 3 material wastage |
| 3 | 1 | 3 material wastage |
| 4 | 2 min. pitch | 2 material wastage |
| 5 | 1 | 2 material wastage |
| 6 | 1 | 1 |
| 7 | 1 | 1 |
| 8 | 2 set out problems | 2 material wastage |
| 9 | 2 min pitch | 1 |
| 10 | 5 | 1 if steel decking |
| 11 | 1 | 1 |
| 12 | 2 mass of tiles over skylight | 1 |

### CEILING TYPES

| Roof Type | Flat | Vaulted |
|---|---|---|
| 1 | 1 | 3 coordinated planning |
| 2 | 1 | 4 not sensible |
| 3 | 1 | 4 not sensible |
| 4 | 1 | 3 coordinated planning |
| 5 | 1 | 3 coordinated planning |
| 5 | 1 | 3 coordinated planning |
| 6 | 1 | 2 coordinated planning |
| 7 | 1 | 4 |
| 8 | 1 | 3 |
| 9 | 1 | 2 |
| 10 | 1 | 5 |
| 11 | 3 | 1 |
| 12 | 5 | 1 |

### ROOF LIGHTS

| Type | Roof Lights | | | | Solar collector |
|---|---|---|---|---|---|
| | Dome | | South light | | (sensible use) |
| | F | V | F | V | |
| 1 | 3 | 2 | 5 | 4 | 1 |
| 2 | 3 | 5 | 5 | 5 | 1 |
| 3 | 3 | 5 | 5 | 5 | 1 |
| 4 | 3 | 2 | 5 | 4 | 1 |
| 5 | 3 | 2 | 5 | 2* | 1 |
| 6 | 3 | 2 | 5 | 2 | 2 |
| 7 | 3 | 2 | 5 | 2 | 2 |
| 8 | 3 | 2 | 5 | 4 | 2 |
| 9 | 3 | 2 | 5 | 2 | 2 |
| 10 | 1 | NA | 3 | NA | 5 |
| 11 | 5 | 2 | 5 | 1 | 3 |
| 12 | 5 | 2 | 5 | 1 | 3 |

*Assumes glazing the gambrel end, otherwise difficult.

# CONVENTIONAL ROOF FRAMES • 46

## Top plate

The top plate is the continuous timber member on top of all walls that are to support the roof structure and transfer the roof mass to the walls. Top plates are required whether the roof is built on timber stud frames, or on brick, block or stone walls.

## Ceiling joists

The ceiling joists are members that fulfil the double function of:
a. providing the structure onto which the ceiling is fixed
b. providing the member that connects or ties the lower ends of rafter couples and therefore prevents the rafters spreading and causing roof sag or collapse.

Ceiling joists are normally horizontal members but can, under special circumstances, be fixed on a slope.

## Rafters

The rafters are members that are fixed on a slope and run in the same direction as the roof slope in a conventionally framed roof. Rafters are generally erected in opposing pairs or couples so that, at their apex on the ridge of the roof, they meet each other directly.

The rafters give the roof its required slope and provide the structure onto which the roof covering is fixed.

Where a rafter meets its supporting top plate it is cut so it rests directly on the top plate. The cut is called a 'bird's mouth cut' because of its resemblance to a bird's open beak.

## Purlins

The purlins are members that are fixed horizontally and perpendicular to the direction of the roof slope. In a conventional roof frame the purlins are used to support the underside of the rafters to allow for the most economic rafter size and to prevent roof sag and collapse. They are given the special name of under purlins because they are fixed under the rafters.

## Struts and props

The struts and props are specifically required in the roof frame to transfer the loads on under purlins and other beams to the load-bearing walls. Struts and props should be adequately supported over studs in timber walls and should have anti-slip blocks where necessary, to prevent them sliding horizontally. Struts and props generally are more efficient the closer they are to the vertical. Angles between struts and the horizontal of less than 45 degrees should be avoided.

## Collar tie

The collar ties supplement the ceiling joists and resist the rafter couples spreading. They are normally fixed halfway up the rafter couple and are located on alternate rafter couples. In roofs with rafter lengths in excess of 5000 mm, additional collar ties may be required at two or more levels. A roof of this scale should be checked by a qualified consultant before and after construction.

## Ridge (beam)

The ridge in a conventional framed roof is required not so much as a structural member but more as a setout member to make sure rafter couples meet at a consistent height, giving a horizontal ridge line free from sags. If the ridge is required to act as a beam (to give strength to the roof) then it should be selected for size more carefully than would be normal.

## Hanging beam

The hanging beam is generally a deep timber beam located perpendicular to ceiling joists. It has the function of reducing the span of the ceiling joists and therefore allows them to be a more economic and consistent section than may otherwise be the case. The hanging beam size must be selected carefully from timber framing tackle and is required to be adequately supported on its ends over load-bearing walls by blocking pieces of the same timber as the ceiling joists with the grain of the block lying horizontally.

The blocking piece must be used in this manner to reduce any problems associated with differential shrinkage between the block and the ceiling joists, which causes ceiling defects. The ceiling joists are fixed to the hanging beam with hoop iron straps, timber battens or purpose-made ceiling dogs.

## Strutting beam

Strutting beams are used in many forms and locations in a conventionally framed roof, but in all cases their function is to support roof members, generally under purlins, where there are no conveniently located load-bearing walls. The strutting beam must transfer the roof load directly to load bearing walls, and must not under any circumstances rest on, or transfer load to, a ceiling joist. Strutting beams cannot double as hanging beams, because they are likely to sag and will therefore cause ceilings to sag. Also, as they are transferring roof loads that are often dynamic due to wind pressure acting on the roof, they may move up and down or side to side, causing the ceiling to move and possibly crack. Struts, props and blocks used to support strutting beams must have their grain direction parallel to the load applied by the beam, as this is generally their strongest direction. Timber shrinks much less along the grain than across it, and any shrinkage of a supporting block would cause the strutting beam to move down and hence allow localised roof sagging.

## Roofing battens

The roof battens are often provided by the contractor supplying and fixing the roof cover. This is particularly common when tiles, shingles, slates or the like are being used but less common for sheet roofing materials like galvanised steel, aluminium, fibrocement or such materials. The roof battens have the obvious function to support the roof cover material, but in many instances they are also an essential part of the roof structure itself, as they effectively tie rafter couples one to the other.

## Soffit bearer

The soffit or eaves bearers, although technically a structural component of the roof itself, are essential for the fixing of eaves linings where a horizontal or boxed eaves condition is required. In wide eaves conditions, that is eaves wider than 600 mm, they can assist in a structural way to reduce eaves sag.

## Fascia board

The fascia board is the member fixed to the ends of the rafters. It has the structural function of restraining and tying the rafter ends, and the general function of enclosing the eaves in conjunction with the soffit lining. The fascia should be of adequate size to resist twisting and should be of a material that will remain sound in such an exposed location.

## Barge board

The barge board is similar in function to the fascia but instead of being located at the ends of rafters, and perpendicular to them, it is fixed parallel to rafters and located on a slope at the gable or skillion end of a roof.

## Hip rafter

The hip rafter is located at an external angle change of roof direction, and its function is similar to that of the ridge beam. That is to say it is used to assist in the location and support of rafter couples as they make the transition from one roof plane to another. Care must be taken to correctly size and support the hip rafter, as they invariably have difficult end support join conditions, particularly at the roof crown.

# 46 • CONVENTIONAL ROOF FRAMES

## Valley rafter

The valley rafter is located at an internal angle change of roof direction, and in many ways is similar to the hip rafter. However, there is the added complexity of supporting the upper end of a valley rafter which, in many instances, may be located partway up the rafter length of the adjacent roof plane and often removed from a structure that can adequately transfer its load to a wall.

## Roof construction timbers

The timber used in roof construction varies considerably from location to location. Traditionally it seems to be generally accepted that Oregon (Douglas fir) is the best timber for a conventional roof frame, as it combines acceptable strength with lightweight and easy workability.

Oregon is not available in all Australian states and, as it is imported from North America and Canada, it is expensive. Alternative timbers include Australian hardwoods (including Victoria, New South Wales, Queensland, Tasmania, together with Jarrah in Western Australia), which make strong but heavy roofs. *Pinus* (grown locally or imported from New Zealand) makes good light roofs, as does Cypress pine (used mainly in Queensland) and Canada pine (Hemlock), which varies in quality.

It is possible to build a conventional framed roof using post form steel sections. However, while a good sound roof can be achieved, the methods of erection would normally be too complex for an OB project, and it seems that for the present this product will be used only in mass-manufactured houses.

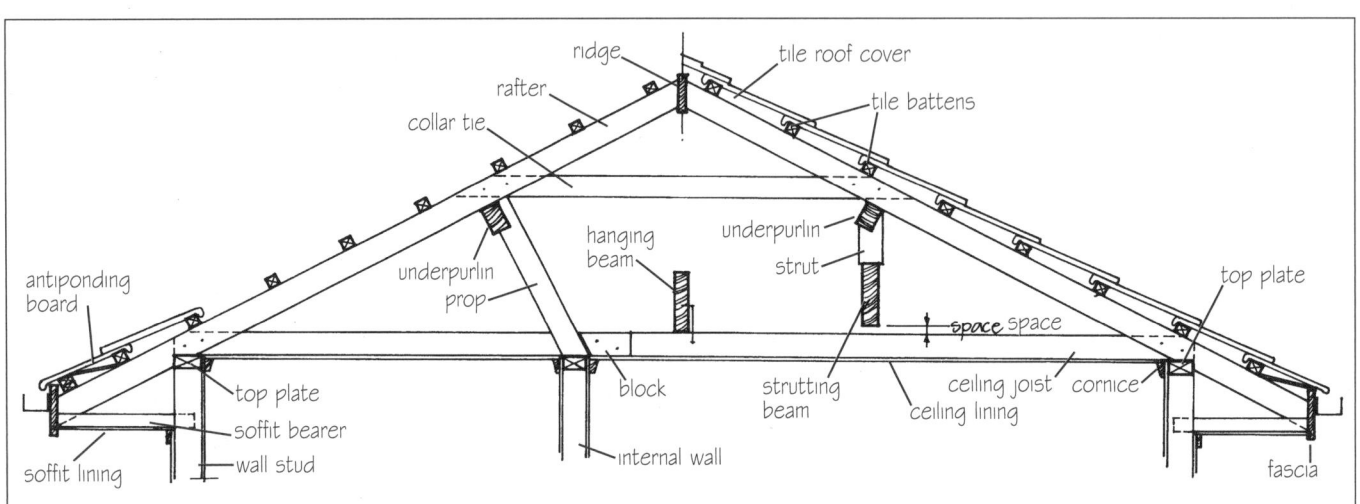

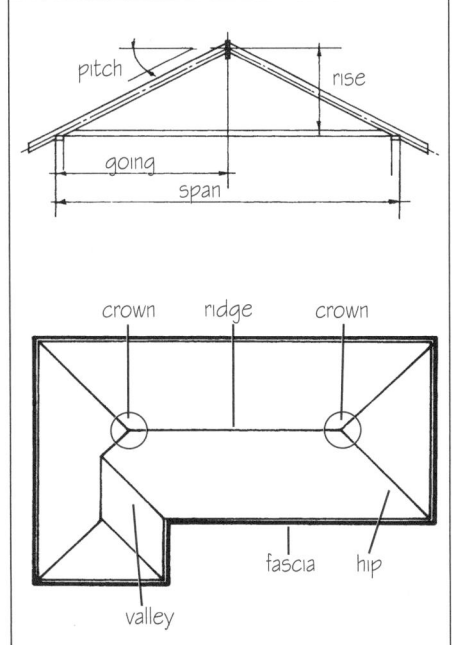

| ROOF PITCH | | TILE ROOF | | SHEET ROOF | |
|---|---|---|---|---|---|
| Degrees | Ratio | T.C. | Conc | Deck | Profile |
| 1° | 1:57.3 | | | | |
| 2° | 1:22.9 | | | | |
| 5° | 1:11.5 | | | | |
| 7° | 1:7.6 | | | | |
| 10° | 1:5.7 | | | | Some types only |
| 12° | 1:4.5 | | | | |
| 15° | 1:3.7 | | | | |
| 17° | 1:3.2 | | Some types only | | |
| 20° | 1:2.7 | | | | |
| 22° | 1:2.4 | | | | |
| 25° | 1:2.1 | | | | |
| 26° ¼ pitch | 1:2. | | | Generally can be used on these roof pitches | |
| 27° | 1:1.9 | | | | |
| 30° | 1:1.7 | | | | |
| 35° | 1:1.4 | | | | |
| 37° | 1:1.3333 | | | | |
| 40° | 1:1.9 | | | | |
| 44°/45° | 1:1 | | | | |
| 50° | 1:0.8 | Special fixings required¶ | | | |
| 60° | 1:0.6 | | | | |
| 70° | 1:0.4 | | | | |

Roof pitches have traditionally been expressed in degrees of angle from the horizontal to the slope—or by ratio. Early roofs in Australia were often 45 degrees or a ratio of 1:1 as this was easy to calculate and construct. Later, but still during colonial times, a roof frame based on the 3/4/5 triangle was used extensively. This allowed relatively simple arithmetic to be used to calculate roof member lengths.

During the bungalow period of 1915 to 1935 the ¼ pitch roof became the norm, that is where the rise of the roof is ¼ of the span or a pitch ratio of 1:2. After WW2 roofs were commonly referred to by angle, but recently efforts have been made by the Australian Standards Association to recommend that roof pitches be stated as a ratio in preference to degrees of angle.

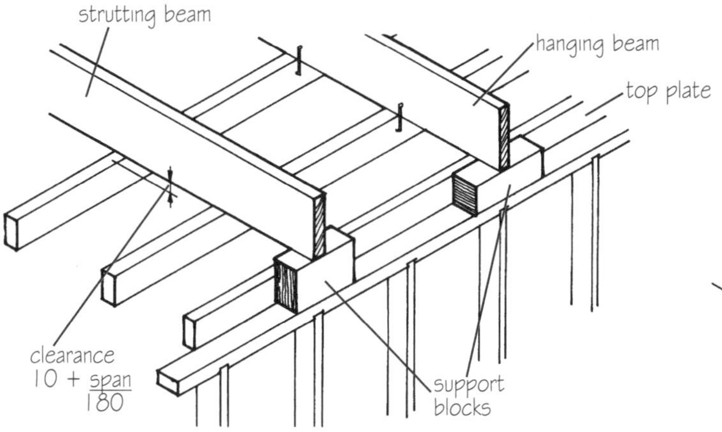

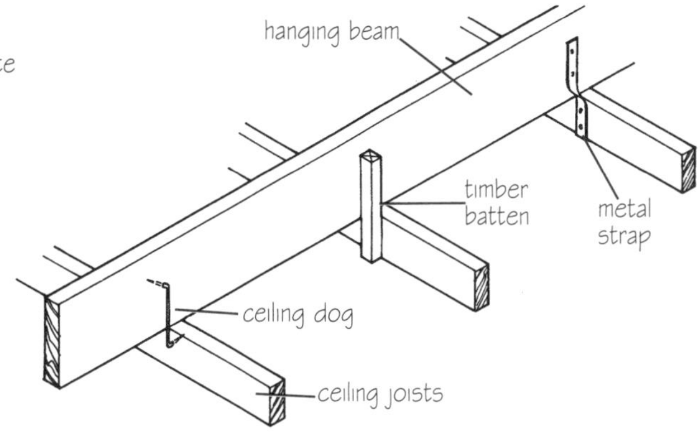

# CONVENTIONAL ROOF FRAMES • 46

Blocks to
Strutting beam have VERTICAL grain
Hanging beam have HORIZONTAL grain

When the end of a hanging beam has to be cut at such an angle to avoid contact with roof covering that the quality of bearing is reduced then:
1. the hanging beam should be adequately bolted to the side of an appropriate rafter
2. a jack joist should be fixed to the under edge of the hanging beam by means of a galvanised steel nail plate

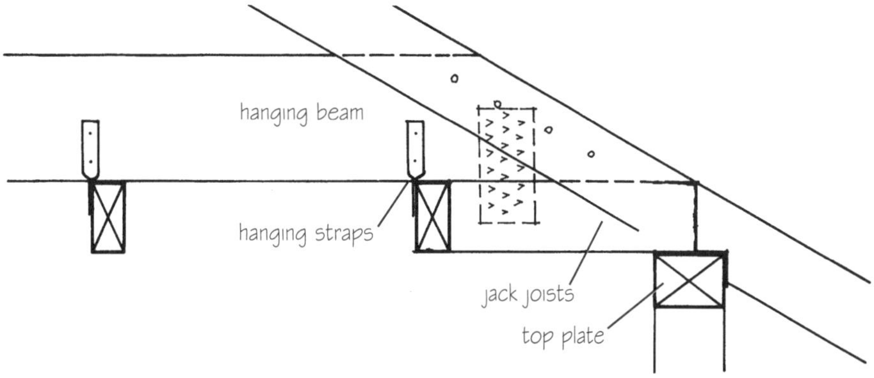

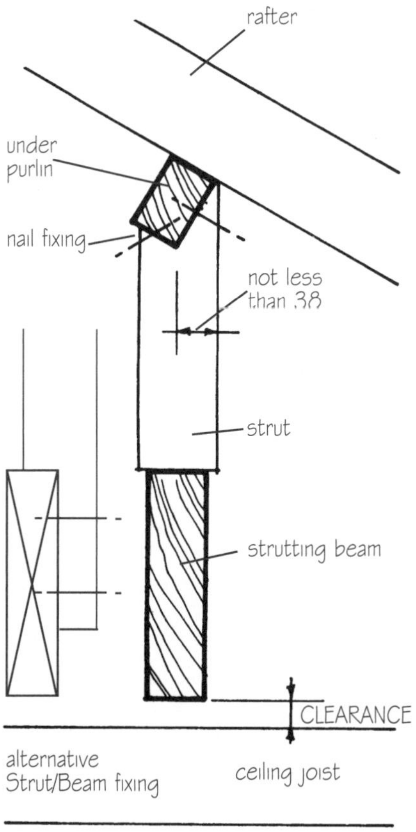

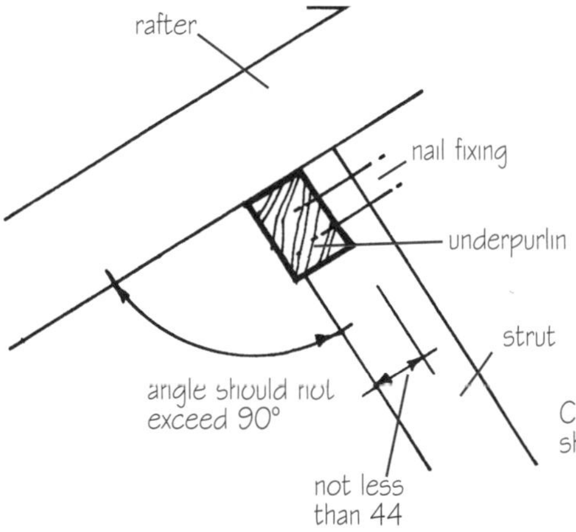

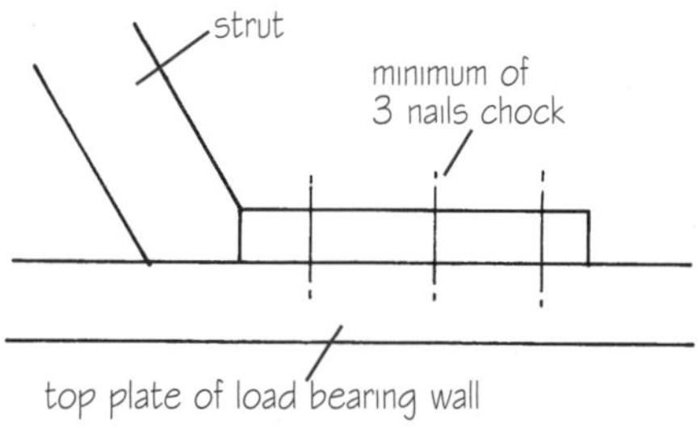

CLEARANCE UNDER STRUTTING BEAM should be:

$$10mm + \frac{\text{span of Beam in mm}}{180}$$

eg: for a strutting beam 4000 long,

then $10 + \frac{4000}{180} = 32mm$

The design deflection of a concealed timber beam is:

$$\frac{SPAN}{180}$$

The design deflection of an exposed timber beam is:

$$\frac{SPAN}{360}$$

# 46 • CONVENTIONAL ROOF FRAMES

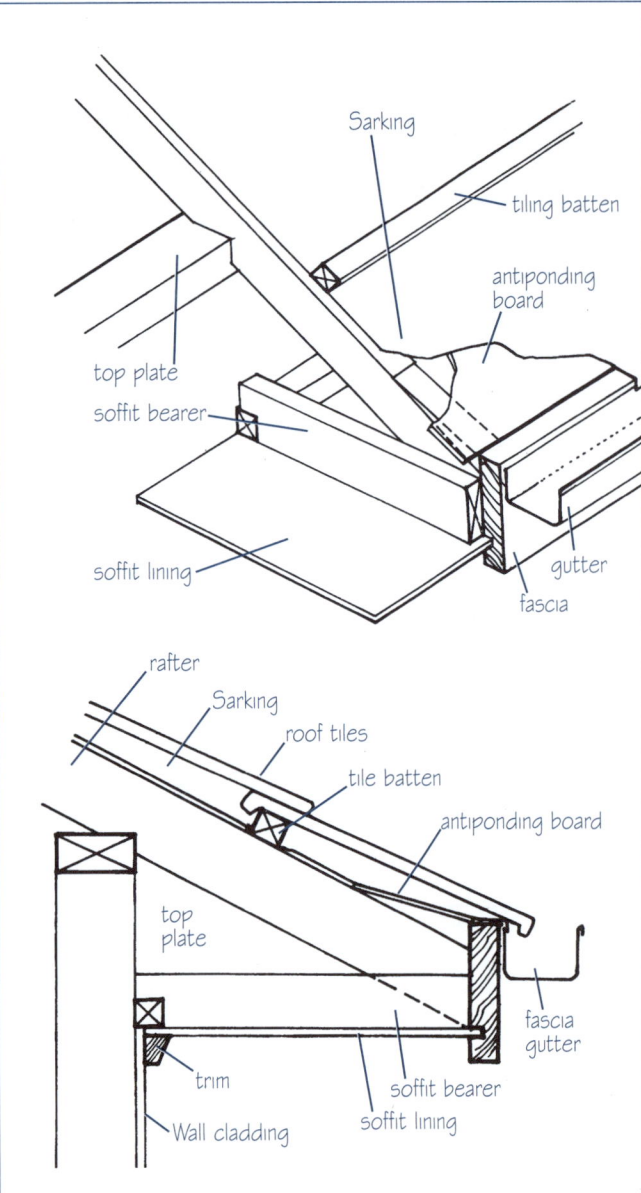

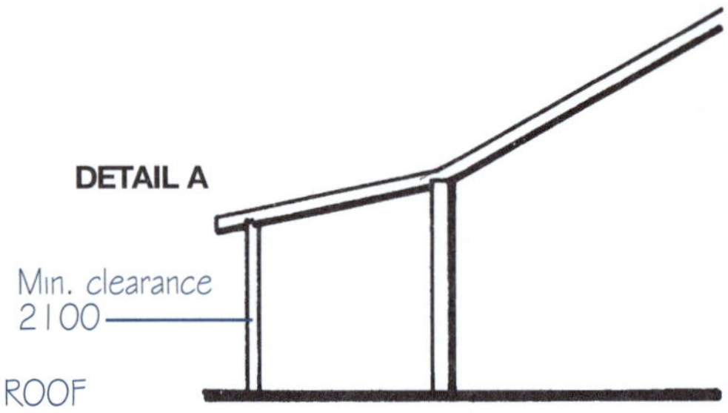

### Sarking
Sarking has been used under low pitched tile and sheet roofs for many years and has a number of sound reasons to recommend its use on all roofs.

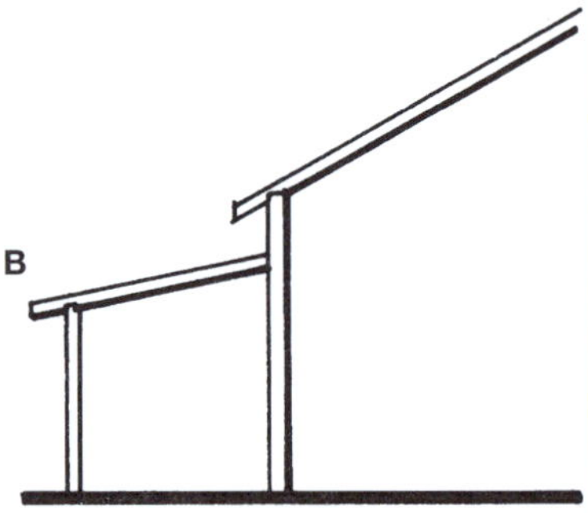

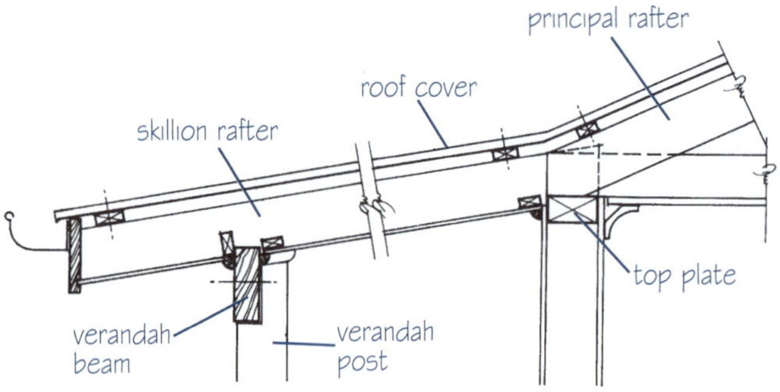

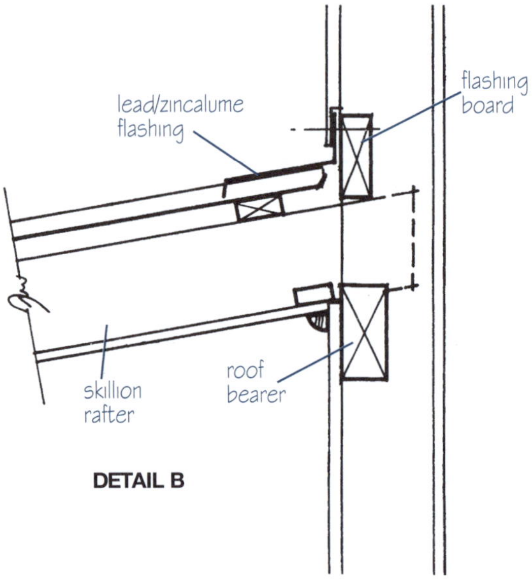

a. The sarking is generally based on a moisture proof building paper and therefore it will give added protection to the house by reducing the likelihood of water damage because a roof tile cracks, and as condensation normally forms below the roof cover and above the sarking, staining of ceilings caused by condensation drips is reduced.

b. Modern sarking papers generally have reflective metallic surfaces which serve as effective heat insulation particularly by reducing the thermal impact of summer heat.

### Anti-ponding board
When sarking is used it is essential that the sarking is carried beyond the roof space over the fascia and into the roof gutter. If this is not done any moisture running down the top of the sarking will be deposited in the eaves space and is likely to cause staining or worse.

To avoid any moisture collected by the sarking ponding behind the fascia board, an anti-ponding board is fixed between the last filling batten and the fascia board on top of the rafters.

**VERANDAH ROOF OPTIONS**

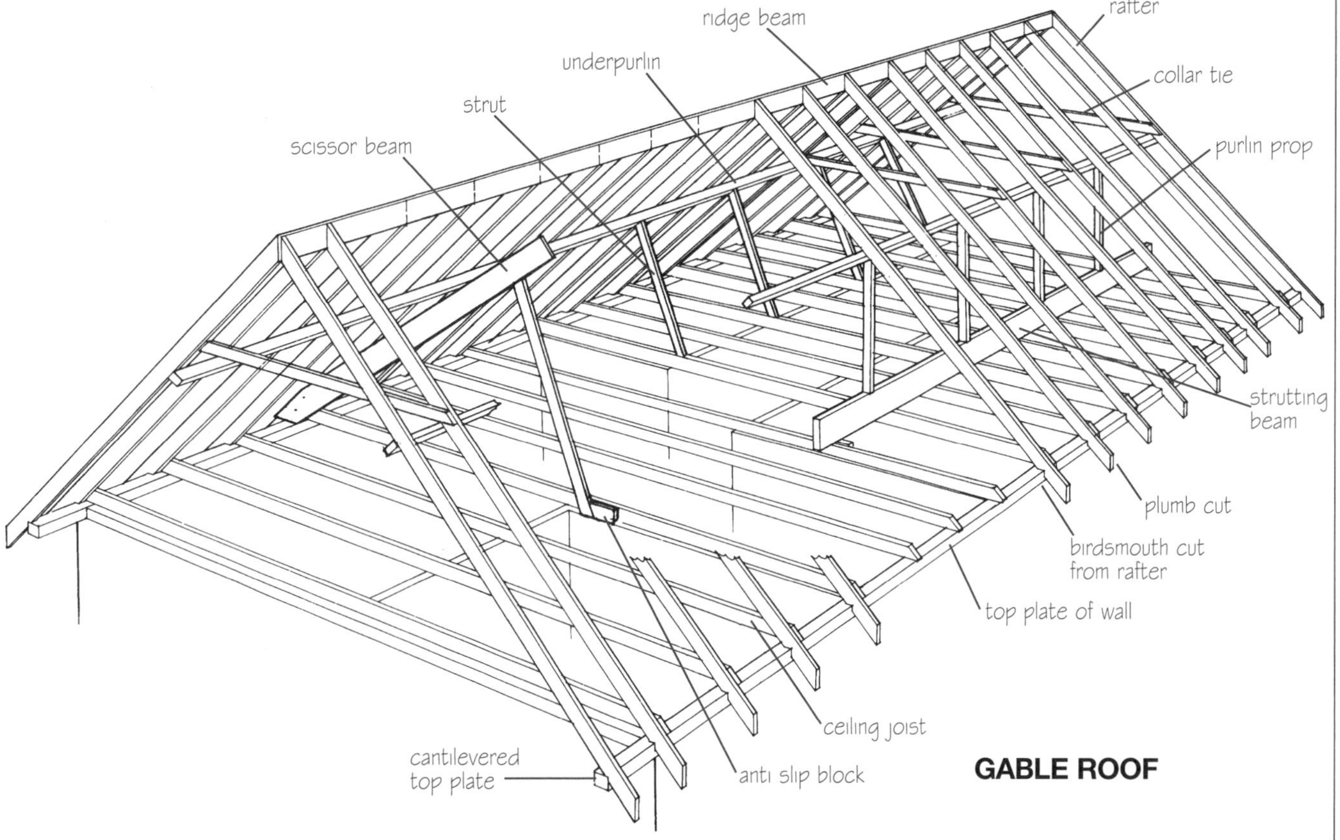

# CONVENTIONAL ROOF FRAMES • 46

**GABLE ROOF**

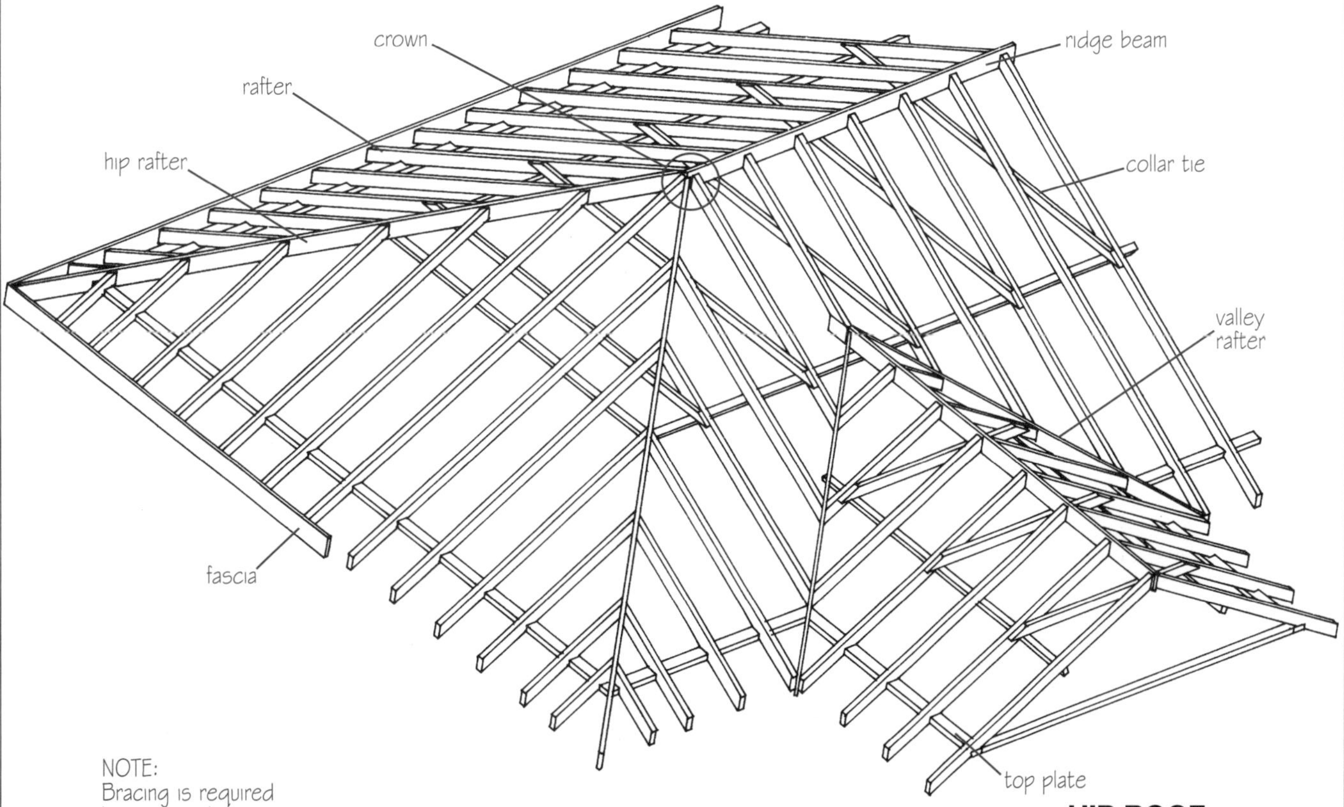

Labels: ridge beam, underpurlin, strut, scissor beam, rafter, collar tie, purlin prop, strutting beam, plumb cut, birdsmouth cut from rafter, top plate of wall, ceiling joist, anti slip block, cantilevered top plate

**HIP ROOF**

Labels: crown, rafter, hip rafter, ridge beam, collar tie, valley rafter, fascia, top plate

NOTE: Bracing is required but is not shown

# 46 • CONVENTIONAL ROOF FRAMES

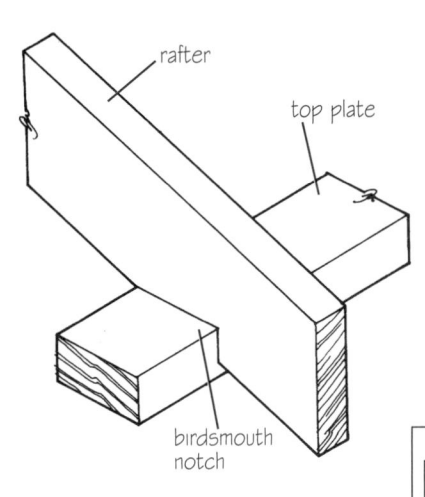

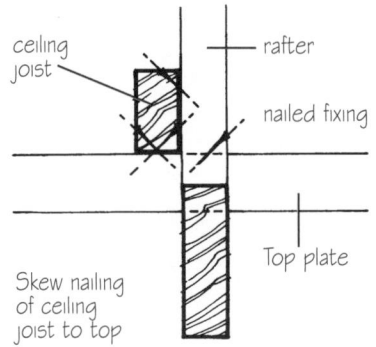

Skew nailing of ceiling joist to top

**The birdsmouth notch**
Rafters are seated on to top plates by using a 'birdsmouth' notch to form a horizontal bearing area in the rafter. AS 1684* requires that the timber that is left in the rafter after cutting the birdsmouth notch should be at least two-thirds of the depth of the minimum rafter section allowed by the standard.

*AS 1684—SAA (light) Timber Framing Code

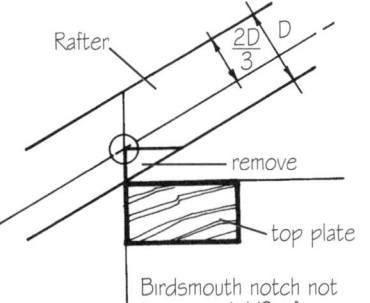

Birdsmouth notch not to exceed 1/3 of rafter depth

### Roof Members Section Size Guide— based on seasoned timber of approximately stress grade F8

Note: These section sizes are to aid in estimating only, and should not be used for construction before checking in the appropriate sections of AS1684—1999 Residential Timber Framed Construction.

| Member | Common Section | Maximum Span | Maximum Spacing | Comments |
|---|---|---|---|---|
| Rafter/Tiles | 90 x 45 | 2400 | 600 | Continuous Spans |
| Rafters/Sheet | 90 x 45 | 2700 | 1200 | Continuous Spans |
| Ceiling Joist | 90 x 35 | 1800 | 600 | Plasterboard |
|  | 90 x 45 | 2400 | 450 | Timber lining boards |
| Purlin/Tiles | 140 x 70 | 2400 | 2700 | Continuous Purlin |
| Purlin/Sheet | 120 x 70 | 2400 | 4200 | Continuous Purlin |
| Struts | 90 x 70 |  |  |  |
| Collar Tie | 90 x 35 |  | 1200 |  |
| Ridge Board | 180 x 20 |  |  |  |
| Hanging Beam | 140 x 45 | 2400 | 3600 | Rafters @3600=140x35 Tiles |
|  | 190 x 45 | 3000 | 3600 | Rafters @3600= 120x35 Sheet |
|  | 240 x 45 | 3600 | 3600 | Ceiling Joists @3600=120x35 |
| Strutting Beam Tiles | 190 x 70 | 3000 | 2400 | Strutting beams must be used with care to achieve a suitable balance between structural requirements and economics |
|  | 240 x 70 | 3600 | 2400 |  |
| Strutting Beam Sheet | 190 x 70 | 3600 | 2400 |  |
|  | 240 x 70 | 4200 | 2400 |  |
| Verandah Beam Tiles | 170 x 45 | 2700 |  | All continuous spans All for up to 2000 wide verandah |
|  | 190 x 45 | 3000 |  |  |
| Verandah Beam Sheet | 170 x 45 | 3000 |  |  |
|  | 190 x 45 | 3300 |  |  |

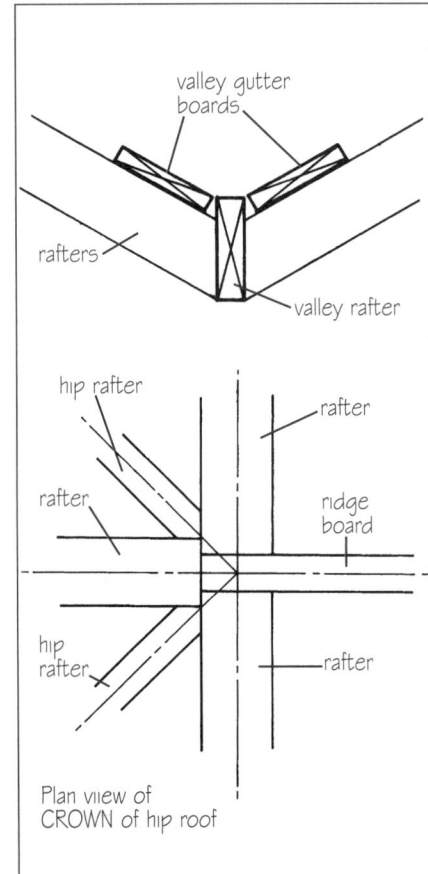

Plan view of CROWN of hip roof

# TRUSSED ROOFS • 47

## The principles of trussed roofs

A trussed roof is one where a series of two-dimensional, triangulated structural frames are used to support the roof load and transfer it to load-bearing walls. The truss is a static mechanical system consisting of one or more triangular frames which, when adequately designed and constructed, make a structural spanning system capable of very good span-to-weight ratio.

Normally roof trusses are arranged in parallel then tied and braced to form a stable three-dimensional structure. In domestic construction in Australia, a particular type of roof truss has been developed to fulfil the local requirements of the home building industry. This type of roof truss has become known as the lightweight gang nail truss, for it is indeed light by historical standards, being made of scantling-size timber members and joined with light gauge, integral rail steel plates.

The lightweight trusses used in residential construction are of particular value to OBs for they allow accurate on-site prefabrication and relatively (compared to conventional roof framing) easy erection. As well as being designed to span from outside wall to outside wall, they also free internal walls from a load-bearing function and therefore allow more flexible initial planning and long-term adaptability.

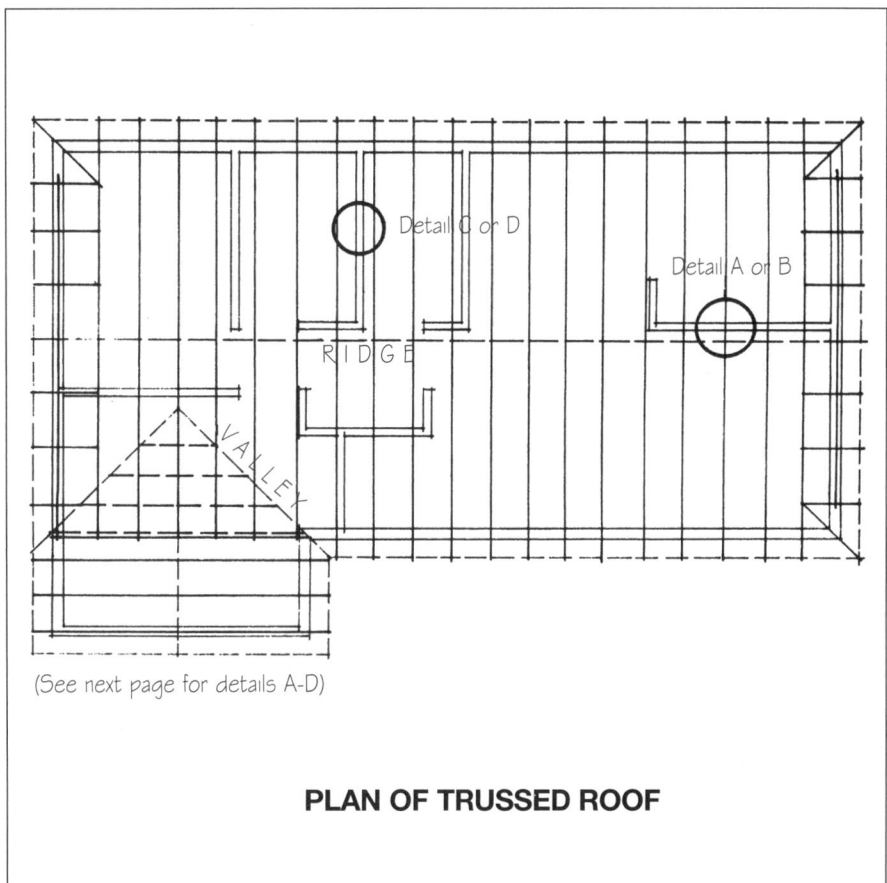

(See next page for details A-D)

**PLAN OF TRUSSED ROOF**

**GENERAL VIEW OF TRUSSED ROOF (GABLE)**

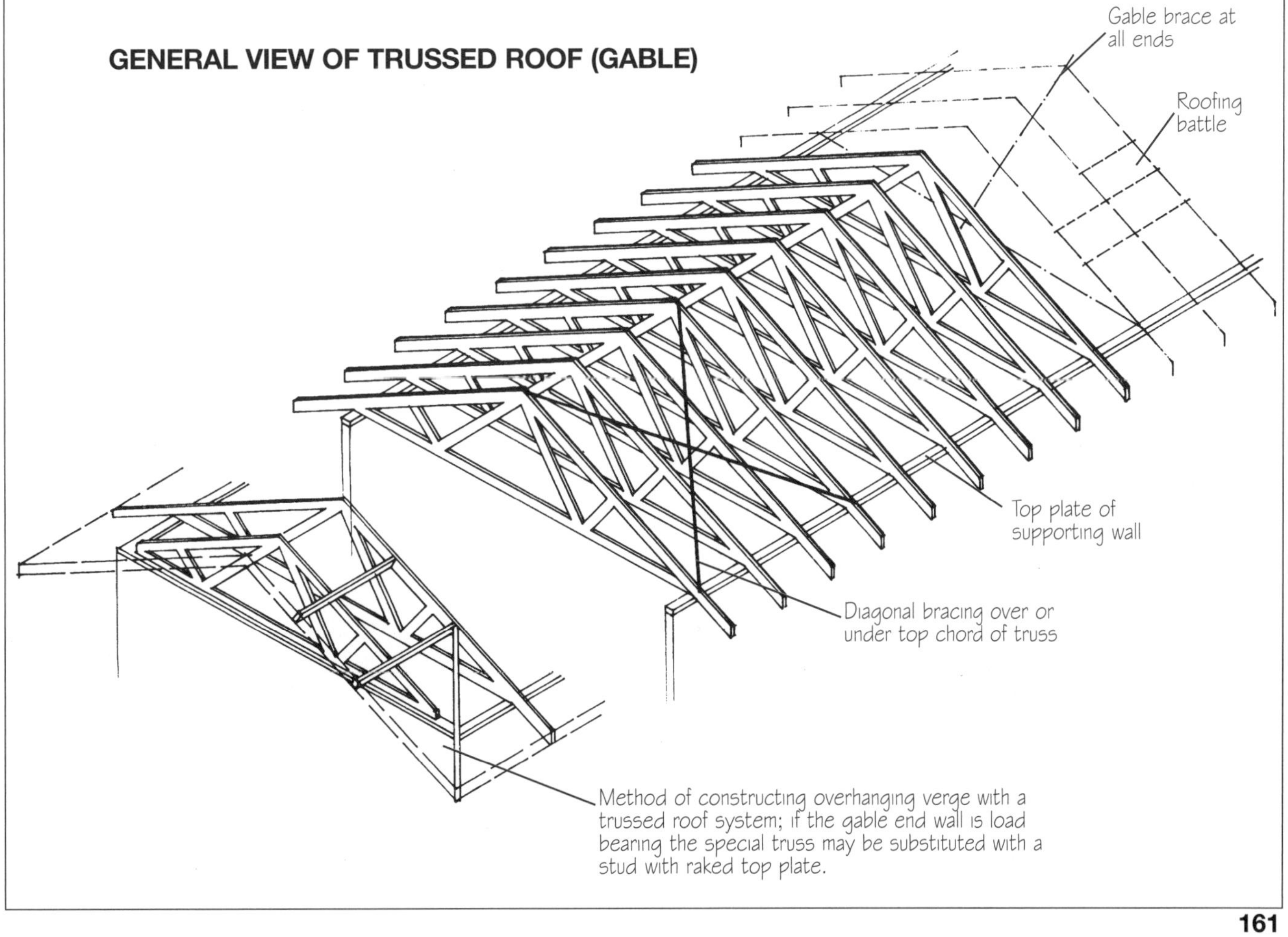

- Gable brace at all ends
- Roofing batten
- Top plate of supporting wall
- Diagonal bracing over or under top chord of truss
- Method of constructing overhanging verge with a trussed roof system; if the gable end wall is load bearing the special truss may be substituted with a stud with raked top plate.

# 47 • TRUSSED ROOFS

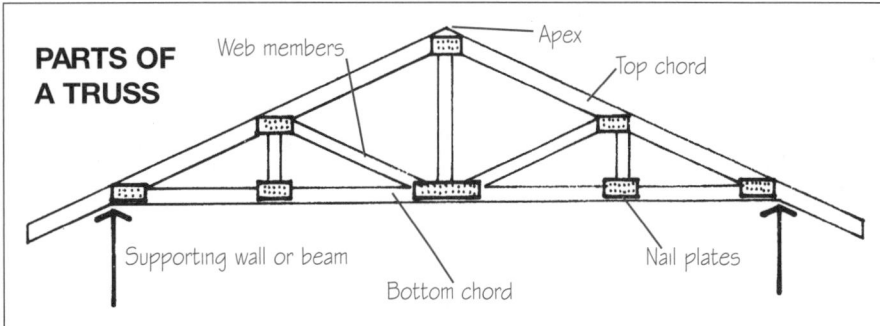

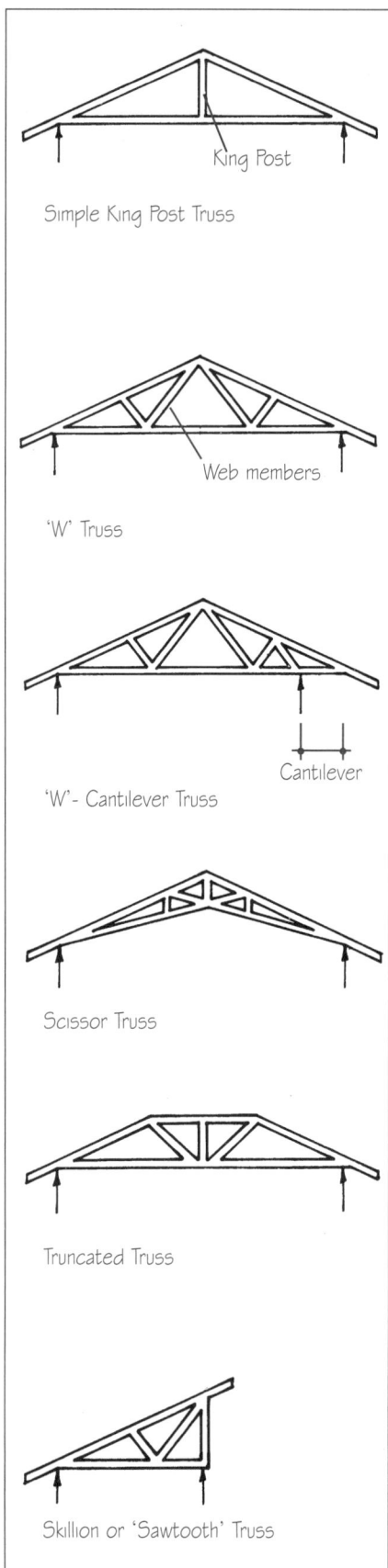

## The parts of a truss

Lightweight roof trusses are constructed from three basic parts:
1. **top chord**—if the truss has a flat top, an uncommon though not unused form of truss, then the top chord may be one piece of timber. However, as most trusses are of simple triangular shape, the top chord is generally two pieces of timber that meet at the apex of the truss. The top chord then becomes the rafters for the roof.
2. **bottom chord**—the bottom chord is normally one horizontal timber member, but can be multi-membered, as in the scissor truss and similar. In all cases the bottom chord in a lightweight trussed roof may be used as a ceiling joist but, if desired, the trusses may be exposed to view and the ceiling fixed over the top chord. Lightweight roof trusses are normally required to be at 600–1200 mm centres and therefore care should be taken if trusses are to be exposed, so as not to have an over-busy appearance.
3. **web members**—are those pieces of timber that take compression and tension loads between the top and bottom chords and complete the triangulation and therefore the stability of the truss. In a lightweight truss the depth of the individual sections may vary to suit specific design conditions but the width of the sections must remain the same if simple jointing is to be used.
4. **jointing**—the chord and web members of the truss are connected at their intersecting points, called panel points, with integral nail plates. These are rectangular sheets of galvanised steel that have been formed to take the appearance of a 'bed of nails'. The nail plates are positioned on each side of the joint and then pressed into the timber to give a cheap, efficient connection.

Note: lightweight timber trusses are generally designed to be supported on walls at the extreme ends of the bottom chord. No additional support or alteration to the design support points should be undertaken without a specially designed or engineeringly modified truss being used.

## Timbers used in roof trusses

Lightweight timber roof trusses can be and are fabricated from a variety of timbers, including, but not limited to, *pinus*, hardwood, Oregon and cypress pine, dependent only on local availability. No timber appears specifically better than another and every truss is engineered to suit a particular set of conditions.

## Truss types

There are basically two types of roof trusses used in residential building:
1. the simple king post truss—can be used for short-span applications up to approximately 6000 mm and has particular application in garages and carports.
2. the 'W' web truss—can be used for spans upto 10,000 mm with ease and is ideally suited to simple gable roof types. Sub types of the 'W' web truss are in common use and include:
   a. the cantilevered 'W' web truss—specifically designed to allow verandahs to be incorporated into a trussed roof house.
   b. the truncated truss—allows a trussed roof to be used with a hipped shape.

Further variations are numerous, and include two forms that have specific and special application:
   c. the scissor truss—used when a vaulted ceiling is designed, but the cost of large section solid timber rafters is inappropriate and unrealistically expensive.
   d. the skillion or sawtooth truss—used both as a lean-to extension for an existing roof or as a roof in its own right, often using the vertical post of the truss to incorporate windows.

## Fixing to walls

When fixing a timber roof truss to a load-bearing wall make sure that the bottom chord of the truss has its designed seating on the support wall. If there is less than adequate bearing achieved, then localised defects in wall and ceiling linings are likely to occur.

If the roof truss has adequate bearing on the supporting wall plate and the trusses in the roof are tied one to the other and adequately braced, the roof is stable in its static condition. The roof needs to be tied down to the wall plate—not to hold it up, but rather to hold it down in windy conditions when aerodynamics cause the roof to lift. To resist the uplift condition a patent three-way metal connector is used to fix the truss This is generally done by side nailing from the top chord to the wall, and normally by nailing into the side of the top plate. OBs should check local requirements for specific design regulations required to

# TRUSSED ROOFS • 47

make roofs safe in high winds, particularly in areas where cyclones are encountered.

Even when the truss is adequately fixed to the top plate this is not sufficient to withstand any more than a slight breeze if the top plate is not able to transfer the uplift to an adequate anchor. Do not take chances with the roof fixing. Have an engineer or the local authorities check your design for adequacy.

## Use of trusses for complex roofs

The project home builders are seemingly able to roof any shape of floor plan with a trussed roof. Many of the floor plans that they build consist mainly of a serpentine exterior wall that forms room shapes with the addition of a minimum number of internal walls.

This may be an economic way to mass-produce housing but the results are generally repetitive and often rely on faux period details to give a semblance of character.

OBs should take care in attempting to replicate the prefabricated wall frames and a truckload of trusses, building system favoured by the project home builders.

In the majority of trussed roof designs all of the roof forces are carried by the external walls. These loads are a combination of mass, uplift due to wind and torsion due to overturning.

The location and design of openings for doors and windows must be carefully considered.

The head of the opening must be capable of transferring the room load to the supports on either side of the opening. These

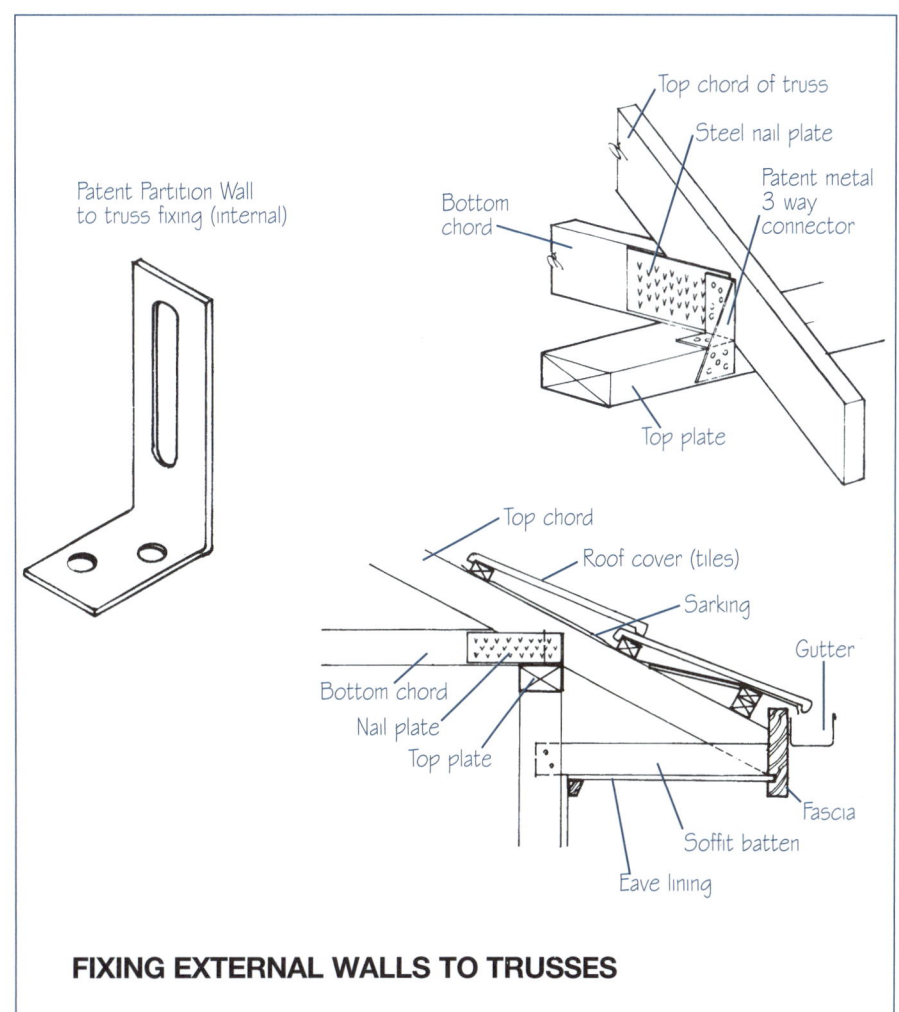

**FIXING EXTERNAL WALLS TO TRUSSES**

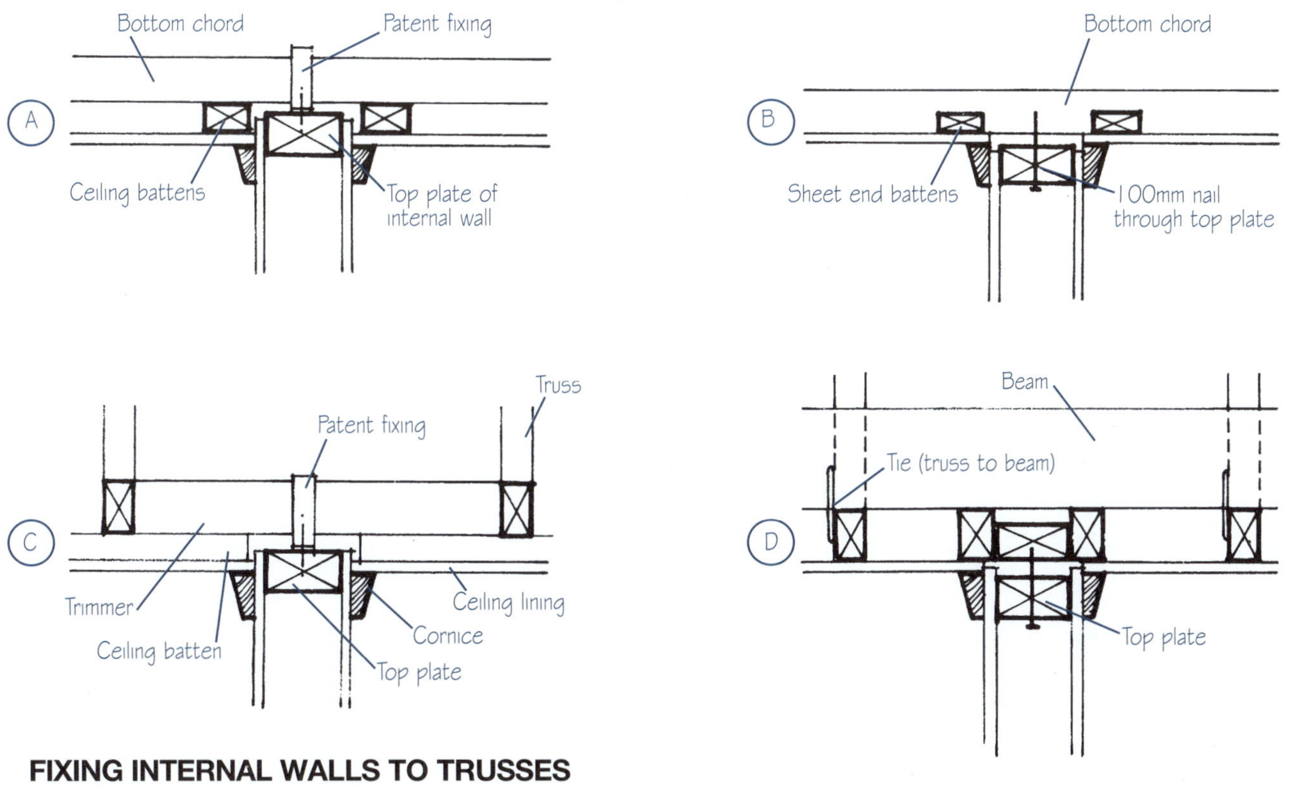

**FIXING INTERNAL WALLS TO TRUSSES**

# 47 • TRUSSED ROOFS

supports then must transfer the load through the footings into the foundation.

The uplift forces on the roof must be resisted by the head of the opening; this generally requires metal straps over the trusses and fixed to the side of the window head. The supports on either side of the opening must be capable of resisting this up force by transferring the load to the footings. Continuous tie downs are needed from the opening beam all the way into the mass of the footings or floor slab. This is necessary as any nailed joint or any simply mortar-jointed brick base wall can be pulled apart, due to the uplift generated by wind.

Wide spans in trusses will also introduce a tendency for the top plate of the head to move in and out during a windstorm. When windows were about 2000 mm wide, rooms average less than 4000 mm wide, walls were 100 mm thick in hardwood and ceiling joists were 100 x 38 at 450 centres, this was of minimum concern. Today a project home may use windows over 3500 mm wide, with trusses at 600 mm centres spanning over 9000 mm, all relying on a 90 mm *Pinus* stud frame. This can be a deadly combination if an opening head begins to shimmy to the extent that thin gauge metal connectors tear – and the roof flies free.

Bracing of the roof, ceiling and wall planes is an essential component of designing a trussed-roof house. The bracing often requires detailed engineering design, as these houses often have so little wall plane area that even bracing every wall may not be sufficient.

The modern trussed-roof house has pushed lightweight timber construction methods too close to their limit, to a point where the component, connection, assembly and system are often interdependent.

## Some Terms Specific to Truss Roof Systems

**Bottom Chord:** Truss member forming bottom edge of truss, generally used to support ceiling.

**Clear Span:** The horizontal distance between the inner faces of the truss supports.

**Gable Truss:** Normal triangular shaped truss

**Girder Truss:** Truss designed to support one or more trusses.

**Hip Truss:** Jack truss specifically designed to make the 45° plan angle required for trussed roofs.

**Jack Truss:** Sawtooth truss with an extended top chord to allow for hip roof designs

**Sawtooth Truss:** A truss with a horizontal bottom chord, a single sloping top chord, a king post to complete the triangle plus required web members.

**Saddle Truss:** Particular gable type truss designed to fix at right angles to the top chord of main roof trusses to make a valley roof connection.

**Station:** The position of a truss measured from the outside face of the end wall. Usually used to describe the position of truncated Girder and Standard trusses in a Hip End.

**Truncated Truss:** A truss that has two sloping top chords joined by another top chord parallel to the bottom chord.

**Top Chord:** Truss members forming the top edges of truss, generally used to support roofing system.

**Waling Plate:** Timber member bolted to the face of a truss to support intersecting rafters or trusses. May also used to support intersecting battens or purlins.

**Web:** The internal members of a truss. Usually only subject to axial loads due to truss action.

### DUTCH HIP END

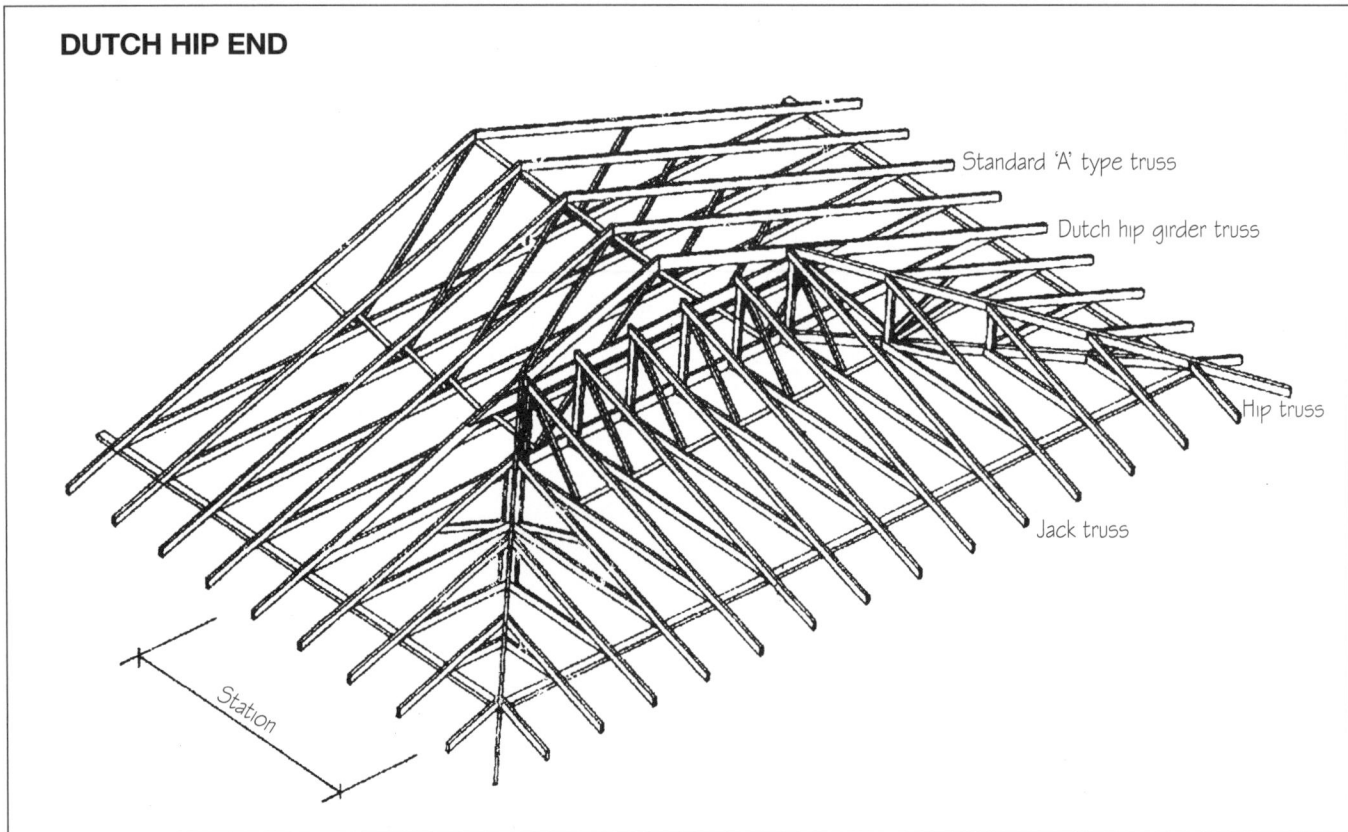

# TRUSSED ROOFS • 47

## HIP END

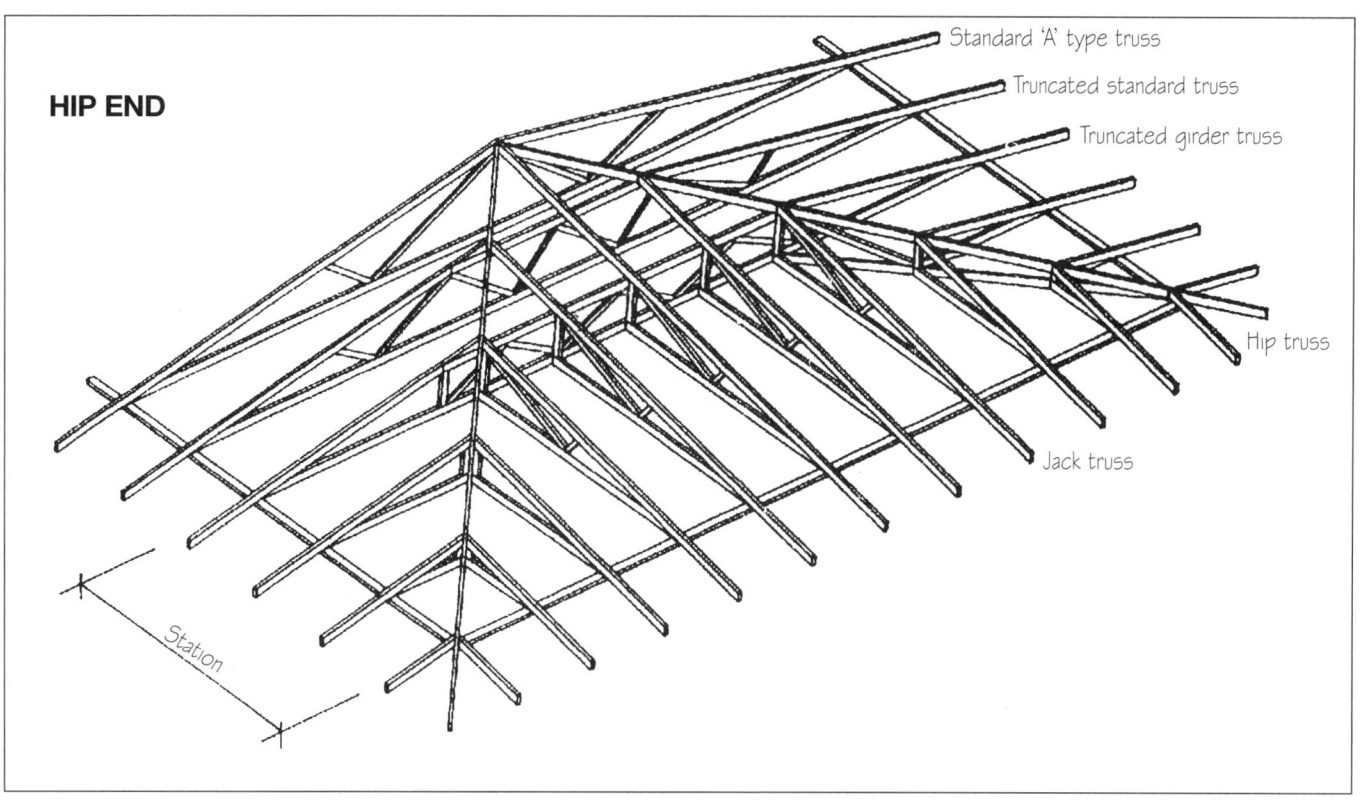

## TIMBER BEAMS

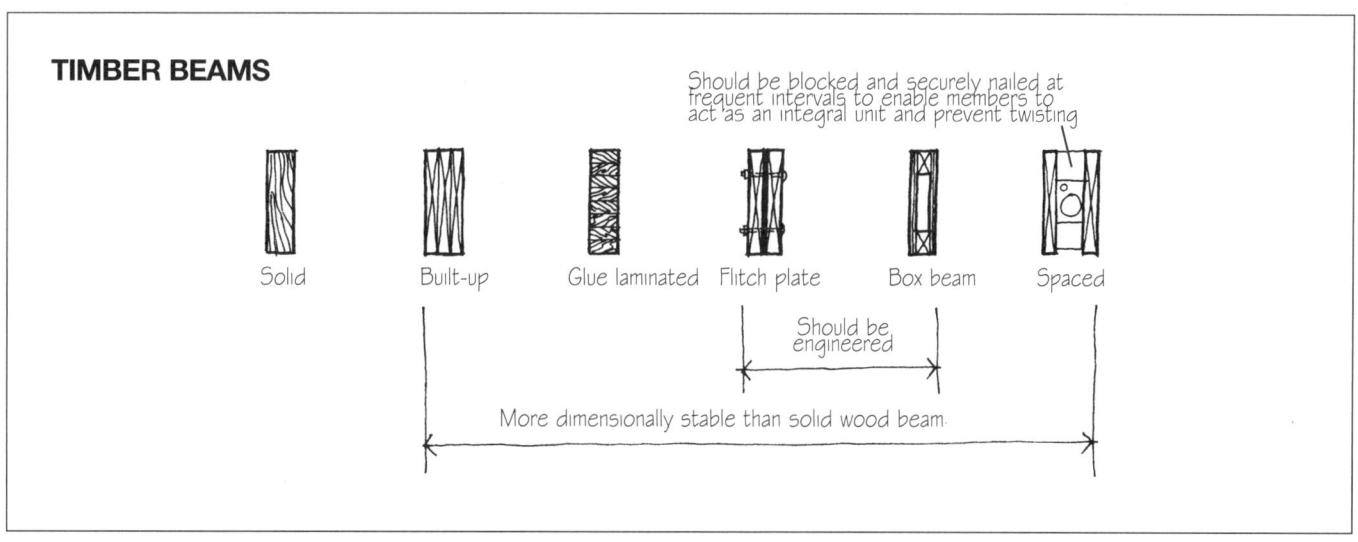

## TIMBER BEAM/COLUMN CONNECTIONS

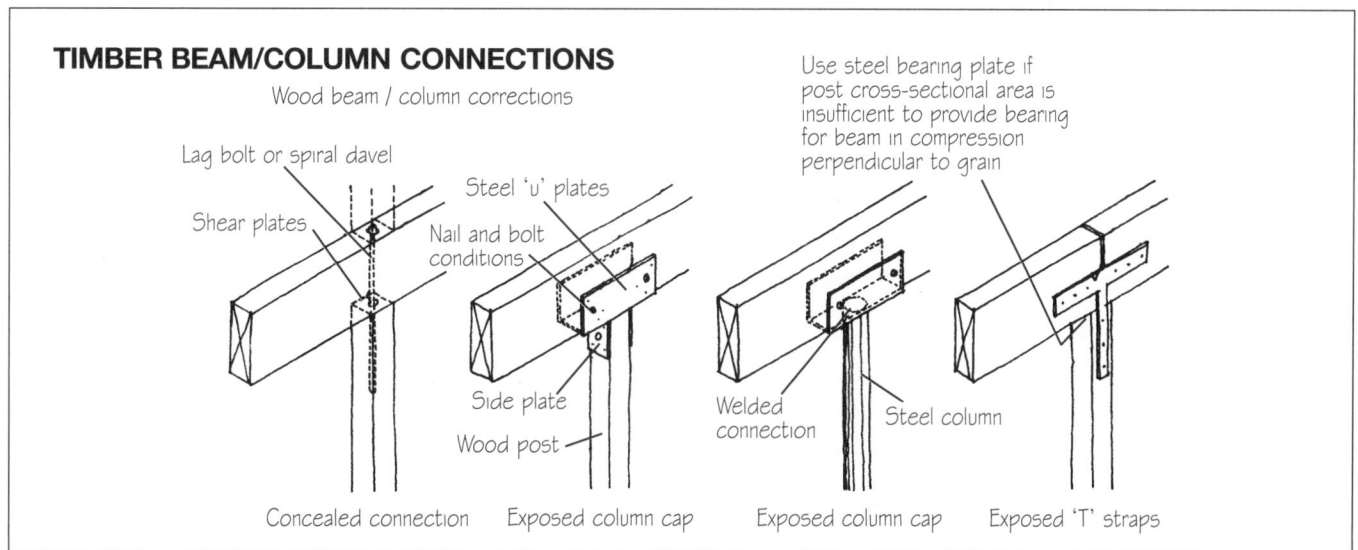

# 48 • ROOFS—FLAT, SKILLION & VAULTED

## Construction

Conventional roof framing is often described as being stick-built—a large number of sticks of timber are cut and placed on-site to create a complex interdependent network of members that give structure and form to the roof.

Trussed roof framing can be described as a combination of two-dimensional, triangulated frames built into a three-dimensional system.

Flat, skillion and vaulted roofs are generally described as post and beam—the structure is built up by erecting walls and posts, then topped by individual beams until the desired shape and structure are achieved.

The post and beam method of construction generally needs more care than either conventional or trussed-roof methods. This is because the system tends to rely directly on gravity to maintain structural integrity, rather than triangulated frames, as in conventional or trussed roofs. The structural components in post and beam construction are individually more critical to the structure as a whole than in the other two methods, and the failure of one component member can often cause the collapse of the whole, a situation less likely to occur in either conventional or trussed roofs.

Vaulted ceiling systems often require the use of open rafter couples, which means the restraining forces supplied by the ceiling joists and collar ties in conventional roofs and the bottom chord in trussed roofs are not provided in the roof frame itself, but rather by walls and beams designed to counter the rafter spreading forces.

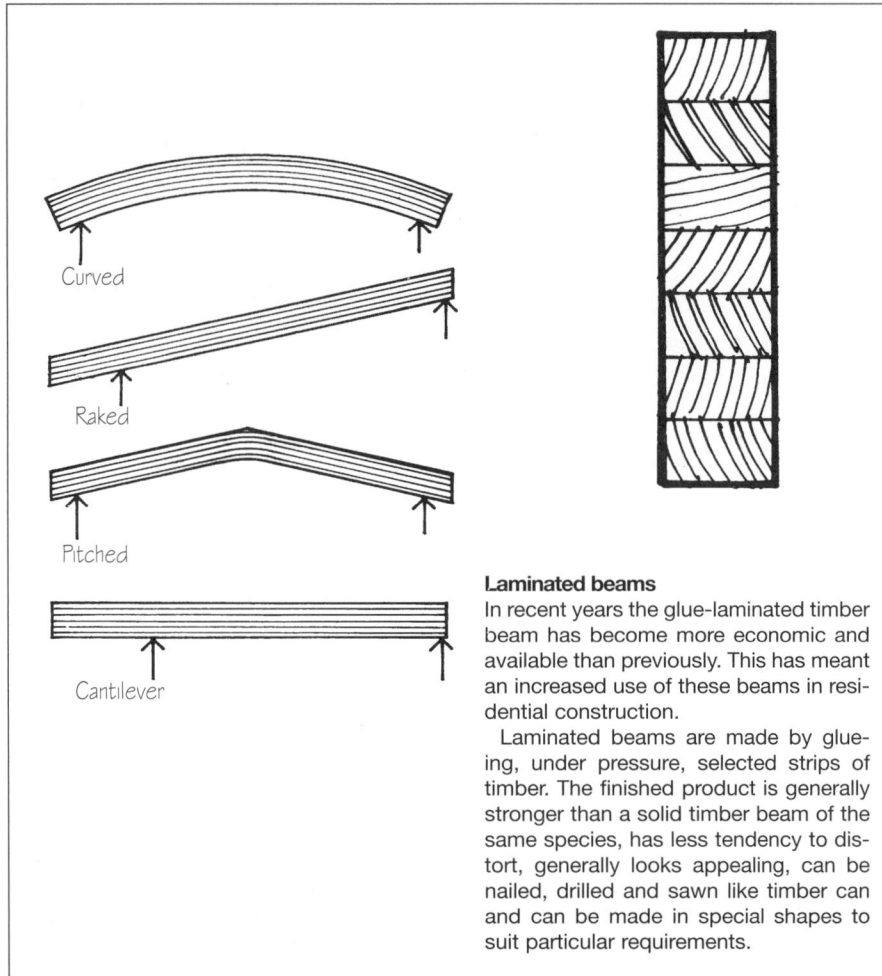

### Laminated beams

In recent years the glue-laminated timber beam has become more economic and available than previously. This has meant an increased use of these beams in residential construction.

Laminated beams are made by glueing, under pressure, selected strips of timber. The finished product is generally stronger than a solid timber beam of the same species, has less tendency to distort, generally looks appealing, can be nailed, drilled and sawn like timber can and can be made in special shapes to suit particular requirements.

## SKILLION ROOF—REFERENCE DRAWING

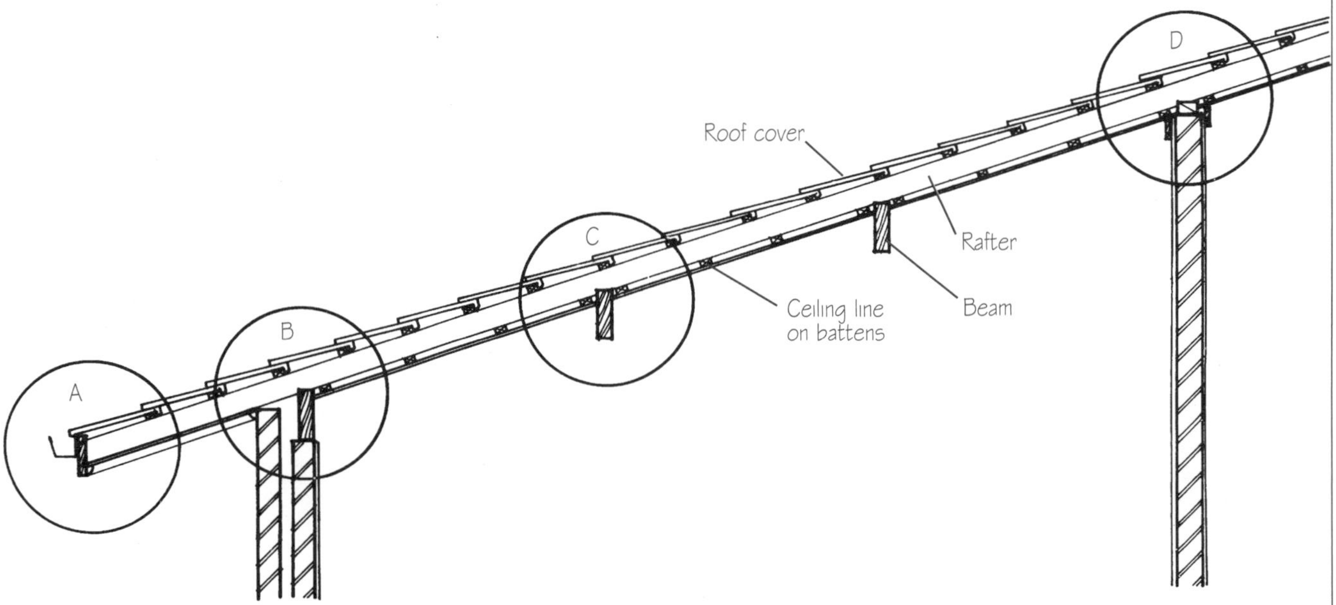

# ROOFS—FLAT, SKILLION & VAULTED • 48

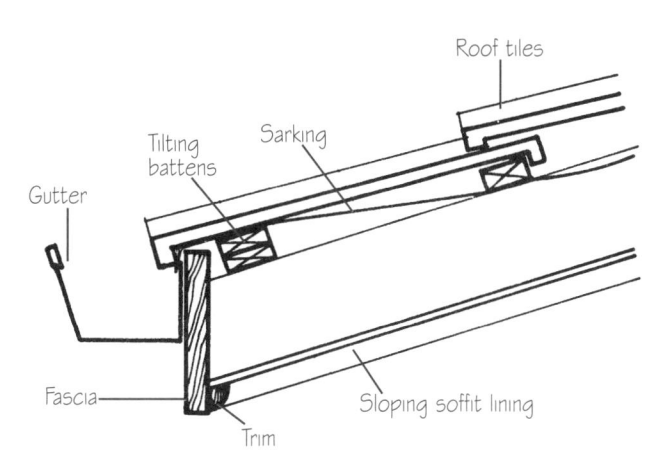

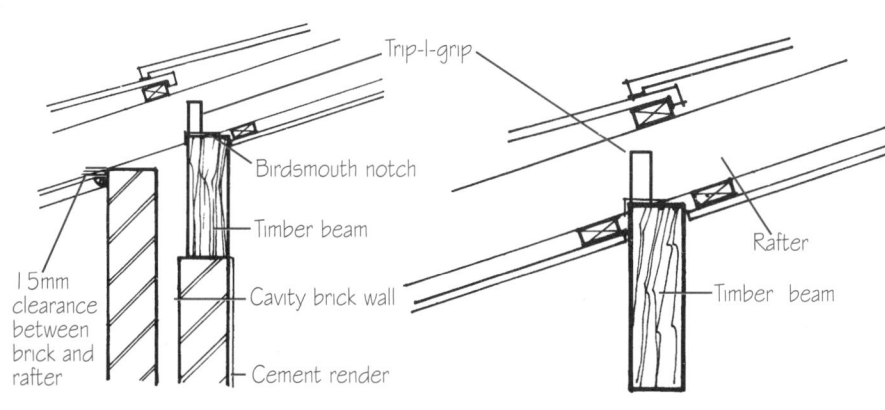

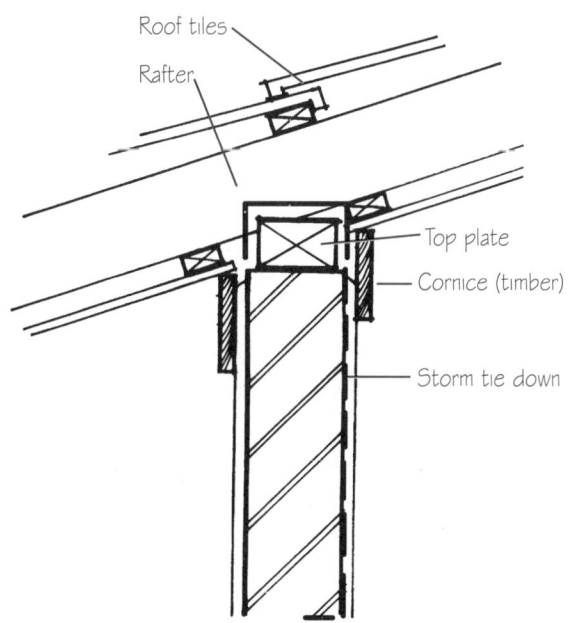

**SKILLION DETAILS**

**A Rafter to external wall junction**
Skillion roof construction generally uses the underside of the rafters as the structure to which the ceiling is fixed. Therefore, detail drawing B, shows this system. Owner builders should note that this detail (B) also applies equally to all raked ceiling conditions.

In this detail (B) the normal top plate has been replaced by a timber beam. This is a system used by some architects where they design a beam that will span the largest opening in the external wall of the house, and then use this beam continuously around the perimeter of the building regardless of whether there is a window, door or wall. Although this is generally used for aesthetic reasons it does mean that brick work is not carried over openings on steel lintels, and that there is a continuous bonding member at the roof pitching line.

**B&C Rafter to beam junction**
Where a rafter crosses an exposed timber beam the rafter must be birds mouth-notched to the beam. Technically these beams are purlins and may be supported by load beaming walls or posts. Adequate fixing between the rafter and the beam is essential not only to achieve a structurally sound building but also because noisy disturbing roof flutter may develop during windy conditions.

The beam is shown to be dressed timber, but sawn timber, steel beams, open web joists, cold formed steel purlins, laminated or boxed timber beams could be used. Location, appearance and design span being the likely determinants for choosing one material over another.

**D Rafter to wall junction**
The rafter to internal wall plate connection is by simple birds mouth notch and adequate nailing or, when required by local conditions, three dimensional steel fixing plates.

Matters requiring particular attention in this detail (D) are:
1. Fixing of the top plate to the wall. This is no particular problem in stud frame construction but can pose major problems in single skin brick wall. If hoop iron straps are used on the face of the brick work these are unsightly in exposed face work and tend to crack any render skin placed over them. There appears to be no common answer to this problem and it is recommended that each case be studied on its individual requirements and a system be devised to satisfy those requirements.
2. The unrestrained height of internal walls in skillion and other raked rafter houses, is sometimes in excess of that which is desirable to achieve a sound structure-particularly where brick walls are used. Walls should be designed, wherever possible, with adequate return corners and cross walls with stability.

# 48 • ROOFS—FLAT, SKILLION & VAULTED

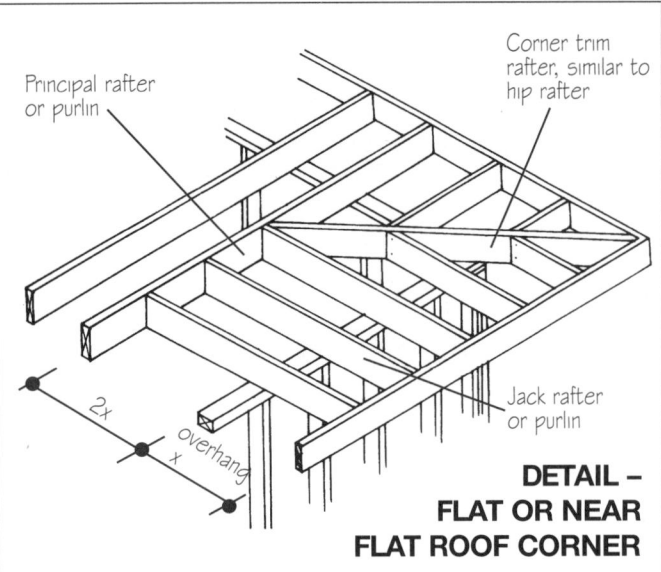

**DETAIL – FLAT OR NEAR FLAT ROOF CORNER**

If a flat or near flat roof requires both eave and verge overhangs then a special corner detail must be applied. This detail is illustrated (above) and it should be noted it does not matter whether it is rafters or purlins that make up the majority of the roof structure, their detail is still applicable. The jack rafters should not cantilever to form the eave or verge by more than one-third of their length and must be fixed to the principal rafters so that uplift is restrained as well as on support of roof load. If exposed rafters or purlins are being used in the house design then this detail must be given careful consideration.

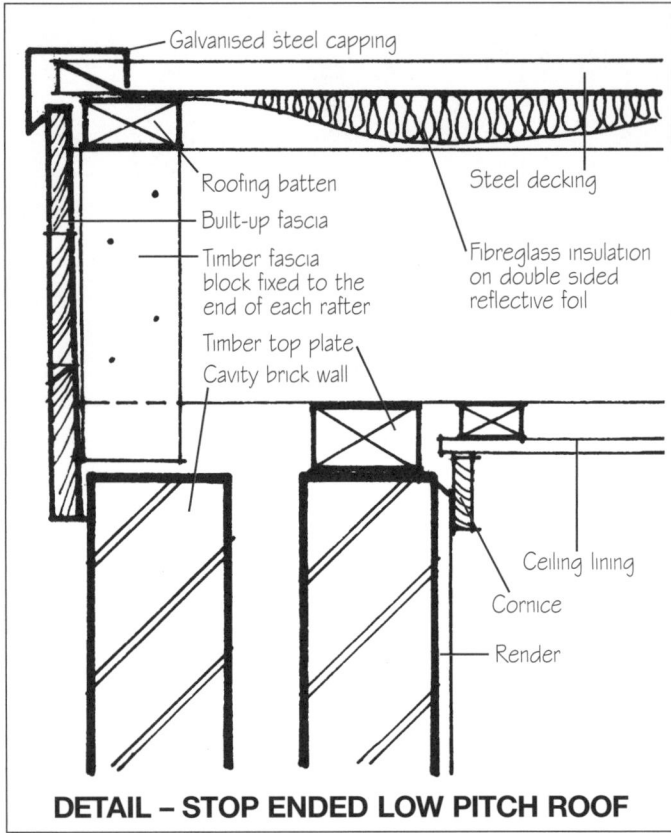

**DETAIL – STOP ENDED LOW PITCH ROOF**

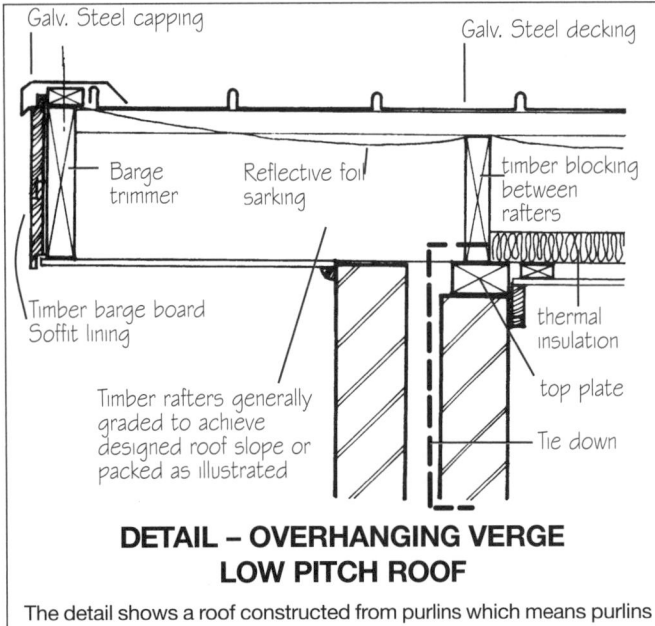

**DETAIL – OVERHANGING VERGE LOW PITCH ROOF**

The detail shows a roof constructed from purlins which means purlins are placed perpendicular to the roof slope. This allows easy cantilevering to give wide verge overhangs.

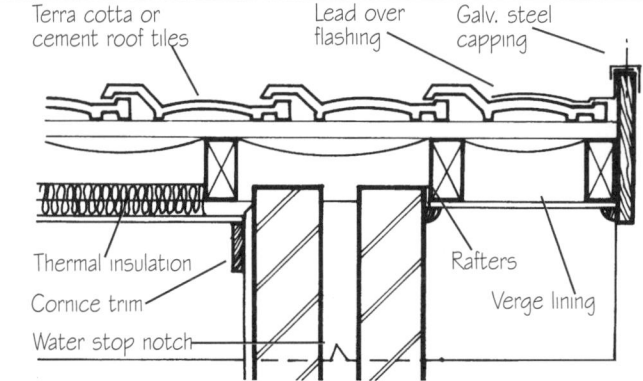

**DETAIL – VERGE CONDITION WITH RAKED CEILING AND TILE ROOF**

It is normal when tiles are used on a vaulted/raked ceiling house that the rafters are carried on large section under purlins that are featured inside the house as exposed beams. Care must be taken in high wind areas to tie the under purlin beams down adequately, for although the beams are very heavy to lift into place, once combined into the roof structure are only a small proportion of the roof mass.

### DEFINITIONS
| | |
|---|---|
| Fascia | board fixed to end of rafters |
| Barge | board fixed to end of purlins |
| Eave | overhang created by extending rafters beyond support wall |
| Verge | overhang created by extending purlins beyond support wall |
| Rafter | principle roof member at the slope of roof |
| Purlin | principle roof member perpendicular to slope of roof |

# ROOF TILES • 49

The majority of Australian houses are roofed with tiles of one type or another, and many people consider that a terracotta tiled roof cannot be beaten for economy and longevity.

There is much evidence to confirm that manufactured roof tiles, whether terracotta or cement, are particularly suited to the Australian climate, which generally does not have high winds or snow loads—the elements most likely to reduce the life of roof tiles.

Other tile roofing materials include: slate, which is a naturally occurring rock; fibrocement shingles, which are slate-like products manufactured in fibre-reinforced cement; timber shingles and shakes, which use sawn or split timber boards; and aluminium or steel pressed tiles, which are metal shaped to mimic a tile, and often covered with coloured sand glued to the metal.

## Tile roofs

Roof tiles can generally be used on roofs of 15 degrees or greater pitch, although some materials and methods of manufacture require 25 degrees or greater pitch.

Normally pitches between 17 degrees and 27 degrees are used, giving a very uniform roofing appearance to many suburbs, particularly when the colours are most likely to be terracotta red, brown or charcoal.

When heavy tiles are used on a roof, steep slopes are actively discouraged. In some Australian states, tilers demand a premium payment for laying tiles on slopes of more than 45 degree pitch, causing many architects who want a steeply pitched roof to restrict the slope to 44 degrees.

House plans are generally required to have right-angled corners if roof tiles are to be used. To tile a roof where the meeting planes of hip or valley intersections do not meet at 45 degrees generally creates messy detailing, often because the manufacturers and tilers believe that tile roofs should be simple, and therefore only produce, or are trained, to that level.

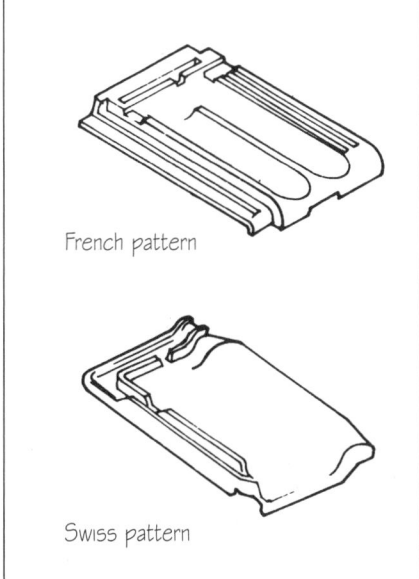

French pattern

Swiss pattern

## Roof tile types

### 1. Terracotta
Terracotta roof tiles are manufactured in many locations throughout Australia. The tile is formed in wet clay then fired to give a hard, long-life roofing product. They come in a number of profiles, including French/Marseillaise, French/modern, Swiss and Roman. Terracotta tiles have a roof life in excess of 50 years in most environments, although salt atmospheres sometimes cause premature fretting.

### 2. Cement
Cement roof tiles (or more technically, concrete roof tiles) are generally modelled on terracotta tiles and for many years were considered inferior and cheap. Modern cement tiles have roof lives approaching those of terracotta, are stable, warp-free and have the same basic profiles as terracotta tiles. The colour on the cement tiles is an applied coating and, although some colours fade quickly (within 5 years), most of the popular red/brown/charcoal colours have long colour retention. Cement tiles are available in some locations in a flat, shingle-like profile.

### 3. Cordovan
Cordovan tiles are those that grace Spanish missions in California. They are available in the main capital cities of Australia on special order and can be purchased in red terracotta or salt-glazed brown terracotta. These look good but should be restricted to areas of low rainfall, as they are not particularly watertight.

### 4. Slate
Slate roofing was used extensively on quality Australian homes during the Victorian era but, as there were few deposits of good quality roofing slate in Australia, the majority of the slate used came from England. Terracotta roof tiles are much cheaper than slate and pushed this material into a very small expensive sector of the roofing market. Slate is a natural stone that can be split into thin sheets. Good quality slate has an almost infinite life in temperate climates, but, as the quality of the slate drops and the solar load increases, the life of the slate reduces proportionally.

### 5. Fibrocement Shingles
Fibrocement shingles generally consist of a multi-notched sheet of fibre-reinforced cement that has been specially coloured and is sometimes textured. The multi-notches make the sheet, when fixed, appear to be a number of shingles (see diagram). This system allows for a saving in labour and material. The material saving is achieved because the notches occur only where the shingle is visible; two layers of multi-notched boards will achieve a waterproof roof, whereas three layers of shakes or individual shingles are required to achieve a waterproof roof.

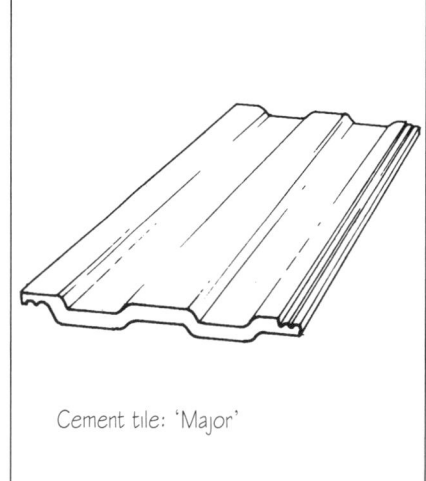

Cement tile: 'Major'

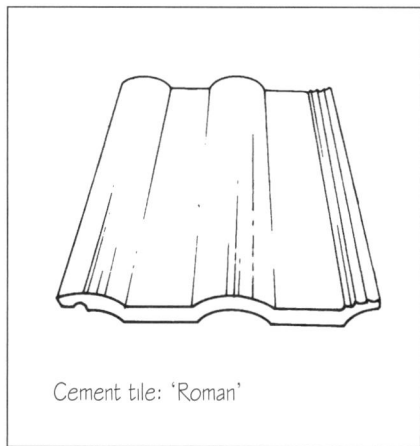

Cement tile: 'Roman'

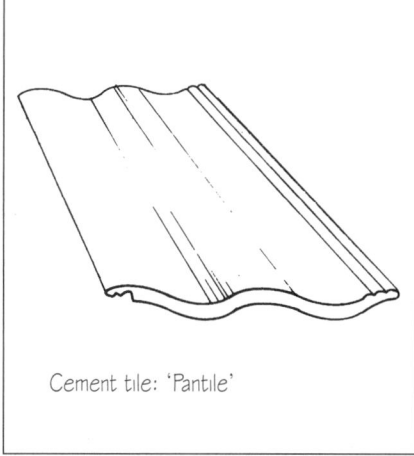

Cement tile: 'Pantile'

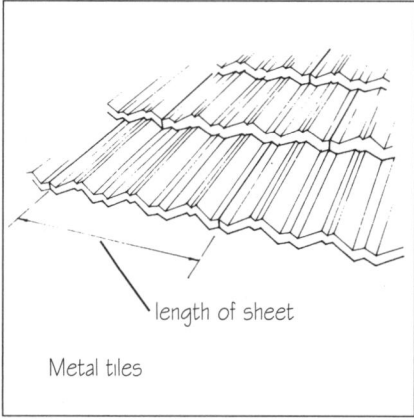

length of sheet

Metal tiles

# 49 • ROOF TILES

### 6. Timber Shingles and Shakes

Timber shingles and shakes are generally made from highly durable timber species that can be split into shakes or sawn into shingles. Although some shingles are still cut from Australian hardwood timbers, the majority of shingles used in Australia are imported from North America.

Shingles have a pleasant rustic appearance a good roof life in ideal conditions. Shingles are expensive to purchase because they must be cut from good-quality timber. They are three layers thick on the roof; each shake has to be individually fitted and fixed and best results are obtained on a roof where sub-roof sheeting has been fixed to the top of the rafters (see diagram).

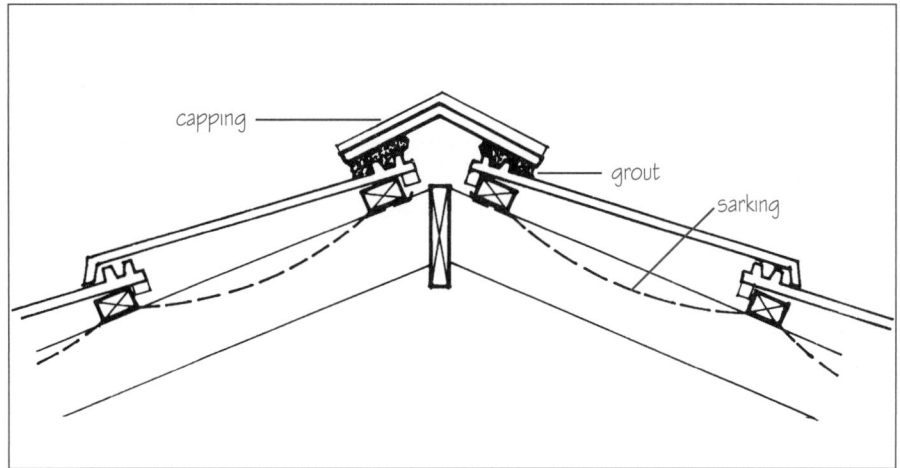

### 7. Metal tiles

Metal roof tiles are generally cold-formed in long lengths of galvanised steel or aluminium. Although some metal roof tiles are sold as pre-painted smooth metal, recent technological advances have allowed some manufacturers of metal roof tiles to coat them after formation with sand-like textured surfaces in various colour combinations. These are not cheap products but they are lightweight compared with cement tiles and can be transported cheaply over long distances, allowing a house owner a tile-like roof, even in extremely remote areas. They are particularly suited to earthquake areas, where the alternative would be profiled metal sheet or clad roofing.

### Sarking

Most manufacturers of roof tile products indicate a minimum pitch for their product without sarking and a different minimum pitch for roofs with sarking.

Sarking is advisable under any roof, particularly if the sarking can perform a dual role:
1. as a secondary protection against moisture penetration due to wind-blown spray, broken tiles, condensation and the like.
2. as a reflective foil surface to reflect the radiant heat of the sun.

If a roof is built in the 'sarking required' pitch zone, OBs should realise that the manufacturer is saying that, without sarking, there is a high chance that the roof will let in water. That being the case, OBs should take care when installing the sarking that it is waterproof and that there are no joints that could leak or accidental holes or tears in the membrane.

Remember, the sarking on a tile roof is under the battens and therefore hard to replace if it is subsequently damaged. It is difficult to install if left out of the initial building.

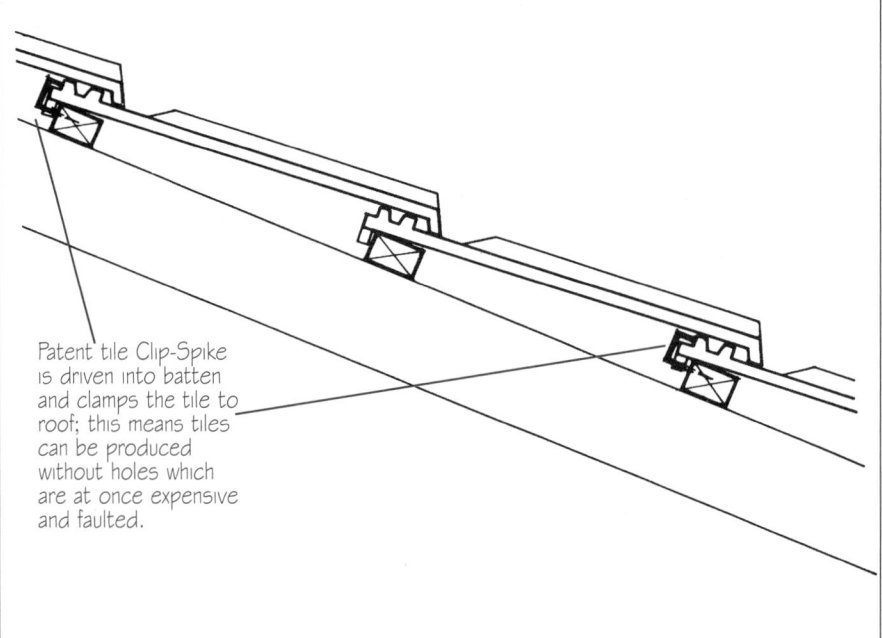

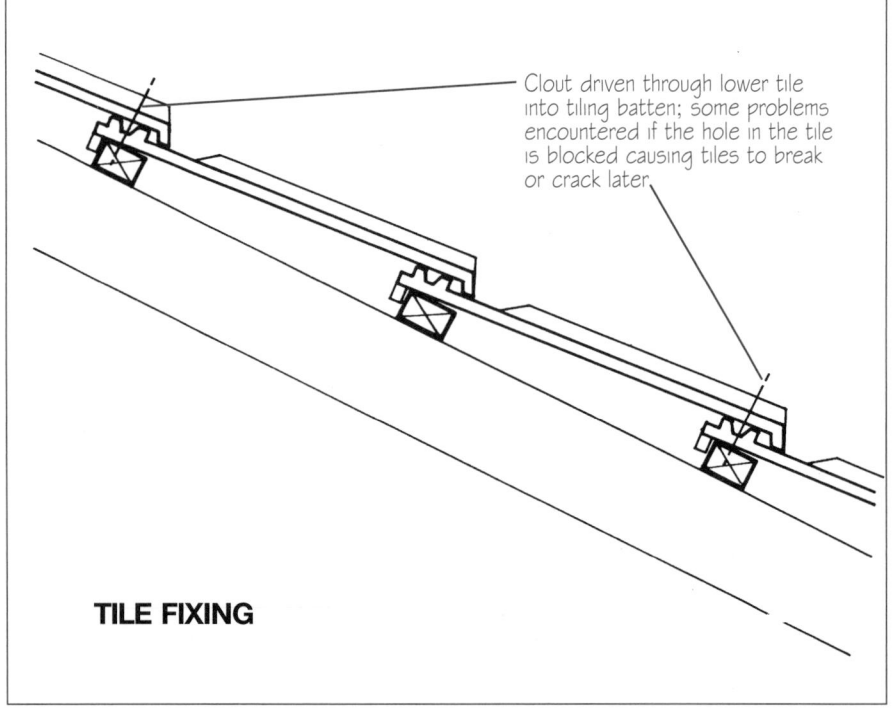

**TILE FIXING**

# ROOF TILES • 49

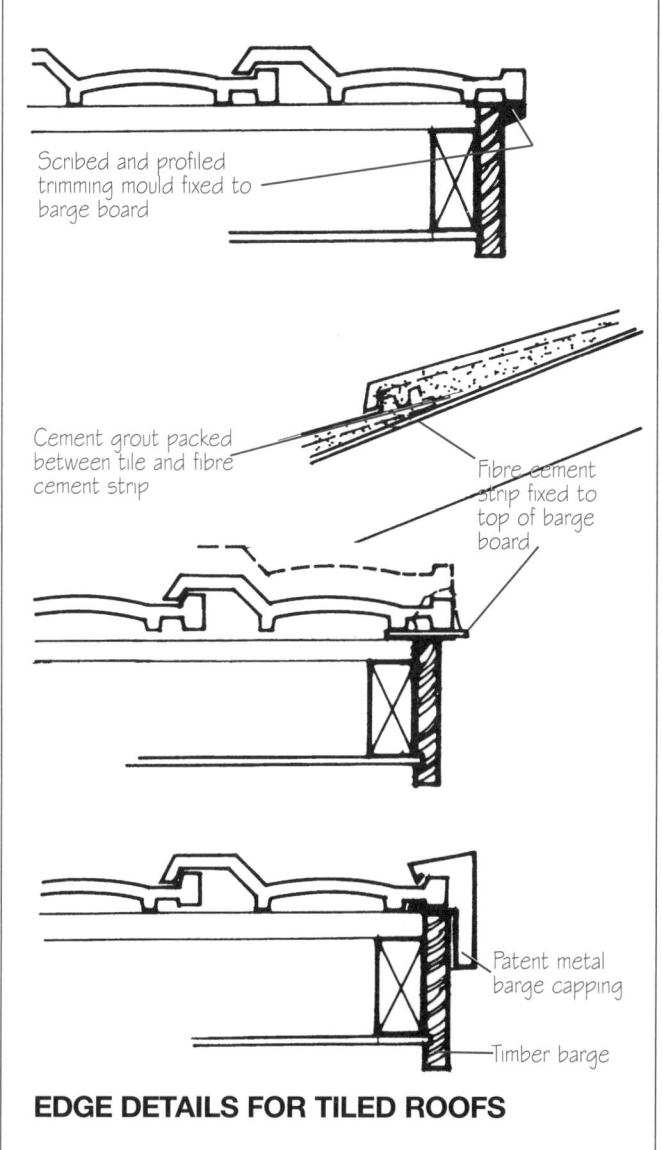

**EDGE DETAILS FOR TILED ROOFS**

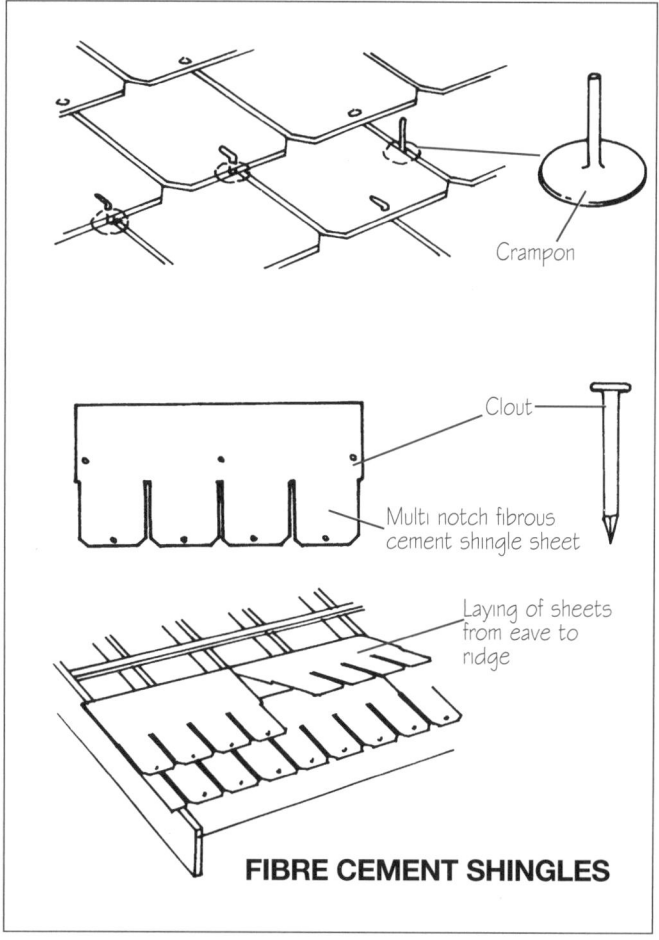

**FIBRE CEMENT SHINGLES**

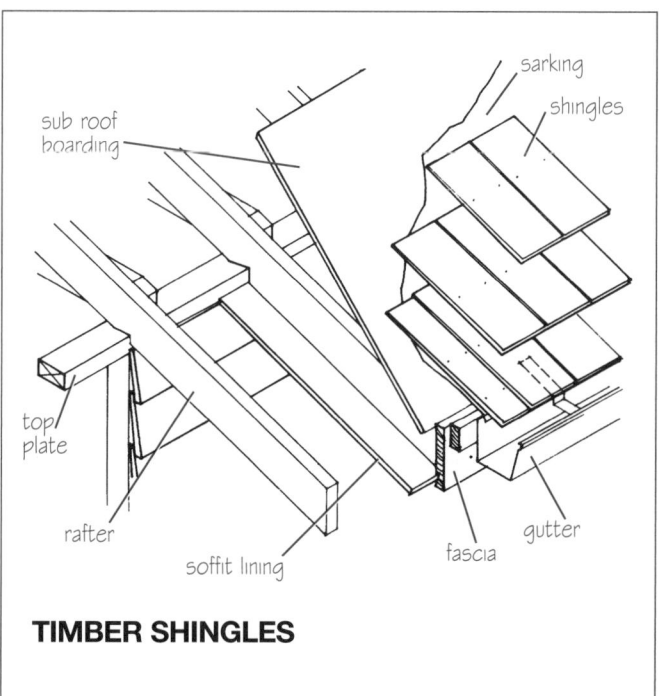

**TIMBER SHINGLES**

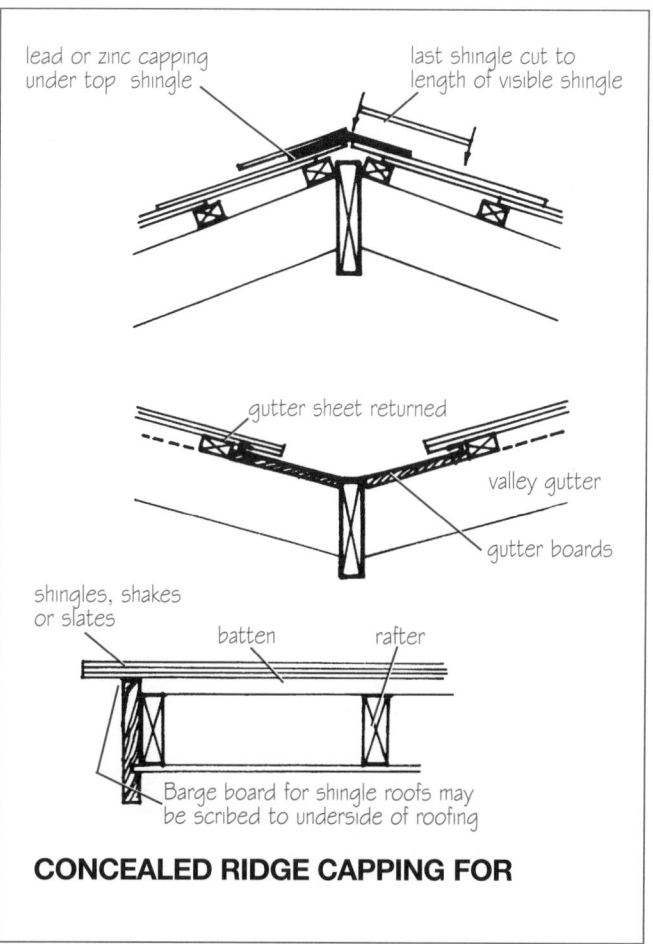

**CONCEALED RIDGE CAPPING FOR**

# 50 • ROOF SHEETS

## Aesthetics

The general acceptance of roof materials, particularly sheet roofs, varies from locality to locality. The original sheet roofing product was galvanised iron and has changed little from when it first arrived from England. Today, the profile of the sheet is almost identical, although the iron is now steel and the hot dipped galvanised (zinc) coating has given way to an aluminium/zinc-alloy coating.

Galvanised iron, or zincalume corrugated Custom Orb, as it is now known, has had three major differences from the original product.
1. It is now available in long lengths, limited only by trucking constraints.
2. It is now available in pre-painted coloured sheets.
3. The fixing methods have been improved to reduce sheet damage and leakage.

Galvanised iron is currently enjoying a period of acceptability as a roofing material for houses and a number of architects, notably in New South Wales, have used it extensively in award-winning houses.

Zincalume coated steel roof sheets are also produced in many different profiles to suit particular applications and appearance preferences. However, the standard corrugated profile still satisfies the widest range of applications—it is still the only profile that lends itself to curving, which allows its use on 'colonial' type bullnose, ogee and draped verandah roofs.

Of the other steel roof sheet products, the steel decking systems, which have deep strong profiles and secret non-puncturing fixings, are of special interest, as these products have been developed in Australia as low pitch (as shallow as 1:50) roof coverings.

Prior to the development of the steel deck products during the 1960s, roofs of very low pitch had to be built in reinforced concrete, which is both expensive and seldom leak-proof, or built-up bitumen felt roofs. Both were developed in cool climate European and US locations and proved to be unsatisfactory in hot Australian summers because they melted and flowed, causing cracks and leaks.

The flat roof can be used to create aesthetically pleasing houses, particularly where a complex floor plan is desired, or required.

Other sheet roof products include aluminium sheets and decks which have not, to date, attracted a significant proportion of the market, mainly because they have mostly been marketed in the commercial industrial building sector and are prone to annoying thermal noise problems.

Corrugated fibrocement sheets have enjoyed some acceptance by architects from time to time, but the short lengths of the sheets and the complex fixing and flashing systems have limited their market penetration The stigma of fibro as being a cheap roofing material has not assisted this good-quality long-life roofing material achieving its potential.

## Suitable roof types

Generally, roof sheets can be used on any roof type, but each particular profile has a particular range of roof pitches in which it is most efficient. OBs should check with local manufacturers to ascertain the most suitable and economic roof sheet for the roof type they are considering.

## Water collection

The run-off from the metal roof sheets during rain is immediate and complete, making both steel- and aluminium-based materials ideally suited to farmhouses or other special circumstances, where the roof water is collected for storage and hence water supply.

Fibrocement products are less suited to collection of water for drinking purposes but for other purposes the water collected from a fibrocement roof is satisfactory.

## Wind

Sheet roof products have often been considered more likely to blow off in high wind conditions than roof tiles or shingles. This problem has, mostly, been because of inadequate fixing of the sheeting to the roof structure, or inadequate fixing of the roof structure to the building as a whole.

OBs considering using sheet roofing in high wind or cyclonic zones, should take care to construct their houses correctly from the footings upwards, to resist the uplifting forces created by wind blowing across a roof—and to fix the roof sheets as recommended by the manufacturer.

The manufacturers of roof sheets generally produce good-quality technical information for their products and often give extensive information on fixings for high wind areas. In some instances special profiles are produced for use in extreme climatic conditions.

## Roof pitch

Most sheet roof products will remain adequately watertight down to a pitch of 73 degrees, and some, particularly the decking profiles, will remain watertight on pitches that may be considered nearly flat: 1:50.

OBs should take care when using very low-pitched roofs in very high rainfall zones. Sometimes the water cannot be shed from the roof quickly enough and the water builds up on the roof higher than the crests of the sheets and therefore increases the chance of water penetration.

## Insulation and sarking

Generally speaking, all sheet roofs should have a waterproof sarking immediately under the roof sheets. The sarking can simply be reinforced kraft building paper, or, more commonly today, reflective foil sarking. The reflective foil improves the roof's thermal performance by reflecting a significant proportion of the solar load falling onto the roof.

The sarking must be adequately fixed. Often the battens or purlins for sheet roofs are widely spaced and poor-quality sarking can flap about in the wind, causing annoying noise and possible tearing of the sarking membrane. Therefore, the sarking purchased should be of adequate quality to do the job being asked of it. Wide spans between supports will require the use of heavy-duty, internally reinforced sarking, or a supplementary support net under the sarking of chicken wire or similar.

The primary function of the sarking under sheet roofs is to reduce the problem of condensation forming on the underside of the roof. This can be a particular problem in low-pitched metal sheet roofs where, because of the low pitch, the condensation drips off the sheet onto the ceiling below rather than running down the underside of the sheet and out of the building. The problem is further compounded because the air space between the ceiling and the underside of the roof sheeting in low-pitched roofs is often too small to allow adequate flow of air and therefore a space of high humidity air is allowed to develop, causing an increased condensation problem.

The sarking, when faced with reflecting foil, will improve the solar load control of the house. In most parts of Australia this reflective foil needs to be supplemented by thermal blanket insulation to reduce heat loss from the house in winter or heat gain by the house in summer.

Every house has its own particular problems in relation to thermal insulation and it is recommended that, in extreme climatic conditions, OBs supplement the general information of this book with information particularly relevant to the locality where they are building, and suitable to the house they are constructing.

In the moderate latitudes, care should be taken not to over-insulate for it is possible to insulate in a way that will create as many or worse problems than it solves.

When thermal insulation is used on a low/flat-pitched roof it is common to place the thermal blanket on top of the reflective foil sarking and then fix the roof sheeting directly on top of this sandwich. This works satisfactorily if the ceiling is very close to the underside of the roof but leaves the ceiling uninsulated, which can cause a lowering of thermal insulation efficiency.

It is generally better to place the thermal blanket insulation immediately over the ceiling sheets and, even if blanket insulation is used directly under the roof sheeting, ceiling insulation should be considered.

A further option is to use rigid board insulation that doubles as ceiling sheet and insulation. There are a number of rigid board insulation/ceiling systems available use a wide range of insulation materials including plastic foams, compressed straw,

# ROOF SHEETS • 50

fibreglass and mineral wools and wood shavings. Most of these boards have been developed as integrated systems, including structural beams in steel or timber, and are assembled on some form of modular grid.

## Colorbond

Most of the roof sheeting manufacturers produce pre-coloured roof sheets in many different profiles and a number of different metals—including high-tensile steel, stainless steel and aluminium.

Generally, the basic-grade material is chosen for houses. The available colours vary from time to time and from place to place—there are special Queensland colours, for example. This means that future matching of roof sheeting colours, should an extension ever be required to the house, may be a problem.

In 2001, BHP, the effective single source of the coloured roof sheeting stock, released three new premium Colorbond products.

**1 Colorbond Stainless**
This allows OBs to build within 100 metres of the breaking surf, according to BHP's promotion. It may also be a sensible selection for a roof over an indoor swimming pool.

**2 Colorbond Ultra Steel**
This one is suitable between 100 and 200 metres of the breaking surf.

**3 Colorbond Metallic Steel**
Where as the Stainless and the Ultra Steel are coated in the traditional Colorbond colours, this range has a metallic finish not unlike that used on motorcars. Interesting challenge to designers but isn't zincalume also metallic, and often rejected by authorities because it is too reflective?

## Building with sheet roofing

OBs will generally be able to use most sheet roof products with little or no training, but there is important information that should be considered for every job.
1. Choose a sheet that suits the design of the house you are building. It is worthwhile taking a look around at houses that have a similar product on their roof, and check that there are no apparent detail or appearance problems.
2. Read the manufacturer's catalogue information thoroughly, and contact the company for advice if your particular application is not clearly illustrated or described.
3. Make a list of all the components you require: the sheets, cappings, flashings, fixings, special tools, and so on.
4. Accurately measure the length of sheets you require, take great care because it is extremely awkward to cut some sheets on-site, and sometimes impossible to add to sheets if they are too short. Accurate measurement will save on-site problems and money.

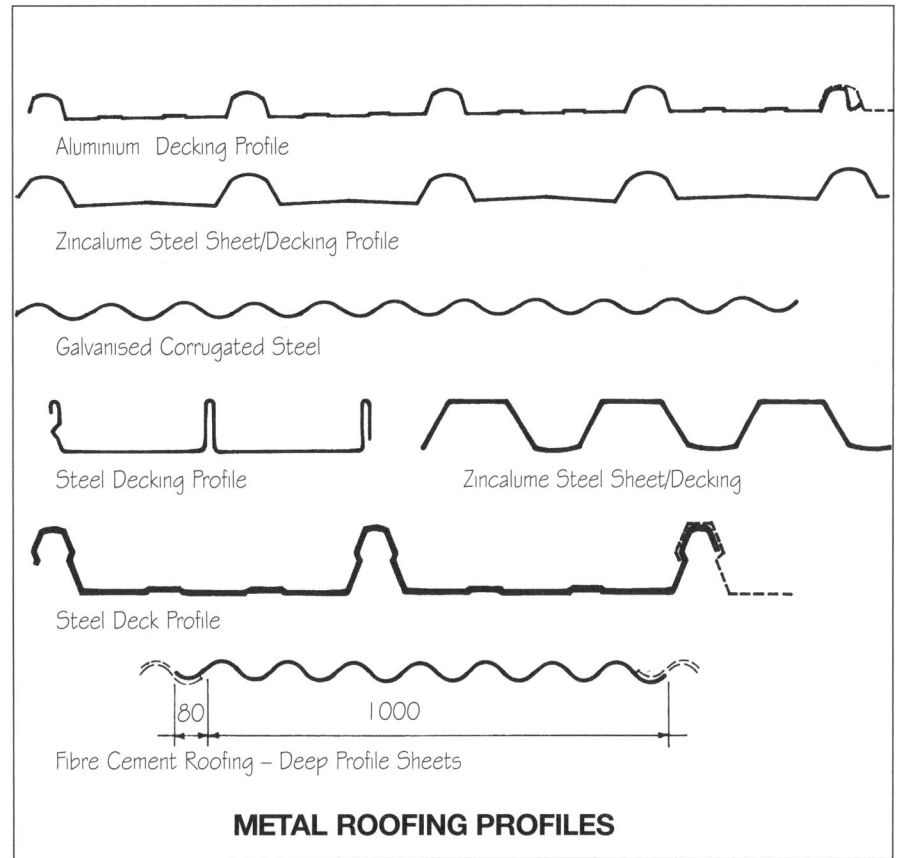

**METAL ROOFING PROFILES**

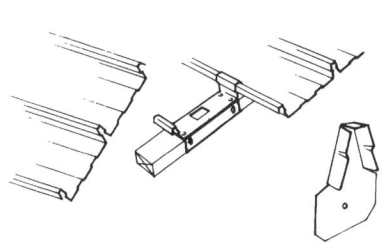

Clip fixed, steel and aluminium deckings are available with secret clip fixings.

This allows for very shallow roof falls (as low as 1:50) because the combination of the deep section of the decking and the complete lack of penetrations means that it is nearly impossible for a steel or aluminium deck roof to leak.

However, great care must be exercised when laying the decking to make sure the high ends of sheets are stop ended and that no small holes are caused by this process and the gutter ends are turned down into the gutter and have the rib ends filled with an end cap.

In high wind areas clipd that fix to the side of the purlin/battens are preferred over the standard nail down type, and special extraa deep deck profiles should be considered.

The manufacturers of steel and aluminium decks produce first class information on erection sequences and methods and all users should make themselves conversant with this information.

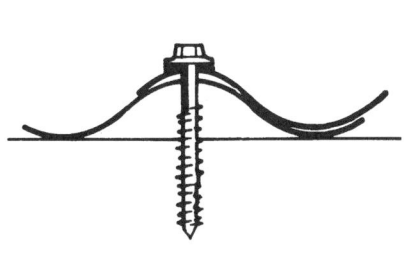

Crest fixed, self sealing screw with neoprene washer on hexagonal head to suit power fixing. A cutting edge is manufactured in the point of the fixing so that the screw drills its own hole through the galvanised steel.

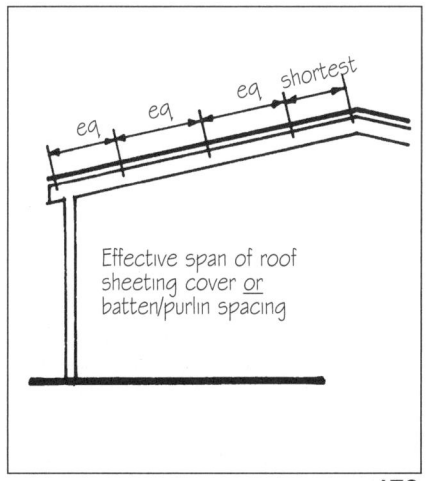

Effective span of roof sheeting cover or batten/purlin spacing

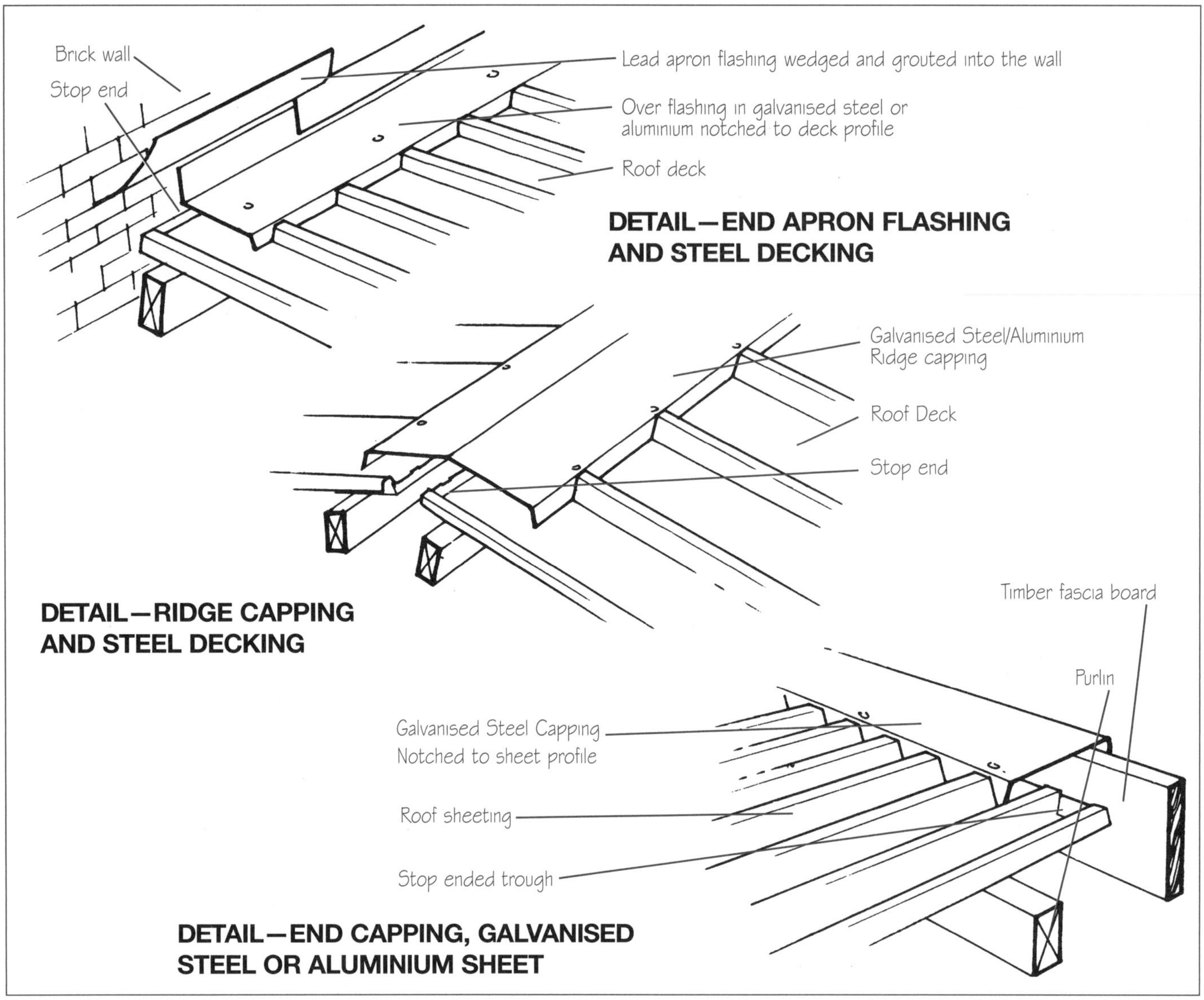

# 50 • ROOF SHEETS

**DETAIL—END APRON FLASHING AND STEEL DECKING**
- Brick wall
- Stop end
- Lead apron flashing wedged and grouted into the wall
- Over flashing in galvanised steel or aluminium notched to deck profile
- Roof deck

**DETAIL—RIDGE CAPPING AND STEEL DECKING**
- Galvanised Steel/Aluminium Ridge capping
- Roof Deck
- Stop end

**DETAIL—END CAPPING, GALVANISED STEEL OR ALUMINIUM SHEET**
- Galvanised Steel Capping Notched to sheet profile
- Roof sheeting
- Stop ended trough
- Timber fascia board
- Purlin

5. Check delivery delay. You should attempt to have the roofing material delivered when it is required; miscalculation or guesstimation of the delivery delay may cost time and money.
6. If the sheeting is ordered in long lengths, make sure there is a space available for the delivery as close to the job as possible; in some cases the truck may be equipped with a crane which can lift the packs of roof sheets directly onto the roof structure, to save as much site handling as possible.
7. On the day(s) you set aside to fix the roof, make sure you arrange enough labour to complete the job, as unfixed sheets lying on the roof can become lethal if a wind springs up. Do not lay sheets on a windy day.
8. Some roof sheets, including steel deck and fibrocement have a number of operations, such as stop ending and mitring, that can be done on the ground before the sheets are hoisted onto the roof. If these tasks are done in advance, a more economic use of time can be achieved on the day you select to fix the roof cover.
9. The chicken wire (if necessary), sarking and insulation can be precut and ready to lay just ahead of the sheeting. This will save time and reduce the risk of wind, water and other accidental damage to these materials, for it only takes one small hole in the sarking to cause future problems.
10. If an exposed beam ceiling is being used, and the ceiling sheets are to be laid over the beams and the roofing fixed over, it is advisable to keep a tarpaulin on the site to protect the ceiling if there is an unexpected shower of rain. Also, where the ceiling is placed first, and for that matter in all roofing operations, provide wide safe planks that can be laid ahead of the roof covering operation to give your roof-fixers a safe platform to work on.

## ACCESSORIES

The accessories for sheet roof are often available preformed and in long lengths. Owner builders will require a combination of side end cappings together with rainwater gutters for even the simplest building. Plus many other optional accessories which include, but are not limited to, apron flashings, ridge capping, valley gutters, box gutters, down pipes, roof lights. All flashings must be assembled with care for often many hours are spent, after the project is complete and the first rain has arrived, searching for small but critical weaknesses in the weather tightness of the roof cover.

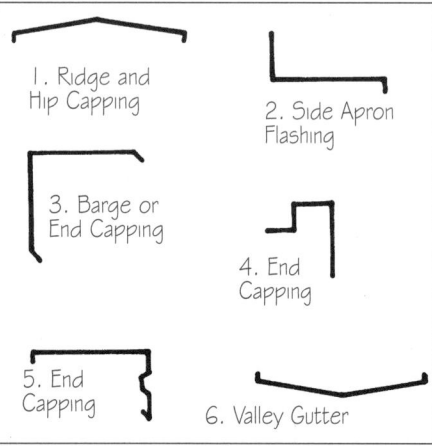

1. Ridge and Hip Capping
2. Side Apron Flashing
3. Barge or End Capping
4. End Capping
5. End Capping
6. Valley Gutter

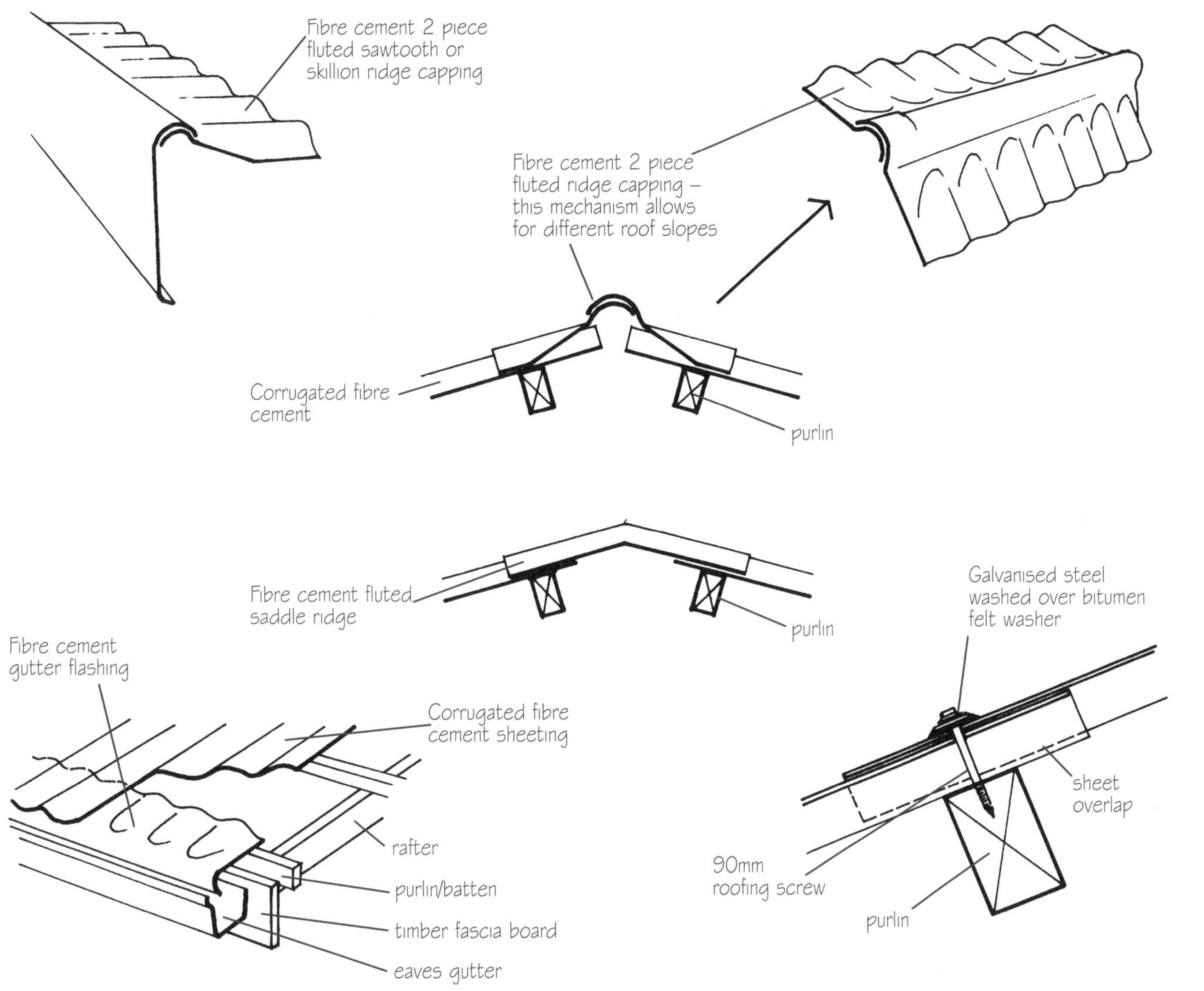

# ROOF SHEETS • 50

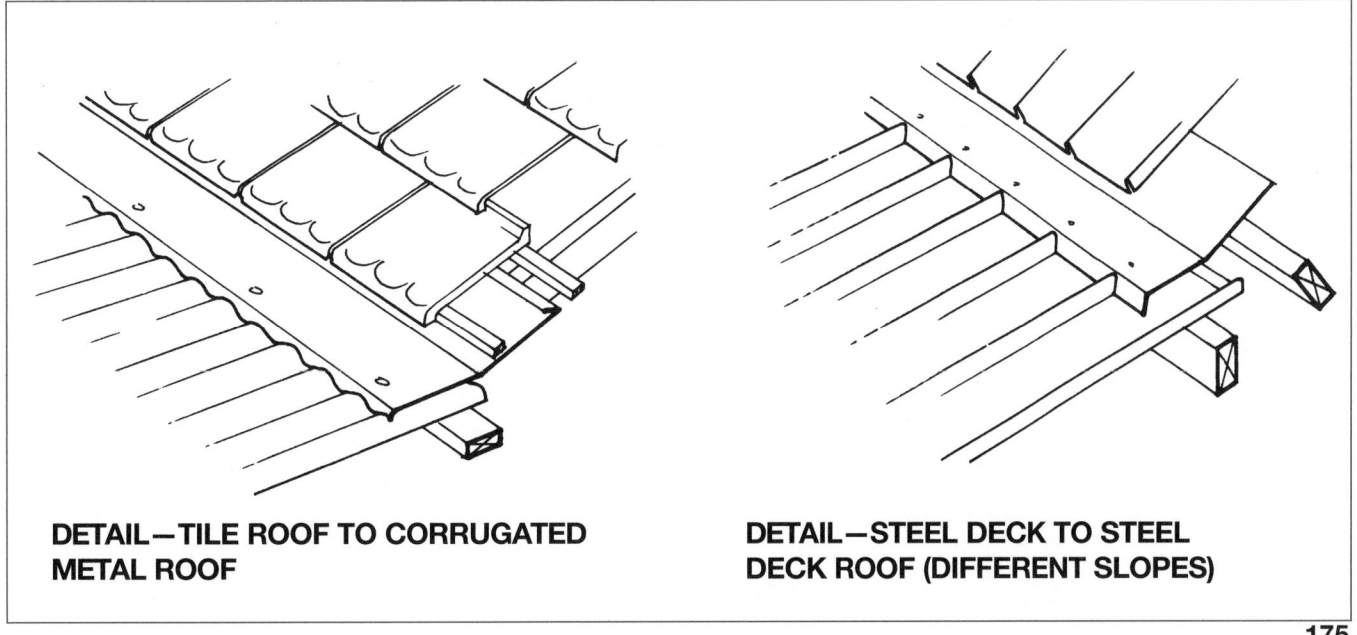

**FIXING CORRUGATED FIBRE CEMENT TO ROOF PURLINS**

**DETAIL—TILE ROOF TO CORRUGATED METAL ROOF**

**DETAIL—STEEL DECK TO STEEL DECK ROOF (DIFFERENT SLOPES)**

# 51 • CEILINGS

When choosing a material and the form of a ceiling, OBs are faced with a wide variety of possibilities, including:
- flush smooth materials
- natural timber materials
- manufactured timber materials
- modular panels that can be fixed:
  flat horizontal
  flat on a slope
  with beams
  on curves
  compound designs.

It is generally true to say that almost anywhere in Australia you can build, horizontal flush-jointed plasterboard ceilings are inexpensive and in common use. Because of this, OBs often want to use anything but plasterboard on their ceilings. This is not always realistic and often simple design aids, like exposed beams, sloping ceilings, wall (ceiling) paper can give quality results.

Timber boards, when used as ceiling lining, are often favoured by OBs. However, care must be taken not to make a ceiling too dark by using dark timber, or too woody by using excess amounts of timber without control. Often OBs will use timber because they think it is easier to erect than plasterboard, which may need a skilled fixer.

There are two problems with this belief:
1. timber can cost as much for the material as does plasterboard for material and labour
2. timber is not as easy to fix as it appears and will only be as straight (or warped) as the joists or battens that support it.

Manufactured timber panels, like plywood, hardboard or prefinished sheets, have limited use in ceilings. Great care must be taken to avoid unsightly sheet joints and, as these materials are generally nail-fixed, nail heads are often visible.

Thicker plywoods of 5–10 mm are useful in sub-roof panelling but, unless good quality sheet material is used most of the thinner/cheaper reconstituted timber sheets and plywoods have a tendency to expand and contract excessively, and may frequently warp and distort.

Modular ceiling panels are normally fixed to some form of metal, or the less common timber grid—which is sometimes a structural system capable of supporting the ceiling and the roof.

When deciding whether to use flat or sloped (or curved) ceilings, a number of considerations should be taken into account. The following checklist will be of assistance.
1. What is the difference in cost between flat and sloped ceilings? (Sloped ceilings are generally more expensive because there is a greater area of ceiling and walls than with a flat ceiling for the same rafter springing line.)
2. What advantage will be gained from using a sloped ceiling? (Advantages in visually higher spaces: clerestory windows, exhaust ventilation and solar conditioning devices could be considered.)
3. What disadvantages will there be in using a sloped ceiling? (Common disadvantages in sloped ceilings are the close proximity of the roof cover to the ceiling; interior and exterior design have to be carefully integrated, often leading to more expense or compromises. It is often much harder to gain structural efficiency in a sloped ceiling system, which cannot take advantage of the static triangular frame that is provided in a trussed roof or a rafter couple and tie roof system.)
4. Is a sloped or flat ceiling better suited to any particular space? (It is wise to check each space and room in a house to determine whether sloped or flat ceilings would be most appropriate. Generally the larger the space the more applicable is a sloped or other consciously designed ceiling.)

## Flush-jointed ceiling

The majority of ceilings in Australia are lined with sheets of gypsum plaster or fibrocement-based products finished with flush-sealed joints.

This requirement for flush-jointing has been a major challenge for manufacturers of ceiling lining boards for, when the first flush-joinable sheets of fibrous plaster came on the market, they were challenging the site-applied lath and plaster ceiling for a market share.

Eventually the fibrous plaster sheet successfully became the dominant ceiling material in use but it remained in final appearance similar to the product it displaced: that is, smooth and jointless.

Fibrous plaster ceilings were also much less messy to fix than lath and plaster and moulded ceilings made available to all home purchasers. The demand for fancy moulded ceilings waned but the smooth flush finish remains the dominant factor in home builder choice of material.

After World War II many new products came on the market to challenge fibrous plaster's dominance of the ceiling market, including hardboard, fibrocement sheet, fibreboard and chipboard. Many of these products were cheaper than fibrous plaster and easy to fix by nailing. Therefore, they were much more able to be fixed by an OB. But all the new products had one major failing: the joints were either exposed or covered with strips; they were not flush, and the public wanted flush joints, so the new products never achieved wide success.

Just as fibrous plaster appeared to have the market redominated, a new product from America began to appear on the Australian market. Plasterboard had arrived—simply a very thin 10 or 13 mm slab of gypsum plaster sheeted on both faces with light cardboard. The fibrous plaster industry did not see the new product as a threat but, slowly, as techniques to use plasterboard in the Australian housing industry improved, so did its acceptance.

Today plasterboard has all but pushed fibrous plaster out of the industry and remains the dominant ceiling and wall lining material used in Australia. Other flush-jointed ceiling sheets are also marketed. The most common of these is made from fibrocement and is used as an alternative to plasterboard, particularly where exposure to moisture is likely. The cement surface of the fibrocement product has a greater resistance to deterioration from water than the cardboard surface of plasterboard.

## Some cornice alternatives

Ceilings are generally trimmed at their wall junctions with a moulded section of plaster or timber called a cornice. Most cornices are manufactured, cut to length, jointed and fixed on-site.

The selection of cornice can significantly change the appearance of a room, and OBs who wish to achieve a particular historical period style, or just have an interesting cornice, should check in their locality for manufacturers who have old fibrous plaster cornice moulds and who will, for relatively small cost, produce complex cornice designs.

Cornices are not essential, and are often omitted, particularly if timber ceilings and/or walls are used However, if cornices are not to be installed on plasterboard ceilings, it is often hard to achieve a sharp crack-free ceiling/wall junction.

## Insulation

Most ceilings should have a blanket of thermal insulation immediately on top of them or integrated with them.

Insulation on top of the ceiling will always be an advantage when the ambient temperature falls below the desired internal temperature of a home. Heat rises to the ceiling and without adequate insulation will pass through and be lost.

Insulation on top of the ceiling may cause a house or room to hold heat in summer, particularly if there are no high-level controlled exhaust vents or windows, or if the windows in the room face directly into a solar heat load (the sun)—when a glass house effect is created and the internal temperature rises above ambient.

Insulation on ceilings is an important part of any efficient energy control system, but it is only part of a system. Care must be taken to design a fully integrated active and passive energy utilisation system before stuffing the ceiling full of a thermal blanket. Use only the thickness of insulation that gives you the desired reduction in thermal conductivity.

# CEILINGS • 51

## BASIC CEILING STRUCTURE

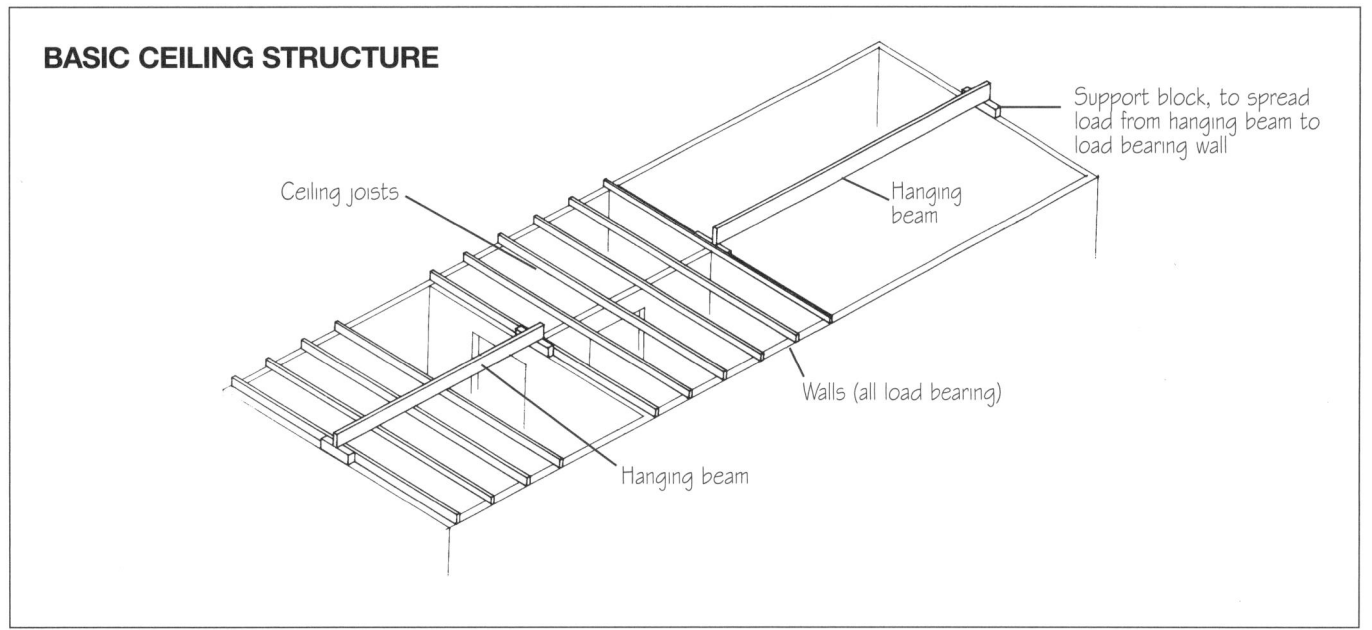

##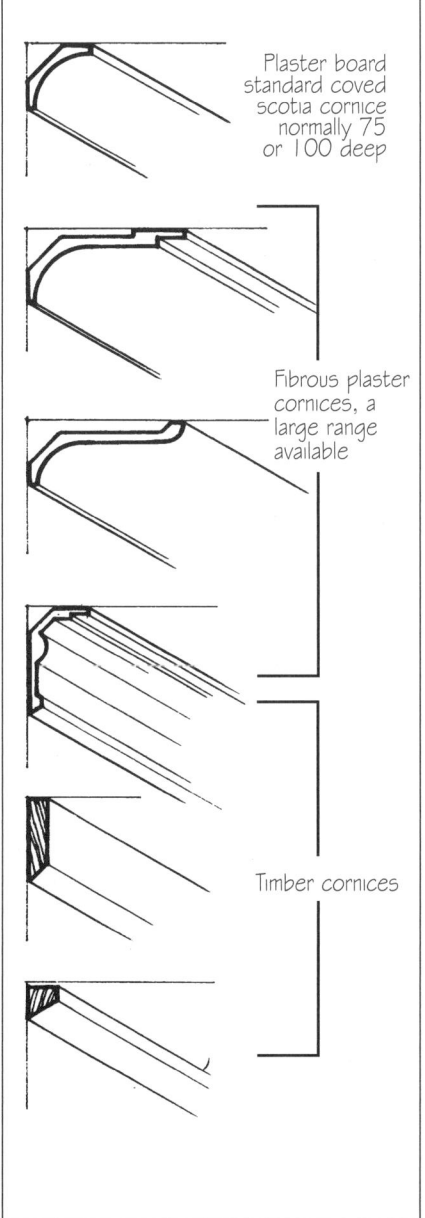

Plaster board standard coved scotia cornice normally 75 or 100 deep

Fibrous plaster cornices, a large range available

Timber cornices

## FLUSH JOINTED SHEET FIXING

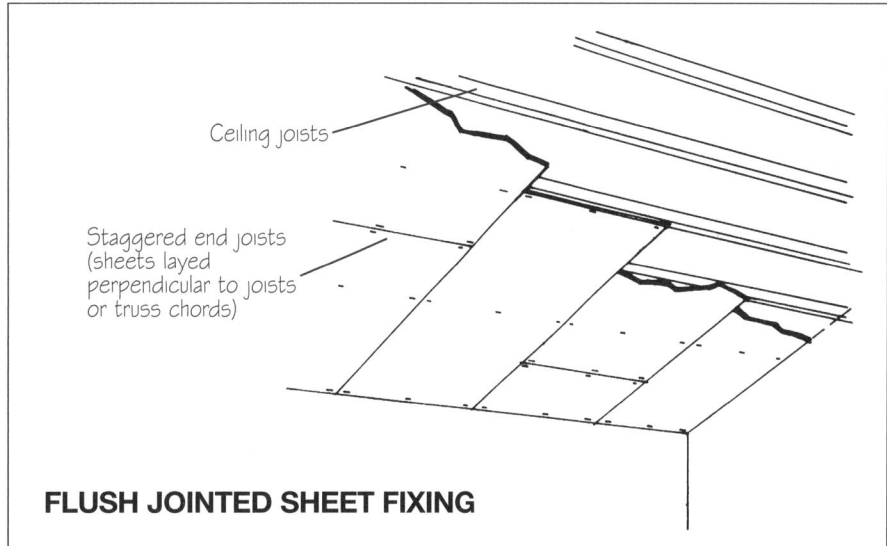

## WALL-CEILING JUNCTION

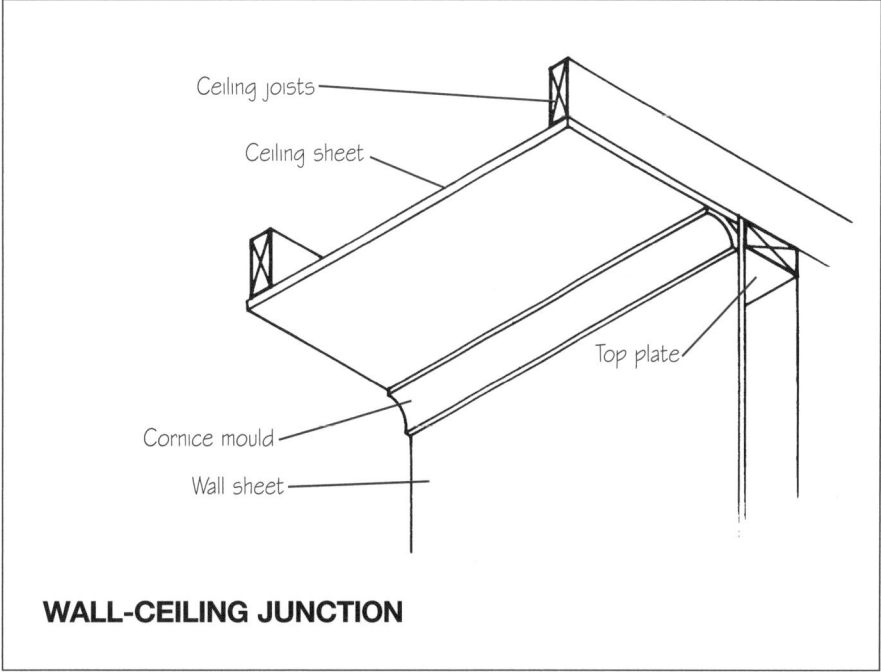

177

# 51 • CEILINGS

**Some examples of timber ceilings**

There are many different kinds of moulded lining boards including shiplap, vee-jointed, double vee-jointed, reeded point and many others with or without tongue-and-grooved joints and secret nailing.

A layer of fire proof kraft building paper should be positioned between the boards and the ceiling joists to control dust penetrating the room below from the ceiling space.

Many timbers are suitable for moulding into lining boards, these include:

- Pinus (Radiata)
- Canada Pine (Hemlock)
- Oregon (Douglas Gir)
- Western Red Cedar
- Red Pine/Redwood
- Pacific Maple
- Jarrah
- Tasmanian Oak/kiln dried hardwood
- Other clear or knotty pines
- Selected Australian hardwoods
- Imported exotic timbers, including Teak, Palm, Oak, Ash, Beech

Timber lining boards are not cheap and owner builder must be careful to select the thickness and quality necessary to satisfy the particular application. Special care must be taken in choosing the thickness of a board to achieve a balance between cost and structural sufficiency. It is often cheaper to use close-spaced joists or battens, which allow thinner lining boards, than to skimp on support structure but have to use thicker boards, or get sagging and distortion.

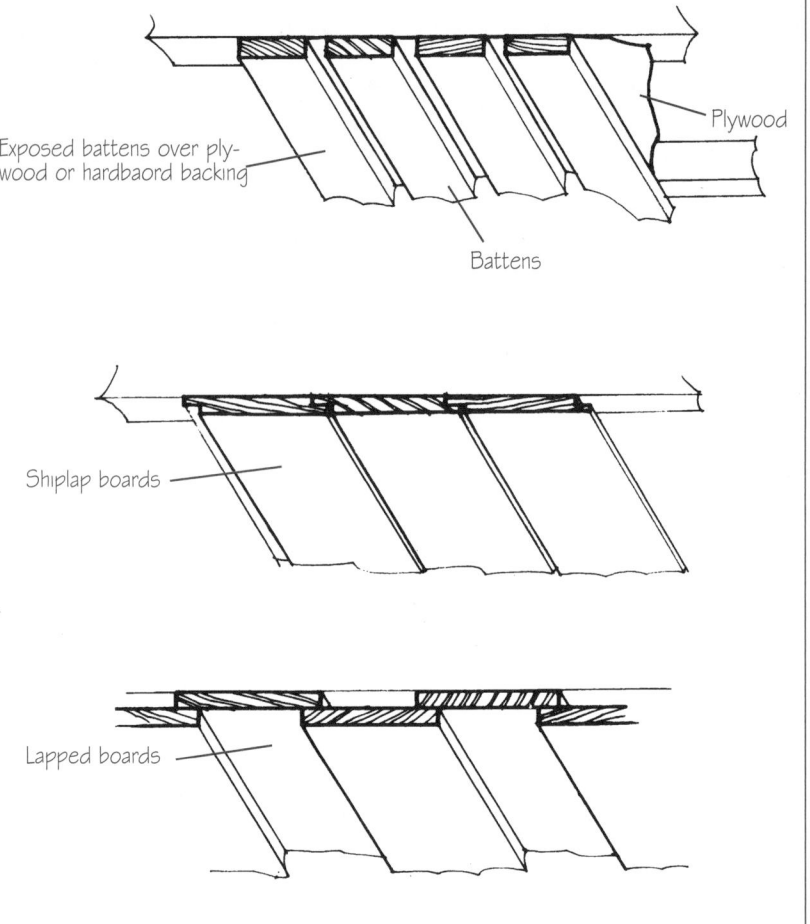

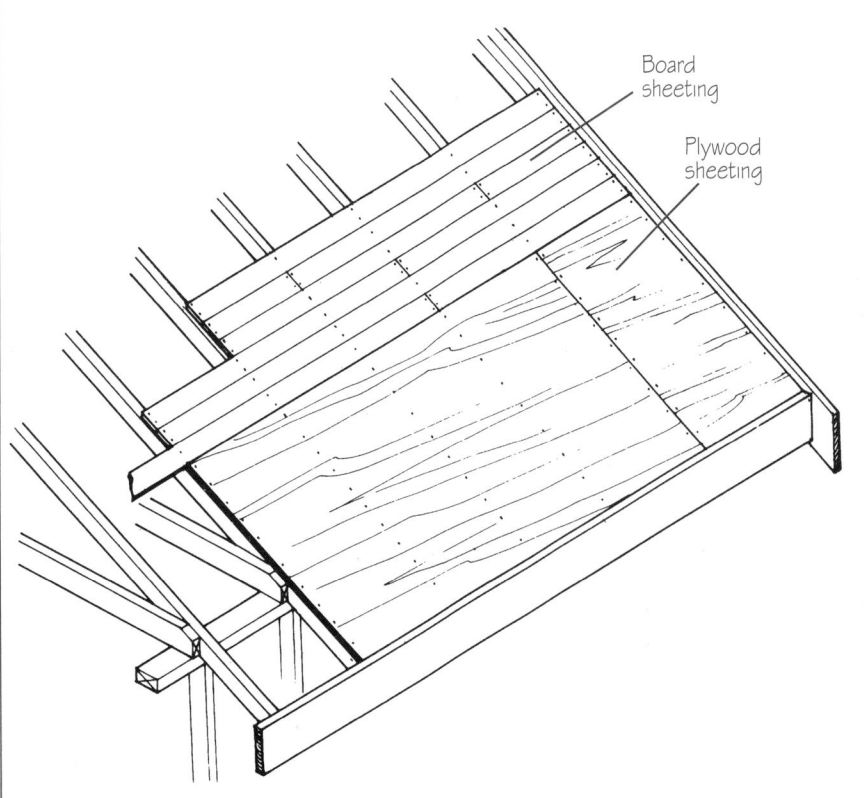

**Sub-roof panelling**

This method of sheeting the roof completely before fixing roof cover is favoured for use in areas of extremely cold climate with high rain and snow falls.

Battens are not needed to fix roof cover and it is most suited for use with timber shingles and shakes.

The diagram shows alternative plywood or boards, but pineboard, hardboard, compressed fibreglass board or similar can be used as alternatives.

Sub-roof panelling can also be used as a ceiling system where the rafters are exposed. When this method is used great care must be taken to protect the boarding from rain until the roof is covered and flashed.

When rafters are exposed in a ceiling care must be taken to lay out the rafters so they are evenly spaced in the rooms below and that odd junctions are eliminated.

# CEILINGS • 51

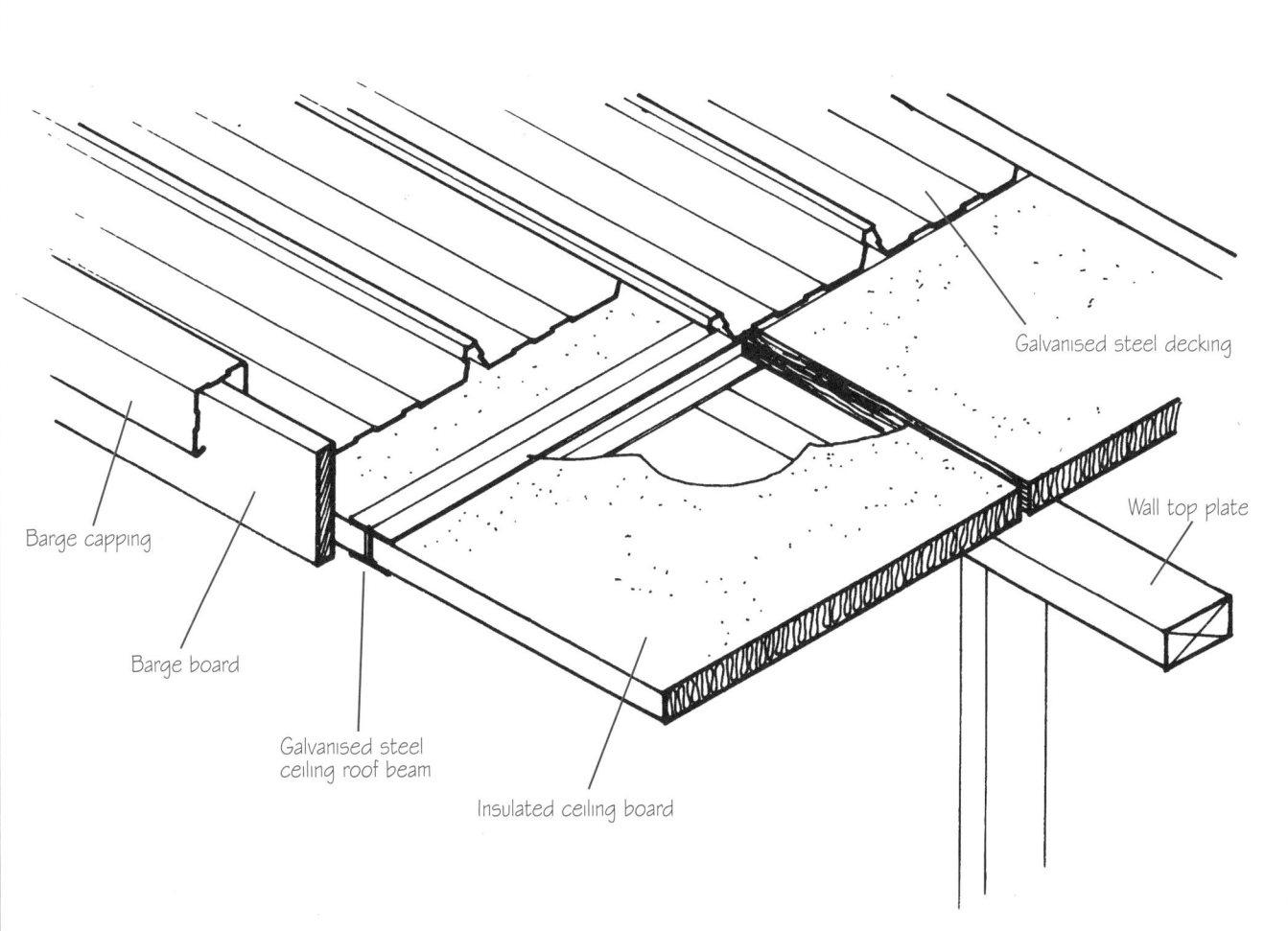

**Patented roofing systems**

There are a number of sheeting systems available in Australia that combine structural components, ceiling system, insulation barrier and roof cover.

The most common system uses a range of formed galvanised steel beams and other assorted interlocking modular components to form the spanning roof structure. The system can span over 6000mm without any supplementary structure. If the building to which the patented roofing system is applied is designed from the ground up, the system is simple and inexpensive for the owner builder to use. Manufacturers produce very good assembly information sheets that should fill any gaps in knowledge.

The patented roof system can be used in conjunction with more conventional roof spanning systems like exposed beams or trusses, or the steel sections can be replaced completely with timber members – which may be more satisfactory in some applications.

The ceiling and insulation system consists of 50mm thick boards of compressed straw sheeted on both sides with heavy paper – similar to plaster board – or some special rigid insulation material such as exposed straw, wood spirals, styrofoam or composite aluminium.

The most common ceiling/insulation panel is the compressed straw type, and it should be noted that the straw is treated to render it fire proof. The board offers high levels of thermal and noise insulation.

The surface of the standard paper faced board can be painted, or the board can be ordered with special face treatments including cork, straw, vermiculite, woodchip and acoustic.

The roof system used with the patented roof system is normally a non-punctured low pitch steel decking. This form of roofing can be installed down to a pitch of 1:50. Other roof cover, including tiles, corrugated steel or shingles, can be used with careful attention to detail.

The patented roof system is extremely useful for owner-builders, particularly those who wish to have a high hands-on component, or those who wish to build in a remote area where access to traditional building materials is limited.

# 52 • SPECIAL ROOF DETAILS

## Roof lights, skylights and clerestory windows

It is common today to incorporate windows into roof designs, and there are a number of manufacturers that produce roof light windows. If it is your intention to incorporate windows in the roof of your house then there are basically two avenues open to you.
1. Use a standard production roof light.
2. Use a normal window as a clerestory or glazing bars on the roof slope.

## Standard production roof lights

There are two common forms of production roof light:
1. The acrylic dome type.
2. The double-glazed pivot frame type.

## Acrylic dome roof lights

The acrylic dome roof lights generally consist of a dome-shaped acrylic roof light fixed into an aluminium or galvanised steel frame. The acrylic dome can be clear or obscure and is some times double-skinned for improved insulation, safety and security.

The metal frame is supplied with flashing kits to suit tiled and sheet roofing systems and can be fixed into sloped or near-flat roofs. Also, the frame is often available with ventilators built into it to satisfy the ventilation requirements of some authorities if the dome is to be used to provide natural light and ventilation to an otherwise windowless room.

Acrylic dome lights are produced by many companies around Australia and each manufacturer has its own special design. Therefore, if an OB is considering using one or more of these products, it is worthwhile having the manufacturer's representative look at your plans and advise on the appropriate type and size required. You may find it prudent to get two companies to give recommendations and a quotation.

The manufacturers generally have good catalogue/pamphlet information available on their products. OBs are advised to consult this information during the planning stage of their proposed house so as to avoid on-site problems.

### Double glazed pivot roof windows

These products are generally imported from Europe and are of high quality, which means that they are relatively expensive. However, as they are generally used to bring natural light and ventilation into an attic space, then their price needs to be compared with the cost of a dormer type roof window. When this is done, their value is apparent.

These windows are purchased with a flashing list appropriate to the roof cover they are fixed through. There is a wide range of 'extras' available for those purchasers who require them, such a sun blinds, insect screens, storm screens (to protect from hail and snow) and remote-opening devices.

The manufacturers produce extensive catalogue/pamphlet information which should studied by intending users before ordering.

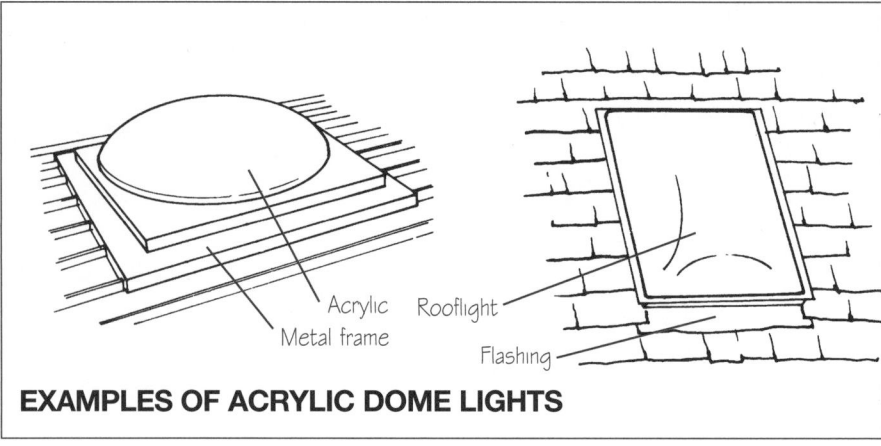

**EXAMPLES OF ACRYLIC DOME LIGHTS**

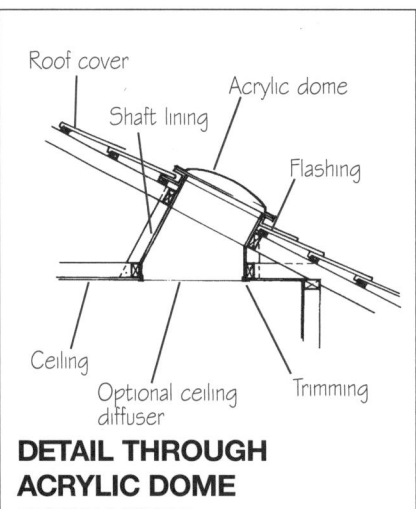

**DETAIL THROUGH ACRYLIC DOME ROOF LIGHT**

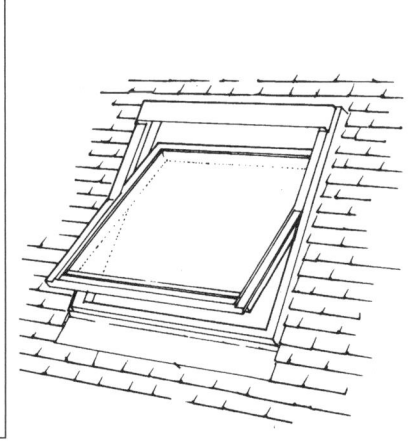

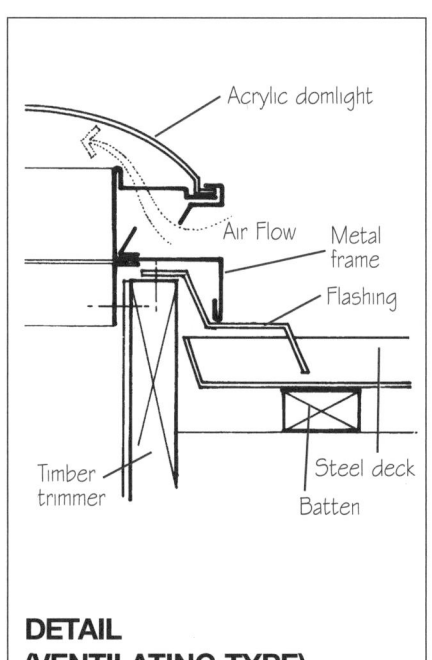

**DETAIL (VENTILATING TYPE)**

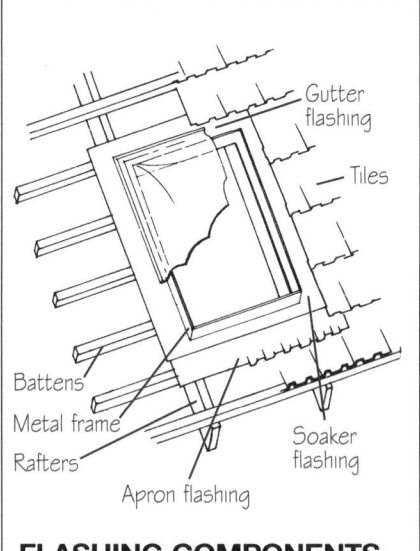

**FLASHING COMPONENTS OF ACRYLIC DOME LIGHT**

# SPECIAL ROOF DETAILS • 52

## Clerestories and glazing bars

The use of normal timber or aluminium window frames and clerestory windows is acceptable, but OBs should take care to:

1. draw to detail a section through the clerestory to accurately plot the size of the roof members and window. Advice from an architect could be beneficial here because, although clerestory windows can be a most attractive feature of a house, poor or inadequate detailing can give many construction and aesthetic problems.
2. make sure any opening sashes can be opened.
3. make sure the windows can be cleaned safely.
4. adequately flash the window frames, because wind-blown rain as it passes over a roof swirls through the air, making clerestory windows more likely to leak than normal wall windows.
5. only use the windows vertically. Few standard timber or aluminium windows will function if they are not fixed vertically.

If a large area of roof requires glazing then proprietary glazing bars may be used. These are extruded aluminium sections designed to be fabricated on site and to carry glass or acrylic (perspex) glazing either vertically or on a slope. The glazing bar section works because it is designed to allow any water not carried by the glazing and that penetrates through the joints to be caught in an integral mini-gutter and run clear of the wall or roof.

Glazing bars have been used for decades in greenhouse and factory construction, but have recently begun to be used extensively in houses, particularly solar/ecology/conservation projects, where greenhouse type rooms are required as part of the natural climate-control system.

Large areas of roof glazing can be troublesome and dangerous if not designed correctly. An OB considering using glazed roofs should consult an expert in the field of solar house design, or at least do some research. Please note that the information used should be applicable to the southern hemisphere, where the midday sun is in the northern sky, and based on a latitude similar to the one where the proposed house is to be built.

## Dormer windows

The vertical window set into a roof with its own small roof and gablet has been used in houses for as long as houses have had windows. In Europe, where the climate tends to be colder than most people would consider comfortable for many months of the year, it is very sensible to build rooms into the roof space, where the warm air from the lower-floor fires rises.

Attics and dormers have not been as popular a design factor in Australian houses for a number of reasons, including the relatively large residential allotments used,

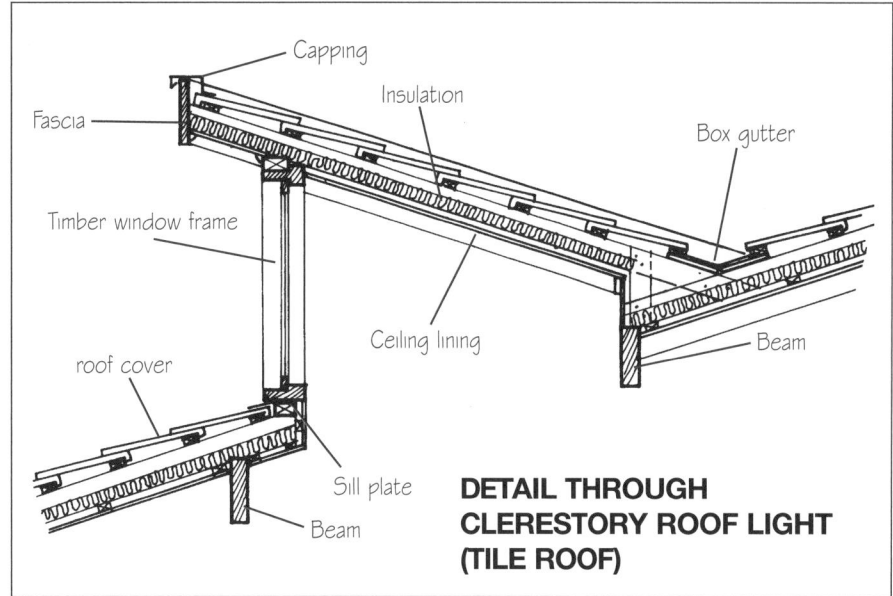

**DETAIL THROUGH CLERESTORY ROOF LIGHT (TILE ROOF)**

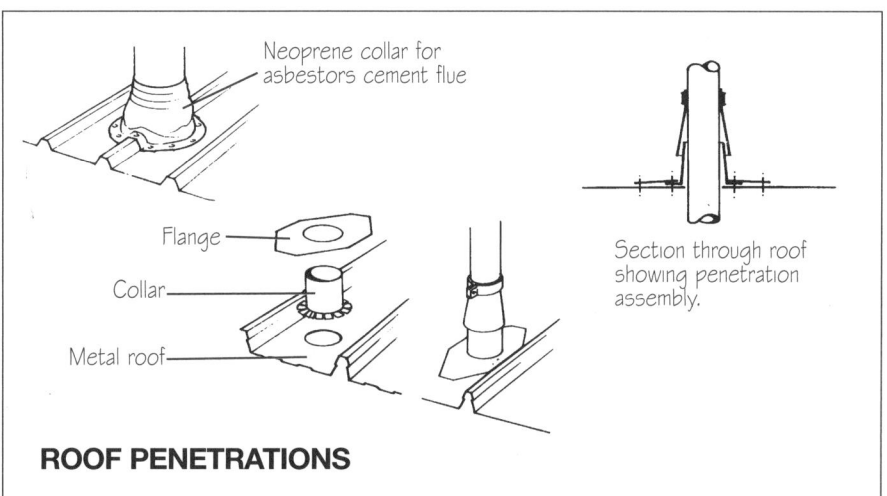

**ROOF PENETRATIONS**

which allow for cheaper and convenient single storey alternatives.

Early colonial houses often did have attics and dormers but by the Victorian period most houses were either single-storey or had full-walled second storeys, and this has continued until today.

Inner-city cottages often display dormer windows. Some of these are original but many were added during the gentrification of the inner-city residential zones.

To make use of attic windows it is normal to have an attic, although some buildings use dormer windows to allow more light into a large space with a vaulted ceiling. Houses with attics are generally built in a traditional manner with a deep joist upper floor structure to provide a trafficable floor, and a steep roof, generally 45 degrees or greater.

The roof is commonly formed by rafter couples, sometimes with collar ties but often without. The attic space suits a gable roof form as this gives a reasonable useable floor area and the gable end wall can often have windows.

The window elevation of the dormer will look best if there is a higher proportion of

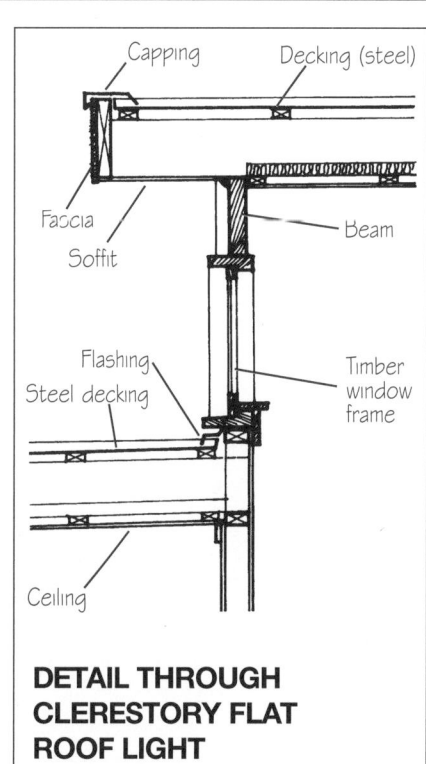

**DETAIL THROUGH CLERESTORY FLAT ROOF LIGHT**

# 52 • SPECIAL ROOF DETAILS

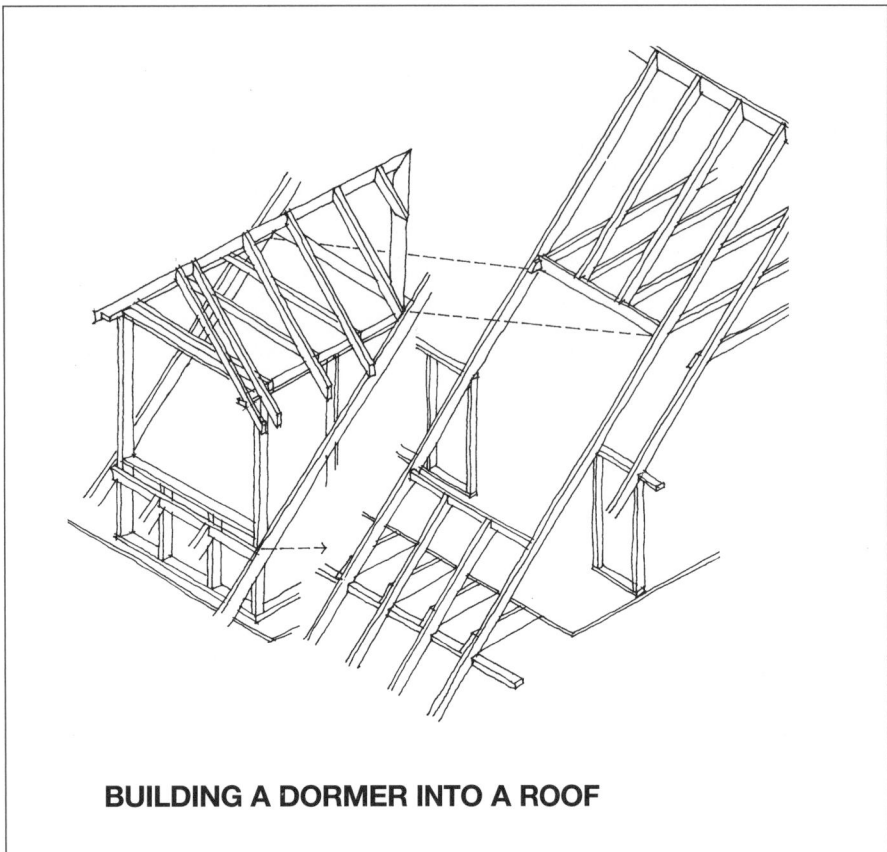

**BUILDING A DORMER INTO A ROOF**

glass to frame, and if the window is higher that it is wide—a proportion of about 3 high to 2.1 wide is a reasonable starting point. The sketch shows the use of studs on flat for the side walls of the dormer; this assists in keeping the bulk of the unglazed portion down.

The sketch shows a flat ceiling in the dormer but a vaulted ceiling is also acceptable. It also shows collar ties close together to suit a ceiling but this is not essential in small roof spans; larger spans will need structural engineering.

In Australia the climate and construction methods preferred by builders tend to make dormer and attic construction uncomfortable during the hotter months of the year and generally much more expensive for a relatively inefficient spatial solution.

Insulation, orientation, ventilation and access need careful consideration—well-designed attics will give a reasonable quality environment in Hobart but there are few places on mainland Australia where they are practical, except for romantic effect.

## Glass roofs

Glass roofs need special knowledge to construct; they need the combination of a designer who understands the limitations of glass and associated products, and a fabricator experienced in constructing safe and leak-proof roofs.

Calling a glass roof leak-proof, or saying that it is a safe-glass roof, is a contradiction in terms. Eventually, someone will design and build a good-looking, glass roof system that will never leak and will resist the impact of heavy hail and falling house bricks—but not yet.

Some architects are skilled in designing glass roofs, but many designers and builders will have limited or no experience in designing in this medium. Specialist companies are often the safest choice—they may even guarantee their work.

Framing systems are made of timber, aluminium and plastic—or combinations of these. Investigate appropriate choices carefully, and try to speak to someone who has had a similar system for a few years.

Glazing can be plain glass or perspex—neither of these is recommended and both are unlikely to conform to current safety codes. Discuss the costs and benefits of alternative specialist products available with a suitable supplier. Some alternatives to begin the search are:
1. **laminated glass**—a glass and plastic sandwich
2. **toughened glass**—treated glass to make it stronger and safe
3. **wire cast glass**—metal wire embedded in glass sheets
4. **Lexan**—super-high-quality, plastic sheeting.

## Penetrations

Other than roof lights, there can be many other penetrations through the roof cover. These include, but are not limited to:
1. flues from solid-fuel heaters
2. flues from gas heaters and hot water services
3. chimneys
4. sewer vents
5. electricity connections
6. flues from oil heaters
7. airconditioning systems
8. solar hot water units.

All of these roof penetrations must be treated with care, as it is generally because of a poor standard of workmanship, poor quality of materials or poor design of flashing to roof penetrations that roofs leak.

Flashing through roofing should be carried out wherever possible to the recommended practice, as set out by the roofing manufacturer and the supplier of the appliance being flued or vented where applicable. A roof plumber is generally skilled in all roof flashing applications and should be employed wherever possible, at least for complex or critical details.

Care must be taken when hot flues pass through the roof that they are not in contact with or adjacent to any material likely to catch fire and endanger the house and its occupants.

Solar hot water systems with the tank on the roof and roof-mounted airconditioning units require special consideration because of their weight and possible vibration.

# SERVICES • 53

The connection of a modern house to utility services is as important as finding the correct foundation, accurately cutting the timber frame or ensuring a waterproof roof. There are many services to consider, many of which are controlled by specific statutory bodies and restricted to licensed tradespersons. These therefore limit the input that can be made by an OB.

The electrician and plumber are the most common of the special licensed tradespeople that need to be employed by OBs. The electrician obviously installs all electrical wiring and fittings, but also is likely to install internal communication wiring, telephone wires, television antenna wiring and thermal/smoke alarms.

The plumber, in most areas, is licensed to install sewer and stormwater drains (note: not all areas require stormwater drains to be installed by plumbers), connect water supply from a town service or a private system, install hot water reticulation systems, gas services and septic disposal systems. In most localities plumbers are licensed to install water systems; most are also licensed sewer drainers and approved septic system contractors. Many are also gas fitters. In some places drainers can be licensed without being tradespersons.

This section deals in detail with the placement of services and the normal methods of identifying them on the drawings. Local authorities have such diverse codes and regulations relating to the installation of services that every OB will have to extend the information provided herein by having detailed discussions with the local authorities responsible. Sometimes the regulations of one authority are interpreted differently by individuals within the authority, and OBs should make sure they talk to the officer most likely to have responsibility for their specific project.

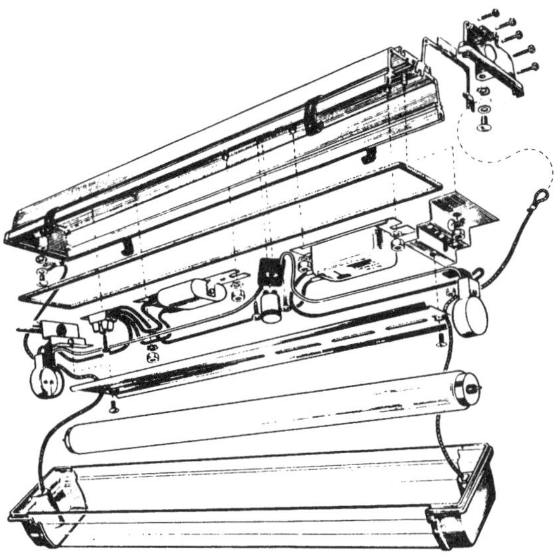

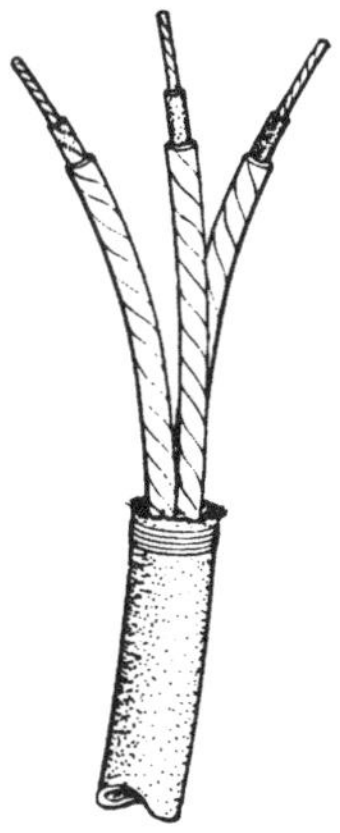

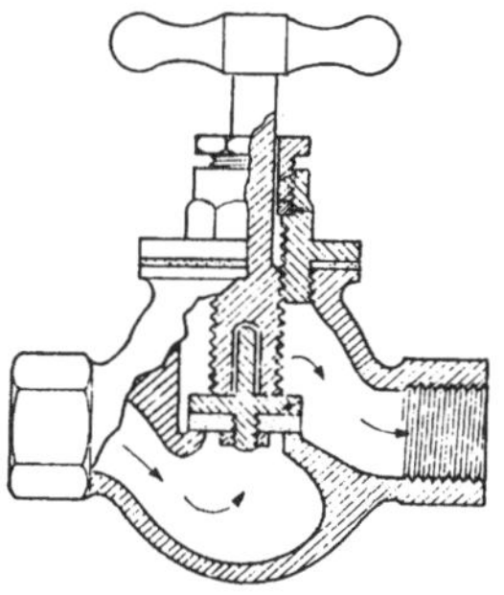

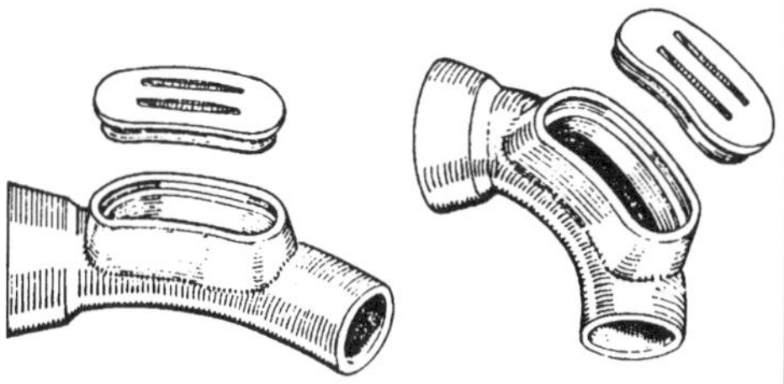

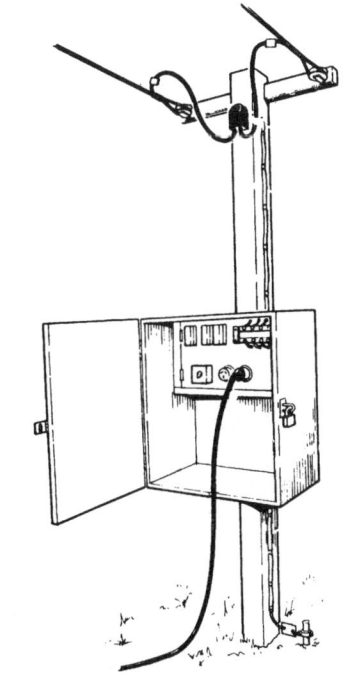

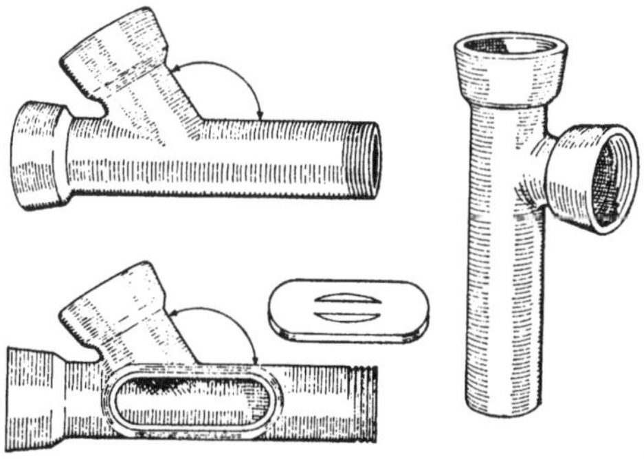

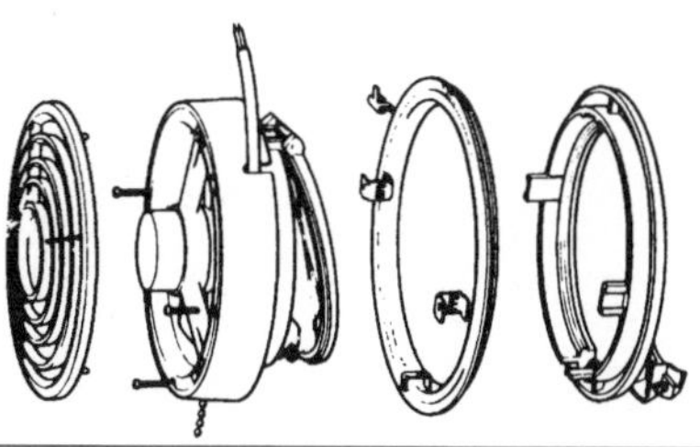

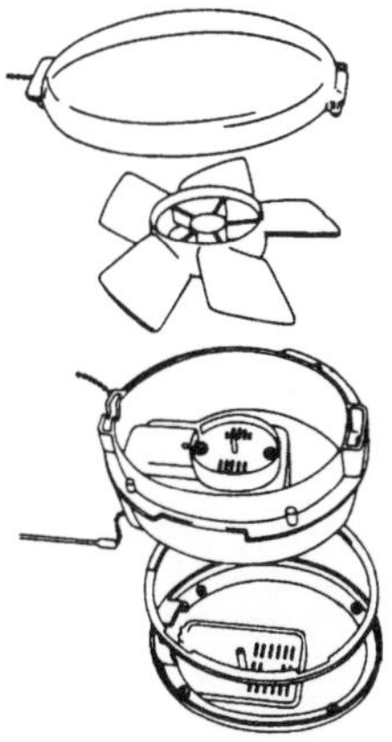

183

# 54 • DRAINER

## Introduction

The drainer is the tradesperson who is generally responsible for the installation of the sewer and stormwater drainage system, as well as the septic disposal system in non-sewered areas. Licensing requirements for drainers vary from state to state but in nearly all locations in Australia the installation of sewer and septic disposal systems requires the employment of a tradesperson licensed with the appropriate local authority. A few locations require the stormwater disposal system to be installed by licensed drainers but this is much less common than the requirement for effluent disposal.

The licensing of drainers for sewer work is normally the responsibility of the local sewerage disposal authority. It is normal that all licensed plumbers are also licensed drainers, but the reverse is not always true, as in some locations a special limited drainer's licence is available. Always make sure the drainer you use is licensed to carry out work under the authority that has jurisdiction.

Septic system installation may also require the employment of a licensed drainer, but in many locations the supervision of septic installations is the responsibility of a completely different authority from sewer installations. It is important to check the drainer's credentials.

Although few authorities require licensed tradespeople be employed for stormwater or sub-surface agricultural drainage, OBs should be aware of the problems that may be encountered and the skills that are required in laying any sub-surface drain to even grades. Most local authorities that do not require the employment of a licensed drainer for stormwater installations will require an on-site inspection of the installation before it is covered with dirt.

## Stormwater drains and disposal

### Introduction

In residential construction the stormwater drainage and disposal system will generally consist of three parts:
1. roof water disposal
2. surface water disposal
3. sub-surface (agricultural) drainage.

### Disposal

There is obviously little point in having a stormwater drainage system if there is nowhere to dispose of the waste water. Therefore, the first task is to establish the best disposal method.

The three main sources of water requiring disposal are identified above. Normally the three separate systems can be combined into one disposal method, but not always. There will be times when it is more convenient, economic or expedient to have multiple disposal points and methods.

The most common stormwater disposal methods are to pipe to a:

1. natural watercourse, eg. a creek, river, dam, or lake
2. constructed watercourse, eg. a channel, culvert or pipe
3. absorption pit or trench
4. expiration bed
5. collection tank(s).

If you intend to dispose of your stormwater into a natural watercourse, a number of special considerations must be taken into account, particularly:

a. Is the waterway under control of a specific authority and is approval required before any stormwater can be disposed of into it?
b. Are there any clean water and/or anti-pollution regulations in force over the waterway, and how does your stormwater conform to these regulations?
c. If the boundaries of the property enclose the waterway, or part of the waterway, generally there is no trouble getting approval to dispose of stormwater. But if the waterway that is to receive stormwater waste is in another person's or crown land then special approvals may be required to gain easement rights.
d. In some areas, stormwater can only be disposed of into natural watercourses if it is first run through a series of pits that incorporate silt traps and debris screens. In other areas, a simple open channel cannot be opened directly into the bank of the natural watercourse.

When designing a stormwater disposal system that empties directly into a natural watercourse always check if there is any authority having jurisdiction over the waterway; and, even more importantly, design the system so that, if the watercourse is running near capacity, your domestic stormwater system continues to operate. It is possible to install a stormwater system that will only work when there is little or no water in the receiving watercourse. Check for the local method to install a flood-proof overflow valve.

If the area in which you are building is blessed with a constructed stormwater disposal system, then you are generally at an advantage over all those people in Australia who are lucky if they are supplied with an open kerb and gutter method of stormwater disposal by the local authority.

Where the local authorities supply a system of channel and/or pipe stormwater disposal, then it is normal that home builders be required to carry all their stormwater to this system for disposal. Before finalising the plans for their house OBs should:

a. check the location of the pipes or channels in relation to their block of land
b. check what easement rights they enjoy and what easement rights others may enjoy over parts of their property, and what restriction this may have on them
c. check the lowest invert level (how deep the main stormwater drain is) and calculate whether all the drains they require can flow to the main (minimum fall is approximately 1:60)
d. discuss with the authority that has jurisdiction over the main stormwater disposal system if there are any problems, like how to make water run up hill. Some authorities require that householders dispose of all site stormwater into the main stormwater system, even if pumps have to be employed. This is a situation that should be avoided by OBs if possible, because pumps are initially expensive and need regular specialist maintenance if they are to give a relatively trouble-free operation.

Absorption pits and transpiration beds normally mean on site disposal. The first method makes use of the natural ground's ability to absorb water. The second method is used when the natural ground will not absorb sufficient water; water is then disposed of into the atmosphere by transpiration from the leaves of selected grasses and by evaporation.

The size, shape and type of absorption system varies significantly throughout Australia in an attempt to meet special local conditions of rainfall and the absorption coefficient of the soil. In areas with sandy soils, site absorption disposal systems work very efficiently, but in soils with rock foundations little or no absorption may be possible. Clay soils generally will absorb water but have a secondary problem: they swell when wet and that can often create adverse foundation conditions and be the cause of foundation/footing movement and consequent building cracking and deterioration. The general rule for absorption disposal systems is to make them as large as practicable and keep them as far downhill from any building as possible.

Transpiration beds are seldom used for the disposal of stormwater but in areas of low ground absorption, supplementary transpiration systems may be considered. In some areas transpiration beds have been used successfully in conjunction with holding tanks that slowly release the stormwater into the transpiration bed, but this is an expensive system to install.

Collecting the stormwater in tanks may not be, technically, a disposal system, but in many parts of Australia water is altogether too precious to dispose of without at least one extra node in the water cycle. Care must be taken to separate clean roof water, which can be used for drinking, from less pure surface water that can be used for washing functions after settlement and filtration.

### Collector drains

For both the roof water and the surface water collection systems a network of sub-surface pipes and pits is required to convey the waste water from the pick-up points (downpipes, grated drains, sumps, and so on) to the disposal point.

The main aim of these networks is to convey the water at the least cost, which generally means that effort should be taken to keep pipe lengths to a sensible minimum, and to control the depth of the pipes to avoid expensive and deep excavation. Drainage pipes are commonly laid in plastic (PVC) pipe but there is still significant use of earthenware, concrete, fibrocement and even galvanised steel and cast iron.

For best results, the stormwater drains should be laid to a fall of around 1 in 50 (that is 30 mm in 1.5 metres); the fall should be even and all intersections properly formed by using specially manufactured junction sections or well-constructed pits or sumps.

It is absolutely essential that the pipes be well supported on a stable bed—sand is the most common—and that the joints are leak-proof. Any leak in a stormwater system can cause erosion of the pipe bed, eventually leading to pipe fracture and failure of the system.

## Sewerage installation

The sewerage system is of pipes, vents, pits, traps, gullies and other assorted fittings that combine to remove the unwanted septic and effluent wastes from houses.

In Australia, a sewerage system is considered to be a system of liquid waste disposal that removes the waste via pipelines to a remote disposal point. An onsite disposal system is generally referred to as a septic system and will be referred to in a separate section.

The sewerage system is broken into three main parts:
1. the sanitary plumbing—generally considered to be the waste pipes that link any sanitary fitting to the under-floor sewer drainage system
2. the house service drains—generally considered to be all the sewer pipes and other fittings that make up that part of the sewerage system linking the sanitary plumbing to the main sewer drainage system and under the control of the local sewerage disposal authority
3. the main sewerage mains and the disposal system.

Most urban areas in Australia have a town- or city-wide, sewerage drainage system and a disposal installation.

### House service

The sewerage drains that run from the underside of the floor to the mains are called the house service drains. Traditionally these drains were laid in earthenware salt glazed pottery pipes, but in recent years more and more authorities have allowed the use of plastic (generally PVC) pipe for sewerage drains. As with the sanitary plumbers' work the installation of house service sewerage drains has been significantly simplified by this trend, but rather than widening the range of people who are allowed to lay house service drains, some sewerage authorities have moved to further restrict the tradespeople licensed to do the work.

It is not easy to lay a set of house service sewer drains but, as most of the work is in fact tied to the simple physical act of digging the trenches in which to install the pipes, it would appear that a more enlightened attitude to OBs wishing to carry out their own house service drainage could be allowed. Inspectors from the responsible authorities could inspect the work just as they do in most cases now, and the design could be undertaken by qualified hydraulic engineers or licensed plumbers.

Like the stormwater installation previously discussed, the main consideration in installing a sewer house service is to find the invert at the sewer authority's point of entry on your property, and excavate the trenches for the pipes to an even fall of around 1 in 50.

Over the last 20 years sewerage authorities in Australia have progressively reduced the number of regulations governing sewer installations and many of the complex over-vented systems of the past have been superseded by new regulations that allow simple single-vent and single-stack installations in many areas.

### The authorities' sewer mains

The authorities' sewer mains generally have little impact on an OB, except in relation to the location of the point of entry to the authorities main' that has been allocated to the property.

However, on the odd occasion when OBs find that the main sewer bisects their property diagonally, a request to build over the sewer may need to be made to the authority. Building over sewers is not the impossible problem it was some years ago, and today most authorities will allow buildings to be built over sewer mains if the property owner takes full responsibility for allowing access by any of the authority's employees to carry out maintenance or the like. However, even those authorities with an enlightened attitude to building over sew-

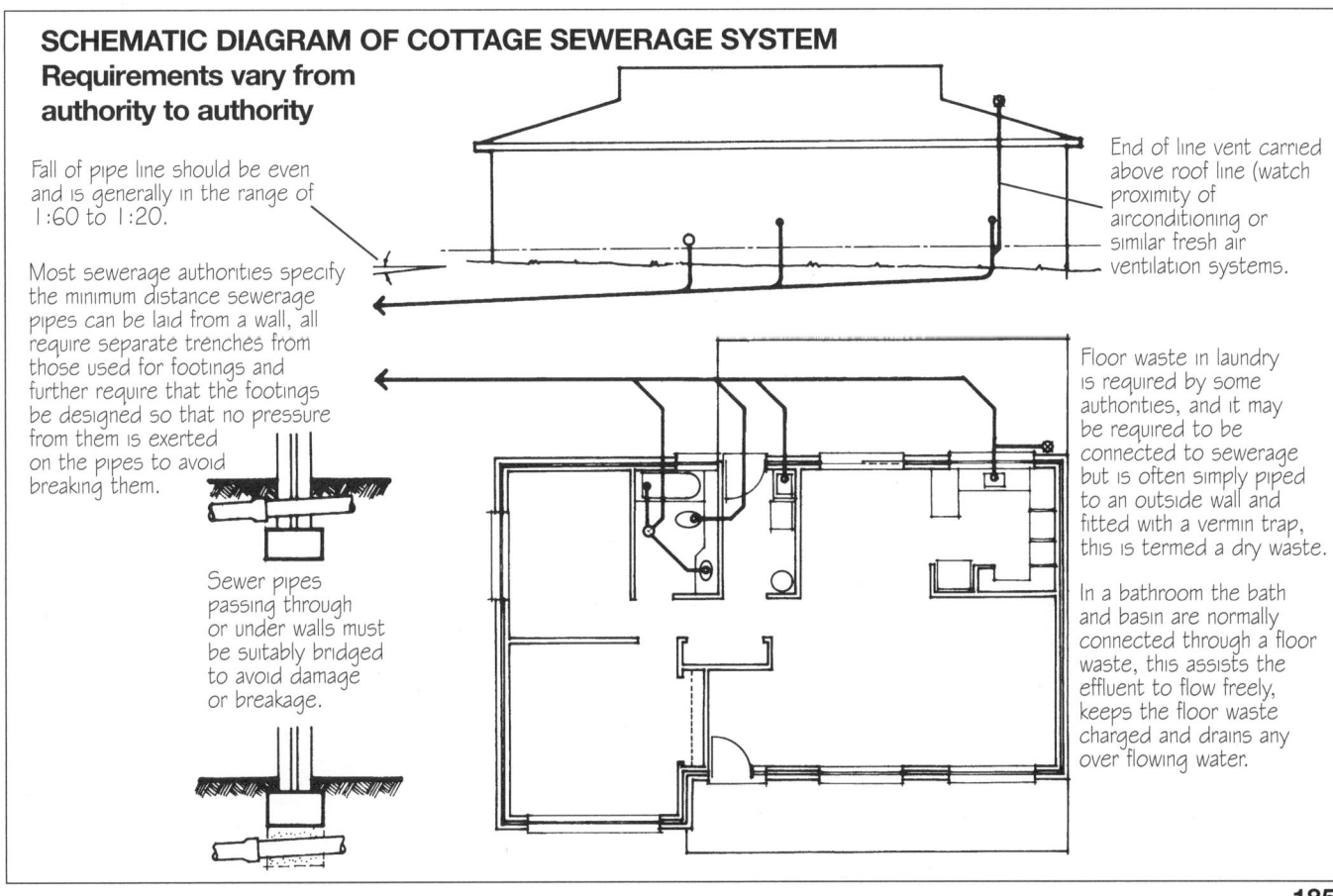

SCHEMATIC DIAGRAM OF COTTAGE SEWERAGE SYSTEM
Requirements vary from authority to authority

# 54 • DRAINER

ers usually find ways of charging such high fees to allow home owners to build over sewers that alternative approaches are more acceptable. No sewer authority appears to have been prepared to give the property owner that 'as of right' approval to build over sewer mains.

Note that sewerage authorities take a very dim view of anyone who allows stormwater to drain into the main sewerage drains and disposal system. Often the pipes are not of sufficient size to contain both sewerage and stormwater and the system may discharge raw sewage in an unsanitary manner. In some areas where there is a high danger of sewage backing up in the pipes, house services may need to be fitted with a reflux valve designed to stop the flow of sewage the wrong way in the pipe.

## Septic tanks and disposal system

In areas of Australia where there is no main sewage disposal system most houses are connected to what is called a septic tank.

What is meant by this is that all the sewage and effluent wastes produced in the house are treated on-site and in most cases disposed of within the confines of the property. When the disposal is not on the site then either the part-treated effluent is held in a holding tank and removed regularly by a mobile tanker for disposal elsewhere, or the part-treated effluent is pumped via small bore pumps to a community sewerage disposal plant. The mobile tanker system is common in areas with undulating topography and a high bedrock sub-stratum. In these areas it is very difficult and expensive to develop a main line sewer system.

Sloping land with high rock sub-strata is generally highly unsuitable for the installation of on-site disposal septic systems. Community treatment works and disposal systems are uncommon in Australia, but there are examples and there may well be many more in the future.

In a septic system we generally think of two separate sub-systems:
1. the sewage disposal system charged with reducing human waste to a low toxicity liquid and disposing of it with the least side effects possible
2. the effluent disposal system charged with getting rid of the large amounts of water generated each day by showers, baths, washing machines and kitchen sinks. This effluent generally contains little toxic material but non-biodegradable chemicals can at times cause some problems. The kitchen sink is generally treated as a special case, as it often contains high amounts of fatty and greasy material, and most authorities require a grease arrester trap between the kitchen sink and any on-site disposal system.

### The sewage disposal system

It is the sewage disposal system that incorporates the septic tank, and in general a pipe leads directly from any toilet, bidet (in most states, although some authorities prefer no bidets at all in septic installations) and/or urinal directly to the septic tank. In the tank, bacterial action reduces the sewage into a liquid that is then disposed of through an absorption system or transpiration system similar to those described under stormwater disposal.

In most states the control of septic tank installations is in the hands of the local government council, which is charged with administering a section of a state Health act specifically relating to septic tank installations—hence the variations in local requirements defy description.

The golden rule on septic tank installations is to ask the local authority first, second and last, then do as they say unless there is special good reason for challenging the requirements.

### The effluent and grey-water disposal system

This system that attempts to dispose of all the water produced by a residence that is not defined as sewage or stormwater; it is the waste water that is becoming commonly referred to as grey water. Until recently it was normal to attempt to dispose of grey water by use of absorption systems or transpiration beds, but grey water can be stored in a holding tank and used as garden water. If such a tank is used (these are not always approved of by local government, especially in urban areas) then an overflow absorption trench system is still recommended. Take care with detergents used in the house, as some of their chemicals are harmful to plants.

In most locations OBs can install their own septic disposal system under the eye of the local health inspector. However, if the system is to be connected to mains sewer in the foreseeable future, the sanitary plumbing and the majority of the house service may have to be installed by a licensed tradesperson.

Check local regulations carefully and gain all necessary approvals before commencing any work. The septic installation may have been shown clearly on the plans and specifications submitted for the building

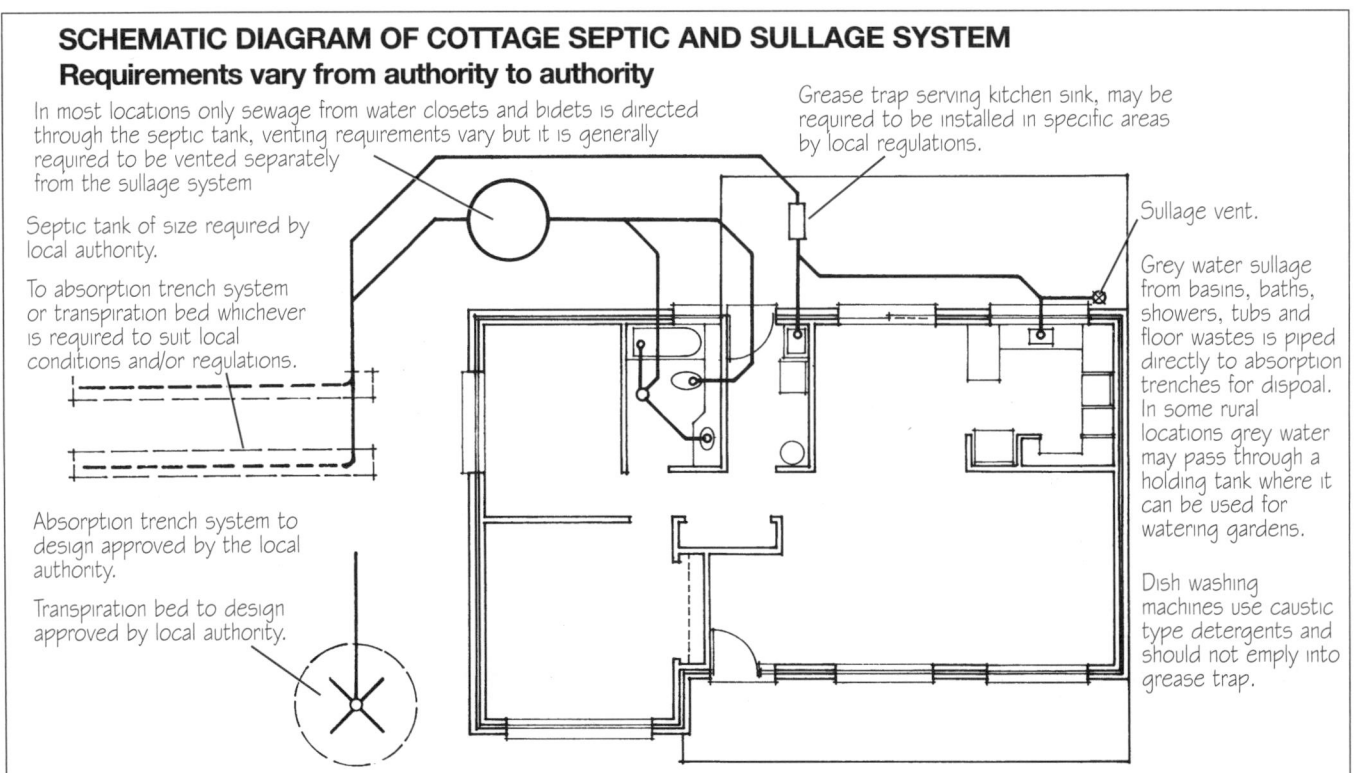

**SCHEMATIC DIAGRAM OF COTTAGE SEPTIC AND SULLAGE SYSTEM**
**Requirements vary from authority to authority**

In most locations only sewage from water closets and bidets is directed through the septic tank, venting requirements vary but it is generally required to be vented separately from the sullage system

Septic tank of size required by local authority.

To absorption trench system or transpiration bed whichever is required to suit local conditions and/or regulations.

Absorption trench system to design approved by the local authority.

Transpiration bed to design approved by local authority.

Grease trap serving kitchen sink, may be required to be installed in specific areas by local regulations.

Sullage vent.

Grey water sullage from basins, baths, showers, tubs and floor wastes is piped directly to absorption trenches for disposal. In some rural locations grey water may pass through a holding tank where it can be used for watering gardens.

Dish washing machines use caustic type detergents and should not empty into grease trap.

# DRAINER • 54

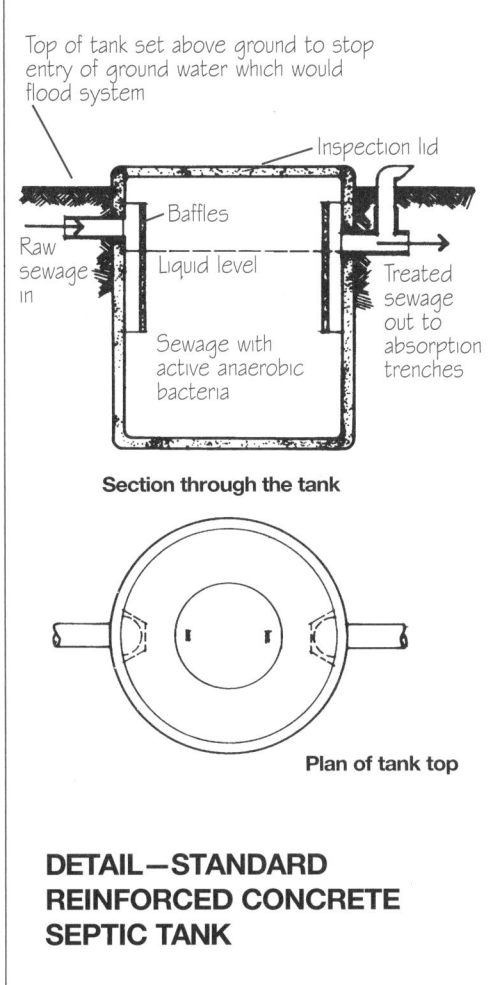

**Section through the tank**

**Plan of tank top**

**DETAIL—STANDARD REINFORCED CONCRETE SEPTIC TANK**

### DETAIL— TRANSPIRATION BED

The size and depth of the transpiration bed should be designed to suit the volume of liquid to be disposed and the local weather conditions. Municipal health departments are generally able to give appropriate information.
Transpiration beds are used in low absorption soils and use evaporation as their main method of disposing of liquid waste.

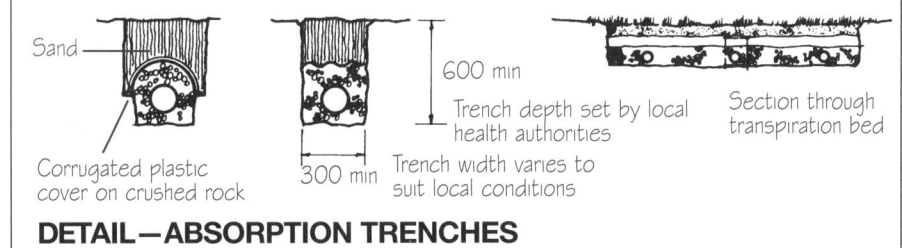

**DETAIL—ABSORPTION TRENCHES**
—For use in septic, sullage and storm water disposal systems

permit, but this does not always mean it was approved with the permit, even if the same department considers both applications. A separate application for the septic tank must be approved.

### Modern disposal systems
Local health authorities often demand sewage effluent disposal systems that allow only potable water or harmless products as on-site disposal material. System acceptance and availability vary, so check which ones are approved. Some systems allow liquid waste to be used as garden irrigation; others produce worm castings—ideal garden compost.

### Sanitary plumbing
The sanitary plumbing includes all fittings within the house that are connected to a water supply and empty into the sewerage drains. The fittings include, but are not limited to:
• water closets
• wash handbasins
• baths
• showers
• laundry tubs
• kitchen sinks
• bidets.

It should be noted that neither clothes washing machines or dishwashing machines are considered sanitary fittings, although the wastewater from them must be disposed of through a sanitary fitting. There is some doubt whether a waste disposal is a sanitary fitting, as the regulations governing the installation of this appliance vary from location to location.

The sanitary fitting is at the boundary between the work of a licensed plumber or a licensed drainer and that of the carpenter or other person who may have some part in the installation of the fitting. The regulations vary from location to location on how much of the installation of sanitary fittings is prescribed work for plumbers/drainers.

The connection of modern sanitary fittings to sewer drains, which are commonly considered to start at the floor level of the building, is generally a fairly straightforward matter of assembling numerous precision

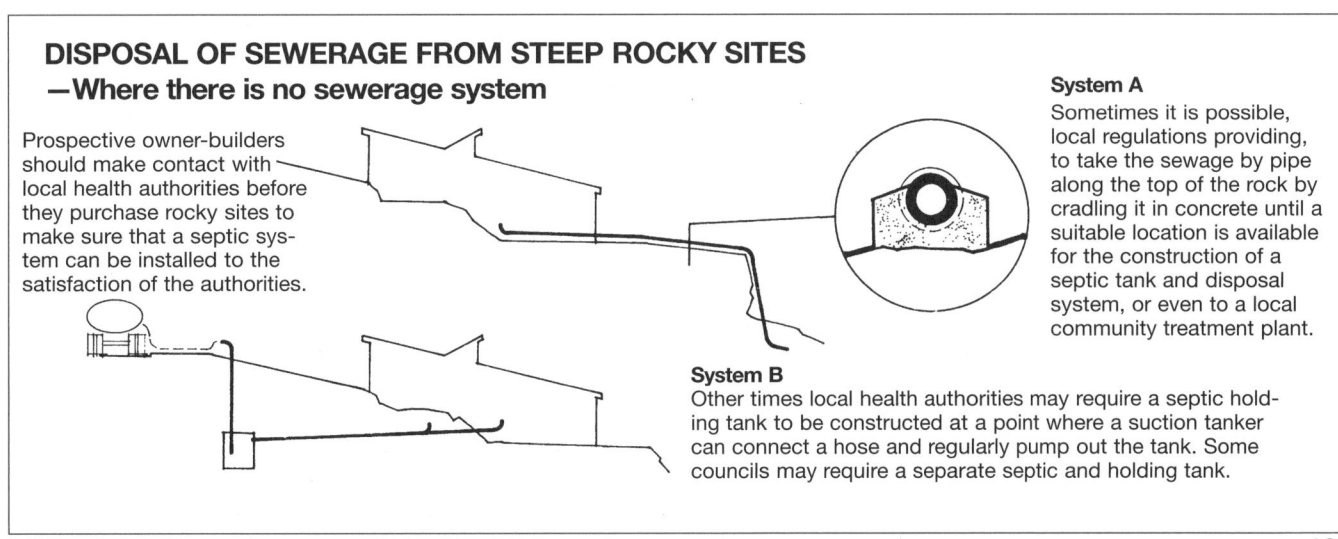

### DISPOSAL OF SEWERAGE FROM STEEP ROCKY SITES
—Where there is no sewerage system

Prospective owner-builders should make contact with local health authorities before they purchase rocky sites to make sure that a septic system can be installed to the satisfaction of the authorities.

**System A**
Sometimes it is possible, local regulations providing, to take the sewage by pipe along the top of the rock by cradling it in concrete until a suitable location is available for the construction of a septic tank and disposal system, or even to a local community treatment plant.

**System B**
Other times local health authorities may require a septic holding tank to be constructed at a point where a suction tanker can connect a hose and regularly pump out the tank. Some councils may require a separate septic and holding tank.

made rigid plastic pipes, traps and junctions together by way of threaded joints, patented connectors or simply glue. Gone are the days when plumbers soldered formed lead pipes together. The mystique and skill vanished a decade ago and it is only the conservativeness of the plumbing trade and the traditions of the rule makers that still demand that only licensed plumbers should carry out sanitary plumbing jobs.

If any of this highly specialised sanitary plumbing is to be exposed in your home, make sure the plumber understands that mixing white pipe with grey fittings and allowing the solvent glue to form blue abstract patterns is not your idea of first-quality tradespersonship.

**Alternative to the s-bend***
Since the development of the 's-trap' water seal, most sanitary fittings have been fitted with one of these. Water seal traps take up a significant amount of space in kitchen cabinets and vanity units, are hard to conceal in pedestal basins and often have to hang through the floor under showers and baths. At least one company (Hepworth) makes an alternative waterless self-sealing waste valve—it conforms to AS/NZS 3500/2/1 and relies on a one-way plastic valve that allows the passage of waste water and seals against foul air entry.

These waterless traps can be used on basins, sinks, bidets, showers, baths, urinals, washing machine wastes, dishwasher wastes and in some applications floor wastes. Currently they are not applicable to flushing toilet bowls.

**Alternative to the toilet cistern***
Few items are as inefficient in houses as the flush tank cistern used on most toilets. They are noisy when they flush, even noisier as they fill, do not allow instant re-flush, are inaccurate in measuring flush volumes, leak too often and generally look unattractive.

The alternative has been the flush-o-metre valve; these have been available since at least the 1920s but, although sometimes used in large buildings, they were seldom seen in houses. The problem was they required a water holding tank in the roof space and all the fittings were expensive heavy-duty chrome-plated brass.

There is a residential unit distributed in Australia by Caroma—the Schell WC Flush Valve—which if in general provided with a length of 25 mm cold water supply, will dual flush—3 or 6 litres—without delay as there is no cistern to be filled.

*OBs should check with local authorities before fitting either of these items.*

# PLUMBER • 55

The plumber is a licensed tradesperson responsible for the installation of the cold and hot water reticulation within the house, plus fixing the roof plumbing, special flashing and the gas service—where applicable.

Almost without exception, in all locations in Australia where piped town water is supplied to a property, there is a requirement that the water reticulation system be installed by licensed plumbers. This requirement is to ensure that the water supply is not polluted by septic fluids entering it.

People who live away from town water supplies are generally unrestricted in that they are normally allowed to install their own water reticulation system if they desire.

## Cold water reticulation

When using town water supply the first job of the plumber is to tap into the main water supply pipe and bring this service on to the property. In most areas, the tapping service is terminated at a water supply authority's water meter.

From the meter the water supply is reticulated to all cold water supply points in the house. The pipe used is generally copper but, in recent times, most water supply authorities have approved the use of special plastic pipes and fittings. There are also a few installations using galvanised steel pipe, but this method of reticulating water is almost obsolete.

Cold water pipes can be run under the house, in the walls of the house and in the roof, wherever is the most convenient. Care should be taken to make sure the plumber saddles the pipes adequately, as a banging pipe buried in a wall can be impossible to fix, when the building is complete, without major structural surgery.

In frost-prone areas remember to insulate any pipes that are exposed to areas that may have temperatures below 0°C.

Many homes are now using single-lever or dual-mixer (quarter- or half-turn) ceramic washer taps. These generally give move even flow control than a standard bouncing washer tap.

The ceramic washer tap does not give as satisfactory anti-syphoning protection as the traditional washer and in some localities it is likely that every ceramic washer will require a normal valve stop cock up-line to provide this protection.

Replacing a traditional tap washer has been a householder chore since this type of tap was introduced, and most Australians pride themselves on being able to carry out this task successfully. The ceramic washers are less likely to fail but, when they do, it is not a home handyperson task—a plumber will be needed and then the tap body may have to be returned to the supplier for repair. To reduce the chance of ceramic washers failing an inline water strainer is recommended—this is a device to remove particulate mater from the water supply and should not be confused with water purifiers. By removing the grit from the water supply, the tiny holes in the ceramic discs that allow the controlled flow of water are less likely to become clogged.

## Water purification

There are some places in Australia where to drink the reticulated water is a health risk. If this is the case a water purification system can be fitted to the drinking water supply. Always make sure that these systems are installed, maintained and used strictly to the manufacturer's recommendations.

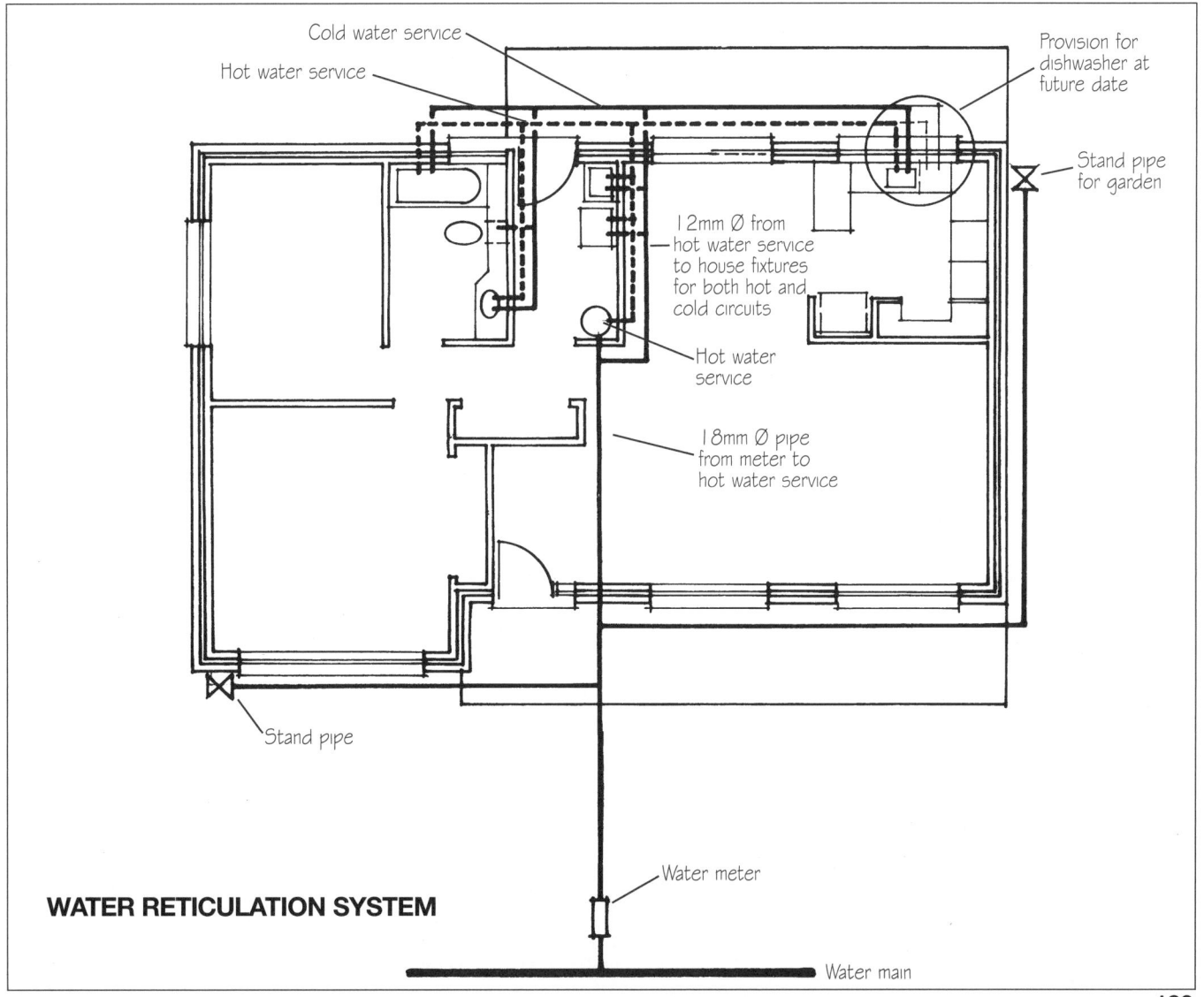

**WATER RETICULATION SYSTEM**

# 55 • PLUMBER

There is a small danger that the filtering system itself can become a breeding area for microorganisms if it is not changed or flushed regularly.

## Hot water reticulation

As with cold water reticulation it is normal for the hot water to be piped in copper pipes, the only difference being that the hot water supply pipes are insulated. Traditional hemp type fibre insulations are still used by some plumbers, but modern synthetic foam insulating is more efficient.

It is possible to use modern polybutylene with hot water systems without insulation in low temperature transfer environments. If piping is exposed to temperatures below freezing then insulation is recommended. Plastic piping sags when it is hot and therefore extra clipping is recommended. Some hot water services tend to heat the pipes close to the water heater to temperatures above the recommended working temperatures for plastic piping; check the manufacturer's recommendations for tempering measures—generally at least a metre of metal pipe between the water heaters and the commencement of the plastic piping installation.

Traditionally, Australian homes had one central off-peak storage hot water service. This is still a very acceptable system under many circumstances, but in modern homes the hot water supply points are spread all over the house. Sometimes the wait for hot water to come can mean a long wait while the static cold water is drawn out of the system. In these cases the delay waiting for cold water can be reduced and the nuisance of the water going cold because someone else in the house turned on a hot water tap can be greatly improved, if two hot water services are supplied instead of one.

## Hot water systems

Efficient and reliable hot water is a critical component of modern houses. Every member of the family demands a hot shower each morning and only the shortest possible delay can be tolerated in the time required before the hot water flows.

The problem is that water heaters use large quantities of energy that is often simply flushed down the drain. With the development of efficient cold water clothes washing systems the majority of domestic hot water is used for bathing. Humans love the feel of hot water when it is cascading over their bodies or providing relaxation in a deep hot bath.

If people believe that non-renewable energy should be conserved there are basically three options:
- use less hot water when bathing
- use less energy to heat the water
- use solar or other renewable energy.

Ideally, effort should be made to incorporate all three into new house designs.

## Selection of hot water system

The selection of a hot water system is a compromise between what the owner would like to have and what they can afford. On one hand users want an adequate supply of hot water with a minimum of inconvenience, but on the other hand they are governed by the initial cost of the system (including installation) and running costs, including for energy supply and maintenance. There are additional factors, such as space limitations, non-availability of a particular energy source or even restrictions by local supply authorities that may influence the selection of the unit.

## Storage type

There are two types of storage units.

**Mains Pressure Units**—systems that operate at mains pressure or close to mains pressure. These units are the most common of the storage types and the cylinder can be located internally or externally—generally they should be in a position where the pressure relief valve and overflow pipe can be maintained and monitored. Mains pressure systems can be heated using electricity, gas, kerosene/heating oil, solar or heat exchanger/heat pumps.

**Reduced pressure or Gravity Feed Units**—once the most commonly installed unit but seldom used today, as their best location is in the roof space of the house; low-pitched roofs and trussed construction methods have limited their appeal. In-roof systems are generally limited to non-flame energy systems. The difference in the low-pressure hot water supply and the mains pressure cold water supply often means variations in temperature during a shower. Modern single-lever taps are not suited to low-pressure systems.

## Continuous flow type

As consumers become more aware of energy conservation issues, gas-fired continuous flow hot water systems have been promoted as using less energy than electric storage units. In 2002 the lower rate charged by electricity supply companies in many regions still made the cost to run an electrical storage unit similar to a continuous gas unit. The purchase price of the units is also similar.

Progressively, the authorities have lowered the temperature at which hot water can be reticulated. This has tended to favour the continuous flow systems as, with the lowering of the temperature at which water can be stored, there is a subsequent fall in the energy being stored in a storage unit. Therefore, the lower energy use storage systems will run out of hot water earlier than the older higher temperature storage units. This has led to the promotion of dual-element storage units that use real-time electricity to maintain a reasonable quantity of hot water in the storage tank—about one-third of the volume of the tank is maintained at full operating temperature.

Check at what rate the booster electricity is being charged, as in some locations it is billed at the off-peak rates (lower price per energy unit) and in other locations the full-peak-time rates apply.

There are some factors to consider when evaluating a continuous flow system. They:
- are effectively restricted to gas flame heating
- heat water on demand
- have a smaller unit than a storage type
- can be fitted internally or externally
- operate at or near mains pressure
- are available in different capacities to suit different household sizes
- can be fitted with temperature control units in a convenient location.

## Selection of energy type and source

The progressive deregulation of energy supply companies means that in many locations electricity and gas can be purchased from the same provider. Many of the companies are resellers and do not own any electricity generators or gas reserves; they often do not own the distribution cabling or pipelines. The price of energy could drop in highly competitive urban locations and escalate in less competitive regions—centralised pricing structures controlled by state government instrumentalities are disappearing. Technically, consumers are allowed to generate their own electricity—assuming they have a suitable energy source and can gain planning approvals.

Larger properties in constant wind flow zones, suitable for the erection of a wind generator or with an adequate strongly flowing watercourse, suitable to drive a hydro-electric generator, can even sell surplus electrical energy back to the distributor.

## Natural gas or liquid petroleum gas (LPG)

Gas is suited to storage or continuous systems, heats or reheats on demand and with correctly sized units there is minimum risk of having an insufficient volume of hot water.

There is little cost advantage when comparing the cost-to-buy and the cost-to-operate figures between the two systems.

Environmental authorities indicate a similar preference for either system if a house is continuously occupied. If a house is not continuously occupied—a weekender—or is under occupied—one person when a storage system has been sized for a family—the continuous system will excel, as no energy is used until a tap is turned on.

Natural gas is a finite resource, although it

could be replaced by manufactured gas from vegetable sources. It has less greenhouse gas impacts than most other fossil fuels. This can mean that using natural gas efficiently may produce less environmental damage than using electricity generated at a coal-fired power station.

Natural gas is normally reticulated through a mains pipeline system and delivered through a meter to a residence. LPG is pipe-distributed in some areas but is often delivered in large steel bottles that are refilled periodically.

The price of natural gas versus LPG per unit of heated water is often difficult to calculate and will vary significantly from place to place.

## Electricity

Electricity is suited to off-peak storage systems rather than continuous systems or on-demand storage units. The cost to operate differential can be as high as 1:2. At a residence, electricity is a clean, fire-safe and odour-free energy but, until it is produced from environmentally friendly and renewable sources, it remains an imperfect choice.

## Solar energy

Research continues to find the most economic and efficient method of using the energy from the sun to heat domestic hot water. Most systems, dependant on location, will provide between 50 per cent and 100 per cent of a residential family's demand for hot water. The ability of a solar hot water system to operate is directly affected by the efficiency of the system, its locality, the time of the year and the ambient temperature of the input supply water. The quality of solar systems varies markedly:, some high quality systems work at high efficiency and have an extended life of over 10 years without major maintenance or parts replacement. Other units are much lower in efficiency and deteriorate significantly in well under 10-years.

Generally solar systems are more expensive than equivalent-capacity storage or continuous systems and although they can provide cost-free hot water, they often need the assistance of an electrical booster element to maintain constant temperature hot water. The combined premium cost of the unit, its maintenance costs, relatively short amortisation period and the cost of booster electricity (often at full-peak rates) means that fewer than desirable numbers of families have been attracted to the system.

## Heat pumps and heat exchangers

This highly efficient system uses the same principal as a refrigerator but in reverse. With the use of a compressor and refrigerant gases (modern gases are ozone-friendly) heat is absorbed from the atmosphere using roof-top panels or out of the ground using tubular coils. It is then piped to a water tank, where the heat energy is transferred to the water—result: hot water. These very efficient systems generally use electricity to drive the compressor motor (heat pump) but it is possible to use any driving force. If wind energy is used to drive the pump, either directly or through an electrical generator and a storage battery, the system moves close to being continuously renewable. The only fossil-fuel input—and therefore potential greenhouse emission—occurs during the manufacture of the components.

Although many environmentalists believe that airconditioning systems in domestic applications are a serious waste of energy the truth is that many Australian households have no other way of maintaining a cool house during humid summer days. The real waste is that the aircon unit actually absorbs heat from the interior of the house and then blows it into the atmosphere at an external condenser coil—if rather than blowing the heat into the atmosphere, the waste heat was transferred to the water through a heat exchanger in a storage hot water system, there would be significantly less energy waste.

## What is an adequate supply of hot water?

This means that in the worst season of the year (winter) hot water should be available at the temperature, rate of flow and quantity to suit particular needs as indicated below.

Few people are in the situation where they can afford a hot water system that will never run out. The majority of us have to adjust our bathing habits to suit the capabilities of the system they can afford.

The number is a maximum where a higher standard of living and/or cold climate conditions apply, the capability of any unit is reduced. Best practice is to install the correct size of unit in relation to the accommodation potential of the size of the dwelling and the standard of living of its occupants. Thus, a heater for a couple in a four bedroom home would be out of proportion to the accommodation potential of the home. This empirical calculation adjusts for element rating as the one hour availability is reduced as the element rating is reduced.

## Gas supply

The five basic types of fuel gas in use in Australia for domestic purposes are:
- town gas
- natural gas
- TLP (tempered liquefied petroleum gas)
- SNG (simulated natural gas)
- LPG (liquefied petroleum gas).

Natural gas is obtained from bores sunk into the earth either on land or at sea. It is primarily methane. Town gas is of variable composition depending on the plant and raw materials used in its production. Originally based on coal, gas coal is becoming less significant as a raw material and reformed gas is becoming more important.

TLP, or tempered liquefied petroleum gas, is a mixture of propane (or butane) and air for distribution by pipe supply from a central source—as with town gas. It is being used to replace town gas, particularly in country areas, due to the ease with which it can be produced by automatic plant operation.

SNG, or simulated natural gas, is a mixture of LPG and air for distribution by pipe supply from a central source. As the name implies, it approximates the characteristics of natural gas and is used as a substitute for natural gas, either before natural gas is used, or as a peak load substitute.

LPG, or liquefied petroleum gas, consists of commercial propane or butane or a mixture of both. These gases are obtained as a by-product from the processing of raw natural gas or from oil refining. The gases can be distributed as a gas or, more frequently, compressed into a liquid under pressure and transported in a pressure vessel.

LPG, when sold under a trade name such as 'Portagas' or 'Shellane', is usually commercial propane. Reticulated gas is produced and distributed by undertakings generally known as the Gas Company, as many of the early undertakings were private enterprises. In Australia today there are several types of ownership of gas companies.

Progressively the Australian states are deregulating energy distribution, including gas supply. In many places companies sell or will be allowed to sell gas to local consumers without having to own the source or means of distribution.

Gas, electricity, telecommunications and in some places even water will become listed marketable commodities that can be bought and sold in the same manner as many other resources. There will be regulations to ensure quality of the product and safety of installations, but the companies will prosper or fail by market forces. There may be greater competition, leading to lower prices for those consumers savvy enough to assess the right choice.

Some companies sell gas by its heat content, as so many cents per kWh, or per therm. Others sell gas by the cubic metre. In this case the calorific value of the gas should be known in order to compare the heating cost with other fuels.

Essential features of gas appliances are:
1. all town gas appliances are required to meet the approval standards of the Australian Gas Association
2. all town gas appliances are required to be fitted with a gas governor, a regulating device that prevents the build-up of gas pressure at the appliance and over-gassing of the burner
3. all gas appliances require an adequate supply of air for operation, the absence of which will produce imperfect combustion and may promote sooting, smells and poisonous carbon monoxide conditions

# 55 • PLUMBER

### Roof plumber
It is not essential for the person who does the roof plumbing to have a licence, but plumbers are the tradesmen who have the knowledge to install the system of gutters, downpipes, cappings, flashings and soakers that make roofs watertight and able to carry away the rainwater. Some parts of the roof plumbing system are installed before the roofer, some parts are after. Getting the plumber to butt up the eaves, valley and box gutters before the roofer arrives is important to remember.

### Flashings
There are flashings in many buildings other than those described under roof plumber. These include purpose-made cranked flashings in cavity brick walls and other special flashings over retaining walls, under doors and windows to the sides of all openings.

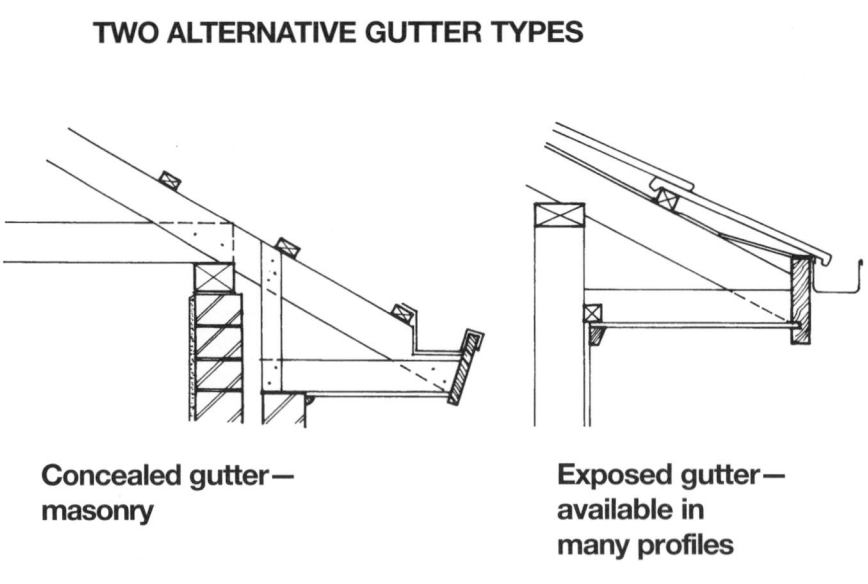

**TWO ALTERNATIVE GUTTER TYPES**

Concealed gutter— masonry

Exposed gutter— available in many profiles

4. all gas appliances release products of combustion that must be removed from the site of operation. Large fixed appliances are installed with a flue to remove such products from the appliance. Smaller appliances may be permitted with unflued installation but this is only allowed where there is adequate ventilation. Inadequate ventilation of an unflued appliance will produce undesirable or dangerous effects such as smothering of the flames, imperfect combustion, condensation on walls and ceilings, smells, and discolouration of painted surfaces or textiles.
5. town gas appliances may operate on tempered LPG but will not operate on natural gas or LPG without modification
6. LPG appliances do not necessarily have a gas governor
7. LPG appliances should be fitted with 100 per cent safety shut-off devices to stop all gas flow should the pilot flame become extinguished.

### Gas plumber/fitter

In most locations where there is a town gas supply, plumbers are required to have a special licence or endorsement on their licence before they undertake any gas installations—obviously for the safety of the occupiers of the house. In areas where there is no town gas supply, and bottled gas is used, it will not necessarily be required that a licensed gas fitter be used for the installation, but it is highly recommended.

If you purchase any gas appliances for installation in your new home, make sure the ones you buy are compatible with the gas supply they are to be connected too. Not all gases are the same and the appliance should be manufactured to suit the gases being used. Before making a commitment with your gas plumber, check with the local gas supply company to see if there are any discount or bonus deals going at that time, for gas companies are often attempting to get people to connect to gas on very favourable terms. They have been known to offer free gas for a period of time, free connection, or even apparently low priced appliances—always check the deals.

| Sizing a hot water system | | | | |
|---|---|---|---|---|
| **Family size** | 1 to 3 | 3 to 5 | 5 to 7 | 7 to 9 |
| **System** | | | | |
| **Gas** | | | | |
| Continuous | 16 L/min | 20 L/min | 24 L/min | 24 L/min |
| Storage | 90 Litres | 170 Litres | 200 Litres | 260 Litres |
| **Electrical** | | | | |
| Off-peak storage | 160 Litres | 250 Litres | 315 Litres | 400 Litres |
| At-peak storage | 40 Litres | 80 Litres | 125 Litres | |
| **Solar** | | | | |
| Collector | 2 sq.m | 4 sq.m | 6 sq.m | 8 sq.m |
| Booster tank | 160 Litres | 300 Litres | 370 Litres | 440 Litres |
| **Heat pump** Individually designed | | | | |
| **Heat exchanger** Individually designed | | | | |

# ELECTRICIAN • 56

In all states of Australia, it is a requirement that all electrical installations be carried out by qualified licensed electricians.

The main reason for this is the obvious danger of contact with live electric wires, but the secondary consideration is the possibility of fire caused by electrical short circuits.

The main grid distribution of electricity in Australia is 50 hertz (cycles per second) alternating current (AC), which is normally stepped down from the high distribution voltages to 240 volts for domestic consumption.

It is normally the responsibility of the electricity supply authority to provide the electricity conductors up to the point of attachment to the house requiring supply. If the distance from the supply authority's mains (whether overhead or underground) is around the normal residential building alignment then this connection is normally carried out without additional charge—but not in all locations.

The home builder's electrician is responsible for all electrical wiring and other installations from the point of attachment, including providing the meter box for the electricity—they use meters supplied by the supply authority (this is true in most areas of Australia, but because of the multitude of supply authorities it is recommended that an OB check with their supply authority for local requirements).

OBs are advised to mark up a copy of the working drawings with the final location of all light points, power points, appliances, heaters, hot water services, fans, airconditioners, computers, telephone systems, faxes, internet connections and any other specific electrical points. This marked-up drawing should be given to the electrician before the commencement of any wiring in the house, and in this way there should be few expensive additional points required after the number of circuits has been established. The electrician should be able to confirm the final cost of the installation before the work is commenced.

If external power points are required ensure their location is communicated to the electrician, as they are often significantly more expensive to install later. It is often simple and inexpensive to install a power point in the meter box, as there is no wiring to install and if the box has a lid it may not have to be an expensive weatherproof outlet.

Remember when marking up the electrical drawing that the electrician needs to know the location and height of all light switches, as well as the height above the floor of all power points and wall lights. Recessed and fluorescent light fittings are commonly installed by the electrician, but make it clear who is to supply them: the electrician or you. Fancy light fittings are not attached by the electrician unless special arrangements are made—many electricians have been held responsible for dropping expensive light fittings and they are therefore wary.

## Temporary supply

It used to be common practice to attach a temporary electricity supply to every building site wherever power was available. In recent years, however, with the trend to underground power supply and an ever-escalating cost spiral, the attachment of temporary power to building sites is becoming less common.

OBs should research whether they absolutely cannot do without temporary electricity supply. If a temporary supply is required then it must be installed by a qualified licensed electrician.

## The electrical rough-in

The electrical installation is carried out in two distinct parts: the rough-in and the fix-off.

The electrical rough-in of a house is carried out when the house has reached the stage in its program where the majority of the structure is in place—but while it is still possible to run the electrical cables in locations that mean they will be finally concealed by the wall and ceiling finishes.

The rough-in wiring consists of placing all the main electrical reticulation cables throughout the house, leaving enough wire for the final fixing of the circuits.

When the walls are sheeted out it is important to remember to mark clearly or pull through any cables that may otherwise become lost behind the ceiling or wall sheeting.

## The electrical fix-off

When the house is nearly complete, just before the painter is due to start, the electrician can fix off the ends of all the cables that were roughed-in previously.

This means that the electrician can connect all the lighting outlets, power points and other electrical fittings. Make sure you let the electrician know the colour, type and configuration of all power outlets and light switches. This is of special importance if a number of lights are switched from the same plate; it can be both inconvenient and frustrating if the switches are not arranged logically.

All permanently wired appliances, heaters and other electrical equipment should be delivered to the electrician before the fix-off to avoid waste of time, return calls and the possibility of incorrect wiring.

When the electrician has completed the installation there is normally a requirement for the supply authority to check the installation before it can be used by the occupants. Remember this condition—if you do not have the electrical installation correctly tested and approved you may be delayed in moving into your new home

## Electricity supply

Electrical generation, distribution and supply are being deregulated in most locations around Australia. Governments, which once owned and controlled every step in the system, are selling the assets and withdrawing with a sizeable financial gain, to allow the market to work out the future. There are new rules being introduced to allow for a more rapid development of renewable electrical energy sources.

Private individuals, small companies and large corporations are able to generate electricity and sell it into the distribution grid. Initially this electricity will have to be absorbed and credited. The long-term aim is to have the majority of Australia's electricity generated by wind, tide, water, solar and other non-polluting, minimum greenhouse effect and fully renewable sources.

Electricity is generated by:
1. thermal generation—using steam power produced by coal or oil, or mechanical energy from diesel motors
2. hydro-electric generation—using the energy from water falling through height
3. wind power
4. organic waste utilisation— from yearly replanted crops such as rice and sugar cane
5. solar power
6. tidal power.

Electric energy is distributed over large distances by raising its voltage at the generating plant and lowering the voltage at the local points of distribution. Electricity in Australia is universally 50 hertz AC and household supply is in one of three forms:
1. single-phase—240 volt, two wires, one of which is at zero potential (the neutral) and the other at 240 volts above and below zero
2. two-phase—415/240 volt, three wires, one the neutral, the other two active and 240 volts above or below the neutral but with 415 volts between them
3. Three-phase—415/240 volt, four wires, one neutral and three actives each 240 volt above and below the neutral and 415 volts from each other.

Declared voltage and cycles may vary from 240 and 50 hertz in some areas. In addition, direct current (DC) supply may be used in isolated areas. It is important to remember that AC appliances do not normally operate on DC and special provision is needed to cope with DC supply.

# 56 • ELECTRICIAN

## SCHEMATIC DIAGRAM OF COTTAGE ELECTRICAL SYSTEM

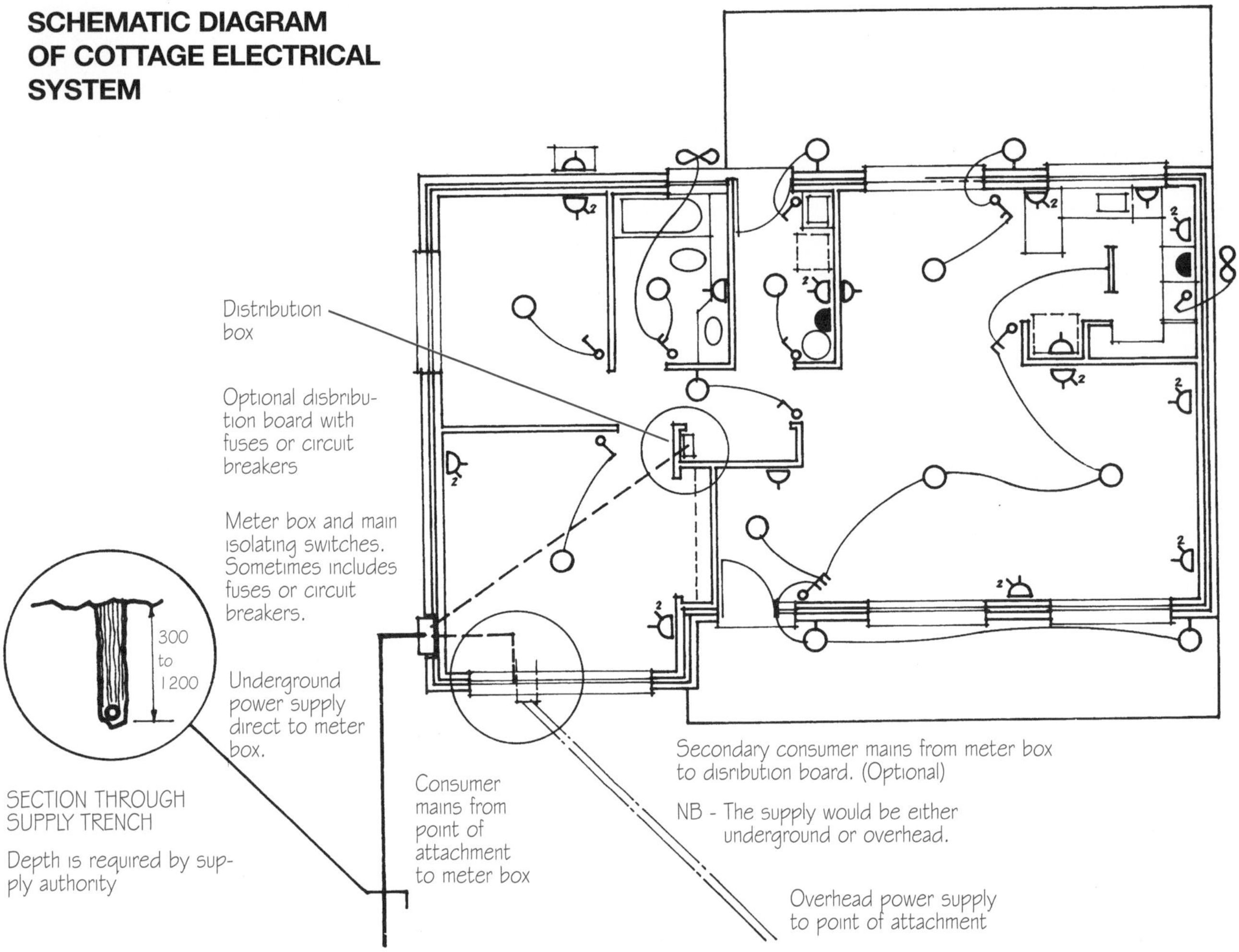

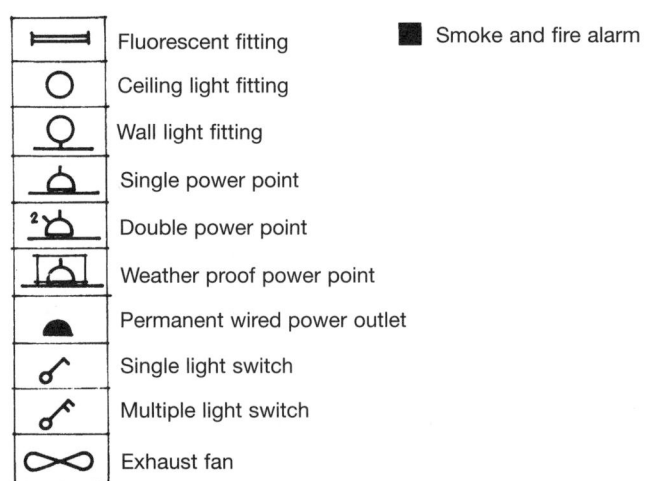

- Distribution box
- Optional distribution board with fuses or circuit breakers
- Meter box and main isolating switches. Sometimes includes fuses or circuit breakers.
- Underground power supply direct to meter box.
- Consumer mains from point of attachment to meter box
- Secondary consumer mains from meter box to distribution board. (Optional)
- NB - The supply would be either underground or overhead.
- Overhead power supply to point of attachment

### SECTION THROUGH SUPPLY TRENCH
300 to 1200

Depth is required by supply authority

## METER BOX

Earth Leakage Protection Device (RCD) is generally located in the switchboard; this protects people from the danger of electrocution if they contact a live wire by cutting the energy flow in an instant.

Take care: Lighting circuits are not always earth leak protected.

Supply meters, number may vary if 'off peak' or three phase supply is required.

Isolating switches, fuses and/or circuit breakers.

## SYMBOLS USED ON PLAN

- Fluorescent fitting
- Ceiling light fitting
- Wall light fitting
- Single power point
- Double power point
- Weather proof power point
- Permanent wired power outlet
- Single light switch
- Multiple light switch
- Exhaust fan
- Smoke and fire alarm

NB  There are many different symbols used on electrical layouts. These used on this diagram are closely based on Standards Association recommendations.

# ELECTRICIAN • 56

## Sale of electricity

Electric energy is measured and sold by the kilowatt hour (kWh). For domestic use one or more tariff methods may apply:
1. standing charge for period plus so much per kWh used
2. block tariff—initial block at high rate per kWh (primary units) then next lot at lower rate and so on
3. special tariffs for specific purposes (eg. off-peak water or space heating)—electricity billing period may be monthly, two monthly or quarterly, depending on the supply authority concerned.

## Controlled energy supply

The cost of producing and distributing energy can be split into components. The demand cost and energy cost. The demand cost covers the capital cost needed to generate the electricity, the energy cost covers the cost of providing the energy, reading meters and so on. As the demand cost is determined by the highest energy demand on the system each month, (this will occur for only 30 minutes on one day in each month) any energy sold which does not contribute to the maximum demand can be sold at a lower tariff based on energy cost without demand cost.

Most supply authorities offer energy as 'off-peak' or 'night rate' energy during the night hours when the demand on the system is very low. Some may, in addition, offer 'it extended off-peak' in which energy is available most of the 24 hours and only shut off to avoid the peak periods. Usually this energy is sold at a higher tariff than off-peak or night rate. In each of these, energy supply is controlled by time switches at each installation set to cut in and out at predetermined times.

## Flexible cord core colours

In the past, the requirements concerning the identification of conductors of flexible cords have provided that the colour green shall be used to identify the earthing conductor. The colours red and black have been used to identify the current-carrying conductor. Recently, in conformity with overseas practice, electrical authorities throughout Australia have agreed to accept an alternative system of colour coding. These are:

|  | Obsolete | Standard |
|---|---|---|
| **Active** | Red | Brown |
| **Neutral** | Black | Blue |
| **Earth** | Green | Green/yellow (striped) |

## Choosing light fittings

There are four basic types of light sources used in Australia.

### Incandescent-tungsten

These are the traditional globes with bayonet or Edison screw bases; they come in clear, pearl or enhanced glass. They are commonly 240 volt—10, 15, 25, 40, 60, 75 and 100 watts and available in standard size, pilot size and in fancy shapes to suit special decorator fittings.

The clear globes are generally used behind a diffuser, pearl globes are used in fittings where the globe is visible, and enhanced globes are often gas-filled, to reduce the hot spot of the filament.

Other enhanced fittings include mirror back spot and flood lamps, and mirror front lamps for use with special reflectors.

It is generally considered that the life of an incandescent light globe is about 1000 hours.

Tungsten globes generally give a warm light but some globes have colour correction to take them into the white zone or even slightly into the blue spectrum.

There are coloured globes for festive events and special colours for special tasks—to make plants grow, to attract insects, to repel insects.

Heat lamps used to keep food warm and to heat bathroom are of the incandescent type.

Most incandescent lights can be connected to a dimmer.

### Halogen-dichroic

These are the small-diameter lights used as downlights in many modern residential lighting designs. These are often powered by transformers that convert 240 volts to 12 volts—ventilated space within a reasonable distance of the light fitting is required to house these transformers. Some newer versions of the dichroic-type downlights are all-in-one units that do not require a separate transformer and can be connected directly to the 240 volt supply.

Other exposed dichroic fittings can be used, even without insulated 12 volt wire supply—contact with a 12 volt wire used to power residential dichroic fittings is not considered dangerous to people—but it is wise never to be casual about any electrical fitting and make sure all work is installed and tested by a licensed electrician.

Special thin halogen lights are made to be mounted within narrow shelves.

Halogen light is in the blue range and there is a slight danger of ultraviolet exposure to people working too close to a fitting.

Most halogen-dichroic installations are dimmable.

### Fluorescent-tubes

The 1200 mm long fluoro tube has been around for a long time and, although the fitting most used in commercial applications, has not attracted many admirers in the residential market. They used to flicker and buzz; these problems have been significantly trained out of modern tubes but the light remains clinical and cool. Choose tube colour carefully to suit its intended task.

A benefit of the fluorescent tube is that it is of high efficiency and provides lots of light for less money. Useful in garages, with some applications in kitchens and bathrooms if used carefully. They are difficult to dim.

### Fluorescent energy-saving compact

This is the long-life low-energy-use light that will fit into a standard light socket—either bayonet or Edison. It is a winner except it is very expensive to buy, even though the laboratories state its long life and low energy use will bring it out in front after a life span of 8000 to 16,000 hours – if this light was used for 8 hours every day it would last for 5 years.

Compacts can be used in downlights and are of value, particularly in kitchens. They are useful in table lamps and some bed lamps. Available in warm and cool colours, they can be used almost anywhere if careful selection is made. One of the problems is whether replacement lights will be available in 5 years; there already have been a number of design changes to some of the products in the market.

Like all fluoro lights these are difficult to dim.

There is a particular type of compact fluoro that will operate at 12 volts. These are worth exploring by those OBs who are environmentally conscious, as they give out amazing light on minute amounts of energy.

# 57 • MAJOR APPLIANCES (WHITEGOODS)

Traditionally, the major appliances in houses were free-standing items, but in more recent times many of these appliances have been designed to be 'built in' to benches, cupboards and walls.

This move to the built-in appliance means that they must be selected and often purchased earlier than was previously necessary, to allow proper detailing of cabinetry and the provision of proper services.

The following table indicates the range of sizes and service connections generally applicable to the commonly available appliances but, with many imported appliances now becoming available in Australia, off-standard sizes or installation requirements may occur.

**Notes on the chart details**

1. Stoves are normally 900 mm high but may vary from 850 to 920 mm. Each must be checked. Stoves are available with eye level second ovens or with the oven and cooktop side by side (the so-called elevated range). Both these types are in the minority.
2. Stoves are normally 600 mm deep but some imported odd-ball or plug-in types may be less. Few are deeper.
3. Cooktop depth is only a consideration when allowing sufficient space under the bench. Remember a drawer cannot be placed immediately below some cooktops.
4. Wall ovens are more or less standardised at 600 mm wide, but there are some special wider types on the market. The opening to receive the oven is different from the fascia panel and must be checked with each individual model.
5. Wall ovens are normally a little less than 600 mm deep but, because of the danger of building in high-temperature appliances, clearances at the rear (and sides) of the oven may require a cupboard 100–200 mm deeper and wider than the appliance itself.
6. Refrigerators are still manufactured at many and varying heights, widths and depths. However, in all cases they are designed with one dimension narrow enough to fit through a standard 820 mm door opening. Take care not to reduce access to the kitchen to under the smallest dimension of the refrigerator selected.
7. Refrigerators using gas energy are uncommon but are still manufactured for use in remote areas where the electricity supply may be unreliable.
8. Refrigerators requiring cold water supply are some of the sophisticated units that incorporate automatic ice making and chilled water dispensers.
9. Freezers are not as easy as refrigerators to fit into close-fitting recesses. Whereas a refrigerator has good air flow over the exposed condenser coils (normally on the back of the unit so it will function efficiently), many freezers have the condensers built into the side and rear skins of the unit so that all faces of the freezer must be exposed to air flow.
10. Dishwashers are normally between 800 and 850 mm in height so that they may be installed under normal 900 mm high kitchen benches. Freestanding units may be up to 900 mm high with the addition of a work surface on their top. This work surface can often be removed if the unit is to be built-in in the future.
11. Dishwashers are normally connected to cold water supply, but there are some units that require either hot, or hot and cold water connections. Take care when installing water connections that all connections and hoses are in good order and well fitted, because a hose failure can fill your kitchen with water while you sleep.
12. Clothes washers normally have a set of taps in the wall immediately adjacent to the machine, take care to check the height of the hob of top loader machines to ensure they are below any wall taps. Front loader washers may be free standing or can be built in under bench tops; ensure the taps suit the machine chosen and note the rear clearance necessary for the hoses, including waste.

Where the symbol 'P' is used in the electricity the appliance is sometimes permanently connected to the electricity supply. Check the supply requirements for the proposed appliance because these may vary, even between different models from the same manufacturer.

## General comments

**Stoves**—are available in gas or electric models. If gas, then care should be taken to purchase a stove compatible with the gas to be used, ie., town gas, natural gas, LPG. Also, many gas stoves need an electrical connection to operate clocks, lights, rotisseries, hob power points, and so on.

Semi-commercial all-in-one stoves are often used in custom-designed cook's kitchens. These are generally wider than the standard 600 mm, ranging from 750 through 900, 1050 to 1200 mm. The larger stoves often have two special-use ovens. A semi-commercial stove will commonly have gas hobs, often with wok, fish, grill or barbecue burners. Their ovens can be either gas or electric, with multiple heat sources, fans and lights. Check what gas, electric and even water connections are needed for the more elaborate of these combo-ranges.

| Appliance | Size range | | | Service connections | | | | |
|---|---|---|---|---|---|---|---|---|
| | Height | Width | Depth | Electricity | Gas | Hot water | Cold water | Sewer drain |
| STOVE | 900(1) | 550–1200 | 600(2) | ✓ P | | | | |
| | | | | | ✓ | | | |
| | | | | ✓ | ✓ | | | |
| COOKTOP | 30–150(3) | 500–750 | 400–600 | ✓ P | | | | |
| | | | | | ✓ | | | |
| | | | | ✓ P | ✓ | | | |
| WALL OVEN | 450–900 | 600–900(4) | 600(5) | ✓ P | | | | |
| | | | | | ✓ | | | |
| | | | | ✓ | ✓ | | | |
| REFRIGERATOR | 850–1800 | 600–1200 | 600–750(6) | ✓ | | | ✓(8) | Optional on some models |
| | | | | | ✓(7) | | ✓ | |
| FREEZER | | | | | | | | |
| -Upright | 850–1800 | 600–900 | 600–750(9) | ✓ | | | | Optional on some models |
| -Cabinet | 750–900 | 600–1800 | 600 | ✓ | | | | |
| DISHWASHER | 800–900(10) | 600 | 600 | ✓ | | ✓ or | ✓(11) | ✓ |
| CLOTHES WASHER | 750–900 | 600–750 | 450–750 | ✓ | | ✓ & | ✓(12) | |
| CLOTHES DRYER –rotary | 700–900 | 500–700 | 450–700 | ✓ | | | | |

# MAJOR APPLIANCES (WHITEGOODS)

**Cooktops**—are becoming more sophisticated and it is now possible to choose from electricity or gas, electricity/gas combined, plus barbecue, deep-frying or bain marie combinations. Some ranges offer modular cooktops to allow purchasers to build up any combination of the above units.

**Wall ovens**—are available in gas, electric or combined gas/electric models. When gas wall ovens are used, take care to install strictly to the manufacturer's and gas authority's recommendations so as to avoid the potential danger of gas leakage into the space surrounding the oven. Wall ovens are available with single or double ovens, with separate grillers, with fan-forced convection and in some cases microwave generators.

**Refrigerators**—most are still free-standing appliances but several built-in units are available for bars and kitchens.

Generally, other than refrigerators with vertical split double doors, the direction of door swing can be ordered to suit particular kitchen layouts. Make sure that this factor is considered.

**Freezers**—both chest and upright, continue to be made but there appears to be a trend by manufacturers to promote upright models—and there are even a small number of built-in units appearing on the market. Note: never put a refrigerator or freezer in a pantry, because these units make as much heat on the outside as they make cold on the inside and are likely to adversely raise the pantry temperature.

**Dishwashers**—are almost standard equipment in all new kitchens, but there are still those who, for whatever reason, do not wish to install one. However, it is strongly recommended in designing any new kitchen that provision be at least made for future installation by way of space, electrical point under the bench, a water point and sewer waste.

Many dishwashers offer the option of fitting door panels to match the kitchen design; details of these should be obtained before commencing fabrication of the kitchen. At least one manufacturer has a drawer-type dishwasher that can add flexibility to kitchen designs.

**Clothes washers**—few are currently built-in but there is a strong trend towards producing the more efficient frontloading machine that is adaptable to below-bench installations.

**Clothes dryers**—can be wall- or floor-mounted to suit special installations. Some are designed to pair with front-loading washing machines and be mounted on top of them. At least one built-in model is available. The bigger dryers work more efficiently if they can exhaust the wet air through a duct to the outside—this also reduces the condensation problems often associated with clothes dryers.

At the premium end of the clothes dryer market the units use condensers to remove the water from the clothes, this means there is no requirement for exhaust ductwork as there are no fans or loose lint. Most of these dryers can be pigeon paired to a matching frontloader washer. There are some semi-commercial dryers available for domestic use—some of these need a 20 amp electricity supply and others use gas as the heating source.

**Microwave ovens**—often placed on benchtops but, progressively, built-in locations are being used. There are benchtop models, benchtop models with build-in kits and fully built-in models available. Progressively, the built-in models are being manufactured in sizes that fit with normal kitchen unit modules. Most microwaves are simple plug-in appliances but some, particularly those that incorporate browning and cooking elements, may require a higher current than provided by a standard 10 amp outlet.

**Coffee makers**—built-in cappuccino machines are made by some manufacturers. These are about the size of a microwave and need to be included at the design stage of the kitchen. They need a water connection and some types may also be connected to a sanitary waste.

**Rangehoods**—are progressively replacing ceiling exhaust fans as the preferred means of extracting cooking vapours from a kitchen. Some are built into the overhead cupboards and either recycle the vapours through filters or exhaust the vapours to the outside of the house. They are often fitted with pull-out slides to extend their exhaust effectiveness. Many rangehoods are fully on display above the cooking range, have built in lighting systems, micro-fine filters, multi-speed multi-fans and complex ducting systems. Rangehoods should be at least as wide as, and extend out as close as possible, to the furthest edges of the cooking range being served. There are also benchtop exhaust systems designed to suit the heavy smoky vapours from barbecue units.

**Water filters**—are generally located under the kitchen sink and consist of a bundle of pipes and removable filter cartridges. Make sure there is a convenient water supply and sufficient room in the cabinets—remember to allow for dodging the kitchen waste pipes. If the unit has a water cooler option then a power point and even more space are required.

**Boiling water units**—really fill the cupboards under the sink. These provide an instant hot beverage service. They need water and electricity supply. With both the water filters and the boiling water units, extra holes need to be allowed in the sink for their spouts. Alternatively, the boiling water unit can be located on a wall over a sink or tub.

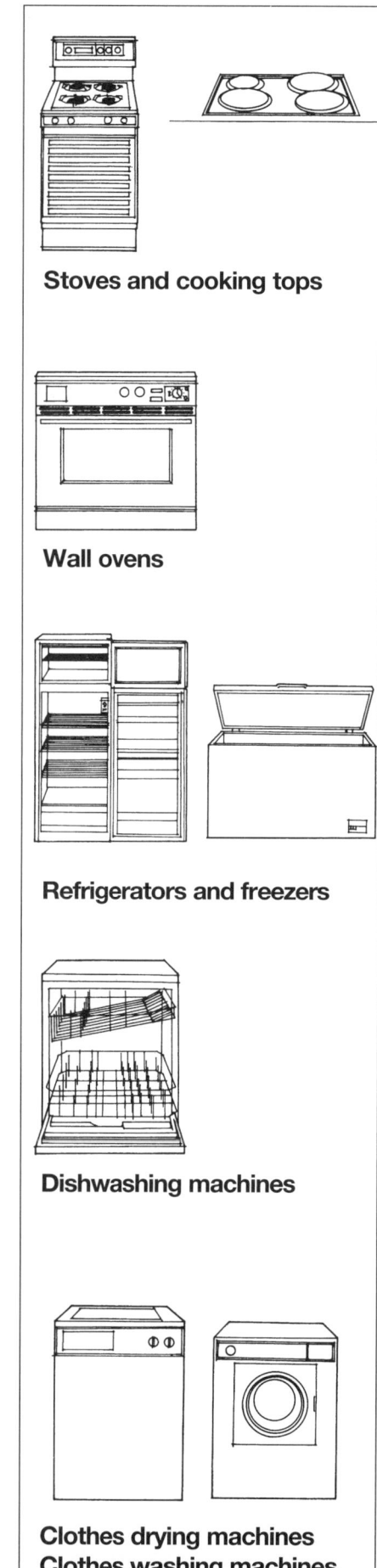

**Stoves and cooking tops**

**Wall ovens**

**Refrigerators and freezers**

**Dishwashing machines**

**Clothes drying machines**
**Clothes washing machines**

# 58 • FIXING OUT

Once the main structure of the house is built then is the time to add all the bits and pieces of trims, cabinetry, joinery, equipment, machinery, appliances and the like that are necessary for the final appearance and functioning of a modern home.

The fixing-out can need the co-ordination of many skills and often a number of specialised tradespeople will be required to work at the same time. OBs who wish to do much of the fix-out themselves should plan the period carefully so that any special tradespeople are contacted well in advance. Any delays in the arrival of tradespeople so late in the project can be extremely detrimental to the completion date. Also, ordering of fix-out materials should be undertaken as early as practicable; shortages in fix-out materials can be a nuisance, particularly if special timber mouldings have been run for the job and may take many weeks to be run again.

The whole area of fixing out buildings has changed significantly over the last decades. As close as the late 1960s and early 1970s the work of fixing out a house was the domain of a skilled tradesperson—now the work has a lower skill component. Most of the fixing out consists of installing prefabricated components into their designated positions in the building. Everything from pre-hung architraved doors, through easy-fit sliding wardrobe doors to complete kitchens are made elsewhere and trucked to the site for fixing, so the role of the skilled carpenter and joiner have all but been eliminated from mass-market housing.

The skilled tradesperson does hang on grimly in the custom- and specially designed house building sector but even here the days of site-fitted joinery seem limited, as the cost to use quality joinery climbs out of reach. The number of apprenticeships being offered in carpentry and joinery is also falling steadily, which means the skilled tradesperson seems destined to disappear. This is a good reason to be an OB, because then at least you can spend the time and effort to fix out your home personally with quality fittings.

Many years are needed to train an all-round carpenter and joiner, training that most OBs will not have, nor could get even if they wanted to. OBs can, however, attend selective courses at technical colleges, building centres or elsewhere to gain sufficient knowledge and skill to fix out their own home. Nothing improves the quality of work as well as practice. If you must practise on your own new home, start in the third bedroom and work towards the living room.

One of the secrets of a quality fix-out is the use of correct, well-sharpened tools. If you do not have a kit of quality tools, do not attempt a fix-out. Check the cost of a kit of good tools and compare them with a quote for the labour to fix out your house. If the tools cost as much as the labour would, then employ the labour, for if you are short by this many tools the chances are you have not had enough practice with the tools required and you will likely do a poor quality job anyway.

## MDF (medium density fibreboard)

Medium density fibreboard (MDF) is a compressed sheet/board material made from *Pinus* and other wood fibres combined with urea formaldehyde or similar glues. It has a much finer texture when worked than particleboard. This product has proved to be ideally suited for the fabrication of cabinets and can be moulded to the shapes required for architraves, skirtings and other mouldings. It is a suitable material for OBs but reasonable care should be taken to avoid breathing in the dust when sawing and working the material, and filter masks should be worn.

Some people are allergic to the gases that are given off by MDF and particleboard. The phenomenon of 'out gassing' will continue for varying lengths of time, depending on the type, amount and strength of adhesive used and the sealing applied to the board. Other materials can 'out gas', including carpet. It would be wise for OBs to read textbooks dealing with healthy and unhealthy building materials.

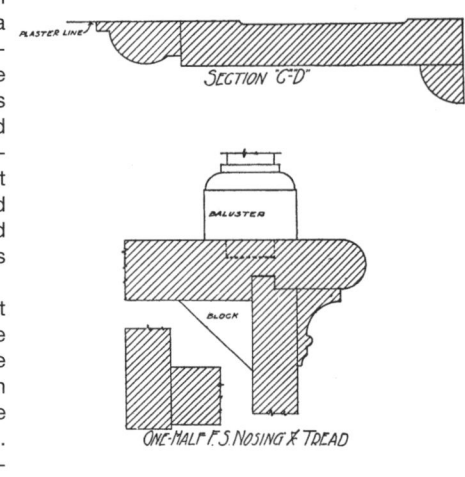

# TRIMS • 59

Trims are those pieces of timber or other material used to tidy up or give character to joints between different building components and materials, as well as in, or around some changes in direction of wall, floor and ceiling sheeting.

The most common trims that are used in houses are:
- skirtings
- cornices
- architraves
- linings
- angles.

## Skirtings

The skirting board is the board fixed to the wall at its junction with the floor. Skirting boards can be shallow, as little as 40 mm, or up to 400+ mm high. Simple skirting boards may be finished with a bullnose, slayed or bevelled moulding. Or they can be very elaborate and moulded. When skirting boards are fixed to the wall they should be set at least 5 mm above the finished floor surface to allow for settlement of the building structure.

When joining external angles for skirtings it is normal to mitre the junction and, when pining the inside corners of skirting, a scribed and saw junction is generally more satisfactory. The skirting board should be thinner (or at least not larger) than the door architraves, otherwise the skirting will stand proud of the architrave wherever they make contact. This problem can be solved by putting a skirting/architrave block at all intersections and, while uncommon today, it was used extensively when elaborately moulded skirtings were the order of the day.

## Cornices

The cornice is the mould used on the junction between the walls and the ceiling. If the walls and/or ceiling are lined with render or plasterboard then the cornice is likely to be moulded in plaster. If the walls and/or ceiling are lined with timber then timber may be chosen for the cornice.

Plaster cornices are generally precast and delivered to the site in long lengths. They can be simple cove sections or elaborately decorated mouldings that may use multiple sections to build up an extremely ornate unit.

Sometimes when lavish cornices are used, moulded ceiling roses may be used to decorate the ceiling.

Cornices should be fixed firmly to the ceiling, but, if possible, they should not be fixed to the wall. This is not always possible or practicable but, where they are not fixed, the ceiling can move independently from the walls and this will reduce unsightly cracking at the cornice-to-wall junction.

## Architraves

The architraves are trims fixed to cover the gap between the wall lining and the frames of windows and doors. The architrave traditionally is a moulded timber section, but sometimes other materials may be used. Recently a number of project homes have begun using hardboards and/or plastic sections.

Many pre-hung doors have factory-fixed architraves, making an OB's job much easier. However, if you fix your own architraves, only fix them to the door or window jamb. Fixing it to the wall is likely to pull on the frame and this can cause the mitred corners to open up.

Some windows have nosings and apron moulds on their internal sills, the traditional method of trimming the bottom of a window. This is advisable for OBs who want to give a degree of authenticity to a period style house.

## Linings

The linings are the timber boards used to make up wall thicknesses, such as door jambs, window reveals or similar applications. The timber used for all linings should be carefully chosen, with a straight grain and should be well seasoned. Any shrinkage or twisting of these members can cause doors to jam as well as look unsightly.

Linings should be well fixed to grounds or furrings and should be at least 15 mm thick, veneered chipboard can be substituted for real timber in many applications but care must be taken not to use this material if there is any risk of dampness, as this product will swell up and disintegrate.

## Angles:

Angles are used to trim wall-to-wall junctions and are often unavoidable in timber lining applications. They can look a well-detailed part of the panelling.

However, angle trims should be avoided where possible in flush wall sheeting applications like plasterboard, but must be used with many fibrecement and hardboard sheets. The angles are often extruded from plastic or aluminium.

## Other trims

Throughout any building there is always a temptation to take the easy way out and cover a sloppy piece of work with a timber moulding or an angle. Avoid the temptation as much as possible, for there is no doubt that most houses look better with the minimum of unnecessary trims.

Where trims are essential, detail them to fit in with the other work in the area and try to make them appear as an essential part of the design, not an afterthought.

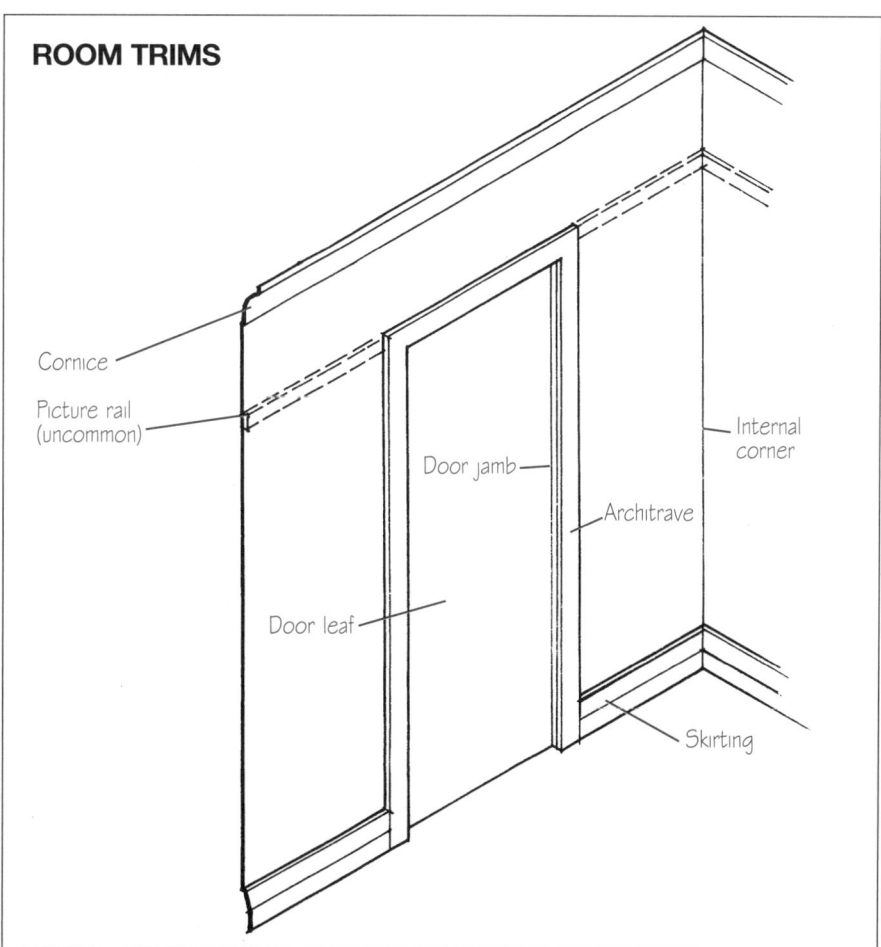

**ROOM TRIMS**

# 59 • TRIMS

## TIMBER MOULDINGS IN COMMON USE

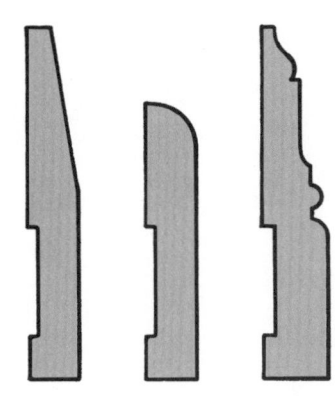

1 Splayed skirting
2 Bullnose skirting
3 Period (various) skirting

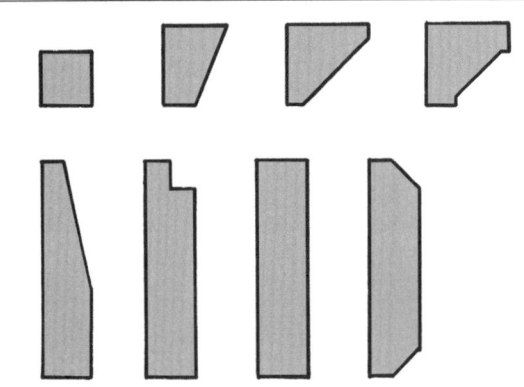

4 Square fillet
5 Splayed fillet
6 Bevelled fillet
7 Sunk bevelled fillet
8 Splayed architrave
9 Rebated board
10 Rectangular section
11 Double bevelled architrave

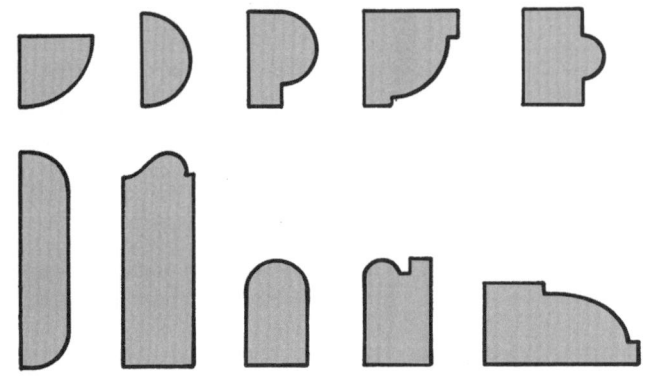

12 Quad(rant) mould
13 Half round mould
14 Bead
15 Ovolo mould
16 Half round bead on block
17 'D' mould
18 Double curve nosing
19 Bull nose mould
20 Quirk
21 Thumb mould

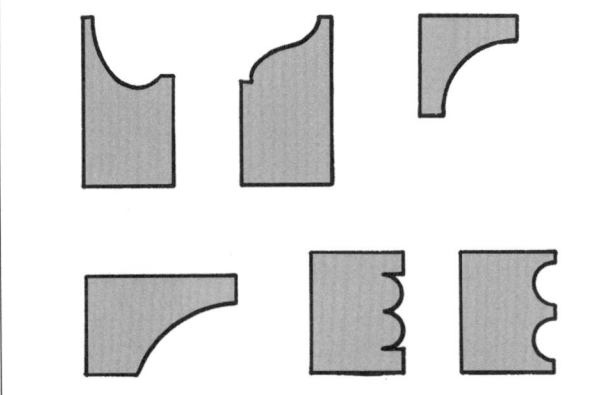

22 Deep scotia mould
23 Ogee mould
24 Cove (scotia or cavetto)
25 Hollow chamfer
26 Reeding
27 Fluting

# KITCHENS • 60

In the years since World War II there has been a revolution in the design, appearance and content of Australian kitchens.

To understand this we should remind ourselves of what kitchens were like 40 years ago.

A typical kitchen then featured a black iron stove—although gas cookers had been available, they were unsophisticated and very expensive to operate.

Stainless steel sinks were beginning to penetrate the market, but enamelled cast iron or china sinks with whitewood or terrazzo draining boards still dominated. A cold water tap was located over the sink and occasionally there was a gas sink-heater on the wall to supply hot water. Central hot water services were uncommon, and generally available only for better-off families.

Food was stored in a pantry, and perishables took their chances in an ice-chest. Refrigerators were still a luxury item.

Built-in cupboards and benches were limited and every kitchen boasted a free-standing dresser. Laminex was only just becoming available. A chromium-plated table with a Laminex top was the centrepiece of most kitchens.

How things have changed! Kitchens are now streamlined and crammed with labour-saving appliances.

The kitchen in most new homes is a well-designed food storage, preparation and hygiene centre.

Efficiency has become the catchword in all kitchen planning and design. Fully integrated, modularised cupboard, bench and appliance components are assembled into an almost limitless array of solutions that can be tailored to suit the needs of the majority of families. The kitchen has moved from a special room in the house to a centrally located facility for the dispensing of food and drink.

The planning of a kitchen today is not simply a matter of grouping a set of storage, cooking and washing facilities around the walls of a room arbitrarily designated the kitchen, but rather the integration of a host of specialised components into a space that has been located to allow efficient operation of kitchen functions and provide direct access to the family eating area.

## Planning

There are many ways to plan a kitchen. The following are the five basic plans used in most Australian homes.

## The one-wall plan

This is not used extensively in Australian homes, although it may be found in smaller cottages and flats.

*Advantages*—space-saving, particularly if located along one wall of the only living room in a small flat. It is cheap to supply services to and can be hidden by a screen if necessary.

*Disadvantages*—limited in length by the available room and the distance a user can comfortably move.

## The galley plan

A fairly common plan, often favoured by architects because of its easy-to-control appearance and flexibility.

*Advantages*—allows for good separation of food storage, preparation, cooking and cleaning functions. With no return corners, all underbench storage is freely and easily accessible.

*Disadvantages*—a tendency to make the kitchen a walk-through passage, with access to the laundry or the like at one end. Also, if the benches are too close together (less than 1.2 metre) it is awkward for more than one person to be in the kitchen at the one time. If, however, the benches are wide apart, then the distances between activity points can become uncomfortable.

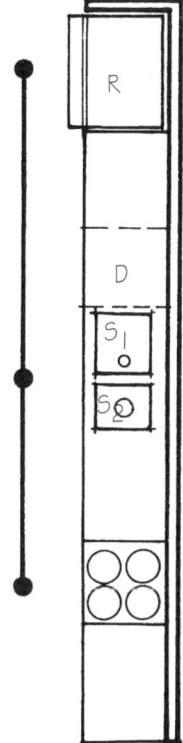

R Refrigerator
C Cooker (with or without oven)
O Oven (optional in all layouts)

S1 Standard sink
S2 Supplementary sink (with waste disposal optional)
D Dishwasher (optional)

**COMMON KITCHEN LAYOUTS**

# 60 • KITCHENS

## The L-shaped plan

This is an older type of kitchen plan not used extensively today but favoured by households that like to have their kitchen table in the kitchen. Sometimes it is called the homestead, or country kitchen plan.

*Advantages*—allows for an eat-in kitchen with central table and, depending on the size of the room, generally allows more than one person to work without getting in each other's way.

*Disadvantages*—if the two bench wings become too long, then the efficiency of the food preparation activity is reduced. The kitchen table becomes an extension of food preparation activity and an overlap with its main use—to eat at—can be achieved.

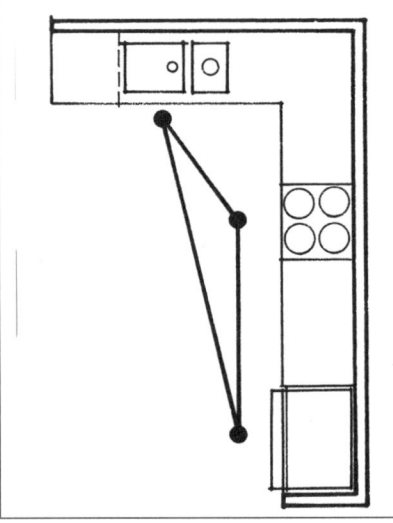

## The U-shaped plan

This is by far the most common layout currently favoured in Australian homes. The kitchen is often located next to a family room and uses one of its benches to separate the two, often by the addition of a servery or breakfast bar to that bench.

*Advantages*—allows for efficient movement and separation of storage, preparation, cooking and cleaning functions. Located in the family room, it offers proximity to family leisure activities.

*Disadvantages*—two dead corners are inherent in the design and either require expensive lazy-susan corner units or some other specially contrived device. Although efficient for one cook, two cooks will often disrupt this efficiency.

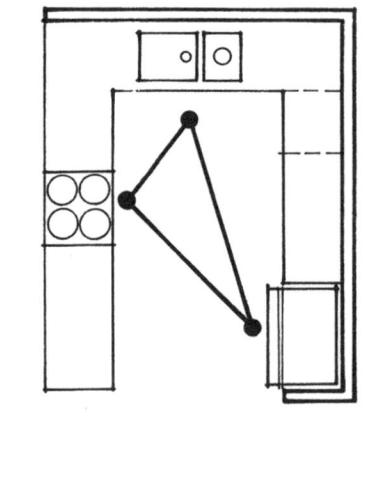

## The island plan

This takes components from the L-shape, U-shape and galley plans and produces what many people believe to be the most flexible and efficient kitchen layout.

*Advantages*—provides for efficient working movement—two or more cooks can operate simultaneously. The island bench can incorporate a breakfast bar, a concept similar to the central kitchen table without some of its disadvantages.

*Disadvantages*—often expensive to build and requires rather a large space to attain a great efficiency.

Note: all these layouts work satisfactorily. Extra appliances, pantries and/or overhead cupboards may be added.

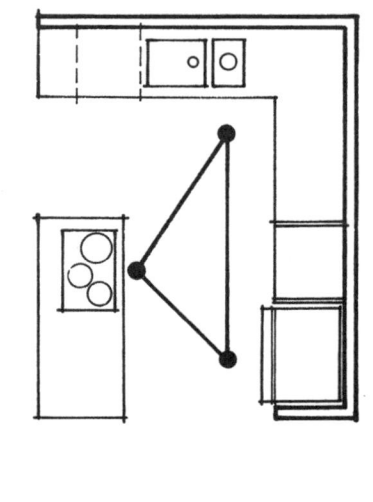

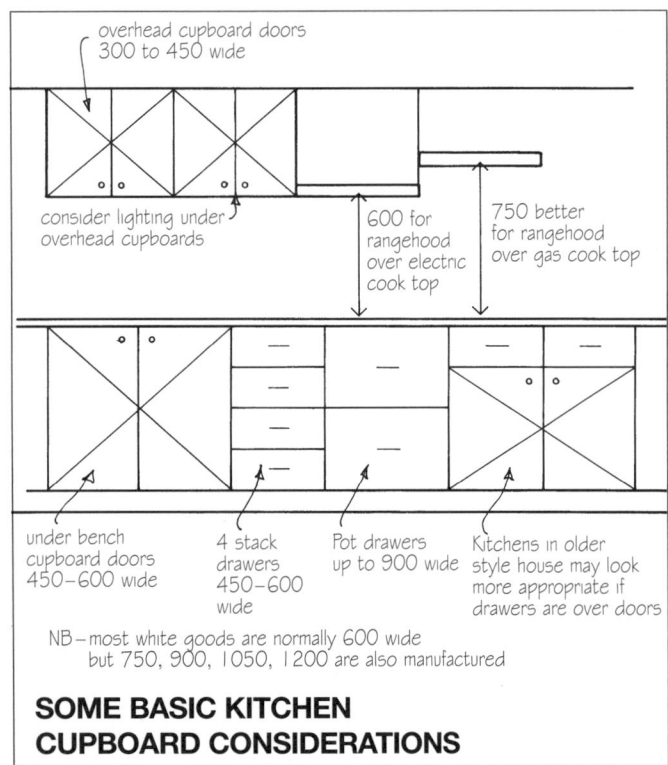

## SOME BASIC KITCHEN CUPBOARD CONSIDERATIONS

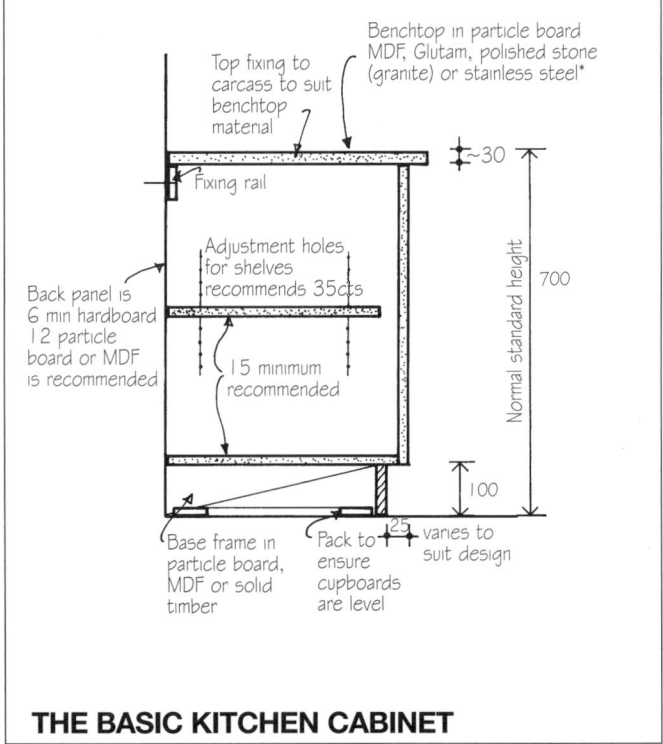

## THE BASIC KITCHEN CABINET

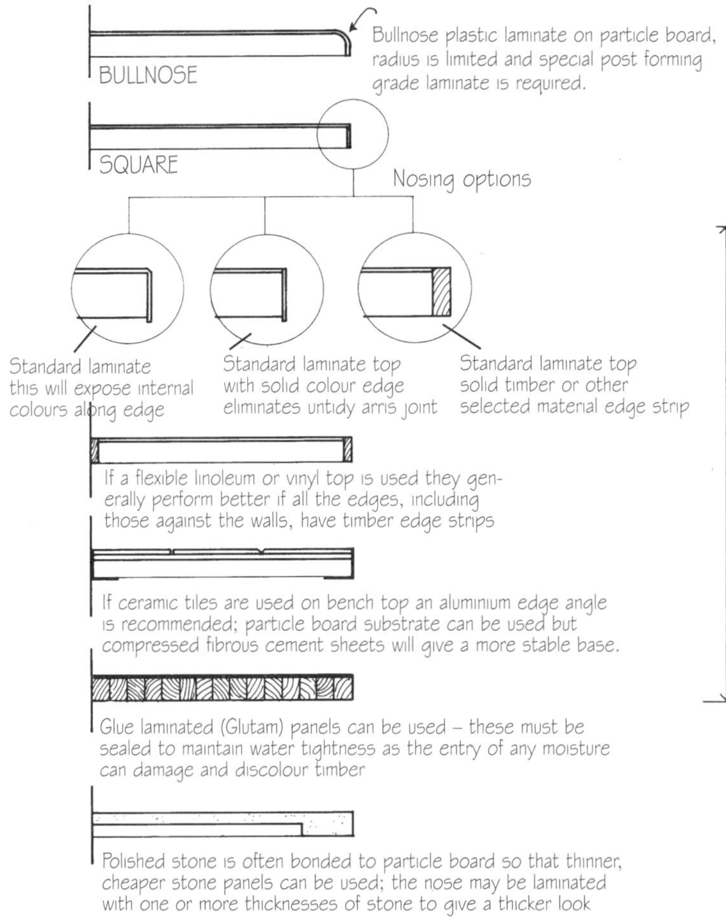

# KITCHENS • 60

## KITCHEN BENCH TOPS

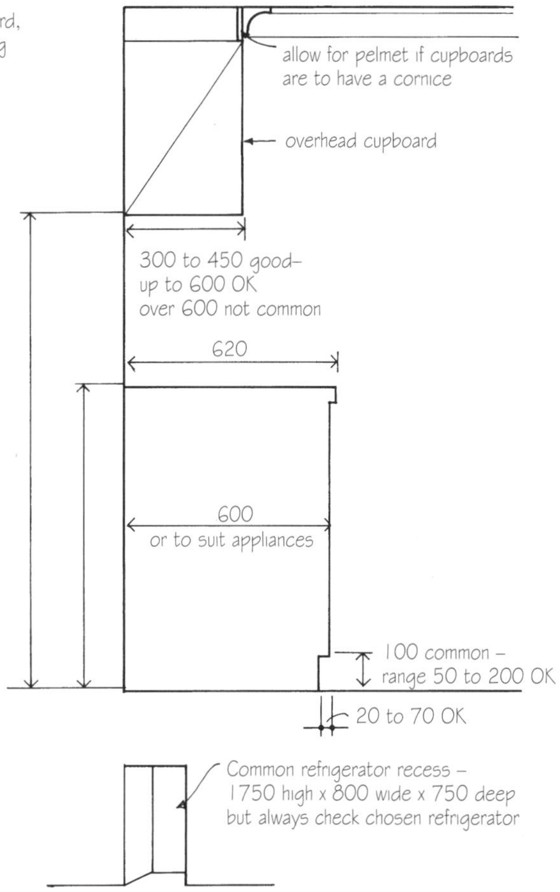

**BULLNOSE** — Bullnose plastic laminate on particle board, radius is limited and special post forming grade laminate is required.

**SQUARE**

Nosing options:
- Standard laminate this will expose internal colours along edge
- Standard laminate top with solid colour edge eliminates untidy arris joint
- Standard laminate top solid timber or other selected material edge strip

If a flexible linoleum or vinyl top is used they generally perform better if all the edges, including those against the walls, have timber edge strips

If ceramic tiles are used on bench top an aluminium edge angle is recommended; particle board substrate can be used but compressed fibrous cement sheets will give a more stable base.

Glue laminated (Glulam) panels can be used — these must be sealed to maintain water tightness as the entry of any moisture can damage and discolour timber

Polished stone is often bonded to particle board so that thinner, cheaper stone panels can be used; the nose may be laminated with one or more thicknesses of stone to give a thicker look

## GENERAL KITCHEN CUPBOARD SECTIONAL DIMENSIONS

- allow for pelmet if cupboards are to have a cornice
- overhead cupboard
- 300 to 450 good — up to 600 OK — over 600 not common
- 620
- 600 or to suit appliances
- 100 common — range 50 to 200 OK
- 20 to 70 OK
- Common refrigerator recess — 1750 high × 800 wide × 750 deep but always check chosen refrigerator

## Appliances

Most modern kitchens are equipped with a wide range of appliances, both fixed and portable. It is important to plan a kitchen so that all the appliances you will ever want can be incorporated. Particular attention should be given to allowing for electric, gas, water and sewer connections.

## Joinery

Appliances are separated by cupboards and benches. These joinery units are generally constructed from melamine-coated chipboard and are designed for multi-functional use. By far the majority of kitchen benches and cupboards are faced with laminated plastic veneers.

## Benches

Today the recommended bench width is 600 mm. This is required for modern inset sink units and is the common depth of most major appliances except refrigerators. The most cost-efficient bench surface is laminated plastic, but solid laminated timber, solid plastic, polished stone—granite, marble, limestone—stainless steel and tiles may be used to achieve a specific appearance or to satisfy a functional requirement.

## Cupboards

The doors on modern cupboards are normally hung on spring-loaded hinges that make door catches unnecessary. A food cupboard (pantry) is a worthwhile addition and is best equipped with a sliding door to allow convenient entry. Do not operate a refrigerator or freezer in a pantry, as these appliances are as hot on the outside as they are cold on the inside. Therefore, the pantry becomes a hotbox, ruining the food stored in it and reducing the refrigerator's efficiency.

## Drawers

Where possible, use top-quality drawer runners. Cheap plastic or timber drawer slides will wear quickly, causing drawers to jam and become expensive to repair.

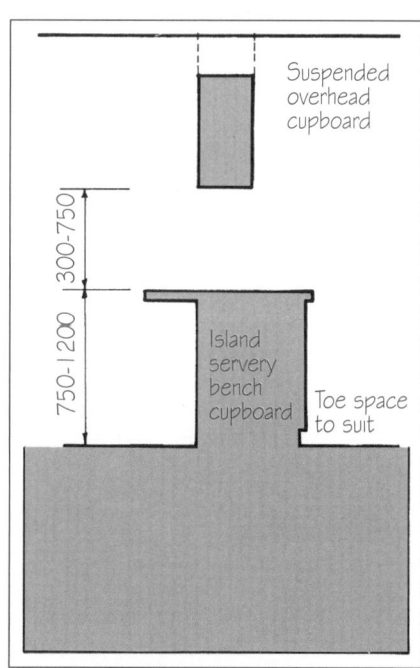

- Suspended overhead cupboard
- 300-750
- 750-1200
- Island servery bench cupboard
- Toe space to suit

# 61 • BATHROOMS

The bathroom used to be a room containing a bath and little else. Today bathrooms contain many more fittings and in some houses, the bath itself has disappeared in favour of the shower.

Australian homes, which would have been unlikely to have more than one bathroom in 1960, are now likely to have at least two bathrooms and often more.

The common arrangement is to have a central bathroom for general use with an adjacent lavatory, a smaller ensuite shower room adjacent to the main bedroom, often with its own lavatory, and a third shower room accessible from outside the house, to be used in conjunction with outdoor activities particularly swimming and gardening.

The trend to more bathrooms in our homes means that often over 15 square metres of floor space is given over to bathrooms. This represents around 10 per cent of the floor area of a 150 square metre home.

In most houses only the kitchen and the bathrooms are thoroughly pre-planned. This has led to a standardisation of the size of fittings and fixtures used, and generally leads to efficiently planned spaces.

The building regulations require that all bathrooms have adequate provision of light and ventilation. In practice, most modern bathrooms are provided with windows much larger than the set minimums, but very often bathroom ventilation is not nearly effective. Bathrooms should have two forms of ventilation, continuous and controllable. The first should be provided by fixed ventilators in or close to the ceiling. These fixed ventilators are essential, as they allow the moisture that condenses on the walls to dry off, thus reducing the growth of mildew moulds that are unsightly and smelly. The second should at least be provided in the form of an opening sash in the window. In addition, many bathrooms are fitted with electric exhaust fans that will mechanically extract vapour from the bathroom.

Some care is required when installing exhaust fans. It is no longer recommended to simply blow the wet air from the bathroom into the ceiling space of the house, as many ceilings are sarked and insulated and the moisture from the bathrooms may lead to a damp mouldy ceiling space. It is much better if the moisture-laden air is exhausted directly to the outside by use of ducts or by placing the fan high on the wall.

The old theory of grouping bathrooms to save money is still valid, but the savings are seldom recognisable—so place bathrooms and toilets for convenience, not necessarily for grouped plumbing. There is however a downside to too widely spread plumbing—the furthest bathroom from the hot water service still has the coldest showers and the longest wait for hot water.

Traditionally Australian bathrooms are 1650–2100 mm wide and 2100–2700 mm long—this is tiny, particularly when filled with a shower, bath, toilet suite and basin. In an effort to make 5 square metre bathrooms comfortable, many families have invested in thousands of dollars worth of expensive tiles and sanitary fittings, for little gain. Whether a bathroom has low-quality or high-quality finishes and fittings, small remains small.

Consider a larger bathroom, 8 square metres or greater, and reduce the opulence factor slightly—the result is likely to be a more comfortable, better looking and easier to clean bathroom. Space is a luxury in a bathroom that will often transcend polished stone and gold plate. Add good natural light, a mirror that does not fog up, a shower with at least two spray heads and a good flow of hot water—the fittings and finished can be modest.

Spa baths come in many shapes and sizes; choose carefully to suit your needs. Spa jets in a standard-type long deep bath will often be the best choice, particularly if only to be used by one person at a time—they work just as well as a normal bath, they often have well positioned jets, they use a reasonable volume of water (they will not run the hot water system to cold), many allow flat floor sections so a shower can be installed over and they are generally easy to build in.

If the spa is as much for fun and recreation, then consider sizes that will accommodate two or more people. Many of these shapes are difficult to incorporate into bathroom designs and some, particularly the smaller-corner type designs, are uncomfortable, so take care. If a large capacity spa is required, consider fitting a water heating feature, as many domestic hot water systems will not have capacity to fill and maintain a suitable temperature.

A standard square shower recess is about 900 x 900 mm. This is big enough for anyone—but 1000 x 1000 mm has less of a Houdini's man in the bottle feel about it, and a 1200 x 1200 mm shower is positively grand. If it is any bigger it may as well fill the whole room. Chop a 45 degree corner off a 1200 x 1200 mm shower to provide a corner door and the bathroom will look good in soap operas, but there is unlikely to be any gain in function.

Some cities, Sydney particularly, have traditionally used hobs to define shower stalls even when they are fully enclosed in glass. With modern wet-seal processes they are generally unnecessary and do little but provide the tiler with a bonus.

Toilet pans are actually quite small in volume, so they do not take up much room. In most bathrooms, the centreline of the pan and cistern should not be closer than 400 mm to a corner, and 450 mm is close to ideal. The dimension from the wall supporting the cistern to any obstruction at the end of the pan should not be less than 1200 mm and 1500 mm is definitely more comfortable.

If there is demand for a pan in a separate room, a space 900 mm wide by 1500 mm long is the minimum desirable; if there is a basin in the room then the room should be widened out to 1100 mm and the length stretched to 1800.

There is now a regulation that requires provision to be made to ensure that a person who collapses in any room containing a toilet cannot jam access to people who attempt to provide assistance. In general this means that if there is less than 1200 mm from the nose of any pan to the arc of the door swing, then the door may have to be undercut, or open outwards, or be provided with hinges or other devices that allow the door to be removed readily to gain access in an emergency. OBs should check the local interpretation of this requirement.

Most water supply authorities in Australia are concerned when bidets are specified. Their concern is that, because in a traditional French-style bidet there is a water supply spray at the bottom of the bidet, if the bidet fills with contaminated water it could be back-siphoned into the drinking water supply.

If this style of bidet is selected, a fully approved anti-siphon hot and cold water supply system must be installed—these are quite expensive and may require regular inspection and re-certification.

There are compromise bidets available where the outlet spray head is above the rim of the bidet; these generally are not required to meet such stringent regulations.

If there is an available wall with appropriate aspect, then windows can add life and value to a bathroom. A reasonably sized clear glass window with appropriate privacy devises is generally more attractive than a traditional obscured glass window. If a bathroom is internal, then a clear glass roof-light window will generally be a more appropriate alternative to an obscure plastic dome at the end of a long, often spider-infested, shaft. In hard to light bathroom, where neither windows nor full roof lights can conveniently be installed, consider the domed plastic and reflector tube type of roof light system. These provide reasonable natural light but little else.

Bathroom storage can be a problem; too often the only storage is in drawers and cupboards in the vanity unit or in a tight shallow mirror cabinet. These satisfy basic storage requirements, but consider fitting a tall cupboard 300 mm deep, about 600 mm wide and in the order of 1800 mm high. This will provide storage in the bathroom for back-up toilet rolls, extra towels, large bottles of spa crystals and shampoo, plus allow space to hide the bathroom scales.

This cabinet can have a mirror door which gives a full length image, brave people can put the mirror on the outside of the door and more modest people can put the mirror on the back of the door.

All mirrors should have lighting placed around or above them to ensure there is a light source on the face and body of the person using the mirror. Central lights in a bathroom are of little value. Dichroic or halogen low-voltage lights work, as does a combination of tungsten and fluoro fittings.

Most bathrooms require mechanical exhaust ventilation to remove steam and odours—some models are very noisy, so choose carefully. Heat lamps provide efficient heating on cold winter mornings but they do rather overstate their presence.

# BATHROOMS • 61

## Construction

The construction of bathrooms has been modified over the years to accommodate changing building methods. No longer are bathrooms always built with solid brick walls and concrete floors; in fact, this method of construction is likely to be the exception rather than the rule.

With the trend to build housing estates in far-flung suburbs, it became obvious during the late 1960s that the traditional concrete floor used in bathrooms was becoming increasingly expensive. Builders wanted a floor that could be laid on to a timber base, like the rest of the flooring in the house. In Victoria, the regulations did not require concrete floors but in New South Wales, the Water Board remained adamant that bathrooms must have concrete floors.

An answer was eventually devised when Hardies released their compressed fibrocement floor sheets. This product was the same thickness as timber flooring boards, and could be fixed directly to a timber floor frame. At the same time it was technically 'concrete'—so the Water Board capitulated. More recently, the use of certain specially manufactured plywoods and chipboards has been allowed for bathroom floors.

Bathrooms have traditionally had tiled walls, a tradition that is still followed today. Tiling onto a solid brick wall poses no special problems—the technology to do this was developed in Roman times—but fixing tiles to a timber-framed wall is not so easy, or, at least, it was not until the necessity to tile bathrooms in brick veneer homes became important. Waterproof wall sheets were tried instead of tiles, but these were generally unattractive and failed to gain market acceptance. Again, it was a fibrocement product that solved the problem.

Today bathrooms in timber-framed brick veneer houses are generally lined with smooth flush-jointed fibrocement (the asbestos has recently been removed) boards that can either be painted or tiled to suit the decor.

The tiling of the sheeted walls is now easily and efficiently carried out using tile adhesives that have been perfected only in the last 10 years. Some of the early adhesives only held the tiles onto the wall for a few years until someone slammed the bathroom door and all the tiles fell into the bath.

Although tiles are the most popular wall covering in bathrooms it is possible to use timber panelling. This, however, normally requires great care in the sealing of the timber, fixing of the panels and a commitment to regular maintenance. Pre-finished impervious wallboard sheeting may also be used. Modern pre-finished sheets are a premium-quality product in most cases, but make sure of the product's quality and suitability for the job before you commit it to your bathroom walls.

The most important part of the construction of a bathroom is the sealing of the intersection of the wall with the floor and the sealing of the internal wall corners of the room, especially if the bathroom is on an upper floor. The corners should be sealed by the use of continuous purpose-made plastic angles, which are bonded and sealed to walls and the floor before the wall and floor tiles are fixed.

The bathroom floor should have a floor waste and the floor must be sloped so that any water will drain towards it. To gain this floor slope it is normal to lay the floor tiles on a cement bed with a graded fall. This means that small tiles, like mosaic, give flexibility in being laid to two-dimensional falls without needing to cut tiles, as is sometimes necessary with larger-area tiles.

Therefore, if you select a large, often expensive, floor tile make sure that the tile layer knows how to lay the chosen tile with the minimum of unsightly cuts.

An alternative to using manufactured tiles such as ceramic, glazed, glass, porcelain and encaustic, are the natural materials, particularly polished stone. These may require a tiler with specific experience in the material to be used, if high-quality results are to be achieved. Take care with stone products, as even very dense granite may not be waterproof—it may not be damaged by water, but it may simply be too porous.

Stones such as marble and limestone are particularly beautiful, but are also vulnerable to reaction with some chemicals in cement and adhesives. Slate can also be used, but the cheaper varieties of this material need to be sealed, as some of them will dissolve in water.

## Fittings

### The bath

Modern baths are either manufactured from steel sheet pressed to shape or moulded in reinforced plastic. Locally made cast-iron baths were withdrawn from the market in 1982. However, for those people who still believe that nothing matches a good cast-iron bath then there are some imported models available, but they are priced in the super-luxury bracket.

Choose the style and colour of your bath carefully, for once the bath is built into the bathroom it is a major job to change it. Remember that the lilac baths that were all the rage in the late 1960s are looking dated today. There are three standard bath sizes 1675 x 750 mm, 1500 x 750 mm and 900 x 900 mm, the last being the shower bath.

### The lavatory pan

All lavatory pans are manufactured in vitreous china and come in a wide range of styles and colours. The flushing cistern may also be in vitreous china but is more likely to be moulded plastic. Cisterns are either exposed in the lavatory space or concealed in the wall. If you wish to have an in-wall type then check carefully to make sure it can be fitted, as there is considerable pre-planning necessary to conceal a cistern inside a brick wall. Some timber-framed walls are too thin to contain the cistern. Generally the pan seat will be moulded in plastic and some models combine the cistern and seat to give a smooth, integrated appearance.

### The shower

The shower is made up of two parts: the base and the screen. Today, most shower bases are built up on-site using the same tiles as used elsewhere on the bathroom floor, but moulded plastic, cast terrazzo and stainless steel shower bases are available. Shower screens are normally fabricated using an aluminium frame and glass or plastic panels.

Watch for hard-to-clean grooves in some aluminium mouldings and check that the glass supplied is of a safety type required by law. Toughened glass frame-less screens can be purchased and, although these are premium products, they can visually enhance a bathroom.

### The taps and spouts

Good taps are made from solid brass, or so the story goes. The trouble, is solid brass taps are expensive and out of the financial reach of many people. Plastics have taken over a large number of the parts used in taps today and, although not ultimately as

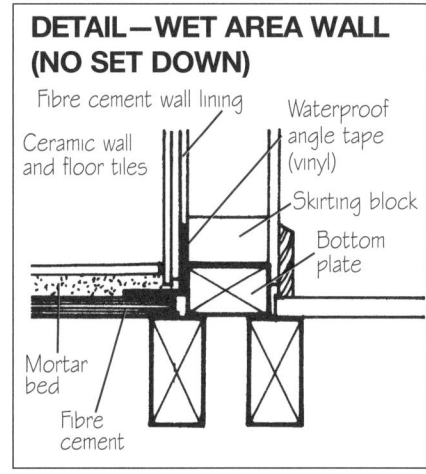

**DETAIL—WET AREA WALL (NO SET DOWN)**

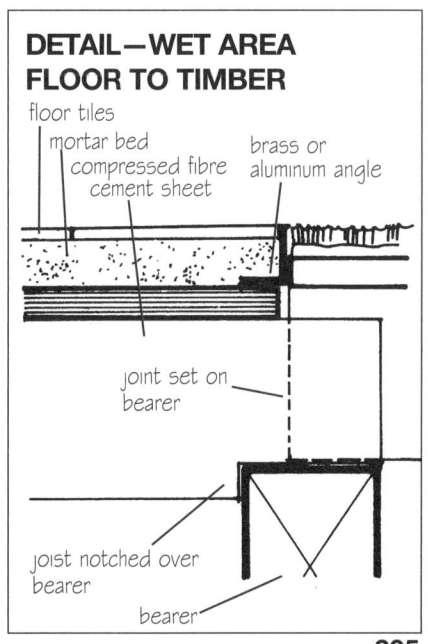

**DETAIL—WET AREA FLOOR TO TIMBER**

205

# 61 • BATHROOMS

durable as brass, at least it can be said that taps are a fashion item and most people will replace them before they wear out, whether brass or plastic.

Bathrooms, en-suite and lavatories are often the most expensive rooms in the house on a dollar-per-square-metre basis and, once completed, are expensive to modify. Therefore, choose the fixed sanitary fittings such as the bath, vanity and lavatory pan carefully.

Modern single-lever ceramic washer tapware should be considered—but remember to specify early to avoid incorrect numbers of holes being provided in basins, or the incorrect in-wall valve body being installed for the shower.

### The basin
Most basins are manufactured in vitreous china. However, some units are being produced in moulded plastic and there are still some porcelain enamel steel basins made. Basins can be used as an individual unit or combined with a bench and storage unit to become a vanity bar. Some vanity bars have the benchtop and the basin moulded in a continuous piece of plastic, thereby reducing the biggest problem of a vanity bar—the joint between the basin and the benchtop. Semi-recessed basins should be considered, as they provide the advantage of being built into a vanity unit but avoid excessively bulky cabinets and are much easier to clean and seal than fully recessed or rim-fitted vanity basins

### The bidet
This fitting is not common in Australia, although considered essential in many parts of western Europe. The installation of some types of bidet requires the provision of a separate hot water service and a disconnected cold water supply. This is justified on the grounds that there is a health risk that waste in the bidet could be siphoned back into the drinking water reticulation system. The cost of installation has effectively kept the bidet out of many Australian bathrooms.

### The spa bath
The spa bath or hot tub is used in many modern bathrooms. They can change the size of the normal bathroom because of their size: small spas are often 1800 x 800 mm and may be as big as 3000 x 2000 mm. There are many different types of spa on the market, some good and some downright poor. It is important not to purchase from brochures. Make sure you see the product in operation.

Some points to watch: if the spa holds a large quantity of water and it doesn't have a reheat facility, your hot water system may not be able to supply sufficient hot water for the spa and the rest of the house as well. Some spas have the spa pump motor built-in with the spa; ask whether the spa you like has easy access to any mechanical part for inevitable maintenance. Remember that some spas hold a large quantity of water; make sure the floor you place it on can support the weight (a spa 2000 x 1200 x 750 mm could have over 2000 kilograms of water in it when full—that's two tonnes).

### The wall cabinet
Wall cabinets and mirrors are an important part of any bathroom and may be purchased 'off the peg' in either plastic, steel or timber frames. If the cabinet is to be recessed into the wall, remember that it can only be recessed the thickness of the wall. In solid brick construction the back of the cabinet becomes the wall on the other side; this can cause problems if a wall is to be rendered.

### The final checklist
Bathrooms, including shower rooms, ensuites, powder rooms and toilets, compete with kitchens as the most expensive room per-unit area—bathrooms often win.

Before designing a new bathroom, sit in an existing bathroom and contemplate your real needs. Make a checklist to determine some priorities.

- Does the main bedroom require an ensuite bathroom?
- Do any other bedrooms need ensuite bathrooms?
- Is a bath needed in the house?
- Should the bath be a spa?
- At the peak ablutions period of the day, how many showers are required?
- Can the bath be used as a shower?
- How many toilet suites are required?
- Can these be located in bathrooms?
- Is there a need for a single-cell toilet?
- Should it have a basin and double as a guest toilet?
- Does anyone in the family want a bidet?
- Do all the bathrooms have to have windows?
- Can some bathrooms make do with roof lights?
- What storage is needed in the bathrooms?
- What lighting is needed in the bathrooms?
- What heating or cooling is needed in the bathrooms?
- How high are the tiles to be on walls?
- Are heated towel rails needed?

## BATHROOMS WITH SPACE

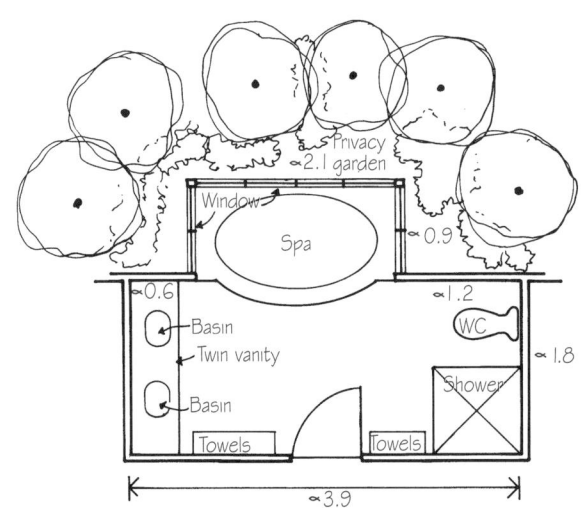

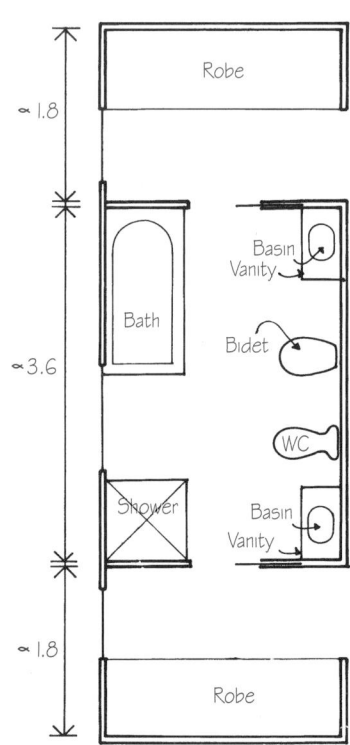

# BATHROOMS • 61

**SOME LAYOUTS TO CONSIDER**

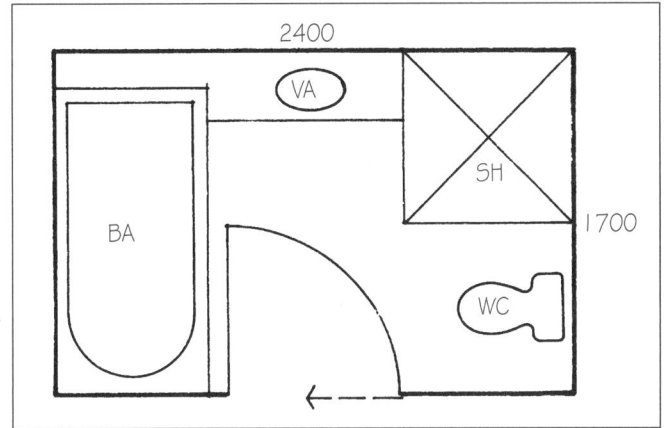
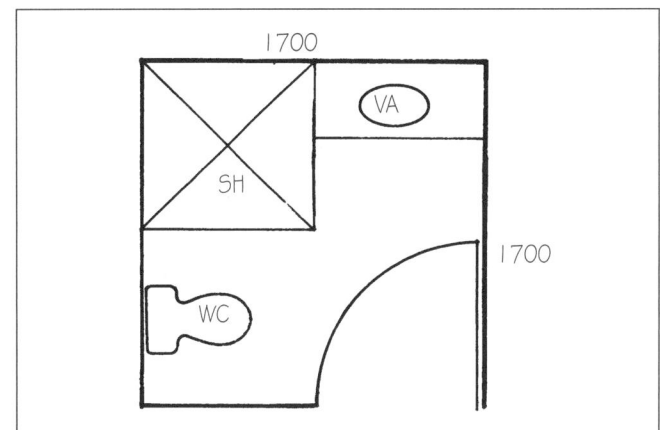
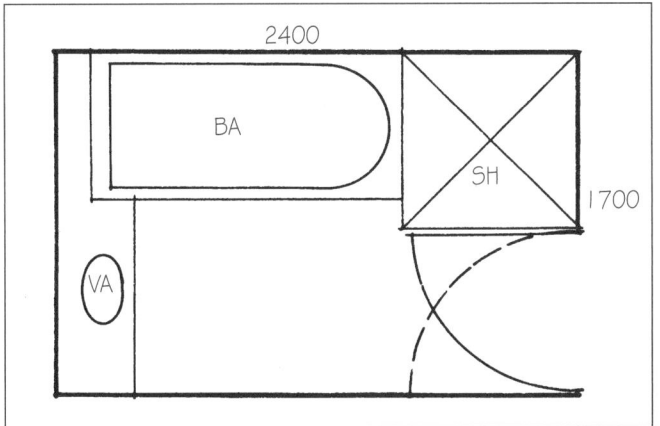
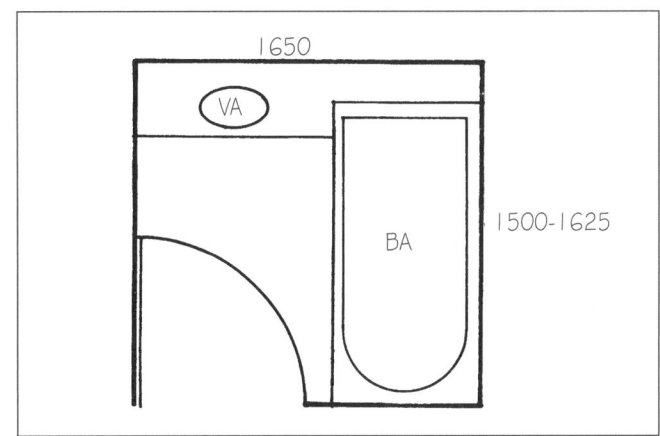
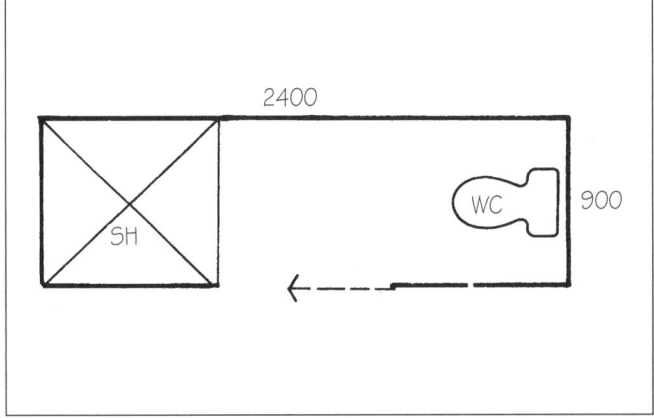
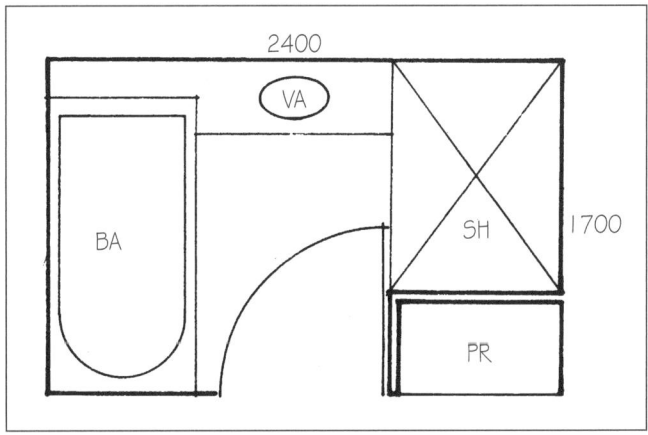
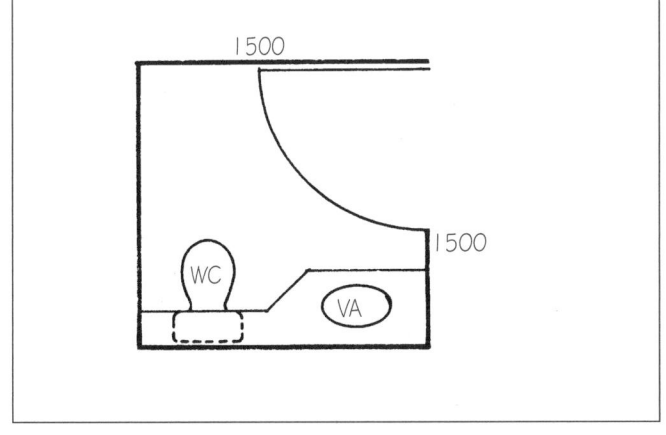
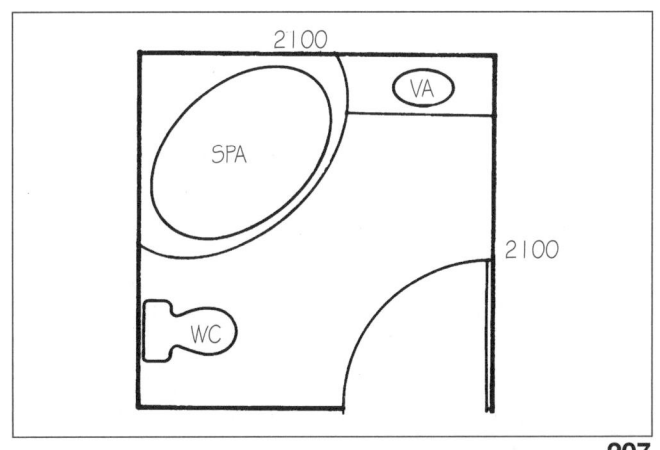

# 61 • BATHROOMS

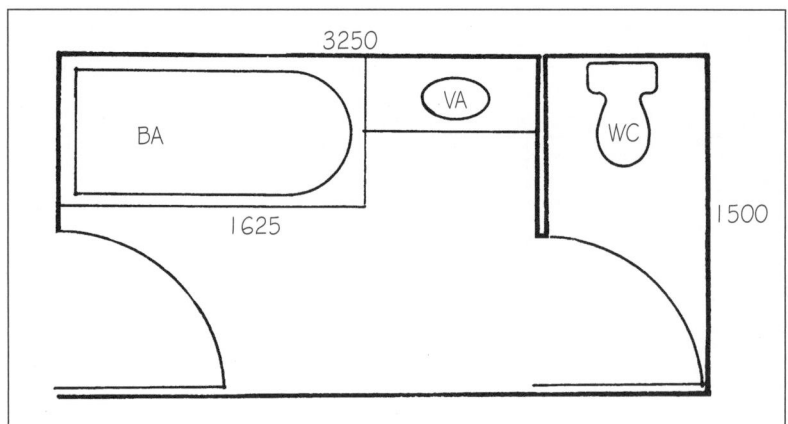

WC—Water closet
VA—Vanity basin
BA—Bath
SH—Shower
PR—Linen press

Note:
Dimensions shown are recommended minimum clear dimensions.

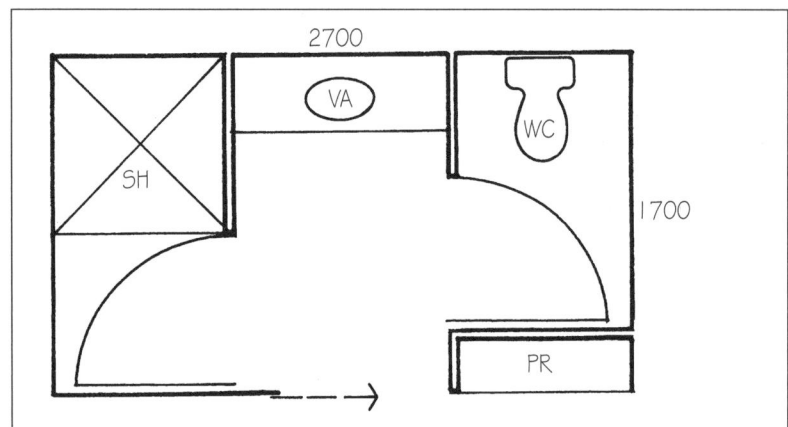

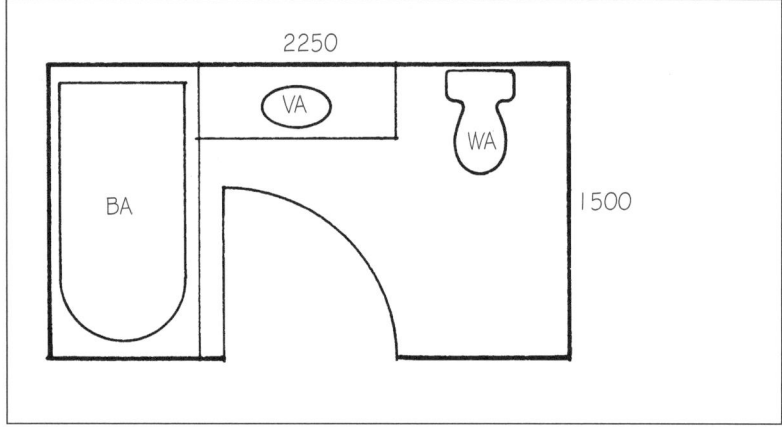

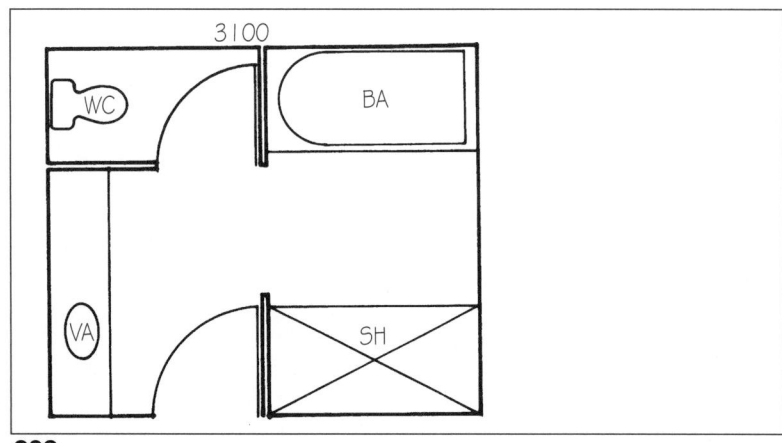

# BATHROOMS • 61

**LAYING OUT FLOOR TILES**

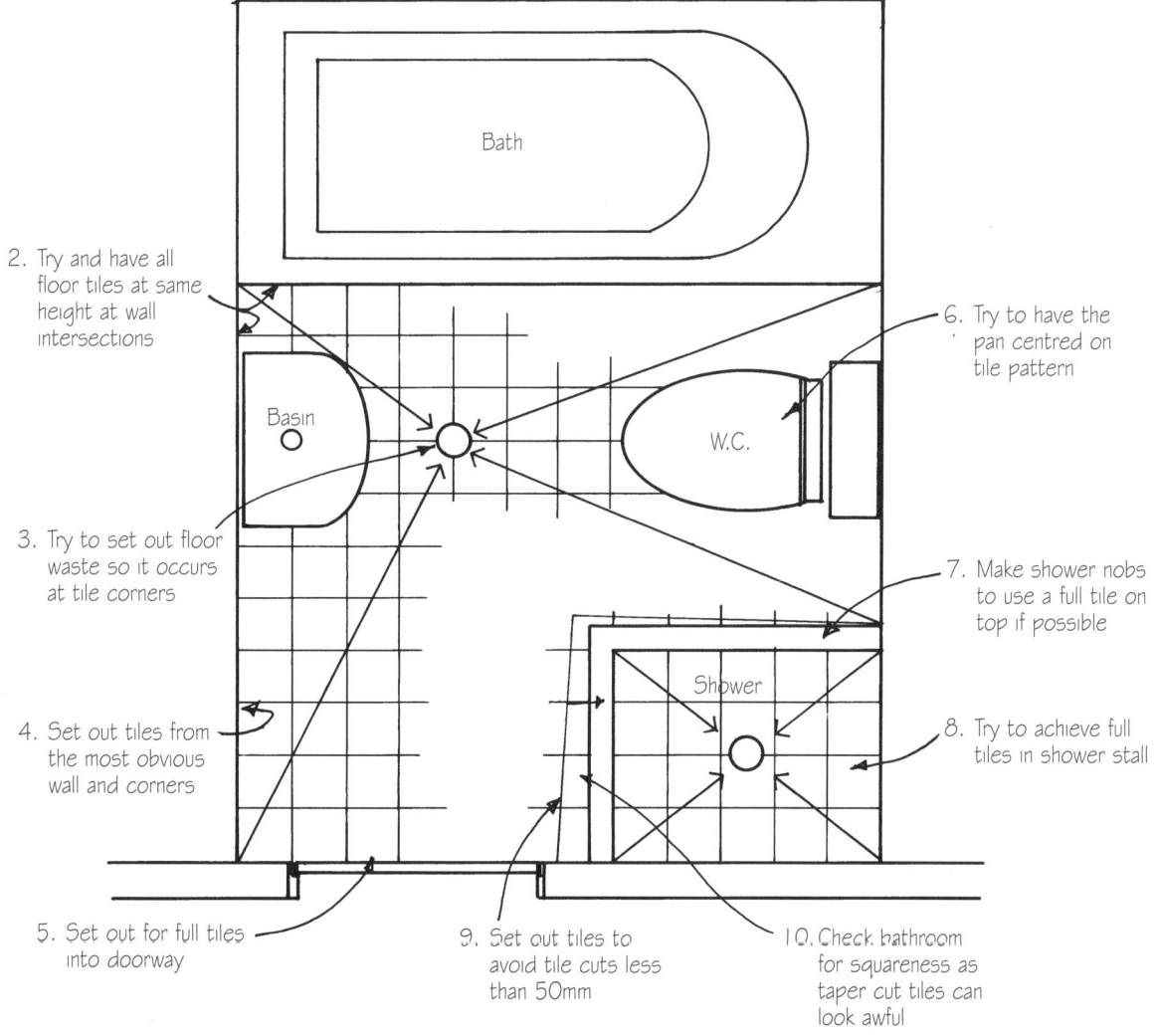

When laying out tiling for a bathroom floor consider these factors:

1. Smaller tiles are easier to fit with sanitary pipes and floor falls than larger tiles
2. Try and have all floor tiles at same height at wall intersections
3. Try to set out floor waste so it occurs at tile corners
4. Set out tiles from the most obvious wall and corners
5. Set out for full tiles into doorway
6. Try to have the pan centred on tile pattern
7. Make shower nobs to use a full tile on top if possible
8. Try to achieve full tiles in shower stall
9. Set out tiles to avoid tile cuts less than 50mm
10. Check bathroom for squareness as taper cut tiles can look awful

# 62 • ROBES AND PRESSES

The provision of robes, presses and other built-in cupboards is considered essential to the average Australian household. Over the years the methods used to build these units have altered significantly, until today they have been reduced to their bones.

The most economic method of building any of these cupboards only requires the provision of a space between two parallel walls, which is then closed off by 2400 mm high metal-framed, hardboard or mirror-panelled sliding doors. These doors are factory assembled and supplied to the site ready to install.

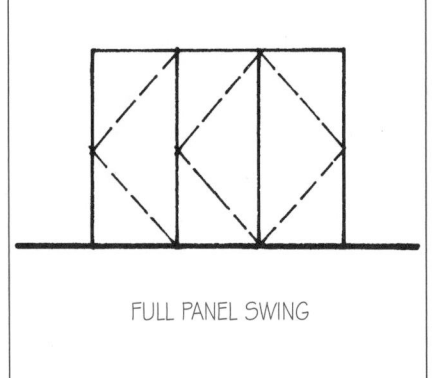

FULL PANEL SWING

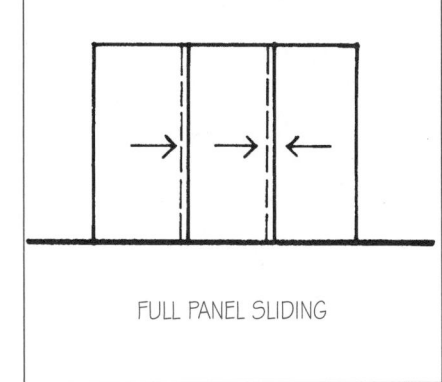

FULL PANEL SLIDING

### Robes

A built-in robe s normally 600mm deep including the thickness of the enclosing doors. A shelf is installed at approximately 1700mm above the floor, which is normally at the same level as the adjacent bedroom and carpeted to match.

Shelving is best constructed in 18mm melamine-faced chipboard, and supported with full-depth vertical panel at approximately 900 centres. Plastic covered or chromium plated 18mm diameter hanging rails give good service.

Any drawer units are best supplied for the wide range of whitewood or melamine covered chipboard units that are available very cheaply from whitewood stores. Specially-made joinery units nearly always cost more and offer no specific benefit.

Doors should, wherever possible, be chosen from standard size leaf widths, 620, 670, 720, 770, 820mm and of either 2030 or 2330mm in height. They can be either simple flush panel units or panelled units or panelled doors. If panelled, a saving can sometimes be achieved if they are ordered with flush backs.

### Linen presses

Linen presses for most families need to be a minimum of two doors wide and with a depth of around 450mm. Shelves should include a false floor about 100mm above the floor and then shelves spaced between 300 and 400mm apart.

### Broom cupboards

A broom cupboard of either one or two doors wide satisfies most family requirements. The bottom shelf should be approximately 1500mm above the floor and other shelves should be 300 and 400mm apart. A depth of 600mm is ideal for a broom cupboard, but units as shallow as 400mm are still useful.
Some families find that the broom cupboard is a good location for the hot water service. If you wish to do this, check that you can do this in your locality and that the hot water service you want can be located in this way.

### Cloak cupboard

Many houses are installing cloak cupboards in or near the front entry. These cupboards can be as narrow as a single door, but should be 600mm deep wherever possible.

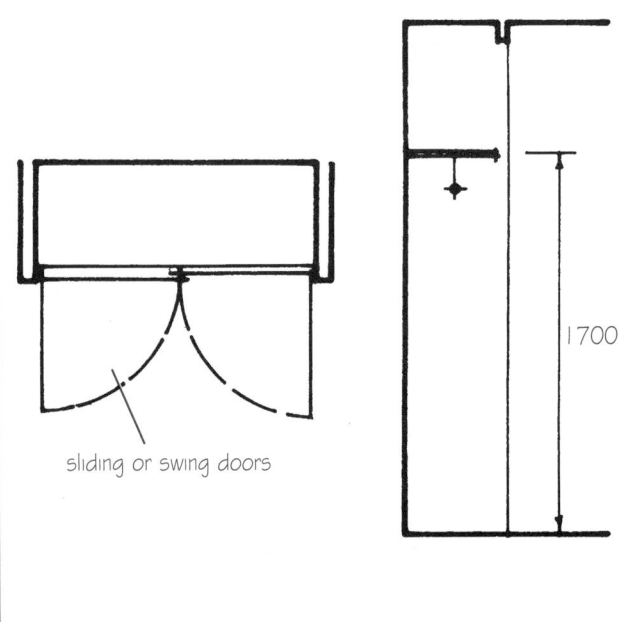

sliding or swing doors

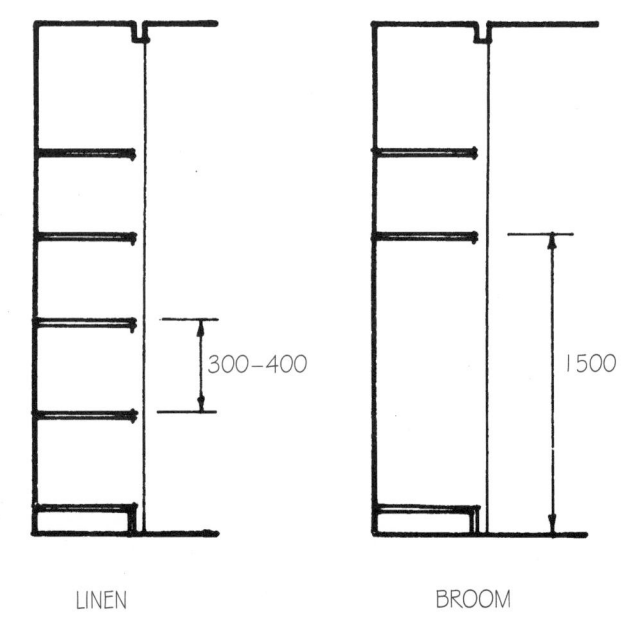

LINEN          BROOM

# ROBES AND PRESSES • 62

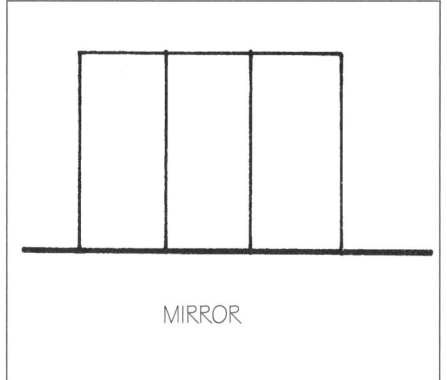
MIRROR

PANEL

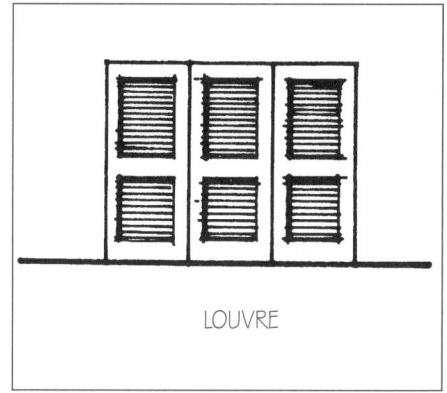
LOUVRE

## Pantries

The traditional pantry has made a major return to the modern kitchen. The most important factor to remember when building a pantry is the need to keep vermin out, but to allow a good flow of ventilating air.

Care must be taken to line the cupboard with a material that will not leave a bad taste in the food stored in it. Melamine-coated chipboard seems to be the currently favoured product. It is easy to clean and relatively odour free.

The door may be either hinged or sliding. A sliding door allows access more easily because the door can be left open when the family cook is busy preparing food in the kitchen. Refrigerators and/or freezers should never be placed in pantries as they generate as much heat as they make cold and therefore the pantry becomes very hot and food can be spoiled.

Allow some deep shelves in the pantry of about 450mm deep to store bulky items, but generally shelves of between 350 and 250mm are the best as they avoid some items being hidden by those in front.

## Walk-in closets

Many homes incorporate walk-in closets in master bedroom suites, and these need to be about 1800mm wide to give efficient three-wall hanging and other storage. The door to these closets is generally best hung to swing outward, or they can use swinging doors.

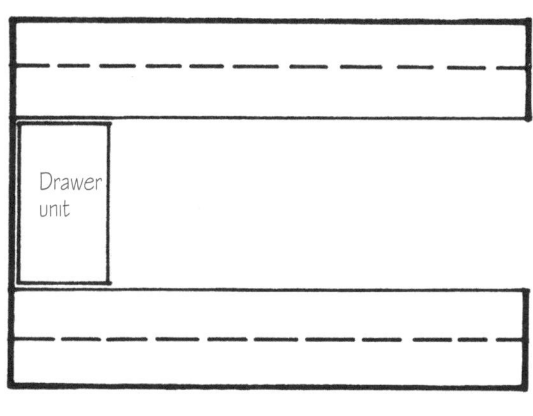

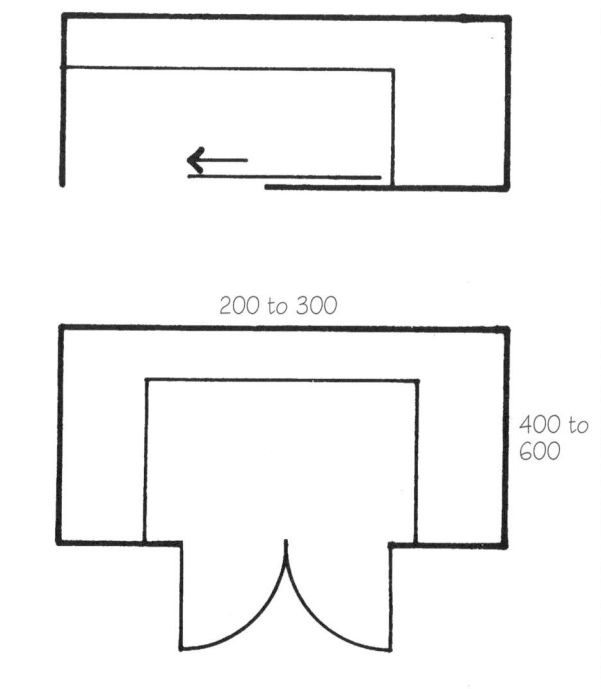

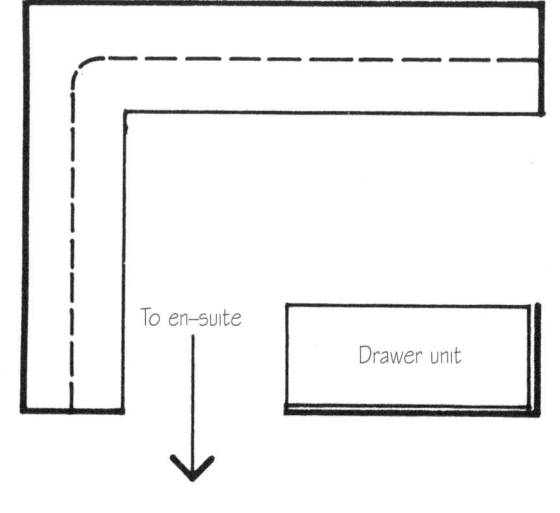

# 62 • ROBES AND PRESSES

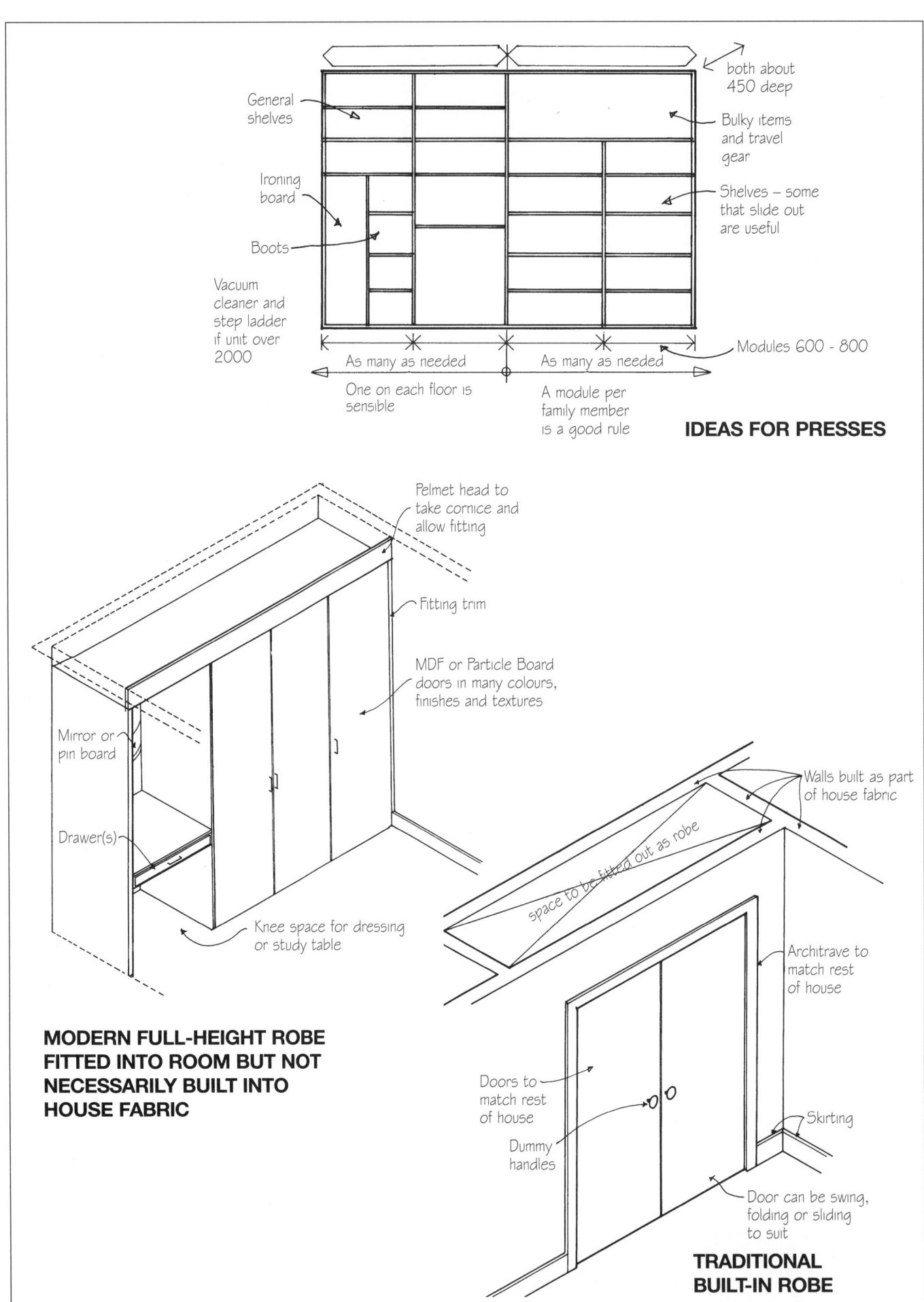

**IDEAS FOR PRESSES**

**MODERN FULL-HEIGHT ROBE FITTED INTO ROOM BUT NOT NECESSARILY BUILT INTO HOUSE FABRIC**

**TRADITIONAL BUILT-IN ROBE**

# FIREPLACES • 63

Fireplaces are not an economical means of heating. Tests indicate that, as ordinarily constructed, they are about one-third as efficient as a good stove or circulator heater.

However, a well-designed, properly built fireplace can:
- provide additional heat
- provide all the heat necessary in mild climates
- enhance the appearance and comfort of the room
- burn as fuel certain combustible materials that otherwise might be wasted.

## Design

A fireplace should harmonise in detail and proportion with the room in which it is located, but safety and utility should not be sacrificed for appearance.

Location of the fireplace within a room will depend on the position of the existing chimney or the best position for safe construction for the proposed chimney. A fireplace should not be located near doors.

Fireplace openings are usually made 600–1800 mm wide. The kind of fuel to be burned can suggest a practical width.

Height of the opening can range from 600 mm for an opening 600 mm wide to 1000 mm for one that is 1800 mm wide. The higher the opening, the more chance of a smoky fireplace. (See table)

In general, the wider the opening, the greater the depth. A shallow opening throws out relatively more heat than a deep one, but holds smaller pieces of wood. You have the choice, therefore, between a deeper opening that holds larger, longer-burning logs and a shallower one that takes smaller pieces of wood, but throws out more heat. In small fireplaces, a depth of 300 mm may permit good draft, but a minimum depth of 400 mm is recommended to lessen the danger of brands falling out on the floor. Suitable screens should be placed in front of all fireplaces to minimise the danger from brands and sparks.

## Authorities and pollution

In some (often inner-city) locations, applications to construct new, solid-fuel open fireplaces will not be approved by the authorities. Restrictions may apply as to when existing fireplaces can be used, and to the volume and type of smoke emitted. The smoke from open fires is often considered a pollutant: it quickly builds up in the atmosphere and combines with other pollutants to create smog. If an OB must have naked-flame heat, there are solid-fuel, high-efficiency heaters that are almost smokeless and are a suitable alternative in restricted areas. Other options include gas log fires—not real, but often quite realistic.

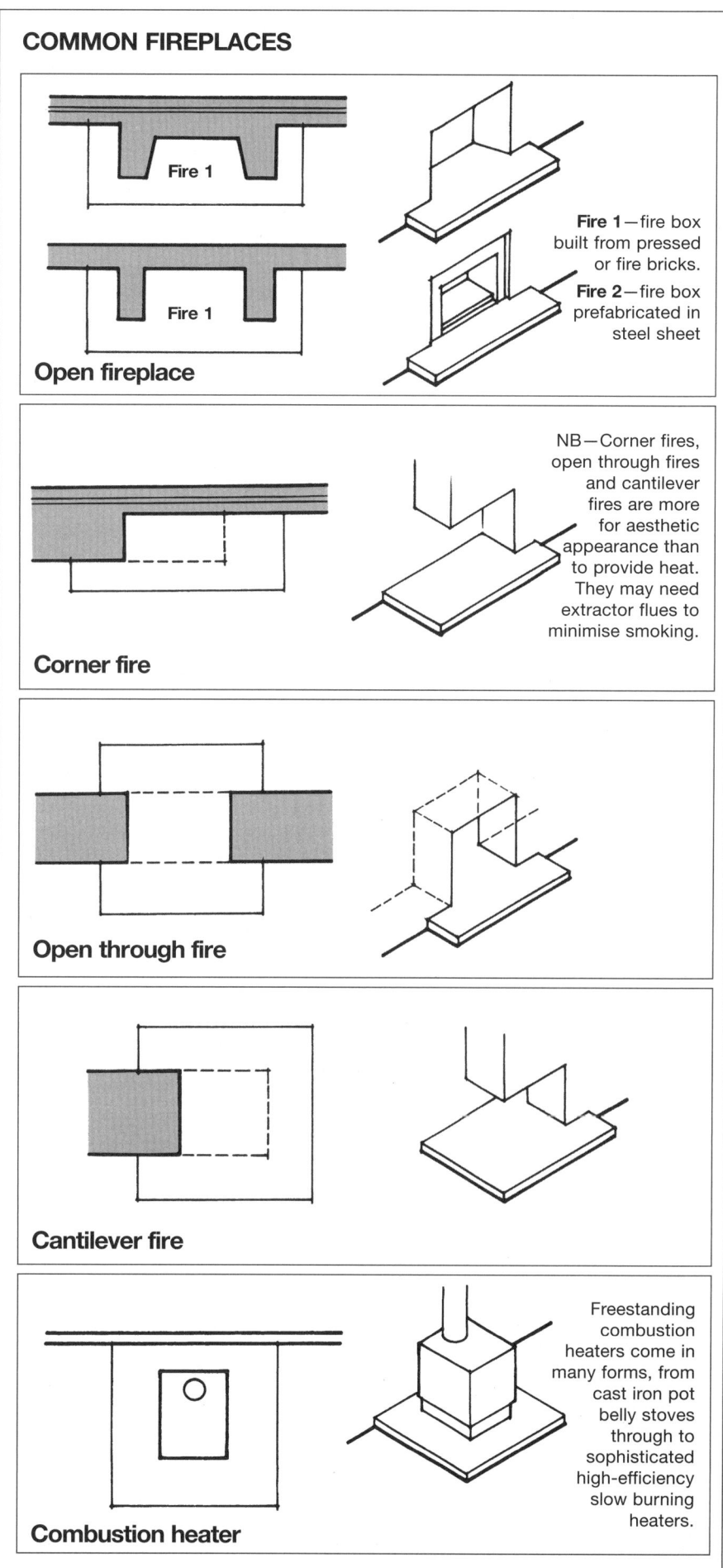

**COMMON FIREPLACES**

**Open fireplace**
Fire 1—fire box built from pressed or fire bricks.
Fire 2—fire box prefabricated in steel sheet

**Corner fire**
NB—Corner fires, open through fires and cantilever fires are more for aesthetic appearance than to provide heat. They may need extractor flues to minimise smoking.

**Open through fire**

**Cantilever fire**

**Combustion heater**
Freestanding combustion heaters come in many forms, from cast iron pot belly stoves through to sophisticated high-efficiency slow burning heaters.

# 63 • FIREPLACES

## OPEN FIREPLACE DETAILS

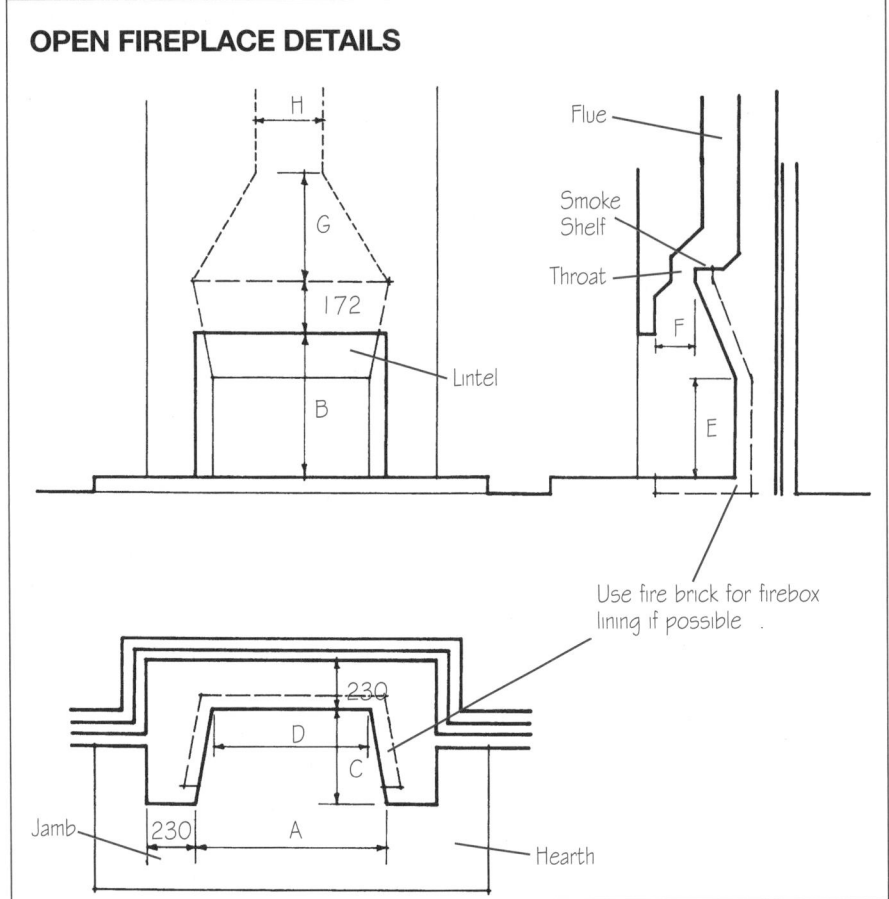

## Construction

Fireplace construction is basically the same regardless of design. The diagram in this section shows construction of a typical fireplace. The accompanying table recommends dimensions for essential parts of fireplaces of various sizes.

## Hearth

The fireplace hearth should be made of brick, stone, terracotta, or reinforced concrete at least 100 mm thick. It should project at least 550 mm from the chimney-breast and should be 600 mm wider than the fireplace opening.

The hearth can be flush with the floor so that sweepings can be brushed into the fireplace or it can be raised. Raising the hearth to various levels and extending in length as desired is presently common practice, especially in contemporary design.

In buildings with timber floors, the hearth in front of the fireplace should be supported by masonry trimmer arches or other fire-resistant construction.

## Walls

Some building codes generally require that the back and sides of fireplaces be constructed of solid masonry or reinforced concrete at least 200 mm thick and be lined with firebrick or material not less than 50 mm thick, or steel lining.

## Jambs

The jambs of the fireplace should be wide enough to provide stability and to present a pleasing appearance.

For a fireplace opening 900 mm wide or less, the jambs can be 300 mm wide if a wood mantle will be used or 400 mm wide if they will be of exposed masonry. For wider fireplace openings, or if the fireplace is in a large room, the jambs should be proportionately wider.

Fireplace jambs are frequently faced with ornamental brick or tile.

No woodwork should be placed within 150 mm of the fireplace opening. Woodwork above and projecting more than 50 mm from the fireplace opening should be placed not less than 300 mm from the top of the fireplace opening.

## Lintel

A lintel must be installed across the top of the fireplace opening to support the masonry.

For fireplace openings 1200 mm wide or less, 12 x 75 mm flat steel bars, 100 x 75 x 6 mm angle irons, or specially designed damper frames may be used. Wider openings will require heavier lintels.

If a masonry arch is used over the opening, the fireplace jambs must be heavy enough to resist the thrust of the arch.

## Throat

Proper construction of the throat area is essential for a satisfactory fireplace.

The sides of the fireplace must be vertical up to the throat, which should be 150–200 mm or more above the bottom of the lintel.

The area of the throat must not be less than that of the flue and the length must be equal to the width of the fireplace opening.

The sidewalls should start sloping inward 125 mm above the throat to meet the flue.

## Damper

A damper consists of a cast-iron frame with a hinged lid that opens or closes to vary the throat opening.

Dampers are not always installed but they are definitely recommended, especially in colder climates.

With a well-designed, properly installed damper, you can:
- regulate the draft
- close the flue to prevent loss of heat from the room when there is no fire in the fireplace
- adjust the throat opening according to the type of fire to reduce loss of heat. For example, a roaring pine fire may require a

| Acceptable dimensions to achieve a good drawing fire | | | | | | | | |
|---|---|---|---|---|---|---|---|---|
| | | | | Vertical back | Throat depth | Smoke chamber | Flue size | |
| Width A | Height B | Depth C | Back D | E | F | G | Square | Circular H |
| 800 | 740 | 400 | 480 | 360 | 220 | 600 | 300 x 300 | 250 |
| 900 | 740 | 400 | 580 | 360 | 220 | 680 | 300 x 300 | 250 |
| 1000 | 740 | 400 | 680 | 360 | 220 | 730 | 300 x 300 | 250 |
| 1200 | 800 | 450 | 830 | 360 | 220 | 940 | 400 x 400 | 300 |
| 1500 | 940 | 560 | 1070 | 400 | 330 | 1140 | 400 x 500 | 380 |
| 1800 | 1000 | 560 | 1370 | 400 | 330 | 1420 | 500 x 500 | 500 |

# FIREPLACES • 63

## CHIMNEY FLUES—DETAIL INFORMATION

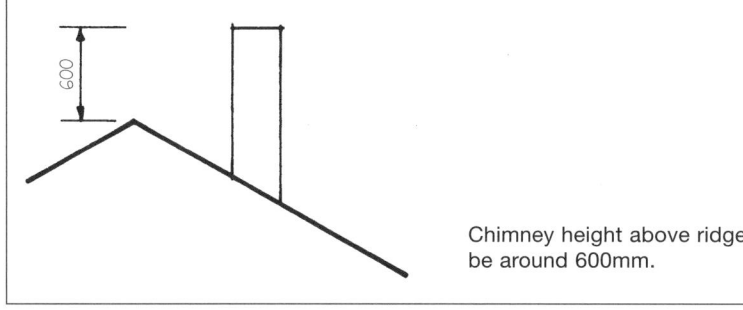

Chimney height above ridge should be around 600mm.

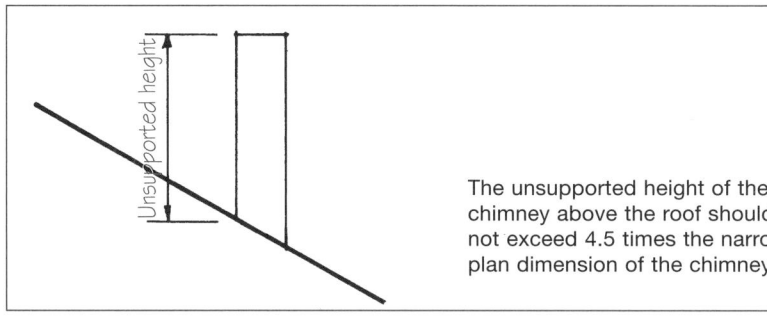

The unsupported height of the chimney above the roof should not exceed 4.5 times the narrowest plan dimension of the chimney.

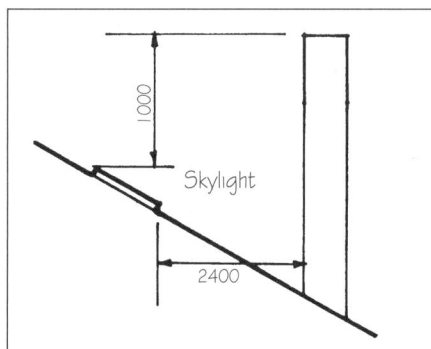

Chimney flues should not be lower than 1000mm above a roof opening. Chimney flues and roof opening must be separated 2400mm.

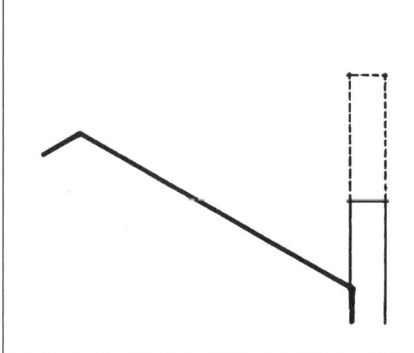

Where practical it is advisable to check the drawing capacity, as the chimney is erected, to achieve the best possible draw.

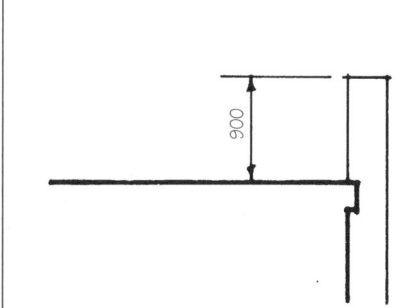

Where practical the chimney for a flat roof house should be at least 900mm above the roof.

full throat opening, but a slow-burning hardwood log fire may require an opening of only 50 mm. Closing the damper to that opening will reduce loss of heat up the chimney.
- close or partially close the flue to prevent loss of heat from the main heating system. When air heated by a furnace goes up a chimney, an excessive amount of fuel may be wasted.
- close the flue in summer to prevent insects from entering the house through the chimney.

Dampers of various designs are available. Some are designed to support the masonry over fireplace openings, thus replacing ordinary lintels.

Responsible manufacturers of fireplace equipment usually offer assistance in selecting a suitable damper for a given size of fireplace. It is important that the full damper opening equal the area of the flue.

## Smoke shelf and chamber

A smoke shelf prevents downdraft. It is made by setting the brickwork at the top of the throat back to the line of the flue wall for the full length of the throat. the depth of the shelf may be 150–300 mm or more.

The smoke chamber is the area from the top of the throat to the bottom of the flue. As indicated, the sidewalls should slope inward to meet the flue.

## Flue

Proper proportions between the size (area) of the fireplace opening, the size (area) of the flue, and the height of the flue are essential for satisfactory operation of the fireplace.

The area of a lined flue 6 metres high should be at least one-twelfth of the area of the fireplace opening. The area of an unlined flue or a flue less than 6 metres high should be one-tenth of the area of the fireplace opening.

## Manufactured fireboxes

Modified fireplaces are manufactured fireplace units, made of heavy metal and designed to be set in place and concealed with the usual brickwork or other construction. They contain all the essential fireplace smoke shelf and chamber. In the completed installation, only grilles show.

Modified fireplaces offer two advantages:
- the correctly designed and proportioned firebox provides a ready-made form for the masonry, which reduces the chance of faulty construction and assures a smokeless fireplace
- when properly installed, the better designed units heat more efficiently than ordinary fireplaces. They circulate heat into the cold corners of rooms and can deliver heated air through ducts to upper or adjoining rooms.

# 63 • FIREPLACES

The use of a modified fireplace unit can increase the cost of a fireplace (although manufacturers claim that labour, materials, and fuel saved offset any additional cost). It should not be necessary to use one merely to ensure an attractive, well-proportioned fireplace; you can create an equally attractive and satisfactory masonry fireplace through careful construction.

Even a well-designed modified fireplace unit will not operate properly if the chimney is inadequate. Therefore, proper chimney construction is as important for these units as it is for ordinary fireplaces.

## Prefabrication fireplaces

Prefabricated fireplace and chimney units—all parts needed for a complete fireplace-to-chimney installation—are on the market.

Such units offer these features:
- a wide selection of styles, shapes, and colours
- pre-tested designs that are highly efficient in operation
- easy and versatile installation—can be installed free-standing or flush against a wall in practically any part of the house
- a light weight
- a lower cost than comparable masonry units.

The basic part of the prefabricated fireplace is a specially insulated metal firebox shell. Since it is light in weight, it can be set directly on the floor without the heavy footing required for masonry fireplaces.

## Chimneys and flues

All fireplaces and fuel-burning equipment such as stoves and furnaces require some type of chimney. The chimney must be designed and built so that it produces sufficient draft to supply an adequate quantity of fresh air to the fire and to expel smoke and gases emitted by the fire or equipment.

A chimney located entirely inside a building has better draft than an exterior chimney, because the masonry retains heat longer when protected from cold outside air.

## Flue size

The flue is the passage in the chimney through which the air, gases and smoke travel.

Proper construction of the flue is important. Its size (area), height, shape, tightness, and smoothness determine the effectiveness of the chimney in producing adequate draft and in expelling smoke and gases. Soundness of the flue walls may determine the safety of the building should a fire occur in the chimney. Overheated or defective flues can be a cause of house fires.

Manufacturers of fuel-burning equipment usually specify chimney requirements, including flue dimensions, for their equipment. Follow their recommendations.

## Height

A chimney should extend at least 900 mm above flat roofs and at least 600 mm above a roof ridge or raised part of a roof within 3 metres of the chimney. A hood should be provided if a chimney cannot be built high enough above a ridge to prevent trouble from eddies caused by wind being deflected from the roof. The open ends of the hood should be parallel to the ridge.

Metal extensions must be securely anchored against the wind and must have the same cross-sectional area as the flue. They are available with a metal cowl or top that turns with the wind to prevent air from blowing down the flue.

## Support

The chimney is usually the heaviest part of a building and must rest on a solid foundation to prevent differential settlement in the building.

Concrete footings are recommended. They must be designed to distribute the load over an area wide enough to avoid exceeding the safe load-bearing capacity of the soil. They should extend at least 150 mm beyond the chimney on all sides and should be 200 mm thick.

If the house wall is of masonry at least 300 mm (cavity wall) thick, the chimney can be built integrally with the wall and, instead of being carried down to the ground, it can be offset from the wall enough to provide flue space by corbelling.

Chimneys in frame buildings should be built from the ground up, or they can rest on the footings.

## Mortar

Brickwork around chimney flues and fireplaces should be laid with cement mortar; it is more resistant to the action of heat and flue gases than lime mortar.

## Insulation

No wood should be in contact with the chimney. Leave a 50 mm space between the chimney walls and all wooden beams or joists (unless the walls are of solid masonry 200 mm thick, in which case the framing can be within 15 mm of the chimney masonry).

Fill the space between wall and floor framing with porous, non-metallic, incombustible material, such as loose cinders. Do not use brickwork, mortar or concrete. Place the filling before the floor is laid, because it not only forms a firestop but also prevents the accumulation of shavings or other combustible material.

Flooring and sub-flooring can be laid against masonry.

Wood studding, furring, or lathing should be set back at least 50 mm from chimney walls. (Plaster can be applied directly to the masonry or to metal lath laid over the masonry, but this is not recommended because settlement of the chimney may crack the plaster.) A coat of cement plaster should be applied to chimney walls that will be encased by wood partition or other combustible construction.

If baseboards are fastened to plaster that is in direct contact with the chimney wall, install a layer of fireproof material between the baseboard and the plaster.

# STAIRS AND ENTRY • 64

## Stairs

Australian houses have traditionally placed their staircases in close proximity to the front entry door - even when there was no particular logic for this. More recently the staircase has move into the living rooms, this could well have been influenced by TV sitcoms where the stair becomes a prop to allow entry and exit from a scene. There is no right place except that the staircase should be fully integrated into the house design as a functional element not just a feature.

First determine the stair that is required, if the floor to floor rise is about 3000 mm then 10 to 18 step rises will satisfy the Building Code of Australia (BCA).

This means the area of the stair measured on the plan, if a straight flight without landings is about 4000 mm x 1000 mm. A stair with a full return half landing will require a floor area of about 3000 mm x 2000 mm on both floors served plus the space to access the stair.

Allow a space of around 8 to 10 sqm, that is about the size of a small bedroom on each floor.

Stairways do provide a fine entry feature but they can be a trap as when visitors arrive there is little opportunity to move up and down in the house without being seen - this can be embarrassing if any member of the household is caught in a compromised state of un-dress.

If the stair is located in or close to the family room, there is generally better privacy but food odours from the kitchen and the sound of the TV and music can spread through the upstairs areas of the house.

Stairways are generally enhanced by having a window to provide natural light and a pleasant aspect but take care that any windows do not provide a view from the street or to the neighbours of an under-dressed family. Even obscured glass or stained glass can be amazingly transparent at night.

Generally the most comfortable and safest stairs are ones that have two straight flights separated by a quarter or half handing. Curved stairs can be elegant alternatives but they are difficult and expensive to construct, they must be carefully calculated to conform to the BCA and can be difficult for the young and infirm to negotiate. Spiral stairs in general do not meet the detailed requirements of the BCA and are very dangerous, particularly if they are used as the main vertical movement system in a busy house.

It is recommended only to consider using spiral stairs as a secondary stair or as access to an attic or similar low traffic destination, always check with the local authorities to ensure the stairs chosen satisfy their particular coded requirements.

Where space is available a full 180° return open newel stair is a very safe choice and provides a quality appearance, equally a stair with two quarter landings - particularly where there are ceiling heights above 3000 mm- can provide a highly suitable solution.

Simple closed newel half landing stairs work well if the stairs flights are about a metre wide. If there is insufficient room for a full half landing then consider using a stair where there is a quarter landing plus a 90° winder set of three treads—this is a reasonable compromise between safety, convenience and space allocation. A full 180° winder return is also satisfactory but is much less safe and is troublesome to less able bodied people.

The BCA rules for stair are regularly amended, always check what is allowed in a particular locality before proceeding to order any stairs. If winders approved satisfy local rules and they satisfy a particular projects requirements then at least attempt to keep the winder angles constant and not less than 30°.

## Entries

The colder the winters the more likely it is for a house to have a formal entry. If the entry has two sets of doors, one from the outside and then one to the interior of the house from an enclosed entry, this is considered to be a weather lock. These are critical in particularly cold zones but they also function well in high temperature, high humidity zones - particularly if the house has two storeys and hot humid air could entry the house then rise and displace cooler air from the top floor.

Most of Australia's capital cities are in climatic zones that offer fair comfort for at least half of the year but during winter and the height of summer ambient temperatures can be uncomfortable. Consider carefully the use of entries and stairs as part of an active internal temperature control system.

### Straight flights

The simplest form of stair is the straight flight. These are built with or without mid-flight landings.

Most localities restrict the number of rises between floors or landings to a maximum of 18. Always check with your local building approval authority or code. Most straight flight stairs are built against a wall. This reduces the impact that the stair flight makes on the room.

Mid-flight landings are traditionally the same width as the flight. Check locally – some regulations will allow lengths of mid-flight landing to be reduced to 700mm.

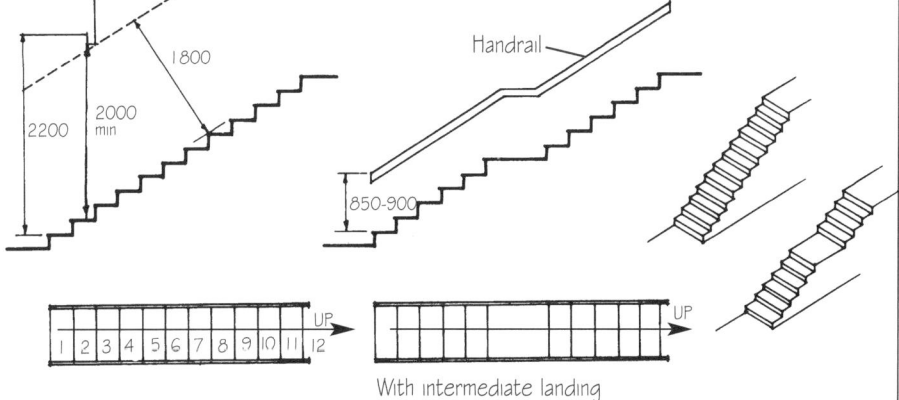

With intermediate landing

### Return flights

The return flight stair is the most common stair layout used in houses, flats and commercial buildings. It may be built with half, quarter or wider landings.

The drawings illustrate the half and the quarter landings; the wider landings can be seen in the illustrations for the 90 degree stair flights.

When a half-landing is used, and if a smooth balustrade transaction is required, then the risers at the half landing are normally off-set as is illustrated.

Return flights may be built free-standing, but are normally built against 1, 2, 3 or 4 walls, depending on the design requirements.

It is possible to combine quarter landings with return winders to achieve a required landing but to maintain a compact plan.

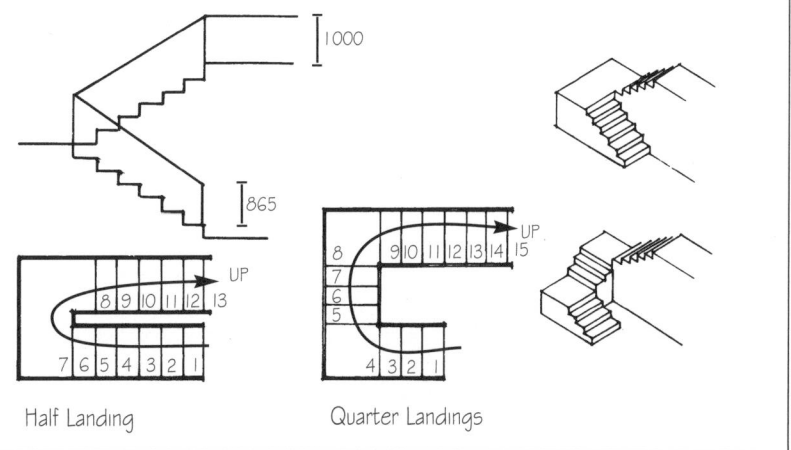

Half Landing   Quarter Landings

# 64 • STAIRS AND ENTRY

## 90 degrees flights

The 90 degrees or right angle flights are used extensively in smaller two storey cottages or town houses where the staircase is exposed in a living area of the house.

The 90 degrees flight is normally built against two walls where they meet at an internal or external 90 degrees corner. It is possible to change direction of the stair flight at a different angle than 90 degrees and 30 degrees, 60 degrees, 120 degrees and 150 degrees are all common.

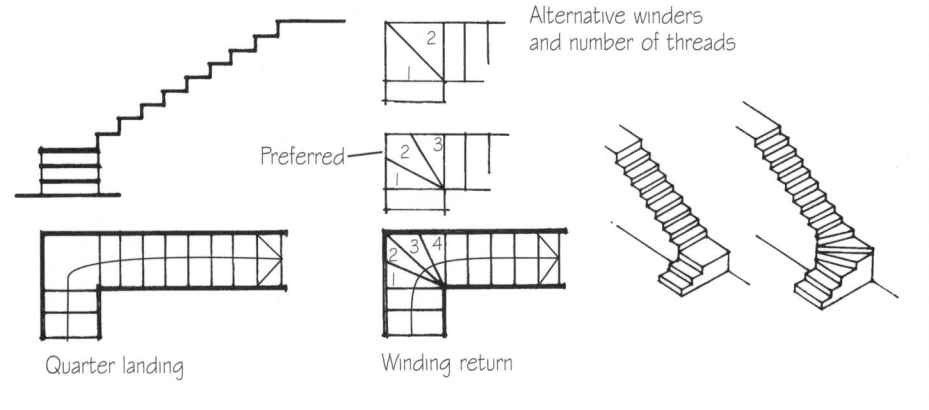

Alternative winders and number of threads

Preferred

Quarter landing

Winding return

## Circular/spiral flights

The true circular stair, which has its treads cantilevered off circular enclosing walls, is uncommon in residential buildings.

Much more common and very popular with the second floor conversion designers and builders is the circular stair which cantilevers its treads off a central post.

These central post stairs generally need less space than other stairs and can be made to be relatively, structurally self-sufficient.

A central post, circular or spiral, stair normally requires 12 or 16 treads per revolution, depending on the required use and local regulations.

Circular stairs are normally 1500 to 2100mm in diameter, the 1500 to 1750mm range have 12 treads per rotation, 1800 to 2100mm have 16 treads per rotation.

Not all building regulations allow spiral stairs. Check your local codes.

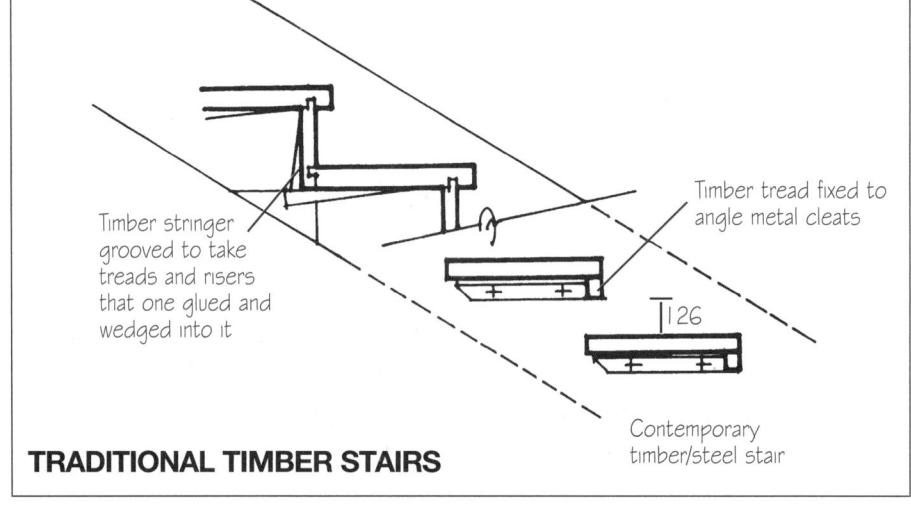

Timber stringer grooved to take treads and risers that one glued and wedged into it

Timber tread fixed to angle metal cleats

Contemporary timber/steel stair

### TRADITIONAL TIMBER STAIRS

### Balustrades and the BCA

Clause D2. 16. The Building Code of Australia now requires that all stairs have a balustrade 865mm above the nosing of any stair that is more than 1000mm above a floor, and that landings have balustrades that are 1000mm high. Note that a balustrade to a stair or a landing cannot have any opening through which a 125mm diameter sphere can pass.

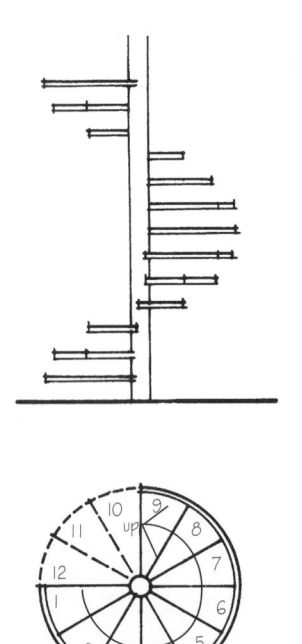

### REINFORCED CONCRETE STAIR

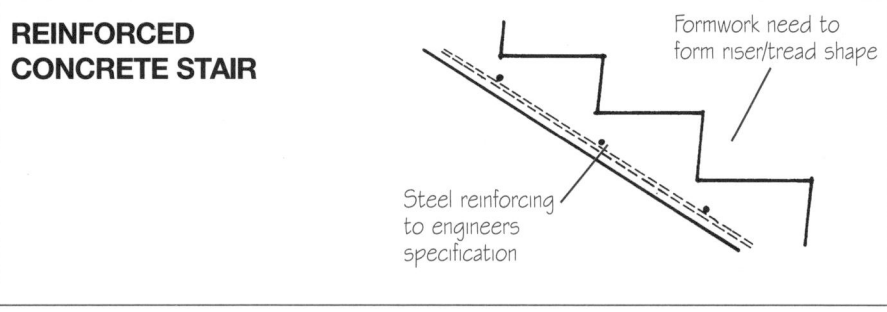

Formwork need to form riser/tread shape

Steel reinforcing to engineers specification

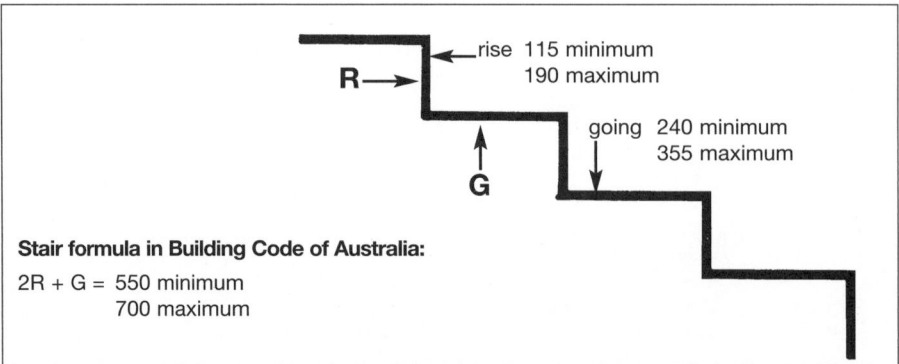

rise 115 minimum
190 maximum

going 240 minimum
355 maximum

**Stair formula in Building Code of Australia:**

2R + G = 550 minimum
700 maximum

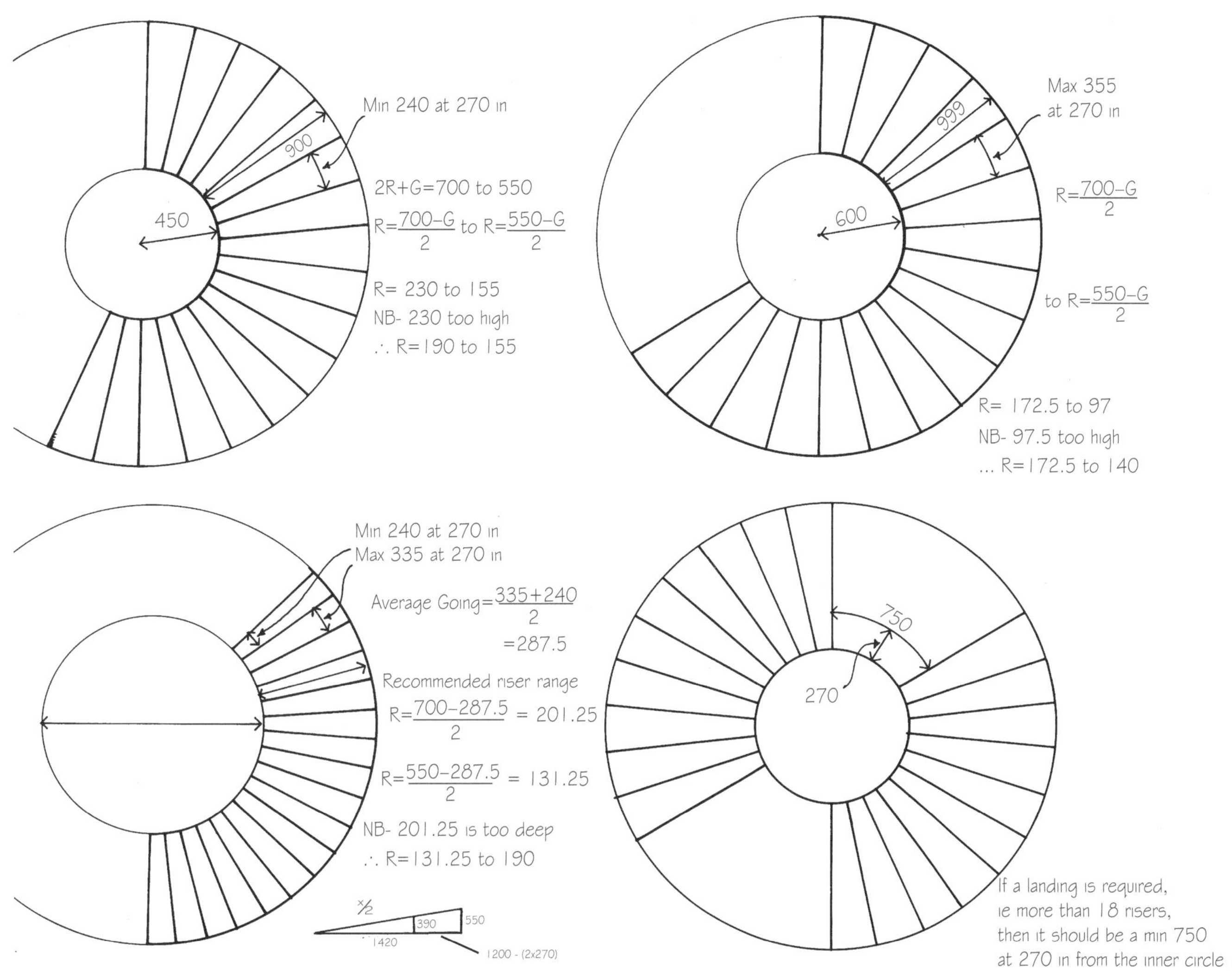

# STAIRS AND ENTRY • 64

## SOME WORKING EXAMPLES OF CIRCULAR STAIRS

Min 240 at 270 in
900
450
2R+G=700 to 550
$R = \frac{700-G}{2}$ to $R = \frac{550-G}{2}$
R= 230 to 155
NB- 230 too high
∴ R=190 to 155

Max 355 at 270 in
999
600
$R = \frac{700-G}{2}$
to $R = \frac{550-G}{2}$
R= 172.5 to 97
NB- 97.5 too high
∴ R=172.5 to 140

Min 240 at 270 in
Max 335 at 270 in
Average Going = $\frac{335+240}{2}$ = 287.5
Recommended riser range
$R = \frac{700-287.5}{2} = 201.25$
$R = \frac{550-287.5}{2} = 131.25$
NB- 201.25 is too deep
∴ R=131.25 to 190

750
270
If a landing is required, ie more than 18 risers, then it should be a min 750 at 270 in from the inner circle

½  390  550
1420   1200 - (2x270)

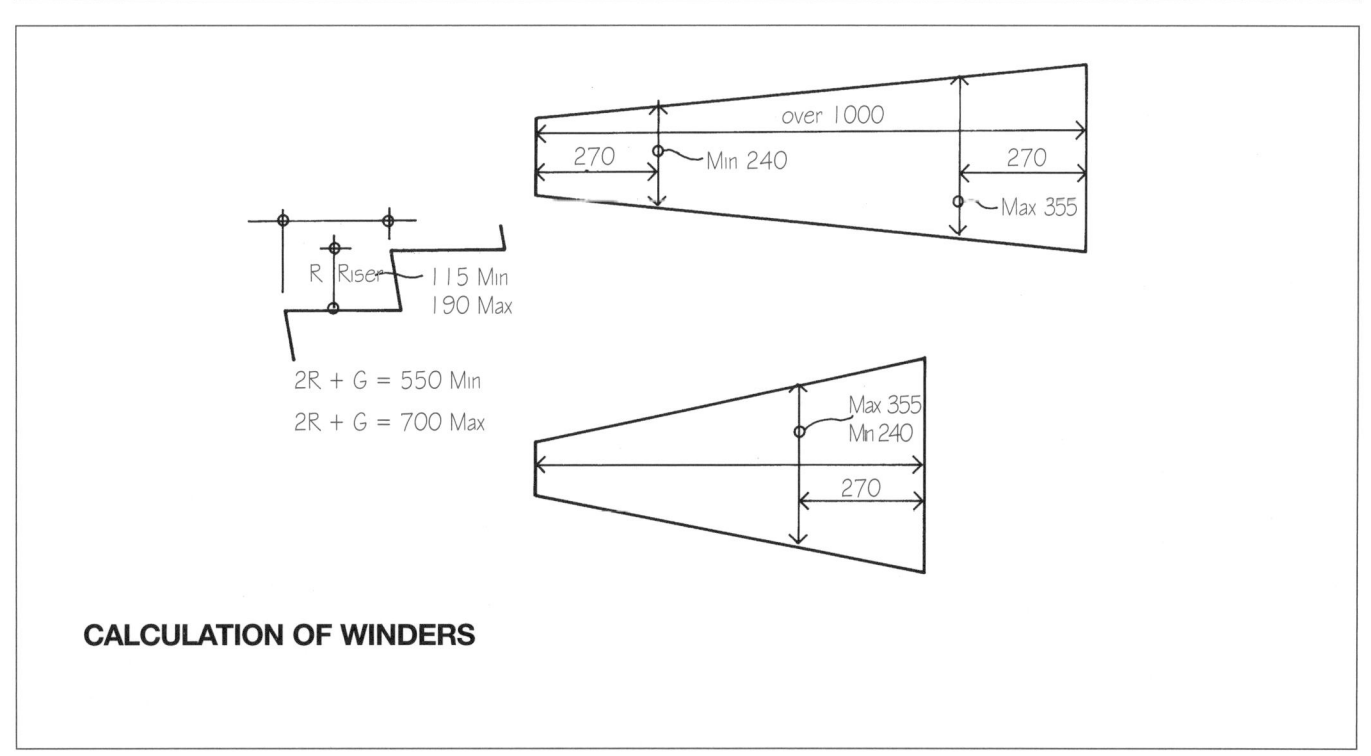

R Riser — 115 Min 190 Max
2R + G = 550 Min
2R + G = 700 Max

over 1000
270 — Min 240
270 — Max 355

Max 355
Min 240
270

## CALCULATION OF WINDERS

219

# 64 • STAIRS AND ENTRY

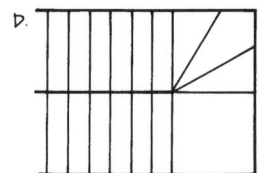
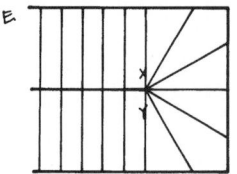
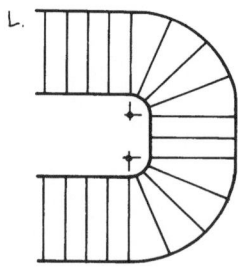
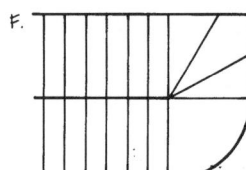
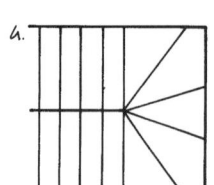
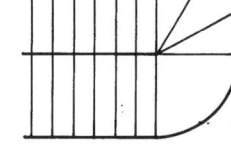
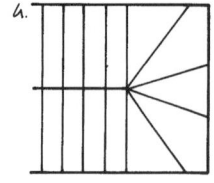
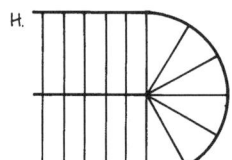
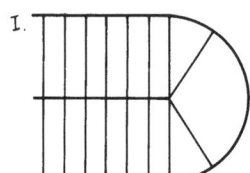
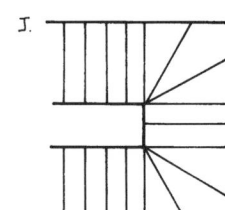
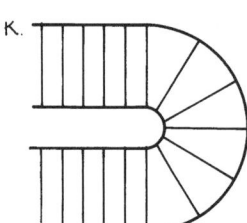

## SOME COMBINED WINDERS

D  Simple combination of quarters landing and three-kite winder set – compact, minimum complexity

E  The very compact six-kite full return winder set – use only if space is critical – vertical drop at post from X to Y in the order of 7 x 180 = 1260

F  Similar to D with added elegance of a quarter-circle landing

G  The five-kite winder set is considered by some to have a more elegant appearance than other winder options – is extremely complex to construct for often minimal gain

H  The six-kite semi-circular winder set – can be stylish if the curve can be expressed

I  The use of winders leading to and from a 120 degree landing - interesting , but can be confusing to casual users

J  Open newel stairs with two sets of three-kite winders

K  Open newel semi-circular six-winder set

L  Open newel stair with winders and three flights

### Stair Readi-Reckoner

| Floor to Floor Height | Minimum Number of Risers | Height of Minimum Riser | Recommended Number of Risers | Height of Recommended Riser | Sum of 2R = G if G = 265 |
|---|---|---|---|---|---|
| 2500 | 14 | 178.6 | **14** | 178.6 | 622 |
| 2600 | 14 | 185.7 | **15** | 173.3 | 612 |
| 2700 | 15 | 180 | **15** | 180 | 625 |
| 2800 | 15 | 186.7 | **16** | 175 | 615 |
| 2900 | 16 | 181.3 | **16** | 181.3 | 628 |
| 3000 | 16 | 187.5 | **17** | 176.5 | 616 |
| 3100 | **17** | **182.4** | **18** | 172.2 | 630/609 |
| 3200 | 17 | 188.2 | **18** | 177.8 | 620 |
| 3300 | 18 | 183.3 | **19** | 173.7 | 612 |
| 3400 | 18 | 188.9 | **19** | 178.9 | 623 |
| Stairs below this line must have a landing and minimum flights of two risers. | | | | | |
| 3500 | 19 | 184.2 | **20** | 175 | 615 |
| 3600 | 19 | 189.5 | **20** | 180 | 625 |

Using the formula 2R + G = 550 to 700 (average 625)
Risers not less than 115 nor more than 190
Note minimum going for a residential stair is 240 but 265 is more comfortable

# CABINET JOINTS • 65

Cabinetry joints are seldom used in on site trades in the house building industry today as much of the cabinet work is either built off site in a factory or is made on the site from materials that have limited amounts of joinery required.

Chipboard and medium density fibreboard (MDF) has revolutionised the cabinet trades by rendering the timber pint almost obsolete. This remarkable product is simply glued and screwed to make pieces of cabinetry that required a fine skill to produce in ladder framing and plywood sheeting less than a few years ago.

This section illustrates a range of timber and chipboard joints that OBs, who wish to do their own joinery will find a useful reference. If any further information on joinery is required then one of the many textbooks on this trade should be consulted.

When joining chipboard the pints should be well glued and screwed. The screws should be of a type specially developed for use in this material. It should be clearly understood that screws fixed into the edge grain of chipboard are very insecure and care should be taken not to load any joint that is not fixed to the face of the chipboard. Use an appropriate patent fixing clip, bracket, plug or the like specifically designed to be used with chipboard.

Chipboard is available in a number of different grades for different purposes so make sure the chipboard you use is compatible for the tasks you require it to perform. Chipboard, ideally, should be cut on a saw bench using a fine-toothed, hardened-tip saw specifically designed for cutting chipboard. Great care must be exercised when cutting melamine-faced chipboard, because the surface coating of this board is very hard and likely to chip and split along the cutting line.

If plain chipboard is used to make joinery doors or the like and is faced with a timber or laminated plastic veneer, it is important to face both sides of the sheet with similar products. If this is not done the sheet is likely to warp, bow and/or twist because of differential moisture content causing one surface to swell while the other remains stable.

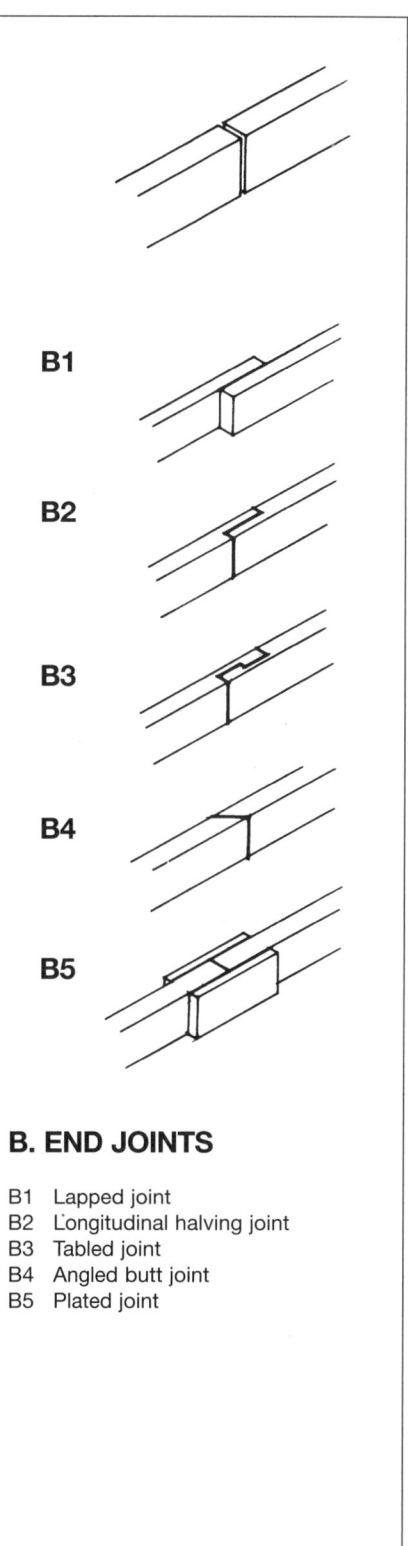

## A. CORNER JOINTS

- A1 Rebated butt joints
- A2 Common housed or trenched joint
- A3 Tongued joint
- A4 Trenched dovetail joint
- A5 Boxed joint
- A6 Angle dovetail
- A7 Common mitred joint
- A8 Mitre and rebate joint
- A9 Lock mitre joint
- A10 Stop mitre joint
- A11 Mitre and tongue joint

## B. END JOINTS

- B1 Lapped joint
- B2 Longitudinal halving joint
- B3 Tabled joint
- B4 Angled butt joint
- B5 Plated joint

# 65 • CABINET JOINTS

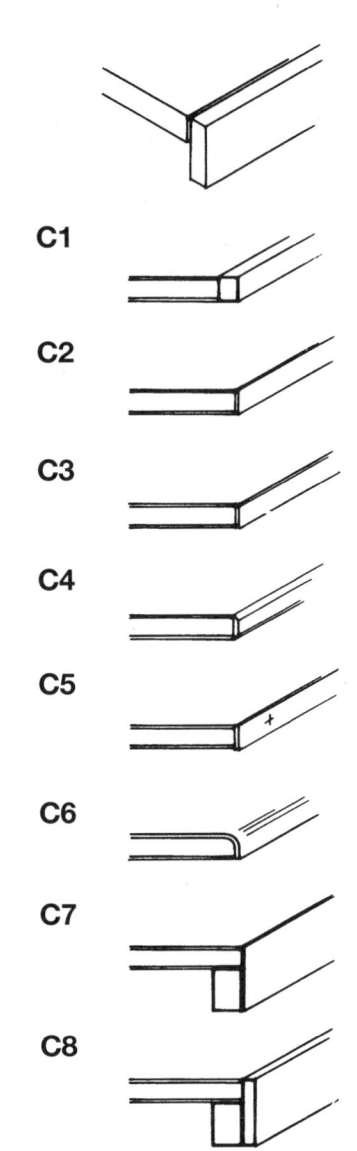

## C. EDGE JOINTS

C1  Timber edge strip
C2  Laminated edge with stopped top veneer
C3  Laminated edge with through top veneer
C4  Moulded edge strip (metal or plastic)
C5  Metal edge strip (heavy duty use only)
C6  Formed edge using post-forming grade laminated plastic
C7  Built up edge in laminate
C8  Built up edge in solid timber

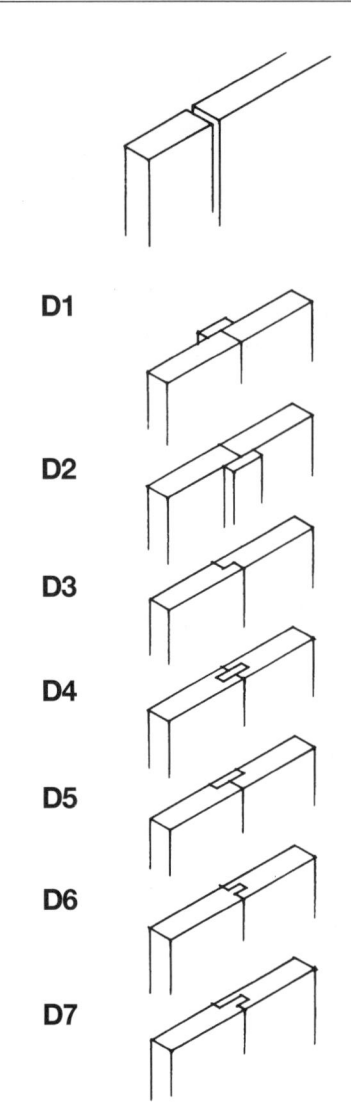

## D. PLANK JOINTS

D1  Batten behind
D2  Exposed batten
D3  Rebated
D4  Loose tongue
D5  Rebated and stripped
D6  Standard tongue and groove
D7  Tongue and groove— secret nailing

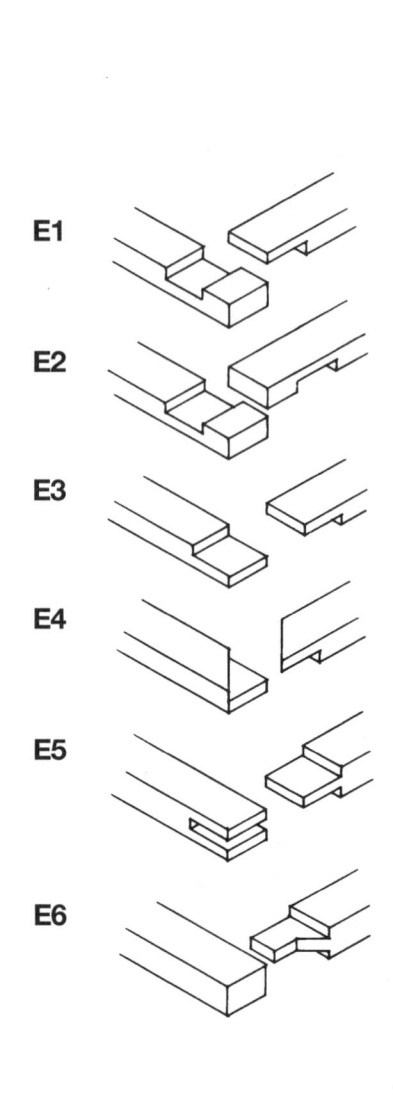

## E. FRAME JOINTS

E1  Junction halving joint
E2  Cross halving joint
E3  Corner halving joint
E4  Mitred halving
E5  Corner mortice and tenon
E6  Corner haunched tenon

A home is not only a house on land; it is many other physical, social and economic manifestations of family life together. This section investigates some of the physical options families add to their homes to personalise them and to improve their comfort. Common optional additions to family homes are:

**Car accommodation** in the form of garages or carports.

**Additional weather protection** in the form of verandahs, pergolas, and awnings.

**Outdoor leisure areas** in the form of decks, patios, terraces and barbecue areas.

**Vegetation** in the form of landscaped areas, lawns and ground contouring.

Some of these options are included in the first stage of the owner-building project while others are added to the home at a later stage. In all cases the owner builder should plan for their inclusion at the earliest possible stages so as to avoid problems in the future.

Few local authorities make the provision of sheltered car accommodation compulsory, but a number of local councils are requiring that adequate parking is provided for the family cars inside the property boundaries.

A debate has raged in recent years over the need for councils to regulate many of the options that families add to their homes. Some authorities believe that there is no good reason for councils to determine whether a simple structure conforms to an often arbitrary system of rules—and that families should be free to do as they like to their homes as long as a reasonable level of safety, soundness and cleanliness can be maintained.

Alas, most authorities still rule by the use of restriction and require families to submit in detail all work that is intended to be carried out to a residence, including in some areas things as simple as on ground paving and internal garden fences or shade houses.

Check the requirements of your local control authorities carefully, remembering that in many cases it is easier to include all the options that you wish to build into your home on the plans that you submit to the council. This way you avoid unnecessary friction with municipal building inspectors and give the authority a chance to refuse approval if they wish.

## External paving

The external living areas of a home can be paved or decked with many materials, the most common of these are covered briefly in this section and include:
- Concrete slabs,
- Concrete pavers,
- Clay pavers,
- Quarry tiles,
- Split slate,
- Stone cobbles,
- Timber decking.

### Concrete pavers

Concrete pavers are generally made by the same companies that make concrete masonry blocks, and are essentially solid moulded concrete blocks specifically made to be used as on ground pavers.

Concrete pavers are manufactured in various thicknesses to suit different applications, 40 mm for footpaths and 90 mm for driveways. The shape of the paver varies from simple square or rectangle tiles to decorative interlocking special shapes and cobblestone look-a-like units.

Concrete pavers are designed to be used on a well consolidated sand base, and when laid to the manufacturer's directions give an attractive paved area.

A special type of concrete paver is manufactured to allow it to be used in such a manner that lawn may be planted in large holes that are manufactured into it—this allows cars to be parked on areas that have the appearance of lawn but the support of paving.

Concrete pavers are available in a range of colours, and in some cases can be used to build other landscaping features; retaining walls, barbecues, steps and garden seats.

### Clay pavers

Clay pavers are much the same as concrete masonry pavers, except that they are made of vitrified clay in the same way as bricks are manufactured.

Clay pavers are generally laid on a sand base, and are available in standard square or rectangular shapes as well as special and interlocking shapes.

Standard bricks can be used as pavers, but the specially manufactured clay pavers generally give a wider range of colours and textures. Pressed and extruded manufacturing processes are used and a range of clinker and glazed finishes are available in some ranges.

Clay pavers range in thickness from about 50 mm through to about 100 mm and the manufacturers will give free advice on the correct paver for your needs.

Clay pavers can be laid on concrete slabs as if they were tiles and some of the thinner examples are ideally suited for this purpose, however if they are used on a suspended concrete slab the capacity of the slab to take the extra weight should be checked.

### Quarry tiles

Quarry tiles are any vitrified clay tile that is made for external application and is designed to be fixed to a concrete slab base (or in some instances a compressed fibrous cement sheet base).

Quarry tiles are normally manufactured in natural burn clay colours, but some are available in fashion colors. They are normally square or rectangular in shape but some are available in hexagonal or other shapes to suit particular architectural styles—Spanish mission or Roman villa.

The concrete slab used as the base for quarry tiles should be finished with a rough surface to allow the tiles to key to the slab and reduce the chance of them coming loose. The quarry tiles are laid on a bed of mortar and therefore, the thickness of the tile plus its mortar bed must be taken into account where setdowns are required to allow the tiles to line up with another surface finish. The tiles vary in thickness from around 20 mm to 40 mm in thickness and a 20 mm to 30 mm bed is generally required.

Quarry tiles are either pressed or extruded, and care must be taken with some of the extruded types as they are manufactured by a process that makes two tiles from one extrusion, this often means the bottom of the tile is very rough and needs skilled fixing to avoid uneven paving.

### Split slate

Split slate makes a very attractive external paving material and many natural shades can be chosen, from warm light beige/browns through to cool dark blues and blacks. Normally the darker the colour the more durable the slate and care should be taken to make sure the slate chosen will give the life expectancy desired.

Split slate is either used in a random irregular form or in a sawn rectangular form; both are equally acceptable dependent only on personal preference. The slate, being a natural stone material that is mechanically split, is not normally of an even thickness and great care must be exercised in its laying to achieve an acceptable near level surface.

Slate is normally laid on a concrete or compressed fibrous cement base, but it is possible to use locally quarried slate in thicknesses that allow it to be laid on a sand base. When this method is used, however, grouting between the slate pieces is often required to avoid excessive erosion of the sand bed.

### Stone cobbles

Stone cobble paving is a particularly attractive paving system, however it is expensive and out of the budgets of many OBs. Natural Australian Blue Stone or Sandstone makes very satisfactory paving but if these stones are not exotic enough special granites and marble like stone cobbles may be imported from Europe or Asia.

Newly quarried or previously used stone is equally acceptable, but care must be taken to cut the cobbles so that the natural grain of the stone runs in the direction which will give the least wear and last for the longest time. A discussion with a stone mason or with the quarry company is highly recommended before undertaking this work.

### Timber decking

Timber decking is a favoured method of building outdoor entertainment areas in some parts of Australia, particularly the hotter climate areas where its surface does not hold heat as long as masonry and the gaps between the boards allows for the breeze to blow through.

Every State has its own preference as to which timber is most suited to decking but there is no doubt that some timbers are more suitable than others. Tallowwood, Redgum and Jarrah are suitable hardwoods and tanolith-treated Pinus is a good all round softwood.

# 66 • OPTIONS

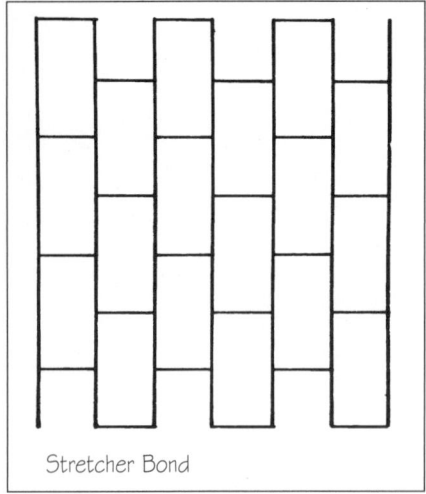
Stretcher Bond

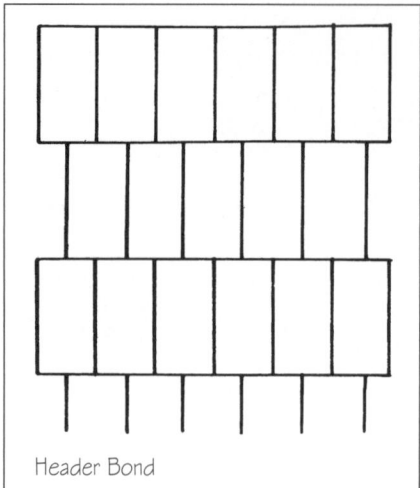
Header Bond

Basket Weave Bond

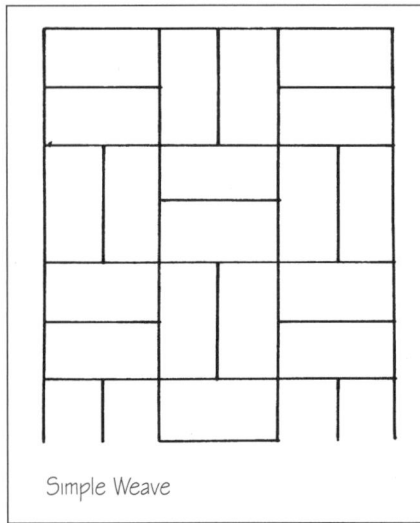
Simple Weave

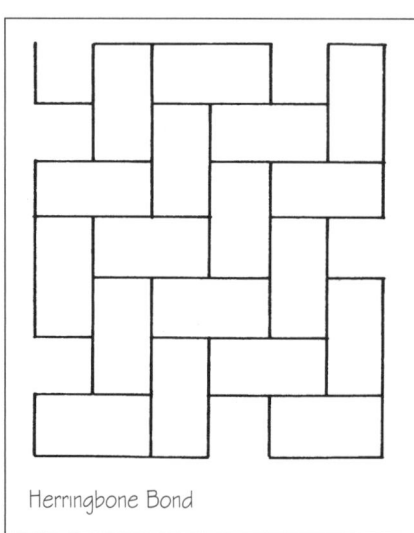
Herringbone Bond

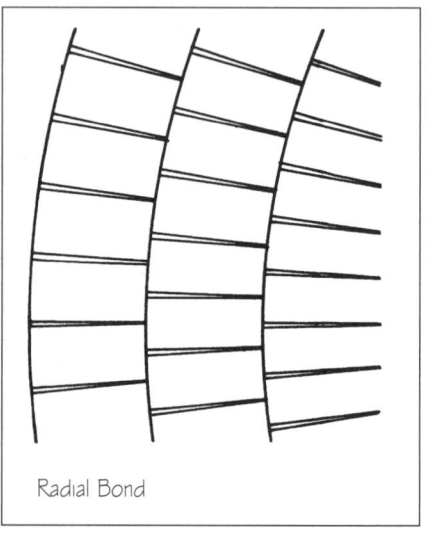
Radial Bond

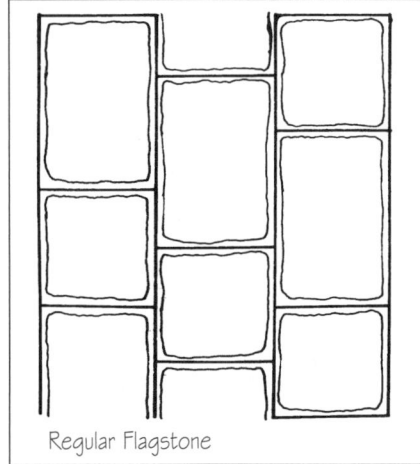
Regular Flagstone

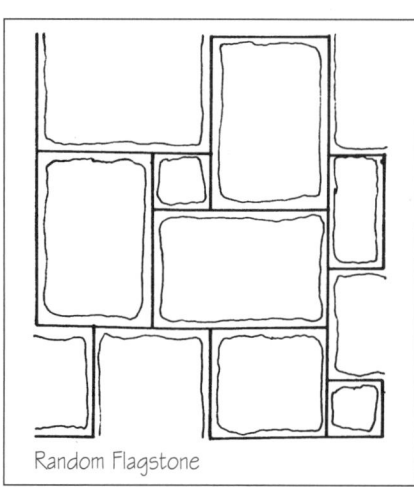
Random Flagstone

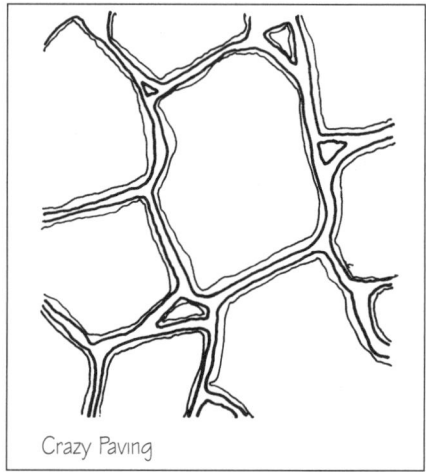
Crazy Paving

**Drainage**
Care must be taken in all paving to ensure that surface water is adequately collected and disposed of. Paving should fall away from any building, even when it is protected by a verandah, with a fall of not less than 1:50. Collection should be via formed gutters and grated drains or pits, with a piped disposal system to keep any excessive water flow away from the building footings and the foundation substrata.

# DOUBLE-STOREY CONSTRUCTION • 67

Double storey construction varies from single storey construction only in small ways. Obviously, a stairway must be added to the building if access to the first floor is to be attained.

Other differences are confined to:
Stronger lower walls,
Upper floor framing.

All else remains basically the same as it was when treated as a single storey construction. The footings may be wider but do not significantly change, the base remains unaltered in most instances, the single storey walls are the same as the upper floor walls, the roof remains the same and there is insignificant change to the services systems.

## Lower floor walls

The only significant change in the lower floor walls to those used in the upper floor is the sizes of the beams over the openings, whether the building is timber framed or masonry, heavier duty beams are needed to carry the load on the upper floor and the added structural weight. In timber frame construction a small increase in the size of the wall framing members used may be required, and this information is freely available in appropriate framing size tables from the Standards Association of Australia (AS 1684) or from tables issued by timber merchant associations based on the standards.

Generally, all walls in the lower frame are load bearing and therefore must be located over base work so that their loads are carried directly to the foundation. In timber framed construction it is possible to not have upper floor internal walls over the lower walls, but only if the house has a roof design that does not place any load from the roof onto any wall not designed to carry it.

Houses with solid brick construction on the lower floor which use timber non-load bearing partition walls in the upper storey may also avoid having to place upper storey walls over the lower walls.

Houses that are solid masonry on the upper and lower storey must, as a rule, position the upper walls over the lower walls. This can be a severe planning constraint, which can only be overcome by using a specially-designed reinforced concrete upper floor which is strong enough to carry wall loads on the span of the slab.

If any beams are used in the upper floor design then the part of the wall that supports the beam end may require heavy frame members, if in timber frame, or a pier or load spreader if in masonry. Take care of landing main floor beams on top of opening beams, this can cause unsightly or potentially dangerous conditions to arise. If this condition is inevitable an engineer should be consulted to determine structural sufficiency.

## Upper floor construction:

The upper floor varies from the ground floor to the extent that it is normally required to

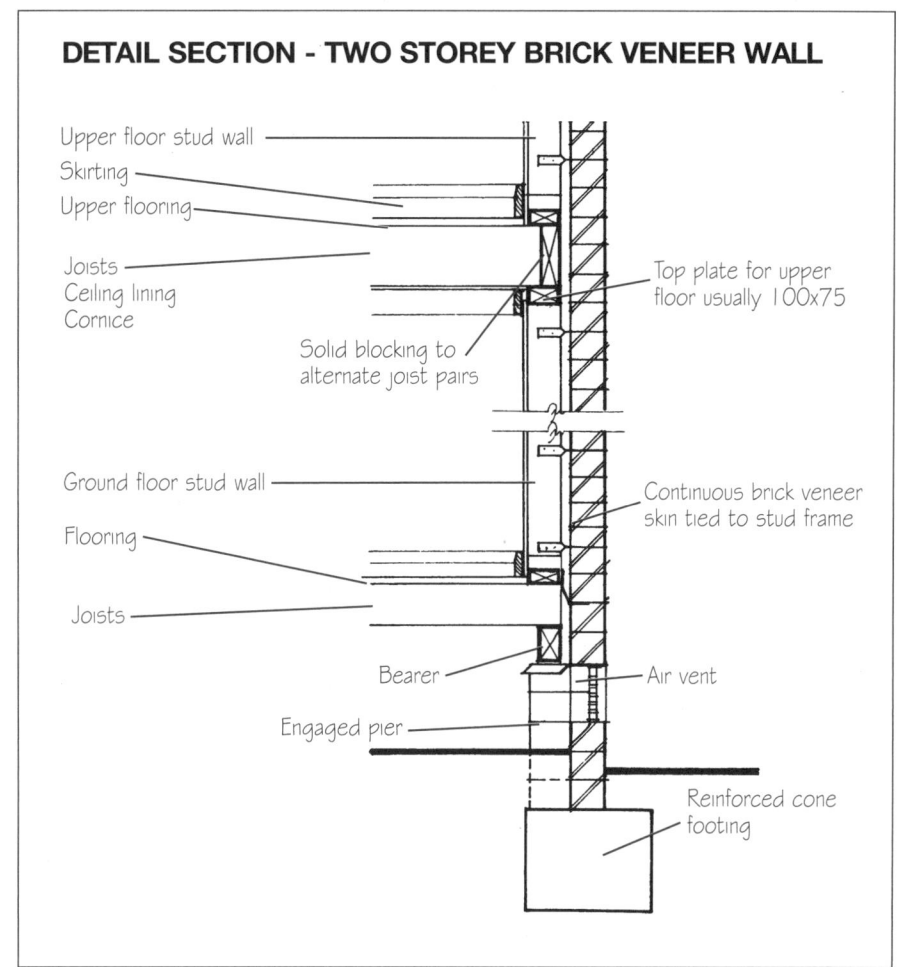

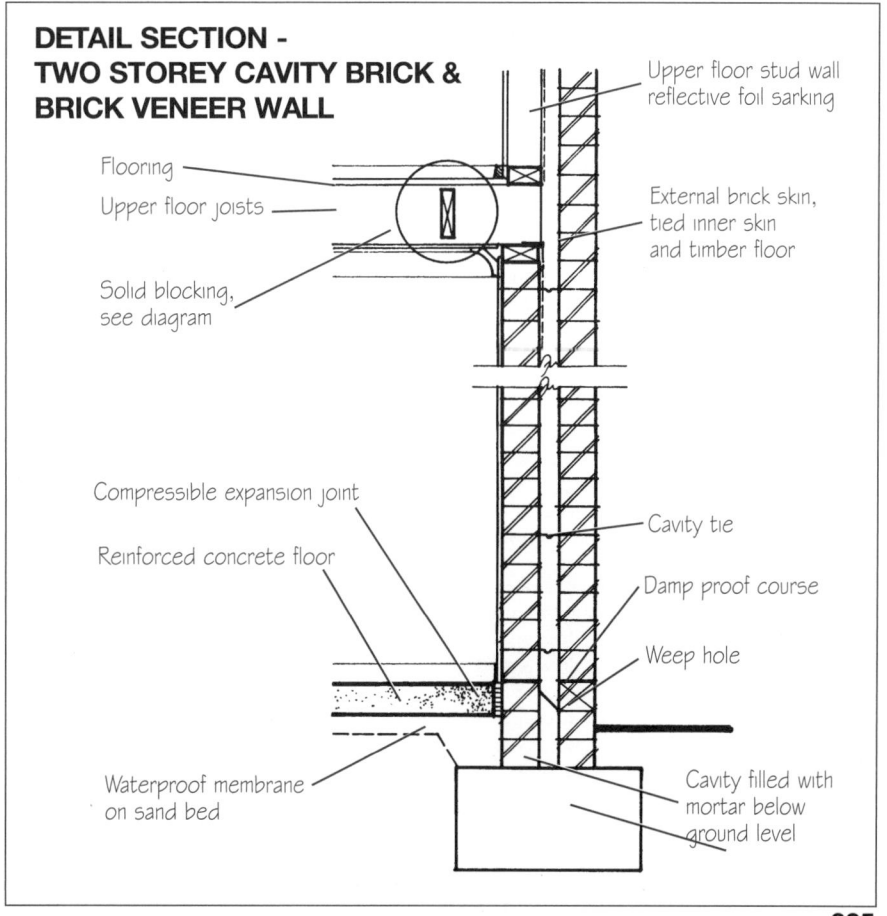

# 67 • DOUBLE-STOREY CONSTRUCTION

bridge a wider span than the ground floor. This means that much greater care must be exercised in the design of the structure to gain the most economic and safe system while retaining good appearance.

Economic spans in residential two storey buildings range to around 4000 mm. Over this span timber sizes become hard to obtain in the quality required and concrete slabs become uneconomically thick and complex.

A room with a width of around 4000 is acceptable for most homes. There are few rooms that require a wider dimension than this, double garages and billiard rooms for instance. A beam on the underside of a garage ceiling is normally tolerable and this solves that problem, but the billiard room may pose a greater problem with a desirable span of nearly 5500 mm.

Ceiling beams are not widely accepted in billiard rooms as ceiling height is often at a premium, there are a number of answers:

Thicken the floor slab over the billiard room.

This means that if the house is built to the minimum ceiling height the billiard room will set the minimum ceiling height for the ground floor.

The billiard table could be placed upstairs.

The billiard room roof could be built in a single storey wing of the house.

This example points out some of the problems of the design of two storey houses where most people want big wide living rooms downstairs and multiple bedrooms and bathrooms upstairs. This is the reverse of the plan that is most easy to construct.

With timber upper floors some services, like drain pipes from an upstairs bathroom, can be hidden in the thickness of the floor framing, this is not always possible but there is a chance that a poorly located toilet can be covered below. If reinforced concrete floors are used then there is no way a wayward pipe can be concealed. There have been cases of sewer traps turning up on living room ceilings.

If you have your heart set on a fine two storey villa, have the plans checked by a building designer or architect—for poor planning integration and span control can lead too often to unnecessary expense and even embarrassing waste pipes.

Upper floor timber frames use deep narrow timber joists, these would fail by screwing over if they are not constrained. The constraints are in the form of blocking or herringbone strutting. These should be spaced at around 1800 spacing along floor joists and between every joist pair continuously.

A timber upper floor frame is from 200 to 300 mm thick.

A reinforced concrete upper floor slab is 100 to 150 mm thick, in most cases.

Waterproofing upstairs bathrooms is essential, particularly shower recesses, as these often allow water to seep through to the floor below. Any repairs are likely to be a nuisance and expensive, much better to be careful when the house is being built.

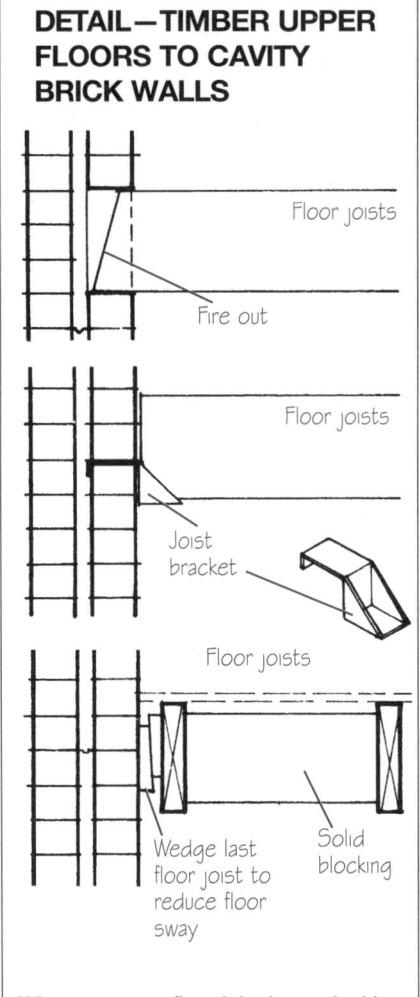

**DETAIL—TIMBER UPPER FLOORS TO CAVITY BRICK WALLS**

When an upper floor joist is used with a cavity brick wall the end of the joist is cut off at an angle to allow it to fall out of the wall if there is a fire in the house rather than pulling the wall down.

A modern alternative to the fire cut is the special joist bracket made from galvanised steel.

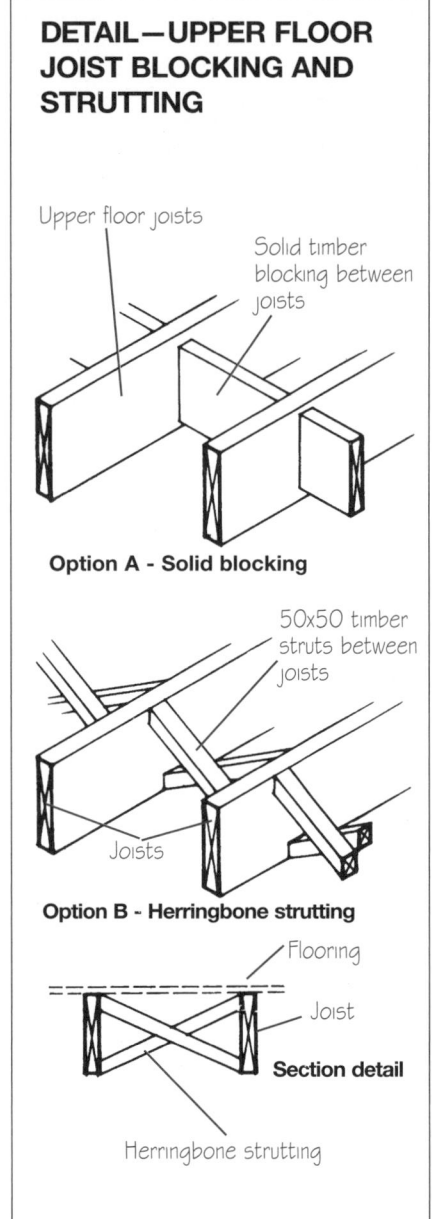

**DETAIL—UPPER FLOOR JOIST BLOCKING AND STRUTTING**

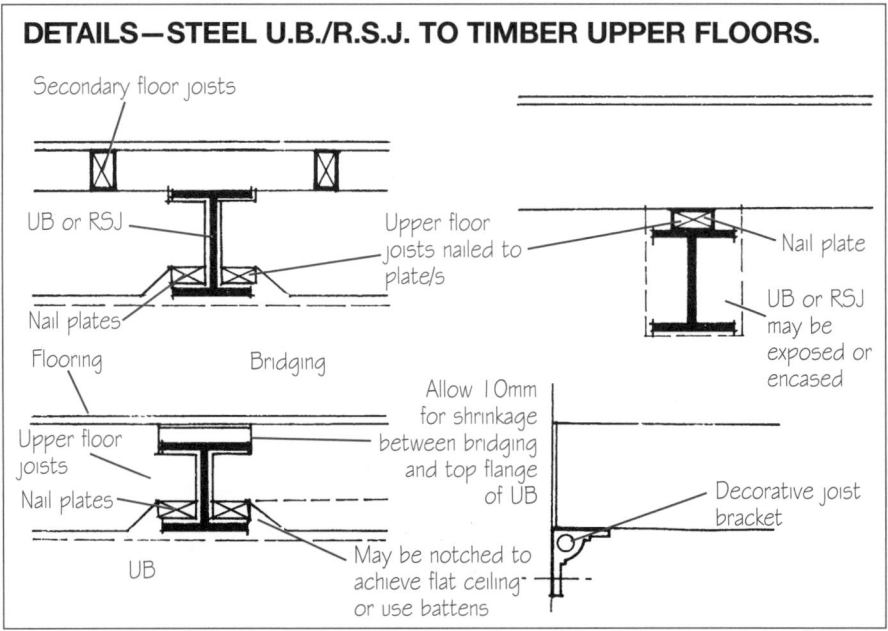

**DETAILS—STEEL U.B./R.S.J. TO TIMBER UPPER FLOORS.**

# DOUBLE-STOREY CONSTRUCTION • 67

**Long-span joists**
When extra long spans are required to support upper floors, beyond the spans of normally available solid timber joists, there are some specially manufactured alternatives available including:

Hyspan Beams—have the appearance of very thick plywood planks. When they are used the plies are in the vertical plane.

Hyne–I–Beams—have solid timber top and bottom flanges with a groove, into which a plywood web is glued.

EdgeBeam LGL—are beams where strips of timber are glued edge to edge to form a vertical stack, with the end joints offset.

HyneBeam and other GluLam Beams—are beams where strips of timber, approximately square in cross-section are glued together.

Posi-Joists—joists that are composites of timber flange chords and formed steel lattice webs gang-nailed on either side.

All these products and others have stretched the economical span of timber construction. OBs must use them strictly as set out in the manuals published by the manufacturers or have an engineer prepare the design.

Long spans generally mean higher end support loads; this could mean that a standard timber stud wall may need extra support or stiffening.

# 68 • CAR ACCOMMODATION

The garaging of the family car is an important part of an Australian home, whether in an enclosed garage or in an open carport.

An area of approximately 6000 mm long by 3000 mm wide is needed to accommodate a motor vehicle; this is a generous allowance for many modern small cars but it is considered a good working basis. Make the area smaller only if constrained by available space or funds.

Most people can drive their family car comfortably through an opening 2400 mm wide by 2100 mm high. These dimensions has become the standard for single garage doors while 4800 mm wide is used in double garages. Whereas the width is unlikely to change, the height of vehicles might with the trend to a wider range of family vehicles. Some off-road and people-mover types may require doors higher than 2100 mm. Garages and carports can be:
• double or single (or triple, etc)
• attached or detached
• adjacent to or under the house.

## Attached or detached

It is generally cheaper to attach a garage or carport to the structure of the house than it is to build it free-standing, depending on space available.

In some states fire separation is required between the car accommodation and the residence. This generally requires the garage to be separated by a distance of around 900 mm, or a masonry wall at least one brick thick. Where fire separation is required on a garage located under the floor of a house, a fire-rated, multi-sheet plasterboard ceiling, or even a concrete floor may be demanded.

Attached car accommodation is more suitable to near-flat sites, and much less suitable on sloped or undulating sites, where it can end up with poor proportions and detract from the appearance of the residence.

If the regulations in your state allow it, a door from the car accommodation directly into the house is most desirable, particularly on a wet and cold winter's night.

## Adjacent or under

Siting the car accommodation adjacent to the house, where both the garage and house are at ground level, or under the house is generally solved by the topography of the building block.

On a sloping site there are many good reasons to build the garage under the house, but take care that any retaining walls are correctly engineered and an efficient sub-surface water disposal system is installed.

In many cases the garage is the biggest room in the house, but all too often its structural sufficiency is downgraded or simply overlooked.

In all garages, the beam over the door(s) is critical. The size of this beam and its method of installation can often determine the appearance of the house. A sagging, cracking, stained, under-sized concrete beam is a lasting reminder of poor planning.

When the garage or carport is a free-standing unit it is essential to make sure that the structure is braced to be rigid in all three dimensions. Rigidity is often hard to achieve in a carport but if it is not achieved, the building is likely to blow over in the first high wind. Remember to brace the roof plane to resist twisting of the whole structure. This is also essential in attached garages and carports.

The best floor for a garage is reinforced concrete laid at a slight slope up to about 1 in 50; this allows any water or other fluid to run to a drain which preferably should be of a continuous type across the entry doors. If the floor is poured after the structure is complete then it can be set to suit the outside access driveway conditions. Make sure that a smooth transition from the outside to the inside is achieved—this is particularly important on a site with difficult access. Crests and hollows should be smoothed out, to avoid the bottom of the car scraping on the driveway when manoeuvring into or out of the garage.

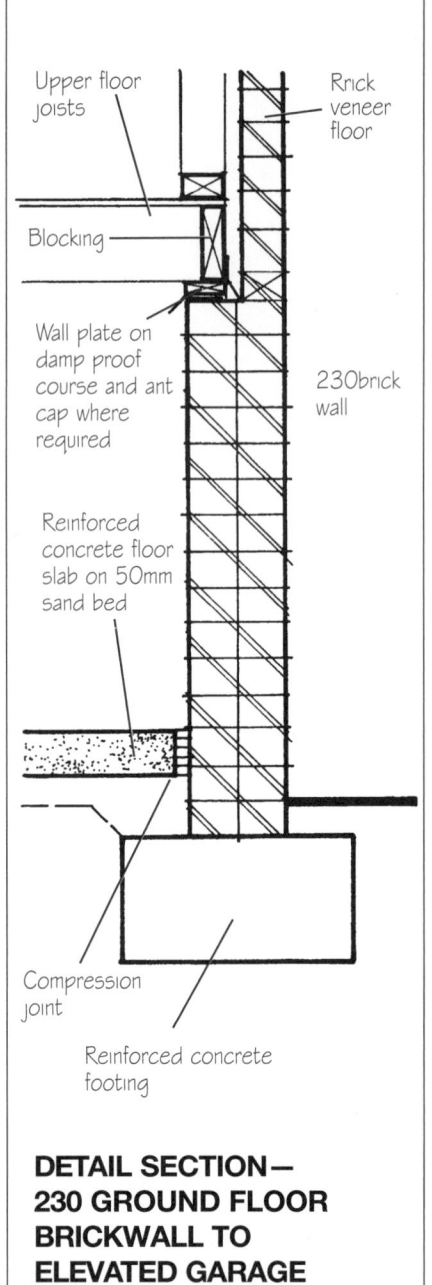

**DETAIL SECTION— 230 GROUND FLOOR BRICKWALL TO ELEVATED GARAGE UNDER DWELLING**

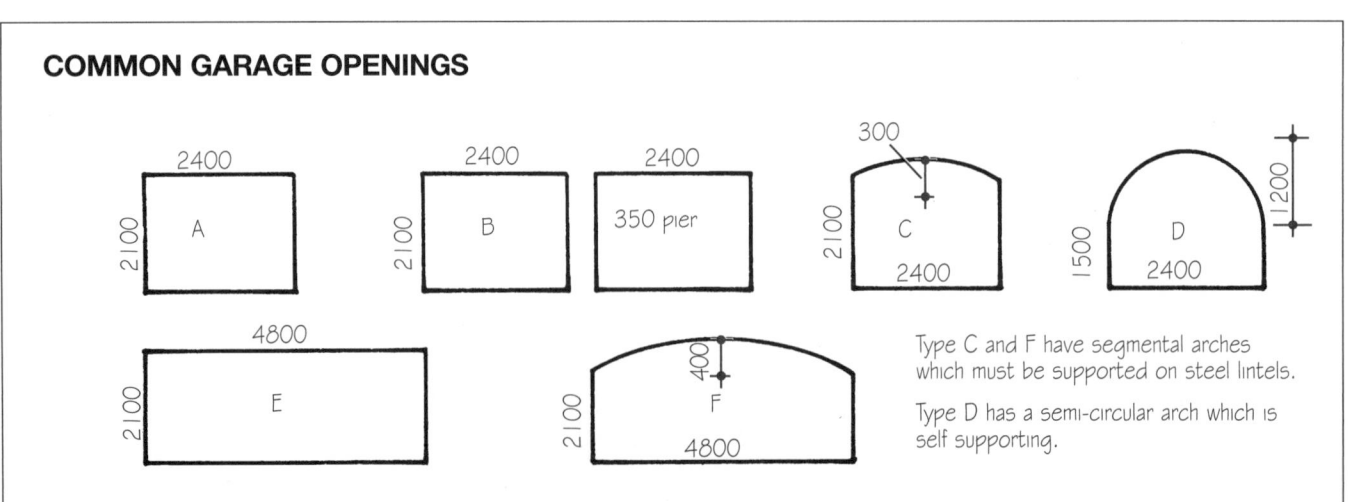

Type C and F have segmental arches which must be supported on steel lintels.

Type D has a semi-circular arch which is self supporting.

# CAR ACCOMMODATION • 68

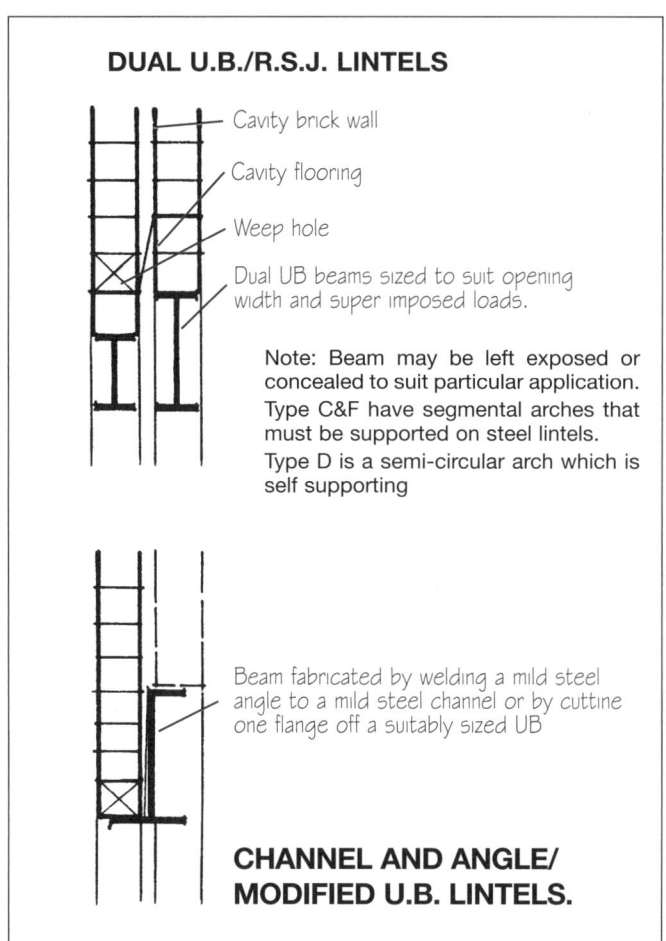

**DUAL U.B./R.S.J. LINTELS**

**CHANNEL AND ANGLE/ MODIFIED U.B. LINTELS.**

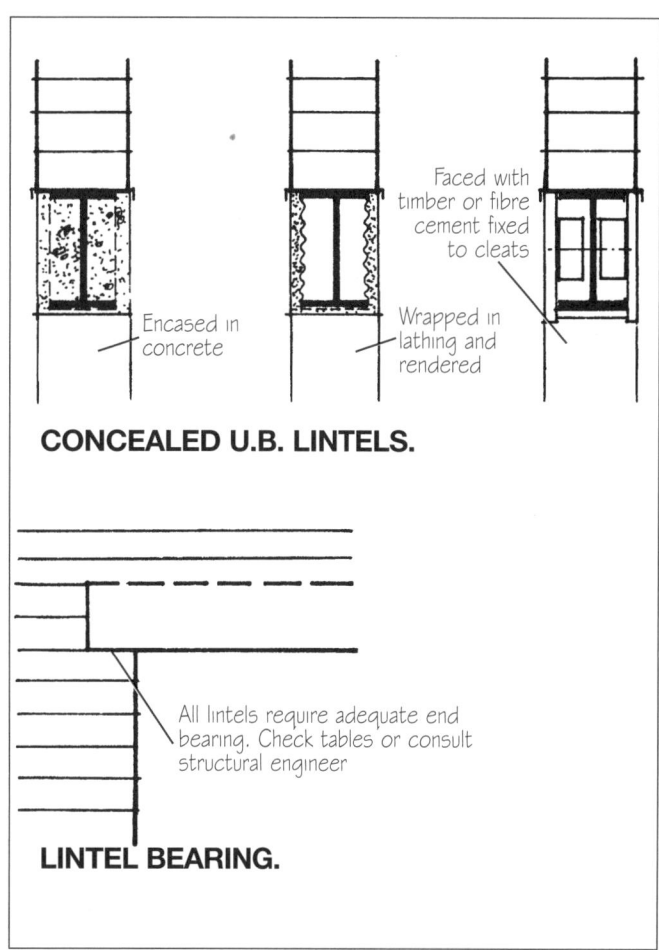

**CONCEALED U.B. LINTELS.**

**LINTEL BEARING.**

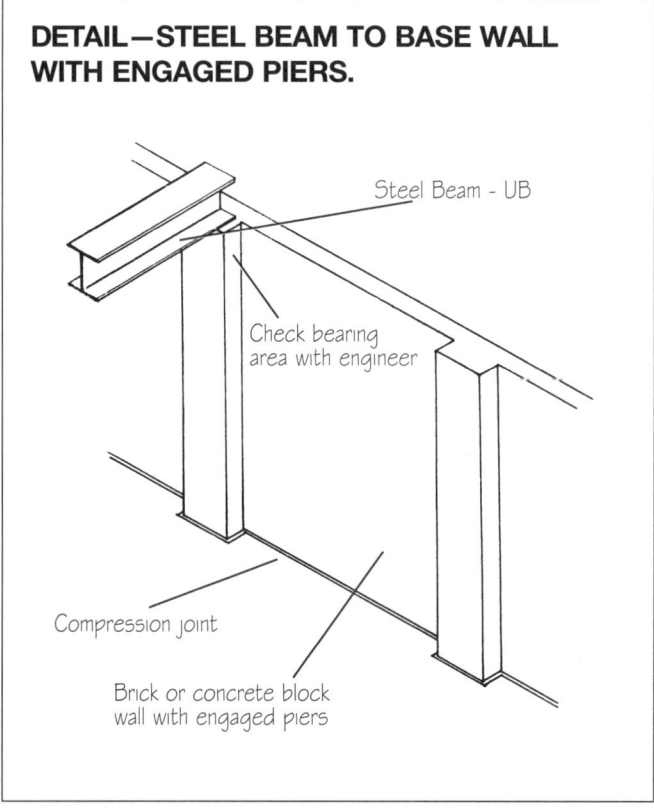

**DETAIL—STEEL BEAM TO BASE WALL WITH ENGAGED PIERS.**

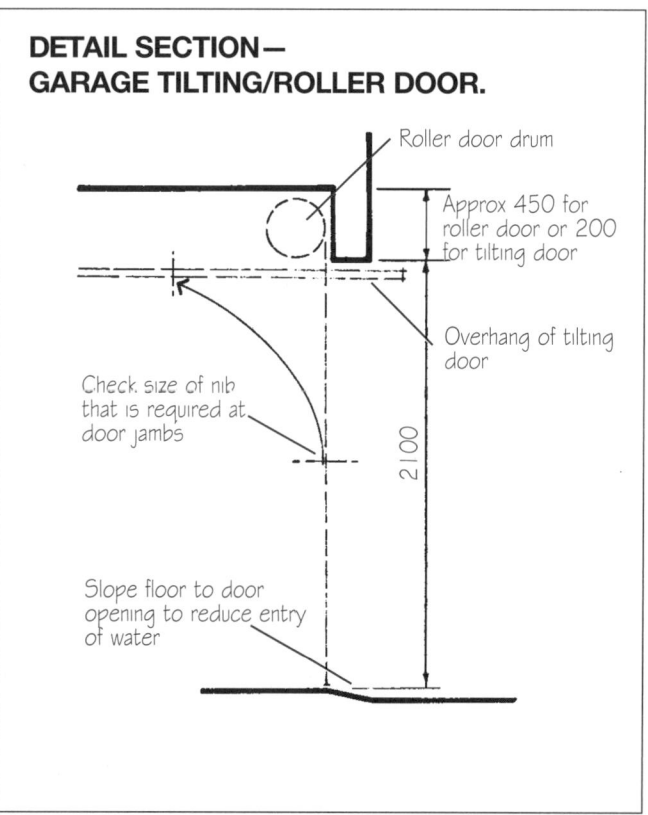

**DETAIL SECTION— GARAGE TILTING/ROLLER DOOR.**

# 69 • THERMAL COMFORT

## Comfort

Climate affects us both physically and emotionally and is therefore a factor of considerable importance in building design. The thermal characteristics of a building are extremely important human comfort. It is at the building stage itself where the demands of climate and psychological comfort must be resolved.

Our comfort within buildings is determined by the heat exchange processes that go on between the building and the outdoor environment. A building gains or loses heat by conduction, convection and radiation. The flow of heat by conduction through walls, floors and ceilings may occur in either direction. Conductive heat losses occur more often in winter, while conductive heat gain will happen in summer. The material composition of the walls, floor and ceiling will determine the rate of conduction.

Heat exchange by convection can occur through building surfaces by the movement of air between areas of different temperatures. For example, movement of air between the outside and inside of a building around doors and windows in winter is a convective heat loss.

The radiation of heat through glass or other transparent surfaces can add considerable warmth to a building. Conversely thermal radiation from interior surfaces to cool exterior walls will influence to a small degree a building's heat loss. The amount of solar heat gain is influenced by window area, building orientation and sunshading.

**In cold weather a house is heated up by:**
1. the sun's rays through radiation
2. people, lights, cooking, washing, drying, hot water systems
3. solar, fossil fuel, electric or space heating and hot water heating.

**In cold weather a house is cooled down by:**
1. radiation to dark sky
2. house walls and glass by conduction to outside cold air
3. cold outside air by convection (infiltration ventilation)
4. humidification (air temperature drops as humidity rises until external heat is applied)
5. ground is a basement or underfloor of house is warmer than the ground
6. drains and flues—heat is lost down the drain or up the flue.

**In hot weather a house is heated up by:**
1. house walls and windows by conduction when outside air is hot
2. hot and/or humid outside air by convection (infiltration ventilation)
3. people, lights, cooking, washing, hot water systems.

**In hot weather a house is cooled down by:**
1. radiation to dark sky
2. house walls and glass by conduction when outside air is cool
3. cool outside air by convection (ventilation)
4. ground by conduction from the crawl space under the house
5. humidification if air is dry.

The rate at which a house gains or loses heat for a given temperature difference (between inside and outside) depends on the surface characteristics of the materials and to a great extent on how well insulated the house is. A well-insulated house will be cosier without any heating in colder weather than a less well-insulated one.

When discussing the rates of heat loss it is normally sufficient to consider conductive losses through the house structure and those through ventilation, rather than go into conduction, convection and radiation separately. Every material has some resistance to flow of heat. This resistance is measured by the 'R' value of the material (R for resistance).

The higher the 'R' value the higher the resistance to heat flow. To find the total 'R'

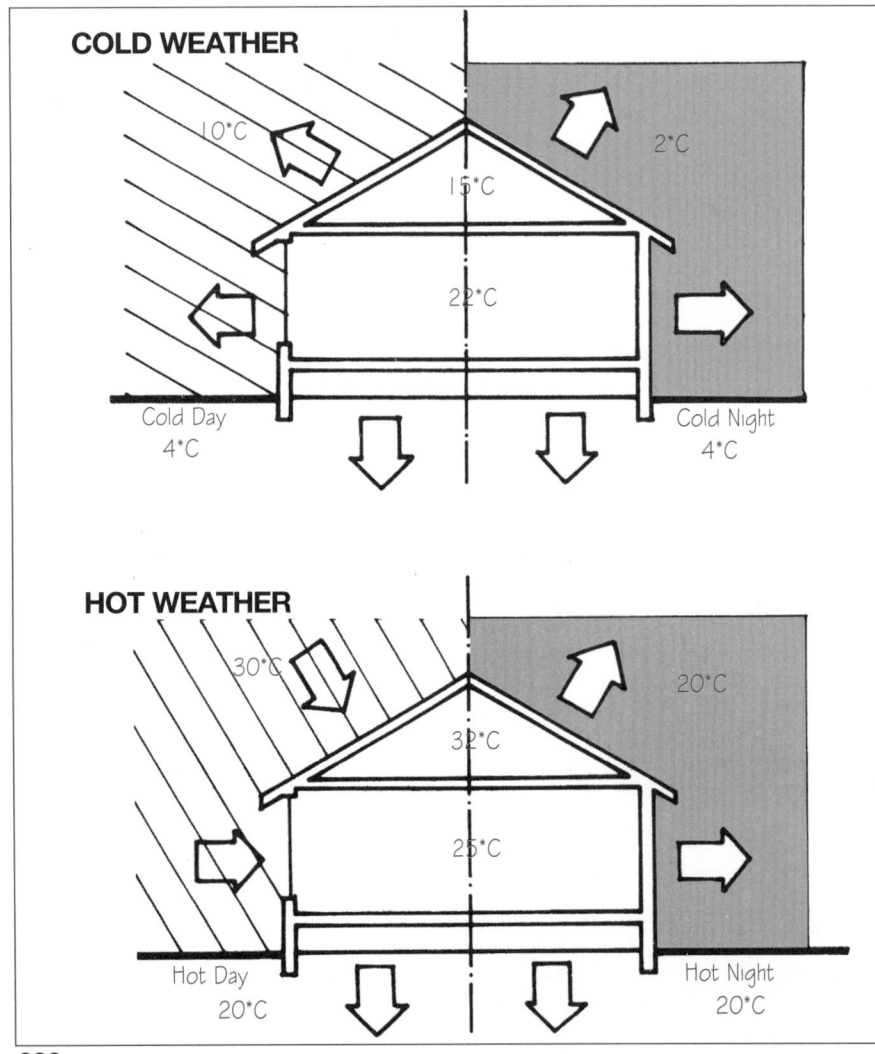

# THERMAL COMFORT • 69

value of a wall one adds up the 'R' value of each component of that wall and its air layer. As summer and winter heat are transmitted through a building differently, the summer 'R' is different from the winter 'R' of many different building components (see chart of 'R' values of materials). The 'R' value of a material is given for one square metre and for one degree temperature difference.

## Insulation

As mentioned earlier, insulation reduces the heat transfer through the building enclosure. There are two types of insulation:
1. bulk insulation—reduces the flow of heat by conduction. They are usually fibrous or cellular particles that enclose small pockets of air. In this way convection currents cannot occur within the layer of insulation.
2. reflective insulation—reduces the flow of heat by radiation.

## Insulation materials

Bulk insulation comes in several forms:
1. batts—consist of fibrous or cellular material bonded to form a lightweight sheet and sized to fit between standard ceiling joists or wall studs
2. blankets—similar to batts but in roll form
3. boards or sheets—made of more rigid materials like polystyrene, polyurethane or strawboard. Some are manufactured as decorative thermal and acoustic tiles. They may have plasterboard or vinyl linings.
4. loose fill—usually comprising granules or fibres of insulating material laid or pumped into position. Because of the nature of this material it is usually laid on a flat surface. Can be made from recycled paper, which should be treated with boric acid as a fire retardant and rodent inhibitor, or from granules of polystyrene or rock wool, fibreglass or exfoliated vermiculite. Loose-fill materials settle in due course, resulting in a decrease in thickness, which means a reduced insulation value.
5. reflective insulation—in the form of metallised foils, usually laminated to bitumen-impregnated paper or glass fibre reinforcing mesh. It usually comes in a roll and can be reflective on one side or both. Commonly referred to as reflective foil or sarking.

### Thermal Resistance of Materials (R Value)

| Construction | Summer heat flow | Winter heat flow | With appropriate insulation added | |
|---|---|---|---|---|
| Pitched roof, tiled with plasterboard ceiling | 0.76 | 0.29 | With reflective foil | Summer 1.81<br>Winter 0.55 |
| | | | With R2 bulk insulation | Summer 2.76<br>Winter 2.29 |
| Metal deck roofing with raked ceiling | 0.43 | 0.38 | With reflective foil and R2 insulation | Summer 3.137<br>Winter 3.067 |
| Timber frame with weatherboards | 0.46 | 0.46 | With foil<br>With R2 bulk insulator | 1.685<br>2.305 |
| Brick veneer | 0.45 | 0.45 | With foil<br>With R2 bulk insulator | 1.517<br>2.457 |
| Cavity brick | 0.51 | 0.51 | | |
| Concrete block (200 mm) | 0.35 | 0.35 | | |
| Single glazed window | 0.166 | 0.156 | | |
| Double glazed window | 0.312 | 0.302 | | |
| Timber floor | 0.39 | 0.43 | With foil | Summer 0.617<br>Winter 1.28 |
| Suspended Concrete Floor | 0.43 | 0.54 | | |
| Steel partition wall | 0.52 | 0.52 | | |
| Mud brick (300 mm) | 0.4 | 0.4 | | |

# 69 • THERMAL COMFORT

## Location of Insulating Materials

| Insulation types | Roof | Walls | Floors |
|---|---|---|---|
| **A. Bulk:** | | | |
| Batts | | For conventional roofs, adjacent to ceiling lining | Between timber studs, for timber frame and brick veneer | Upper floors or floor high off ground |
| Blankets | | " " | " " | " " |
| Boards/sheets | | Concrete roofs—on top surface with waterproof membrane above | | |
| Loose fill | | " " | " " | " " |
| **B. Reflective** | | | | |
| Foil both sides | | | On outside of studs, for brick veneer | |
| Foil one side | | On roof rafters, foil side down, as top side gets dusty, and is a hazard during construction | | |
| 2 sheets of foil with air space between | | For conventional roofs, adjacent to ceiling lining | | |

# FLOOR SURFACE FINISHES • 70

## Natural finishes

### Hardwood
Hardwood is available in strip or block form, prefabricated 'mosaic or parquet tiles'. As a material it is resilient, good-looking, warm and hard-wearing. It improves with age and it can be sanded and resealed. The disadvantages of hardwood include shrinking, dampness and swelling; some forms such as parquet can be expensive. A polyurethane seal will bring out the colour and the grain. Waxing a hardwood floor will only make it slippery.

### Softwood
From pine and fir trees. This is a common timber finish. Usually in the form of strips timber, these boards can be tongue and grooved or simply butt jointed. The material can be sanded and sealed. Rich colours will show up the grain and look attractive or, alternatively, boards can be coloured with pigmented translucent wood sealers, with water-based stain then sealed, or with paint. This timber does not have a great number of disadvantages; its primary one is that as a soft material, it is susceptible to impact marks. The timber should be sealed with resinous or polyurethane sealer.

### Hardboard
Can be used as an inexpensive surface covering in a similar manner to plywood or chipboard. There are light and dark hardboards or they can be stained. Oil-tempered hardboards cannot be sealed but can be waxed to give a pleasant finish. Hardboards have the advantage of being inexpensive and make a useful short-term floor, which could later have carpet or other material laid over it. It is also good for making a level base over older or inaccurately laid floors. Its disadvantages are that it is not hard-wearing, it may stretch somewhat, will rot if water gets underneath and will stain if the sealer is allowed to wear off.

### Plywood
As well as being a good sub-floor, plywood can be used as a surface covering and comes in sheets in both tongue-and-groove form and butt-jointed form. It can be used directly on timber framing instead of boarding and it comes in various timber finishes.

Unsealed or presealed, it can be stained. It has the advantages of being comparatively inexpensive and attractive. It is more expensive than softwood boarding but quicker to lay in larger sheets. Its disadvantages are that, like softwood, it is not very impact-resistant. If the stain wears off, the ply will stain and deteriorate.

### Chipboard/particleboard
Made from various sizes of wood chips mixed with urea formaldehyde resin and bonded under pressure. Like plywood, it can be used as a surface covering or as a finish material. Its chief advantages are that it is inexpensive, reduces sound transmission and can be made to look attractive. The disadvantages for non-sealed chipboard are that it is very porous and stains or discolours considerably, so a seal should not be allowed to wear off.

Also referred to as particleboard, it is available in moisture-resistant and weather-resistant varieties. It can be used for carcass construction in kitchen cupboards and resists quite high moisture environments, particularly if melamine-faced, without deterioration. This is the most common material used for timber floor panelling. It is normally in 900 mm wide sheets with tongue-and-groove edges. Floor sheets can be used under carpet, flexible floor sheets and wet area tiling, and is recommended as a sub-floor under quality timber boarding.

### MDF (medium density fibreboard)
Made from wood fibres compressed with ureaformaldehyde resin or equivalent to form rigid boards varying in thickness from about 5 mm through to about 35 mm. Available in a standard form, a moisture-resistant form and a high-density form. This is a very flexible product, mostly used for cabinetry and solid-core doors. Takes paint well with correct preparation and is an ideal base for timber and laminated plastic veneers. Tends to sag more than chip or particleboards when used for shelves.

### Quarry tiles
Made of clay pressed into tiles and fired at high temperatures. Tiles can be handmade or machine-made; some have a non-slip finish. The advantages are that they are tough, waterproof and impervious to almost all household liquids. Disadvantages are that they can be cold, noisy and rather tiring for feet.

### Mosaic tiles
Made of glass silica or ceramic. A small tile, approximately 25 mm square, glazed or unglazed, semi-glazed or fully vitrified, and usually mounted on paper to make them easier to lay. Advantages are as for ceramic tiles. Disadvantages are that they are cold, hard and noisy.

### Bricks and pavers
A good finish both indoors and out. Most types of brick make a good, hard-wearing floor, and can be sealed with wax or polyurethane sealers to prevent staining. Their disadvantages are that they are hard and fairly noisy.

### Marble
A hard-wearing, natural material available in many colours. Advantages are that they are durable and long lasting. Disadvantages are that they are cold and hard under foot. An expensive material.

### Travertine
A marble that is fissured and requires special filling if it is to be used for flooring in wet areas such as bathrooms, as the fissures collect dirt and grime. Advantages are as for marble. Disadvantages are that they are expensive, cold, hard under foot need initial sealing.

### Terrazzo
Marble chippings set in cement. Plain or coloured then ground smooth. Available in tiles or large slabs or laid in situ. Advantages are that it is hard-wearing and maintenance-free. it also has a good visual effect. Disadvantages are that it is fairly hard, noisy and can sometimes be slippery.

### Slate
A dense non-porous stone that varies in colour. Looks best in square or rectangular shapes rather than random pieces. Available in sawn or polished finishes. Advantages are that it is beautiful material, easy to care for and exceptionally hard-wearing. Disadvantages are that it is expensive, fairly noisy, cold, hard and can be slippery when wet.

### Stone
Sandstone, granite or limestone slabs. Also cast stone made from chippings mixed with cement. Advantages are that it is extremely hard-wearing. Disadvantages are that it is hard, cold and noisy.

### Cork
Made from cork granules and natural or synthetic binders compressed and baked. Natural colour earth tones. Advantages are its good insulation qualities, that it is warm, resilient and quiet. it is also hard-wearing and if sealed, easy to maintain. Disadvantages are that with relatively light wear, no grease or heavy splashing, looks attractive if left natural, but manufacturers recommend seal or polish. The tiles chip or crumble at the edges and fade in strong sunlight.

## Synthetic finishes

### Vinyl
Available in both sheet and tile form. The better quality versions have a cushioned backing. Good vinyl flooring is as expensive as medium quality carpet. Available in a wide range of colours, patterns and textures. Advantages are that it is waterproof and resistant to oil, fats and most domestic chemicals. A textured finish shows up marks less than plain finish and is slip-resistant. Cushioned vinyl is quiet, resilient and warm under foot. Disadvantages are that unbacked vinyl over concrete can be hard and cold. Vinyl tiles need very accurate laying both for pattern continuity and quality finish.

### Linoleum
Made from ground cork, wood flour, linseed oil and resins pressed onto a jute or hessian backing. Comes in different grades and thicknesses. Available in sheet or tile form in a good range of plain colours or marble and patterned. Advantages are that it is hard-wearing and thicker types have high-resistance and are warm under foot. Disadvantages are that it is inclined to rise, peel, or rot if water gets underneath. it is also sensitive to alkalies.

# 70 • FLOOR SURFACE FINISHES

### Carpet
Carpet is resilient and easy to fit but presents cleaning problems. It will readily attract dirt through static electricity. Available in a range of both natural and synthetic fibres and a range of pile types, namely cut pile, loop pile and cut and loop pile. Synthetic carpets may be hard-wearing.

### Epoxy resin
Not strictly a sheet material but rather a plasticised liquid either laid with a trowel as a screed or applied in a self-levelling liquid form. Suitable only for concrete sub-floors, is very hard-wearing and comes in a wide range of colours. Is expensive for a small area, but is available in a non-slip finish.

### Rubber
Often a combination of natural and synthetic rubber, available in sheets or tiles in plain colours or mottled with inlaid patterns. Advantages are that it is hard-wearing, resilient, quiet and waterproof. Disadvantages are that it is difficult to lay and reacts badly to grease, fruit juices and spirits. Rubber heels from shoes make black marks. Synthetic rubber is now used to a greater extent than natural rubber.

### Metal tiles
Can be matt, polished or patterned. An expensive material that is noisy, but hard-wearing. Noise can be reduced by gluing insulation board to the underside of the tiles. Withstands extremes of heat, making a good surround for work areas. It is hygienic and easy to clean, but copper tiles are susceptible to scratches and stains from acid fluids.

### Plastic laminate
Made from several grades of paper with a finish paper either plain, coloured or textured and bonded with natural or synthetic resins. They come in a wide range of colours and patterns. A laminate is glued to a sub-surface with an adhesive. It can be contoured in a special press, is easy to cut, simple to clean and withstands a certain amount of heat, knocks and scratches.

### Washable wallpapers
Vinyl wall coverings are made from PVC. They are water- and steam-proof, which means they are ideal in kitchens and bathrooms. Because they are impermeable, they must be stuck down with a fungicidal adhesive so mould does not form underneath. Washable papers have a transparent plastic coating that makes them water-resistant rather than waterproof. They are not as tough as vinyl but have a pleasant matt surface unlike the sheen of vinyl.

### Paint
Oil-based paints are more durable than water-based emulsion ones. In general, the glossier the paint the more hard-wearing it is. Gloss or satin finishes are available in a myriad of colours with any number of additives.

## Surface Finishes

| Material | Substructure | | Uses | | |
|---|---|---|---|---|---|
| | Timber Frame | Masonry Concrete | Floor | Walls | Ceilings |
| Hardwood | ✓ | ✓ (parquetry) | ✓ | ✓ | ✓ |
| Softwood | ✓ | | ✓ | ✓ | ✓ |
| Hardboard | ✓ | | ✓ | ✓ | ✓ |
| Plywood/Chipboard | ✓ | | ✓ | ✓ | ✓ |
| Quarry Tiles | ✓ (fibrous cement base) | ✓ | ✓ | | |
| Ceramic Tiles | ✓ (fibrous cement base) | ✓ | ✓ | ✓ | |
| Mosaic Tiles | ✓ (fibrous cement base) | ✓ | ✓ | ✓ | |
| Bricks, pavers | | ✓ | ✓ | | |
| Marble | | ✓ | ✓ | ✓ | |
| Travertine | | ✓ | ✓ | ✓ | |
| Terrazzo | | ✓ | ✓ | ✓ | |
| Slate | ✓ (fibrous cement base) | ✓ | ✓ | ✓ | |
| Stone | | ✓ | ✓ | ✓ | |
| Cork | ✓ (hardboard base) | ✓ | ✓ | ✓ | ✓ |
| Vinyl | ✓ | ✓ | ✓ | | |
| Linoleum | ✓ | ✓ (hardboard base) | ✓ | | |
| Carpet | ✓ | ✓ | ✓ | ✓ | |
| Epoxy Resin | ✓ | ✓ | ✓ | | |
| Rubber | ✓ | ✓ | ✓ | ✓ | |
| Asbestos Cement (profiled sheet) | ✓ | | | ✓ | ✓ |
| Insulation Board/Tiles | ✓ | | | ✓ | ✓ |
| Metal Tiles | ✓ | ✓ | | ✓ | ✓ |
| Plastic Laminate | ✓ | | | ✓ | |
| Washable Wallpapers | ✓ | ✓ (render) | | ✓ | ✓ |
| Paint | ✓ | ✓ | ✓ | ✓ | ✓ |

### Notes

**1. Materials**
The following list of surface finishes and materials are finishes that would be applied to a sub-structure such as timber frame and masonry materials and does not necessarily include structural materials that would have an integral finish.

**2. Timber frame structure**
This refers either to a timber framework, be it floor frame, wall frame or ceiling structure, to which the finish material is applied, or to a timber frame over which a standard material, such as plasterboard, has already been applied and to which the finish material will be applied.

**3. Masonry concrete sub-structures**
Includes concrete slabs, brickwork, concrete blocks and the like.

# PAINTING • 71

## Paint as a finish

Painting is a simple way of protecting, preserving or decorating a variety of surfaces. Certain building materials, require painting not only for their decorative aspect but also to protect their surfaces from corrosion or decay. Others require painting for reasons of sanitation and cleanliness. The chief materials requiring preservation are timber and steel. If left untreated, all softwoods and most hardwoods suffer deterioration from the constant changes of temperature and moisture in the air. All internal plaster and timber surfaces of a building require periodical redecoration to keep them clean and sanitary, for the dust they harbour or form themselves may constitute breeding places for insects or bacteria.

Painting is generally carried out using a sequence of paints (the combination of coats) that will vary in complexity in relation to the needs of the project. Any painting system is only as good as the surface to which it is applied and the quality of the materials used.

## Surface preparation

Before painting any surfaces the following factors must be considered. The surfaces must be:

1. free of dust, flaking materials and loose particles
2. free of grease or oily deposits
3. free of rust and corrosion
4. dry
5. as smooth as possible (although paints for rough-sawn materials are available).

## New surfaces

1. Timber surfaces should be sanded smooth. For clear finishes, apply first coat before stopping the punched nail holes with a similarly coloured filler. If a painted surface is required, stopping should be carried out after priming.
2. Plasterboard should have all cracks filled with a suitable filler, sanded flush, then sealed with an alkali sealer.
3. Masonry and cement surfaces should be left at least one month for each 25 mm thickness of concrete. If this is not possible, etch the surface with a phosphoric acid solution then neutralise before sealing with a general purpose primer.
4. Steel surfaces must be completely free of all rust and scales from milling. All grease and oil must be removed by a metal cleaner. Loose rust may be brushed off the surface and the surface treated with a rust converter. To protect the metal from rust attack a cold galvanising compound should be applied to fresh metal surfaces, but the surface to be treated must first be spotless.
5. Galvanised iron and zincaneal should be thoroughly dusted, all corrosion removed, and painted with galvanised iron cleaner before finishing. Aluminium should be lightly abraded to provide a key for subsequent painting. New galvanised iron should be allowed to weather for several months before painting.
6. Plaster (set) may cause problems unless the surface layer is first dissolved with a weak phosphoric acid solution then left to make sure the surface is completely dry.

## Old surfaces

1. Exterior timber should be checked for any repairs needed. Replace rotting timber and loose nails. The surface should also be checked for blisters, flaking paint or chalky surfaces. These should be removed entirely and spot primed with pink or white primer. Alternatively, paint may be softened with a blowtorch and scraped off. Difficult areas may be stripped using commercial liquid stripper. Careful washing with mineral turpentine and subsequent sanding will give a new surface to work on. Soundly bonded old paint may be sanded to a matt surface.

   Clear finishes can be applied to an interior timber after sanding. Oil or stain finishes merely require brushing with a stiff brush and recoating.
2. Plasterboard needs no stripping. Chalking paint should be lightly sanded then painted with a binder.
3. Masonry surfaces can be burned off or wire brushed to prepare the surface. If enamel paint is used on internal areas an alkali sealer will be needed to stop alkali attack on the paint.
4. Metal surfaces can be prepared by power wire brushing, which is the easiest method. Paint stripping or burning may also be used.

## Paint types

1. **Oil paint and oils**—interior and exterior finishing paint available in gloss, semi-gloss, satin, textures.
2. **Water paints**—mostly of the oil-in-water type and have flat drying interior or exterior finishes, although satin and gloss plastic paints are also available.
3. **Lacquers**—generally thinner-based and usually sprayed on. Brushing lacquers are available for furniture finishes. The use of solvents is critical for fine finishes.
4. **Shellacs**—methylated spirit-based furniture finish that comes ready-mixed or as flakes that have to be dissolved in methylated spirits. The use of shellac requires many coats and then requires the art of French polishing.
5. **White washing and lime wash**—an exterior finish often used on fences or bagged brick work.
6. **Stains**—may be water-, oil- or spirit-based. The latter is difficult to use, as it dries very quickly, causing patchiness and is not recommended for the non-professional. Stain is rubbed into the work well and then the excess is wiped off.
7. **Primers, undercoats and sealers**—interior and exterior undercoats for oil paints and enamels. Different varieties are available for wood, metal and masonry.

## Applications

1. **Brush and rollers**

   The brush remains the principal tool for paint application. The reasons for this are:
   a. no other method of paint application is available that gives comparable versatility and economy in operation
   b. no other tool or equipment gives effective application and is readily transportable
   c. a cleaner or healthier method of application is not yet available.

   A wide range of brushes caters for all classes of work and it is to the painter's advantage to use the right brush.
2. **Spray painting**

   *Advantages*—Because of its speed of application large surfaces can be covered quickly by spray painting. The spray application of lacquers and other quick-drying finishes reduces the drying time between successive coats, so the job may be completed in hours rather than days. Perforated screens, grilles, louvres and uneven surfaces to be painted are frequently inaccessible or uneconomic for brush use. Spray painting is preferable for these. A more uniform film can be applied by spraying than brushing. By using various adjustments on a spray gun, considerable control can be exercised over the delivery of the spray in both density and shape.

   *Disadvantages*—Where there are no large surfaces and the colours have to be changed, it is often more costly to do the work by spray gun than by brush or roller painting, even when using a small plant. In spraying, considerable loss of material can result when finely atomised paint is delivered with the gun held some distance from the surface. Wind or draft can carry away a high percentage of fine particles before they reach the surface.

   The loss of paint may not exceed 10 per cent, but frequently in house painting the circumstances for spraying are far from favourable. In brushing the painter relies largely upon a sense of touch in judging the right amount of paint to be placed on the surface and thereby guards against any excess that may produce sagging or wrinkling. In spraying the sense of touch is missing. This is one reason why the spraying of slow-setting and slow-drying finishes such as oil and enamel is not usually attempted.

## Painting as a trade

As an OB, it is important to recognise the link between plastering, rendering and

# 71 • PAINTING

painting as trades. Normally, after the plasterboard liner or cement renderer have finished their trade, the builder would move in and sand back the surface, making it clean and ready for the painter. As an OB this becomes your responsibility. The reason for this is that it is not normally part of the painter's job to prepare the surface for painting. If you wish your painter to do this, advise say so beforehand and your request will be reflected in the quotation for painting.

Every consultant is free to charge whatever the market will bear, so the first question for an OB to ask is: can we see your fee schedule?

## SURFACE FINISHES

### Surface—exterior

| | Type of finish | Paint system | Application Brush | Roller | Spray | Clean up Solvent | Water |
|---|---|---|---|---|---|---|---|
| **Timber:** Painted (trim, doors, windows, weatherboard, eaves) | Gloss | 1 coat primer<br>1 undercoat<br>1-2 coats enamel | ✓ | ✓ | ✓ | ✓ | |
| | Matt | 1 coat primer<br>2 coats acrylic | ✓ | ✓ | ✓ | | ✓ |
| **Timber:** stained/natural (trim, doors, windows, weatherboard, eaves, fences) | Natural<br>Opaque Matt<br>Stain | 3 coats polyurethane<br>3 coats acrylic<br>3 coats stain | ✓<br><br>✓ | ✓<br><br>✓ | ✓<br><br>✓ | ✓<br><br>✓ | ✓ |
| **Masonry:** (brick, fibre cement, cement render, stucco concrete block) | Gloss | 1 coat primer<br>1 coat undercoat<br>1 coat enamel | ✓ | ✓ | ✓ | ✓ | |
| | Matt | 1 coat primer<br>2 coats acrylic | ✓ | ✓ | ✓ | | ✓ |
| **Concrete:** (paving, floors, footpaths, etc.) | Gloss | 2 coats polyurethane<br>2 coats paving paint | ✓<br>✓ | ✓<br>✓ | ✓<br>✓ | ✓<br>✓ | |
| **Metals:** Ferrous metal | Gloss | 1 coat general purpose metal primer or cold galvanising<br>1-2 coats enamel | ✓ | ✓ | ✓ | ✓ | |
| | Matt | 1 coat general purpose metal primer or cold galvanising<br>1-2 coats acrylic | ✓ | ✓ | ✓ | | ✓ |
| Galvanised steel | Gloss | 1 coat galvanised iron primer<br>1-2 coats enamel | ✓ | ✓ | ✓ | ✓ | |
| | Matt | 1 coat galvanised iron primer<br>2 coats acrylic | ✓ | ✓ | ✓ | | ✓ |
| Roofing | Gloss | Contact paint companies | ✓ | | ✓ | ✓ | |

### Surface—interior

| | Type of finish | Paint system | Brush | Roller | Spray | Solvent | Water |
|---|---|---|---|---|---|---|---|
| **Timber:** (trim, doors, windows, furniture, cupboards) | Gloss | 1 coat undercoat<br>1-2 coats enamel | ✓ | ✓ | ✓ | ✓ | |
| | Matt | 1 coat undercoat<br>1-2 coats acrylic | ✓ | ✓ | ✓ | | ✓ |
| **Timber:** stained/natural (trim, doors, windows, cupboards, furniture) | Polyurethane<br>Gloss<br>Matt | 3 coats polyurethane<br>3 coats clear enamel<br>3 coats stain | ✓<br>✓<br>✓ | | ✓<br><br>✓ | ✓<br>✓<br>✓ | |
| **Plasterboard:** (cement render, plasters) | Gloss | 1 coat alkali sealer<br>1 coat undercoat<br>1 coat enamel | ✓ | ✓ | ✓ | ✓ | |
| | Matt | 2 coats acrylic | ✓ | ✓ | ✓ | | ✓ |
| **Masonry:** (brick, concrete block, cement render, fibre cement) | Gloss | 1 coat filler<br>1 coat alkali sealer<br>1 coat enamel | ✓ | ✓ | ✓ | ✓ | |
| | Matt | 1 coat filler<br>2 coats acrylic | ✓ | ✓ | ✓ | | ✓ |
| Concrete/timber floor | Gloss | 2 coats paving paint/<br>2 coats polyurethane<br>(1 pack—light duty)<br>(2 packs—med./heavy duty) | ✓<br><br><br>✓ | ✓<br><br><br>✓ | ✓<br><br><br>✓ | ✓<br><br><br>✓ | |
| **Metal:** Ferrous | Gloss | 1 coat general purpose metal primer<br>1-2 coats enamel | ✓ | ✓ | ✓ | | |
| Galvanised/Zincanneal | Matt | 2 coats galvanised iron primer<br>1-2 coats acrylic | ✓ | ✓ | ✓ | | ✓ |

# PAINTING • 71

| PAINT TYPE | Readily Available | Readily Available with special formulas | Special purpose paints |
|---|:---:|:---:|:---:|
| **EXTERIOR FINISHES** | | | |
| Gloss Enamel (Alkyd) | ✓ | ✓ | ✓ |
| Semi-gloss Enamel | ✓ | ✓ | ✓ |
| Matt Enamel | ✓ | ✓ | ✓ |
| Gloss Acrylic | ✓ | ✓ | ✓ |
| Semi-gloss Acrylic | ✓ | ✓ | ✓ |
| Flat Acrylic | ✓ | ✓ | ✓ |
| Roof Paint – for water collection | ✓ | ✓ | ✓ |
| Roof Paint - not for water collection | ✓ | ✓ | ✓ |
| Aluminium Paint | ✓ | ✓ | ✓ |
| Oils/Stains/Preservatives | ✓ | ✓ | ✓ |
| Acrylic Timber Finishes | ✓ | ✓ | ✓ |
| Clear Finishes | ✓ | ✓ | ✓ |
| Polyurethane | | ✓ | ✓ |
| Marine Systems | | ✓ | ✓ |
| Paving & Floor Paints | ✓ | ✓ | ✓ |
| Texture Finishes | | ✓ | ✓ |
| Lime Wash | ✓ | ✓ | ✓ |
| **INTERIOR FINISHES** | | | |
| Gloss Enamel (Alkyd) | ✓ | ✓ | ✓ |
| Semi-gloss Enamel | ✓ | ✓ | ✓ |
| Matt Enamel | ✓ | ✓ | ✓ |
| Gloss Acrylic | ✓ | ✓ | |
| Semi-gloss Acrylic | ✓ | ✓ | |
| Flat Acrylic | ✓ | ✓ | |
| Polyurethane | ✓ | | ✓ |
| Floor Polyurethane | | ✓ | ✓ |
| Clear Finish | ✓ | ✓ | ✓ |
| Lacquer | | | ✓ |
| Oils/Waxes/Stains | | ✓ | ✓ |
| Vinyl Paints | ✓ | | ✓ |
| **PURPOSE MADE PAINTS** | | | |
| Heat Resistant Paints | | | ✓ |
| Fire Retardant Paints | | | ✓ |
| Clear Silicone Paints | ✓ | | ✓ |
| Bituminous Paints | ✓ | ✓ | ✓ |
| Metal Cleaners | ✓ | | ✓ |
| Cold Galvanising | | | ✓ |
| General Purpose Metal Primers | ✓ | | ✓ |
| Galvanised Iron Primers | | ✓ | ✓ |
| Etch Primers | | | ✓ |
| Zinc Chromate Primers | | ✓ | ✓ |
| Wood Primers | ✓ | ✓ | ✓ |
| Undercoats | ✓ | ✓ | ✓ |
| Chemically Resistant Paints | | | ✓ |
| Alkali Sealers | ✓ | ✓ | |
| Oil Sealers | ✓ | ✓ | |
| Stain & Bleed Sealers | ✓ | ✓ | |
| Anti-Moulds | ✓ | | ✓ |
| Surfacers/Putties/Fillers | ✓ | ✓ | ✓ |
| Solvents | ✓ | | ✓ |
| Strippers | ✓ | | ✓ |

## HOW MUCH PAINT DO WE NEED?
Number of litres

| | WALLS — Length of room | | | | | | | CEILING/FLOOR — Length of room | | | | | | |
|---|---|---|---|---|---|---|---|---|---|---|---|---|---|---|
| Width of room | 2400 | 3000 | 3600 | 4200 | 4800 | 5400 | 6000 | 2400 | 3000 | 3600 | 4200 | 4800 | 5400 | 6000 |
| 2400 | 4 | 4 | 5 | 5 | 6 | 6 | 6 | 1 | 1 | 2 | 2 | 2 | 2 | 2 |
| 3000 | 4 | 5 | 5 | 6 | 6 | 6 | 7 | 1 | 2 | 2 | 2 | 2 | 3 | 3 |
| 3600 | 5 | 5 | 6 | 6 | 6 | 7 | 7 | 2 | 2 | 2 | 2 | 3 | 3 | 3 |
| 4200 | 5 | 6 | 6 | 6 | 7 | 7 | 8 | 2 | 2 | 3 | 3 | 3 | 4 | 4 |
| 4800 | 6 | 6 | 7 | 7 | 8 | 8 | 8 | 2 | 2 | 3 | 3 | 3 | 4 | 2 |
| 5400 | 6 | 6 | 7 | 7 | 8 | 8 | 8 | 2 | 3 | 3 | 3 | 4 | 4 | 5 |
| 6000 | 6 | 7 | 7 | 8 | 8 | 8 | 9 | 2 | 3 | 3 | 4 | 4 | 5 | 5 |

# 72 • BUILDING MATERIAL PROPERTIES AND USES

**ANALYSIS CHART**
Construction element.
Materials suitable for that element.
The available forms of that material as industry produces them.
The use of that form of material as:
A. Structural use.
B. As a finished material

—major advantages, significant disadvantages.
The suitability of those forms of material:
(i) for a subcontractor to use under contract to an OB, and
(ii) for OB to use as a 'hands on' material.

Code:  1 Most suitable
       2 Suitable
       3 Reasonable
       4 Unsuitable
       5 Not viable

| Construction Element | Materials Suitable | Available Forms | Use Struc. use | Use Finish use | Advantages | Disadvantages | Suitability For OB Use Sub-cont. | Suitability For OB Use Hands on |
|---|---|---|---|---|---|---|---|---|
| Footings | Concrete | Cast in place | ✓ | | Can be reinforced | Access for trucks | 1 | 2 |
| | Masonry | | ✓ | | Simpler for small jobs | Discontinuous member, not as strong as concrete | 2 | 4 |
| Subsoil drain | Masonry | Terra cotta pipes | | | If locally available | Multi-unit, multi-step installation process | 2 | 4 |
| | Plastic | Straight pipe | | | Generally available, simple to use | Some types crush easily | 2 | 2 |
| | | Corrugated/flexible | | | | | 1 | 1 |
| Surface drains | Concrete | Cast in place | ✓ | | Availability | Skilled installation | 2 | 3 |
| | | Precast units | ✓ | | | | 3 | 4 |
| Retaining walls | Masonry | Bricks | ✓ | ✓ | Easy to lay | High height: thickness ratio | 2 | 4 |
| | | Concrete blocks (reinforced) | ✓ | ✓ | High strength if reinforced | Multi-stage construction | 2 | 4 |
| | | Stone | ✓ | ✓ | Good appearance | Expensive, hard to get in some locations | 3 | 4 |
| | Concrete | Cast in place | ✓ | ✓ | High strength | Skilled labour | 2 | 5 |
| | | Precast units (inc. interlocking) | ✓ | ✓ | Flexibility | Skilled labour | 2 | 2 |
| | Timber | Interlocking units | ✓ | ✓ | Flexibility | Will eventually decay | 3 | 2 |
| Ground to floor systems | Masonry | Bricks | ✓ | ✓ | Flexibility | | 1 | 2 |
| | | Concrete blocks | ✓ | ✓ | Flexibility | | 2 | 2 |
| | | Stone | ✓ | ✓ | Appearance Only | Availability, labour intensive | 4 | 4 |
| | Concrete | Cast in place | ✓ | ✓ | High strength | Skilled labour | 2 | 4 |
| | | Concrete stumps | ✓ | ✓ | Flexibility | Skilled labour | 1 | 1 |
| | Timber | Posts/poles | ✓ | ✓ | Local Availability | Heavy to handle, indefinite life | 3 | 2 |
| | Metal | Steel posts/columns | ✓ | ✓ | Prefabrication high | Skilled labour, on-site fabrication | 3 | 3 |
| Floor framing | Timber | Sawn | ✓ | | Cheap, forgiving | Susceptable to biological attack | 1 | 1 |
| | | Glue laminated units | ✓ | ✓ | Wide span | Expensive | 1 | 1 |
| | | Timber trusses | ✓ | ✓ | Wide span | Inflexible | 3 | 2 |
| | Metal | Steel beams | ✓ | | Wide span | Expensive, heavy handling | 2 | 3 |
| | | Steel channels | ✓ | | Available secondhand | Connections | 2 | 3 |
| | | Galvanised SHS/RHS | ✓ | | Strong and termite free | Few compared to timber | 1 | 1 |
| Floor finish | Timber | Strip flooring | ✓ | ✓ | Flexible appearance | Skilled labour | 1 | 3 |
| | | Chip board flooring | ✓ | ✓ | Cheap, fast | Must be covered, joints | 1 | 1 |
| | | Plywood | ✓ | ✓ | Strong, fast | Must be covered | 1 | 1 |
| | | Fibre cement | ✓ | | Decay free | Heavy, expensive, hard to cut | 1 | 2 |
| Concrete floor slabs | Concrete | Cast in place | ✓ | ✓ | Strong, durable, flexible | Requires engineering design, skilled labour to finish slab | 1 | 3 |
| | Composite | Precast T beams Fibro form panels Insitu wearing slab | ✓ | | Economic method of suspended floor | Complex to set out | 2 | 3 |

# BUILDING MATERIAL PROPERTIES AND USES • 72

| Construction Element | Materials Suitable | Available Forms | Use Struc- use | Use Finish use | Advantages | Disadvantages | Suitability For Owner Builder Use Sub-cont. | Suitability For Owner Builder Use Hands on |
|---|---|---|---|---|---|---|---|---|
| | | Permanent Steel formwork Insitu wearing slab | ✓ | ✓ | Good finish to underside no form-work to remove | More expensive than formed concrete | 1 | 2 |
| Walls 37-45 External wall structure | Masonry | Bricks (structural grade) | ✓ | ✓ | Cheap, flexible, lasting | | 1 | 2 |
| | | Concrete blocks | ✓ | ✓ | Cheap, flexible, modularised | Detailing | 1 | 2 |
| | | Stone | ✓ | ✓ | Appearance | Expensive, skilled labour | 4 | 4 |
| | | Mud bricks (Adobe) | | | Cheap | Finding right soil | 5 | 1 |
| | Timber | Sawn framing | ✓ | | Flexible, forgiving | Shrinkage | 1 | 1 |
| | | Dimensioned framing | ✓ | | | | 1 | 1 |
| | | Posts/poles | ✓ | ✓ | Local available, difficult sites | Inflexible handling, elaborate detailing | 4 | 3 |
| | Metal | Structural steel | ✓ | | Strength | Inflexibility, unforgiving | 2 | 2 |
| Internal wall structure | Masonry | Bricks | ✓ | ✓ | Sound proofing, strength | Must be rendered at least one side | 1 | 3 |
| | | Concrete blocks | ✓ | ✓ | Sound proofing, strength, face both sides | Detailing | 1 | 2 |
| | | Stone | ✓ | ✓ | Appearance strong, sound proof | Thickness, expensive | 4 | 4 |
| | Timber | Sawn framing | ✓ | | Flexible, forgiving | Shrinkage, low sound proofing | 1 | 1 |
| | | Dimensioned framing | ✓ | | Clean, easy to handle | | 1 | 1 |
| | | Composite panels | ✓ | ✓ | Modularised | Modularised (limiting) jointing | 3 | 2 |
| External wall cladding (and finishes) | Masonry | Cement render | | ✓ | Appearance | | 1 | 4 |
| | | Glass blocks | ✓ | ✓ | Thermal efficiency | Expensive | 2 | 4 |
| | | Brick (veneer) | | ✓ | Lasting appearance | | 1 | 3 |
| | | Stone (veneer) | | ✓ | Appearance | Expensive skilled labour | 3 | 4 |
| | | Simulated masonry | | ✓ | Cheap | Detailing structure | 3 | 2 |
| | Timber | Weatherboards | | ✓ | Range of profiles, appearance | Maintenance | 1 | 2 |
| | | Plywood | | ✓ | | Appearance, detailing | 2 | 2 |
| | | Shingles/shakes | | ✓ | Appearance | Expensive | 3 | 3 |
| | | Hardboard planks | | ✓ | Durable, ready to paint, cheap | | 1 | 1 |
| | Metal | Prefinished | | ✓ | | Problems in refinishing, susceptible to damage | 1 | 5 |
| | Sheet products | Fibre cement sheets | | ✓ | Cheap | Detailing, jointing | 1 | 1 |
| | | Fibre cement | | ✓ | Cheap | Detailing, jointing | 2 | 2 |
| | | Plastic panels | | ✓ | 'potential use' | Untried under Australian conditions | | |
| | | Composite tiles/sheets | | ✓ | | | | |
| Interior linings (and finishes) | Masonry | Cement render | | ✓ | Flush finish on masonry | Skilled labour, wet trade | 1 | 4 |
| | | Simulated masonry | | ✓ | Appearance | | 3 | 2 |
| | Timber | Lining boards | | ✓ | Appearance, flexible | Expensive | 2 | 1 |
| | | Plywoods | | ✓ | Appearance | Jointing, fixing | 2 | 2 |
| | | Particle board | | ✓ | Solid, good sound proofing | Jointing, fixing | 2 | 2 |
| | | Shingles | | ✓ | Appearance | | 3 | 3 |
| | | Hardboard sheets, planks | | ✓ | Cheap, ready to paint | Jointing, buckling | 1 | 1 |
| | | Canite | | ✓ | Cheap | Damages easily, jointing | 1 | 1 |
| | Sheet products | Plasterboard | | ✓ | Cheap, flush, forgiving | Susceptible to moulds | 1 | 2 |
| | | Fibrous plaster | | ✓ | Cheap, flush, forgiving | Not commonly available | 1 | 3 |
| | | Plastic sheets | | ✓ | Alternative to tiles | Jointing, fixing | 1 | 1 |
| | | Fibre cement | | ✓ | Cheap, water resistant | Unattractive | 1 | 1 |
| | | Fibre cement flush jointed | | ✓ | Flush finish, durable | | 1 | 3 |

# 72 • BUILDING MATERIAL PROPERTIES AND USES

| Construction Element | Materials Suitable | Available Forms | Use Struc-use | Use Finish use | Advantages | Disadvantages | Suitability For Owner Builder Use Sub-cont. | Suitability For Owner Builder Use Hands on |
|---|---|---|---|---|---|---|---|---|
| Windows | Timber | Sliding, awning, double hung | | ✓ | Appearance, flexibility | Maintenance | 1 | 2 |
| | Metal (aluminium) | | ✓ | | Durable limited range | Inflexible, | 1 | 2 |
| Doors | Timber | External - glazed | | ✓ | Appearance, design flexibility | Expensive | 2 | 2 |
| | | - paneled | | ✓ | Suitability | | 2 | 2 |
| | | Internal - glazed | | ✓ | | Expensive | 1 | 2 |
| | | - panelled | | ✓ | Flexibility | | 1 | 2 |
| | Metal | External - glazed | | ✓ | Appearance, relatively inexpensive | Limited size range | 1 | 2 |
| Roof structure | Timber | Sawn | ✓ | ✓ | Flexible | Spans limited by economics and size availability | 1 | 3 |
| | | Dimensioned Prefabricated truss | ✓ | | Large span, relatively low cost | Less flexible than stick built | 1 | 1 |
| | Metal | Prefabricated truss | ✓ | | Large spans | Coverings and fittings | 2 | 2 |
| | concrete | Cast in place | ✓ | ✓ | Durable, walk on | Expensive, hard to waterproof, need high quality control, skilled labour | 2 | 4 |
| | | Precast units | ✓ | ✓ | Durable, walk on | Expensive, hard to waterproof, need high quality control, skilled labour | 2 | 3 |
| Roof cover | Masonry | Stone tiles (slate) | | ✓ | Appearance | Expensive, heavy | 2 | 4 |
| | | Concrete tiles | | ✓ | Durability | Inflexibility, heavy | 1 | 4 |
| | | Terracotta tiles | | ✓ | Durability, appearance | Inflexibility, heavy | 1 | 4 |
| | Metal | Steel decking | (✓) | ✓ | Low pitch, long life | Expensive | 1 | 1 |
| | | Steel sheet | | ✓ | Flexibility, lightweight | | 1 | 1 |
| | | Aluminium sheet | | ✓ | Long life | Thermal noise, corrosion | 2 | 2 |
| | | Metal roof tiles | | ✓ | Lightweight | Due to dissimilar metal connections | 2 | 3 |
| | Sheet products | Fibre cement shingles | | ✓ | Appearance | Fixings | 2 | 3 |
| | | Composite shingles | | ✓ | Appearance | Expensive, fixings | 2 | 3 |
| | Timber | Shingles/shakes | | ✓ | Appearance | Expensive, fixings | 3 | 4 |
| Ceiling structure | Timber | Dimensioned | ✓ | | Flexible, cheap | Spans, deflection | 1 | 2 |
| | | Sawn | ✓ | ✓ | | | | |
| | Metal | Steel channel | ✓ | | Interlocking, modular | Modular | 2 | 1 |
| | | Patent system | | | | | | |
| Ceiling finish | Timber | Lining boards | | ✓ | Appearance | Fixings | 2 | 2 |
| | | Plywoods | ✓ | ✓ | Appearance | Fixings | 2 | 2 |
| | | Particle board | ✓ | ✓ | Smooth surface, durable | Heavy | 2 | 2 |
| | | Hardboards/Canite | | ✓ | Cheap | Deflection, jointing | 1 | 2 |
| | Sheet products | Plasterboard | | ✓ | Cheap, flush finish | | 1 | 2 |
| | | Fibrous plaster | | ✓ | Cheap, flush finish | | 1 | 3 |
| | | Fibre cement (flush joint) | | ✓ | Cheap, flush finish | | 2 | 3 |
| | | Composite sheets | | ✓ | Modular, appearance finished | Modular | 2 | 1 |

# BUILDERS • 73

Generally, OBs will not be required to employ a builder, but it is worthwhile understanding the position that a builder takes in the house building process.

In Australia the majority of house builders are initially trained as carpenters and then go on to become builders either by calling themselves builders or by doing further study, generally part-time at a technical college.

This is the traditional description of the Australian house builder. But this description is changing. Now many builders have never been tradespeople but have gone directly to a technical college, college of advanced education or university to study building, not only as a process but also as a business.

As the states legislate control over the house building industry, the educational standard of builders is slowly being increased. However, education alone does not make a good builder. Much of the work revolves around the ability to co-ordinate a loose-knit band of subcontractors into a team that produces a house, on time and of good quality, and make sure that a profit margin is maintained.

The building industry is one of the few in which competitive tendering is still expected. Few of us would consider getting competitive tenders on our new car. We may shop around for the best price—but try writing to a car dealer asking them to send a tender price on a car, then asking them to hold the price for a couple of months while you think about it, then when you pick it up not pay for it for 30 days plus 7 days to process the cheque and withold 5 per cent of the purchase price for three months against any possible defects that may occur.

Few, if any, car dealers would agree to such terms but we commonly ask builders to work under these conditions.

Builders have to be financial jugglers. On the one hand, the purchaser of the home will attempt to delay payment for as long as practicable to reduce mortgage interest payments—and at the same time avoid paying the builder for any rises in labour and materials that could not have always been predicted.

On the other hand, the suppliers of labour and materials to the builder require payment on the best possible terms. In the case of some suppliers, particularly brick companies, advance payment is demanded and it takes a strong builder to refuse to pay a group of bricklayers in cash by lunchtime on a Friday.

This is not an argument for home purchasers to be kind to their builders and agree to all demands, but it does point out the imbalance that often occurs in building contracts.

OBs seldom employ builders, for this would normally defeat the purpose of being an OB. Many of the savings that can be achieved by the OB are simply in the removal of the cost of the builder's administrative role and profit margin. How much is to be saved by removing the builder from the building process and what, if any, costs are associated directly with that removal?

Builders normally aim to recover the cost of their time spent on a project doing administrative work, plus gain a profit that allows them to remain in business and keep ahead of inflation and other outgoings that reduce their income. In general terms, the builder is normally looking to achieve 20–30 per cent on top of the cost of materials and labour.

A significant proportion of their gross profit is expended to service normal business expenses—bank overdraft, accountancy charges, mistakes made in the original estimates, vehicle and plant wear and tear, and inflation—but not limited thereto. Builders who achieve a net profit of 10–15 per cent of contract value are considered to be successful. This can be equated with the fact that most one-person builders can complete about $300,000 to $900,000 worth of contracts per year and therefore expect to have a taxable income of between $30,000 and $135,000 per year.

It is this 10–15 per cent that OBs can aim to save, plus a small proportion of the remainder of the builder's gross profit. Many of the charges included in this section will remain, even for the most resourceful OB.

The jobs that an OB would undertake if there was no builder employed would include, but would not necessarily be limited to, the following:
1. setting out the job
2. Providing sheds and amenities
3. Erecting hoarding and cross over
4. Arranging temporary services
5. Negotiating with and employing (sub)contractors
6. Paying (sub)contractors, every week
7. Appeasing inspectors from the council and every other bureaucratic authority
8. Ordering and arranging delivery of materials and supplies
9. Co-ordinating the job progress
10. Cleaning up after (sub) contractors
11. Learning a new language—the building jargon
12. Fixing any shortfall in deliveries immediately
13. Controlling the weather
14. Collecting taxes
15. Dealing with Departments of Fair Trading
16. Solving demarcation problems
17. Writing contracts that protect the OB but are binding on the (sub)contractor
18. Answering instantly any question asked by any person about any aspect of the job.

Although OBs normally shun using builders, there are occasions when OBs may find it opportune to use one. The most common of these occasions is when an OB employs a builder to construct the proposed house up to the stage termed 'lock up'. This means that the builder is employed to do the site works, the structural frame, the cladding and roofing and install the doors and windows, or in other words do the minimum of work to achieve a building that is weather- and security tight.

By using a builder to lock up stage the OB reduces the time delays often associated with weekend building and improves the control of the subcontractors on-site who would otherwise be unsupervised during the working week. Also the OB is then relatively unhampered by poor weather conditions in bringing the house to completion.

Further, many lending sources are still nervous about lending money to untried OBs and the employment of a builder to bring the house to lock up can often soften the conditions of borrowing imposed on OBs. OBs may find that there are still significant savings to be made by employing a builder to achieve lock up, for if they can borrow long-term, lower interest money from the lending sources for this part of the work and even mortgage their land, they have the opportunity to use their saving or short-term borrowing to complete the house with minimum delay.

## The case for using a Builder

Note: In most states of Australia, OBs have been more and more restricted over time. Once it was a right, now it is seen as a special and very limited privilege granted by governments.

OBs will often be required to gain a special licence to build or modify their own home and will be limited as to how many times and at what frequency they can build a house for themselves.

Other restrictions include taking out home owners warranty insurance policied to protect future purchasers of the house from faulty work practices and structural failures.

In some states the tradespeople all have to be licenced or in some way registered. They will often have to provide warranty insurance for the work they do and the OBs will be responsible to keep these policies and pass them on to purchasers, if the house is sold to another party during the period of the warranties.

And don't forget all goods and services attract GST, so most OBs will need an ABN if they want to collect the tax credits. See an accountant and insurance broker before committing to be an OB.

# 74 • CITY CONDITIONS

**Wind direction**

| City | am | pm |
|---|---|---|
| Adelaide | NE | SW |
| Brisbane | SW | NE |
| Canberra | NW | NW |
| Darwin | SE | NW |
| Hobart | NNW | W |
| Melbourne | N | S |
| Perth | E | SSW |
| Sydney | W | NE |

**Temperatures (°C)**

| City | Jan | July |
|---|---|---|
| Adelaide | 29 | 15.5 |
| Brisbane | 30 | 20 |
| Canberra | 29 | 11 |
| Darwin | 32 | 30 |
| Hobart | 21 | 11 |
| Melbourne | 25 | 13 |
| Perth | 29 | 16 |
| Sydney | 26 | 17 |

**Rainfall (mm)**

| City | Jan | July |
|---|---|---|
| Adelaide | 19 | 63 |
| Brisbane | 146 | 47 |
| Canberra | 52 | 40 |
| Darwin | 410 | 0 |
| Hobart | 46 | 55 |
| Melbourne | 46 | 49 |
| Perth | 10 | 180 |
| Sydney | 98 | 123 |

## Bushfire considerations

The easiest way to avoid having a house burnt to the ground during a bushfire is not to build in the bush. This appears obvious today, 5 January 2002, as I sit in my study in the Blue Mountains surrounded by the smoke of bushfires raging less than 5 kilometres away.

### Starting with the design

How much of the bush can be cleared around the house? It is recommended that a minimum cleared zone of over 15 metres is maintained; 40 metres of clearing is a significant improvement but it is not until a clearance of at least 100 metres is achieved that the risk is reasonably reduced.

It is of course possible to design a residence that would remain intact after a bushfire passes through. Any people in the house may not fare so well, as it is harder to fireproof people and provide them with breathable air while an extreme-temperature wild fire consumes all the oxygen.

Plan your house to have minimal flammability by taking the following steps:

- Select building plan shapes that will avoid trapping debris and fire, in nooks and crannies.
- Select a roof shape that is easy to keep free of debris, with gutters that can be easily kept clean. In very high-risk areas, removing gutters may be an option.
- Select windows that can at least have premium wire gauze on all openings.
- If possible, fit metal roller or hinged shutters over windows.
- Use the strongest glass available; double glazing, toughened or laminated glass may give better protection from radiant heat and flying debris than plain float glass.
- Avoid timber decks attached to houses, particularly those with open perimeters.
- Select incombustible materials wherever possible.
- Select sub-floor ventilators fitted with sliding covers.
- Fit a proprietary permanent-spray sprinkler drenching system.
- Think carefully about all design decisions that are made.

A timber-framed house, on a concrete slab or full-brick base, with full wall insulation and a plasterboard firecheck barrier directly under non-combustible cladding, roofed with a simple low-pitched metal roof with, sealed eaves and a combined sisalation/fibreglass insulation blanket between the sheeting and the rafters is likely to perform better than a brick veneer cottage.

Although a tiled roof, brick veneer cottage seems to have a totally incombustible exterior, it is highly vulnerable if there are any open vents or gaps that can let super-hot air or flames into the wall cavity to ignite the timber wall studs that are then inaccessible to firefighting. The eaves of the brick veneer cottage are often thin fibro sheets that can disintegrate in a super-hot environment and allow fire to gain access to the roof space, setting fire to the timber roof structure. Many tiled roofs do not even have sisalation under the tiles; this can mean that the roof can act like a flue and the fire quickly burns through the tile battens. The tiles then fall through, making even bigger flue holes, and the fire becomes uncontrollable.

Anyone considering building in a bushfire prone area should get all the latest information on bushfire-proof house design from the local council authorities, the fire brigade, the CSIRO and similar bodies. It is advisable to consult an architect who specialises in the field of bushfire proofed house design.

# THE GREEN SUPPLEMENT • 75

## What is the perfect house?

Is it the house with street appeal that makes other people envious? Is it the house that accommodates all the family needs? Is it the most expensive house in the suburb with the highest real estate values?

It is unlikely to be any of these and there will never be a perfect house; but it is important that every house should meet the requirements of its occupants as closely as possible.

This supplement provides guidance that will assist everyone to achieve the best, most economic, environmentally friendly and sustainable house for them.

## Appearance

It is generally accepted that single-family houses should appear responsive to the local urban texture. This does not mean all houses should look the same, but that reasonable effort is made to design them to look as though they belong in their urban neighbourhood. This protects the local and individual household property values.

Why is appearance considered a primary factor in assessing the quality of a house? It is the street elevation that is the first visual contact made with a house, and its worthiness or not is often assessed at this time.

Returning after work, school or play; a welcoming appearance reconnects people to their home and family.

## Comfort

A home should respond positively to its occupants. It should feel fresh, be cooling when it is hot outside and feel warm, snug and safe when it is cold, windy and wet outside.

In most residences, comfort begins by locating the windows in the most suitable positions to take advantage of the local conditions. Living rooms and bedrooms should generally have a northerly aspect. An easterly aspect will provide early morning sunlight but less direct daytime sunlight.

Windows facing west will gain afternoon sun and in many locations this aspect should be avoided, or at least should include active shading. However, there are areas in the southern latitudes or high country altitudes where the warming setting western sun is cherished.

South elevations of houses will receive no sun in the winter months and only glancing sun early in the morning and late in the evening in summer. Locations where there are cold winters will generally shy away from windows to the south, unless that is where the view is. In locations with temperatures over 20 degrees for the majority of the year, windows facing south may be advantageous.

Windows provide daylight, ventilation and solar access. They open to allow fresh air to enter and foul air to escape. It is important to design window locations and configurations to maximise cross ventilation. The best result is achieved where the window openings are low on the inlet side and high on the outlet side. If it is difficult to achieve cross ventilation airways, then controllable ceiling vents with breeze-assisted cowls will allow used air be displaced by incoming fresh air.

Carefully designed eaves or similar awnings can control sunlight falling on north-facing windows. The solar heating gain can be stored in the building if it is absorbed into an efficiently located thermal mass. For example a concrete floor with insulation between the slab and the ground provides excellent thermal mass. Internal brick or concrete block walls will provide similar results.

ACC products and other lightweight building materials have low thermal mass but high thermal insulation. They can reduce heat transfer but have low heat storage capacity

East and west facing windows need active external screens, blinds or louvers to control sun penetration.

The low elevation of the sun in these sectors limits the effectiveness of eaves and similar projecting elements.

Many builders use double or even triple glazing to reduce solar gain and to insulate windows. This is an expensive, complicated and often imprecise way to achieve sun control and thermal efficiency. Instead, consider active external sun screening in combination with insulated internal drapes.

Many modern houses have extreme areas of glazing. It is hard to resist large panorama windows where there is a breathtaking view, but consider designing the window for the view and resist taking glass right down to floor level, when a sill height of 450mm or more could be achieved without diminishing the view. It is relatively easy to insulate a wall to R4.0 insulation value or higher, but it is much harder to achieve this level of insulation even with expensive multi-glazed windows and thermal breaks in the frames.

After designing an effective window and natural ventilation system, the next important consideration is to design suitable and appropriate insulation in the external walls, under the roof and within the ground level floor. There are a number of computer programs available to design passive insulation systems for residences. The best of these programs are location sensitive, often using postcodes as a reference. Insufficient or poorly installed insulation can reduce the comfort of the interior of a house. There is also little to be gained from over insulation.

In many parts of Australia, a correctly insulated house where the windows have opening sashes, are logically located and properly shaded, is all that is needed for a basically comfortable environment, where the temperature is seldom less than 15 degrees or above 30 degrees. Steady breezes and a humidity below 70% RH are generally comfortable for most people.

In very cold climatic areas central heating may be required, but in many locations an enclosed locally-warmed living room is sufficient if combined with sensible clothing.

Tropical locations have their own special considerations. People either have to adjust to high temperature and high humidity climates or they shut everything down and turn on air-conditioning. Air-conditioning generally works by consuming energy to move energy and is unsustainable, i.e. to absorb heat energy from one place and transport it to another to release it.

It is important to be comfortable, but it is incumbent on all households to minimise the amount of energy used and carbon released in achieving suitable comfort levels.

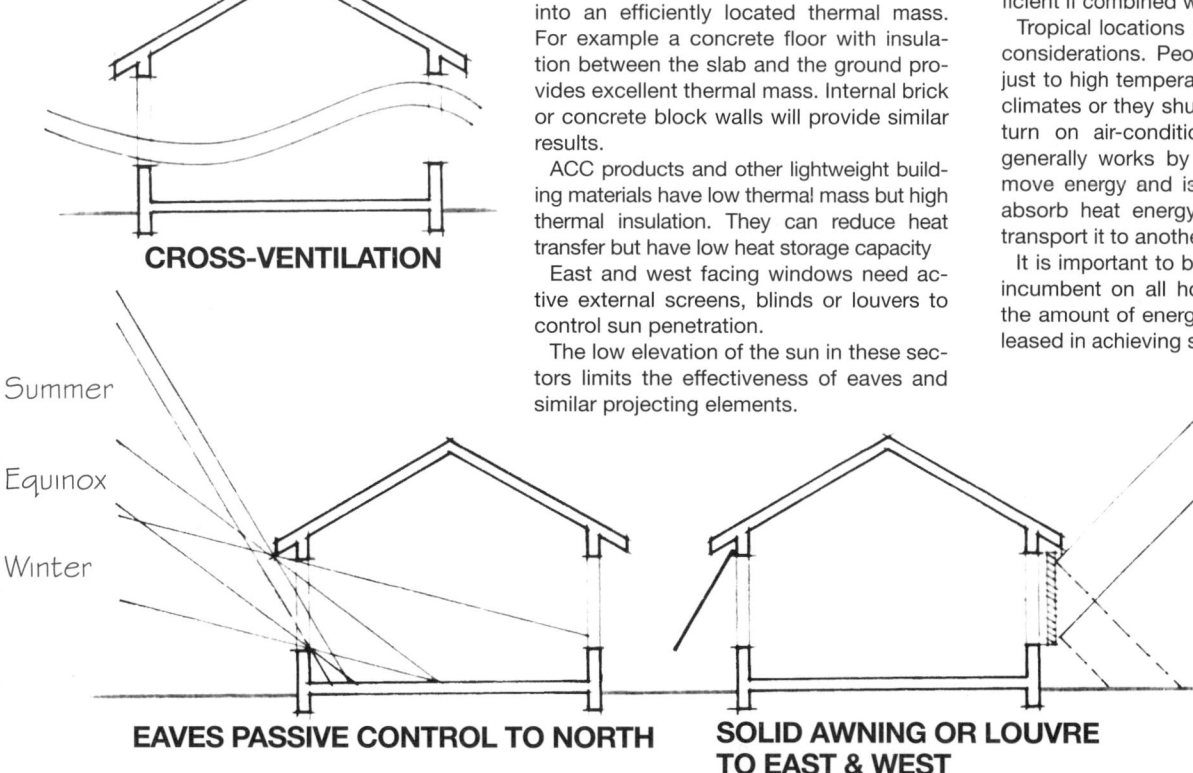

**CROSS-VENTILATION**

**EAVES PASSIVE CONTROL TO NORTH**   **SOLID AWNING OR LOUVRE TO EAST & WEST**

# 75 • THE GREEN SUPPLEMENT

## Ecologically sustainable

Our aim should be to minimise our impact on the Earth's available resources. Houses, particularly in developed societies, often draw heavily on these resources. Carbon is the element that allows life to exist and it follows that understanding and fostering good carbon transfer policies underpins the ecological sustainability of our planet and the future of humankind.

If, as our ancestors did, we constructed our dwellings from the materials immediately available in the surrounding environment, there would be little or no impact on the Earth's ecology. Any stone used would simply be rearranged close to where it was found and any timber used would be cut down but the natural cycle of the timber eventually rotting and returning to the earth is maintained.

Dwellings constructed by utilising only the energy of human and animal muscle are seldom more than basic shelter. When metals and ores were extracted, refined and forged into tools; energy from carbon-rich fuels was transferred into the metal and the waste carbon released into the atmosphere as carbon dioxide.

For centuries the carbon dioxide was recovered from the atmosphere by trees and other vegetation and the carbon cycle maintained a sustainable balance. Now the carbon cycle is significantly out of balance, there is insufficient vegetation to absorb the carbon gases in the atmosphere and the Earth's ecology is struggling to cope.

How can owner builders help rebalance the ecology? Although an individual's impact on ecological sustainability may appear small, it is the combined waste of billions of people that is the problem. Every individual has a responsibility throughout their lives to aim to maintain or improve the Earth's ecological balance.

Owner builders can contribute by:
- Using materials that require a minimum of carbon-derived energy to manufacture and transport them.
- Constructing buildings with the longest possible lives. This includes reusing and adapting existing structures rather than demolishing and building new.
- Considering the building's orientation to achieve the best value from the sun, breezes and other micro-climatic factors.
- Choosing all materials, fittings and fixtures for their longevity, durability and, where appropriate, their ability to be re-used or recycled.
- Thinking, researching, considering and applying logical thought to all decisions needed to provide a home. It is difficult to always make rational decisions, but careful consideration is a step to an ecologically sustainable future.

## Environmentally efficient

The key factor in environmental efficiency is to mitigate against factors that reduce the quality of the natural environment.

Reducing reliance on energy from greenhouse gas emitting power stations is a critical step in reducing environmental damage.

Designing or adapting houses in a manner that lowers reliance on mains power electricity is a sensible starting point.

Other considerations should include:
- Greater use of natural daylight through placement and sizing of windows.
- Choosing lighting systems that have low energy usage.
- Choosing human comfort systems to provide sensible levels of warming or cooling comfort in conjunction with sensible clothing.
- Choosing appliances and home entertainment systems with the lowest available energy consumption that will provide the desirable level of efficient operation. Take care that the ratings are based on reality; low energy ovens are not always efficient cooking appliances.

Waste heat is the major by-product of electrical energy consumption no matter what the source of the energy. So-called green energy systems still contribute to global warming by simply adding heat to the atmosphere. Having a personal solar photovoltonic or wind generation system should not be an excuse to indiscriminately waste energy.

The quantity of mains distributed reticulated potable water consumed by households is of great concern in many localities. It is sensible that all water used should be treated with respect, but it is not necessarily always true that the less water that passes through an individual residence's water meter the better the environmental result.

Clean water is important for hygienic household environments, for washing and preparing food, for washing clothes and for irrigating gardens. It was a major contributor to the reduction of epidemic diseases so rampant right through into the early part of the 20th century.

Recommendations and regulations requiring the reuse and harvesting of rain, storm and grey water is laudable in areas where there is a demonstrable shortage in the capacity of the local catchment or river system to deliver sufficient clean water. Take care and seek reliable advice on safe efficient methods of cleansing, sanitising and detoxification of dirty water. Get it wrong and disease and even death can result.

Liquid household waste has been a major environmental problem for humankind since the beginning of urbanised societies. During that long history there has been a constant search for the suitable conveyance and disposal of stormwater, sullage and sewage.

The modern systems of underground pipes conveying the liquid waste to treatment and safe disposal areas is only about 150 years in development and even today not all urban areas even in a developed country, like Australia, are connected to the most efficient collection, treatment and disposal systems.

There are sections of society, including some university academics and public authorities, that are promoting on-site at home treatment of liquid waste.

This is promoted as an alternative to the authorities collecting the liquid waste, treating it, warranting its quality and reticulating it back to the households from where it came.

The collection and treatment of roof-collected rainwater presents few problems. If treated by basic filtering and sanitisation, it can reach safe potable standards.

Even the collection of stormwater from hard-standing and garden beds can be stored and used for irrigation, toilet flushing and some cleansing purposes, with the minimum of treatment – some care is needed to avoid infectious bacteria being misted into breathable vapour, however.

Grey water or sullage is the low-grade dirty water from bathrooms and laundries; this generally contains predominantly grit and detergent residues. Some cleaning products contain phosphates and other chemicals, but grey water generally needs little treatment before it can be used for toilet flushing and garden irrigation. If it is not efficiently treated and filtered it may flush the toilet but leave difficult to clean scum marks. It is recommended that if grey water is used for garden irrigation it is not misted and that the acid/alkaline balance of the soil is tested regularly and adjusted if required.

Black water or sewage is liquid containing human body waste products plus cooking residues, particularly fats, oils and sugars. This liquid often contains live bacteria, virus, toxins and corrosive or volatile chemicals. Reasonably effective on-site disposal systems for sewage have been available for many decades; few treat the sewage to the point beyond where it is relatively safe for an on-site absorption or transpiration system. Most of these systems will eventually contaminate the disposal bed area and often leach out into underground watercourses and even onto natural extra-urban areas.

Complete on-site treatment of sewage through to potable water is possible but the systems required are; expensive, complicated, demand careful inspection regimes, regular maintenance and can still fail.

A complete on-site water harvesting, treatment and recycling system requires multiple tanks, multiple pumps, multiple treatment units and a complex network of pipes and valves.

In some localities in Australia new homes are required to collect rainwater and to recycle grey water – always check with local authorities to ensure access to the most recent regulations.

Whenever, a water collection and reuse system is to be installed, always get profession advice – be wary of marketing and sales promoting the best systems.

Utilising today's technologies it is probably possible to develop house designs that are comfortable, ecologically sustainable and environmentally efficient – but it is much harder to provide all these factors and achieve a universally acceptable appearance.

Check real estate advertisements:
- They have a picture to show the best possible street appeal.
- They list the number of bedrooms and bathrooms.
- They numerate the car spaces in garages, carports and on-site.

Unless demanded by law they seldom provide potential buyers with information on the comfort, ecologically sustainable or environmental efficiency of a house.

It is easy to promote the reality that:
**Unless houses lower their impact on the natural environment and are constructed to a high degree of sustainability; the future of human kind is endangered.**

It is difficult to gain universal acceptance of this truth at an individual level.

Australians have over 200 years of equating housing quality by evaluating:

### Location
*Best suburb*
*Best street*
*Prestige location*
*Panoramic views*
*On the high side of the street*
*Close to shops, schools, etc*
*Close to buses, trains, etc*

### Appearance
*Quality street appeal*
*Georgian, Victorian, Federation, Californian Bungalow, Art Deco, etc*
*Ultra modern, Fresh …*
*Set in an established garden*

### House size
*Large family home*
*Over 25 squares\*, 275 square metres*
*3 Living rooms, 4 Bedrooms, 2½ Bathrooms and 2 car Garage*
*Separate in-law accommodation*

\*Imperial measurement – One square = 100 square feet or approximately 9.3 square metres

### Lot size
*Quarter acre block\**
*Over 1000 square metres*
*Room for a pony*
*Room for a tennis court and/or swimming pool*

\* Imperial measurement – A quarter acre is about 1000 sq metres

### Potential resale value
*Most sought after area*
*Potential for higher density.*
*(With Council approval)*
*Renovator special, Needs some TLC*

They may be prepared to sacrifice size for what is considered to be a better location with views or prestige.

They are much less likely to give up any of the above for a smaller house with high environmental and sustainability credentials, even if the annual cost of maintenance and utilities is a significant saving.

No one wants to be a family with a house that is untested in the market and often constitutes a high proportion of a family's assets. House purchasers/owners are by nature conservative and, as a rule, avoid purchasing radical cutting-edge houses.

What follows are detailed methods of incorporating efficiencies and sustainability into houses without requiring a radical change to acceptable housing measurables.

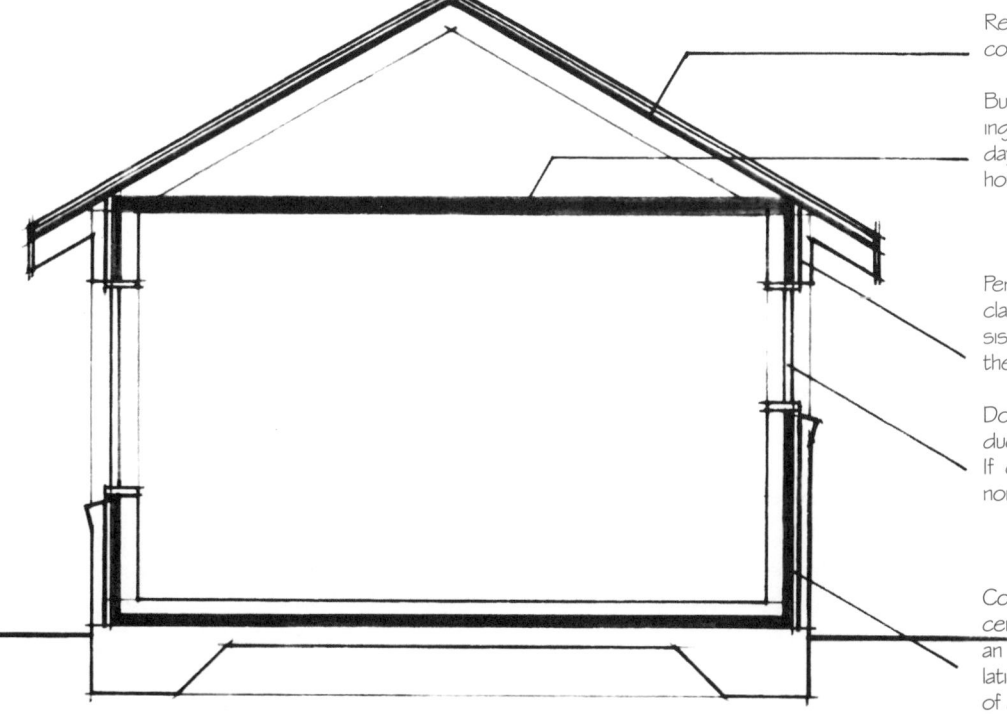

## THE SIMPLE INSULATION GUIDE
Always ensure ratings for insulation suit the local climate conditions and the construction of the house

# 75 • THE GREEN SUPPLEMENT

## 1. Credits for using existing buildings

What should be assessed prior to purchase?

### Orientation

Aim for a building that has good exposure to the northern sun or at least has the potential to be converted to provide a principal room window with this orientation.

In most Australian locations, avoid buildings where rooms face hot summer orientations into the late afternoon or daytime living rooms that only have windows facing south.

A gentle slope away from the house to the north – east is advised. Avoid dense evergreen planting that obscures the sun, but choose deciduous trees that provide shade in the summer months but allow the sun to shine through in winter.

Easterly breezes, at least along the eastern seaboard, are generally moderate and cooling. Become acquainted with the local microclimate as a matter of high importance, this is essential in ensuring the best natural ventilation is achievable.

- Natural environments rely almost exclusively on the sun for warm.
- Natural cooling requires air movement.
- Natural breezes that have passed over water are the most favoured.

Some families will compromise their search for a naturally advantageous orientation to buy a house with a seductive view.

No matter how good the view, if the orientation is poor the house is likely to be uncomfortable, unsustainable and eventually unsatisfying

Gently sloping, sunny and well-ventilated land wins every time over land that is steep, shaded and stagnant.

### Adaptability

Not every house will immediately be ideal for orientation or liveability but often a sturdy moderately-sized house in a desirable location is likely to be adapted to meet a family's requirements.

The rules are simple:

- Keep as much of the original fabric of the house as possible.
- Avoid excessive demolition.
- Build just as much as is needed to satisfy the family's needs.
- Renovate for improved environmental efficiency and ecological sustainability.
- Improve the appearance of the house because in the future when the house is to be sold on to another owner it is likely to improve the chances of a sale. Quality of design, neatly executed details and appropriate finishes should always be part of any building project.

It is always pertinent to have any house that is being considered for purchase checked for defects and compliance with local regulations by a qualified building surveyor and a pest inspector. It is also recommended that if a house is being purchased for adaption, a suitably experienced architect should provide advice on its potential prior to any purchase contract commitment being entered into.

### Structure

The structural walls of most Australian houses are either; load bearing masonry or lightweight framing with a weather excluding cladding. Most roofs are framed. Floors are generally limited to either framed and sheeted or reinforced concrete.

The oldest masonry houses in Australia are in New South Wales and Tasmania where they are mainly constructed from squared coursed sandstone rubble with dressed ashlar freestone being reserved for the street façade in all but the most expensive houses. Other states used other stone that was availably locally, particularly blue stone basalt in Victoria and a dense shale Bluestone in South Australia.

These were durable houses and some are over 200 years old and a significant proportion are over a century old. Most are still in sound condition. Their thermal mass means they provide slow temperature variability and a sense of wellbeing. From an environmental perspective they contain few carbon-credits, as most of the energy expended in their construction was human sweat. Renovating and upgrading for modern living can reverse this low impact status because houses of this age often need new floors, new roofing, replastering, extensive window frame maintenance or replacement, up-graded plumbing, safety up-grading of the electrical wiring, the installation of thermal insulation, repainting and decorating

When the supply of locally-available stone became expensive compared with the availability of low cost machine made bricks, then brick became the most common masonry. Mass manufactured bricks altered the carbon cycle significantly, high use of fossil-based energy is required to quarry the clay, power the plant and fire the kilns.

Where stonewalls and early brick walls were bedded on lime mortars that used relatively low amounts of energy in its manufacture, progressively the mortar used in brick walls became cement based. Cement manufacture requires high-energy inputs and is a major atmospheric polluting process.

Brick buildings have similar thermal mass when compared to stone buildings but progressively modern construction methods have allowed larger window opening. It is a sound decision to maintain an existing brick building rather than add to its energy debt by using extra energy in demolition and reconstruction.

Where any stonework is to be demolished there is almost always a useful use for it on-site and bricks bedded in lime mortar can be manually cleaned and reused.

Bricks bedded in cement mortar are very difficult to clean but there are now many organisations that will crush and recycle old brickwork (also stone and concrete waste) into useful building aggregates.

When stone or brick construction was not suited due to availability, cost or climatic factors, timber-framed houses were generally the alternative. Although there are examples of timber buildings over 100 years old, they are likely to have many defects and require extensive maintenance. Timber framed buildings if in sound repair are generally more adaptable than masonry buildings.

In most localities in Australia framed houses require thermal insulation to achieve all season comfort. Few Australian houses were adequately insulated before 1970 and many even after that date have less than adequate insulation.

The need to check older houses for insulation or the ability to easily retrofit insulation requires careful inspection and pre-purchase consideration. They often have low thermal mass, particularly older houses with timber floors, adding thermal mass into the existing enclosed building wall cavities is likely to be counterproductive but by introducing insulated concrete floors or internal masonry walls in additions should be considered.

There is almost always an environmental and ecological impact associated with adapting any existing timber framed house and these impacts should be evaluated carefully. Saving the maximum amount of existing building fabric is essential to gain maximum green credentials.

Australia may or may not have invented the brick veneer house, but it has been the dominant construction system for over 50 years. Only in recent years since the beginning of the 21st century has the BV begun to loose its dominance. So much of the housing stock is BV that it will be the most likely type of house many owner builders will attempt to convert. Watch out for:

- Trussed roofs that span from outside wall to outside wall. These houses require very specific load-bearing and bracing systems. Never cut through a truss without an engineer's approval.
- Complex concrete floor slabs with unseen cracks under the carpet particularly at counter-flexure points.
- Leaking drains under the floor slab. Cutting up a raft type slab to repair a drain is fraught with many unforeseen dangers.
- Poorly engineered cut and fill site levelling, that results on the house being partly-supported on stable ground and unstable fill.
- Older style aluminium windows with broken corner joints, missing seals, non-existing flashings and thin glass. These windows have almost no thermal insulation value.
- Porous cement-based roofing tiles without sarking.

# THE GREEN SUPPLEMENT • 75

- The majority of mass-market BV project homes will have factory-assembled timber stud frames. These are relatively easy to modify. However, there are also BV homes with metal stud frames which can be difficult to pick with a simple visual inspection because all the studs are concealed. Metal frames are unlikely to be as simple to modify as a timber stud wall, but with reasonable effort should not hold too many restrictions.
- Fibrous cement eaves linings and some wall sheeting may contain asbestos. It is always wise when buying any house, and critically important if alterations are to be undertaken, to have an asbestos audit carried out.

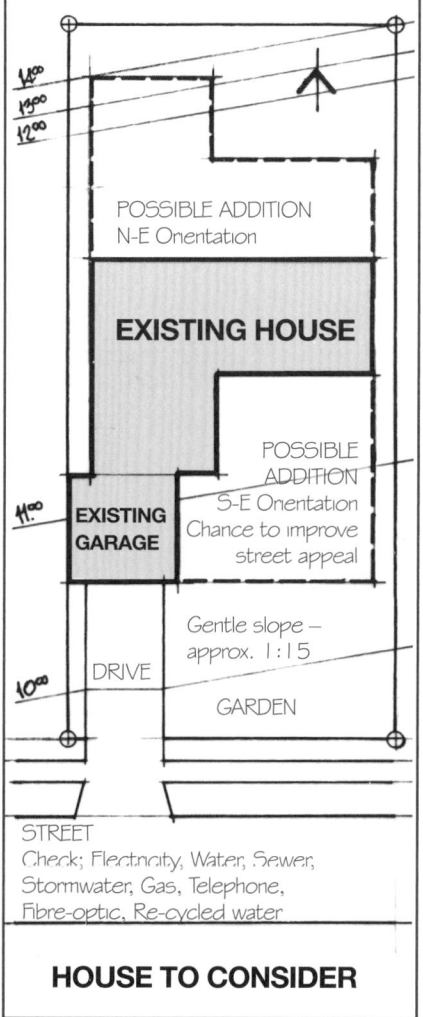

## Liveability

The house that will cost the least to adapt is likely to be the one that is the size required and has a plan layout close to your needs.

The urban belief that the worst house in the best street is the way to make a real estate killing is almost always a myth and is of little consequence if the house is to be occupied by a purchasing family.

It is poor planning to adapt a house without sensibly considering whether there is a market for the house after it is adapted.

In an emergency would a quick sale return all the costs associated with the project plus a worthwhile profit for the effort and resources expended?

The simple aims of liveability include:
- The living spaces suit an owner's uses.
- There are bedrooms for every family member who will live in the house plus if required extra sleeping space for guests.
- The kitchen suits the cuisine of the family.
- The bathing facilities are efficient and suit the family.
- Main rooms have windows that face the most suitable solar exposure for the house's location.
- The habitable rooms have controllable through ventilation.
- The occupants have sensible levels of privacy.
- The house is thermally comfortable and suited to the prevailing weather conditions.

## Condition of wiring, plumbing & drainage

Wiring is the essential component for the distribution of electricity.

Check with the electrical supply authority as to the adequacy of the consumer mains connected to the house.

Ensure that the location of the consumer mains and their point of attachment meet current requirements.

Investigate the current requirements, recommendations, bonuses and credits available if an on-site photovoltaic or wind driven electrical generation system is installed. Both these systems need to be carefully designed to suit specific locations; it is sensible to be well briefed on these matters before committing to a house purchase. Engaging a consulting engineer or a similarly qualified person from a reputable installation company to provide an on-site inspection and report is recommended. An incorrect decision at this juncture could lead to expensive installation requirements or even mean that an on-site electrical generation system cannot be installed.

Renew, upgrade or replace the switchboard, remove any obsolete fuses and circuit breakers and replace with current fittings.

Have an electrician check that the conductors are correctly sized, properly insulated, are adequately earthed and are free of any leakage to earth.

Remove all redundant electrical cabling, this will ensure an open ended cable cannot be accidentally re-connected to power with potentially fatal consequences.

Repair and upgrade during any adaption works, this is the most economic and least disruptive time to carry out the work.

Add new or upgraded circuits for any likely alterations to plant and equipment.

Plumbing is the pipes that supply household water from the street mains or from an on-site storage system.

Check if there is a working water meter and note down the reading at the time of taking possession of the property. If the meter is complicated, take a photograph of the dials or digital display. At an appropriate time turn off all the taps on the property, including the supply to the hot water service. Then monitor the water meter, if the readout shows water is being used there are three likely reasons:

1. Not all the taps were turned off or there are leaking taps, check and re-monitor.
2. There is a cracked pipe or loose joint somewhere in the system, these can be out of sight under the ground. Engage a plumber with the equipment to check for leaks.
3. Someone has connected a T piece into the water supply after the meter and is stealing water.

If none of these three reasons apply, then contact the water supply authority to have the meter checked.

Turn on and off every tap on the property, to make sure they all work. Ensure the flow rate at every tap, spout or showerhead meets local requirements. This can be carried out by timing how long it takes for a fully turned on tap, to fill a bucket of known capacity. If the flow rate exceeds the local requirements then the outlet will require modification or a new tap fitted.

Check the flow rate and temperature of water at hot water outlets. If these exceed the local regulation requirements adjustments may be required.

If there is a reticulated garden irrigation system installed, ensure it is of a type accepted by the local authorities. In some areas no fixed irrigation attached to mains supply is allowed. If irrigation from mains water supply is allowed it is still worthwhile to have the system checked to ensure that water is not wasted.

For the occupant's safety; check that a fully operational backflow protection valve is installed between the potable water supply and the irrigation system. This is required to ensure that the irrigation pipes cannot slowly seep contaminated water containing garden chemicals and manure into the drinking water supply.

Where there is a treated water reticulation system for garden watering and irrigation purposes, make sure that there can be no cross contamination between the potable and grey water systems. Pipes and taps used in a grey water system are required to be colour coded.

Generally a more relaxed code of practice has been used where properties use rainwater tanks or ground water for their water supply.

It is advisable that before completing purchase of any property with a water supply other than a mains supply, a consulting hydraulic engineer or similarly experienced person be engaged to check and report on the capacity and purity of the water in the system.

# 75 • THE GREEN SUPPLEMENT

## Sewerage

This is the system of fittings and pipes that convey the waste liquids, other than rain or stormwater, away from the property through underground mains or to an on-site disposal system.

There are still many old style, minimum treatment, septic tank systems operating throughout Australia. These are systems where liquid waste, including human waste, is partially treated in a holding tank and then slowly released into the ground to be absorbed or transpired by selected vegetation in to atmosphere. There are few areas where these systems comply with the local codes and it is recommended that they be upgraded with the minimum of delay to a high efficiency treatment system.

Where there is a septic system, there is also likely to be a grease-arresting trap to catch the greases and oils flushed down kitchen sinks. The removal of greases and oils from the wastewater ensure higher efficiency treatment. The installation of a dishwasher using high caustic cleansers and very hot water seriously reduces the efficiency of grease traps. Grease traps need to be cleaned out at regular intervals; qualified operators should carry out these services.

The majority of urban areas are connected to underground sewer mains. These mains convey the sewage liquid to a disposal site. The majority of sewer authorities treat the raw sewage before it is released onto specially-developed agricultural areas or into a river, bay or the ocean.

The quality of the treatment varies from locality to locality and varies from basic treatment to some modern plants that have the capacity to produce treated effluent that is suitable for irrigating parks and gardens. A few plants are capable of treating the sewage to the point where the liquid leaving the plant is purer than some main reticulation water supply systems.

In some localities refreshed water is reticulated for garden watering and external yard cleansing. A significant roll out of potable refreshed water is progressing but it requires expensive plant and extensive infrastructure upgrade.

There are sewage treatment systems for individual households. These systems are expensive to purchase, install and maintain and very few are suited to being retro-fitted to existing sewer drains.

Where the sewer system and the sullage systems are separately collected, treatment of the grey (sullage) liquid waste is relatively easy and safe to install. Some jurisdictions may require a grey water collection, storage, treatment and reuse system if an application is made to significantly alter or enlarge a residence.

## Stormwater and rainwater

Stormwater and rainwater are often confused.

Stormwater is all the water that falls on a property or flows through a property due to precipitation.

Rainwater is commonly defined as the water that falls as rain on the roof of the property or other specifically designated areas suited to the collection of rainwater that can be easily treated and used as a potable supply.

Collection and disposal systems for stormwater and rainwater vary extensively from place to place. In locations with no reticulated water supply, most residences will have a system of tanks to collect and store rainwater. Where there is reticulated watersupply the use of tanks has traditionally been influenced by local requirements.

In areas where the reticulated water is hard (has a high alkaline content) tanks often supplement the mains supply to improve the taste of drinking water and to provide a water supply that will suds up when used with soap or detergent. Other urban areas for many decades were almost tank free. This was particularly true in areas where the water supply authority had the power to regulated and limit the installation of tanks and to increase its revenue by charging for the mains supply at a water meter.

In recent decades authorities have had difficulty supplying mains water in the quantities expected and have moved to require new dwellings to be fitted with rainwater collection and storage systems. This requirement is also in place in some locations when extensive alterations or additions are carried out to an existing property.

Collecting and storing rainwater for use on a property where it falls makes sense and most people support the concept, but take care. The system should be carefully designed and fitted with a suitable cleansing system. It should be regularly maintained and kept in correct working order.

There is no point in collecting rainwater if it is not put to use. Incorporating it into the water supply with a system that calculates the maxima/minima range for storage verses use is essential if the best water use profile is to be achieved and the minimum amount of water is drawn from the mains. Part of this system is active human intervention – only turn on a tap when water is essential, limit the flow rate and the time the taps is turned on.

Stormwater that flows off the non-roofed areas of a property is less likely to be collected, as it may be polluted with fertilisers and other chemicals from gardens and oils and material brought on to the site by motor vehicles.

It is important to check what the local requirements are for rainwater and stormwater collection before purchasing a house for alteration and addition. This is particularly important in urban fringe bushfire prone areas, where the combined need to provide a large volume of passive water storage to fight bushfires plus rainwater tanks, grey water storage, treatment and reticulation systems and even a black water treatment system can require a considerable area of land and add significantly to the cost of development.

## Where to look for lead and asbestos

### Lead

Lead is a metallic element and has been one of the most commonly used construction materials throughout history. It is, however, a dangerous poison that can lead to diseases in humans that shorten lives.

Few products today contain lead, although some pure lead sheet is still used for special roof flashings where no other material is as suitable.

Until the introduction of plastic-based paints, most external and many internal enamel paints contain lead. A significant amount of these lead-based paints are still attached to older houses, particularly those constructed before the 1970s.

If the paint is in good condition there is minimal danger, but old lead paint is often cracked and flaky, although some may be concealed under later coats of acrylic paint. If lead-based paint is suspected and it should to be removed, contact a reputable paint supplier and determine the most efficient method to ensure safe, dust-free removal.

Paint manufacturers often tint modern primer/sealer paints to have a reddish colour. These primers are lead free but they mimic traditional red-lead primers that were used for generations to protect external woodwork. If a house dates back before 1970 take care, scrape back a small area of the surface paint carefully through to the bare timber and collect the scrapings. Have the scraping tested for lead content. If lead is found then proceed with great care to eliminate any chance of breathing in or ingesting any of the lead-based paint. Never remove paint containing lead by burning or heat treatment. Also, take care when using any volatile paint remover, read instructions very carefully and use only within the recommendations.

Some very old houses may still use lead pipes in the water supply. These should be replaced as soon as is practical. The more likely location of lead in the water supply is in the solder used to join pipes and to make joints in galvanised steel or copper tanks and other vessels. The risk to health from lead soldered joints may be low but not every authority agrees. It is recommended that at the very least all lead in the water supply should be progressively removed and replaced with materials considered to be unlikely to affect human health.

Sheet lead was frequently used to make roof and other weatherproofing flashings. Unless the water from the roof flows over

# THE GREEN SUPPLEMENT • 75

the lead and is then used for human consumption, there is unlikely to be any threat to the health of occupants. If lead has to be handled it is recommended that disposable gloves are worn or at least hands are very carefully cleansed after any contact.

## Future considerations

How far into the future should be considered when a house is being designed or purchased? Too many owners do not undertake enough time to consider and plan the future of the house they will own.

Once families had a house or sometimes a number of houses that stayed within the family and were passed on from generation to generation. This is less likely to be the case today and the family home has become a tradable asset.

There are many factors that need to be taken into account when deciding on home ownership. The most important consideration is likely to be how quickly and what return can be achieved if a house has to be sold.

It is likely that many families still believe that a house is a whole of life purchase; this is where houses can be confused with homes. A home is where the family lives; a house is a piece of real estate.

This may appear harsh, but a home will generally be happier and less stressed if the financial burden of home ownership is within the family's capacity.

Can any of the following be predicted?
- Divorce
- Income reduction
- Requirement to move
- An unplanned child
- Twins/triplets
- Widow/widower in-laws
- Guessing when an adult child will leave
- Guessing when an adult child will return
- Supporting adult child and spouse.

Plan the family's real estate purchase carefully and there should be enough flexibility to maintain a home suited to changing needs.

## Likely remaining life

It is important when purchasing, altering, extending or constructing a house to determine the likely life of the building and what level of maintenance will be required. The table on p250 provides a basic reference to the likely life of some of the major components of a house, with and without regular maintenance.

## Asbestos

This is probably the most dangerous material that has ever been used in building products. It is proven that asbestos fibres can become trapped in lung walls, causing irritations that can lead to mesothelioma, a deadly form of lung cancer.

For generations, asbestos fibre was combined with cement to make hard brittle building boards. Asbestos cement (AC) sheets can be found in nearly every Australian house that was constructed between the early 20th century until the 1970s and even later in some places. It can be found as wall sheeting inside and outside houses, as ceiling lining, as soffit lining and as eaves lining. There is even dense, high tensile sheeting designed to be moisture resistant flooring under tiled wet area floors.

It is highly recommended that a specialist consultant or certified contractor should check all fibrous cement products that are to be removed from a building and recommend the appropriate handling and disposal methods. The rules applying to the removal and disposal of AC products vary slightly from place to place but for personal and community safety it is always a wise decision to employ an approved contractor to remove and dispose of the material.

AC was also sold as a corrugated roofing material. These roof sheets, regardless of age, are almost certain to contain asbestos. Another use was in the manufacture of roofing shingles.

Some older heating and cooking appliances may contain asbestos insulation and it was often used to lag hot-water/steam pipes.

Check for it as the backing in electric meter boxes and switchboards and anywhere else heat insulation properties were required.

AC can even be found underground as it was sometimes used to make stormwater and sewer pipes.

It is generally considered that painted AC sheeting is reasonably safe to retain if it is regularly painted and no dust is allowed to escape into the atmosphere.

It is difficult not to recommend its complete removal from family homes, this is an insidious fibre that can reside in a human lung for decades before the setting of its cancerous time bomb – for which there is little treatment and the results are too often fatal.

## THE QUESTIONS TO ASK WHEN CONSIDERING A HOME PURCHASE

| | | | | | | | | |
|---|---|---|---|---|---|---|---|---|
| Are renovations required? | Y | N | How long will they take? | | | How much will they cost? | $ | |
| Are any major alterations or additions required? | Y | N | How long will they take? | | | How much will they cost? | $ | |
| Are there likely to be any additions to the family? | Y | N | Babies? | Y | N | Is extra space required? | Y | N |
| | | | In-laws? | Y | N | Is extra space required? | Y | N |
| | | | Children's Partners? | Y | N | Is extra space required? | Y | N |
| Can the proposed house meet all these needs? | Y | N | Would an alternative house be required? | Y | N | When would the alternative be needed? | | |
| Can any house meet all future needs? | Y | N | Would it be too large in the future? | Y | N | Could a large house be hard to sell? | Y | N |
| Think ahead; plan the family's future needs? | | | In the next 5 years | | | | | |
| | | | In 5 to 10 years | | | | | |
| | | | In 10 to 15 years | | | | | |
| | | | In 15 to 20 years | | | | | |
| | | | In 20 to 30 years | | | | | |
| | | | In 30 to 40 years | | | | | |
| | | | At active retirement | | | | | |
| | | | Less active retirement | | | | | |

# 75 • THE GREEN SUPPLEMENT

| COMPONENT | LIFE IF NOT MAINTAINED | LIFE IF WELL MAINTAINED | Always treat any asbestos material as required by local regulations. |
|---|---|---|---|
| Stone/brick footings | 50 to | 50 to 100+ | Soil moisture controlled. |
| Concrete footings | 75 to 125 | 75 to 125+ | Care required in reactive clay soils. |
| Concrete raft slab | 25 to 75 | 50 to 100 | Soil moisture controlled care required in reactive clay soils. |
| Brick piers | 25 to 50 | 50 to 100 | Check, re-plumb, re-point & re-pack. |
| Concrete stumps | 25 to 50 | 25 to 75 | Check, re-plumb & re-pack. |
| Steel sub-floor posts | 10 to 50 | 25 to 75 | Check fixings, re-plumb & check for rust, clean and reprotect. |
| Timber floor frames | 10 to 50 | 25 to 100+ | Use only durable hardwoods or treated timber. Check annually for termite attack and dry rot. Ensure well-ventilated dry subfloor space. |
| Light steel floors joist | 25 to 50 | 50 to 75+ | Check fixings & check for rust, clean and re-protect. Use proprietary systems s to manufacturers recommendations. |
| Suspended concrete floor | 25 to 75 | 50 to 100+ | Check for and correct any water entering slab. Detect and treat any visible rust or surface bulges indication rust. |
| Cavity brick construction | 50 to 75 | 50 to 150 | Check for cracks requiring foundation/footing rectification and joint re-pointing. Old walls may have sand/lime mortar that has deteriorated re-point with similar mortar – DO NOT USE A CEMENT MORTAR. |
| Brick veneer | 25 to 50 | 50 to 75+ | Check for cracks requiring foundation/footing rectification and joint re-pointing. |
| Cladding on stud framing | 10 to 25 | 25 to 50+ | Treat any asbestos material as required by local regulations. Remove and replace all rotting weatherboards. Prepare and re-paint at the intervals recommended by the paint manufacturer. |
| Flat roof, timber frame | 10 to 25 | 25 to 50 | Ensure adequate ventilation, flat roofs are prone to fungi/mould attack with potential material damage and health concerns. |
| Pitch roof propped frame | 25 to 75 | 50 to 100+ | Traditional roof frames should be checked regularly and any deterioration of members or joints corrected without delay. |
| Pitch roof truss frame | 25 to 50 | 25 to 75 | These have highly stressed small timber sections. Regular inspections and corrections are recommended. |
| Metal roofing | 10 to 25 | 10 to 50 | Most metal roofs are steel based and can rust, regular inspection and rust treatment at damage areas is essential. |
| Tile roofing | 15 to 35 | 25 to 100 | Terra-cotta tiles can last over 100 years if carefully maintained, check after ever storm. TC tiles can deteriorate from the underside in seaside locations. Cement tiles lose their colour coating between 15 and 30 years, they can be recoated. |
| Aluminium windows | 10 to 30 | 25 to 75 | Aluminium windows vary extensively in quality. There is little effective maintenance, so good quality is essential for long life. |
| Timber windows | 10 to 25 | 25 to 100+ | There are cheap timber windows made for the mass housing market, regular cleaning and painting will extend their life. Windows, using durable timber species will require less maintenance and last longer. |
| Plasterboard wall/ceiling | 10 to 20 | 20 to 60 | Damp houses, poor quality painting and rough treatment significantly reduce the effective life of plasterboard. Untreated moulds will attack the paper facing. |
| Interior paint | 5 to 25 | 20 to 75 | Careful preparation, sealing, undercoating and quality top-coats, combined with regular cleaning and recoating to the manufacturers instructions will extend paint life. Paint colours are fashion so a short life may be acceptable. |
| Exterior paint | 5 to 20 | 20 to 60 | Careful preparation, sealing, undercoating and quality top-coats, combined with regular cleaning and recoating to the manufacturers instructions will extend paint life. |
| Water supply | 25 to 50 | 30 to 70 | High-grade tubing and joints are the essential components of a long life water system. |
| Taps | 5 to 15 | 15 to 30 | To extend the life of ceramic washer taps, fit a grit removal strainer in the water supply. |
| Hot water service | 5 to 10 | 5 to 20 | The cost is a good indication of how long a HWS will last. |
| Solar HWS | 5 to 10 | 5 to 20 | The quality of the solar collector is critical. Always follow the manufacturers maintenance instructions. |
| Sewer Drainage | 20 to 40 | 25 to 75 | Modern sewer pipes are durable and resistant to root invasion. It is advisable to keep separation between tree roots and pipes. Clear blockages immediately, compacted blockages can damage pipes during clearing. |
| Stormwater drainage | 20 to 40 | 25 to 75 | Generally stormwater pipes will be a lower quality than sewer pipes and highly susceptible to crushing if close to the surface. Any crack attracts roots and terminal damage can result in a very short time. |
| Electrical wiring | 25 to 50 | 25 to 75 | Most electrical wiring is of high standard, the weak points are cheap switches, powerpoints and fittings. |
| Electrical light fittings | 5 to 15 | 10 to 25 | If they are cheap they are very likely to have very short lives. A quality fitting can last for many decades, take care of changes in technology that shorten the life of fittings. |
| Fixed electrical appliance | 5 to 10 | 10 to 25 | Read the manufacturer's maintenance instructions and apply if the maximum life is to be achieved. |
| Kitchen cupboards | 5 to 15 | 15 to 35 | The board quality and connecting hardware determine the life of kitchen cabinets. The board quality and connecting hardware determine the life of kitchen cabinets. Many high quality cabinets could last past 35 but are likely be replaced by a new trendy fashion before then. |
| Sanitary-ware | 10 to 20 | 15 to 35 | Vitreous china sanitaryware will last for more than 30 years with reasonable cleaning, adjustment and maintenance but may be replaced by a new trendy fashion before then. Initial cost and place of manufacture is often a good indicator of likely life. |

## THE GREEN SUPPLEMENT • 75

### Likely maintenance cost impact over the life of components

| COMPONENT | 0 to 25 years | 26 to 50 years | 51 to 75 years |
|---|---|---|---|
| Stone/brick footings | Seldom required | Seldom required | $500+ /lin.m. |
| Concrete footings | Seldom required | Seldom required | Seldom required |
| Concrete raft slab | Seldom required | Nil to major | Nil to demolition |
| Brick piers | Seldom required | Seldom required | $250 to $500 per pier |
| Concrete stumps | Seldom required | $200 to $400 per stump if not maintained | $200 to $400 per stump |
| Steel sub-floor posts | Seldom required | $300 to $700 per post if not maintained | $300 to $700 per post |
| Timber floor frames | $300 per inspection per year $300 to $1500 per sq. m. if not maintained | $300 per inspection $300 to $1500 per sq. m. if not maintained | $300 per inspection $300 to $1500 per sq. m. if not maintained |
| Light steel floors joist | Seldom required | $300 per inspection $350 to $1750 per sq. m. if not maintained | $300 per inspection $350 to $1750 per sq. m. not maintained |
| Suspended concrete floor | Seldom required | Seldom required $300 per inspection Damage repair about $500 per bulge | Seldom required $300 per inspection per year Damage repair about $500 per bulge |
| Cavity brick construction | Seldom required | Seldom required | Seldom required |
| Brick veneer construction | Seldom required | nil to say $1000 | nil to say $2000 |
| Clad stud framing | $25,000 | $50,000 | $75,000 |
| Flat roof, timber frame | Seldom required | nil to say $2000 | nil to say $5000 |
| Pitch roof propped frame | Seldom required | Seldom required | Seldom required |
| Pitch roof truss frame | Seldom required | nil to say $500 per sq.m | nil to $1000 per sq.m. (total replacement) |
| Metal roofing | nil to say $400 per sq.m. | nil to say $400 per sq.m. | Approx. $400 / $600 per sq.m. |
| Tile roofing | Seldom required | Seldom required for terra cotta Approx. $250/ sq.m. to refinish cement tiles | nil to $500 per sq.m. |
| Aluminium Windows | Seldom required | Likely replacement wide price variation | Probable replacement wide price variation |
| Timber windows | Seldom required | Likely replacement of lower quality wide price variation. $500 per window per decade for painting | Probable major repairs or replacement wide price variation. $500 per window per decade for painting |
| Plasterboard wall/ceiling | Seldom required | Some re-sheeting likely @ $1000 per room | Probable re-sheeting likely @ $1000 per room |
| Interior paint | $15,000 per 25 year period | $15,000 per 25 year period | $15,000 per 25 year period |
| Exterior paint | $10,000 (unpainted brick) to $25,000 (WB) per 25 year period | $10,000 (unpainted brick) to $25,000 (WB) per 25 year period | $10,000 (unpainted brick) to $25,000 (WB) per 25 year period |
| Water supply | Seldom required | Seldom required | Expect minimum $5000 through to complete re-plumbing $20,000 |
| Taps | Nil to $2000+ | $2000+ | $2000+ |
| Hot water service | $2000+ | $4000+ | $4000+ |
| Solar HWS | $2000+ | $4000+ | $4000+ |
| Sewer Drainage | Seldom required | $5000 | $5000 |
| Stormwater drainage | Seldom required | $4000 | $4000 |
| Electrical wiring | Seldom required | nil to $10,000 | $10,000 |
| Electrical light fittings | $3000 | $4000 | $5000 |
| Fixed electrical appliance | $10,000 | $12,000 | $15,000 |
| Kitchen cupboards | $25,000 | $50,000 | $50,000 |
| Sanitary-ware | $3000 | $3000 | $3000 |
| Tiling | $2000 to $5000 | $5000 | $5000 |
| Carpets | $3000 to $10,000 | $6000 to $20,000 | $6000 to $20,000 |

- All prices are indicative of 2011 costs; they may be significantly less for a well built well maintained property or much higher in a house that is in poor condition or has be extensively upgraded, particularly if there is a high fashion component.

# 75 • THE GREEN SUPPLEMENT

## Windows

Upgrading, replacing, renewing, modifying or enlarging windows in an existing house requires careful consideration.

If the existing opening in the building fabric can be reused then the task is generally reasonably cost effective, unlikely to affect the building structure and less likely to contravene local authority planning requirements.

If the house is of framed construction then enlarging window openings is relatively easy to achieve but it is always wise to have an architect, structural engineer or building surveyor inspect the building structure to ensure that any new window head lintels and lintel abutments are capable of supporting all likely superimposed loads.

Masonry houses pose many more difficulties if a window opening is to be widened. Modern diamond saws allow stone, bricks and concrete to be cut, this can be expensive and messy.

Careful planning is required including:
- Make sure the style of window desired is available locally.
- Is the window to be made off-site and delivered as a unit or are there components to be assembled on-site?
- Do the window manufactures install the window frame?
- Do the window manufactures include the glass and glazing in their quote?
- Do the window manufacturers provide a full flashing kit with their windows?
- What site preparation is required for the window installers?
- What is the exact date of delivery and installation?
- Is the window delivered fully glazed?
- If on-site glazing is required, is this done on the same day the frame is installed or is there a delay requiring temporary weatherproofing and security?
- Do the window installers reinstall or renew all trims, including any lining and architraves?
- Do the window installers make good any damage made to external or internal wall surfaces, including any repainting?

Make sure there are no water pipes or electric cables in the space where the opening is to be increased. If any are found they must be altered by certified/licensed trades people.

Support the load above where the opening is to be made as instructed by a certified building professional.

Carefully remove any existing windows, all trims, lining and cladding to be reused, then demolish progressively using the correct tools and equipment, and wearing appropriate safety clothing.

If there is a time delay between opening-up the wall and the installation of the new windows remember to have considered a temporary weather and security scheme.

Window lintels for wide openings are often very heavy and may required specialist lifting equipment, make sure it is ordered and on-site when required.

The process may appear arduous but a new window correctly located in an existing house can vastly improve liveability and will often add real value to the house when it comes time to sell.

## COMPARING WINDOWS ON EXISTING HOUSES

| Window location, types, efficiencies and deficiencies |||||||||
|---|---|---|---|---|---|---|---|---|
| | North | | East | | South | | West | |
| Large | 10 | | 6 | | 6 | | 0 | |
| Medium | 8 | | 8 | | 6 | | 4 | |
| Small | 6 | | 6 | | 6 | | 6 | |
| None | 0 | | 4 | | 6 | | 8 | |
| | | | | | | | | |
| Timber | 8 | | 8 | | 8 | | 8 | |
| Aluminium | 6 | | 6 | | 6 | | 6 | |
| Efficient | 10 | | 10 | | 4 | | 10 | |
| | | | | | | | | |
| Clear glass | 4 | | 5 | | 6 | | 4 | |
| E-Glass | 7 | | 8 | | 4 | | 7 | |
| Double glazed | 10 | | 10 | | 8 | | 10 | |
| | | | | | | | | |
| Openable | 10 | | 10 | | 10 | | 10 | |
| Cross-ventilated | 5 | | 5 | | 5 | | 5 | |
| Fixed | 3 | | 3 | | 3 | | 3 | |
| | | | | | | | | |
| No eaves | 0 | | 0 | | 0 | | 0 | |
| Wide eaves | 10 | | 0 | | 0 | | 0 | |
| Ext. blinds | 6 | | 8 | | 0 | | 10 | |
| Totals | | | | | | | | |
| Grand total | | | | | | | | |
| Check local regulations for window requirements in new houses and alterations and additions. This table is useful to compare one house to another, the higher the score the better the solar access and utilisation. |||||||||

# THE GREEN SUPPLEMENT • 75

## Sustainable water supply

There was a time when water supply authorities tempted homeowners to use as much water as they needed to have long hot showers and lush green lawns.

Times have changed and just about everywhere in Australia city water is becoming a scarce and ever more expensive commodity.

When considering the purchase of a house it is important to check the available water supply and all the services and systems connected to it or supporting it.

| | |
|---|---|
| RW | Rainwater tank |
| P | Pump |
| GW | Grey-water tank |
| T | Grey-water treatment |
| SET | Sewer treatment |
| SU | Sullage treatment |
| RE | Recyclable effluent |
| SW | Stormwater & recycled effluent tank |

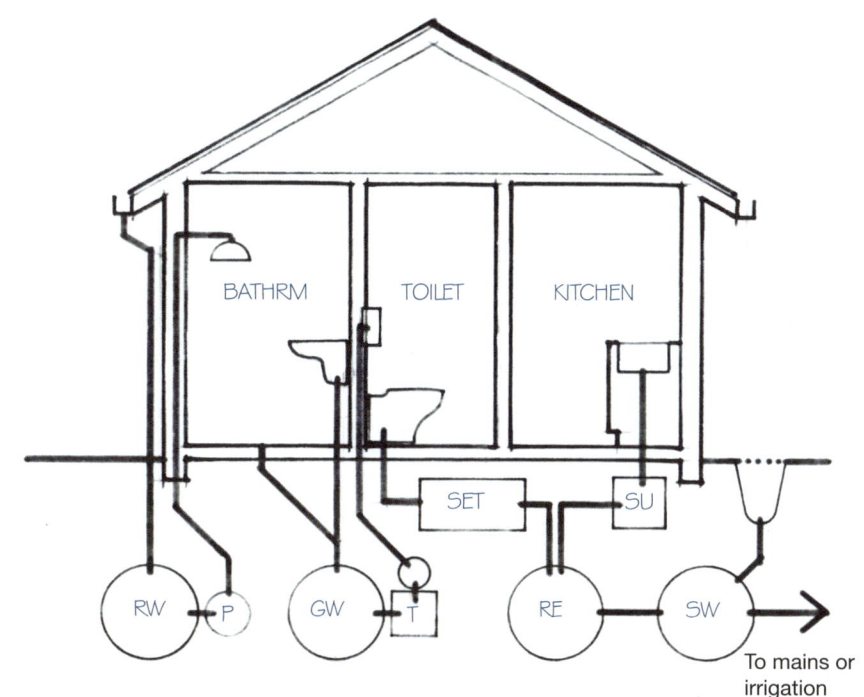

**SCHEMATIC DIAGRAM OF ON-SITE LIQUID WASTE TREATMENT SYSTEM**

**Large** is when the total window area averages < 5 sqm/room
**Medium** is when the total window area averages < 3 sqm/room
**Small** is when the total window area averages <1 sqm/room
**None** is when the total window area averages >1 sqm/room
**Timber** framed
**Aluminium** framed
**Efficient** framed means window frames with R ratings < 5
**Clear glass** is window glass including float, laminated & toughened
**E-glass** is window glass with low heat transmission
**Double-glazed** is two sheets of glass in all window panels
**Openable** means double hung, awning, sliding, etc.
**Cross-ventilated** means there are two opening windows in the room on opposite or adjacent walls
**Fixed** means no opening sashes in the windows
**No eaves** means no horizontal shade projection over windows
**Wide eaves** means at least 600 wide horizontal projections
**External blinds** means adjustable external shade devices to limit the amount of sun on the windows

| Water use comparability table | Yes | No | Score |
|---|---|---|---|
| Rainwater tanks connected to garden irrigation | 10 | | |
| Rainwater tanks connected to toilet flushing system | 15 | | |
| Stormwater harvesting and storage- garden | 10 | | |
| Grey water collection, treatment and toilet flushing | 15 | | |
| | | | |
| All taps rated to current recommended flow restriction | 10 | | |
| More that half of all taps rated to recommended flow limits | 5 | | |
| No taps rated to current recommended flow limits | | 5 | |
| | | | |
| Over 500 sqm of garden watered by aerial spraying from mains | | 25 | |
| Less than 500 sqm but more that 250 sqm of garden sprayed | | 15 | |
| Less than 250 sqm of garden aerial sprayed from mains | | 5 | |
| No garden sprayed from mains | 10 | | |
| Garden irrigated by site harvested water | 20 | | |
| No irrigation required | 25 | | |
| | | | |
| Modern solar / gas coupled hot water system | 10 | | |
| Heat pump hot water system | 5 | | |
| Instantaneous gas hot water system | 5 | | |
| Traditional gas or electric storage hot water system | | 5 | |
| A comparative indicator score | | | |

Use this table only to compare houses one to the other it is not definitive in determining a maximum grade environmentally sustainable house.

A score of 50 would indicate a house that has given high consideration to sustainability; a score of zero or negative indicates significant work is required to reach a reasonable sustainability level.

# 75 • THE GREEN SUPPLEMENT

## Quality of insulation

Most authorities consider in-roof above ceiling the most effective place to install insulation. In this location, if correctly rated and carefully installed, homes are provided with a system that should reduce the influx of heat during summer and limit heat escaping during winter. In-roof insulation should be placed on or as close to the top-side of ceiling as possible; take care not to cover any electrical cables as these can overheat and even cause fire.

Where possible, reflective foil insulation should be installed immediately below roof tiles or metal roofing sheets.

Walls should be insulated where winter heating is required but is much less critical in the hot tropics.

In cold winter locations insulation to floors is highly desirable and there are many products in the market to assist homeowners to increase the insulation of existing houses. Not all these systems are as efficient as they advertise. Choose carefully and check with authorative sources if there is any uncertainty.

Insulation should always be designed to provide comfort and energy saving benefits. Always ensure that location, orientation, heating, cooling, and thermal mass are considered.

## Quality of open space

**Is the garden arable?
Are the plants in the garden flourishing?**

If a garden is to be used to grow vegetables to supplement the family table then the quality of the soil should be assessed. If there is any concern, at least buy a soil testing kit and if there is further uncertainty employ a qualified horticulturalist to provide an assessment.

Are there any likely detrimental impacts from climate change?

If a property is close to the ocean then check with the local authorities for estimates of sea level rises. These estimates will be come more accurate with the passage of time as more reliable data can be modelled.

Climate change remains difficult to accurately assess, but most information sources suggest that the extreme climate zones are likely to increase in their extremes. Hot zones will become even hotter and wet zones will become wetter. Temperate zones are less likely to have extreme change but most climate models indicate there will be a progressive general increase in temperatures.

## Quality of ventilation

**Ventilation to houses is generally provided through open able windows.**

Evaluate all the opening sashes for:
- Ease of opening
- High ventilation value, low draft impact
- Weather impact, particularly in areas with high temperatures and wet weather
- Security if left open when the house is unoccupied or overnight
- Cross-ventilation potential
- Insect screening in areas with nuisance insects

Check also for reversible ceiling fans, fixed ventilators and exhaust systems in bathrooms, laundries and kitchens.

Are there signs of damp?

Rising damp seems to concern many people and this is correct in many instances. Take care that what appears to be dampness at the internal base of walls is really rising damp, in many cases the water could be entering a wall cavity due to damaged or nonexistent flashing around window and door openings.

Rising damp can be eliminated but it will sometimes need expert consideration to eliminate the source of the water, to upgrade sub-floor ventilation, to improve under floor drainage and to introduce new or upgraded damp proof coursing. Beware of companies offering cheap easy fix solutions; they often only deal with part of the problem or cover up the symptoms while leaving the root cause untreated.

When checking a house for purchase look for these tell tale signs of dampness:
- Damaged roof tiles or roof sheets.
- Loose flashings and cappings on roofs.
- Grey/green/black spotting on eaves linings and in bathroom, laundries and kitchens.
- The smell of bleach in bathrooms.
- Musty smells when sub-floor access doors are opened.
- Water staining along internal windowsills indicating that there could have been excessive condensation.

Damp houses are sick houses. If it is not clear there is a suitable effective method to eliminate dampness they will not provide a healthy family environment.

## 2. Credits for using salvaged materials

### Determining the impact of processing salvaged materials

#### Masonry

Natural stone is often salvageable and can be reused many times. There is a controllable impact of using too much hydrocarbon fuelled equipment to lift, transport and manoeuvre heavy pieces of stone. Used carefully it has high weather protection, structural stability and thermal retention properties.

Reused bricks are a good option, but take care in most areas as bricks built into walls since 1950 use a high cement content mortar that is very difficult to remove. It is generally better to use hard burnt, machine pressed bricks manufactured before 1950 and laid in lime rich mortar that can be cleaned from the bricks using hand tools.

Stone, bricks, concrete blocks and reinforced concrete are being crushed and recycled as gravel suitable for hard standing, filling and in certain instances as aggregate in concrete.

There is a high-energy input required but it is considered by most authorities to be a lesser impact that quarrying virgin stone for crushing.

If masonry waste is not recycled it will become useless landfill, reducing further this limited resource.

#### Metal sheets

The predominant sheet metal used in buildings is steel, and it generally has a zinc rich coating (galvanising or zinc-alume) or a factory applied bonded colour coating (Colorbond).

Aluminium has also been used in some building sheets and copper, lead and pure zinc sheets can be found in some historic buildings and more recent prestigious buildings.

Rolling corrugations or other deformations into the sheet profile generally stiffens sheet metal. The most common steel sheets are corrugated and used as roofing and in some cases walling, although there are many other profiles including clip fixed steel decking.

Eventually all metal sheets will oxidise if they are exposed to the weather. Traditionally old style galvanised iron (steel) sheets have been reused in progressively more rustic ways until they are eventually eaten away by rust. Modern Colorbond sheets tend to be in very long lengths that if removed carefully can be reused.

To ensure the maximum life of a roof it is still generally sensible to use new sheets and to send old sheets to be recycled.

If there are historic corrugated sheets on a heritage or period house the sheets may be able to be removed and remanufactured. This could mean cutting off a section of the ends and sides of sheets that have untreatable rust, filling and treating old nail holes and having the sheets regalvanised.

#### Timber

Most timber can be reused or recycled. The limits are timber that has fugal damage, insect attack or has already been reconstituted with chemical bonding agents, for example chipboards and fibreboards.

The limits to timber reuse are often economic. It is considerably more expensive to remove timber from an existing building, transport it to a suitable plant where it is denailed, recut, resized, regraded, visually sorted and marketed, than it is to simply cut down a tree and saw it to suitable sizes.

# THE GREEN SUPPLEMENT • 75

Only some species are economic to reuse; these include Douglas Fir (Oregon) and timber that have attractive grain structure and suitable durability to be used as flooring. Most Australian hardwoods (eucalypts) become too hard with age to be worked with hand tools. Plantation timbers (mostly Pinus Radiata) overdry and deteriorate very quickly when exposed to weather, and they are often visually unattractive.

Some hardwoods can be reused in glulam beams but this is a limited market.

Salvage yards play an important part in timber reuse by allowing components and assemblies, such as fireplace surrounds, doors, windows, panels, planks, mouldings and even staircases to reuse in new locations.

Depending on location, timber can contain insects – generally termites or borers – or fungi spores that can lead to deterioration of the carrier timber and other timber in contact or close proximity.

There are some potential chemical toxin dangers associated with recycled timber including high arsenic content in some early types of treated pine. Other timbers, particularly those that were close to contact with the ground, may have been coated with creosote and residual paint that could contain lead. Use only timber that is clean and free of contaminants.

## Roof tiles

Roof tiles can be reused but take care as most tiles that are removed from existing buildings have either expended most of their expected life or will not provide sufficient life to cover the cost of removal, transportation and refixing.

There are two main materials used to manufacture roof tiles, clay and cement. Both of these materials can be crushed and reused as aggregates or as gravel substitutes in pathways.

Specialist recycling yards may keep supplies of roofing tiles of specific profiles, finishes, colours and manufacturers – these can be of particular benefit if an existing roof is to be extended and near identical tiles would be appropriate.

Faux slate-style shingles often contain asbestos and if they are not able to remain undisturbed they should be removed and disposed of by specialist companies.

## Doors and window frames

If soundly made from durable timber, timber door and window frames should be able to be refurbished and reused.

Old framed and panelled timber doors were often made from timber species ideally suited to the task, many of these species are no longer available.

These doors will always be valued and if correctly maintained will provide generations of use. All doors removed during building demolition or renovation, unless badly damaged, can be reused – if not required on the site of the works then they should be redistributed through a building materials salvage yard.

Traditional timber windows used in buildings up to the 1960s and in some cases later than that, were generally hand fabricated from the best quality selected timber. Although there is often rot in the sill members and in the smaller frame sections, old timber windows if they have been sensibly maintained and regularly painted can be reused with a reasonable certainty for at least another generation of life.

Some of the older windows will be glazed with textured glass, leadlight and even stained glass. These windows can be very rare and should be preserved wherever possible as many of the textures and stained glass colours are no longer available and cannot be economically reproduced. Installing an outer toughened or laminated glass panel over the old glass will reduce the chance of damage due to mower stones and wayward cricket balls, and it will increase the insulation value of the window significantly.

With advanced ordering and careful measuring double glazed panels can be ordered and installed in most timber window frames.

Aluminium window frames date from the 1950s and have generally been improved progressively to the currently available insulated framed models with comfort glass or double-glazing.

For most of the 20th century aluminium window frames were inefficient and often poorly manufactured. Few aluminium windows from this period can be reused and those that can often do not perform to currently required insulation standards. If the windows in a house are poor quality, the aluminium and the glass should be recycled and replaced with modern efficient frames and glass.

Steel framed windows were fabricated from the 1920s through the 1960s when they were out marketed by lower cost, no paint required aluminium frames.

The style of some houses, particularly those from the Art Deco period rely significantly on the appearance of steel framed windows, particularly those containing curved glass.

Steel frames can still be sourced in major cities and replacement galvanised steel window frames with double glazing are available to special order.

Plastic window frames have been marketed since the 1960s but have never made a serious market penetration in Australia.

Early types were highly susceptible to ultra-violet damage from sunlight. Plastic windows benefit from being well sealed, low conductive and are often factory double-glazed. They should be considered if there is proof of longevity and an extensive warranty guarantee to back the manufacturers claims.

## Metals

There are four metals commonly found in houses:
- Steel and aluminium both can be reused and recycled and
- Copper and lead both highly recyclable.

Neither steel or aluminium lose their strength due to age, however rust and other corrosion can damage the material and reduce its strength.

Steel is particularly suitable for reuse as it is relatively easy to refurbish, reform, adapt, weld and bolt. If a second-hand suitable size section of steel is available then there is no reason to use new steel.

Aluminium is often less suitable for reuse but as it is a highly suitable recyclable material all scrap should be used to make new products. Where possible new product made from recycled aluminium should give preference over products fabricated from virgin aluminium.

In some older houses cast iron or wrought iron components may remain, these are likely to date from the 19th and early 20th centuries and are unlikely to have a modern equivalent.

If any iron components are no longer required for the project where they have been discovered they should be sent to an old building material resale yard as there is nearly always a market for building components made of iron. Only if the components are damage beyond repair should they be considered for recycling.

The third metal to be found in most houses is copper and copper alloys. These are used extensively for water piping and the majority of electric cabling has copper conductors.

All copper products are recyclable and often have a significant market value.

Lead was used extensively in roof and wall flashings and all though it may have deteriorated with age it is still a highly recyclable material. There are dangers associated with handling lead. Seek advice and be careful.

# 75 • THE GREEN SUPPLEMENT

**Cabinetry**

Over time, fitted cabinets have progressively replaced freestanding furniture units, particularly in kitchens, bathrooms and bedrooms.

How do modern cabinets compare with furniture cabinets and traditionally site made built-in closets and kitchen units?

Historic wardrobes were often made from selected timbers, were French polished but they did not provide for coat-hangers which have only come into favour over the last century. This caused them to be phased out in favour of deeper modern wardrobes; these were often unitised for on-site assembly, an early form of flat pack, or were built-in as the house was constructed. These wardrobes were mostly made with basic pine or hardwood frames with low toxicity plywood sheeting. They remain as appropriate today as they always were.

Progressively natural timber framing and plywood panelling was replaced with particleboard sheets that combined structure and sheeting into one. Early particleboards contained high levels of urea-formaldehyde, a potential carcinogenic. In warm dry environments the urea-formaldehyde out gassed and is considered to be a contributor to sinus conditions and other diseases.

Over time, the out-gassing reduces to acceptable readings but old chipboard wardrobes should be checked particularly if they predate the now ubiquitous white melamine coating.

Pre-modular kitchen cabinets, i.e. those less than 600 deep and not constructed to contain modern standardised appliances, are often demolished and trashed as they seldom meet current ideas for the model kitchen.

The modern practice of changing cabinets when appliances malfunction or simply drop out of fashion is one of the current problems in evaluating the environmental credentials of houses.

The problem is compounded because very few of the materials used in the construction and fitting out of a modern kitchen are reusable or recyclable. Some appliance manufacturers offer return to manufacturer systems where old appliances are disassembled and the component parts recycled. This is barely sustainability window dressing, when all the old cabinets, polished stone, designer tiles and many of the appliances go to landfill.

Fashion kitchens go with fashion food where appearance is always changing and waste is supported … are these excesses sustainable?

**Electrical lights and fittings**

Major steps forward have been made in the efficiency of light fitting, electrical fittings and appliances. Few older houses in the potential renovation category are likely to have fully upgraded electrical systems.

Working through the maze of material flooding the market on the most energy efficient lights and appliances is often baffling.

The days of a single 240v incandescent light hanging in the middle of a room should be a thing of the past, but so should a multitude of low voltage dichroic down lights and flashing fluorescent tubes.

Consider LED light fittings as they use less energy, generate little heat and can be connected to a low voltage circuit.

It is important to assess the new developments carefully, and here it is difficult to know whom to trust as all parties have a vested interest, but careful comparative study and excessive claims will assist in eliminating the poorest performers.

## 3. Credits for using recycled materials

### The impact of recycling

**Crushed masonry aggregate**

This is a high-energy process that that uses less energy than quarrying and crushing. If a natural gravel is used, then there is likely to be a further energy saving advantage but gravel quarries leave big holes in the ground.

**Refinished ashlar stone**

Recutting blocks of building stone may seem natural and sustainable, but if electrical or combustion energy sources are used to recut, dress and move the stone then some of this advantage is lost.

**Re-sawn, dressed and moulded timber**

Although close to the same amount of energy is used to reprocess timber as would be used for newly felled timber, there is a real advantage in saving valuable carbon absorbing forest trees. Even a pinus plantation sucks carbon out of the atmosphere.

**Redressed slate and roofing shingles**

This is a sweaty process, even today most redressed slates and shingles are reworked manually. The slates are likely to have arrived in Australia from a quarry on the other side of the Earth many years before and some could even have arrived a ballast in windjammers. Very worthy.

**Moulded plastic products from reprocessed waste**

There are many reprocessed plastic products being marketed as timber substitutes. Reprocessing of plastic into these products uses relatively low energy input; but the whole of life energy used to manufacture the original plastic product, that is required so that it can become waste and be available to be reborn as an environmentally sensitive recycled product, could be more environmentally damaging that using a piece of plantation grown timber.

| Tabulating the benefits | | | 20-50% demolished material | 50-100% demolished material |
|---|---|---|---|---|
| Masonry | Bricks | Reuse | *** | ***** |
| | | Recycle | ** | ** |
| | Stone | Reuse | *** | ***** |
| | | Recycle | * | * |
| Timber | Reuse | Scantling | ** | **** |
| | | Mouldings | ** | **** |
| | Recycle | | * | ** |
| Roof Tiles | Reuse | | ** | *** |
| | Recycle | | * | * |
| Metal Sheets | Reuse | | ** | *** |
| | Recycle | | * | * |
| Doors Frames | Reuse | | *** | **** |
| Window Frames | Reuse | | *** | **** |
| | Plus Up-glaze | | **** | ***** |
| Cabinetry | Reuse | | * | ** |
| Electric Lights | Reuse | | * | * |
| | Plus Up-grade | | ** | ** |
| Electrical Fittings | Reuse | | * | * |

The table is relative guide only indicating where reuse pays significant benefits; it is not a definitive guide to any of the authoritative environmental rating systems.

# THE GREEN SUPPLEMENT • 75

Be careful of products marketed as environmentally sustainable, low energy, ecologically friendly or carbon neutral. At least compare the claims to a tree.

If three trees are planted for every tree harvested for construction timber on a 25-year cycle, it is possible that within 50 years 13 trees have been harvested and 27 trees remain. After 75 years, 40 trees have been harvested and 81 trees remain and after a century 121 trees have been harvested and over 240 trees remain … is this carbon neutral? If the first tree was used to construct a well-designed house it is likely it is still holding up floors, walls and roofs a century on.

### Is recycling always the best solution?

It will not always be the best solution but it is always worth considering. If there is a limestone quarry near a cement plant beside a hydro-electric powerhouse on a continually flowing river close to the ocean with no other demands on the water supply and there is a quarry-able supply of clean gravel suitable to make concrete blocks within a few kilometres of a building site; then this could be as good or better a solution than transporting cement over long distances to combined with recycled crushed bricks and old concrete product to make concrete blocks. This just to indicate that it is necessary to check that the processing is local, low energy and non-polluting before finally concluding that recycling is best.

## 4. Credits for using re-processed materials

Chip and fibreboard products from salvaged timber bring a minor ecological gain. The product is less likely to be a strong as a new product and there could be detrimental impacts from cleaning up the salvaged product and then adding a potentially toxic adhesive back into the recycled product.

Aluminium fittings from reprocessed scrap aluminium in most cases will be an environmental advantage, taking into account the vast amounts of energy required to mine, process and manufacture new aluminium product and the fact that aluminium is one of the easiest materials to remanufacture. However, if a less energy-consuming material was chosen, then there could have been a higher environmental advantage.

Glass is effectively fused sand. Grind it up and it becomes sand again. Whether new or recycled, glass remains sand with energy added.

Insulation products from cellulose fibre and glass waste should be fit for the task. This means with the correct R, having longevity, being incombustible, being non-toxic and vermin resistant. Whether it is recycled from old paper or old glass if it does not have a high safe, sound and sanitary factor it is not practical.

Reprocessed plasterboard panels from waste gypsum faced with reprocessed paper is of some value if there is a local reprocessing facility. This is a product with a high relative density, so transportation distances are a major contributor to its environmental impact.

Reuse, reprocessing and recycling have been common for generations. There has been waste but every tonne of waste taken to landfill by a builder constitutes loss of profit.

- Reduction of waste is good.
- Reduction in the embedded energy per useable volume of buildings is good.
- Determining the impact of reprocessing is essential if there is to be real lowering environmental impact.
- Not all recycling, reprocessing and remanufacturing systems are clean or green; there are times when a well-chosen new product will have a better environmental outcome.

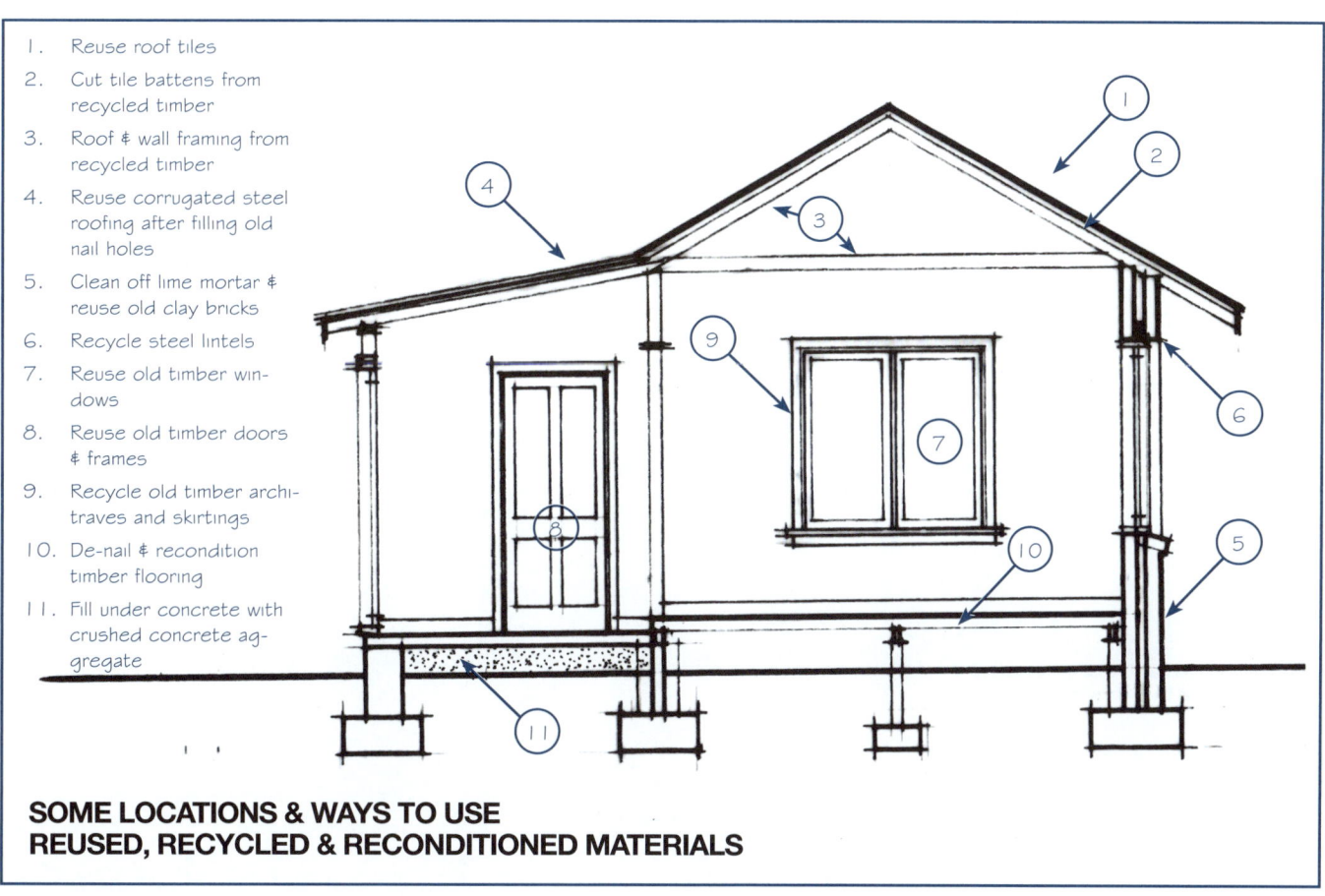

1. Reuse roof tiles
2. Cut tile battens from recycled timber
3. Roof & wall framing from recycled timber
4. Reuse corrugated steel roofing after filling old nail holes
5. Clean off lime mortar & reuse old clay bricks
6. Recycle steel lintels
7. Reuse old timber windows
8. Reuse old timber doors & frames
9. Recycle old timber architraves and skirtings
10. De-nail & recondition timber flooring
11. Fill under concrete with crushed concrete aggregate

**SOME LOCATIONS & WAYS TO USE REUSED, RECYCLED & RECONDITIONED MATERIALS**

# 75 • THE GREEN SUPPLEMENT

## 5 Credits for using water efficiency measures

### Water used to produce building materials

### Water used on-site during construction
Evaluate where water must be used during construction and attempt to eliminate waste.

### Water that will be consumed during the lifetime of the building
The cost of water is going to increase significantly over the decades to come. Unless the predictions are massively incorrect, most Australian cities are facing major limitations on their reticulated water supplies. Currently there are bonuses available to householders to harvest rainwater from the roofs of houses and to store it in tanks to provide alternative water for tasks where pure potable water is not essential. Stored rainwater, with minimum treatment, can be used to water gardens, flush toilets and even do the family clothes washing.

Progressively, as more sophisticated and affordable treatment and purification systems are developed, householders will add stormwater and grey water into the mix and further reduce their reliance on mains water supply.

On-site treatment of sewage (black water) is unlikely to be the norm but expect to see local neighbourhood treatment plants that can treat waste liquids from domestic premises and reticulate it back to the household who provide it in the first lace, for garden irrigation.

Locations with regular consistent rainfall are the easiest to convert to low wastewater use. Expect to see in the future local water authorities harvesting the municipal stormwater systems and even paying householders who are able to harvest more water than they need from their properties to supply it in to a collector system.

The on-site, neighbourhood, local, regional and even national system of water harvesting infrastructure will grow slowly at first but as the demand for and the price of water continues to rise, novel and radical water reticulation systems will emerge.

Owner builders will of course meet all the rules and regulations in force when they construct their new homes. They should also study the emerging trends, particularly in their region, to be prepared to benefit from them.

Low flow taps act as an advance warning of the likely restrictions to come, but the future does not have to be as dire as some are predicting. Sensible water use in conjunction with integrated water harvesting can provide Australian cities with adequate water supply.

## 6 Credits for using energy efficiency measures

### Energy used to produce building materials
It remains difficult to accurately calculate the amount of energy expended in the production and distribution of building materials. There are some guidelines being developed by research institutes and other authorities but these are often difficult to access and complex to apply.

Taken simply, materials that require large amounts of energy to produce, have to be transported long distances – particularly overland – and make up a major proportion of the volume and/or mass of the building should be avoided if there is a suitable lower energy alternative available.

### Energy used on-site during construction
The amount of energy used to construct a house is relatively low and in many cases is lower per unit of time than the likely energy use of an occupied house. However, it is wise to avoid wasting energy and the area where this can be actively reduced is in the ordering of materials because every load taken to landfill will contain an energy component that is a total waste.

### Energy that will be consumed during the lifetime of the building
During the life of a building the largest energy use will be in maintaining comfort, food, hygiene and leisure for the occupants.
- Design lighting systems to suit tasks and to use efficient luminaires.
- Choose heating, ventilation and cooling systems carefully.
- Choose food cold storage and cooking systems carefully.
- Choose water-heating systems carefully.
- Avoid high energy media systems if possible; these can be a hidden drain on electrical energy.

## 7. Credits for maximising the utilisation of green energy systems

Are the materials manufacturers using solar or other green energy systems in the manufacturing processes?

When would a site-specific energy generating system be more efficient, sustainable and environmentally responsible than using mains power over the assessable effective life of the on-site system?

## 8. Credits for maximising sustainability concepts

### What place is there for radical, novel and traditional building materials and systems?
New environmental home building books often feature radical building shapes and materials.

These individualised constructions are interesting and are useful comparisons but are seldom as environmentally or ecologically advanced as they are portrayed, and are unlikely to be suited to the building allotments available to most families.

Odd shaped buildings are too awkward to sit comfortably on standard rectangular building lots and achieve little if any comfort advantage. Earth-covered dwellings are really only suited to perfectly oriented gently sloping large holdings.

In some localities stabilised earth – pise and adobe – houses can be suitable and there are places where straw/hay bale construction has recognisable advantages.

The reason that these shapes, methods and materials are used is often cited as a way of avoiding the high levels of embedded carbon and possible toxic effects associated with highly processed modern building materials.

For most families radical design is not the answer. Home owners are generally conservative, as are the town planning rules that limit non-traditional streetscapes.

There is a real shortage of residential land in most developed cities; pressure is being brought to bear by authorities to increase residential densities and to reduce unsustainable impacts on urban infrastructure.

It will become more and more a luxury to live in a house with a private garden. The future will require a higher level of sharing. If society is sensible urban residences will be carefully designed to balance the needs of comfortable living with low environmental and ecological impacts, combined with building materials and construction methods that provide long sustainable whole of life solutions.

This is not radical or novel. Georgian and Victorian urban-scapes still exist, sometimes after hundreds of years. If correctly oriented, brick-built terraced housing has a low maintenance nearly infinite life. An amazing fact is that the brickwork was generally in lime rich mortar that allows them to be disassembled with hand tools and the bricks reused with potentially no impact on the atmospheric carbon. Even the internal structural and decorative timbers and the artisan made doors and windows are fully reusable with a minimum of reworking.

If a house remains in the use of one occupier for an extended period of time there are many real benefits. However many Australian families believe that the most effective way to increase personal wealth is to utilise the apparent capital gain in their houses. Is this sustainable and is it socially, environmentally and ecologically responsible?

# ENHANCED ACCESS AND LIVABILITY • 76

**Introduction to Enhanced Access and Livability**
The following information is based on information published by Livable Housing Australia.

It recommends that all new dwellings should be designed and constructed in a manner that ensures that a home is suitable for occupants of all ages and abilities to live with the maximum comfort and flexibility that can be reasonably provided. (*Livable Housing Australia, 2017*).

It should be:
- Easy to enter
- Easy to navigate
- Designed for purpose and future adaption

## BENEFITS FROM IMPROVED ACCESS AND FLEXIBILITY

**Benefits for Parents and Children**
Homes that are planned to provide suitable access ways throughout the external and internal spaces, will improve ease of movement for parents who are manoeuvring pushers and reduce the likelihood of tripping and falling injuries to children.

**Benefits for family members suffering an injury or after medical or surgical treatments**
Easy access through the home and its accessible facilities ensures a safe and comfortable environment to occupy during recovery.

**Benefits for senior family members with reduced mobility and senses**
The provision of step free access from the property entry through the living, sleeping, bathrooms and utility spaces of the home, using wide doors and corridors that provide access for mobility devices. The extra facilities in bathrooms and toilets ensure convenience and safety.

**Benefits for family members living with disability and their carers**
The provision of extra space beside beds, in bathrooms and in toilets allows carers to provide assistance to people living with a disability or illness.

**Assessing the extent of improvements to access and flexibility**
Livability design enhancements to homes improve the comfort, safety and enjoyment of all people of who occupy a home throughout its lifetime.

Attempt to provide the highest possible level of easy access when designing a home that will assist in providing generations of occupiers with a suitable environment through all stages of their lives.

Livability recommendations may increase the initial cost of a home but providing for easy upgrading to a higher level of access and livability in the future should be considered.

**What should be included in every home design**
Every new home design should evaluate the needs of all family members when considering their immediate and longer term accommodation requirements.

**Step free access route into the home**
Every pathway from the street footpath to the entry door to the home should be free of any steps and endeavour to have a slope of not more than one in fifteen (1:15) with flat areas no more than nine metres apart.

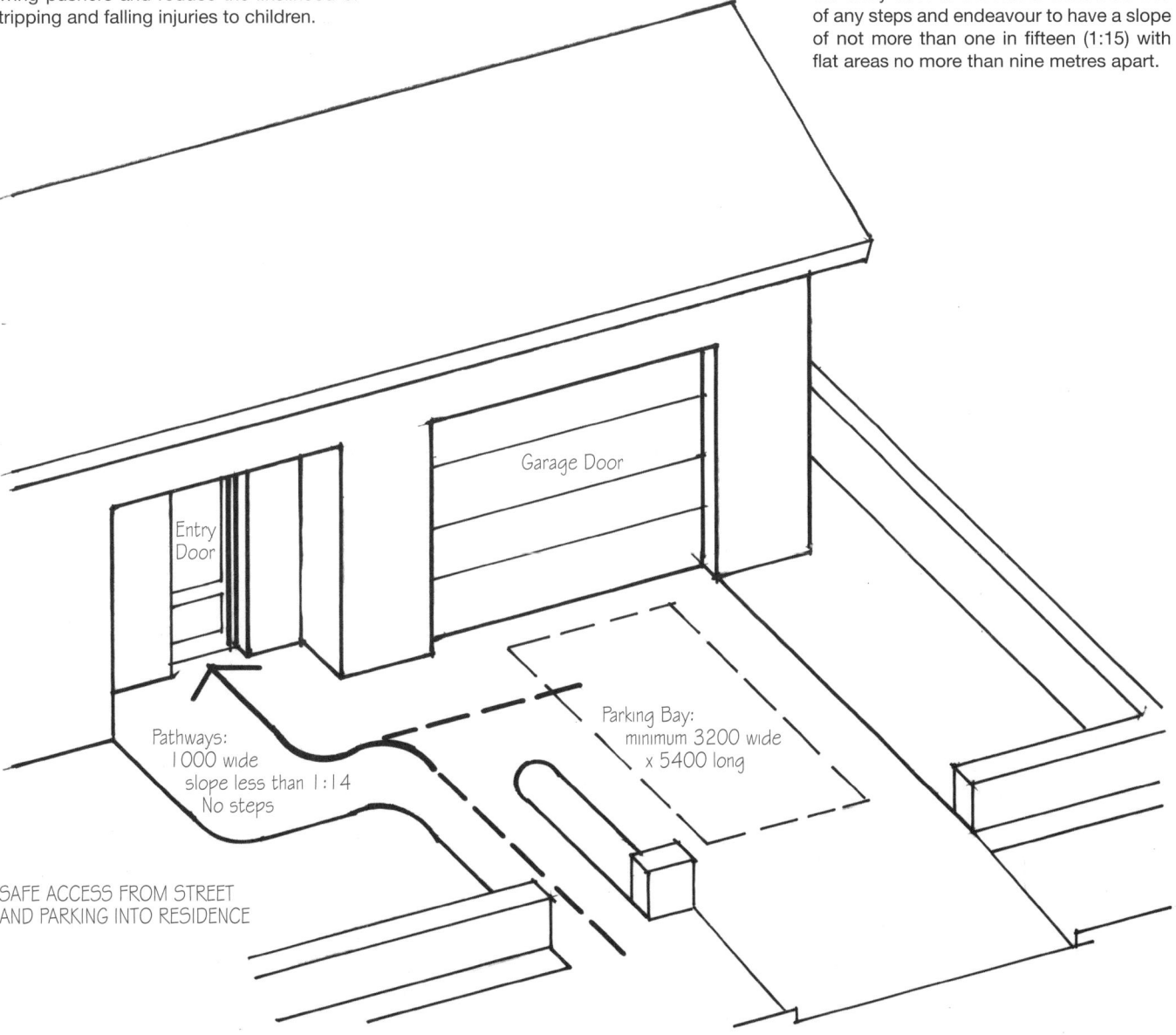

SAFE ACCESS FROM STREET AND PARKING INTO RESIDENCE

# 76 • ENHANCED ACCESS AND LIVABILITY

### Suitable Corridor and Door widths
Where possible corridors should be at least one metre wide and doorways should have at least 820mm clear opening.

The operation and location of doors should provide a lever handle nominally 900mm above the floor.

### Easy access toilet on main living level
All new homes should have an easy access toilet on the main living level of the house. There should also be an easy access toilet on all floor levels containing bedrooms.

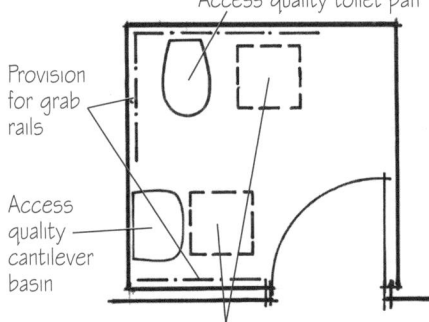

EASY ACCESS TOILET
Recommended size approx. 4 sqm. (2000 x 2000)

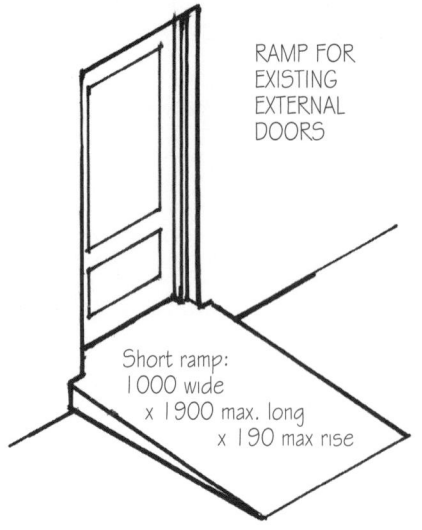

RAMP FOR EXISTING EXTERNAL DOORS
Short ramp: 1000 wide x 1900 max. long x 190 max rise

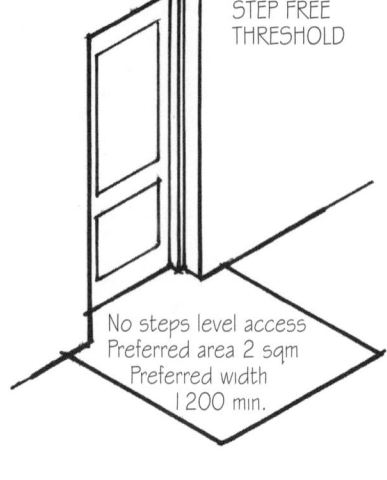

STEP FREE THRESHOLD
No steps level access
Preferred area 2 sqm
Preferred width 1200 min.

### Level access to shower
It is advisable to provide hob free access into all showers.

### Provision to install support rails in toilets and bathrooms
Where possible install suitable grab and handrails in toilets, over baths and in shower cubicles. If the installation of these rails is not required by any member of the family immediately, it is important to ensure the wall areas where these devices would be attached are of sufficient strength to allow their attachment at a later time.

### Suitable continuous handrails to stairs
All stair flights should be provided with a continuous handrail. Handrails improve the access and safety of users; they should be constructed in a manner that allows people to have continuous uninterrupted support for the full length of every stair flight.

### What should be included for family members with identified specific requirements

### Assessing the specific requirements
When designing a family home these important questions that should be answered, assessed and responded too. These include, but are certainly not limited to, family members with specific disabilities;
- Blindness and limitations to vision
- Deafness and other hearing limitations
- Reduced mobility
- Chronic illness
- Lowered cognitive ability
- Requiring continuous care

## ADDRESSING THE REQUIREMENTS

**What should be included to ensure long term suitability of access and flexibility in a home**

### Access to Home from Property Entry

### Step free access for Pedestrians
A pathway not less than one metre (1000mm) wide, 1200mm is preferred, should connect the point of entry into the property to the entrance into the home. The path should as far as practical have:
- No steps
- A firm surface
- An even slip resistant surface
- Crossfall less than 25:1000
- Ramps no steeper than 1:14
- Landings 9 metres apart

In existing homes, a short ramp ideally over 1000mm wide can be provided at the entrance.

With a maximum of 1:10 slope and no longer than 1900mm.

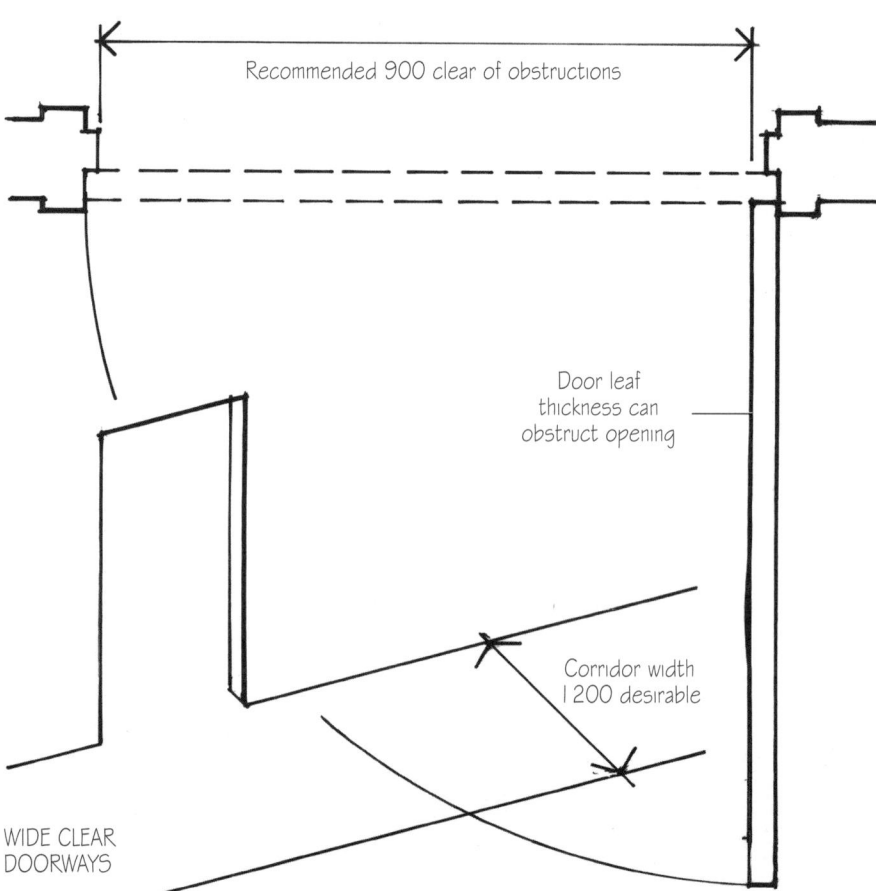

WIDE CLEAR DOORWAYS
Recommended 900 clear of obstructions
Door leaf thickness can obstruct opening
Corridor width 1200 desirable

# ENHANCED ACCESS AND LIVABILITY • 76

**Step free from Vehicular Space**
If a car parking space is at least 3200mm wide and 5400 long, with a surface slope not exceeding 1:40 for concrete and 1:33 maximum for bitumen it can be included in the access pathway.

**Entrance into Home**

**Requirements**

**Step free threshold**
The entry door threshold should not exceed a transition level change of more than 5mm.

**Suitable external landing**
An external level landing area should be provided at the entrance door, with a minimum area 900mm x 900mm. If possible, use a landing with a minimum dimension of 1200mm with a surface area of 2 square metres.

**Wide doorway**
A standard door is nominally 820mm wide but when rebated doorstops and the thickness of the door are considered the unobstructed clear opening, it is likely the opening is 755mm (820 – 15 -15 -35). This may be satisfactory for an able body person but can be too limiting for a person using a mobility device or wheelchair.

It is recommended that a clear opening of at least 850 is provided and where possible choose a door to provide 900 clear opening.

If the home is designed to provide for a people who are infirm or requires mobility devices, ensure there are satisfactory unobstructed routes from the entry door to living spaces, cooking spaces, a bedrooms and a bathroom and toilet.

Include suitable door widths and consider their ease of operation.

**Vehicle Parking**
Consideration should be given to allowing people with a physical disability to drive a motor vehicle or be a passenger in a motor vehicle that can be conveniently parked with easy access into their home.

This may not be reasonably provided where there is no vehicular access to the property, the property has a particularly steep slope or there is insufficient space.

**Wide parking space**
A level vehicle parking space of at least 3200mm wide is recommended to allow sufficient space for a person with a disability to easily enter and leave the vehicle, and to have room to move around the vehicle, particularly if mobility devices are required.

**Space for mobility device**
A minimum length of 5400mm is recommended for any parking space. Consideration should be given to lengthening the space, by up to 3000mm if a wheeled mobility device is required to enter the rear of an accessible vehicle.

When a vehicle provides roof top racks for wheelchairs and similar mobility devices, or there is a high lift rear hatch to allow mobility devices to enter the rear of a vehicle, it is recommended that a clear height of at least 2500 is provided over the parking space.

If a parking space is to be used by any person with a physical disability it should have at least roofed weather protection.

**Allowable slope**
Depending on the firmness and surface texture of the parking space a slope of 1:40 is desirable, if this cannot be achieved no slope should exceed 1:33.

**Doors and Corridors**
Many homes have 820mm door leaves, that often provide less than 770mm clear opening. Corridors could be as narrow as 900mm and have right angle changes of direction. These dimensions may be considered suitable for able bodied people but can reduce or limit access for people with disability.

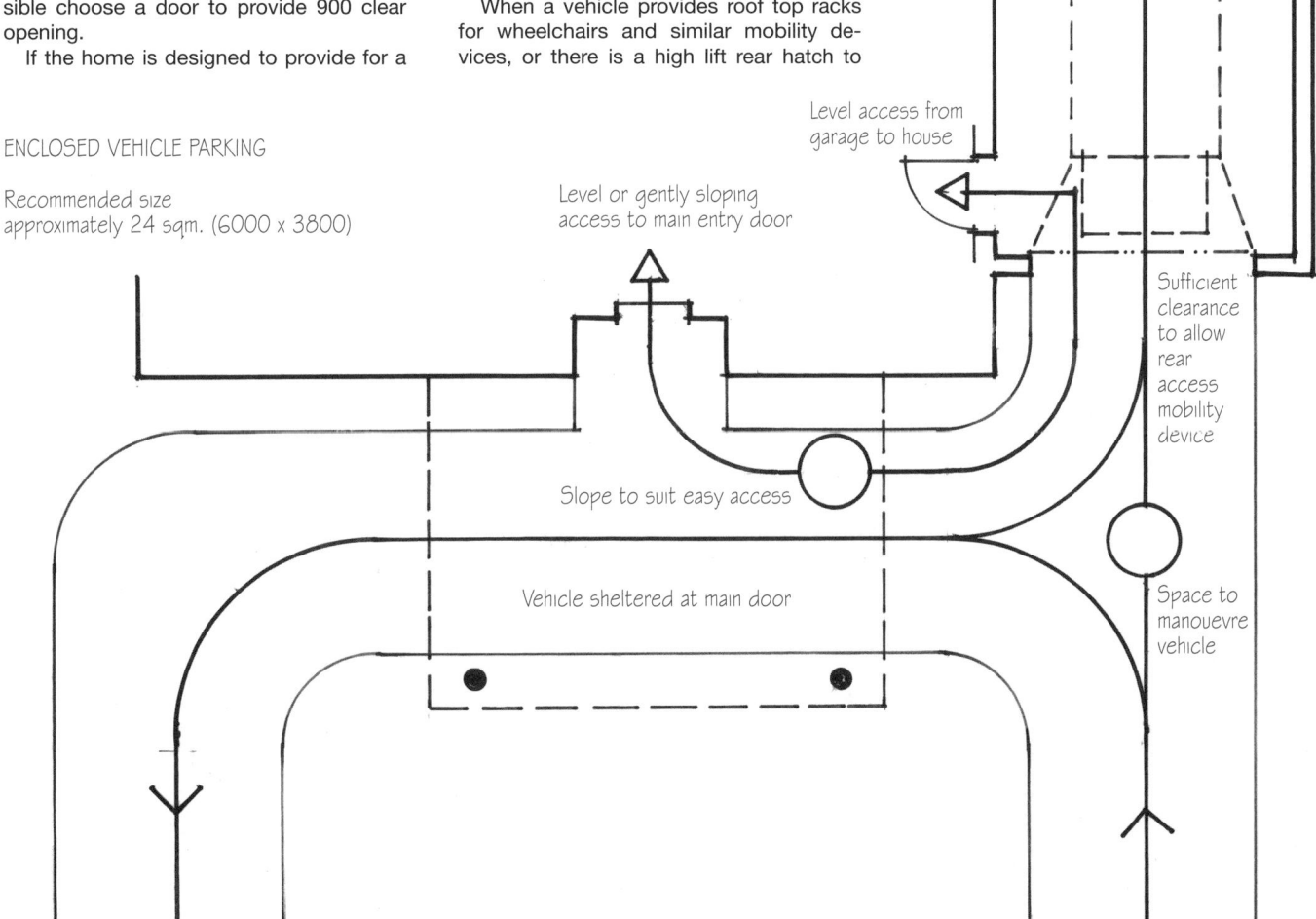

ENCLOSED VEHICLE PARKING

Recommended size approximately 24 sqm. (6000 x 3800)

# 76 • ENHANCED ACCESS AND LIVABILITY

### Wide corridors
It is recommended that all corridors and passageways should have at least 1000mm clear width. Main corridors should be 1200mm clear wherever possible and only short linking passages be reduced to 1000mm. It is advisable that all corridors with right angle corners should be 1200mm wide.

### Wide step free doorways
Ideally all doorways should have 900mm clear opening width. Doors into bathrooms and toilets may have narrower opening width, but doors with at least 820mm clear openings should be considered as the minimum achieved.

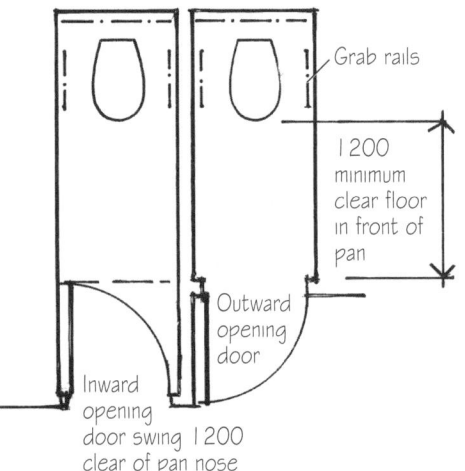

AMBULATORY TOILET
Recommended size 900 wide
Grab rails
1200 minimum clear floor in front of pan
Outward opening door
Inward opening door swing 1200 clear of pan nose

### Toilet
There should be an ambulatory equipped toilet provided in an easily accessible location in all homes.

### Located on Ground Floor
In most homes the ambulatory toilet should be on the ground floor, but some designs may determine another location would be more suitable.

### Suitable Clearances –

### Toilet only cubicle
Every toilet in a separate cubicle should have a clear width of 900mm and have a 1200mm clear space in front of the pan with an outward opening door. If an inward opening door is required, the door swing should not intrude into a 1200mm clear space. This clearance ensures access can be gained if a person has collapsed inside the toilet cubicle.

In existing homes changing the door hinges to a type that allows the door to be removed from the outside the cubicle should be considered.

### Toilet and basin cubicle
To achieve a high level of accessible convenience for the greatest level of abilities a toilet cubicle with sufficient space beside the pan for a person to transfer from a mobility device on to the pan is desirable.

The cubicle should have clear unobstructed easy access from the door to the manoeuvrable space around the pan.

The wash hand basin should be located where there is clear space to allow a mobility device to be stand directly in front of the basin and for the basin to be of a type suitable in height and projection to suit a person using a mobility device.

At least one toilet in every home should be provided with support rails to allow infirm people of low ambulatory ability to use the toilet conveniently or at the very least make provision allowing their installation in the future.

The National Construction Code (NCC) has guidelines for the design of toilets suitable for people requiring specific accessibility and although these guidelines may not be mandatory in a family home, they assist in ensuring suitable solutions.

### Shower Rooms
Every home should have a room which contains a shower and a wash hand basin that provides easy access for people of all abilities.

The room may have other facilities including baths, vanity units, mirrors.

Although significant care is required to achieve a suitable functioning design, it should not limit the appearance of the room.

### Located on Ground Floor
In most homes a shower room should be on the located on the ground floor, but some designs may determine another location would be more suitable.

### Level step free access
There should be a no step threshold at the doorway into the shower room. The floor should be as close to level as permitted by the provision of an appropriate slope to clear surface water. The showering area must be a continuation of the adjacent floor without any hobs to obstruct access.

### Showering area and access
The showering area should be at least 900mm x 900mm. Where possible it is recommended that there is a minimum dimension of 1000mm and a showering area of 1.5 square metres with an adjacent clear space of over 2 square metres if carer assisted showering is required.

A fixed seat or stable in shower chair should be considered if the shower is to be used by infirm people.

### Slip resistance
It is important that the floor drains efficiently and the floor surface is impervious and of a suitable slip resistance. Grated trench drains around the edge of the showering area will limit the spread of water onto the surrounding floor further reducing the danger of slipping.

### Grab rails
The shower area should be located in a corner where possible to allow two walls that can have supporting grab rails attached.

### Shower screens
Shower curtains are acceptable and are the safest and most convenient where carer assisted showering is required. Curtains are easy to move out of the way to allow carer access and to allow support rails to continue along the space adjacent to the showering area to improve the support mobility and safety of infirm people.

If glass or other rigid materials are used to reduce water escaping from the showering area, they should be hinged to open outwards from the showering area and be designed to ensure stable support rails are provided to allow safe movement into and out of the showering area.

### Allowance for grab rails in toilets and bathrooms
The NCC provide information and details on the location and sizes for support rails and grab rails. This information should be used where appropriate in toilets and bathrooms where infirm people require access.

If support and grab rails are not installed in a new or upgraded toilet and bathroom, it is strongly recommended that the areas of walls where they may be required to be fixed at a later time, be constructed of appropriate material or be reinforced to ensure stability when they are installed.

### Stairways
Stairways are an impediment to easy access throughout a home by infirm people. It is recommended that every home is designed so that there is a level floor storey, that can be easily access from the outside by people of all abilities.

This encapsulated in the following:

An easy access level floor, containing a bedroom, with an accessible shower room, and access to a suitably equipped kitchen, laundry and living spaces.

This is a minimum requirement to ensure reasonable comfort for a person living with a disability.

### Location
Stairways should be located for the maximum convenience and safety of all occupants. They should; be well lit during the day and at night, should have a clear space at the top and bottom of a flight that is at least the width of the stair, and a clear level approach area of at least 1200mm long in the direction of the flight.

A stair flight against a structural wall allows for future modification to the stairway to incorporate access equipment.

# ENHANCED ACCESS AND LIVABILITY • 76

### Straight Flights
The NCC provided the mandatory information on the design and installation of residential stairways, including the numbers of risers in allowed in any flight and the formula to determine a suitable going to rise ratio of every step, these are the minimum deemed to comply regulations.

To provide the easiest, most comfortable and safe access, consideration should be given to providing straight stair flights, between floors and landings. Winder and kite shaped treads and curved flights should be avoided.

### Handrails
Continuous easy grip handrails should be provided on both sides of all stair flights, around all landings and projecting sufficiently at the top and bottom of flights to aid visibility and provide support when commencing to use a stairway.

Even in steps joining floors separated by 1000mm or less benefit from the inclusion of handrails.

### Dimensions
The NCC provides clear information on the heights required for balustrades and handrails, these are prescribed codes and must be complied with.

The minimum going for a tread in a residence is 240mm, this is considered as only just appropriate for an able bodied person. In a commercial building the minimum allowable going is 250mm and this a more appropriate minimum for all stairways.

A going greater than 250mm should be considered, in many applications 265mm is a more suitable dimension.

The accepted formula for calculating stairway riser and going dimensions is, two times the riser height plus the tread going expressed in millimetres with a resulting sum of between 550 and 625. Then calculate the the number of risers suitable for the vertical height between the two floors to be served by the stair.

No flight can exceed 18 rises between floors or landings.

For example; if the height between the lower floor FFL and the upper floor FFL is 2750mm, is divided by the allowable height range for the riser (115mm to 190mm), then the choice is between 15 to 18 risers of (15 x 183.3, 16 x 171.9, 17 x 161.8 & 18 x 152.8).

From this calculation 16 or 17 risers appear to offer the most suitable ratio, when checked in should not be less than 550 or more than 625. (2R + G = 550 to 625. The height of all risers in a stairway must be even and need to be carefully determined by calculating the height between finished floor level (FFL) on the lower floor and the finished floor level on the upper floor by applying the 2R + G formula, 2x161.8+265=588.6 and 2x171.9+265=608.8.

Therefore, both 16 and 17 risers in the stairway flight would be satisfactory.

A flight width of 1000mm is recommended and the across flight measurement between handrails should exceed 900mm

### Safety
All risers are required to be enclosed, the treads must be slip resistant and the nosing must be in a contrasting colour, generally white, black or metallic non-slip strip 50mm wide.

### Kitchens, Laundries and Utility Spaces
It is especially difficult to design a kitchen that is suitable to be used by people of all abilities.

Kitchens designs should provide sufficient flexibility for layouts to be modified over time to suit the changing requirements of users.

### Layout
An important consideration to allow for future modification to the layout of kitchens is to continue the floor covering under cabinets and appliances, particularly those that could be altered or renewed.

The dimensions of many built-in appliances have been standardised over time, particularly ovens & dishwashers, this allows ease of upgrading and change of location.

### Workspace clearances
The movement spaces in kitchens should have clearance between benches of not less than 1200mm, a wider clearance of 1550mm should be considered if people using mobility devices require access to the space.

### Bedroom at Entry Level
Where possible every home should have an easy access bedroom on the entry level of the home.

### Bedroom layout
The room should be an absolutely minimum of 10 square metres, measured inside the skirtings and other protrusions and be clear of all permanent obstructions including built-in wardrobes and dressing tables.

### Bedroom dimensions and clearances
It is recommended that the bed provided should be no less than 1250 wide (double bed), a 1500 wide bed (queen size) is recommended.

Larger beds can be used but they may reduce the ease of access for an infirm person.

The height of the mattress above the floor should as far as practical should be level with the height of a wheelchair seat.

The bedhead should be able to be located against a wall that measures over 4100mm wide between obstructions.

This width is the sum of 1000mm clear space, a 1550mm Queen bed and a wide manoeuvring space of 1550mm = 4100mm

Beds are generally 2050mm long and it is recommended that a space at the end of the bed exceeds 1000mm.

When an extra allowance of nominally 610mm is added to provide for wardrobes along one wall.

The appropriate internal dimensions within the finished wall faces for an easy access bedroom are:

4710 (4100 + 610) x 3050 (2050 + 1000) = 14.4 square metres.

### Electrical switches and power points
All electric switches operating light fittings or power-points should be a consistent locations throughout a home.

STRAIGHT STAIR FLIGHTS

Handrail starts and finishes beyond first and last nosing

Continuous handrail with clear finger space from walls

Mid flight landings over 750 in direction of travel

Handrail starts and finishes beyond first and last nosing

# 76 • ENHANCED ACCESS AND LIVABILITY

**Location of Light Switches**
It is recommended that light switches should be located between 900mm and 1100mm above the floor and on the wall face close to the jamb, inside the room containing the light fitting at the opening edge doors.

**Location of power points and other electric and electronic connection points.**
In living rooms and bedrooms, it is recommended that all wall outlets are at east 300mm above the finished floor level.

They should be located to minimise their obstruction by furniture.

In wet areas, kitchens, bathrooms, laundries and utility rooms, it is recommended that all 240volt power outlets are at least 900mm above the floor and at the regulatory separation from all taps, basins, sinks, baths and showers.

Outlets above benches should allow a clearance of 200 above the bench where practical.

There will be other switches and power points with specific tasks located throughout homes, care should be taken to locate these where they can be easily and safely accessed.

**Electric Switch Type**
It is recommended that the switch mechanism has a simple toggle or rocker operation of a style easy to locate and operate by touch.

In some applications contrasting colour or glow in the dark escutcheon plates should be considered.

**Door Handles**
All door handles should be easy to grasp levers or D-pull handles.

**Location**
It is recommended that door handles, to all doors accessing the home and rooms should be located between 900mm and 1100mm above the floor.

(Note: the height of door handles should be the same as the height of light switches.)

**Type**
Entry door: avoid nob type handles and consider quarter turn downward rotation lever handles.

It is recommended that the entry doors to a home should be fitted with dead bolts.

Many of these type of locking devices require keys to be inserted and turned to release the locking bolt. If there are members of the home occupants that have reduced capacity to insert and turn a key to release a dead bolt, it is recommended that advice is sought from a security lock firm. A push button lock operation pad or an electronic lock system may provide a mechanism suitable for people with limited dexterity.

Internal hinged doors; use quarter turn lever latches.

Internal sliding doors: avoid recessed flush mounted handles that require fingertip operation, wherever possible set sliding doors with sufficient leaf projection in the open position to allow the use of D-pull style handles.

Note: Avoid fiddley privacy snibs that are combined with many door handles and use an easy to operate separate privacy latch with an emergency entry override mechanism.

**Living spaces manoeuvrable space**
When fitting out and furnishing, living rooms, dining rooms and family rooms ensure a circular area can be provided to allow easy circulation when mobility devices are used.

**Recommended dimensions**
Movement space within a living space should not be less than 900mm wide.

If a wheelchair or motorised mobility devise is to be used clear routes through the living spaces should be in the order of 1100mm and wider at changes of direction.

Wheelchairs and motorised mobility devices often require a circulation space of 1800mm to 2300mm diameter to comfortably exercise a complete forward rotation.

**Windows**
All windows in living spaces and bedrooms should allow occupants to look through the window particularly; where there is view, to observe children at play and to adjust the ventilation of the occupied space. It is advisable to select windows that can be operated by a person seated in a wheelchair.

**Sill height**
The clear viewing level of windows in living spaces and bedrooms should not be higher than 1000mm.

Note: this allows a 900mm high sill line and a 100mm allowance for the window frame.

**Operation**
In living spaces, the operating mechanism to open and close windows should be between 900mm and 1100mm above the floor and have handles and action requiring the lowest possible effort to open, shut and secure.

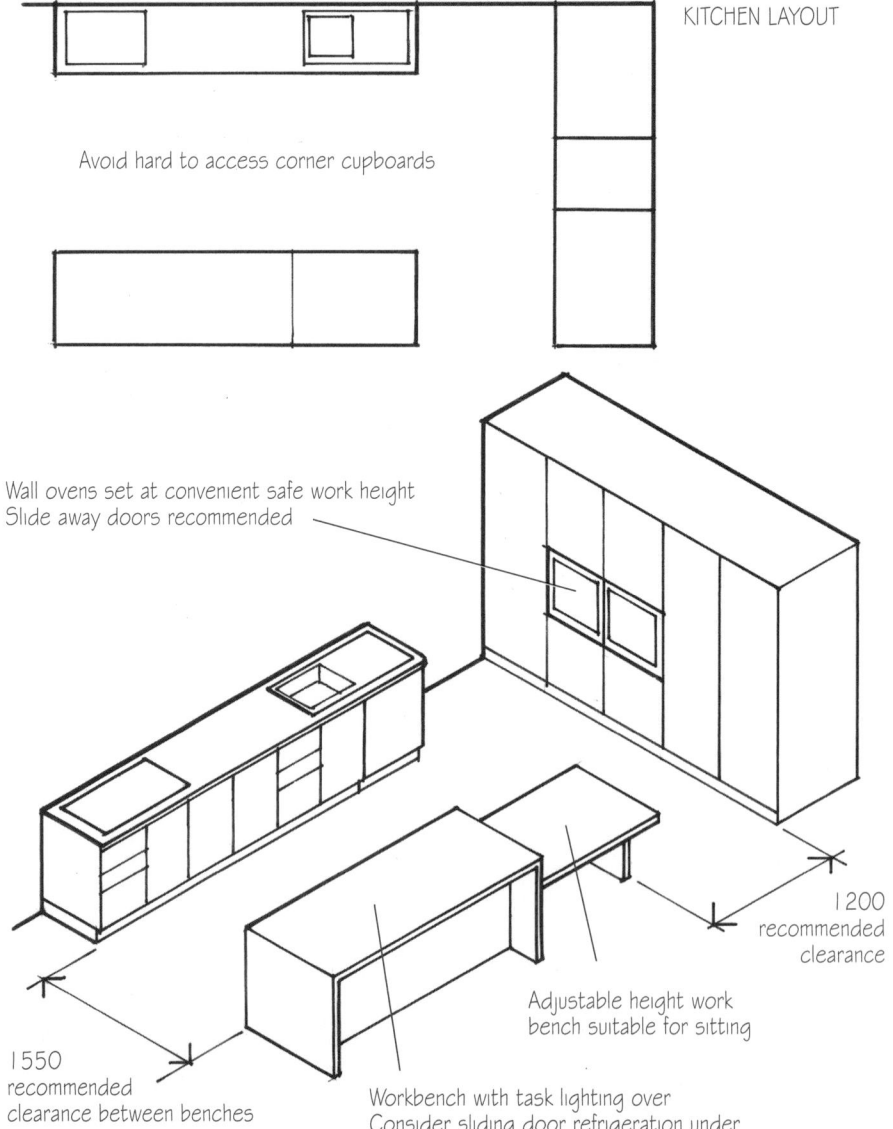

KITCHEN LAYOUT

Avoid hard to access corner cupboards

Wall ovens set at convenient safe work height
Slide away doors recommended

1200 recommended clearance

Adjustable height work bench suitable for sitting

1550 recommended clearance between benches

Workbench with task lighting over
Consider sliding door refrigeration under

# ENHANCED ACCESS AND LIVABILITY • 76

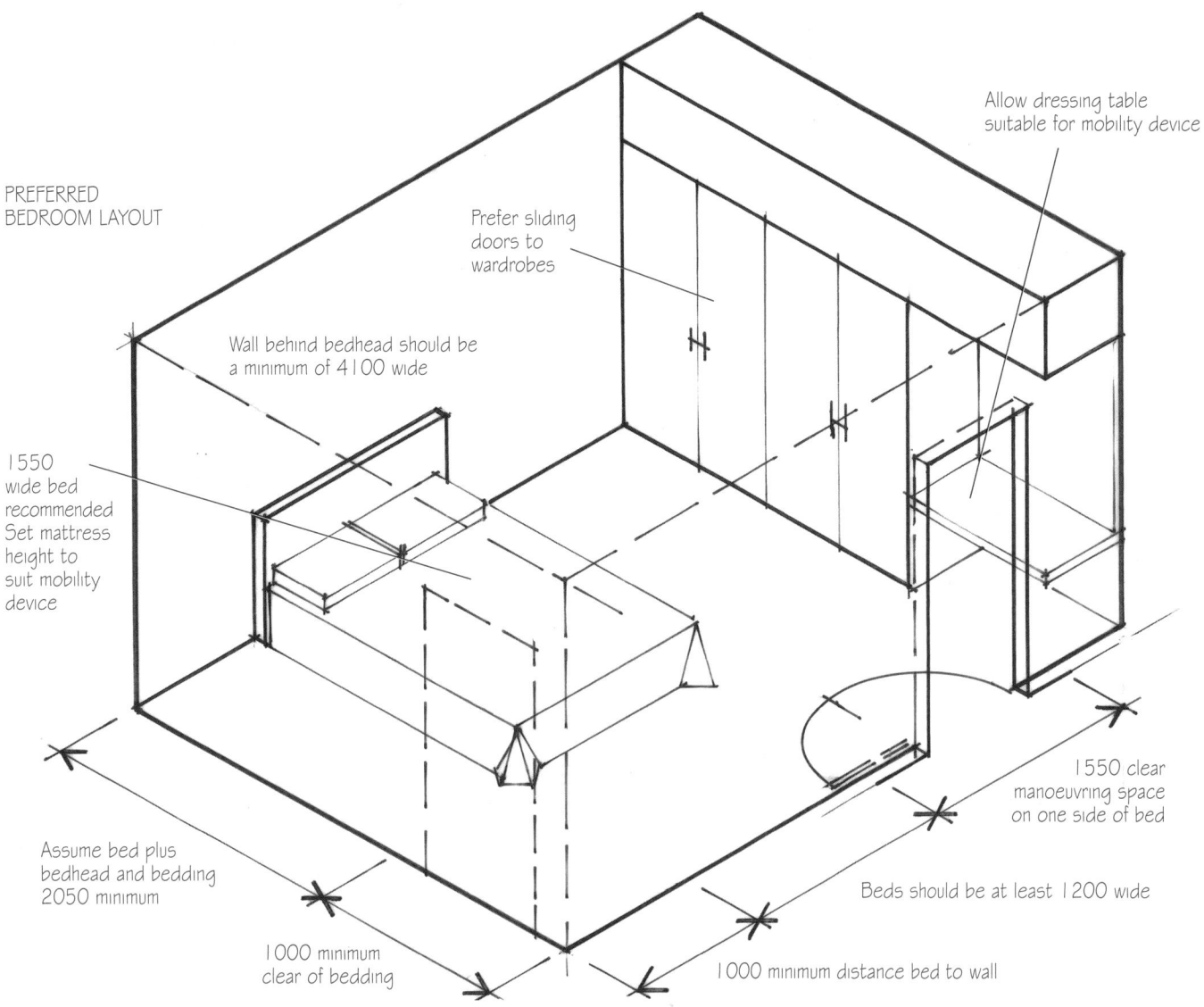

PREFERRED BEDROOM LAYOUT

Prefer sliding doors to wardrobes

Allow dressing table suitable for mobility device

Wall behind bedhead should be a minimum of 4100 wide

1550 wide bed recommended Set mattress height to suit mobility device

1550 clear manoeuvring space on one side of bed

Assume bed plus bedhead and bedding 2050 minimum

Beds should be at least 1200 wide

1000 minimum clear of bedding

1000 minimum distance bed to wall

### Coverings
Where window coverings are provided for privacy or solar control it is advisable to use easy winding mechanism, a power operated system operated by buttons or a remote activation device.

Pull apart drapes and chain operated roller or Roman blinds can be difficult for infirm people to operate and should be avoided on windows to rooms that are likely to be occupied by people with restricted movement abilities.

### Flooring
All floors in homes should have a firm slip resistant surface with no unnecessary changes in level.

Short ramps of no greater slope than 1:14 can be used, longer ramps are better if they do not exceed a slope of 1:20.

Any change level between different flooring materials should be limited to a maximum of 5mm.

Loose rugs that can trip a person or slip from under them should be avoided in spaces that are used by people who are infirm, have reduced mobility or require to use a mobility device.

### Access between floors in a multi-floor dwelling
New homes should be designed to limit the need for people with restricting disabilities to have to move between storeys in homes.

### Stairways
If stairways are required in a home consider designing the home in a manner that would allow occupants with restricting disabilities to have adequate living spaces, bedrooms, kitchen and bathroom on the entry level of a home.

Care should be made to ensure that all occupants of a home have the ability to interact without undue restriction or limitations.

### Lifts
There are many lifts available that can offer small one to two person capacity, these can use less than one square metre in area on each of the floors serviced.

These lifts can be installed to provide level access and can accommodate one person using a walking frame or on a fold down seat.

Slightly larger lifts that can accommodate one wheelchair. Some of these designs are suitable for forward entry and exit for a standard wheelchair.

Other larger lifts can accommodate motorised mobility devises and an accompanying carer. These lifts will require two to three square metres of space on all floors served. These larger lifts may require space for the operating system. In some jurisdictions they will require licencing and regular maintenance.

### Stair lifts and Inclinators
Stair lifts and inclinators have limited use and should be avoided in new homes. They should only be considered if no other more efficient mechanism is suitable.

### Ramps
The use of ramps inside homes is generally unsuitable.

Consider a ramp used to join two floors that have 2700mm height separation between their finished floor levels.

The ramp would need a horizontal dimension of over 40 metres, calculated at a slope of 1:14 with two level landings along the travel length. (2.7 x 14)+(2 x1.2)= 40.2m.

# 77 • BUILDING IN BUSHFIRE PRONE AREAS

## Introduction

During the summer of 2019-2020 bushfires raced through every state of Australia with catastrophic loss of life and homes. Whole communities were left wondering how this could happen.

The fire services throughout Australia had been continually upgrading their equipment and the fire fighters are considered to be well managed, trained and experienced.

The factors that appear to have contributed to this extreme fire season are:
- A long period of drought conditions preceding the bushfires allowing vegetation to lose moisture and become more prone to supporting combustion.
- A combination of high temperatures, strong winds and no rain provided an environment when combined with the effect of the drought was ideal to support bushfires.

Other factors were significant in starting fires and high among these was the phenomenon of dry lighting, this is when there is a thunderstorm without rain and the lighting strikes set fire to vegetation. These strikes are often in inaccessible locations reducing the opportunity to control the resulting fire before it elevates into a bushfire.

Most of these factors are expected components of the Australian environment and the impacts were considered reasonably predictable. The 2019-2020 bushfires did not conform to historic expectations, the fires were fiercer, faster and more widespread than previous bushfire events.

The different between 2019/2010 and other early summer periods were significant changes in the climate conditions.
- Temperatures exceeded historic predictions
- Winds were drier and stronger than predicted

The critical factor of the increasing temperature of the Earth's atmosphere, has led to unpredictable weather conditions and higher likelihood of extreme bushfire events.

The toll of the 2019-2020 bushfires in Australia:
- people were killed or severely injured
- vast tracts of native forests were severely damaged
- many native wildlife species were significantly impacted
- productive agricultural areas were damaged
- large proportions of livestock were lost

## Impacts on the natural and built environment

This toll has a significant impact on the Australian economy, it will take many years to restore the damage and replace the losses.

The increased intensity of bushfires burning close to human settlements has increased to likelihood of loss of infrastructure, buildings and injury and death of people.

Television shows scenes of raging fires leaping along the treetops, whipping across grasslands and peoples' homes and livelihoods being consumed. Looking at the often smoking images of residential areas after the fire has passed what can be observed is that homes were burnt to the ground leaving piles of scorched concrete, metal and masonry material. The sentinel of a forlorn brick chimney surrounded by twisted corrugated roofing steel was the all too common image of the aftermath of a bushfire.

Extensive media time was dedicated to the brave firefighters and families facing walls of scorching flames with whatever water source was available. Sometimes successfully but too often losing the battle to save the building and even losing their life in these futile attempts to control the awesome power of nature.

## What is wrong with the homes built in bushfire zones?

Few homes can independently resist a bushfire and even with professional firefighters defending with water pumps and hoses homes still explodes in flames… too many homes do not use construction methods and materials that can withstand the fierce heat of a raging bushfire or even resist hot embers driven from a remote fire front.

There is a significant variety of information in the public domain providing considered advice of suitable design and construction systems for homes in bushfire zones. Some of this information is flawed and some even contradicts the information contained in Australian Standard AS3959:2018 *Construction of Buildings in Bushfire Prone Areas* and specifically required in the *National Construction Code – Volume Two* (Building Code of Australia).

Anyone considering constructing a home in any area where there is even the lowest risk of a bushfire event should access these two documents. At the time of writing, AS3959 and the NCC Volume 2 (this volume is for residential construction) could be purchased from Standards Australia (SAI Global) for about $250.

Following the devastating bushfires during the Summer of 2020, knowing and understanding the content and application of the requirements of these two documents is essential for owners, designers and builders of homes in bushfire zones.

It is important to review the recommended methods contained in these two documents, and even if the property can be calculated to have a low risk consider applying the highest level of protection that can be afforded. A few days of neglect in landscape maintenance or a sudden unpredicted climatic event can move the bushfire risk from moderate to extreme.

### HISTORIC INFORMATION

Before using or applying complex assessment tools to determine the likelihood of a property having a high annual exposure to bushfire risk, check the following:

Has there been a bushfire in the local area surrounding the property where a home is proposed to be built?

Follow up with a study of:
- When the most recent **bushfire** occurred?
- Where did it start?
- How did it start?
- Was the local fire part of a wider bushfire event?
- Was there property damage and loss?
- Were there sufficient firefighting resources available?
- Was the bushfire quelled effectively?
- Has likelihood of the fire source starting another bushfire been reduced?
- Has the fuel load, both natural and constructed, been removed or reduced?
- The collection and assessment of this information will only provide a starting point in understanding the frequency and dangers associated with chance of bushfires in the locality.

When after assessing the local historic information and relating it to the current conditions it is considered that, though there may still be a chance of a local bushfire; good design and construction methods may reduce the risk to an acceptable level.

At this stage it is important to check with the local government authorities and the firefighting services to ascertain the laws, codes, regulations and recommendations that apply to the construction of a home on the chosen property location.

Consider also that approximately 95% of homes lost during bushfires are built within 150 metres of bushland.

## Contemporary information

Determine the Bushfire Attack Level, (BAL) this is required as part of land use approval and construction approval in most Australian jurisdictions where properties are located in zones considered to have any likelihood of bushfire threat.

The environment and climate are changing continuously, it is important to consider what information is required even if the likelihood of bushfire risk appear low, if there is any risk then consider also:
- Potential changes in the local climate
- Potential and pending changes to the landscape
- Suitability and efficiency of firefighting water supply
- Adequate access for firefighting vehicles and teams
- Appropriate escape route: through the property under threat and from the property to a safe place

# BUILDING IN BUSHFIRE PRONE AREAS • 77

## Australian Standards

### Australian STANDARD – AS 3959:2018

**Construction of Buildings in Bushfire Prone Areas**
'This Standard specifies requirements for the construction of buildings in bushfire-prone areas in order to improve their resistance to bushfire attack from burning embers, radiant heat, flame contact and combinations of the three attack forms'
AS 3959:2019

## Definitions explained, condensed & summarised

### AUTHORITY
These are individuals, agencies or statutory authorities that have the power to require compliance with the laws, ordinances, codes and standards that assess, inspect, approve and certificate the land use and construction of buildings in a bushfire zone.

### BUSHFIRE
Unplanned out of control fires burning through the vegetated landscape.

### BUSHFIRE-PRONE AREAS
Any area that has been, is now or is likely to be subject to bushfire attack

### BUSHFIRE ATTACK LEVEL (BAL)
This is a formula to provide usable measurement of the likely severity of a building's exposure to ember, heat and flame attack. BAL is used to determine the effective bushfire resisting ability of materials and methodology used to construct any building. The numerals allocated to the six BAL are the radiant heat generated by the bushfire impacting on the surfaces of a building, expressed in kilowatts per square metre, this is the measurement of the level of attack that the construction requirements outline for the BALs should be able to resist.

### FIRE DANGER INDEX (FDI)
This is a formula that endeavours to give a value to an amalgam of; the likelihood of a fire starting, its rate of spread, its intensity and how difficult it would be to control. It takes into account, air temperature, humidity, wind speed and various climatic impacts.

### FLAME ZONE (FZ)
BAL-FZ indicates that any building on a property with this bushfire attack level FZ would be directly surrounded by and engulfed in flames and be unlikely to survive

### FIRE RESISTANT LEVEL (FRL)
This is a system that provides an indication of the fire resistance of building elements. FDI indicates the ability of a specific element to: maintain structural adequacy/ material integrity/ thermal insulation during a specific controlled heat event, expressed in minutes.

An FRL of 120/60/30 indicates it maintains structural adequacy for two hours, maintains material integrity for an hour and achieves the required thermal insulation for half an hour (30 minutes).

### BUSHFIRE RESISTING MATERIALS
There are materials that can resist the full intensity of a bushfire even if within a fire zone (FZ). If used correctly concrete, bricks and some natural stone will maintain structural adequacy, material integrity and a high level of thermal insulation for longer than the expected time period a bushfire would take to pass.

Other materials may be adequate if used with high levels of diligence and to high levels of on-site quality control. These include some species of timber, some timber treatments, gypsum based board products (plasterboard), certain glass products, roofing tiles and some composite materials.

Incombustibility is only a single factor of a material exposed to the ferocity of a bushfire, many steel and aluminium elements will fail to be structurally adequate at temperatures lower than those experienced during a bushfire. There are many examples of aluminium liquifying due to the heat generated by bushfires.

The fire resistance of timber products is specifically covered in Appendix F of AS 3959.

### BUSHFIRE RESISTING SHUTTERS AND SCREENS
Bushfire resistant shutters are easy to operate coverings to the exterior of windows and doors that require protection during a bushfire event. They must cover the whole of the window including the framing and need to have a suitable FRL to suit the BAL. Essentially, they should be designed and constructed to provide the same protection as the exterior walls of the home. Window and door screens where required by the BAL or by local regulations are to reduce ember attack, they are not simply standard insect screens- use only screens with fire resistant frames and heavy duty security screen grade metal mesh.

### CLASSIFIED VEGETATION
This is a very complex process of identification and assessment it is an essential component for the calculation of BAL. AS 3959 in Section 2 includes information and tables to assist in understanding *Classification of Vegetation*.

The first step in identifying vegetation classes is to determine the general form of the vegetation, AS 3959 uses the following heads: Forest, Woodland, Shrubland, Scrub, Mallee/Mulga, Rainforest, Grassland and Tussock Moorland.

### DECKING AND VERANDAS
These are the trafficable surfaces of all decks, verandas, steps, ramps and landings. These surfaces are an important consideration when assessing the ability of a home to resist bushfires as any flammable material that congregates at the junction of the horizontal surface of the trafficable area and the wall of the house requires specific attention in the selection of materials and design.

### DOORS AND DOOR FRAMES
- Essentially the door and frame should be fire resistant and there must not be any through gaps at the jambs, heads, sills or at the meeting edges of multiple door units. This generally means engineered edge seals, solid or planted reveals.

### EFFECTIVE LAND SLOPE
- The effective land slope, flat, upward or downward from the home combine with the classification of the vegetation in determining the effective BAL.

### EMBER, HEAT AND FLAME ATTACK
- Ember attack is when smouldering or flaming debris is blown towards the home with the ability to ignite the building or any other combustible material in direct or close contact with the building.
- Heat attack is where the radiant heat being emitted from the bushfire is at a high enough temperature to cause parts of the building to ignite.
- Flame attack is when there flames from the fire come into direct contact with homes or other buildings.

### EMBER GUARDS
- These are coverings that have fine enough openings and of sufficient fire resistance to stop the passage of embers passing through building surfaces into the spaces or cavities beyond.

### FOLIAGE COVER, OVERSTOREY, UNDERSTOREY
- Foliage cover indicates the effective solar shade provided by foliage.
- Overstorey is the highest vegetation canopy in an area of vegetation this would generally be expected to the tallest trees.
- Understorey is all the vegetation below the overstorey and including small trees, shrubs, herbaceous growth, grasses and dead vegetable matter.

### WINDOWS AND OTHER GLAZED ELEMENTS
- These are all elements that allow light to enter a building, it includes all glass and other translucent material, the frames that support the translucent elements and all sealing materials and hardware.

# 77 • BUILDING IN BUSHFIRE PRONE AREAS

**NON-COMBUSTIBLE**
- These are materials, assemblies and components that have been assessed and determined to be non-combustible by AS1530.1 or the NCC.

**NATIONAL CONSTRUCTION CODE – VOLUME TWO**
- The National Construction Code supersedes the Building Code of Australia and all other State or Local Government Area building regulations. It has the highest level of precedence. Although the NCC is the one code for all Australia there are some specific State codes included.
- The NCC-V2 will also call up certain Australian Standards as requiring to be complied with as if they were in the NCC. Specifically recognised is, *Australian Standard AS 3959, Construction of buildings in bushfire prone areas.*
- The NCC has three volumes, volume 2 deals with residential buildings both single tenancy and multi-tenancy dwellings – free standing and attached.
- Under certain statutory land use plans where development is of land that is designated to have; as of right, exempt or complying privileges as approval to construct a home may only require a LGA or private certified building surveyor to certify that the proposed home complies with the NCC.

**LAND USE APPLICATIONS, CONDITIONS AND COMPLIANCE**
If the property where it is intended to construct a home is likely to be in a bushfire prone area it is advisable to always check with the planning department at LGA to ensure that all application and compliance conditions can be met.

Many LGA will require a two stage process to gain approval to build a home on bushfire prone land:

Expect to be required to lodge a development application, this requests the LGA to issue land use permission. They will often require substantial information to be attached to the application, for example to provide information on how the property owner will satisfy certain performance criteria, including for example:
- How will the building resist the impact of bushfire?
- How will the vegetation on the site be modified and or managed to mitigate against the impact of bushfire to the degree required.
- How will an accessible and reliable water supply be provided to the degree necessary to wet down building surfaces and potentially hazardous vegetation prior to the fire front and to extinguish fires after the fire front.
- How will vehicle access and egress be provided that will allow firefighting apparatus to gain safe access to the property for the likely level of the bushfire threat.
- How will the proposed development serve to lessen the risk to the proposed development and to the surrounding development.
- How will the natural environment on and adjacent to the proposed development be maintained in an appropriate ecologically sustainable manner

Preparing a land use application for a property in a designated bushfire prone area can be a complex process and beyond the capacity of inexperienced applicants. Consider engaging a design team with sound local knowledge and experience in achieving positive results in achieving development approval with achievable conditions and requirements.

Experienced designers and consultants should be able to ensure that any approval gained has achievable conditions, a poorly prepared development application may receive a permit but has conditions that are too difficult to achieve.

**NCC-VOL 2: V2.7.2 Buildings in bushfire Prone Areas**
- This is a relatively short section within the NCC and although it deals with the special requirements under this code for *buildings in bushfire prone areas,* it is important to understand that all other requirements of the NCC must be complied with to gain development permits, construction approvals and have the approval to occupy the completed building.

## Site selection

**INTRODUCTION**
Site selection is made up of many parts from the site appearing to be an attractive location where to live through to the practicalities and costs associated with its development as a home site.

If the site is in a bushfire prone area then there are a number of considerations required prior to proceeding, critically:
- Will the controlling authorities allow a home to be built on the site?
- Can a home be designed to meet all the requirements of the authorities and the occupiers?
- Can the design be constructed for the available budget?

To assess the responses to the questions, follow these steps:

**ALLOWABLE LAND USE**
In some locations there may be home building sites available with exempt or complying status, this allows homes to be constructed without the requirement of the LGA or equivalent authority issuing a site specific land use permit. The requirements of the National Construction Code (NCC) will apply and assessment, approval and certification by a registered building surveyor or equivalent is mandatory. If the site requires a Bushfire Attack Level (BAL) Assessment Certificate then the requirements of certain Australian Standards, specifically AS 3959: Construction of Buildings in Bushfire Prone Land are mandatory.

In most locations the local environmental/land use plans will indicate zones or areas where there is known bushfire risk. Land identified as having bushfire risk will often have detailed limitations, restrictions and controls over the use of land, even to the extent of not allowing homes to be constructed.

**REQUIREMENTS FOR A LAND USE/ DEVELOPMENT APPLICATION**
The items to be submitted to an LGA for a land use permit will vary significantly between jurisdictions, what follows is an indicative list of the likely requirements:
- A scale regional location plan of the propose site location
- A dimensioned scale local plan of the surrounding neighbourhood showing all roads, watercourses and vegetated areas (generally within 200 metres of proposed buildings).
- A dimensioned scale plan of the site including:
- Vehicular access to site, suitability for firefighting vehicles
- Buildings remaining on site
- Buildings to be demolished
- Proposed buildings – homes and outbuildings
- Existing fencing layout
- Proposed fencing layout
- Existing vegetation onsite
- Proposed vegetation layout including plant species
- Accurate plans of existing site contours
- Accurate plans and section of proposed site contours
- Firefighting water supply, reticulated and storage
- Specific firefighting systems and appliances
- Proposed construction documents
- Floor plans
- Exterior elevations
- Cross-sections
- Specifications for labour and materials
- Annotation on drawings as required for BAL and AS 3959
- Plumbing and Drainage Plan
- Photographs of existing site conditions
- Details of micro-climatic conditions

**CHECK THE LIKELY CONDITIONS ON A LAND USE PERMIT**
If a land use permit/development approval is required by the LGA then there can be specific conditions attached to the permit, these will vary significantly from location to location, but expect condition under headings similar too:
- The applicable BAL and requirements for risk assessment certificates or similar.
- The state, local & statutory authorities having jurisdiction over the development.
- Requirement for access from a public road for use by firefighting vehicles.

# BUILDING IN BUSHFIRE PRONE AREAS • 77

- Requirement for reticulated water supply or onsite water storage.
- Requirements for electricity supply to the property.
- Requirements for reticulated and bottled gas supply
- Limit on use of combustible gas piping.

## CHECK THE CONSTRUCTION REQUIRED BY THE BUSHFIRE ATTACK LEVEL

The National Construction Code provides the information that governs the quality of buildings constructed in Australia and National Construction Code – Volume Two focuses on the construction requirements for residential buildings.

In a bush fire prone area, the NCC – Vol 2 must be read in conjunction with Australian Standard AS 3959, where specific construction requirements are set-out for designated Bushfire Attack Levels (BAL).

It is possible that the Local Council will have officers who can offer assistance in understanding the requirements of BALs, but in many location it will be a certified Building Surveyor or a Principal Certifying Authority that will assess compliance with the NCC and AS 3959. These are private individuals who are licensed by governments to ensure that all buildings are constructed to the appropriate codes.

Private building surveyors are engaged by the applicant who has approval to use land for the construct a building. They are paid by the applicant for their work, this includes any preliminary discussions and all inspections and certification required throughout of the construction period to the issuing of the certificates of completion and occupation.

It is important to realise that the role of these building surveyors is to ensure all regulations affecting the construction of a building must be achieved. It is an offence punishable at law, to attempt in any way, to influence a building surveyor to allow construction of a lesser quality building than that required by the NCC, AS 3959 and any other law, code or standard required by legislation.

## BAL-LOW

This is the Bushfire Attack Level that is considered to not warrant specific construction requirements greater than those covered in the National Construction Code – Vol 2.

In many parts of Australia determining that the BAL is LOW actually means that it is not BAL-12.5 or greater. Importantly if the property on which a home is to be built has any chance of being endangered by a vegetation fire then it is reasonable insurance to construct the home at least to the requirements set out in AS 3939: Construction Requirements for BAL-12.5.

Established suburban areas within Australian Capital Cities and major Regional Cities cannot be assumed to be free of bushfire threat. There are Fire Danger Indices (FDI) published for areas that can be assessed to have specific levels of risk.

## BAL 12.5

This level provides basic protection from ember attack.

The 12.5 kW/sqm heat flux is calculated assuming that the site has been cleared of all flammable material.

If there is any build up of flammable material on the site the ember attack protection could be significantly reduced and the BAL increased.

## BAL 19

This level is where protection from; ember attack, igniting local debris and a reasonable protection from radiant heat flux up to19kW/sqm.

Protection is not provided from direct wind fanned flames, that level of protection is not calculated to be available until BAL 40.

## BAL 29

This level is better that BAL 19 and increases the protection from radiant heat from 19kW/sqm to 29kW/sqm, **but** it still is not calculated to withstand direct wind fanned flames.

## Note on determining BAL for peace of mind

Using the recommended building location information and construction methods outlined in the Bushfire Attack Levels, BAL 12.5 to 29 are useful in locations where there is known information about previous bushfire events, where there is significant separation from any chance of flame attack and preferably where there is mains water supply, adjacent access for fire fighting vehicles and trained firefighters.

There is a real danger that houses constructed to achieve BAL 29 will not survive a bushfire if it is flaming around the property and the water supply and the firefighting skills of the people trying to save it are not fully up to the task.

If there is any doubt that the house to be constructed is clear of any chance of wind driven flames attack during a bushfire event, seriously consider using the siting and construction calculations provided for under BAL 40.

## BAL 40

Only at this level do the calculations indicate reasonable protection of property from:
- Ember attack
- Burning debris
- Exposure to a high level of radiant heat 40kW/sqm
- **Some** protection from a flaming fire front.

If there is real and potential risk of an extreme bushfire event close to any site where a house is to be constructed the advice contained in AS3959 and in NCC – 2 (residential) should be followed.

Seek advice from the local authorities, especially the local fire brigade and if in doubt seek assistance from a qualified and experience consultant architect, engineer or registered building designer.

## BAL FZ

Constructing a house in a FIRE ZONE is not for the faint hearted, these are locations where catastrophic bushfire events are **likely** to happen and where the services needed to fight bushfires are severely limited or not existing.

It is not a place to build a house but there are always people who will gamble the odds ... this book offers no advice.

## The basics of bushfire resistant design

Appropriate design can be crucial when protecting your house from bushfire. Below basic checklist to consider when designing a house in a bushfire prone area.

### Roofing

It is critically important that the roofs are designed to avoid potential valleys where embers can be trapped.

Ideally roofs should be pitched, with a slope either side.

Valleys should be eliminated or at least kept to a minimum.

Metal or fibre cement roofs are suitable, the fixing method selected should be designed to reduce distortion in sheets that could open gaps that could allow flame to enter directly or to be drawn up through any gaps in the eaves. Steel is the preferred metal for roof sheeting and should be of a gauge and stiffness to reduce distortion. In some higher BAL zones, the whole of roofs should be sheeted over the rafters with dense fibrous cement fire rated sheets.

Tiled roofs should be fully sarked to prevent embers from entering, the sarking should not be able to support combustion and should be continuous over ridges and hips, all joints should be sealed. Note: it is important to understand that the sarking is generally under the tile support battens. Using oversized battens in fire resistant material should be considered in higher BAL zones.

Timber shakes and shingles are not recommended.

### Verandas

All exposed timber should be of a suitable fire resistant variety or treated with fire retardant. Under floor spaces should be sealed to prevent embers entering and there should be no gaps between decking boards.

Consider constructing veranda and decking frames in a manner that they are independent from the main structure of the house and can be sacrificed during an extreme bushfire event without allowing damage to the main building structure.

No vegetation or debris should be allowed to accumulate on or under any veranda decking.

### Framing

Timber or light steel framing in floors walls and roofs should be separated from any

# 77 • BUILDING IN BUSHFIRE PRONE AREAS

flame or superheated air by appropriate fire rated cladding and fireproof insulation.

It is important to eliminate any opportunity for flames of superheated air from entering at a low level of the walls, around windows and doors and in particular through projecting eaves. The chimneying effect of brick veneer walls are a particular concern it is critical to eliminate base wall ventilators or to fit them with fireproof shutters.

### Gutters

Leaf debris in gutters is where sparks and embers from a bushfire can set buildings alight.

Avoid unnecessary gutters particularly valley gutters. Install non-combustible gutter guards on eaves gutters.

In very high risk bushfire zones consider installing systems that allow water to fill the gutters during a bushfire event

### Windows and Patio Doors

Timber window and patio door frames should be fabricated from solid rebated, fire resistant treated timber and fitted where possible with non-combustible shutters. Non-combustible shutters may protect a house from damage as a bushfire front passes through. However, relying of shutters can provide a false sense of security and increase the risk to occupants who stay within a shutter house during a bushfire

The perimeters of window frames should be adequately sealed to eliminate fire or superheated air to enter the building or to set fire to the structural frame adjacent to the windows. Many of the standard beading and flashing systems offered by window manufactures are not suitable for houses in bushfire zones.

Opening window sash frames must have the same fire resistance as the main window framing members and have high value heat resistant permitter seals. Opening windows and doorways should be installed with high-strength steel insect proof mesh screens. The screens can be woven mesh, expanded mesh or perforated panels. It is critical that they are framed and fixed in a manner that provides adequate resistance against flying debris during bushfire events.

The external sheets of windows in the bushfire zone should be sheeted with toughened glass of an appropriate rating. All frames whether timber, steel or aluminium should be rated to withstand the heat and temperature of the calculated exposure time of the bushfire.

It is critical that a specialist inspection is undertaken after any bushfire event to upgrade or replace any and window and door components that have be damaged or weakened by the fire event.

### Orientation

Numerous factors in the immediate environment can make a difference when selecting the orientation of your house. It is advisable to avoid building on the top of hills or north facing slopes. Always check with the local bushfire authorities on the most appropriate locations on the property to construct a house.

### LPG Tanks

LPG tanks should ideally be installed away from the building, vents on LPG tank located next to houses, should face away from any buildings or flammable material.

### Walls

Timber frames to be clad in weatherboards or similar boards, should be separated from the timber structural frame by a layer of fire rated sheeting. The weatherboards should be of fire resistant timber or an appropriately rated fibrous cement cladding product.

Profiled steel sheets can be used in a similar manners, with profiled sheets it is critical that the ends of the sheets are mechanically sealed to ensure fire and heated air cannot get behind the sheeting and set fire to the framing. Take care with vertical profiled steel sheets as the fluted form of these sheets may become a series of chimney tubes allowing high temperature fire and air to flow up the interior of the sheeting setting fire to the wall framing.

The use of aluminium and PVC sheeting is not recommended in bushfire zones as its resistance to the high temperatures of a bushfire is not sufficient.

### Water Tanks

On properties not connected to mains water tanks with and adequate content of water are required, even if it involves purchasing water.
Fire brigade fittings on tanks are advisable and are mandated in some locations.

An internal combustion pump for pumping water is necessary, to ensure the pump will still operate without electricity.

## The questions remain ...

### Can homes be located in areas where there are bushfire threats?

The risk of loss of building and associated property damage can be mitigated ... but not eliminated.

Bushfire events are seriously unpredictable and the choice to construct a home in an identified bushfire risk zone cannot be recommended.

There are many localities, even within developed residential zones, where catastrophic wildfire events have swept through and destroyed whole residential areas.

Wherever, the choice of location, carefully evaluate the level of bushfire risk and then select the building form and materials to mitigate the risk to an acceptable level.

Be fully conversant with all the regulations, codes and standards applicable to the location of the proposed home.

Then apply this knowledge wisely, seeking assistance from appropriate authorities, raising the built-in property protection to the highest affordable level, to reduce the margins of error.

Ensuring that there are appropriate property repair and maintenance regimes in place to avoid deterioration of properties ability to resist bushfire attack.

Understand that bushfires are all consuming and any weakness in buildings defences and conflagration can quickly and disastrously follow.

### Can families living in bushfire threat areas confidently survive a bushfire event?

The risk of losing lives during a catastrophic bushfire event is highly unpredictable and unsatisfactorily resolved.

There are too many questions that cannot be answered with certainty:

- Will the home provide adequate shelter from catching fire?
- Including how much warning time is needed to prepare the property to resist the fire?
- Will suffocating smoke enter the home?
- Will there be sufficient cool oxygen air to breath?
- Is there danger from trees falling onto the house?
- Is there a safe escape route?
- Including what circumstance could negate the safety of the escape route?
- Is there sufficient fire fighting equipment to quell a bushfire?
- Including how many people are needed to adequately fight the impacts of bushfires?
- How long can an adequate supply of firefighting water be maintained during bushfire events?
- If people sustain injuries, burns or are overwhelmed by smoke are:
- Suitable treatment resources available?
- Likely opportunity for safe evacuation available?
- Can critical predictions be evaluated that demonstrate, the stay and fight choice is sufficiently reliable to warrant its application and that all occupants will survive the ordeal unscathed?

Unless these questions can be answered with certainty, and all people, particularly those who are young or infirm have an almost assured chance of surviving the inferno, then the stay and fight option is mute.

If there is no other choice available than to build homes in bushfire prone zones, then constructing house and adjusting the landscape to resist the fire impact without the intervention of active firefighting is the logical choice.

### AND

During the information search carried out to provide these comments on the risks of building in bushfire prone areas it became apparent that there is a lack of consistency

# BUILDING IN BUSHFIRE PRONE AREAS • 77

in much of the information that purports to provide validated information on home design and construction methods suited to resist bushfire attack.

The National Construction Code and the Australian Standards are the two most important Australia wide sources of information.

The State bushfire authorities provide sound information and when supplemented with information and outreach provided by their regional and local offices are a highly proficient source.

Most Local Government Authorities, in bushfire prone localities will have qualified officers with sound local knowledge.

Other publications may provide useful information but there are some that have doubtful credentials, take care.

If the codes, regulations and laws appear complex and difficult to interpret, investigate if there are specialist architects, consulting engineers or bushfire consultants with sound local knowledge. Consider engage them to ensure the best possible understanding and application of the requirements to building in bushfire prone zones is achieved.

| Radiant Heat Flux | Likely Effects | Approximate Distances |
|---|---|---|
| >29 – 110 kW/sqm | Flame zone | 0 – 20 metres |
| 29 kW/sqm | Ignition of most timbers without piloted ignition (3 minutes exposure) [Level 3 construction] during the passage of a bush fire. Toughened glass could fail | 20 metres |
| 19 kW/sqm | Screened float glass could fail [level 2 construction] during the passage of a bushfire | 27 metres |
| 12.5 kW/sqm | Standard float glass could fail [level 1 construction] during the passage of a bushfire. Some timbers can ignite with prolonged exposure and with piloted ignition source (e.g. embers) | 40 metres |
| 10 kW/sqm | Critical conditions. Firefighters not expected to operate in these conditions although they maybe encountered. Considered to be life threatening < 1 minute in protective equipment. Fabrics inside building may ignite spontaneously with long exposures | 45 metres |
| 7 kW/sqm | Likely fatal to unprotected person after exposure for several minutes | 55 metres |
| 4.7 kW/sqm | Extreme conditions. Firefighters in protective clothing will feel pain (60 seconds exposure) | 70 metres |
| 3 kW/sqm | Hazardous conditions. Firefighters expected to operate for short period (10 minutes) | 100 metres |
| 2.1 kW/sqm | Unprotected person will suffer pain after one minute exposure – non fatal | 140 metres |

# 78 • GLOSSARY

**AHD (Australian Height Datum):** The agreed reference point (datum) above and below which all land profiles are measured. This system is essential if major interconnected drainage systems are to work efficiently. Land survey drawings showing levels or contours should indicate if they used AHD or an assumed datum as a reference.

**Allowable slope:** limitations that apply to slopes; slopes of 1:20 can be negotiated by most people, slopes not exceeding 1:14 are considered to be the maximum slope that infirm people can negotiate even with a mobility device. In public buildings there are standards describing design parameters for ramps and slopes. Residences are less constrained by regulations but ease of access for all likely occupants and visitors is recommended.

**Asbestos:** A grey/blue naturally occurring fibrous mineral used in fibrous cement board and insulating products for many decades until the 1980s. It has since been determined to be highly carcinogenic and must be removed from buildings in an approved manner.

**Australian Standards:** The Australian Standards Association is the body that prepares and monitors quality standards and codes of practice used by industry. There are many Standards applicable to the building industry; some have effective compulsory status by being referred to in laws, for example, AS 1684 the Timber Framing Code and AS 3959 Construction of Buildings in Bushfire Prone Areas.

Many standards are referred to the specifications; some are specific to a sub-trade and effectively unreadable to lay people, while others are critical to achieving an assured level of quality in labour and manufactured products. Some municipal libraries and most university libraries have copies of the Australian Standards.

**Australian STANDARD – AS 3959:2018 – Construction of Buildings in Bushfire Prone Areas:** 'This Standard specifies requirements for the construction of buildings in bushfire-prone areas in order to improve their resistance to bushfire attack form burning embers, radiant heat, flame contact and combinations of the three attack forms' AS 3959:2019

**Authorities/Consent Authorities:** Taken to mean any government, semi-government, quasi-government or non-government organisation given power to demand certain authorised requirements of people working in the building industry. Often supported by Laws, Ordinances, Plans, Codes, Regulations, etc. Some authorities are required to assess applications to carry out work and give consent where appropriate.

**BAL-LOW:** This is the Bushfire Attack Level that is considered to not warrant specific construction requirements greater than those covered in the National Construction Code – Vol Two.

**BAL 12.5:** This level provides basic protection from ember attack.

**BAL 19:** This level is where protection from; ember attack, igniting local debris and a reasonable protection from radiant heat flux up to19kW/sqm.

**BAL 29:** This level is better that BAL 19 and increases the protection from radiant heat from 19kW/sqm to 29kW/sqm.

**BAL 40:** At this level do the calculations indicate reasonable protection of property from; ember attack, burning debris, exposure to a high level of radiant heat 40kW/sqm *and some* protection from a flaming fire front.

**BAL FZ:** A recognised FIRE ZONE, these are locations where catastrophic bushfire events are *likely* and where the services needed to fight bushfires are severely limited or not existing.

**Balustrades:** Fencing that surrounds stair voids, inter-floor openings or changes of level. Balustrades are closely controlled by the NCC.

**Barges:** Boards that follow the slope of roofs to finish the ends of roofing battens or purlins.

**Bricklayer:** Person who erects brick or concrete block walls. See also entry for Mason.

**Brick Veneer:** Ubiquitous Australian house-building technique; combines light-framed structural walling enclosed by a single skin of brickwork. Allows speed of erection and relatively low long-term maintenance cost.

**Building Approval:** The letter, conditions, drawings and other documents, granting permission to erect, extend or alter a building, from the relevant authority. Often referred to as stamped plans. See also entry for Construction Certificate.

**Building drawings:** Generally accurate, to-scale drawings of site plans, floor plans, elevations, sections, services and details. Used to submit to authorities for approval and permits. When combined with a tender (quotation), conditions of agreement and a specification, they form a major component of a building contract. Also called working or contract drawings.

**Bushfire-prone areas:** Any area that has been, is now or is likely to be subject to bushfire attack.

**Bushfire:** Unplanned out of control fires burning through the vegetated landscape.

**Bushfire Attack Level (BAL):** This is a formula to provide usable measurement of the likely severity of a building's exposure to ember, heat and flame attack. BAL is used to determine the effective bushfire resisting ability of materials and methodology used to construct any building. The numerals allocated to the six BAL are the radiant heat generated by the bushfire impacting on the surfaces of a building, expressed in kilowatts per square metre, this is the measurement of the level of attack that the construction requirements outline for the BALs should be able to resist.

**Bushfire authority:** These are individuals, agencies or statutory authorities that have the power to require compliance with the laws, ordinances, codes and standards that assess, inspect, approve and certificate the land use and construction of buildings in a bushfire zone.

**Bushfire resisting materials:** There are materials that can resist the full intensity of a bushfire even if within a fire zone (FZ).

**Bushfire resisting shutters and screens:** Bushfire resistant shutters are easy to operate coverings to the exterior of windows and doors that require protection during a bushfire event.

**CAD (Computer aided drawings/documentation):** The process of using computer programs to prepare drawings that can be plotted out on paper.

**Cement:** A grey/white powder that combines with water to set to permanent rock hardness; the critical ingredient of concrete. Also refers to some bonding agents that set irreversibly hard.

**Changes of level:** where there are changes in floor finishes a change of level my be unavoidable, it is recommended that these changes of level do not exceed 5mm.

**Chipboard & Particleboard:** Boards made by combining woodchips (mostly *Pinus*) with glue (mostly urea-formaldehyde) under heat and pressure. Boards range in thickness from a few millimetres to about 35 mm, and in size from 1800x900 to 3600x1800. Used mainly for cabinetry and flooring.

**Classified vegetation:** The process of identification and assessment of vegetation it is an essential component for the calculation of BAL.

**Colorbond:** This trade name has effectively become the generic name for the process of bonding coloured surfaces onto profiled steel sheeting. Predominantly used for roofing and walling products but potentially has a wide range of applications.

**Complying Development:** The process that allows LGAs to approve house building, alterations or additions applications, by comparing a published list of requirements and restrictions against the submitted application documents and drawings—if the submission complies then a permit will be issued. The application of complying development regulations varies widely from place to place, commensurate with the extent an LGA feels the need to control development.

**Concrete:** A mixture of cement, crushed rock, sand and water; easy to mix, mould,

# GLOSSARY • 78

form and finish when wet and fluid, but sets rock hard.

**Construction Certificate:** In some states, building approvals have been replaced by the issuing of a construction certificates, which is the acceptance that the application, drawings and other submitted documents have met necessary criteria for approval. No specific conditions are attached to a construction certificate. In some states construction certificates can be suitably qualified and registered by private certifiers. See also entry for PCAs.

**Contracting:** The process carried out when two parties —one offering to carry out a task, the other accepting the offer—agree for a consideration to proceed to complete the task.

In building contracts, the building owner is generally called the Principal, the builder is called the Contractor and the consideration is the Contract Sum. Contracts do not have to be in writing; casual/verbal agreements made between OBs and tradespeople or suppliers can be enforceable, particularly if the agreement has been observed by a third party.

**Cornices:** Moulding at the junction of ceilings to walls. Also, a decorated expression to the edge of an exposed parapet coping.

**Corrugated iron/steel:** Available for generations; roof sheeting with a curved, sectional profile of equal troughs and crests. Often called galvanised iron, though today is mostly Colorbond steel.

**Cost Plus:** Generally refers to contracts where the builder agrees to charge all materials and labour at cost–plus—a fee for providing construction administration and project management services.

**DCPs (Damp proof courses):** Horizontal waterproof barriers built into brick or masonry courses between a potential source of water and a location where water penetration would be detrimental. DPCs are commonly located above ground but below the floor structure in the base walls of buildings. Modern DCPs are of either a plastic material or an aluminium cored sandwich. If the DPC fails then water can rise from the ground into the wall above the floor; this is the phenomenon called rising damp.

**Demolition:** The act of pulling a building or part of a building down.

**Design drawings:** Prepared by an architect or other designer; presented to a client to assess and communicate design ideas and concepts. Range from simple freehand sketches to scale plans to three-dimensional CAD images; used as tools towards agreement on the final planning and appearance of a new building, addition to a building or building renovation. Usually need to be re-drawn as working, building or contract drawings before submitted to approval authorities or to form part of a building contract.

**Development Application:** The application required in most jurisdictions to gain approval from the appropriate authorities – often the LGA – to carry out new or changed land use developments.

**Double-hung windows:** Traditional windows that slide up and down; also called sash windows. Developed in the mid-19th century, they have two sliding panels that overlap at a mid rail; each sash has a cord attached that goes over a pulley to a counter-balance weight hanging in a boxed jamb. The sashes can be slid to any location and will remain fixed in that location by the counter-balance weights. Modern double-hung windows often use different sash balances utilising friction spirals or guides.

**Drainer:** Person who installs drainage pipes and other fittings. In many locations, drainers are required to gain a trade licence.

**Easy access:** where the access to, within, and from a home can be achieved with the minimum limitation to movement.

**Eaves:** Horizontal projection of roof structures beyond the wall lines of the building.

**Effective land slope:** The effective land slope, flat, upward or downward from the home combine with the classification of the vegetation in determining the effective BAL.

**Electrician:** Person who installs electric cabling and other fittings. Electricians are often required to gain a trade licence.

**Elevations:** Drawings that show the flat to-scale projections of the faces of a building.

**Ember guards:** The coverings that have fine enough openings and of sufficient fire resistance to stop the passage of embers passing into the spaces and cavities of buildings.

**Energy conservation:** The conscious effort to reduce the use of electricity, gas and other energy sources. Achieved by using energy-efficient appliances and turning off energy-using appliances when they are not required.

**Enhanced access:** the considerations required to ensure a home is suitable for people of all abilities and all ages.

**Environmental Planning Acts:** Town planning was once a series of maps with some written descriptions nominating the approved uses to which land could be put. Modern urban planning schemes use fewer maps, and focus on ensuring that the environmental qualities of a locality and the wider community are enhanced by, or at least minimally harmed by, any land use development.

**Facias:** A board fitted horizontally at the ends of rafters normally supporting the roof gutters.

**Fibrocement:** Fibre reinforced cement material made from inert fibres bonded into an autoclave-cured cement. Available as flat, profile and textured sheets, pipes and moulded shapes. Suitable for roofs, external cladding, internal linings, floors and drainage systems. Once reinforced with asbestos cement, but now with less dangerous fibres; it should always be used strictly to manufacturer recommendations.

**Fibrous plaster:** Flat or moulded sheets of gypsum plaster cast on polished concrete moulds and reinforced with an inert fibre. Used 1920s–1970s and still manufactured in limited amounts.

**Fire Danger Index (FDI):** This is a formula that endeavours to give a value to an amalgam of; the likelihood of a fire starting, its rate of spread, its intensity and how difficult it would be to control. It takes into account, air temperature, humidity, wind speed and various climatic impacts.

**Fire Resistant Level (FRL):** This is a system that provides an indication of the fire resistance of building elements. FDI indicates the ability of a specific element to: maintain structural adequacy/material integrity/ thermal insulation during a specific controlled heat event, expressed in minutes.

**Fixing out:** The process of fixing internal fittings and fixings to the building fabric.

**Flame Zone (FZ):** BAL-FZ indicates that any building on a property with this bushfire attack level FZ would be directly surrounded by and engulfed in flames and be unlikely to survive.

**Foliage cover:** Foliage cover indicates the effective solar shade provided by foliage.

**Foliage cover overstorey:** Over storey is the highest vegetation canopy in an area of vegetation this would generally be expected to the tallest trees.

**Foliage cover understorey:** Understorey is all the vegetation below the overstorey and including small trees, shrubs, herbaceous growth, grasses and dead vegetable matter.

**Footings:** In Australia, the part of the building that carries the mass of the structure onto the foundations. Can be made of stone, brick, timber, steel or concrete. Reinforced concrete is generally considered the most suitable and durable. In some countries, footings are called foundations, so take care when using an international text.

**Formwork:** The support structure to mould reinforced concrete.

**French Doors:** Generally a pair of matching hinged glass doors, sometimes referred to as French casements.

**Geo-technical engineer:** A specialist consultant who investigates and reports on substrate ground conditions.

**Grasp levers or D-pull handles:** all handles that open doors, cupboards and drawers should operate by pushing, pulling or sliding a projecting D-shape handle or by

# 78 • GLOSSARY

a downward quarter turn of a lever handle. Knob type handles are not recommended.

**GST (Goods and Services Tax):** A tax payable to the government on every purchase of materials, goods, labour and services associated with building or altering a house. OBs are required to keep records of all GST transactions.

**Hardboards:** Building sheets about 4 mm to 8 mm thick, made from bonded tempered wood fibres. The common trade name is Masonite. Can also be used as a weatherboard alternative, for example, Weathertex.

**Hardwood:** Timber milled from two different tree groups. Hardwoods traditionally came from deciduous trees—oak, ash, maple, apple. These trees are not native to Australia but the local eucalypt trees produce similar timber have taken that nomenclature in Australia. Examples of Australian hardwood include Tasmanian Oak, Victorian Ash and Jarrah.

**Head:** The support over an opening; a lintel, or the top of a door or window frame.

**Heritage:** Representative items from a previous time. Many houses have been listed as important examples of architectural, social or cultural heritage. If a house is listed as a heritage item, there may be limits to alterations allowed to its fabric, and it is unlikely that the responsible authority will allow its demolition.

**Heritage consultants:**
  **Adviser:** Generally, a private consultant employed to give advice to an LGA. May be accessible to the public but often limited by the LGA's agenda.
  **Architect:** Generally, a qualified architect who has developed particular interest in heritage matters.
  **Consultant:** Generally, a person who advises people about alterations or additions to heritage buildings, particularly those listed by LGAs and subject to conservation requirements.
  **Officer:** Generally, an employee of an LGA who is responsible for the implementation of heritage guidelines and controls.

**Heritage qualifications:** There is currently no national registration of heritage consultants; there are, however, some local groups purporting to monitor qualifications and experience of heritage consultants. It is important to check that any authority requiring heritage inputs recognises the person preparing the inputs—this can make the process of finding a truly independent heritage consultant difficult.

**Heritage reports:** When an owner wants to add to, alter or demolish a heritage-listed building it is likely that the responsible authority will require a heritage report prepared by an authorised heritage consultant. Consultants are obliged to prepare full and unbiased reports, regardless of whether or not the report supports approval of the proposed changes to the heritage item.

**Hydraulic engineer:** An engineer who understands fluid dynamics and designs water supply and waste liquid disposal systems.

**Internal linings:** Internal sheeting or boarding to internal walls.

**Jamb:** Vertical members at door- and window-to-wall interfaces.

**Laminated plastics:** Layers of paper bonded by thermo-plastic material; generally about one millimetre thick. It is the durable material that revolutionised kitchen surfaces in the 1950s and remains a critically important material where flat colourful durable surfaces are required. Often described by tradename, for example, Laminex, Formica, Abet Laminati.

**LEPs (Local Environmental Plans):** LEPs may have different names from place to place but are essentially the control documents that prescribe the performance requirements of all property developments within a defined area. The documents are generally text-based, though many modern versions are profusely illustrated.

**LGAs (Local Government Authorities):** The collective term for municipal governments; shires, municipalities and cities.

**Livability design:** the thoughtful design of homes to ensure maximum comfort for all occupants regardless of their ability.

**Lump sum firm price:** When a contract is agreed to have a set price for the whole of the contract. Importantly, the contractor (the builder) agrees to provide the work described in the contract but is generally not obliged to provide prices to the principal (the owner) for individual parts of the labour, materials supply or the sub-contract amounts.

**Mason:** Person who builds walls from stone.

**MDF (Medium density fibreboard):** Boards made by combining woodfibres (mostly *Pinus*) with glue (mostly urea-formaldehyde) under heat and pressure. The boards range in thickness from a few millimetres to about 35 mm, and in size from 1800 x 900 to 3600 x 1800. Used mainly for cabinetry.

**Metric:** Measurements used by the Australian building industry are modern metric. Dimensions on building drawings are shown in millimetres but the mm symbol is not used, for example, 2.345 metres is shown as 2345. Land survey plan dimensions and reduced levels may be shown as decimal metres, for example, 2.345 and again no m symbol is used. Note: centimetres (cm) are never used by the Australian building industry, these are reserved for human dimensions and for fabric dimensions.

**Mobility device:** any device that improves the mobility of people recovering from or living with disability or infirmness of movement.

**Mullion:** An internal vertical member in a window frame.

**National construction code:** The National Construction Code supersedes the Building Code of Australia and all other State or Local Government Area building regulations. It has the highest level of precedence.
  The NCC calls up certain Australian Standards as requiring to be complied with as if they were in the NCC. Specifically recognised is, *Australian Standard AS 3959, Construction of buildings in bushfire prone areas.*
  The NCC has three volumes, volume 2 deals with residential buildings both single tenancy and multi-tenancy dwellings – free standing and attached.

**Natural materials:** Generally, this classification is reserved for materials that are made from naturally occurring materials without the addition of chemicals or complex industrial processes. For example, timber is a natural material, chipboard is not. Stone is a natural material, simple fired clay bricks are probably a natural material but a silica-lime brick is definitely not.

**Non-combustible:** These are materials, assemblies and components that have been assessed and determined to be non-combustible by AS1530.1 or the NCC.

**Particleboard:** Sheet board material also called chipboard. Made from wood chips glued together with an adhesive. The sheets range in thickness from a few millimetres to 30+ mm, and in size from about 900 x 2400 through to about 1800 x 3600. They are used extensively for cabinetry and flooring uses.

**PCAs (Principal Certifying Authorities):** People who are approved by legislation to grant a prescribed range of development and building approval certificates. Some are direct employees of LGAs, others are private organisations. There is extensive variation in the enabling legislation from state to state.

***Pinus*:** The name used in Australia to identify timber sourced from plantation grown *Pinus radiata* (Monterey Pine).

**Pitched Roofs:** These roofs have an obvious slope. The slope can either be annotated as a ratio, for example, 1:4, or as degrees, for example, 20°. Many older roofs were simply equal ration—1:1, 45°; others were 3–4–5 triangles for easy calculation—3:4, or so called quarter pitch—the roof rises one unit for every 4 units of total span: this is a ratio of 1:2 not 1:4.

**Plans:** Loosely, these are the drawings used by a builder to construct a house. More accurately, they are the to-scale drawings of the downward perpendicular view of the walls and rooms, cut through at about 1500 above the floor.

**Plasterboard:** A wall and ceiling lining

# GLOSSARY • 78

product manufactured with a gypsum plaster core sheeted on either face with kraft type paper. Commonly 10, 13 or 16 mm thick and a very economical product that is incombustible, it can be flush finished and painted easily. Often referred to by its trade names, for example, Gyprock.

**Plumber:** Person who carries out piping, valving and other work associated with water supply and liquid waste removal. It is a requirement in most jurisdictions that water supply and sewerage connections be carried out by a licensed plumber. Stormwater drainage is usually carried out by plumbers but this is seldom a mandatory requirement.

**Plywood:** Boards made from three or more veneers of timber set with the grain direction alternating by 90 degrees. The veneers are bonded with glues ranging from animal cuticle hot glues, through common PVA glues to complex special-purpose chemical adhesives. Plywoods vary from being highly susceptible to moisture to highly waterproof boards. They can be decorative or functional in appearance.

**Prefabrication:** The process of carrying out a building or fabrication activity that is completed away from the final site of the works. Wall and roof frames are commonly prefabricated; cabinets are mostly made from fitting together prefabricated carcasses.

**Pressed bricks:** Bricks made by pressing clay or a clay/shale mix in a mould the size of the required brick. This process has been used since the beginning of the 20th century and is still an important method.

**Prime Costs:** The term used when a monetary allowance is made for a purchasable item in a building contract—the allowance is adjusted to the real invoice price of the item when it is purchased, the owner is credited or debited for the difference. Prime costs (PCs) should only be used when exact specification at the time of signing the contract is difficult, and only for items that will have a clear purchase price. PCs should not be used for labour allowances, as these should be included with all other builder costs in the contract price.

**Provisional Sums:** The term used when a monetary allowance is made for a purchasable item in a building contract requiring supply and labour components—the allowance is adjusted to the real quotation price of the work when it is ordered, the owner is credited or debited for the difference. Provisional sums (PSs) should only be used when exact specification at the time of signing the contract is difficult and only for items that will have a clear quotable price. They are materials and labour allowances but should not include other builder costs, as these should be in the contract price.

**Quarter turn:** wherever possible restrict the movement of lever door handles and lever taps to a rotational travel of 90 degrees. Some taps may rotate clockwise, anti-clockwise and in a upward arc, these provide an appropriate movement for most people. Note; when the water starts to flow from a tap it should be of a temperature that will not scald a person's hands, it should only be able to be moved to a hotter temperature flow after the water has commenced flowing.

**Ranch style:** The home style that suits OBs; single storey, two rooms deep and as long as is required to suit the owner's needs. Long, low and simple; ranch style houses need an area of about 25 x 10 metres to suit a full four-bedroom family house.

**Rectangular and Square Hollow Sections:** A development of traditional galvanised circular steel tubing (piping). They are manufactured by forming a tube from flat steel plate stock and then continuously welding the joint to make a seamless tube. Often pre-galvanised, they are easy to cut and fabricate compared to more traditional steel sections, and provide an alternative to many timber sections.

**Retaining Walls:** These are vertical or near vertical structures that support changes in ground level. Mostly used to retain excavated banks or filled benching. It is recommended that unless of modest proportion, retaining walls—particularly those on property boundaries, supporting the ground around basements or restraining a cut and fill excavation— should be designed by a qualified engineer.

**Rise and Fall:** A clause in a contract that allows the contract price to automatically rise or fall by linking it to specific material and labour indices, such as the consumer price index. Favoured by builders during periods of high inflation.

**Rising Damp:** This is where water from the ground rises up through the walls of a house to allow dampness and mould to develop on inside walls. Generally confined to masonry walls and is generally caused by a breakdown of the damp-proof course and/or excessive ground water caused by poor sub-soil drainage.

**Sarking:** The use of a membrane barrier between a potential source of moisture and a place where the moisture could be detrimental. Commonly used under roof tiles and sheets, where a silver surfaced paper is used to provide some radiant heat reflection. Recommended behind weatherboards, where a fire resistant sarking can significantly enhance a building's resistance to external fire risk.

**Scantling:** The collective term used to denote all the general building timber used in house framing, traditionally the 4" x 2" (fourbetwo) now shrunk to 90 x 45.

**Schedules:** Lists or charts of information required to communicate specific requirements for purchase, application or activity. Common schedules include colour schedules, prime cost schedules, window schedules, lighting schedules.

**Sections:** When used to describe a drawing, it means a drawing that has apparently been cut through to reveal the internal elevations of the building and its structural fabric.

**Services:** There are many services utilised by a modern residential home, including but not limited to electricity, water supply, sewerage, telecommunications, gas, stormwater and cable TV. Most of these need cable or pipeline connection to mains services but there are alternative systems that can make household utility services independent of mains infrastructure.

**Shadow diagrams:** A product of the concerns of environmental planning regulations. Local planning approval authorities often require plan diagrams to show the shadows generated—at various times of the day and at various times of the year—by new houses or additions to existing houses. A complex code of solar access requirements is then consulted to ensure that neighbouring properties have a fair share of the available sun. Generally shadow diagrams are only required of built forms, and trees are not considered.

**Sill:** The bottom member of a window or door frame.

**Skillion roofs:** These are the single direction slope roofs, sometimes referred to as lean-to roofs or shed roofs. There can be more than one skillion roof on a building; the key is that rafters do not meet at a continuous ridge.

**Solid masonry:** There is some confusion between what is solid masonry and what is cavity wall construction. In general, solid masonry construction is where stone, brick or block construction is used for the majority of the walls of a building, from the footings to the roof. Cavity walling is really only a devise used in solid masonry construction to eliminate the entry of rainwater into the habitable interior of the house.

**Specifications:** The written component of building documents. The working drawings generally provide accurate quantitative information about a building project, whereas the specifications provide the qualitative information.

**Statements of Environmental Effects** or **EISs:** Most environmental planning instruments require a written statement of the likely impact of proposed developments on the environment. A statement of environmental effects is generally about 2 to 3 pages long whereas an environmental impact statement, which is often for larger or highly sensitive projects, is often 20 plus pages. Both of these documents require landowners to furnish information on environmental impacts and to propose mitigation methods.

# 78 • GLOSSARY

**Steel decking:** Deep section-formed galvanised steel sheets, originally developed as a long span, low pitch, no piercing roofing material, for example, Kliplok or Speedek. It is still available for this use but has evolved into a highly successful permanent concrete formwork system, for example, Bondek.

**Steel fabricator:** Person who supplies, cuts, forms, welds and generally erects structural steel building elements.

**Step free access:** entry, the recommendation that entries, exits and travel through homes is not compromised by requiring the negotiation of any steps.

**Stile:** This is the vertical edge frame member of a door or window sash.

**Structural engineer:** Consultant who makes computations, designs structural elements and provides necessary drawings and other documents to allow structural elements to be fabricated or constructed. Can also prepare reports of structural sufficiency and supervise specific on-site engineering fabrication and placement tasks.

**Structural steel:** Generally in the form of hot rolled sections produced as Universal Beams (UB), Universal Columns (UC), Channels (PFC) and Angles. This steel can be welded or drilled and bolted. Alternative steel sections include tube stock; circular (CHS), rectangular (RHS) and square hollow sections (SHS), these are often a suitable replacement for timber and are used extensively in modern subfloor structures. There are also medium and light gauge cold formed galvanised steel sections, developed for industrial buildings and used regularly by house builders—most of these sections are either channel or top-hat shapes, but many special shapes are fabricated.

**Sub-floor framing:** The sub-floor frame generally consists of beams in two directions; the lower beams, called bearers, supported on piers or stumps, were traditionally 100 x 75 hardwood F11+ (4" x 3") and were spaced about 1800 (6'0"). On top of the bearers were 100 x 50 (4" x 2") hardwood floor joists at 450 (18") centres. The days of access to high durability hardwoods are limited and sub-floor frames are now often built from treat *Pinus* or a proprietary galvanised steel system.

**Subcontractors:** Traditionally, people who carry out work for a contractor to assist in completing part of a larger contract. In the Australian building industry, subcontractors have become all the tradespeople who provide services to builders, whether or not a main contract exists. OBs actually have a contract directly with the trade labour they use but the term subcontractor is used loosely to identify this service provider.

**Superintendence:** This is the role of a person administering a contract. A highly responsible job often undertaken by the architect but can be done by anyone approved by the principal and the contractor. Every contract gives different roles to a superintendent and must be read carefully by all parties.

**Suppliers:** Person or firm that supplies goods.

**Surveys:** Surveying is about accurately measuring and recording real property. Land surveyors can confirm and peg out parcels of land, accurately locate buildings and other features in relation to boundaries, locate services, take site levels and mark contours. Surveys of the interior of buildings are normally carried out by the architect or builder; these can be very important when designing alterations and additions and assist in avoiding costly mistakes.

**T&G Flooring:** When boards have a groove down one edge and a tongue down the other. When the flooring is laid the tongue in one board engages with the groove in the adjacent board. This connection strengthens the floor, reduces drafts and limits the entry of bugs and dust through the floor.

**Tendering:** The process by which a number of potential contractors submit proposals or prices to carry out a project. Every tenderer is given the same information and all tenders are judged only against this information. Most tenders have a specific closure time and no late tenders are accepted. All valid tenders are kept in sealed envelopes until they are opened by the per-son/s designated to assess them.

**Termites:** Cream-coloured insects about the size of a common ant; they have big teeth. An active colony can severely damage a house within a few weeks. They have traditionally been controlled through galvanised steel caps at the top of piers/stumps; this method is still highly effective. For a period in the 1980s, chemical flooding was used, but the chemicals proved to be very nasty to human health, so the search began for highly efficient mechanical systems. Grits, meshes and barriers are available and must be used strictly to the manufacturer's instructions.

**Terracotta:** The red/orange fired clay used to make traditional roof tiles.

**Title referencing:** The system, mostly based on the Torrens system, that gives a specific allotment of land a unique reference number.

**Toggle or rocker switches:** easy to operate electrical switches should be used throughout homes.

**Trusses:** Triangulated structural frames. The majority of project homes have used this roofing system since the 1970s. Light, strong and cheap; computerised factories can churn out trusses accurately and rapidly.

Care must be taken to ensure the load-bearing points, generally at the truss ends, are adequately supported and braced. Houses with trussed roofs generally carry the entire roof load on two parallel walls, and the internal walls are non-load-bearing. This means that upper floor conversions on single storey trussed roofed houses are too complicated—there is no available support for the floors, walls and roof of the addition, and the trusses cannot be cut through.

**Vaulted roofs/ceilings:** Roofs with sloped ceilings—often fixed to the underside of the rafters.

**Verges:** At the edge, generally used to describe the sloping edge of a roof.

**Waste Management Plans:** In some areas (and the trend is growing), LGAs require a waste management plan to be submitted at the time a development application is lodged. Applicants must commit themselves to a management plan to dispose of all building waste, and to agree to the long-term location of garbage cans and worm farms. Until waste management plans are approved, no construction can commence.

**Weatherboards:** Strips of timber about 200 wide x 25 thick used to clad the outside of houses. The boards are applied horizontally and have been used in abundance on houses since first settlement of Australia.

**Wire-cut bricks:** Bricks made by extruding wet plastic clay or clay/shale mix through a rectangular hole, the plan size of a brick, and then using a wire to cut bricks of the correct height. These bricks commonly have three or more holes through them.

**Wrought and cast iron:** Before steel was widely available, many building elements were made from wrought or cast iron. Wrought iron is made by heating and hammering iron into shape; items are particularly resistant to corrosion, particularly durable and often worth retaining. Cast iron is made by heating iron to its liquid state then pouring it into sand mould; used to create functional or decorative items such as balcony balustrades on terrace houses. Cast iron products are still made, but many look-alike products are cast aluminium.

# INDEX

AAC see autoclaved aerated concrete (AAC)
abba houses 20
absorption trenches 91
access for firefighting vehicles 266
access
    see site—access
acrylic dome roof lights 180
administration
    contract 30–31
    project 68–69
    site 60
agricultural drainage 90–91, 95
allowable slope 261
aluminium claddings 124
aluminium windows 16, 140, 141, 147
ambulatory equipped toilet 261
American balloon frame 19
angles 199
anti-ponding board 158
appearance of house
    deciding on 25
    green supplement, 243
appliances 65, 196–97
appropriate escape route 266
arches 132
architects
    advantages of using 24
    briefing 30
    choosing 29, 32
    fees 30, 240
    role 29
    specification, preparing 48
architectural drawings 37
architecture, domestic 11–12
architraves 199
area, of site 77
AS 4000/4305 59
asbestos 15–16, 123, 248-251
Australian Standard AS3959:2018 266
authorities, government and private sector 33–36
autoclaved aerated concrete (AAC) 15, 65, 120–22
BAL 12.5 269
BAL 19 269
BAL 29 269
BAL FZ 269
BAL-LOW 269
balloon framing 19
barge board 155
base, building 80
basins, bathroom 206
bathrooms 204–209
    construction 205
    planning 47, 204, 207–208
    storage 204
baths 204, 205, 206
batts 128
bellcote roof 153
bidets 204, 206
billiard rooms 226
birdsmouth notch 160
blindness/limitation to vision 260
blue board 126
bluestone 128, 136
boiling water units 197
boundaries, site 76
box frame construction 19
boxed gable roof 153
brick veneer construction 19–20, 117–19
bricklayers 61
bricks
    floor 233
    history 15, 17
    mud 15
    sizes 117
    supplies 64
    types 117, 128
brief, design, developing 24–25
bubble diagrams 25–26
budgeting 25
builder licensing legislation 34
builders 241
building alignments 77
building approval, gaining 10, 35
building construction, history 18–20
building costs and benefits 22
building departments 33
building designers 24
building drawings 39–47
building materials
    control 68
    discounts 64
    history 15–17
    reusable 64
built-in cupboards 210
bush rocks 128
Bushfire Attack Level (BAL) 266

bushfire considerations 242
bushfire resistant design 269
bushfire resisting materials 267
bushfire resisting shutters and screens 267
bushfire zones 266
bushfires 266
buying a house
    costs and benefits 22
    factors to consider 21–22
cabinetmakers 62
cabinets
    joints 221–22
    plan drawings 47
    supplies 64, 65
CAD (computer-aided design) 37
car accommodation 228–29
carpenters 61
carpet 234
cash discounts, for labour 57–61
ceiling joists 155
ceiling plans, reflected 46
ceilings 152, 176–79
    insulation 176
    plasterboard 176
    timber 178
cement, history 15, 17
certifiers, private 31, 59
chimneys 216
chipboard 15, 221, 233
city weather conditions 242
claddings, sheet and board 123–27
classified vegetation 267
clerestory roof 154
clerestory windows 181
clothes dryers 197
clothes washers 197
coffee makers 197
cold water reticulation 189
collar tie 155
Colorbond 173
columns 96
comfort, of house, 243
completion, achieving 72
computer-aided design (CAD) 37
concrete
    autoclaved aerated (see autoclaved aerated concrete (AAC))
    history 15, 17
    ready-mix supplies 64
concrete slab flooring 102–108
concreters 61
conite system 126
Construction Certificate 10, 34, 35
construction details, drawing 39, 45
Construction of Buildings in Bushfire Prone Areas 266
consultants 29–32
continuous base walls 97
continuous footings 87
continuous handrails to stairs 260
contracting 58–59
contracts
    cost plus 57
    legal requirements 62
    rise and fall 57
    standard form 62
cooktops 197
copyright, plans 24
cork flooring 233
cornices 176, 199
corrugated iron and steel 16
cost control sheet 53–56
cost plus contracts 57
costing 52–56
costs
    blow-outs 27
    control 53–56, 68
    to satisfy authorities 35–36
    councils, local 33
    questions to ask 34
covenants 76
critical path program 68–71
cross ventilation, 243
cruck construction 18–19
cupboards 210
cut and fill 89
damp-proofing, base brick work 97
deafness 260
decking, timber 223
deliveries, supplies 69
Department of Fair Trading 62
design drawings 37–38
designers, building 24
designing the home 24–27, 30
development application
    documentation required 30
    lodging 30, 34
    preparing 35

development approval
    document 34
    gaining 35
discounts, cash 57–58, 61
dishwashers 197
documentation 30
dome roof lights 180
door widths 260
doors 149–51
    external 149
    folding 149–50
    French 149
    internal 149
    patio 149
    schedule 51
    supplies 65
doors and corridors 261
dormer windows 181–82
double glazing 148
double storey construction 225–27
drafting services, fees 240
drainage
    agricultural 90–91, 95
    stormwater 33, 184–85
    sub-surface 90–91
    weep holes 95
drainer suppliers 64
drainers 61, 184–85
drawings
    architectural 37
    building 39–47
    design 37–38
    electrical 193
    section 44
    structural engineer's 45
    working 35
dryers, clothes 197
dual occupancy 21
Dutch gable roof 153
easements 77
easy access 259
easy access toilet 260
eco housing 20, 244
effective land slope 267
electrical cable
    colour coding 195
    supplies 65
electrical engineers 32
electrical fix-off 193
electrical rough-in 193
electricians 62, 193–95
electricity supply 33, 74, 193, 195
elevation views 37, 39
elevations, exterior 44
ember guards 267
ember, heat and flame attack 267
energy efficiency 258
energy supply companies 190
energy supply, controlled 194
engineering departments 33
engineers
    electrical 32
    geo-technical 32
    hydraulic 32
    mechanical 32
    structural 31, 240
enhanced access 259
entries, house 217
environmental statements 35, 244
epoxy resin 234
equipment hire 64
excavations 64, 81
excavators 61
existing building
    credits for using 246
    costs of extending 22
factors to consider 21
factors to consider 21–22
family conference 23
fascia board 155
faux-Federation house styles 20
fees, professional 240
fibrocement 15, 65, 123
fibrous plaster 17
filled sites 89
finishes
    schedule 50
    surface 233–34
Fire Danger Index (FDI) 267
firefighting water supply 266
fireplaces 213–16
fixing out 198
fixing timber 64
Flame Zone (FZ) 267
flashings 192
flat roof 154, 166–68
floating boulder 89
floating slabs 102
floor plans 37, 39, 43

flooring
    concrete slabs 16, 102–108
    steel 100–101
    timber 64, 98–99
flues 215, 216
fluorescent energy-saving light fitting 195
fluorescent tube light fitting 195
footing plan 39
footings 85–89
    continuous 87
    integrated 88
    isolated 85
    pier and beam 88
formwork 64
foundation stability 18
frame construction 18–20
freezers 197
gable roof 153
galvanised iron 172
gambrel roof 153
garage 228–29
gas appliances 191–92
gas plumber/fitter 192
gas supply 33, 75, 191–92
    liquid petroleum 190–91
    natural 190–91
geo-technical engineers 32
geology, site 76
glass
    history 16
    window 140, 147–48
glass roofs 182
glazing bars 181
glazing, window
    double 148
    obscured 148
    safety 148
Goods and Services Tax (GST) 52, 58, 59, 61
granite 136
green supplement 243-258
grey-water disposal 186, 243
ground to floor support systems 96–97
GST
    see Goods and Services Tax
halogen-dichroic light fittings 195
hanging beam 155
hardboard 15, 123, 233
hardware 64
hardwood 233
heat exchange processes 230–31
heat exchangers 191
heat pumps 191
heritage advisers 31
heritage statements 35
hip and valley roof 153
hip rafter 155
hip roof 153
hiring equipment 64
Holmesglen system 126
home-owner warranty 35
hot water reticulation 190
hot water systems 190–92
hourly rates, tradespeople's 57
hours of work, restricted 58
hydraulic engineers 32
identification of tradespeople 59
incandescent-tungsten light globes 195
Information to Tenderers 48–49
insulation
    ceilings 176
    materials 231
    roofs 172
    simple guide and drawing, 245
    supplies 65
insurance policies 35, 36
integrated footings 88
interior designers 31
interior linings 138–39
iron, history 17
ironstone 136
isolated footings 85
isolated piers 96
jerkin head roof 153
joinery 221–22
joists, long-span 227
    kitchens 201–203
    plans 47, 200–204
kitchens, laundries and utility spaces 263
labour control 68
labourers 61–62
laminate, plastic 234
landscaping plans 46
lavatory pans 204, 205
lead 248-9
leadlight 140, 141
legal obligations 59

277

# 79 • INDEX

legal obligations 59
    subcontractors 60
legislation builder licensing 34
lending authorities
    specifications required by 48
    use of builders 241
level access to shower 260
levelling methods 83–84
liability, public 36
light fittings 65, 195
limestone 128, 136
liners 61
linings, interior 138–39, 199
linoleum 233
lintel beams 128
Livable Housing Australia 259
local councils
    see councils, local
Local Government Authority 10, 34
location, site 75
long service leave contribution fund 34, 60
long-span joists 227
man made materials 16
marble 233
masonry construction, solid 18, 128–37
materials
    see building materials
MDF (medium density fibreboard) 139, 198, 221, 233
mechanical engineers 32
metal products, history 16
metal tiles 234
microwave ovens 197
mobility devices 259
money control 68
mortars 117
mosaic 233
    quarry 223, 233
    supplies 65
    roof cement 15, 169
    cordovan 169
    fibrocement shingles 169
    metal 170
    slate 169
    supplies 64
    terracotta 15, 169
mosaic tiles 233
National Construction Code 266
Natspec Domestic 48
NCC Vol 2 266
needs, design, chart 23
neighbours
    notifying 35
    rights 77
non-combustible 267
orientation, site 76, 243
Owner Builder's Permit 34, 35
owner-building
    future in Australia 10
    history 8–10
paint 234, 235
painters 62
painting 235
pantries 211
particleboard 15, 233
patented roofing systems 179
pavers 223, 233
paving, external 223–24
payment, method of 57, 58, 60
pier and beam footings 88
piers, isolated 96
plan service 24
plan views 37
planning
    final 26
    preliminary 25
plans 37–9
plasterboard 17, 65, 138–39, 176
plasterboard fixers 61
plastic claddings 124
plastic laminate 234
platform floor 98
platform framing 19
plumbers 62, 189–92
plumbing suppliers 64
plywood 15, 233
pole construction 97
ponding, anti-ponding board 158
posts 96
Powerpanel 122
pre-design 23
pre-payments 57
preferred bedroom layout 263
price blow-outs 27
price control 68
pricing 52
prime cost (PC) items 65

Principal Certifying Authorities (PCAs) 10
project homes
project managers 31
public liability 36
purlins 18, 19, 155
quality assurance 49
quality of vs OB homes 8
quarry tiles 223, 233
raft slabs 88, 102
rafters 155
rainfall
    by city 242
    effect on site 74
rainwater tanks 243
rangehoods 197
recyled materials 256-7
reduced mobility 259
reduced mobility 260
refrigerators 197
regulations, building 10
reinforced concrete block 94
    stabilised banking 92
    tanking 95
    water pressure, reducing 94
reinforcing steel 64
renderers 61
resort-condo house styles 20
retaining walls 92–95
    backfill 95
    brick 93
    cantilever 92
    drainage 95
    failure 93
    gravity 92
    inclined 92
retention fund 60
ridge (beam) 155
rise and fall contracts 57
robes and presses 210–12
rock floaters 89
rocks, bush 128, 136
roof lights 152, 180
roof plan 39
roof plumber 192
roof sheeting 152, 172–75
    insulation 172
    sarking 172
    supplies 64
roof tiles
    see tiles, roof roofers 61
roofing battens 155
roofs 152–75
    concrete 152
    conventional roof frames 155–60
    flat 154, 166–68
    framing 18, 65
    glass 182
    insulation 172
    membrane 152
    patented roofing systems 179
    penetrations through 182
    sheeted 64, 152, 172–75
    skillion 166–68
    stick-built 152
    tile 152, 169–70
    timber used 156
    trussed 152, 161–65
    types 153–54
    vaulted 166–68
    wind, effect of 172
Royal Australian Institute Architects 32
rubber 234
safety gear 65
safety glazing 148
salvaged materials 254-5
sandstone 15, 128, 136
sanitary fittings 65, 206
sanitary plumbing 187–88
sarking 158, 170, 172
scaffolding regulations 33
scale, drawing to 39
scantling timber 64, 110
schedules 50–51
section drawings 44
section views 37–38, 39
septic disposal system 186–87
services 183
    availability to site 74–75, 76
    plan 39
setbacks 77
sewer line 74
sewerage authorities 33, 36
sewerage system 185–86, 248
sheet and board claddings 123–27
    fixing methods 124
sheet roofs
    see roof sheeting
shingles

fibrocement 169
history 15
timber 170
shower recess 105, 204, 205
shower rooms 262
silt traps 90
site
    access 75, 76
    amenities 76
    analysis 74–77
    clearing 81
    filled 89
    plan 39, 42
    preparations 81–83
    services available to 74–75, 76
    setout 81–83
skillion roof 154, 166–68
skirtings 199
skylights 152, 180
slate 15, 169, 223, 233
slip resistance 262
slope types 78–79
slope, site 74
soffit bearer 155
softwood 233
soil types
    cohesionless 93
    cohesive 93
solar energy 191
solid masonry construction 128–37
spa baths 204, 206
space for mobility device 261
space required 204
specifications
    definition 48
    manufacturers' 48
    preparing 35
    requirements for 48
    value of 49
stairs 47, 217–20
stairways 262
standards
    AS 4000/4305 59
    quality assurance systems 49
    specifications 48
statutory insurances 36
steel
    flooring systems 100–101
    history 17
    windows 140
steelworkers 61
step free access 259
step free threshold 260
stone 233
stone cobbles 223
stonework 136–37
stormwater drainage
    disposal system 184–85
    requirements 33, 74–75
stoves 196
    structural engineers 31
    drawings 45
    fees 240
struts and props 155
strutting beam 155
stumps 96
styles, architectural 11–12
sub-surface drainage 90–91
subcontracting labour 62
subcontracting process 61
subcontractors 60–63
    finding 60
    payment of 60
substrata, site 74, 76
suitable clearances 262
sunrise and sunset, direction of 74
superannuation, contractors 60
superintendence of contract 30–31
suppliers 64–65
suppliers 64–65
supplies 64
support rails bathrooms 260
support rails toilets 260
surface finishes 233–34
survey plan 39
survey, check 35
surveyors 31
suspended concrete slabs 105
taps 205–206
tax
    GST 52, 58, 59, 61
    registering with ATO 36
tax invoices 61
telephone service providers 33, 75
temperatures, by city 242
Tenderers, Information to, document 48–49
tendering 57–58

termite proofing 97, 107
terrazzo 233
thermal comfort 230–32
thermal resistance of materials 231
tilers 62
tiles
    floor and wall
    metal 234
timber floor frame 98–99
timber framing 19
timber shingles and shakes 170
timber
    ceilings 178
    decking 223
    fixing, supplies 64
    flooring 64, 98–99
    history 17
    lining boards 139
    mouldings 200
    roof 156, 162, 170
    scantling 64, 110
    weatherboards 123
    windows 140
time control 68
time limits for work 57
toilet and basin cubicle 262
toilet cistern, alternative to 188
toilet pans 205
top plate 155
town planners 31
town planning departments 33
tradespeople, identification 59
travertine marble flooring 233
trees, on site 74
trims 199
trussed roofs 152, 161–65
valley rafter 156
valuers 31
vaulted roofs 166–68
vegetation, on site 74
ventilation, 243
vinyl 233
walk-in closets 211
wall cabinet, bathroom 206
wall frames
    metal 117
    supplies 65
    timber assembly methods 110–14
    post and beam construction 114
    scantling materials 110
wall ovens 197
wallpapers 234
walls
    function 109
    retaining (see retaining walls)
    wardrobes 210–12
washing machines 197
water efficiency 258
water filters 197
water purification 189–90
water reticulation
    cold 189
    hot 190
water supply 74, 253
    sustainable, 253
water supply authorities 33, 36
water units, boiling 197
weather conditions, by city 242
weather, effects on construction 72, 74
weatherboard 123
weep holes 95
wet areas, flooring requirements 99
wheelchair 260
whitegoods 196–97
wide clear doorways 260
wide parking space 261
wind direction 74, 242
wind, roofing considerations 172
windows 140–48
    aluminium-framed 16, 140, 141, 147
    clerestory 181
    dormer 181–82
    leadlight 140, 141
    reusing 252
    schedule 51
    steel 140
    supplies 65
    timber 140
workers' compensation policy 36
working drawings, preparing 35
workspace clearances 263
zincalume 172
zoning, site 76

# NOTES

# NOTES